Recording Studio Design

Recording Studio Design

Third Edition

Philip Newell

NEW YORK AND LONDON

First published 2003
This edition published 2013 by Focal Press
70 Blanchard Road, Suite 402, Burlington, MA 01803

Simultaneously published in the UK by Focal Press
2 Park Square, Milton Park, Abingdon, Oxon OX14 4RN

Focal Press is an imprint of the Taylor & Francis Group, an informa business

Library of Congress Control Number: 2011935549

ISBN 13: 978-0-240-52240-1 (pbk)
ISBN 13: 978-0-240-52241-8 (ebk)

Contents

About the Author

Philip Newell entered the world of music directly from school in 1966, at the age of 17. His first job was as an apprentice in audio electronics, during which time he studied radio and television servicing at Blackburn Technical College, in England. However, he soon gave up his apprenticeship when offered a job as sound and light operator in a local ballroom, owned by the Mecca entertainments company. His work was well-liked, and he was gradually moved to larger ballrooms within the Mecca chain, finally arriving at the Orchid ballroom in Purley, just south of London, which was then one of the largest ballrooms in the country.

These were the days when musical groups did not travel with their own public address systems. They tended to rely on the house systems, and usually the house sound engineer as well. So the Orchid, being such a prominent ballroom, was a natural choice of venue for many of the famous musical artistes of the time. *It was just part of his normal work as the resident sound engineer for Philip to be working with artistes such as Booker T and the MGs, Junior Walker and the All Stars, Eddie Floyd, Arthur Conley, Sam and Dave and many other stars of the Stax/Motown era, as well as groups such at The Who, The Small Faces and other British rock groups, many of which he would later meet again, either in recording studios or whilst making live recordings.*

By the age of 21, Philip Newell knew a lot of musicians, and some had asked him to put together small 'demo' studios (the forerunners of today's project studios) in which they could work, principally, on their song-writing. One such studio, Majestic, in Clapham, south London, began to grow out of all proportion during its construction, finally opening in late 1970 as a quite large, professional studio. However, its control room, much larger and more absorbent than most control rooms of the day, was not well received. The more usual rooms were heavily influenced by broadcast control rooms, and their specifications were quite rigid. Recording staff also tended to be quite conservative. Philip's attempt to build a control room that *he* thought was more accurate than many other control rooms did not see much use. The owner decided that the control room should be reduced in size, brightened up acoustically, and fitted with a proprietary stereo monitor system in place of the custom four-channel system. At this juncture, Philip went to work for Pye Records, in London's West End, and would not attempt anything on the lines of Majestic for another 20 years, although he never lost faith in the concept of highly damped rooms.

Pye was a large studio complex with two studios, two mix-only rooms (reduction rooms, as they were then known), three disc-cutting rooms, two tape duplication rooms and a room for compiling the eight-track masters for the tape cartridges then used in many motor cars. Pye also had a mobile recording unit, and this appealed very much to Philip's love of live music events. His experience of music on-stage made him an obvious candidate for the mobile recording crew. Until late 1971 he was working in the studios, principally as a maintenance engineer, and on the mobile recording unit as a '*Jack of all trades*'. Mobile recordings were then very much a team effort. During this time at Pye records, they built an articulated mobile recording vehicle, chiefly designed by Ray Prickett, the technical manager of the studios. This was used to record many live concerts, with artistes such as The Who (again), The Faces, Free, Emerson Lake and Palmer, Traffic and many other famous groups of that era.

However, the studio's administration manager was beginning to take exception to the length of Philip Newell's hair, *and* his tendency to wear multi-coloured boots. The ultimatum 'get your hair cut, or else ...' resulted in Philip accepting an offer as chief engineer at Virgin Record's almost completed Manor Studios, near Oxford, where the wearing of long hair and multi-coloured boots was almost *de rigueur.* Within weeks he was recording a solo album for John Cale (ex Velvet Underground) with musicians such as Ronnie Wood, now with the Rolling Stones.

Nevertheless, the 'call of the wild' (mobile recording) was still a strong pull, and much spare time was spent putting together a mobile recording vehicle in a corner of the Manor's 35 acre (15 hectare) grounds. For reasons still unclear, Richard Branson (Virgin's chairman) took exception to this, but made an unusual offer, which was tantamount to 'Give me all your equipment in exchange for me financing the building of the world's best mobile recording studio – of which you will be 20% shareholder – or you are fired'.

The choice was not very difficult to make, so Philip soon began plans for the Manor Mobile – destined to be the world's first, purpose-built, 24-track mobile recording studio (using Ampex's pre-production MM 1100 24-track tape recorder) in January 1973. By the end of that year there was so much work that the Manor Mobile Ltd bought the Pye Records mobile recording vehicle. Around this time, Tom Newman, the managing director of the Manor Studios, left Virgin, and Philip Newell, at the age of 24, found himself technical director of a newly-formed recording division of Virgin Records.

1975 saw the rebuilding of the Manor Studios, with Tom Hidley, the then chief of Westlake Audio. During the same year, Philip also spent months working with Mike Oldfield on his *Ommadawn* album, which was re-mixed into quadrophonics in the newly completed 'surround' control room at the Manor. Shortly after he re-mixed the classic *Tubular Bells* into four-channel surround; a mix which was re-released in 2001 as one of the first Super Audio Compact Discs (SACDs).

In 1978, again with Tom Hidley, Philip led the Virgin team who built The Townhouse, in London. In 1979, he was back on the road again, as front of house engineer for Mike Oldfield's 45-musician extravaganza which toured Europe. But, not only was he doing the front of house mixing, he was also producing the recording of the live album, *Exposed*, which was a gold disc, on advanced orders, before it even reached the shops.

During eleven years with Virgin, Philip was involved in a mountain of recordings, both in the studios and with the mobile recording units. He produced artistes such as Gong and Mike Oldfield (producer or engineer on six of his albums), recorded The Warsaw Philharmonic Orchestra; The Duke Ellington Orchestra; Hawkwind; Led Zeppelin; Don McLean; Captain Beefheart; Jack Bruce; Dizzy Gillespie; The Small Faces; Ben E. King; The Buzzcocks; XTC; Nana Mouskouri; The Motors; Jim Capaldi; Stevie Winwood; The Band; Patti Smith; Queen; Can; Tangerine Dream; Steve Hillage; Alvin Lee; The Royal Philharmonic Orchestra – not to mention church organs; English brass bands; fairground organs; Welsh male-voice choirs; Scottish pipes and accordions; gospel choirs; The Edinburgh Festival Choir – the list goes on. The great lesson learned from this variety of recordings, plus an enormous number of long-forgotten recordings, was that a great recording usually begins with great musicians. What goes into the microphones is much more important than what a recording engineer can do with the mixing console.

As Philip Newell was later to say: 'The thing that I found most disappointing about being a recording engineer was the lack of correlation between the effort put into the job and the success of the results. I could work extremely hard, using all my skill and experience, trying to get a half decent recording from a group of mediocre musicians, or I could sit with my feet on the desk, pushing up a fader with one finger, and record an absolutely fantastic guitar sound from Dave Gilmour or Jimmy Page'. This no doubt contributed to his almost total departure from the recording industry in 1982. Virgin was also getting to be much more 'big business' and bureaucratic, which was not well suited to Philip's somewhat free-spirit, so he sold his shares in the company and invested more in his seaplane fleet, which he had begun in 1979. This had been largely in connection with Richard Branson's purchase of Necker Island, in the British Virgin Islands, and on which they were planning to build a tax-haven recording studio. However, the collapse of the pound sterling on the foreign exchange markets, the very high spending by the Virgin group on other projects, and the election of Margaret Thatcher, who greatly reduced the higher tax rates in Britain, all conspired to squash the idea of the Caribbean studio.

However, it was perhaps the 'call of the wild' again, which drew Philip to the wide-open spaces of the world of float-planes and flying-boats. He flew in many air-displays, and also for cinema and television work (and even a BBC radio programme), and by 1982 was a flying instructor, and an examiner on certain types of small seaplanes. However, without the income from the music business to support it, it was difficult to keep these operations afloat; both in the physical and financial senses. In 1983, he sold everything, and returned to music to produce an album for Tom Newman, the co-producer of *Tubular Bells*.

In 1984, he met Alex Weeks, who had a company called Reflexion Arts, specialising in the sale of very expensive gold and silver flutes. In the same year, Philip had been asked to design a studio for Jacobs Studios, in southern England, so he joined with Reflexion Arts to begin a studio design division, and Jacobs 'Court' studio was their first endeavour together. He then designed a range of monitor systems under the Reflexion Arts name.

In 1986, he realised that he needed further, specialised help in the design of a more advanced range of monitors, and sought help from the Institute of Sound and Vibration Research (ISVR) at Southampton University in the UK. He had come into contact with the ISVR quite coincidentally, via flying. His aerodynamics colleagues in Southampton University's Department of Aeronautics and Aerospace, where he was making enquiries about horn design with specialists in trans-sonic (i.e. *through* the speed of sound) wind tunnel construction, shared a building with the ISVR. These investigations drifted him across to the ISVR acoustics department, where he sponsored a 3-year doctoral research programme which eventually led to Keith Holland's AX2 horn, somewhat revolutionary in its time (1989), and which is still used in some of the current Reflexion Arts monitor systems.

The connection with the ISVR continues, where Philip has sponsored a number of students at undergraduate, Masters and doctoral research levels. He was once heard to say to the owner of a school of recording engineering, who taught at the school but had never himself been a professional recording engineer: 'The big difference between us is that students pay you to teach *them*, whereas I pay students to teach *me*'.

Philip Newell left Reflexion Arts in 1988, but has remained in close contact with them since the late Alex Weeks passed the company to new owners in 1991. It now operates from Vigo, Spain, and has clients around the world. In 1992 he moved to Spain, where he has lived since, though he is rarely home. During one period of time, between late 1992 and early 1994, he spent one night at home in 18 months. Philip has now worked, in one capacity or another, in 34 different countries. He is a member of the Audio Engineering Society, a Fellow of the UK Institute of Acoustics and a member of the Seaplane Pilots Association.

His work now involves the designs of studios for music recording, film mixing, television shooting stages, concert halls, multi-use halls, music clubs, rooms for voice recording, discothèques, screening rooms, rehearsal rooms and occasionally he also gets involved in industrial noise control. From time to time Philip still also makes recordings. He has designed hundreds of rooms, and written around a hundred articles for magazines on the subjects of music recording and aeronautical issues. He has also written around 30 papers which have been presented at Audio Engineering Society (AES) and Institute of Acoustics (IOA) conferences, and has also contributed technical works to their journals.

On occasions he is called upon to give talks at colleges, institutes, universities and learned societies, and has done so in the UK, Holland, Spain, Russia, Serbia, Ukraine and the USA, to students of music, recording technologies, and engineering acoustics. This is his seventh book, following on from *Studio Monitoring Design*, *Recording Spaces*, *Project Studios*, the first and second editions of this book, and *Loudspeakers*, co-written with Keith Holland.

On a more personal note, Philip is a member of British Mensa, and the League Against Cruel Sports. The latter is something very dear to his heart, as cruelty of any kind, to any living creature, is something that he abhors. He greatly dislikes 'doing business', and tends to become very personally involved with his designs and constructions. Consequently, he can sometimes be quite abrasive. He can also be volatile and highly explosive, but he tends to cool down as quickly as he blows up. Philip has never suffered fools gladly, even if they were his paymasters [hence not being good at business – P.N.], and it has taken him a long time to understand that not everybody can be as totally committed to the work as he is. However, he has always had a lot of respect for people who try hard and want to learn, whether they succeed or not.

Acknowledgements

Dr Keith Holland; lecturer in acoustics at the Institute of Sound and Vibration Research, with whom I have worked closely for over 20 years. His work is evident throughout this book, and he was responsible for about 150 of the measurements presented in the figures. It is an honour to have him so deeply involved in a work such as this.

Sergio Castro MIOA, who assembled the entire artwork for the book and who was responsible for the graphics for around 200 of the figures. He has been a close friend and colleague ever since I designed his studio, Planta Sonica, in Vigo, Spain, in 1985.

Janet Payne, who not only provoked me into writing this book, but who undertook to put onto a word processor literally thousands of pages of manuscript. For anybody who knows what my handwriting is like, the enormity of the task will be self-evident.

Tim Goodyer and David Bell, for Chapter 17. It can be difficult for a person such as me to enthuse about something to which they cannot 100% commit themselves, but it would have been unjust to write a half-hearted chapter, or to ignore the contribution that Live-End, Dead-End control rooms have made to the recording industry. David is a deft exponent of the technique, and it was courageous of him to step into the lion's den and make his contribution. He deserves great respect as an honest, sincere and capable man.

Professor Jamie A. S. Angus of Salford University, UK, for allowing me to copy some of the figures from her chapters in the book *Acoustics and Psychoacoustics* (co-written by Professor David Howard), and for many stimulating conversations on the subject of this work.

Julius Newell MIOA, MInstSCE, for the hard work involved in making the measurements for Figures 19.15 to 19.17, and for countless hours of discussion.

Melanie Holdaway, Janet Payne's sister, for handling the overloads when the typing schedules became excessively pressurised.

Beth Howard, at Focal Press, for keeping faith as this book grew and grew.

Catharine Steers, for asking for the second and third editions.

Alan Perkins, for a thorough reading of the proofs.

Eliana Valdigem AMIOA, for her help with many experiments and measurements, and also for her general support and encouragement.

And finally, to all the people who have worked so hard on the construction of my designs, without whose diligence and effort the end results would not have achieved their success.

My sincerest thanks to all of them.

Philip Newell

Preface

The intention of this book is to make accessible to many people involved in the daily use of recording studios information which is locked away in many textbooks. The majority of people working in modern music recording studios have not had the necessary formal education in mathematics, acoustics and electronics to make the textbooks appear as anything other than cold print. Largely, also, the days are gone when the majority of studio staff received formal training in the studios themselves, spending years learning under the watchful eyes of previous generations of recording engineers.

This book is not intended to replace the textbooks, but to accompany them, in order to put many of the principles which they define into the context of modern recording studios, in a way which may help to give more meaning to the bare facts. The practical examples given cannot cover the almost infinite range of possible combinations of techniques, but if the examples can be well understood, then they should help the reader to interpolate the data sufficiently to have a reasonable ability to determine for themselves the likely outcome of other approaches. Inevitably, in a book of this size, there will be a certain amount of overlap and repetition. However, where this occurs, it has been left in for reasons of clarity, emphasis of importance, or for the ability of a chapter to stand alone, without the need for unnecessary cross-referencing. Whilst the language used is as plain as possible, there is an extensive glossary at the end of the book to help to explain any unfamiliar terms, and whilst only a minimum of simple mathematics is involved, nevertheless the contents of the book are intended to be as rigorously factual as possible.

Philip R. Newell
Moaña, Spain, 2003

Preface to Second Edition

When this book was originally being discussed with the publisher, a book of about 80,000 words was proposed. However, once work began, it soon became apparent that in order to deal with the concepts to a depth which most previous books on the subject had not achieved, the original estimate for the size of the book had been greatly misjudged. The book grew and grew, up to a point where the then commissioning editor had to decide whether the original marketing proposals would still be valid, and whether a book of such a size was still viable. Fortunately, she kept her faith in the idea, but a halt was called when the word count was approaching 200,000.

Once the book was released it was generally very well received, but numerous readers commented on certain omissions of details that they would have found useful, such as how to make sound isolating doors, as one example. The book was reprinted in 2004, and then in 2005, and even twice more in 2006. In fact, sales had been continuing at a steady pace since the first publication. Focal Press suggested that perhaps the book could stand an enlargement sufficient to incorporate the items that some readers had requested, and also to cover the subjects of more research and developments that had taken place in the four years since the first publication.

In this Second Edition, apart from more on the subject of doors, more material has been added on floated floors, as well as on air-conditioning and climate control. New work has been incorporated on the strengths and weaknesses of digital signal processing as a means of room correction, and more has been added on the use of multiple sub-woofers for room mode cancellation. New sections have also been added on the design of rooms for cinema soundtrack mixing, along with more on the perception of frequency responses in rooms of different sizes and modal activity. Again, in response to reader's requests, sections have been added on rooms for the recording of the spoken voice, and rooms for sound effects. Finally, three entire chapters have been added to the end of the book, dealing with foldback, electrical supplies and analogue interfacing. It is hoped that these new additions will substantially augment the usefulness of the book as a work of reference.

Philip Newell
Moaña, Spain, 2007

Preface to Third Edition

Four more years have now passed, and changes in the industry have taken place at an alarming rate. CDs are no longer the general reference, and high fidelity, uncompressed surround formats have almost disappeared. Further comments have also arrived from readers, and the cinema industry is moving from film to digital formats. A considerable number of updates and additions have therefore been made for this Third Edition. More information has been added on floated isolation systems, and the discussion about film studios, in Chapter 21, has been significantly expanded, including a more thorough assessment of the limitations of 'room' equalisation, which in fact applies to all rooms. New data has also been included on wide-band absorber systems in Appendix 1, along with many other expansions and clarifications throughout the book.

However, in order to avoid confusion between the various editions when referring to the books, the changes have been made in such a way as to preserve as far as possible the same section and figure numbers from the previous editions.

Philip Newell
Moaña, Spain, 2011

Introduction

The development of sound recording studios advanced steadily from the 1920s to the 1980s almost entirely in the hands of trained professionals. By the mid-1980s the professional studios had achieved a high degree of sophistication, financed by a recording industry which drew its money principally from the record, film and advertising industries. These client industries were themselves mainly professional industries, and were accustomed to paying professional prices for professional services.

By the late 1980s, much less expensive recording equipment of 'acceptable' quality (at least on the face of it) became available on an increasing scale, and the imminent arrival of domestic/semi-professional digital recording systems was soon to lead to an 'explosion'. This saw the sound recording studio industry fragment into a myriad of small facilities, which severely damaged the commercial viability of many of the larger studios. It broke up huge numbers of experienced teams of recording personnel, and consequently much of the generation-to-generation know-how which resided in many of the large professional studio complexes was lost.

This boom in the number of small studios spawned a worldwide industry supplying the necessary technology and equipment, but the whole recording studio industry has since become ever more dependent upon (and subject to the wishes of) the manufacturers supplying its equipment. It has largely become an industry of recording equipment operation rather than one based on the skills and knowledge of traditional recording engineering. So much recording is now software-based, and so many people in the modern industry are now largely self-taught, that only a relatively few people out of the total number involved in music recording have, or will ever have, experienced the benefits that a really well-designed studio can offer.

Clearly, things will never be as they were in the past, but although many great advances are taking place in recording technology, some of the basic principles are just as relevant now as ever they were. Good recording spaces, good monitoring conditions, good sound isolation and a good working environment are still basic requirements for any recordings involving the use of non-electronic instruments, which means most recordings, because voices also come under the 'non-electronic instruments' heading.

The general tendency nowadays is to think of the equipment first. Many so-called recording studios are in fact no more than several piles of rather sophisticated equipment set up in any reasonable room that will house them. Many owners realise all too soon after the inauguration of their 'studios' that there is more to recording studios than they first thought. The real needs become all too obvious, which then often leads to some trial and error, and sometimes very wildly misguided attempts to convert their already-purchased, unsuitable space into what they think that they really need.

The sad fact is that there are now enormous numbers of bad studios producing recordings of very arbitrary quality. As this situation spreads with the growth of the less professional industry, many standards are being eroded. The norms of the industry are being set by the mass market, and no longer so much by the skilled professionals with their valuable knowledge of what *can* be achieved, which seems to be a pity.

It is all the more a pity because modern technology *and* the knowledge passed down through the generations can together reach previously unattainable levels of excellence. What is more, the cost is not necessarily prohibitive. Rather it is ignorance which is the enemy, because the cost of doing things badly is often no more than the cost of doing things well. People waste an incredible amount of money by their errors, and lose much valuable income by not being able to offer the first class results which they *should* be able to achieve from their investments.

When The Townhouse studios were completed in London in 1978, the two studios had cost around one million pounds sterling (about 1.4 million euros) and were staffed by two recording engineers, five assistants and five qualified maintenance engineers. The cost of each studio per hour was around £85, which probably relates to something more like £300 (€400) in 2011 money. Few sane people would now spend such a sum of money (inflation adjusted, of course) on a comparable facility. Almost nowhere in the world would it be possible to charge such an hourly rate for music-only recording. We therefore need to be realistic in our approach to modern day designs. Nevertheless, the good news is that with the developments in the recording equipment, the advanced nature of new acoustic materials and techniques, and a much greater understanding of psychoacoustics compared with what was known 30 years ago, we can now achieve comparable, and in many ways superior results to those which were achieved in the original incarnation of the classic Townhouse, and for much less money than ever before.

The financial pressure on recording studios is great. Competition is fierce, and what was once seen as a genuine industry is now often seen more as a glorified hobby. Where banks used to finance many studio projects, large and small, they are becoming unwilling to do so in the twenty-first century. The recording industry is often seen to be unstable, with ill-conceived ideas and a poor track record of adequate professionalism. Banks may still finance the purchase of buildings, which they can sell if the studio fails commercially. Leasing companies may be interested in supplying recording equipment, which they will continue to be the true owners of until such time that the lease is paid in full. However, few organisations will risk the financing of the acoustic control structures that actually define a professional studio. This is simply because if the studio does fail commercially, the labour costs involved in the construction are lost. Furthermore, most of the materials used will not be recoverable in any way that would enable them to have any resale value, and the demolition costs of the heavy, space-consuming acoustic work can be considerable if the next occupiers of the building require it in its 'unmodified' state. The lack of available financing for the acoustic work is one reason why it is often now not afforded its rightful attention. Somewhat unfortunately, the neglect of this one critical aspect of the studios can be a prime reason for their failures to perform, either musically or commercially. Many studio owners and operators are beginning to see this, and it *is* being realised that much of what was once considered an essential part of all serious studios is *still* an essential part of all serious studios.

What this book will now discuss are the fundamentals of good studio acoustics and monitoring, in a language that will hopefully be recognisable and accessible to the people who may well need the information that it contains. It will deal with the basic principles, their application in practical circumstances, and the reasons for their importance to the daily success of recording studios. Because of the importance of good acoustics to the success of most studios, and because of the financial burden which failure may impose, getting things right first time is essential. This applies equally to studios large and small.

It is being presumed that the majority of readers will be more interested in how these things affect their daily lives rather than wishing to make an in-depth study of pure acoustics. Bibliographies at the end of most of the chapters will point interested readers to other publications which may treat the specific subjects more formally, but inevitably we will have to begin with a couple of chapters which set out a minimum of the fundamental principles involved, in order that we can proceed with at least *some* of the basic concepts firmly in mind.

CHAPTER 1

General Requirements and Common Errors

This chapter lays out the fundamental requirements of premises for professional recording purposes, including: common underestimation of need for good isolation; avoidance of disturbance from plant and equipment noises; influence of location on isolation requirements; consideration of artistic needs; control room monitoring basics; types of buildings to avoid; and the need for adequate space and building strength.

1.1 The General Requirements

Some of the things that set a professional recording studio apart from a personal studio are listed below:

1. The ability to work during the chosen hours of use (in many cases 24 hours per day) without disturbing, or being disturbed by, anything or anybody in the local community.
2. The studio should be able to record musicians without delays or impediments to the needs of the musical performance.
3. Studios should inspire confidence in all the personnel involved in any recording.
4. The achievable quality of recording should not be limited by the inadequacy of the studio design or installation. Even a modest studio performing optimally may well outperform a much more elaborate one that has been poorly conceived and installed.
5. The studio should always provide an adequate supply of clean, fresh air, in a temperature and humidity-controlled environment. (See Chapter 9.)

So now, let us look at these points in some more detail.

1.2 Sound Isolation and Background Noise Levels

In the enthusiasm that often accompanies the idea to build a recording studio, the lack of experience of the people involved often leads to a tendency to fail to realise the need for good sound isolation. In far too many cases, people believe that they can work around most of the restrictions which poor isolation imposes. This is a dangerous attitude, because once it is realised that the compromises severely restrict the success of the studio it is often too late or too financially burdening to make the necessary changes. The result is often either a ceiling placed on the ability of the studio to develop, or financial ruin. In 2001, European banks reported bad debts on over 20,000 studio project loans, and this has made things difficult ever since. Optimism must be tempered by reality.

Isolation is a two-way problem. The most obvious need for isolation is to prevent sound escaping from the studio and disturbing any noise-sensitive neighbours. Almost everybody realises that repeated disturbance of neighbours is probably going to lead to complaints and, if nothing is done about it, cause the closure of the studio. Conversely, noises from the local community activity entering the studio can disrupt recordings and disturb the creative flow of the artistic performances. Sound isolation also sets the dynamic range limit for a studio. This latter point is very important in a professional recording situation, but it is often woefully under-appreciated.

1.2.1 From the Inside Out

If a studio only has an effective isolation of 40 dB, then any sounds above 75 dBA in the studio will risk annoying neighbours. The resulting 35 dBA reaching them would certainly be considered a potential noise nuisance, at least if the studio were to be used after 10 pm and was sited in a residential area. For example, one cannot turn down the volume of a drum kit. Playing quietly is no solution, because it produces an entirely different tone quality to playing loud. Realistic drum levels are more in the order of 110 dBA, so 75 dB of isolation (the 110 dBA SPL [Sound Pressure Level] of the drums minus the 35 dBA acceptable to the neighbours) would be a basic requirement, though this could be reduced at low frequencies, as will be discussed in Chapter 2.

Many people decide that they can mix in the control room at night in rooms with reduced isolation, in the belief that they can work with the monitor volume controls reduced below their daytime levels. It soon becomes apparent that if the studio is to be used commercially, it is usually the clients, not the studio owners, who decide at what level they wish to monitor. If they cannot work in the way that they wish or need to work, they will perhaps look elsewhere when planning their next recordings. In addition, when the ability to monitor at higher levels is denied, low level noises or distortions may go unnoticed, only to be heard at a later date. This may result in either the work having to be done again or the bill for the wasted session going unpaid.

Even more disturbing (see next chapter and Figure 2.1 for reasons), mixing at a relatively quiet SPL of 75 dB is at the lower end of the preferred range for music mixing, because it is already descending into a region where the ear is less sensitive to the upper, and especially the lower, frequency ranges. Mixes done at or below this level may tend to sound excessive in bass when reproduced elsewhere at higher SPLs, as would often be the case. Therefore, mixing at a low level so as not to annoy the neighbours is not really a professional option.

It is true that for a voice studio for publicity or radio recording (and especially when the end-product is not likely to be listened to from an audiophile perspective), 40 or 50 dBA of isolation and a 75 dB maximum operating level may suffice, but such conditions would certainly not be suitable for music recording. In conditions of poor isolation, frustrating moments of lost artistic inspiration can be frequent, such as when a good take is ruined by an external noise, or when operating level restrictions deny the opportunity to do what is needed when the moment is 'hot'. Professional studios should be ready for whatever the musicians reasonably require, because capturing the artistic performance is the prime reason for their existence.

1.2.2 From the Outside In

Background noise levels of below 20 dBA (or NR20 or NC20 as variously used) were the norm for professional studios. In recent years, cost constraints on air-conditioning systems, together with the appearance of ever more computer disc drives in the control rooms, have pushed these levels higher. These problems will also be discussed in later chapters, but background noise levels above 25 or 30 dBA in either the studio rooms or the control rooms seriously begin to encroach on the recording operation. Twenty dBA is still optimal.

Most musical instruments have been designed to have sufficient loudness to be heard clearly over the murmur of a quiet audience, but if the background noises in a recording room exceed around 30 dBA there will be a tendency for the extraneous noises to enter the microphones with sufficient level to degrade the clarity of some recordings. Much important low-level information in the tone of an instrument or voice may then be masked by the noise. In the control rooms, we should reasonably expect a background noise level at least as low as that of the recordings. Otherwise, when monitoring at life-like levels similar to those produced by the instruments in the studio, one could not monitor the background noise level on the recording because it would tend to be masked by the higher background noise level in the control room. The number of so-called recording studios which now have 50 dBA or more of hard disc and cooling fan noise in the control room, with monitoring limits of only 90 dB SPL, is now reaching alarming proportions. That represents a monitoring signal-to-noise ratio of only 40 dB. It is absurd that many such studios are promoting their new, advanced, 24-bit/96 K recording systems as part of a super-low-noise/high-quality facility, when the 100 dB + signal-to-noise ratio which they offer cannot even remotely be monitored. One cannot trust to luck and call oneself professional.

1.2.3 Realistic Goals

The previous two subsections have outlined the basic reasons why good sound isolation is required in recording studios. The inside-to-outside isolation is usually dominant, as few studios are sited next to neighbours producing upwards of 110 dBA. As the 30 dBA region is reasonably close to the limit for tolerance of background noise by either the neighbours or the studio, it is principally the 110 dBA or so produced in the studio that dictates the isolation needs.

Of course, a well-judged choice of location can make life easier. Siting the studio in the middle of nowhere would seem to be one way of reducing the need for so much isolation. However, the owners must ask themselves if their clients are likely to travel to such a remote location in commercially viable numbers. Furthermore, one should be wary of other likely problems. One expensive studio was located in a place with little sound isolation because it was so remote from any neighbours. Three months of unseasonably strong winds and heavy rain almost drove them to ruin because of the weather-related noise entering the studio. At great cost, improved sound isolation had to be added after the studio had been completed, which proved to be far more expensive than it would have been had it been incorporated during the initial construction of the studio.

It is client convenience which often drives studio owners to locations in city centres or apartment buildings. Convenient for the clients they may be, but high property prices and/or high isolation costs often cause the owners to look for premises which are too small. Often there is simply no room for adequate isolation in their chosen spaces, even when very expensive techniques are employed. This subject will be dealt with in greater depth in Chapter 2.

1.2.4 Isolation versus Artistry

Artistic performance can be a fragile thing. Curfews on what can be done in the studio and during which hours can be a source of great problems. No matter how clearly it is stressed that the working hours are 10 am to 10 pm, for example, the situation will always arise when things are going very well or very badly, where a few extra hours of work after the pre-set deadline will make a good recording great or perhaps save a disaster. In either case, using a studio where this flexibility is allowable is a great comfort to musicians and producers alike, and may be very much taken into account when the decision is made about which studio to use for a recording.

1.3 Confidence in the System

A professional studio should be able to operate efficiently and smoothly. Not only should the equipment be reliable and well maintained, but also all doubts should be removed as far as possible from the whole recording process. This means that a professional studio needs recording rooms with adequately controlled acoustics and a monitoring situation which allows a reliable assessment to be made of the sounds entering the microphones. This latter requirement means reasonably flat monitoring systems are needed, in control rooms that allow the flat response to reach the mixing position and any other designated listening regions of the room. The monitoring systems should also have good transparency and resolution of fine detail, uncoloured by the rooms in which they are placed or by the disturbances caused by the installed recording equipment. Where doubt exists about the monitored sound, musicians may become insecure and downhearted, and hence will be unlikely to either feel comfortable or perform at their best.

The decay time of the control room monitoring response should be shorter than that of any of the main recording rooms (dead isolation booths may be an exception), otherwise the recording personnel may not know whether the decay that they are hearing is a part of the recording or a result of the monitoring environment. This subject can arouse many strongly opinionated comments from advocates of some older control room design philosophies, but the fact remains that adequate quality control monitoring can be difficult to perform in rooms with typically domestic decay times.

When recording personnel and musicians realise that they can trust that what they are hearing is what the audiophiles will hear in good conditions, it tends to give them more confidence. Despite the fact that very many people now listen to downloaded, data-reduced music via mobile telephones and ear-buds, the majority of musicians still want their recordings to sound good on top quality systems. It is a question of artistic and professional satisfaction. Confidence is often lacking in an insecure artistic world, so anything which can boost it is much to be valued. Small loudspeakers are effectively *de rigueur* in all studios these days, both as a mixing tool and as a more domestic reference. This is a very necessary requirement, as one obviously wants to know what the likely result of a mix will be in 95% of the record buyers' homes. Nevertheless, it still seems to be incumbent on a professional studio to be able to provide the means to monitor the full range of a recording. Those paying fortunes for their super hi-fi systems will not then be disappointed, as they would be when buying poorly monitored recordings that could have been so much better if only the recording studio had had better monitoring. The large monitors are also necessary for a good, full frequency range, *quality control* assessment of the basic recordings, even if they are not to be used at the mixing stage, but this will be dealt with in much more detail in Chapter 19. If there is any one thing that disgraces so much of the 'less than professional' part of the recording industry it is the widespread use of appalling monitoring conditions.

A further point for consideration, although a detailed discussion is outside of the scope of this book, is that it should still not go without mention that nothing really inspires more confidence in a recording process than the participation of an experienced and knowledgeable staff.

1.4 The Complete System

A recording studio is a system, just as a racing car is a system. No haphazard combination of high-quality gearbox, engine, wheels, tyres, axles and chassis will guarantee a well-performing car. The whole thing needs to be balanced. The same principle applies to recording studios. A hugely expensive, physically large mixing console, with large flat surfaces will tend to dominate the acoustic response of a small control room. In such situations, even when using the flattest monitors available, there is little chance of achieving a flat response at the listening position(s) in a small room. When studio equipment outgrows the rooms as the studio expands, the results usually suffer.

Studios should also be well ventilated, with good stability of temperature and humidity, otherwise musicians can become uncomfortable and instruments can vary in their tuning. Correcting the tuning later by electronic means is not a professional solution to any of these problems, because if the problems exist at the time of the recording they will almost inevitably affect the performance negatively. In fact, speaking about negativity, perhaps we should look at some of the typical things that many prospective studio owners get wrong, or misunderstand most often.

1.5 Very Common Mistakes

In an enormous number of cases, prospective studio owners' purchase or lease premises which they consider suitable for their studio before calling in a studio designer or acoustical expert. They often realise that there could be potential problems, but they believe that they can talk their way around any difficulties with neighbours. They invest considerable money in building something which they deem to be suitable for their needs, and then only call in specialists once the whole thing has been completed but the neighbours refuse to 'see reason'.

Acoustics is not an intuitive science, and many people cannot appreciate just how many 'obvious' things are, in reality, not that obvious at all. It is a very unpleasant experience for acoustics engineers to have to tell people, who have often invested their hearts, souls and every last penny in a studio, that the building simply is not suitable. Unfortunately, it happens regularly. The problem in many of these cases is that the buildings are of lightweight construction and the neighbours are too close. The three things most important in providing good sound isolation are rigidity, mass and distance. Lightweight buildings are rarely very rigid, so if the neighbours are close, such buildings really have nothing going for them except cheapness. Even if there is space to build internal, massive, floated structures, the floors may not be strong enough to support their weight because the buildings are only of weak, lightweight construction. In many cases, such premises will have been purchased precisely because they *are* inexpensive; perhaps they were all that could be afforded at that time, which often also means that the money for expensive isolation work is not available. The cost of massive isolation work in a cheap building will obviously be greater than a smaller amount of isolation work in a more sturdily constructed building, and usually the overall cost of the building and isolation work will be cheaper in the latter case.

An actual set of plans for the isolation work in a rather unsuitable building in southern Spain is shown in Figure 1.1. It is sited in the ground floor garage of an apartment building. Initial tests with bass and drums in the proposed studio, *after* it had been purchased, produced 83 dBA in a neighbour's bedroom. This would have meant trying to sleep with the equivalent of a loud hi-fi system playing in the adjoining room. The almost absurd quantity of required sound isolation work eventually reduced the noise level in the bedroom to around 30 dBA, but the cost was not only financial; much space was also lost.

1.5.1 The Need for Space

Space is also something which many potential studio owners underestimate. Whilst it is not universally appreciated just how much space can be consumed by acoustic isolation and control measures, it is still alarming that so many studio owners buy premises in which the rooms, when empty, have precisely the floor area and ceiling height that they expect to be available in the finished rooms. The owners of the studio shown in Figure 1.1 were very distressed when they saw their space being eaten up by the acoustic work. They could only breathe easily again when they realised that the isolation was adequate and that the relatively small remaining space had an open sound in which they could make excellent recordings. They eventually had to market the studio on its sound quality, and not on its size; which on reflection was perhaps not a bad idea. The studio became very successful.

If prospective studio owners can consider space in a *new* building *before* it is completed, then access by the acoustics engineer to the architects can usually provide some remarkably inexpensive solutions. Concrete, steel and sand are relatively cheap materials, and most structures can cope with supporting a lot of extra weight if this is taken into account at the planning stage. What is more, results are more easily guaranteed because the precise details of the structure will be known. Old buildings often lack adequate plans, and the acoustic properties of the materials used are often unknown. Hidden structural resonances can thwart the results of well-planned isolation work, so it is often necessary to err on the safe side when trying to guarantee sound isolation in old buildings, which usually leads to more expense.

Obviously, though, what we have been discussing in the previous few paragraphs require long-term investments. Many start-up studios are underfinanced, and the owners find themselves in short-lease premises in which the acoustic treatment is seen as a potential dead loss when the day comes to move. These people tend to be very resistant to investing in acoustics. Not very much can be done to make serious studios in such premises, certainly not for high-quality music recording, though exceptions do exist.

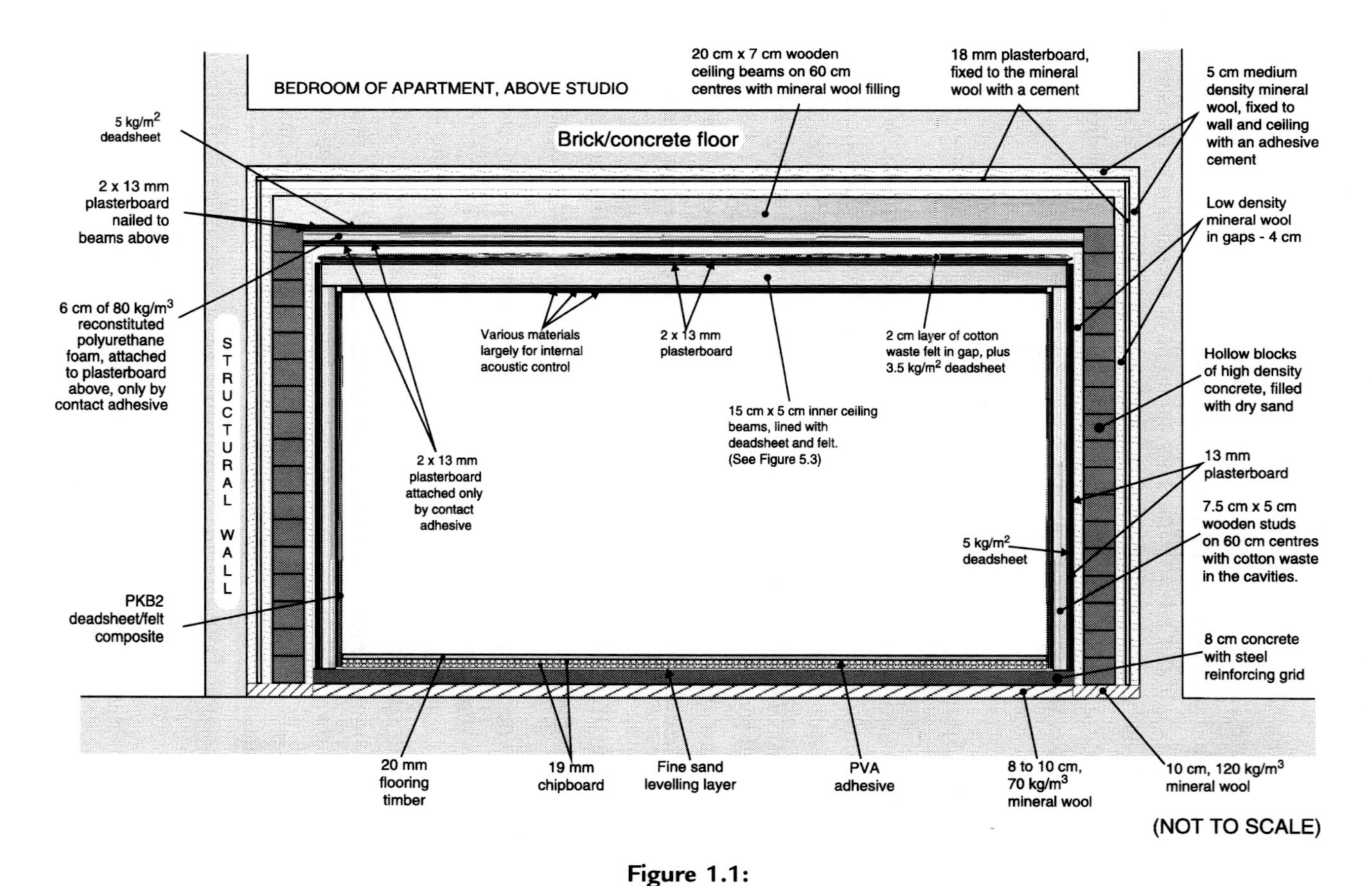

Figure 1.1:
Triple isolation shell in a weak domestic building.

1.5.2 Height

It is very difficult to make a good quality studio, free of problematical compromises, in a space with inadequate height. Control rooms require height because of the need to avoid parallelism between the floor and the ceiling. At low frequencies, all suitable floors are reflective, so the ceilings must be designed such that monitor response problems are not created by the vertical room modes. As will become apparent in later chapters, all forms of suitable treatment for the ceilings are wavelength dependent. So, if a metre is needed for the ceiling structure, and 20 cm or so for a floated floor, then to maintain a ceiling height of 2.5 m within the room, something approaching 4 m will be needed in the empty space before construction.

In the studio rooms, microphones placed above instruments, as often they must be, will be far too close to a reflective boundary unless adequate height is available in the room. Again, with less than 4 m of height to begin with it becomes very difficult to achieve the acoustics necessary to make a flexible, high-quality recording room. Six metres is a desirable height for an area in which a music studio is to be built. Less than 3 m makes the construction of an excellent studio almost impossible.

Experience has shown that if less than 3.5 m of height is available before treatment; the best that can be achieved are rooms of either limited flexibility or idiosyncratic sound. Obviously many rooms *are* built, these days, in spaces with much less height than optimum, but few of them could truly claim to have a 'first division' response. The lack of ceiling height in the chosen spaces is one of the most common errors made by prospective studio owners when acquiring premises.

1.5.3 Floor Loading

In general, sound isolation systems are heavy. The details of why and to what degree will be dealt with further in Chapter 3. There is no simple weight per cubic metre figure for typical isolation, but as an example, an adequately isolated room of 10 m×6 m×4 m in a residential building could easily contain 40 tonnes. On the 60 m^2 floor, this would mean an average loading of around 700 kg/m^2 (or around 150 pounds per square foot in imperial measure). This is more than a general light industrial loading, and much more than a domestic loading, and it is made worse by the fact that the weight is not evenly distributed. There may be areas beneath the lines of dividing walls, such as between the control room and the studio, where loads of 4 or 5 $tonnes/m^2$ may be present.

This is simply often not appreciated by people looking for suitable studio premises. Figure 1.2 shows the steelwork in a reinforced concrete fourth floor of an apartment building in Mallorca. Despite looking quite complex, it was not very expensive to make. Luckily, the prospective studio owner had taken advice from an acquaintance and bought a return air ticket to send to an acoustics engineer to enable him to meet the architect of the building

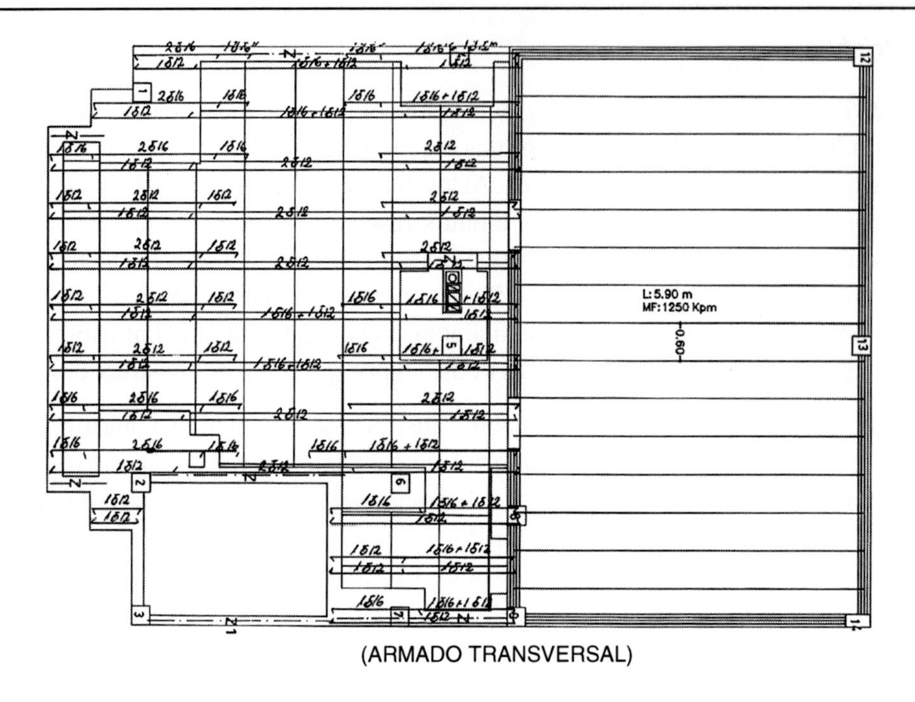

FORJADO PLANTA
PISO 4

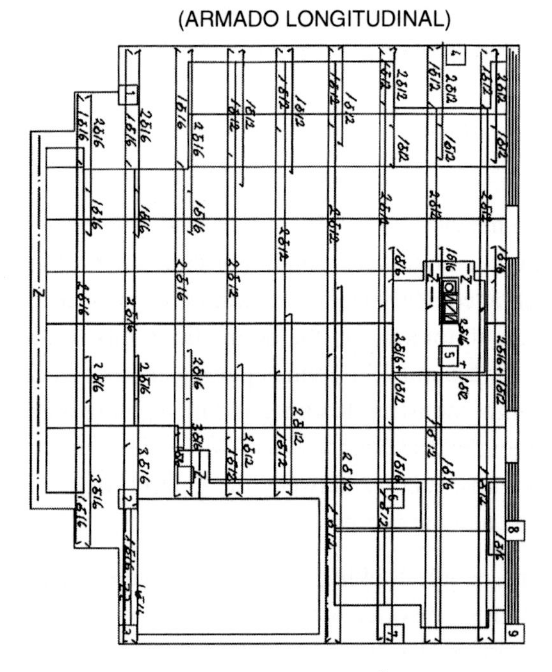

Transverse and longitudinal steelwork

Figure 1.2:
Reinforcing steelwork for supporting a studio on the fourth floor of an apartment building.

before construction began. The floor in Figure 1.2 *can* carry 40 tonnes, and the proposed studio eventually went into operation without problems. Had the owner not had the foresight to consult an acoustics engineer, and had he begun the internal isolation work without the required knowledge, then the studio could have been forced to close soon after opening due to poor isolation to the rest of the buildings. What is more, and in fact worse, the owner could have *tried* to provide sufficient isolation, only for the floor to collapse with perhaps fatal consequences.

The underestimation of the need for adequate floor strength and rigidity is a very common error made by prospective studio owners. What makes the situation worse is that in many cases the buildings that have weak floors often also have weak walls and weak ceilings, which make them the very buildings that *require* the heaviest isolation, which of course they cannot support. Obviously, therefore, they are not suitable as recording studios unless they are without neighbours and in areas of very low external noise, but as previously mentioned, the weather can then cause problems. The lowest floor of a solidly constructed building is clearly a better option.

The requirements for, and the cost of, the sound insulation/isolation can therefore be very much influenced by the nature of the structure of the building and its situation *vis-à-vis* noise sources and noise sensitive neighbours. The cost difference between needing 50 dB or 70 dB of isolation is very great, so if an appropriate building and location can be chosen, even if it is more expensive to buy or lease, it may still work out cheaper when the cost of the entire studio is fully appreciated.

If the things mentioned in this chapter are given due consideration at the very early stages of studio planning, then many problems can be avoided. In addition, if many things are *not* duly considered, problems in the realisation of the studio can be so deep-seated that they may have to be lived with for its working lifetime. Such problems can severely limit the potential for upgrading the studio to suit new ideas or a higher standard of recording. There is no doubt that a comprehensive knowledge of what one is seeking to achieve is a good starting point in almost any form of construction, but when choosing studio buildings it is especially important.

1.6 Summary

The general requirements of a studio should be carefully thought about before a location is chosen.

Good sound isolation is essential, and many people greatly underestimate its importance.

One cannot work more quietly at night time and expect to achieve the same results as working at normal SPLs.

Noisy electromechanical systems, such as ventilator fans, disc drives and air-conditioning units should not be allowed to disturb the recording or monitoring environments. Background noises above 30 dBA are not acceptable for professional use.

Choice of location can greatly simplify sound isolation requirements, but convenient access for the clients may drive studios into more noise sensitive areas. In the latter case, costs must be expected to rise. Potential earnings, on the other hand, may also be greater.

An undisturbed recording environment may be essential for achieving great artistic performances.

Control room and monitor system decay times should be shorter than the decay times in the principal studio (performing) rooms. Otherwise, monitoring environment decay may mask the performing room decay, and make the recorded ambience very difficult to assess.

Large and small monitor systems tend to be needed, each for different reasons.

It is best to seek expert advice *before* choosing a building in which to site a recording studio, because acoustics is not an intuitive science.

Lightweight, inexpensive buildings rarely make good studios. Buildings should also be considerably larger than what is needed solely for the interiors of the finished rooms. Isolation and acoustic control work can be space consuming. Adequate height is also beneficial. Old buildings often have hidden problems, so the prediction of conversion costs can sometimes be difficult to assess accurately.

Adequate low frequency isolation can often require the use of considerable quantities of heavy materials. These need not be expensive, but the question often arises as to whether a given building can support the weight.

CHAPTER 2

Sound, Decibels and Hearing

Important aspects of hearing sensitivity and frequency range. An introduction to the decibel in its various applications. The speed of sound and the concept of wavelength. Relation between absorbers and wavelength. Sound power, sound pressure, sound intensity. Double-distance rule. The dBA and dBC concepts. Sound insulation and noise perception. Aspects of hearing and the concept of psychoacoustics. The sensitivity of the ear and the differences of perception from one person to another. The effect on the perception of loudspeakers *vis-à-vis* live music.

2.1 Perception of Sound

That our perception of sound via our hearing systems is logarithmic becomes an obvious necessity when one considers that the difference in sound power between the smallest perceivable sound in a quiet room and a loud rock band in a concert is about 10^{12} – one to one-million-million times. A rocket launch at close distance can increase that by a further one million times. The ear actually responds to the sound *pressure* though, which is related to the square root of the sound power, so the *pressure* difference between the quietest sound and a loud rock band is 10^6 – one to one-million, which is still an enormous range.

When a pure tone of mid frequency is increased in power by ten times, the tone will subjectively approximately double in loudness. This ten times power increase is represented by a unit called a bel. One tenth of that power increase is represented by a decibel (dB) and it just so happens that one decibel represents the smallest mean detectable change in level that can be heard on a pure tone at mid frequencies. *Ten* decibels (one bel) represents a doubling or halving of loudness. However, the terms 'pure tone' and 'mid frequencies' are all-important here. Figure 2.1 shows two representations of equal loudness contours for human hearing. Each higher line represents a doubling of subjective loudness. It can be seen from the plots that at the frequency extremes the lines converge showing that, especially at low frequencies, smaller changes than 10 dB can be perceived to double or halve the loudness. This is an important fact that will enter the discussions many times during the course of this book.

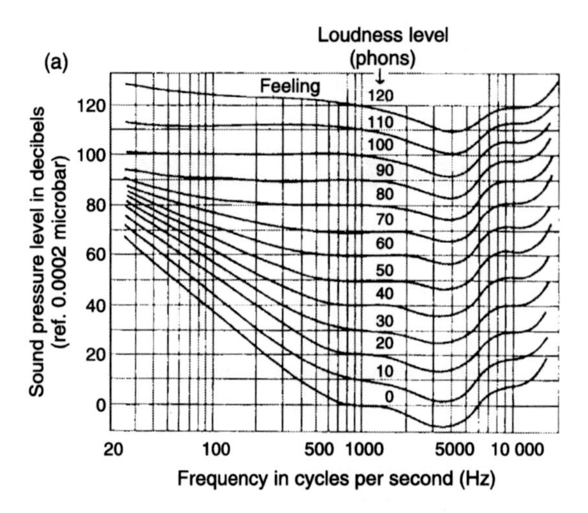

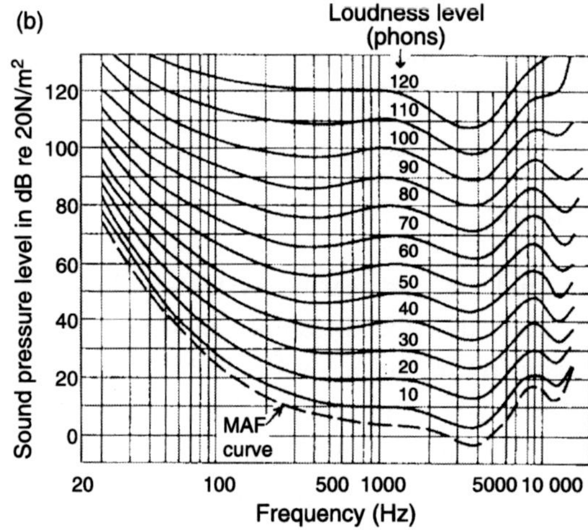

Figure 2.1:

(a, b) Equal loudness contours: (a) The classic Fletcher and Munson contours of equal loudness for pure tones, clearly showing higher relative levels being required at high and low frequencies for equal loudness as the SPL falls. In other words, at 110 dB SPL, 100 Hz, 1 kHz and 10 kHz would all be perceived as roughly equal in loudness. At 60 dB SPL, however, the 60 phon contour shows that 10 kHz and 100 Hz would require a 10 dB boost in order to be perceived as equally loud to the 1 kHz tone; (b) The Robinson–Dadson equal loudness contours. These plots were intended to supersede the Fletcher–Munson contours, but, as can be seen, the differences are too small to change the general concept. Indeed, other sets of contours have subsequently been published as further updates, but for general acoustical purposes, as opposed to critical uses in digital data compression and noise shaping, the contours of (a) and (b) both suffice. The MAF (minimum audible field) curve replaces the '0 phons' curve of the older, Fletcher–Munson contours. The MAF curve is not absolute, but is statistically derived from many tests. The absolute threshold of hearing varies not only from person to person, but with other factors such as whether listening monaurally or binaurally, whether in free-field conditions or in a reverberant space, and the relative direction of the source from the listener. It is therefore difficult to fix an absolutely defined 0 dB curve.

It is the concept of the doubling of loudness for every 10 dB increase in sound pressure level that fits so well with our logarithmic hearing. A street with light traffic in a small town will tend to produce a sound pressure level (SPL) of around 60 dBA, whereas a loud rock band may produce around 120 dBA (dB and dBA will be discussed later in the chapter). The sound pressure difference between 60 and 120 dBA is one thousand times, but it is self-evident that a loud rock band is not one thousand times louder than light traffic. If we use the 10 dB concept then 70 dBA will be twice as loud as 60 dBA, 80 dBA four times as loud (2×2), 90dBA eight times ($2 \times 2 \times 2$), 100 dBA 16 times ($2 \times 2 \times 2 \times 2$), 110 dBA 32 times ($2 \times 2 \times 2 \times 2 \times 2$), and 120 dBA 64 times as loud ($2 \times 2 \times 2 \times 2 \times 2 \times 2$). The concept of a loud rock band being 64 times as loud as light traffic is more intuitively reasonable, and in fact it is a good approximation.

The concept of 1 dB being the smallest perceivable level change only holds true for pure tones. For complex signals in mid frequency bands it has been shown that much smaller level changes can be noticeable. Indeed, Dr Roger Lagadec, the former head of digital development at Studer International, in Switzerland, detected in the early 1980s audible colouration caused by amplitude response ripples in a digital filter at levels only just above ± 0.001 dB. However, whether he was detecting the level changes, *per se*, or an artefact of the periodicity of the ripples, may still be open to question.

Perhaps it is therefore important to note at this early stage of the chapter that many so-called facts of hearing are often wrongly applied. Tests done on pure tones or speech frequently do not represent what occurs with musical sounds. Traditionally it has been the medical and communications industries that have funded much of the research into hearing. The fact that it is a different part of the brain which deals with musical perception to that which deals with speech and pure tones is often not realised. One should be very careful when attempting to apply known 'facts' about hearing to the subject of musical perception. They can often be very misleading.

2.2 Sound Itself

Sound is the human perception of vibrations in the region between 20 Hz and 20 kHz. 'Hz' is the abbreviation for hertz, the internationally accepted unit denoting cycles per second, or whole vibrations per second. The abbreviation 'cps' for cycles per second is still to be found in some older publications. (In some very old French texts, half-vibrations [zero-crossings] per second were used, with a consequent doubling of the frequency figure.[1]) Figure 2.2 shows a graphic representation of a cycle of a sine wave. It can be seen that the pressure cyclically moves from compression to rarefaction and back to compression. The number of times which each whole cycle occurs in a second is known as the frequency. Hence, a frequency of 200 Hz denotes that 200 cycles occur in any given second.

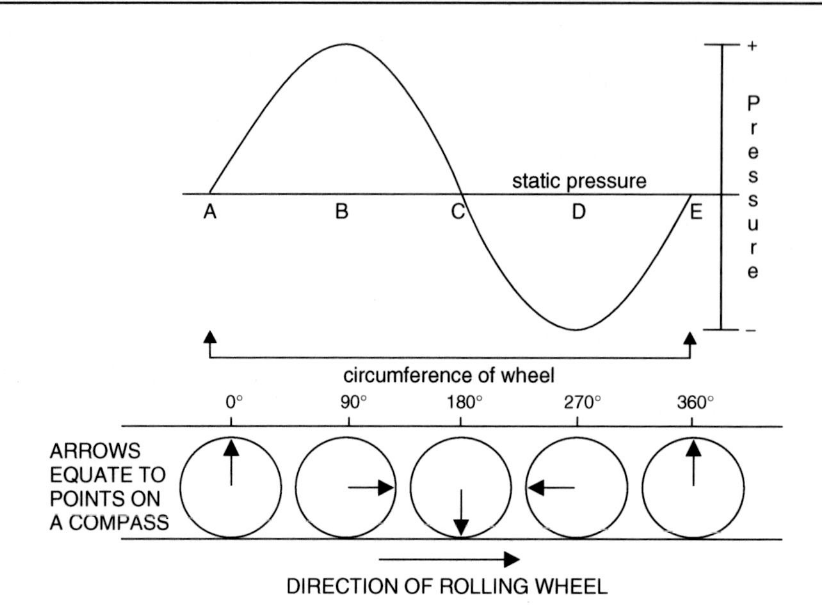

Figure 2.2:
Sine wave; amplitude and phase. If the circumference of the wheel is equal to the period of the sine wave (A to E), then as the wheel rolls, a line drawn radially on the wheel will indicate the phase angle of the associated sine wave. This is why phase is sometimes denoted in radians — one radian being the phase angle passed through as the wheel advances by its own radial length on its circumference. Therefore, $360^\circ = 2\pi$ radians (i.e. circumference $= 2\pi \times$ radius). 1 radian $= 360^\circ/2\pi = 57.3^\circ$ approximately.

The compression and rarefaction half cycles represent the alternating progression of the pressure from static pressure to its peak pressure, the return through static pressure and on to peak rarefaction, and finally back to static pressure. The whole cycle of a sine wave can be shown by an arrow placed on the perimeter of a rolling wheel whose circumference is equal to one wavelength. One complete revolution of the wheel could show one whole cycle of a sine wave, hence any point on the sine wave can be related to its positive or negative pressure in terms of the degrees of rotation of the wheel which would be necessary to reach that point. One cycle can therefore be described as having a 360° phase rotation. The direction of the arrow at any given instant would show the phase angle, and the displacement of the arrowhead to the right or to the left of the central axle of the wheel would relate to the positive or negative pressures, respectively. The pressure variation as a function of time is proportional to the sine of the angle of the rotation of the wheel, producing a sinusoidal pressure variation, or sine wave.

Another way of describing 360° of rotation is 2π (pi) radians, a radian being the length on the circumference of the wheel which is equal to its radius. In some cases, this concept is more convenient than the use of degrees. The concept of a sine wave having this phase component leads to another definition of frequency: the frequency can be described as the rate of change of phase with time.

An acoustic sine wave thus has three components, its pressure amplitude, its phase and time. They are mathematically interrelated by the Fourier transform. Fourier discovered that *all* sounds can be represented by sine waves in different relationships of frequency, amplitude and phase, a remarkable feat for a person born in the eighteenth century.

The speed of sound *in air* is constant with frequency (although not in most other substances), and is dependent only on the square root of the absolute temperature. At 20°C it is around 344 m per second, sometimes also written 344 m/s or 344 m^{s-1}. The speed of sound in air is perhaps the most important aspect of room acoustics, because it dominates so many aspects of acoustic design. It dictates that each individual frequency will have its own particular wavelength. The wavelength is the distance travelled by a sound at any given frequency as it completes a full cycle of compression and rarefaction. It can be calculated by the simple formula:

$$\lambda = \frac{c}{f} \tag{2.1}$$

where

λ = wavelength in metres
f = frequency
c = speed of sound in air (344 metres per second at 20°C)

For example, the wavelength of a frequency of 688 Hz is:

$$\lambda = \frac{344}{688} = 0.5\text{ m} = 50\text{ cm}$$

The wavelength of 100 Hz is:

$$\lambda = \frac{344}{100} = \frac{3.44}{1} = 3.44\text{ metres}$$

To give an idea of the wavelength at the extremes of the audio frequency band, let us calculate the wavelengths for 20 Hz and 20 kHz:

20 Hz $\qquad \lambda = \frac{344}{20} = \underline{\underline{17.2\text{ metres}}}$

20 kHz $\qquad \lambda = \frac{344}{20{,}000} = \underline{\underline{0.0172\text{ metres}}}$
or 1.72 cm

Note the enormous difference in the wavelengths in the typical audio frequency range, from 17.2 *metres* down to 1.72 *centimetres*; a ratio of 1000:1.

There is also another unit used to describe a single frequency acoustic wave, which is its wave number. The wave number describes, in radians per metre, the cyclic nature of a sound in any given medium. It is thus the spacial equivalent of what frequency is in the time domain. It is rarely used by non-academic acousticians, and is of little general use in the language of the studio world. Nevertheless, it exists in many text books, and so it is worth a mention, and helps once again to reinforce the concept of the relationship between sound and the world in which we live.

So what has all of this got to do with recording studios? A lot. An awful lot!

Sound is a wave motion in the air, and all wave motions, whether on the sea, vibrations through the earth, radio waves, light waves, or whatever else, follow many of the same universal physical laws. The spectrum of visible light covers about one octave, that is, the highest frequency involved is about double the lowest frequency. Sound, however, covers 20 – 40 Hz, 40 – 80 Hz, 80 – 160 Hz, 160 – 320 Hz, 320 – 640 Hz, 640 – 1280 Hz, 1280 – 2560 Hz, 2560 – 5180 Hz, 5180 – 10,360 Hz, 10,360 – 20,720 Hz... 10 *octaves*, which one could even stretch to 12 octaves if one considers that the first ultrasonic and infrasonic octaves can contribute to a musical experience. That would mean a wavelength range from 34.4 *metres* down to 8.6 *millimetres*, and a frequency range not of 2 to 1, like light, but more in the order of 4000 to 1.

The ramifications of this difference will become more obvious as we progress in our discussions of sound isolation and acoustic control, but it does lead to much confusion in the minds of people who do not realise the facts. For example, as a general rule, effective sound absorption will take place when the depth of an absorbent material approaches one quarter of a wavelength. Many people know that 10 cm of mineral wool (Rockwool, Paroc, etc.), covering a wall, will absorb sound. Speech frequencies, around 1000 Hz, with a wavelength of around 34 cm will be maximally absorbed, because the quarter wavelength would be 8.5 cm, so the 10 cm of absorbent would adequately exceed the quarter wavelength criterion. Indeed, all higher frequencies, with shorter wavelengths, would also be maximally absorbed. However, 100 Hz, with a quarter wavelength of 85 cm (a wavelength of around 3.4 m), would tend to be almost totally unaffected by such a small depth of material. Nearly a one metre thickness would be needed to have the same effect at 100 Hz as a 10 cm thickness at 1 kHz.

Figure 2.3 shows a medium sized anechoic chamber. The fibrous wedges are of one metre length, and they are spaced off from the walls in such a way that augments their low frequency absorption. The frequency having a four metre wavelength (one metre quarter wavelength) is 86 Hz. With the augmentation effect of the spacing, the room is actually anechoic to around 70 Hz. This means that all frequencies above 70 Hz will be at least 99.9% absorbed by the boundaries of the room.

Contrast this to a typical voice studio with 10 cm of 'egg-box' type, foam wall coverings. For all practical purposes, this will be very absorbent above 800 Hz, or so, and reasonably absorbent at 500 Hz, or even a little below, but it will do almost nothing to control the fundamental frequencies of a bass guitar in the 40 – 80 Hz range. This is why a little

Figure 2.3:
The ISVR anechoic chamber. The large anechoic chamber in the Institute of Sound and Vibration Research at Southampton University, UK, has a volume of 611 m^3. The chamber is anechoic down to around 70 Hz. Below this frequency, the wedges represent less than a quarter of a wavelength, and the absorption falls off with decreasing frequency. The floor grids are completely removable, but are strong enough to support the weight of motor vehicles.

knowledge can be a dangerous thing. A person believing that thin foam wall panels are excellent absorbers, but not realising their frequency limitation, may use such treatments in rooms which are also used for the recording or reproduction of deep-toned instruments. The result is usually that the absorbent panels only serve to rob the instruments of their upper harmonics, and hence take all the life from the sound. The totally unnatural and woolly sound which results may be in many cases subjectively worse than the sound when heard in an untreated room. This type of misapplication is one reason why there are so many bad small studios in existence. The misunderstanding of many acoustic principles is disastrously widespread. (Section 5.11 deals with voice rooms in more detail.)

So, is mineral wool absorbent? The answer depends on what frequency range we are considering, where it is placed and what thickness is being used. The answer can vary from 100% to almost zero. Such is the nature of *many* acoustical questions.

2.3 The Decibel: Sound Power, Sound Pressure and Sound Intensity

The decibel is, these days, one of the most widely popularised technical terms. Strictly speaking, it expresses a power ratio, but it has many other applications. The dB SPL is a defined sound-pressure-level reference. Zero dB SPL is defined as a pressure of 20 micro pascals (one pascal being one newton per square metre) but that definition hardly affects daily life in recording studios. There are many text books which deal with the mathematics of decibels, but we will try to deal here with the more mundane aspects of its use. Zero (0) dB SPL is the generally accepted level of the quietest perceivable sound to the average, young human being.

Being a logarithmic unit, every 3 dB increase or decrease represents a doubling or halving of power. A 3 dB increase in power above 1 W would be 2 W, 3 dB more would yield 4 W, and a further 3 dB increase (9 dB above 1 W) would be 8 W. In fact, a 10 dB power increase rounds out at a 10 *times* increase in power, and this is a useful figure to remember.

Sound *pressure* doubles with every 6 dB increase, so one would need to quadruple the sound power from 1 W to 4 W (3 dB + 3 dB) to double the sound *pressure* from an acoustic source. Our hearing perception tends to correspond to changes in sound *pressure* level, and it was stated earlier that a roughly 10 dB increase or decrease was needed in order to double or halve loudness. What therefore becomes apparent is that if a 10 dB increase causes a doubling of loudness, and that same 10 dB increase requires a ten times power increase (as explained in the last paragraph), then it is necessary to increase the power from a source by 10 times in order to double its loudness.

This is what gives rise to the enormous quantities of power amplifiers used in many public address (PA) systems for rock bands, because, in any given loudspeaker system, 100 W is only twice as loud as 10 W. One thousand watts is only twice as loud as 100 W, and 10000 W is only twice as loud as 1000 W. Therefore, 10,000 W is only eight times louder than 10 W, and, in fact, only 16 times louder than *1 W.* This is totally consistent with what was said at the beginning of the chapter, that a rock band producing 120 dB SPL was only about 64 times louder than light traffic, despite producing a thousand times more sound pressure. Yes, one million watts is only 64 times louder than 1 W – one million times the power produces one thousand times the pressure, which is 64 times louder at mid frequencies. The relationships need to be well understood.

Sound intensity is another measure of sound, but this time it deals with the *flow* of sound energy. It can also be scaled in decibels, but the units relate to watts per square metre. Sound waves in free space expand spherically. The formula for the surface area of a sphere is $4\pi \times r^2$ or $4\pi r^2$. (Free air is often referred to as 4π space, relating to 4π steradians [solid radians].) The surface area of a sphere of 1m radius (r) would thus be 4×3.142 ($\pi = 3.142$) $\times 1^2$ or about 12.5 m^2. The surface area of a 2m radius (r) sphere would therefore be $4\pi \times 2^2$ or $4 \times 3.142 \times 4$, which is about 50 m^2. So, every time the radius doubles, the surface area increases by four times.

Now, if the sound is made to distribute the same power over an area four times greater, then its *intensity* (in watts per square metre) must reduce by a factor of four. For the same total power in the system, each square metre will only have one quarter of the power over the surface of the 2 m radius sphere than it had on the surface of the 1 m radius sphere. One quarter of the *power* represents a reduction of 6 dB (3 dB for each halving), so the sound *pressure* at any point on the surface of the sphere will reduce by a half every time that one moves double the distance from the source of sound in free air. This is the basis of the often referred to 'double-distance rule' which states that as one doubles the distance from a sound source in free air, the sound pressure level will halve (i.e. fall by 6 dB).

However, an important point to bear in mind about the double-distance rule is that it only applies in spaces where the sound wave is free to expand without restriction as it propagates. This is a point which is often not fully appreciated. At the other extreme, if we take the case of a speaking tube, as used for communication in old ships, for example, there is no expansion of the wave as a person speaks in one end and the wave propagates to the other end. What is more, if the tube has rigid walls and a smooth internal surface, the absorption losses may be very small indeed. The French scientist, Biot, in the nineteenth century, found that he could hold conversations, even in a low voice, through the water pipes of Paris over distances of 1 km or more.[1] The free-field and the smooth bore tube are two extremes of wave propagation, and in the latter case the double-distance rule most certainly does not apply. As wave expansion is restricted, the reduction in level with distance diminishes. How the fall-off with distance behaves for sounds being generated by loudspeakers in rooms is a complex issue, and will probably be frequency dependent. It can depend on the directivity of the source, whether it is a simple source or an array, and the nature of the room acoustics.

In terms of annotation, sound pressure level is represented by 'SPL', and sound power level is usually written 'SWL'. The sound intensity is usually annotated 'I'. From a point source in a free-field, the intensity is the sound power divided by the area, or, conversely:

$$\text{Sound power} = \text{sound intensity} \times \text{area (or SWL} = \text{IA)}$$

The sound intensity therefore falls with the sound pressure level as one moves away from a source in a free-field. In a lossless system, they would both reduce with distance, even though the total sound power remained constant, merely spreading themselves thinner as the surface area over which they were distributed increased.

2.3.1 The dBA and dBC Scales

Reference has already been made in earlier sections to dB SPL and dBA. Essentially, a sound level meter consists of a measuring microphone coupled to an amplifier which drives a meter calibrated in dB SPL, referenced to the standard 0 dB SPL. The overall response is flat. Such meters are used for making absolute reference measurements. However, as we saw from Figure 2.1, the response of the human hearing system is most definitely not flat. Especially at lower levels, the ear is markedly less sensitive to low and high frequencies than it is to mid frequencies. It can be seen from Figure 2.1 that 60 dB SPL at 3 kHz will be very audible, in fact it will be over 60 dB above the threshold of hearing at that frequency, yet a tone of 30 Hz would be inaudible at 60 dB SPL; it would lie on the graph below the 0 phon curve of 'just audible'.

Consequently, if a flat reading of 60 dB was taken on an SPL meter measuring a broad-band noise signal in a room, then 25 dB of isolation would be needed at 3 kHz if the neighbours were not to be subjected to more than 35 dBA. However, at 30 Hz, *nothing* would need to be done, because the sound at that frequency would not even be audible *in* the room, let alone outside of it. Hence, providing 25 dB of isolation at 30 Hz to reduce the outside level to less than 35 dB SPL unweighted (i.e. a flat frequency response) would simply be a waste of time, money, effort and space.

To try to make more sense from the noise measurements, an 'A-weighting' filter is provided in the amplifier circuits of most SPL meters, which can be switched in when needed. The 'A' filter is approximately the inverse of the 40 phon curve from the equal loudness contours of Figure 2.1, and is shown in Figure 2.4. The effect of the A-weighting is to render the measurement at any frequency (at relatively low SPLs), subjectively equal. That is to say, a measurement of 40 dBA at 40 Hz would sound subjectively similar in level as a measurement of 40 dBA at 1 kHz, or 4 kHz, for example. This greatly helps with subjective noise analysis, but A-weighting should never be used for absolute response measurements such as the flatness of a loudspeaker. For such purposes, only unweighted measurements should be used.

Also shown in Figure 2.4 is the 'C-weighting' curve. This roughly equates to the inverse of the 80 phon curve of Figure 2.1. C-weighting is more appropriate than A-weighting when assessing the subjectivity of higher level noises, but as it is the nuisance effect of noise that is usually of most concern, the C-weighting scale is less widely used. There is also a 'B-weighting' curve, but it is now largely considered to be redundant, especially as no current legislation makes use of it.

One must be very careful not to mix the measurements because there is no simple way of cross-referencing them. In the opening paragraphs of this chapter reference was made to a loud rock band producing 120 dBA. Strictly speaking, the use of the dBA scale in the region of 120 dB SPL is a nonsense, because the A curve does not even vaguely represent the inverse of the response of the ears at the 120 phon loudness level, shown in Figure 2.1, which is almost flat. The 120 phon level actually equates better to an unweighted response, but, as has just been mentioned, one cannot mix the use of the scales. It would have been more accurate to say that a very loud rock band at 120 dBA would, *in the mid frequency range*, be 64 times louder than light traffic in the street of a small town. The low frequencies may, in fact, sound very much louder. This fact affects the thinking when calculating any required sound isolation, which can also be rated in dBA.

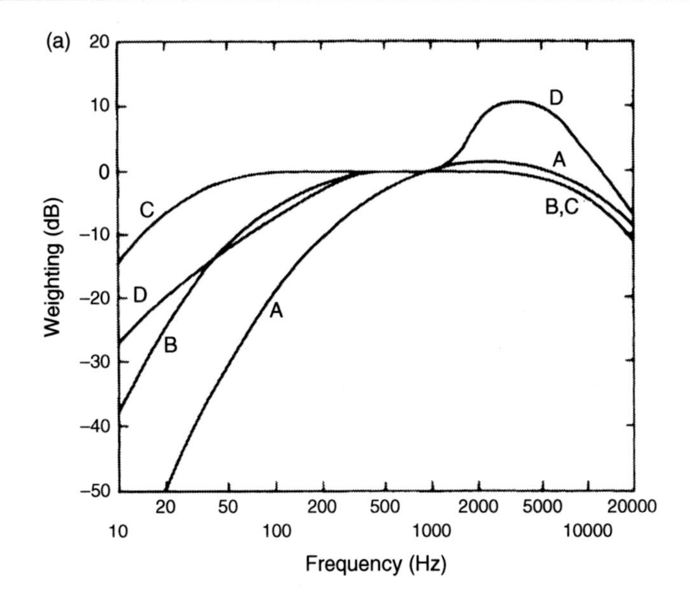

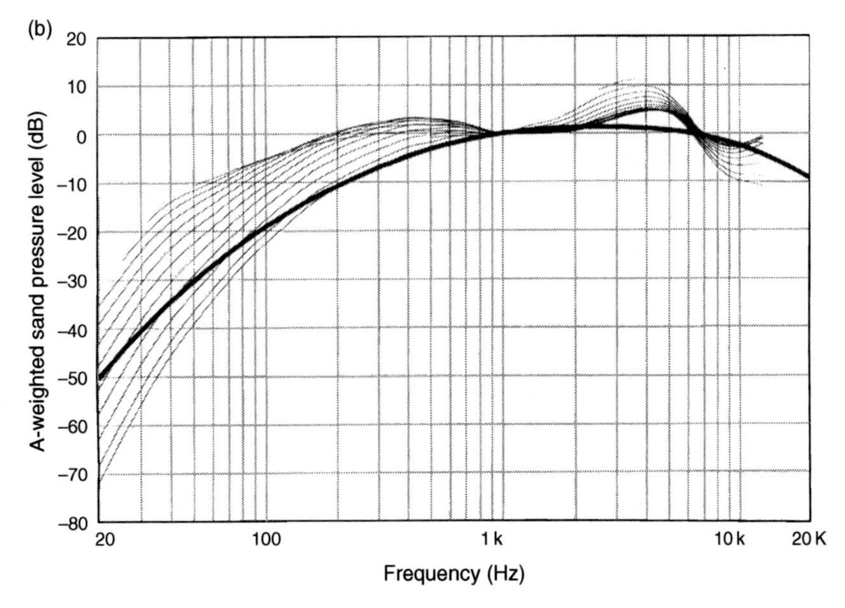

Figure 2.4:
(a) A-, B-, C- and D-weighting curves for sound level meters; (b) Inverse equal loudness level contours for a diffuse field for the range from the hearing threshold to 120 phon (thin lines), and A-weighting (thick line). The above figure was taken from *On the Use of A-weighted Level for Prediction of Loudness Level*, by Henrik Müller and Morten Lydolf, of Aalborg University, Denmark. The paper was presented to the 8th International Meeting on Low Frequency Noise and Vibration, in Gothenburg, Sweden, in June 1997. The work is published in the *Proceedings*, by Multi-Science Publishing, UK.

There are also measurements known as Noise Criteria (NC) and Noise Ratings (NR) which are close to the dBA scale. They allow for reduced isolation as the frequency lowers, in order to be more realistic in most cases of sound isolation, although they are perhaps more suited to industrial noise control than to recording studio and general music use. Where very high levels of low frequencies are present, such as in recording studios, 60 dB of isolation at 40 Hz may well be needed, along with 80 dB of isolation at 1 kHz, in order to render both frequencies below the nuisance threshold in adjacent buildings. An NR, NC or A-weighted specification for sound isolation would suggest that only 40 dB of isolation were necessary at 40 Hz, not 60 dB, so it would underestimate by 20 dB the realistic isolation needed. Not many industrial processes produce the low frequency SPLs of a rock band.

If all of this sounds a little confusing, it is because it is. The non-flat frequency response of the ear and its non-uniform dynamic response (i.e. the frequency response changes with level) have blighted all attempts to develop a simple, easily understood system of subjective/objective noise level analysis. Although the A-weighting system is very flawed, it nevertheless has proved to be valuable beyond what could ever have been academically expected of it. However, all of these measurements require interpretation, and that is why noise control is a very important, independent branch of the acoustical sciences. Failing to understand all the implications of the A, C and unweighted scales has led to many expensive errors in studio construction, so if the situation is serious, it is better to call in the experts.

For example, an 'acoustic' door may have an isolation rating of 35 dBA. This may be an NC, NR or dBA rating. Many people have made the mistake of believing that if there is 90 dB SPL of music in a room, then fitting such a door would reduce the level beyond it to no more than 55 dB SPL, which they may deem to be acceptable. However, after paying a considerable amount of money for the door and its fitting, they may find that there is still much more than the expected 55 dB on the other side of the door. This is because the nominal 35 dB of the door follows a roughly dBA curve (see Glossary if necessary) and so at 60 Hz it may only provide an isolation of 15 dB or so.

Because of the complexity of the frequency and level relationships shown in Figure 2.1, if the music in the room was only at 70 dB SPL to begin with, the low frequency isolation of the door may reduce the bass to acceptable limits, due to the fact that the levels would be down to the threshold of inaudibility. This can be seen on the equal loudness contours, where they squeeze together at low frequencies. On the other hand, if the music was at 100 dB SPL, the provision of only 15 dB of isolation at 50 Hz may be totally inadequate. The subject of doors is discussed further in Section 8.8, but it should be mentioned here that the manufacture of doors with an isolation of 35 dB at all frequencies would not be a very practical proposition. The laws of physics would not permit it to be so. At low frequencies, therefore, the isolation provided by 'acoustic' doors cannot be expected to be as high as the nominal rating.

2.4 Human Hearing

The opening section of this chapter, although making many generalisations, hinted at a degree of variability in human perception. By contrast, the second section dealt with what is clearly a hard science; that of sound and its propagation. Acoustic waves behave according to some very set principles, and despite the interaction at times being fiendishly complex, acoustics is nevertheless a clearly definable science. Likewise in the third section, the treatment of the decibel is clearly mathematical, and it is only in its application to human hearing and perception that it begins to get a little ragged. It is, however, the application that gets ragged, not the concept of the decibel itself. Eventually, though, we must understand a little more about the perception of sound by individual human beings, because they are the final arbiters of whatever value exists in the end result of all work in recording studios, and it is here where the somewhat worrying degree of variability enters the proceedings in earnest.

The world of sound recording studios is broad and deep. It straddles the divide between the hardest objective science and the most ephemeral art, so what the rest of this chapter will concentrate on is the degree of hearing variability which may lead many people to make their own judgements. These judgements relate both to the artistic and scientific aspects of the recording studios themselves, and the recordings which result from their use.

The human auditory system is truly remarkable. In addition to the characteristics described in earlier sections, it is equipped with protection systems which limit ear drum (tympanic membrane) movement below 100 Hz. These are to prevent damage, yet still allow the brain to interpret, via the help of body vibration conduction, what is really there. The ear detects pressure half-cycles, which the brain reconstructs into full cycles, and can provide up to 15 dB of compression at high levels. By means of time, phase and amplitude differences between the arrival of sound at the two ears, we can detect direction with great accuracy, in some cases to as little as a few degrees. Human pinnae (outer ears) have evolved with complex resonant cavities and reflectors to enhance this ability, and to augment the level-sensitivity of the system.

Our pinnae are actually as individual to each one of us as are our fingerprints, which is the first step in ensuring that our overall audio perception systems are also individual to each one of us. When we add all of this together, and consider the vagaries introduced by what was discussed in Section 2.3.1, we should be able to easily realise how the study of acoustics, and audio in general, can often be seen as a black art. Hearing perception is a world where everything seems to be on an individually sliding scale. Of course, that is also one reason why it can be so fascinating, but until it is sufficiently understood, the frustration can often greatly outweigh the fascination.

Psychoacoustics is the branch of the science which deals with the human perception of sound, and it has made great leaps forward in the past 25 years, not least because of the huge amount of money pumped into research by the computer world in their search for 'virtual' everything. Psychoacoustics is an enormous subject, and the bibliography at the end of this chapter suggests further reading for those who may be interested in studying it in greater depth. Nevertheless, it may well be useful and informative to look at some concepts of our hearing and its individuality with respect to each one of us, because it is via our own hearing that we each perceive and judge musical works. From the differences which we are about to explore, there can be little surprise if there are some disagreements about standards for recording studio design.

What follows was from an article in *Studio Sound* magazine, in April 2000.

2.4.1 Chacun a Son Oreille

In 1896, Lord Rayleigh wrote in his book, *The Theory of Sound* (Strutt, 1896, p. 1) 'The sensation of a sound is a thing *sui generis*, not comparable with any of our other senses … Directly or indirectly, all questions connected with this subject must come for decision to the ear, … and from it there can be no appeal'. Sir James Jeans concluded his book *Science & Music* (Cambridge University Press, 1937) with the sentence 'Students of evolution in the animal world tell us that the ear was the last of the sense organs to arrive; it is beyond question the most intricate and the most wonderful'.

The displacement of the ear drum when listening to the quietest perceivable sounds is around one-hundredth of the diameter of a hydrogen molecule, and even a tone of 1 kHz at a level of 70 dB still displaces the ear drum by less than one-millionth of an inch (0.025 micrometres). Add to all these points the fact that our pinnae (outer ears) are individual to each one of us, and one has a recipe for great variability in human auditory perception as a whole, because the ability to perceive these minute differences is so great.

In fact, if the ear was only about 10 dB more sensitive than it is, we would hear a permanent hiss of random noise, due to detection of the Brownian motion of the air molecules. Some people can detect pitch changes of as little as 1/25th of a semitone (as reported by *Seashore*). Clearly there is much variability in all of this from individual to individual, and one test carried out on 16 professionals at the Royal Opera, Vienna, showed a 10:1 variability in pitch sensitivity from the most sensitive to the least sensitive. What is more, ears all produce their own non-linear distortions, both in the form of harmonic and inter-modulation distortions. I had one good friend who liked music when played quietly, but at around 85 dB SPL she would put her hands to her ears and beg for it to be turned down. It appeared that above a certain level, her auditory system clipped, and at that point all hell broke loose inside her head.

Of course, we still cannot enter each other's brains, so the argument about whether we all perceive the colour blue in the same way cannot be answered. Similarly we cannot know that we perceive what other people hear when listening to similar sounds under similar circumstances. We are, however, now capable of taking very accurate in-ear measurements, and the suggestion from the findings is that what arrives at the eardrums of different people is clearly not the same, whereas what enters different people's eyes to all intents and purposes is the same. Of course, some people may be colour-blind, whilst others may be long-sighted, short-sighted, or may have one or a combination of numerous other sight anomalies. Nevertheless, what arrives at the eye, as the sensory organ of sight, is largely the same for all of us. However, if we take the tympanic membrane (the ear drum) as being the 'front-end' of our auditory system, no such commonality exists. Indeed, even if we extend our concept of the front-end to some arbitrary point at, or in front of, our pinnae, things would still not be the same from person to person because we all have different shapes and sizes of heads and hair styles. This inevitably means that the entrances to our ear canals are separated by different distances, and have different shapes and textures of objects between them. Given the additional diffraction and reflexion effects from our torsos, the answer to the question of whether we all 'see the same bluc' in the auditory domain seems to be clearly 'no', because even what reaches our ear drums is individual to each of us, let alone how our brains perceive the sounds.

There is abundant evidence to suggest that many aspects of our hearing are common to almost all of us, and this implies that there is a certain amount of 'hard-wiring' in our brains which predisposes us towards perceiving certain sensations from certain stimuli. Nevertheless, this does not preclude the possibility that some aspects of our auditory perception may be inherited, and that there may be a degree of variability in these genetically influenced features. Aside from physical damage to our hearing system, there may also be cultural or environmental aspects of our lives which give rise to some of us developing different levels of acuteness in specific aspects of our hearing, or that some aspects may be learned from repeated exposure to certain stimuli.

It would really appear to be stretching our ideas of the evolutionary process beyond reason, though, to presume that the gene pairings which code our pinnae could somehow be linked to the gene pairings which code any variables in our auditory perception systems. Furthermore, it would seem *totally* unreasonable to expect that if any such links did exist, that they could function in such a way that one process could complement the other such that all our overall perceptions of sound were equal. In fact, back in the 1970s, experiments were carried out (which will be discussed in later paragraphs) which more or less conclusively prove that this could not be the case. Almost without doubt, we do not all hear sounds in the same way, and hence there will almost certainly always be a degree of subjectivity in the judgement and choice of studio monitoring systems. There will be an even greater degree of variability in our choice of domestic hi-fi loudspeakers, which tend to be used in much less acoustically controlled surroundings.

Much has been written about the use of dummy head recording techniques for binaural stereo, and it has also been frequently stated that most people tend to perceive the recordings as sounding more natural when they are made via mouldings of their own pinnae. In fact, many people seem to be of the opinion that we all hear the recordings *at their best* if our own pinnae mouldings are used, but this is not necessarily the case. It is true that the perception may be deemed more *natural* by reference to what we hear from day-to-day, but it is also true that some people naturally hear certain aspects of sound more clearly than others. In my first book,[2] I related a story about being called to a studio by its owner to explain why hi-hats tended to travel in an arc when panned between the loudspeakers; seeming to come from a point somewhere above the control room window when centrally panned. The owner had just begun to use a rather reflexion-free control room, in which the recordings were rendered somewhat bare. On visiting the studio, all that I heard was a left-to-right, horizontal pan. We simply had different pinnae.

In a well-known AES paper, and in her Ph.D. thesis, the late Puddie Rodgers[3,4] described in detail how early reflexions from mixing consoles, or other equipment, could cause response dips which mimicked those created by the internal reflexions from the folds and cavities of different pinnae when receiving cues from different directions. In other words, a very early reflexion from the surface of a mixing console could cause comb filtering of a loudspeaker response which could closely resemble the in-ear reflexions which may cause a listener to believe that the sound was coming from a direction other than that from which it was actually arriving. Median plane vertical perception is very variable from individual to individual, and in the case mentioned in the last paragraph, there was almost certainly some source of reflexion which gave the studio owner a sensation of the high frequencies arriving from a vertically higher source, whilst I was left with no such sensation. The differences were doubtless due to the different shapes of our pinnae, and the source of reflexions was probably the mixing console.

In the 'letters' page of the October 1994 *Studio Sound* magazine, recording engineer Tony Batchelor very courageously admitted that he believed that he had difficulty in perceiving what other people said stereo should be like. He went on to add, though, that at a demonstration of Ambisonics he received 'a unique listening experience'. Clearly to Tony Batchelor, stereo imaging would not be at the top of his list of priorities for his home hi-fi system, yet he may be very sensitive to intermodulation distortions, or frequency imbalances, to a degree that would cause no concern to many other people.

Belendiuk and Butler[5] concluded from their experiments with 45 subjects that 'there exists a pattern of spectral cues for median sagittal plane positioned sounds common to all listeners'. In order to prove this hypothesis, they conducted an experiment in which sounds were emitted from different, numbered, loudspeakers, and the listeners were asked to say from which loudspeaker the sound was emanating. They then made binaural recordings via moulds of the actual outer ears of four listeners, and asked them to repeat the test, via headphones, of the recordings made using their own pinnae. The headphone results were very similar to the direct results, suggesting that the recordings were representative of 'live' listening. Not all the subjects were equally accurate in their choices, with some, in both their live and recorded tests, scoring better than others in terms of identifying the correct source position.

Very interestingly, when the tests were repeated with each subject listening via the pinnae recordings of the three other subjects in turn, the experimenters noted, 'that some pinnae, in their role of transforming the spectra of the sound field, provide more adequate (positional) cues ... than do others'. Some people, who scored low in both the live and recorded tests using their own pinnae, could locate more accurately via other peoples' pinnae. Conversely, via some pinnae, none of the subjects could locate very accurately. The above experiments were carried out in the vertical plane. Morimoto and Ando,[6] on the other hand, found that, in the horizontal plane, subjects generally made fewer errors using their own head-related transfer functions (HRTFs), i.e. via recordings using simulations of their own pinnae and bodies. What all the relevant reports seem to have shown, over the years, is that different pinnae are differently perceived, and the whole HRTF is quite distinct from one person to another. All of these differences relate to the different perceptions of different sound fields.

Studio Monitoring Design (Newell, 1995)[2] also related the true story of two well-respected recording engineers who could not agree on the 'correct' amount of high frequencies from a monitor loudspeaker system which gave the most accurate reproduction when compared to a live cello. They disagreed by a full 3 dB at 6 kHz, but this disagreement was clearly not related to their own absolute high frequency sensitivities because they were comparing the sound of the monitors to a live source. The only apparent explanation to this is that because the live instrument and the loudspeakers produced different sound fields, the perception of the sound field was different for each listener. Clearly, all the high frequencies from the loudspeaker came from one very small source, the tweeter, whilst the high frequency distribution from the instrument was from many points – the strings and various parts of the body. The 'highs' from the cello radiated, therefore, from a distributed source having a much greater area than the tweeter. Of course, the microphone could add its own frequency tailoring and one-dimensionality, but there would seem to be no reason why the perception of this should differ from one listener to another.

So, given the previous discussion about pinnae transformations and the different HRTFs as they relate to sounds arriving from different directions, it does not seem too surprising that sound sources with spacially different origins may result in spectrally different perceptions for different people. Tony Batchelor's statements about his not being able to appreciate stereo, yet readily perceiving Ambisonic presentations of spacial effects, would seem to be strongly related to aspects of the sound field. He and I would no doubt attach different degrees of importance to the horizontal effects of stereo if working on a joint production. Martin Young (the aforementioned studio owner) and I clearly had different vertical perception when panning a hi-hat; and the two well-known engineers could not agree on a natural high-frequency level in a live versus loudspeaker test.

The implications of all this would suggest that unless we can reproduce an accurate sound field, we will never have universal agreement on the question of 'the most accurate' monitoring systems. Add to this a good degree of personal preference for different concepts of what constitutes a good sound, and it would appear that some degree of monitoring and control room variability will be with us for the foreseeable future. Nevertheless, the last 20 years have seen some very great strides forward in the understanding of our auditory perception systems, and this has been a great spur to the advancement of loudspeaker and control room designs. Probably, though, we have still barely seen the tip of the iceberg, so it will be interesting to see what the future can reveal.

To close, let us look at some data that has been with us for 40 years or more. The plots shown in Figures 2.5 and 2.6 were taken from work by Shaw.[7] Figure 2.5 shows the average, in-ear-canal, 0° and 90° responses for ten people. Note how the ear canal receives a very different spectrum depending upon the direction of arrival of the sound. Figure 2.6 shows the ten individual sources from which Figure 2.5 was derived. The differences from person to person are significant, and the response from one direction cannot be inferred from the response from a different direction.

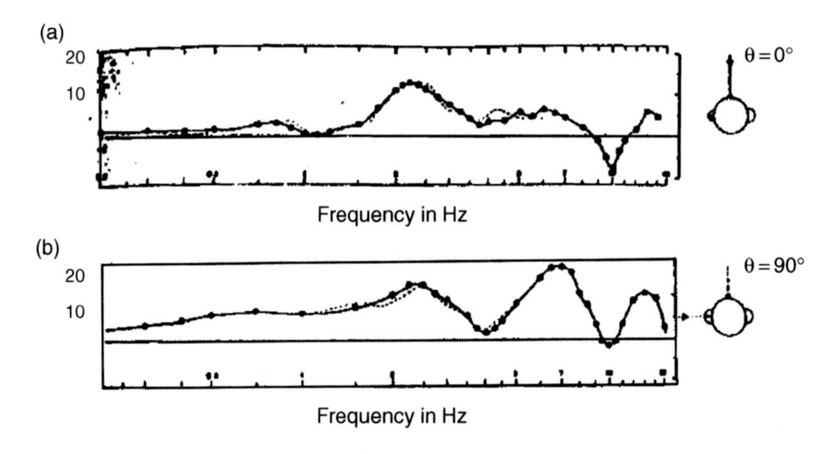

Figure 2.5:
Shaw's data showing the ratio (in dB) of the sound pressure at the ear canal entrance to the free-field sound pressure. The curves are the average of ten responses. (a) Shows the average response for a sound source at 0° azimuth and (b) shows the average response for a sound source at 90° azimuth.

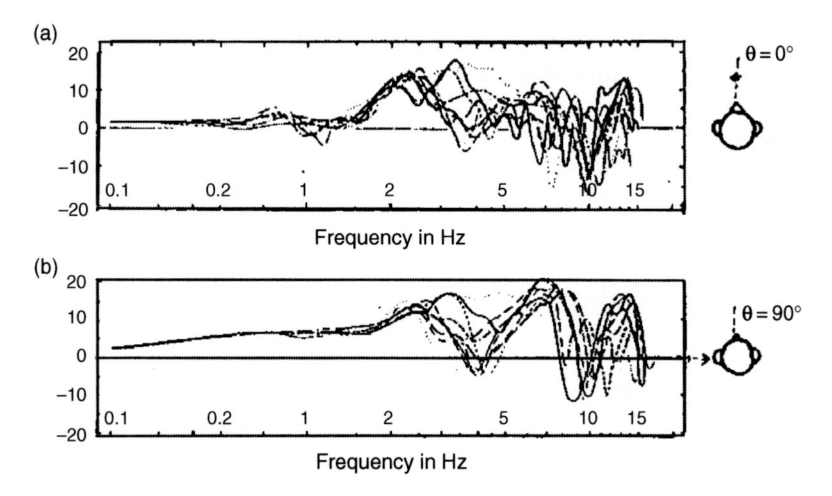

Figure 2.6:
Shaw's data showing the ratio (in dB) of the sound pressure at the ear canal entrance to the free-field sound pressure: (a) shows the response for each of the ten subjects for a sound source of 0° azimuth; (b) shows the response for a sound source at 90° azimuth.

Given these differences, and all of the aspects of frequency discrimination, distortion sensitivity, spectral response differences, directional response differences, psychological differences, environmental differences, cultural differences, and so forth, it would be almost absurd to expect that we all perceive the same balance of characteristics from any given sound. It is true that whatever we each individually hear is *natural* to each one of us, but when any reproduction system creates any imbalance in any of its characteristics, as compared to a natural event, the aforementioned human variables will inevitably dictate that any shortcomings in the reproduction system will elicit different opinions from different people *vis-à-vis* the accuracy of reproduction. As to the question of whether it is more important to reduce the harmonic distortion in a system by 0.02%, or the phase accuracy of 5° at 15 kHz, it could well be an entirely personal matter, and no amount of general discussion could reach any universal consensus. Indeed, Lord Rayleigh was right; the sensation of sound *is* a thing, *sui generis* (*sui generis* meaning 'unique').

2.5 Summary

Hearing has a logarithmic response because the range of perceivable sound pressures is over one million to one. It is hard to imagine how we would hear such a range of pressure levels if hearing were based on a linear perception.

One decibel approximates to the smallest perceivable change in level for a single tone. At mid frequencies, 10 dB approximates to a doubling or halving of perceived loudness. Doubling the loudness requires ten times the power.

The speed of sound in air is constant with frequency (but this is not so in most other substances). The speed of sound in any given material determines the wavelength for each frequency.

The spectrum of visible light covers about one octave, but the spectrum of audible sound covers more than 10 octaves. This leads to wavelengths differing by 1000 to 1 over the 20 Hz to 20 kHz frequency range – 17.4 m to 1.74 cm.

How absorbers behave towards any given sound is dependent upon the relationship between size and wavelength.

The decibel is the most usual unit to be encountered in measurements of sound. Sound *power* doubles or halves with every 3 dB change. Sound *pressure* doubles or halves with every 6 dB change. Our hearing system responds to changes in sound pressure.

Sound *intensity* is the flow of sound energy, and is measured in watts per m^2.

Sound decays by 6 dB for each doubling of distance in free-field conditions, because the intensity is reduced by a factor of four on the surface of a sphere each time the radius is doubled.

The dBA scale is used to relate better between sound pressure measurements and perceived loudness at low levels.

The dBC scale is more appropriate for higher (around 80 dB) sound pressure levels.

For equipment response measurements, or any absolute measurements, only a flat (unweighted) response should be used.

Sound isolation need not be flat with frequency because our perception of the loudness of the different frequencies is not uniform.

Our pinnae are individual to each one of us, so to some extent we all perceive sounds somewhat differently.

Psychoacoustics is the branch of the science which deals with the human perception of sound.

The displacement of the ear drum by a sound of 0 dB SPL is less than one one-hundredth of the diameter of a hydrogen molecule.

Some people can detect pitch changes of as little as one 1/25th of a semitone.

Due to the physical differences between our bodies, and especially our outer ears, the sounds arriving at our ear drums are individual to each one of us.

Some pinnae perform better than others in terms of directional localisation.

The differences in perception between different people can give rise to different hierarchies of priorities in terms of what aspects of a sound and its reproduction are most important.

Perception of musical instrument, directly or via loudspeakers, can be modified by pinnae differences; so the loudspeaker that sounds most accurate to one person may not be the one that sounds most accurate to another person.

References

1 Tyndall, John, *On Sound*, 6th edn, Longmans, Green & Co., London, UK, p. 67 (1895)
2 Newell, Philip. *Studio Monitoring Design*, Focal Press, Oxford, UK, and Boston, USA (1995)
3 Rodgers, Carolyn Alexander, Ph.D. Thesis, Multidimensional Localization, Northwestern University, Evanston, IL, USA (1981)
4 Rodgers, Puddie C. A., 'Pinna Transformations and Sound Reproduction', *Journal of the Audio Engineering Society*, Vol. 29, No. 4, pp. 226–234 (April 1981)
5 Belendiuk, K. and Butler, R. A., 'Directional Hearing Under Progressive Improvement of Binaural Cues', *Sensory Processes*, 2, pp. 58–70 (1978)
6 Morimoto, M. and Ando, Y., 'Simulation of Sound Localisation', presented at the Symposium of Sound Localisation, Guelph, Canada (1979)
7 Shaw, E. A. G., 'Ear Canal Pressure Generated by a Free-Field Sound', *Journal of the Acoustical Society of America*, Vol. 39, No. 3, pp. 465–470 (1966)

Bibliography

Borwick, John, *Loudspeaker and Headphone Handbook*, 3rd Edn, Focal Press, Oxford, UK (2001)
Davis, Don, and Davis, Carolyne, *Sound System Engineering*, 2nd Edn, Focal Press, Boston, USA (1997)
Howard, David M. and Angus, Jamie A. S., *Acoustics and Psychoacoustics*, 4th Edn, Focal Press, Oxford, UK (2009)
Strutt, John William, 3d Baron Rayleigh, *The Theory of Sound*, 2nd Edn, Vol. 1, London (1896). Republished by Dover Publications, New York (1945 – still in print, 2011)
Toole, Floyd E., *Sound Reproduction – the acoustics and psychoacoustics of loudspeakers and rooms*, Focal Press, Oxford, UK (2008)

CHAPTER 3

Sound Isolation

The weight of air. Interaction of the air vibrations with room boundaries. Reflexion, transmission and absorption. Reflexion and absorption as a means of isolation. The mass law, damping and decoupling. Frequency dependence of isolation needs. Level dependence of isolation needs *vis-à-vis* the non-uniform ear response. Floor, wall and ceiling isolation. Weight considerations. Material densities. The journey through a complex isolation system. Considerations regarding impact noises. Matters influencing studio location choice. The behaviour of mass/spring floor systems and the characteristics of fibrous and cellular base layers.

3.1 Vibrational Behaviour

Sound waves in air can be remarkably difficult to stop. Air is a fluid of considerable substance. It can support 500 tonne aeroplanes and blow down buildings. In fact, it is much heavier than most people think. On Earth, at sea level, air weighs about 1.2 kg per cubic metre, which is a very good reason for pumping any unnecessary excess of it out of aeroplanes (depressurising). This process is partially carried out to reduce the pressure differential stresses on the aircraft fuselage when flying at altitude in air of lower pressure (and density), but the aircraft also use less fuel by not having to carry excessive quantities of air over their entire journeys. A jumbo jet can actually reduce its load by around half a tonne by depressurising to an equivalent pressure altitude of around 8000 feet (2500 m). Air is also a rather springy substance, which makes it useful in air-pistols, shock absorbers and car tyres.

When any material is immersed in a vibrating fluid, it will itself be set into vibration to some degree. The characteristics of the resulting vibration will depend on the properties of the material, especially as, unlike air, the speed of sound is frequency dependent in many materials. The 'laser gun' sound sometimes heard from railway tracks when a train is approaching is due to phase dispersion within the steel tracks. The high frequencies travel faster than the low frequencies when the wheel flanges excite the rail modes, and so arrive first, with the lower frequencies following later.

An acoustic wave when travelling through air will continue to expand until it reaches a discontinuity, which may be a solid boundary or a porous material. It will then encounter a change in acoustic impedance that will cause some of the energy to be reflected, some to be absorbed, and some to be transmitted beyond the discontinuity. The degree of reflexion will be determined by the lack of acoustic permeability of the boundary and the degree of internal losses that will tend to convert the acoustic energy into heat. A rigid, heavy, impermeable structure, such as could be made from a 1 m thickness of reinforced concrete with a well-sealed surface, would reflect back perhaps 99.9% of the incident energy. Conversely, a wall of 1 m wedges of open-cell foam or mineral wool would present a very gradual impedance change which would allow the acoustic wave to enter with minimum reflexion, perhaps *absorbing* 99.9% of the incident energy. Mechanisms such as internal friction, tortuosity, and adiabatic losses convert acoustic energy into heat. The internal friction results in losses as the materials, on a molecular or particulate level, rub together. The energy required for these motions is taken from the acoustic energy.

Tortuosity relates to the degree of obstruction placed in the way of the air particles as they try to move under the influence of the acoustic wave. Air, in some circumstances, can act like a rather sticky, viscous fluid, and the viscous losses are increased as the degree of tortuosity increases. The fibres of a medium density mineral wool, for example, present a very tortuous path for the air vibration to negotiate. Remember, though, that there is no net air *flow* associated with these vibrations. The air particle motion is very localised, and is dependent upon the amplitude and frequency of the vibration. Adiabatic losses occur when the heat of compression and the cold of rarefaction, which both normally contribute to the sound propagation (see Chapter 4), are conducted away from the air by the close proximity of a porous medium in which the air is dispersed.

Because the above losses are proportional to the speed with which a particle of vibrating air tries to oscillate within the material, their absorption coefficients tend to be greater as the frequency rises due to the particle movements being more rapid. In addition, as the acoustic wave propagation must stop and reverse when it reaches a solid boundary, the particle motion will tend to zero when the boundary is approached. The absorbent effect of fibrous materials is therefore greater when the materials are placed some distance away from the reflective boundary. The particle velocity is greatest, and hence fibrous absorption is greatest, at the quarter and three-quarter wavelength distances from the boundary, on the *velocity anti-nodes*, which coincide with the pressure nodes of an acoustical wave. (See Chapter 4.)

Wave motion can be thought of like a swinging pendulum. When the pendulum is at its maximum height, the movement is zero, and when the movement is at its maximum velocity, the pendulum is swinging through its point of minimum height. The height is therefore at its maximum when the speed is at its minimum, and *vice versa*. Similarly, in an acoustic wave, the velocity component is at its maximum when the pressure component is at its minimum, and *vice versa*, and the energy in the propagation is continuously passing from one component to the other. This will also be discussed further in Chapter 4, because absorption will be seen to be more relevant to acoustic control than to isolation. When absorbent materials are used in isolation systems, they tend to be used more as mechanical springs and for acoustic damping.

One reason for this is that fibrous/porous absorbers in general have little effect at low frequencies except in great depth and away from a reflective boundary. When we are considering sound isolation for recording studios, we tend to be considering treatment principally at low frequencies and near to boundaries, because preventing the acoustic energy from bass guitars and bass drums entering the boundaries is what we are largely seeking to achieve.

Once acoustic waves are allowed to enter a structure, things may become highly unpredictable. The disturbance will travel through the materials of the structure at their characteristic speeds of sound, sometimes with very little loss, then re-radiate into the air from the vibrating surfaces with unfathomable phase relationships that can produce 'hot spots' of sound energy in some very unexpected places. The speed of sound in solids is usually much greater than in air. For example, in concrete it is around 3500 m/s, in steel over 5000 m/s, and even in liquid water about 1500 m/s. In woods, the speed of sound can be dependent on the direction of propagation. Taking the case of beech, sound propagates along the grain at around 4500 m/s, yet at only just over 1100 m/s across the grain. These speeds are all approximate because the materials are somewhat variable in nature – fresh water and seawater, for instance – and there can be different speeds at different frequencies.

3.1.1 Relevance to Isolation

If isolation is the goal, then it is the degree of transmission through the boundaries that is important. From this point of view absorption and reflexions are often lumped together, because isolation will be achieved either by reflecting the energy back from a boundary or by absorption within it. Reflexion is by far the most important control technique, because absorbent walls would need to be of enormous thickness to be effective at low frequencies – several metres at 20 Hz, for example. In Chapter 2 it is mentioned that an anechoic chamber with 1 m wedges would absorb around 99.9% of the sound power down to 70 or 80 Hz or so, but such absorption is minimal in terms of isolation. Ninety-nine point nine per cent of the power being absorbed leaves only one part in 1000 of the original energy, but still only represents a 30 dB reduction (10 dB for each 10 times reduction). For measurement purposes, this renders reflexions insignificant in most cases, but 30 dB of isolation around a drum kit and bass guitar could still leave over 80 dB SPL on the other side of a 99.9% absorbent wall. To isolate by 60 dB we cannot allow more than one part in *one million* of the sound power to escape. Ten-metre thick walls of 70 kg/m^3 mineral wool may suffice at 20 Hz, but it is hardly a practical solution, especially when one realises that the ceiling would need the same treatment.

3.2 Basic Isolation Concepts

There are essentially four aspects to sound isolation; mass, rigidity, damping and distance. Taking the last one first, if we can get far enough away from noise sources and our neighbours, then we will have solved our isolation problems. At least that seems to be rather obvious, but obvious things in acoustics are rather rare. The characteristics of mass, rigidity and damping are a little more complex.

All other things being equal, it takes more energy to move a large mass than a small one. Consequently, a large mass subjected to an acoustic wave will tend to reflect back more energy than a small mass, because it has more inertia. It has more acoustic impedance because it has a greater tendency to impede the path of the wave. However, if the mass is not rigid, and has a tendency to vibrate at its natural frequencies, energy may be absorbed from the acoustic wave that can set up resonances in the structure. Once the whole mass is resonating, its surfaces will be in movement and will act as diaphragms, re-radiating acoustic energy. If this mass were a wall, then the outer surface would selectively re-radiate the sound which was striking the inner surfaces. Isolation would therefore be dependent on the degree of the freedom of the mass to resonate.

If the mass were perfectly rigid, then resonances could not occur, because vibration implies movement, and infinite rigidity precludes this. Theoretically, of course, if a sealed room were made from an infinitely rigid, lightweight material, then because it could not vibrate it would be soundproof unless the whole thing could be set in movement *en masse*. In the latter case, the inertia of the air inside the room would resist the motion of the shell and set up pressure waves from the boundaries. Unfortunately, lightweight infinitely rigid materials do not exist, so the only way we can normally achieve high degrees of sound isolation over short distances is by the use of highly rigid, massive structures.

3.2.1 Damping and the Mass Law

A great influence on the ability of any structure to provide sound isolation is that of damping. Damping is the degree to which a propagating wave within a material or structure is internally absorbed, normally by the conversion of the vibrational energy into heat. The damping of a material or structure can also be achieved to some degree by the addition of a damping material to its surface – Plasticine on a bell, for example. An acoustically very 'lossy' (highly damped), massive structure can in many cases, for the same degree of isolation, be less rigid, because the passage of the vibrational waves through the structure is severely attenuated before the waves can re-radiate from another surface.

Limpness, to some degree, achieves the same end as rigidity – the inability of the structure or material to vibrate sympathetically – but unsupported limp materials are incapable of forming a structure – they have inertia, but no stiffness. This leads us to partitions that are essentially controlled by the mass law, which roughly states that when the inertia of a panel, rather than its stiffness, is the dominant principle for sound transmission loss, that loss increases by 6 dB for each doubling of the mass per unit area, and by 6 dB for each doubling of frequency. At least, that is, for plane waves at a given angle, this is why the mass law is only an approximation to normal circumstances.

3.2.2 Floating Structures

In practice, the means by which isolation is usually achieved is by mechanically decoupling the inner structure of a room from the main structure of the building. As has been discussed, isolation by pure absorption is very unwieldy – rooms made from 10 m thick mineral wool walls and ceilings are not an option. As no lightweight super-rigid structures are readily available, then neither is rigidity alone a practicable solution. To some degree, mass is used, but if it were to be used alone, then it too can become unrealistic in its application. For example, let us presume that a concrete block wall of 20 cm thickness and 40 dB isolation were to be augmented, by mass alone, to achieve 60 dB of isolation at low frequencies. The mass law would add at the most around 6 dB of isolation for each doubling of the mass per unit area, though the increasing rigidity of the more massive structure could tend to raise the isolation to above the 6 dB mark. Under ideal circumstances, which may not be realised in practice, doubling the thickness to 40 cm would yield 46 dB. Doubling this to 80 cm would result in 52 dB, and doubling this yet again to 1.6 m would still only provide 58 dB of isolation. We would therefore require walls of around 2 m thickness to achieve 60 dB of low frequency isolation if we were to rely on increasing the mass alone, and even this may be compromised by internal resonances within the structure.

The practical answer to the isolation problem lies in decoupling the inner and outer structures. This can be achieved by many means, such as steel springs, rubber or neoprene blocks, fibrous mats, polyurethane foams and other materials. There have even been cases of rooms floated on shredded car tyres, and even tennis balls, but the problem with tennis balls is that the air will gradually leak out over time. Re-inflatable air bags use the same principle, and this is another practical solution sometimes used.

The floating relies on the mass/spring/mass resonant system as shown in Figure 3.1. If the first mass is set in motion, the force exerted on the spring will be resisted by the inertia of the second mass, and above the resonant frequency of the system the spring will be heated by the vibrational energy. This converts the acoustic energy into thermal energy. Such isolation systems work down to about 1.4 times the resonant frequency of the system, below which the decoupling ceases to become effective. The resonant frequency is a function of the mass that is sprung and the stiffness of the springs. Increasing the mass and decreasing the spring stiffness both tend to reduce the resonant frequency.

The effect of this decoupling of the masses is to render the two systems (floated and structural) acoustically independent. Hence if we had a structure of 20 cm sand-filled concrete blocks, with 40 dB of low frequency isolation, we would only need an internal floated structure with 20 dB of isolation in order to achieve 60 dB of total isolation. This may be achieved by

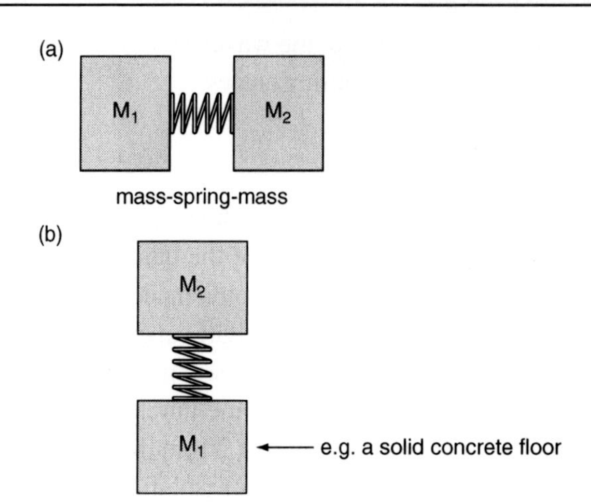

Figure 3.1:
(a) A resonant system; (b) If we now 'earth' M_1 we will have a good representation of a floated floor, where M_1 becomes of almost infinite impedance, such as would be the case for a heavy base slab on solid ground. If mass M_2 were to be set in motion, say by pressing down on the mass and releasing it, the system would oscillate at its resonant frequency until all the energy was dissipated. The length of time for which the system oscillates depends on the resistive losses – the damping. If the mass M_2 were caused to vibrate at a frequency above the resonant frequency of the system, the spring would effectively decouple the vibrations from the earthed mass, M_1.

internal floated walls of 10 cm thickness, such as of sand-filled, concrete blocks, spring isolated from the floor, with a 5 cm mineral wool lined air space between the two walls. Such a system, as depicted in Figure 3.2, would achieve in a total thickness of 35 cm what could only be achieved by 2 m thickness of blocks that were mechanically connected, at least down to the resonant frequency of the floated system. Below that frequency, the isolation would begin to fall, and the advantage would pass back to the 2 m solid wall.

The way that this works is that the internal wall attenuates the sound from inside the room by 20 dB, which is the resultant level on the outside of the inner wall. The air between the walls acts as a spring and it is not capable of efficiently pushing or pulling the much greater mass of the inert outer wall. If the walls were in contact, the masses would be reasonably comparable, so the vibration in the inner layers would have relatively little trouble progressing through the structure. In the isolated wall system, it is only the sound pressure which has already been attenuated by 20 dB which impinges on the outer wall, which then attenuates the sound by another 40 dB before reaching the outside of the building, thus achieving the 60 dB of isolation. Figure 3.2 shows the inner wall floated on a layer of high-density mineral wool, which is an economical solution. However, a drawback is that as more weight is loaded on it with each layer of blocks, it compresses significantly. With a wall of 4 m height it my go down from 10 cm to 6 cm, or thereabouts. This means that the whole room needs to be built with one entire layer

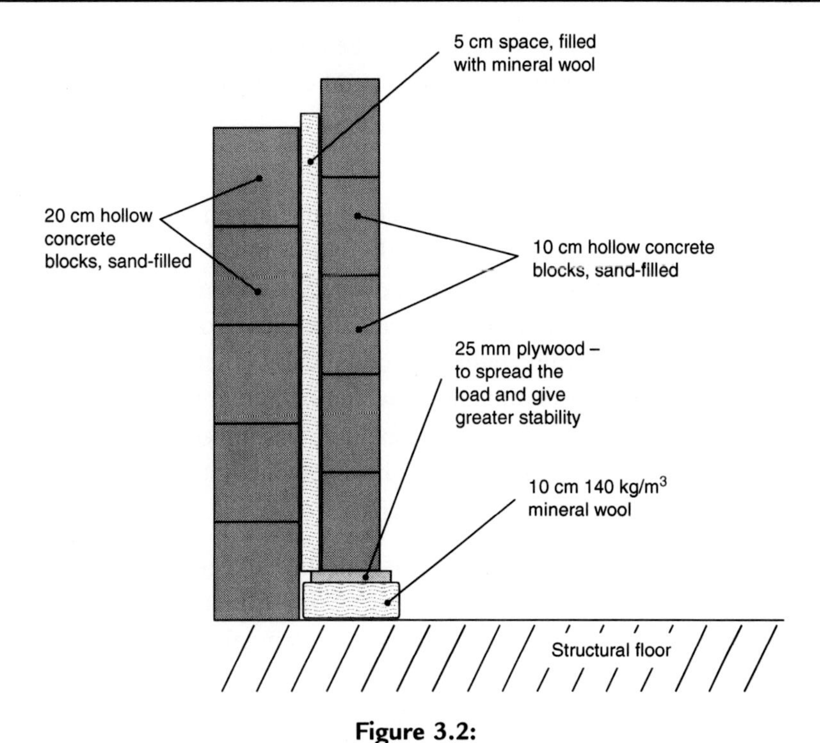

Figure 3.2:
Floating a wall — an earthed mass-spring-mass system in practice.

of blocks at a time because if one whole wall is erected first, the first layers of blocks of the adjacent walls will not line up with those of the first wall which has already compressed the mineral wool below it. It is therefore essential when using mineral wool to make all the walls at the same time, row by row, so that the mineral wool compresses at the same rate. Until any lintels are in place above any doors or windows, some free-standing sections of wall may need to be supported by props, in order to keep them vertical.

As an alternative, there are other isolation materials available made from materials such as neoprene and rubber. Sylomer and Clempol are typical examples, but the precise type of each material needs to be selected according to the total load which they will support. Somewhat confusingly, various manufacturers give the working loads in kg/m^2, lb/ft^2, lb/in^2, kPa, kg/cm^2, N/mm^2 and other units of force. This can be totally confusing for non-specialists, and it is certainly not helpful for specialists, either. Nevertheless, they are all directly convertible.

Materials such as rubber and neoprene would typically be used in thicknesses of about 15 mm, and so would not compress as much as mineral wool during the construction of the walls. This allows a little more flexibility in the number of rows which can be laid at one time in each section of wall, because the compression range per row will be much less, but care should still be taken not to work too unevenly or hot-spots may be formed in some areas whilst other areas may be bridged over. It is, therefore, still best to build the rooms one entire row at a time.

3.2.3 Floating System Choices

The choice of what to use to float the inner structure depends on the mass to be floated and the lowest frequency to which isolation is needed. Two identical studio rooms, one for the recording of bass guitar and one for the recording of speech, both requiring the same degrees of isolation within the frequency range of their intended usages, would need rather different suspension systems if they were to be built in economically efficient ways. The former room would be much more expensive to build, and it would be a total waste of money to make a room such as this if so much low frequency isolation was unnecessary. For simplicity, let us make a comparison of two such rooms in which we are only interested in the sound leakage in the outward direction to a neighbouring room.

Taking the vocal recording room first, most of us will have experienced how the sound of speech can travel through walls. Noisy neighbours talking till late in the night whilst we are trying to get some sleep, either at home or in a holiday hotel, for example. This problem is so easy to solve that it is a disgrace that so many buildings have been constructed with so little thought about the problem. However, if a huge number of people want to go away on holiday as cheaply as possible, it is not too surprising that the hotels that they stay in will have been *built* as cheaply as possible. You often get what you pay for, but not much more.

People talking relatively loudly are likely to produce sound pressure levels in the order of 80 dB SPL (or the same dBA as there is little low frequency content) at the boundary of a room, a couple of metres from the source of the sound. (The speech frequencies, whose lower frequencies begin around 100 Hz, are little affected by the low frequency roll-off of the dBA weighting.) The deepest tones of some male voices extend down to 80 Hz or so, but they tend to be weak in level, and even with a little attenuation through a wall soon reduce to levels that become imperceptible. A glance at Figure 2.1 will reveal that at low frequencies at low levels the equal loudness curves close together. They do so to such a degree that only 5 dB or less reduction in SPL will produce a halving of subjective loudness as opposed to the 10 dB a mid frequency would require to produce the same effect. Therefore, 20 dB of isolation at 70 Hz and 50 dB of isolation at mid frequencies would render speech all but inaudible above a quiet background noise in an adjacent room. Figure 3.3 shows how this could typically be achieved. A person recording drama in an otherwise residential building would be unlikely to make the neighbours aware of their activities if such a treatment of the recording room were to be undertaken.

A musician wishing to practice playing bass guitar at home, in the same building as our drama studio, would not be so lucky. In this case, the fundamental frequencies of some 5-string basses can go down as far as 35 Hz. Enthusiastic playing could produce 100 dB SPL or more

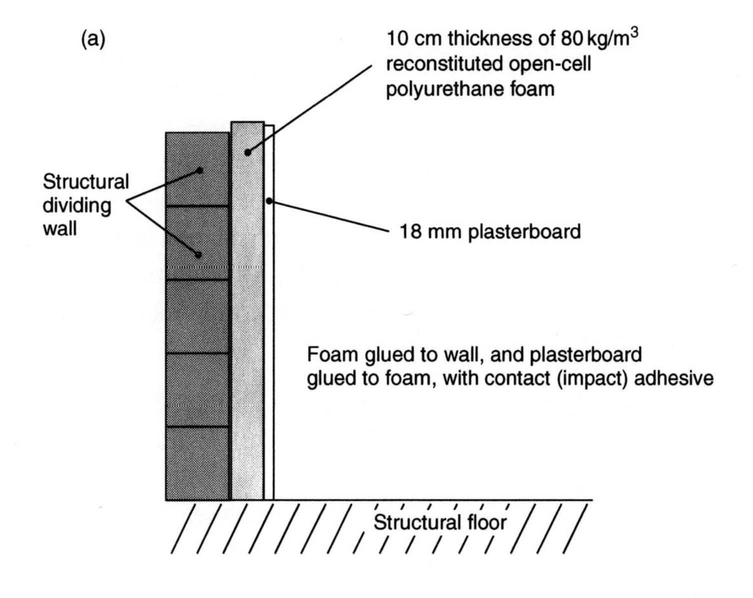

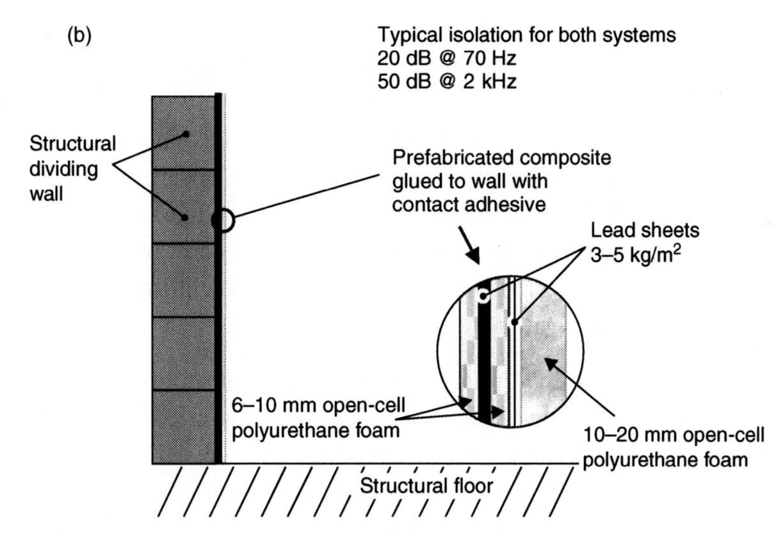

Figure 3.3:
Isolation treatments for direct application to walls: (a) With reflective surface to room; (b) With absorbent surface to room. Typical isolation for both systems: 20 dB at 70 Hz; 50 dB at 2 kHz.

if it was for recording, where the sound of amplifier compression at high level can become an integral part of the sound. What is more, the relatively sustained notes of a bass guitar can exist for a sufficient period to excite strong structural resonances in an acoustically non-specifically designed construction, such as an apartment building.

From Figure 2.1 it can be seen that below about 55 dB SPL, 35 Hz is inaudible. If the bass guitarist were playing at a level of 80 dB at that frequency, then whether the room had 30 dB of isolation or 40 dB of isolation would make no difference, because in either case the resultant leakage at 35 Hz would be inaudible. Thirty dBA is usually considered to be an acceptable noise level for sleeping, and at 35 Hz, 30 dBA would actually be about 70 dB SPL (the 30 phon level in Figure 2.1) so in practice, even 10 dB of isolation at 35 Hz would suffice if the intention was not to annoy the neighbours. At 500 Hz, if the guitar were also producing 80 dB, then 50 dB of isolation would be the minimum needed in order not to annoy the neighbours with more than 30 dB of 'noise'.

If the musician increased the volume by 30 dB, so that both the 35 Hz and 500 Hz components of the bass guitar were producing 110 dB, 30 dB extra isolation would be needed at both 35 Hz and 500 Hz to reduce the 110 dB down to a tolerable 30 phon (the 30 dBA curve equivalent) for the neighbours. However, because of the ear's tendency to increase its sensitivity to low frequency level changes as the frequency descends, the provision of only 15 dB of extra isolation at both frequencies would not have an equal subjective effect. At 500 Hz, the 15 dB excess leakage would somewhat more than double the subjective 'noise' level for the neighbour (10 dB doubles loudness at mid frequencies) which, during the daytime and early evening, may be tolerated. Not so at 35 Hz, though. Figure 2.1 shows that at 35 Hz, 70 dB SPL lies about the 30 phon level. Increasing the SPL to 85 dB (the 15 dB extra leakage) now places 35 Hz on the 60 phon level. In other words, it will have subjectively doubled in loudness from the 30 to 40 phon curves, doubled again between the 40 and 50 phon curves, and doubled yet again between the 50 and 60 phon levels. The result would be an eight times increase in apparent loudness, which the neighbours simply may not tolerate.

What this means is that if the drama group begin to scream and shout and produce 30 dB more sound during the day, then they may well get away with an increase of isolation of only 15 dB in the mid and low frequencies, as they produce little low frequencies. However, the bass guitarist would not remain a friendly neighbour if the volume were to be increased by 30 dB with only 15 dB extra isolation being provided. What is worse is that low frequency isolation in structures is much less effective than high frequency isolation, because of the aforementioned mass law. (See Section 3.2.1.) Therefore, not only will the bass player need more isolation, but also the extra low frequency isolation needed will be much heavier than the equivalent mid frequency isolation. In a domestic building, the basic structure may simply not support the necessary weight of isolation materials, and so being a good neighbour whilst playing a bass guitar at 110 dB SPL may simply not be possible unless the other occupants of the building are deaf.

In practice, it may simply not be possible to provide adequate isolation for somebody who wants to play a bass guitar at 100 dB in a residential building if they are not on solid ground and where a very heavy internal structure can be built. Conversely, it may be possible for a drama group to make 100 dB of noise in a similar residential building without causing any nuisance after only a small amount of isolation work. The circumstances are very frequency dependent.

This variability in hearing sensitivity with level causes so much confusion for non-specialists in the understanding of sound isolation requirements. Simple figures are not possible, as everything must relate to frequency and level. In effect, only *curves* of isolation requirements can be specified, dependent on the frequency range to be isolated and the levels to which they either will produce sound, or will need to be isolated from it. The seemingly simple question of 'How many decibels of isolation do I need?' usually has no simple answer.

Figure 1.1 shows an actual construction that was necessary to prevent bass guitars at around 110 dB from annoying the neighbours in a bedroom above. Looking at this and Figure 3.2 will show clearly the degree to which the isolation of low frequencies at high levels can get wildly out of hand in inappropriately chosen buildings. The room shown in Figure 1.1 weighs about 40 tonnes, and has a floor area of about 27 m^2, so had it been on anything other than the ground floor – and on *solid* ground, with no basement – then it would not have been feasible to construct it in any normal domestic building. The only exception would have been if acoustic engineers could have got to the architects before the building was constructed, in order to suggest the type of reinforcement shown in Figure 1.2. If the premises were to be used for commercial recording, as was the case in Figures 1.1 and 1.2, then such isolation work might just be realistic, but purely for rehearsal it would normally (in fact in almost all cases) be economically out of the question. What often fails to be appreciated is that the sound waves do not know why they are being produced, so the isolation requirements do not change according to the use or the economic viability of their constraint. It is remarkable how many people believe that if they 'only' want to make the sound for personal fun the isolation will somehow cost less than if they were making the sound for commercially serious recording.

3.3 Practical Floors

Most floor isolation systems are based on the mass/spring/mass system shown in Figure 3.1, and some typical examples are shown in Figure 3.4. In each case the floated mass and the material used as the spring are different. Such options are necessary in order to achieve the required isolation, which may vary not only in the overall amount needed but also in the lowest frequency to which it must be effective. In general, however, the low frequency isolation reduces as the floated mass reduces, such as from (a) to (d) in Figure 3.4. For whatever requirements, the floor needs to be tuned so that its frequency of resonance is about half of that of the lowest frequency to be isolated. Also, whatever 'spring' is being used must be operating in its optimum load range. This concept is explained in Figure 3.5.

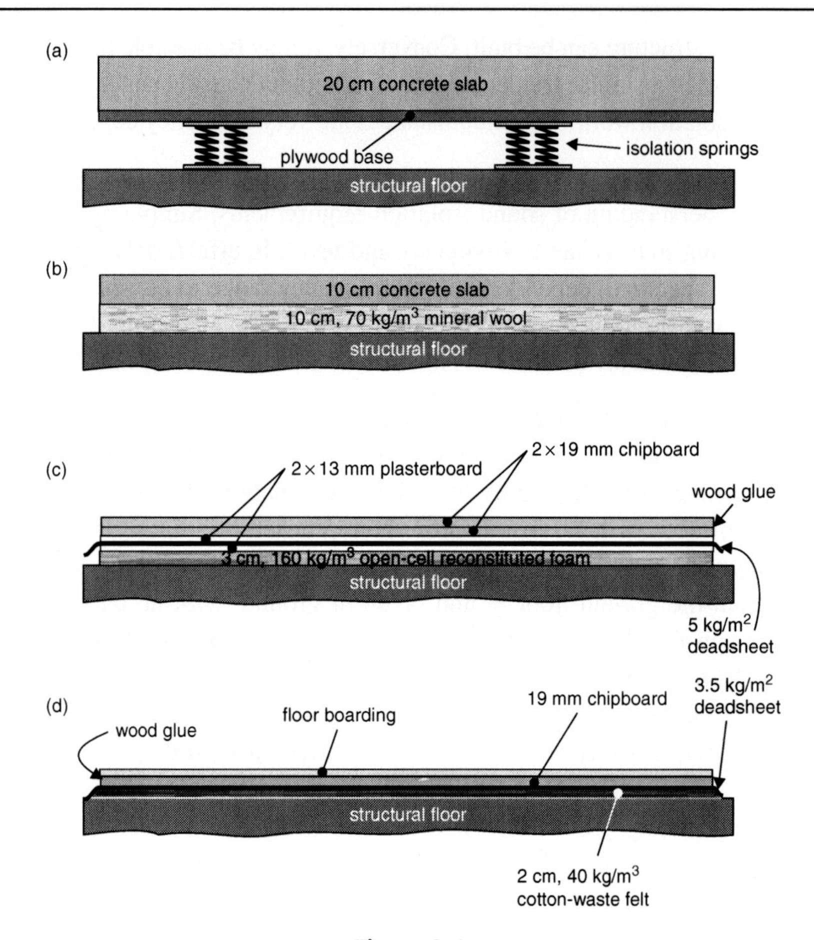

Figure 3.4:
A selection of floor-floating options.

It can therefore be seen that the springs cannot be chosen without an accurate knowledge of the weight that they will be supporting. In turn, the weight cannot be calculated until the structural requirements and the lowest frequency of required isolation are known. The choice of isolating springs may also depend on the floor which supports it; because obviously the springs which are spaced out over the surface of the floor present much higher point-loads on the supporting floor than do the 'blanket' coverings of fibrous materials or polyurethane foam panels (compare Figures 3.4 (a) and (b)). Figures 3.6 and 3.7 show the practical problems which may be encountered if the floor floating spring material is inappropriately chosen, even though the isolation achieved may be entirely adequate. These problems are not restricted to lightweight foams, however, as there have been cases of rooms built on 40 cm concrete slabs, and floated on very low frequency steel springs, which have tilted alarmingly when a heavy mixing console has been placed in an off-centre position when the rooms were equipped ready for use.

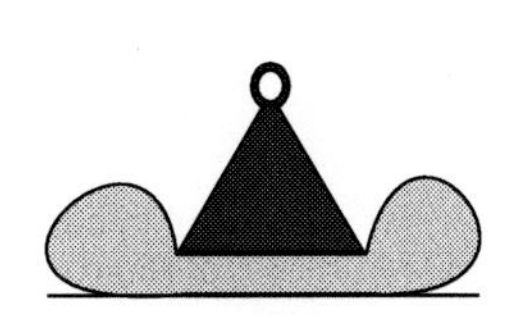

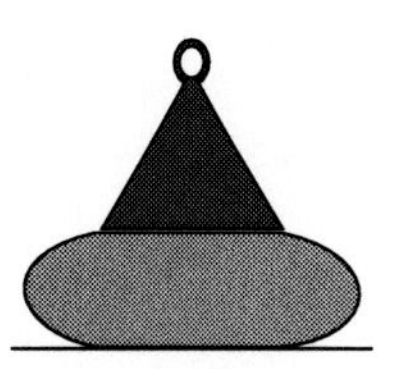

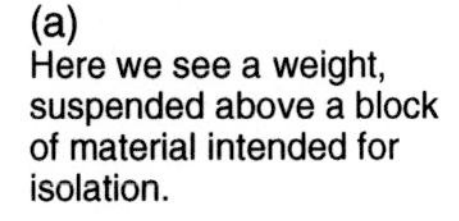

(a)
Here we see a weight, suspended above a block of material intended for isolation.

(b)
In this figure, the weight is resting on the block, but the block shows no sign of compression. Its stiffness must therefore be very great, effectively making a rigid coupling between the weight and the ground. Vibrations in the weight (if, for example, it was a vibrating machine) will be transmitted to the ground.

(c)
In the above example the isolation material is too soft, and the weight has compressed it down to a very thin layer. Its density and stiffness will thus also increase, and the effect will be an almost rigid coupling with the same effect as in (b).

(d)
Above, the weight can be seen to have compressed the isolation material to about half its original thickness. The system is at rest due to the equilibrium being found between the gravitational down-force on the weight and the elasticity of the material.
Vibrations in the weight will be resisted by the mass of the floor and the elasticity of the isolation material. Much of the vibrational energy will be turned into heat by the internal losses in the isolation material, and hence will not be transmitted into the ground.

Figure 3.5:
Float materials need to be in the centre of their compression range for the most effective isolation.

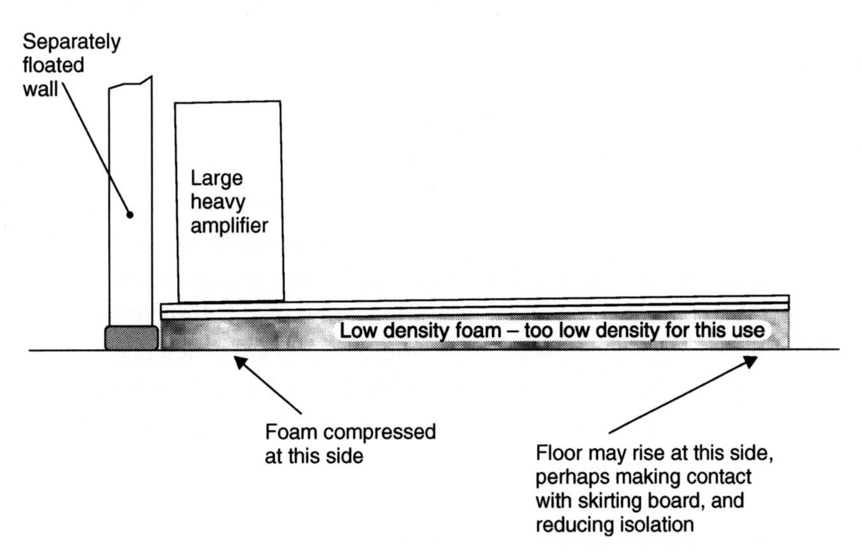

Figure 3.6:
Effect of uneven floor loading on low-density float material.

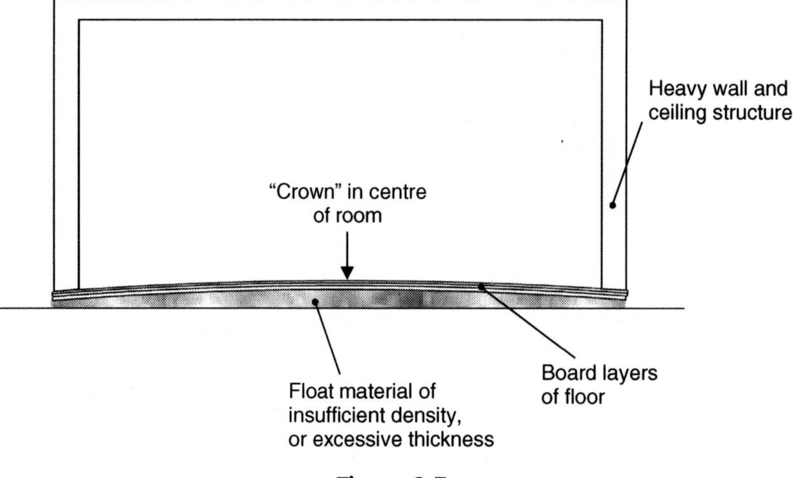

Figure 3.7:
Crown effect.

Vibration conduction via the ground can be considerable, but it is so dependent on the nature of the local terrain that it may be very difficult to predict or describe. Even a very massive rock and concrete floor can be set into vibration by a bass guitar amplifier in contact with it, so in all but the rarest of cases, a floated floor is mandatory for a professional recording studio. This is even more necessary if there is a potential for vibrations to enter the studio, such as may be the case with main roads or railways in the vicinity, or neighbouring factories with heavy, vibrating machinery.

As previously discussed, the low frequency isolation in terms of both the cut-off frequency and the degree of isolation will be dependent upon the type of recording to be undertaken. The isolation system shown in Figure 3.4(a) is typical of the large studios for full frequency range recording. The 20 cm concrete slab weighs about half a tonne per square metre, and the springs are chosen to support this weight in their mid-compression range, when the whole system will probably have a resonance of 5 to 8 Hz. The bases of the spring mounting frames (or boxes) will be likely to have an area of about 400 cm^2, or 0.04 m^2. If there were one spring for every 1 m^2, then the load on each 0.04 m^2 base would be 500 kg, which equates to a pressure of 12.5 tonnes per square metre immediately below the bases. Things like this must be taken into account when designing floor-floating systems. If such spot loads cannot be supported, then something must be done to spread the load. Likewise, the floated slab itself must usually have some load spreading, to avoid the tops of the springs punching through it. However, this is usually automatically a part of the construction, because the slab will need a continuous base over its entire area, above the springs, before it is cast. (See the plywood base in Figure 3.4(a).)

Whether the floated slab supports walls around its perimeter is largely dependent upon the weight of the perimeter walls. For a heavy concrete block isolation wall structure, such as shown in Figure 1.1, it would be typical to float it separately from the floated floor, and only to build the lighter weight acoustic control shell walls directly on the floated slab. Too much weight around the perimeter could lead to the crown effect shown in Figure 3.7 unless extra spring reinforcement were to be placed close to the edges of the slab. It is definitely preferable, however, with heavy structures, to float the walls separately on their own spring base, which makes the calculations and the construction much easier. It is also more flexible should modifications need to be made later. Furthermore, the heavy isolation walls may also have to carry the weight of a ceiling structure, so the downloads below the isolation wall can easily reach 5 or 10 tonnes per square metre, which is excessive in terms of being supported on the edges of a floated slab. In fact, there have been cases where a situation such as that shown in Figure 3.7 has occurred, and where the floated layer has been a reinforced concrete slab which has subsequently cracked open under the load. It is best to restrict the load on a floated slab to the internal acoustic control structures.

Figure 3.4(b) shows a 10 cm reinforced concrete slab floated on a mineral wool base 'spring'. Using this technique, the load is evenly distributed over the entire supporting area. It is relatively simple to cast tubes in this type of floor to carry cables, even though the concrete is relatively thin, as shown in Figure 3.8. There can be advantages and disadvantages to this type of floating because, depending on circumstances, the continuous blanket of mineral wool can transmit more energy into the base slab by virtue of the greater area of contact, or it may improve matters due to the extra damping of the base slab. Certainly, if the floated floor is not on solid ground, such as if it were above a basement, the more even weight distribution of the continuous spring material may be advantageous structurally, even if not acoustically. In general, the discrete spring or elastomer block system of floating is acoustically superior to the fibrous blanket method. The discrete mountings allow an air cavity to exist between the slab and the structural floor, which can improve the isolation. Any such gap is usually damped with a light, uncompressed filling of mineral wool.

Figure 3.4(c) shows a composite floor made from multiple layers of heavy and limp materials, floated on a reconstituted polyurethane foam base. (The polyurethane foam is the sponge [open cell] type, not the aerosol type of closed cell expanding polyurethane foam, which has entirely different properties.) This system is useful where wet-work (cement) is impracticable. Such a floor sandwich would weigh about 50 kg/m^2, which is about one fifth of the weight of a concrete slab of 10 cm thickness. For stability, and to avoid the effect shown in Figure 3.6, the foam spring layer would typically be 3 cm of 120 to 160 kg/m^3 sponge. This type of floated floor exhibits less low frequency isolation than concrete floors, but can achieve what is often an adequate degree of isolation

(a)

(b)

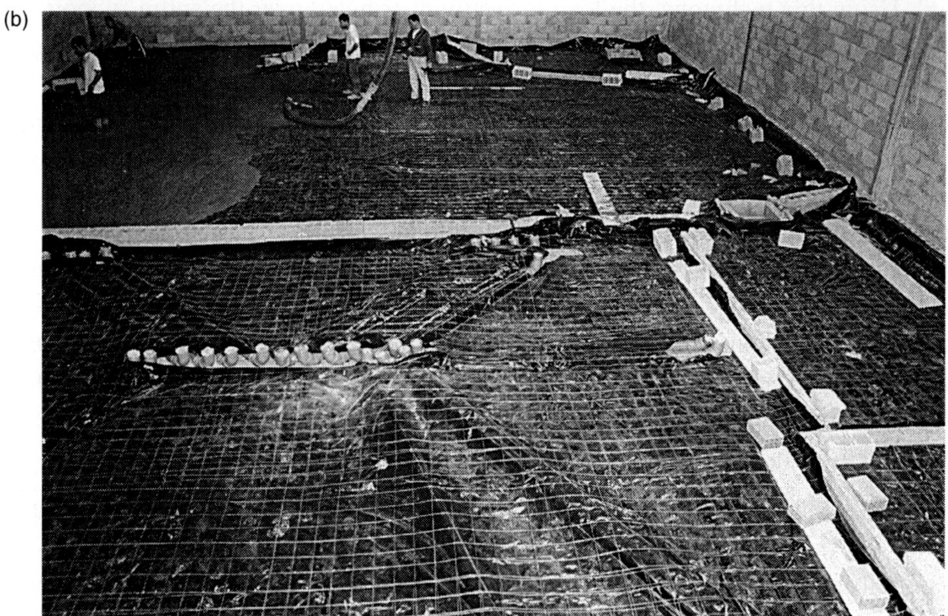

Figure 3.8:

Cable tubes: (a) If the 10 cm of mineral wool is composed of two layers of 5 cm, the upper layer can have channels cut into it to allow the tubes to be lowered on to the lower layer, thus positioning the tube half in the mineral wool and half in the concrete; (b) The tubes and mineral wool are covered in plastic sheeting and steel reinforcing mesh. The tube mouths can be seen protruding. They are blocked with mineral wool plugs prior to the pouring of the 10 to 15 cm concrete slab; (c) Detail of procedure.

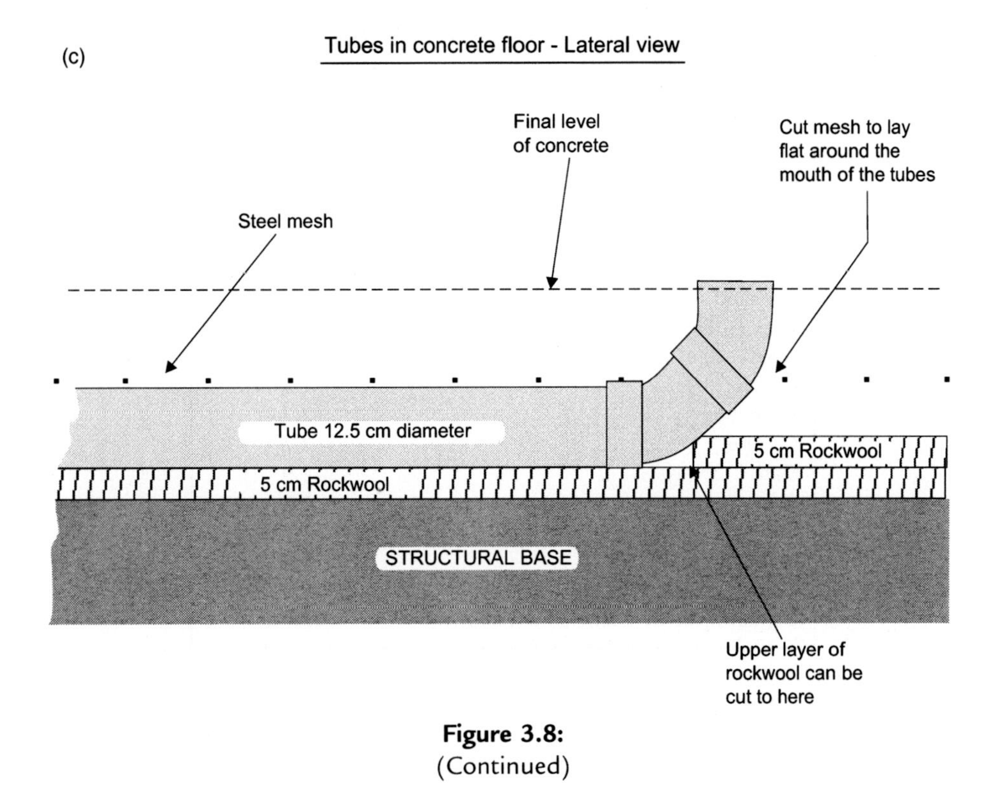

Figure 3.8:
(Continued)

in circumstances where the floor loading is restricted. Many domestic buildings may only have total floor loading capabilities in the region of 120 kg/m^2, which would be incapable of supporting even a 5 cm floated concrete slab. The density of concrete is typically around 2000 to 2500 kg/m^3 so a 1 m square of 10 cm concrete slab would weigh around 250 kg. The typical light industrial loading for a building would be in the order of 400 kg/m^2 (or 100 lb/ft^2 in imperial measure). Table 3.1 shows the densities of some typical acoustic materials, plus a few others to provide some points of reference.

Figure 3.4(d) shows a very light floating system. It consists of a wooden floor surface glued on top of a 19 mm layer of chipboard, floated on top of a composite of a 3.5 kg/m^2 deadsheet and a 2 cm layer of cotton-waste felt. Deadsheets will be discussed further in Chapter 4 (see also Glossary), but they are essentially heavy, limp membranes used as damping layers. This type of floor is principally used to reduce the effect of impact noises from entering a structure. They are particularly useful above a weak ceiling where impact noises from footsteps could enter a structure and be rather difficult to treat from below, especially when the room below has only limited height, and hence not much room for treatment.

Table 3.1: Density of materials

Material	Density
Cork	0.25
Pine	0.45
Typical hardwoods	0.60–0.75
Plywood	0.60–0.70
Plasterboard	0.75
Chipboard	0.81
Water	1.0
Dry sand	1.5
Brick (solid)	1.8
Glass	2.4
Concrete	1.8–2.7
Aluminium	2.7
Granite	2.7
Slate	2.9
Steel	7.7
Iron	7.8
Lead	11.3
Gold	19.3
Osmium (the densest element known)	22.4

The figures are in grammes per cubic centimetre, but they also represent tonnes per cubic metre.

Figure 3.9 shows a range of isolators used for structural floating. It should now be apparent that the range of possibilities for floor floating is enormous, which is just as well because the range of different circumstances requiring floated floors is equally enormous.

3.3.1 Floors on Weak Sub-Floors

The situation of floor floating is complicated enormously when the structural floor is itself significantly resonant. Such can typically be the case when a ground floor studio is sited over an underground car park, or when studios are stacked one above the other in the same building. If the site under consideration as a potential studio building also has low ceilings, then the sound isolation problems can become impossible to solve without major reconstruction of the building. Isolation of floors and ceilings takes up space, as a considerable thickness of materials is needed in each case.

When structural floors are weak, it is very difficult to prevent low frequency noises (such as those caused by a bass drum on a floated floor) from passing through the spring layer and exciting the resonances in the floor below. If the floor is resonant, then it only requires a relatively small amount of energy to excite the natural frequencies of the structure. In many cases, the only way to overcome such difficulties would be to support another floor, on steel beams, above the main floor, but this can only be done in cases where the walls are

(Taken from the publicity material of Eurovib Acoustic Products Ltd. [central group], Fabreeka UK and Mason UK Ltd.)
[www.eurovib.co.uk / www.fabreeka.com / Mason P.O. Box 190, GU9 8XN, UK]

Figure 3.9:
Range of isolation mounts.

strong enough to support the extra weight and when there is sufficient headroom to accept the loss of height. In the following section, these problems will arise again when we look at the problem of low headroom and weak ceilings, and it will also be discussed in Section 3.8.

Essentially, a fundamental requirement for high isolation, unless large amounts of money or space are to be consumed, is that the entire outer shell of the chosen building should be massive, rigid and well damped. Steel, and cast concrete, often fail to provide the last of these three requirements; they can have characteristic resonances that weaken the isolation. However, steel reinforced concrete floors can be excellent when they are sufficiently strong, as the steel rods encased in the concrete can be effective in providing damping, but the situation can be improved further by the addition of damping agents to the wet concrete mix if the floors are to be cast from new. 'Concredamp', for example, is one proprietary brand of such a compound.

3.4 Ceiling Isolation

Reference, once again, to Figure 1.1 will show a range of techniques used for ceiling isolation. The need for this seemingly excessive collection of techniques was precipitated by the somewhat absurd decision by four Andalusians to site a recording studio, for 24 hours per day use by rock bands, directly under the bedroom of an unsympathetic neighbour in a building of unsuitably weak structure. Although, for clarity, the drawing is not exactly to scale, neither is it too far from scale. It does give, therefore, a reasonably realistic view of the amount of space lost to isolation and acoustic control.

Such a construction was only possible because the original space had (barely) adequate height – about 3.5 m – allowing about 1 m for isolation and acoustic control before the inner fabric ceilings at 2.5 m. There is currently a tendency towards smaller and smaller studios as the perceived size requirement comes down by using hard disc-based recording and mixing systems. However, the size of human beings and acoustic wavelengths is not coming down, and as virtual studios do not replace conventional ones, there is a limit to how small one can go. The other force that is bearing down on studio size is an economic one. In many cases, as the real price of recording equipment of good sound quality comes down, an enormous number of people now seem to expect that the cost of recording acoustics should somehow fall accordingly. It is unlikely to do so, though; simply because economics have no bearing on the speed of sound or any other physical law, and it is the physical laws of acoustics that determine whether a room is well isolated or not – and also whether it is going to have good internal acoustics or not. There will be more on this subject in the following chapters, but there is definitely a tendency for people to underestimate the need for adequate ceiling height in order to make a good studio.

Given a starting height of only 3 m, it is almost impossible to install adequate ceiling isolation if the floor above is either noise sensitive or liable to radiate noises that can be prejudicial to the operation of the studio. The only exception is if the floor is truly massive, but this may only be of use in constraining sound within the studio. In the reverse direction, the vibrations in a floor that has considerable mass can be extremely difficult to stop from entering the space below unless an isolated ceiling can be employed which is suspended from floated walls, and not from the ceiling above. In Figure 1.1, an isolated plasterboard inner ceiling, above the acoustic control shell, can be seen suspended from a 6 cm layer of polyurethane sponge (reconstituted foam) which is in turn glued to a plasterboard/deadsheet/plasterboard sandwich fixed below wooden beams. These are supported on the floated concrete walls. In addition, to augment the transmission loss and to prevent resonances between the upper plasterboard layer and the plasterboard suspended from the upper ceiling, a mineral wool infill was used between the beams. There is therefore no physical contact between the structural ceiling and the intermediate isolation ceiling except via the air, and ultimately the ground, but this contact is decoupled by way of the suspension materials between the floated concrete isolation walls and the main structure. Furthermore, on the underside of the structural ceiling (the floor of the bedroom), a layer of 5 cm medium density mineral wool had been attached by a cement. Below that, also by means of a cement, a layer of 18 mm plasterboard had been suspended. A further isolation layer of two 13 mm plasterboards had been laid on top of the inner isolation structure.

One reason why so many different materials and suspension systems were used was to avoid any coincident resonances that can result from using too much of too few materials. Any resonances inherent in any structure or material will tend to reduce the isolation, perhaps seriously, at the resonant frequency. By using a variety of isolation techniques, the resonances will tend to occur at different frequencies, and hence any weak spots will be covered by the other material combinations in the different layers. A brief 'ride' on a sound wave attempting to propagate through the isolation ceiling may be usefully informative, as we can discuss the isolation mechanisms involved as we come across them.

3.4.1 A Trip Through the Ceiling

Before reading this subsection, it may be worthwhile photocopying Figure 1.1 and keeping it visible during the discussion. The description of the passage of an acoustic wave through the ceiling system should be informative because it will encounter the many different types of obstacles that have been built into this ceiling in order to try to achieve the desired isolation. An understanding of something of the feel for these isolation concepts is essential, because experience and knowledge of the practical application of these principles are perhaps the only real way of deciding on the most effective combination.

Even great academic acousticians acknowledge that 'the theoretical analysis of the sound transmission through double leaf partitions is far less well developed than that of single leaf partitions, and that consequently greater reliance must be placed on empirical information'.[1] Clearly, the ceiling construction that we are describing here is even more complex than a double leaf partition, so if any readers get any bright ideas about using their computers to solve these complex interactions – forget it! There are simply no programs to deal with this sort of thing, and even experts need to use a lot of discretion when using what limited computerised help they have. In the words of Fahy: 'The reason is not hard to find; the complexity of construction and the correspondingly large number of parameters, some of which are difficult to evaluate, militate against the refinement of theoretical treatments'.[1] Bearing this in mind, please forgive the following wordy treatment of the subject.

An acoustic wave originating in the recording space shown in Figure 1.1 and travelling in the direction of the ceiling would first pass through the lower layers of the inner ceiling, whose purpose is primarily for controlling the character of the sound within the room itself. The wave would next encounter the two layers of plasterboard fixed on top of the support beams of the inner acoustic shell. Plasterboard consists of a granular textured plaster core, covered with a layer of paper on each side to provide a degree of structural integrity. It is more resistant, less dense and less brittle than pure plaster sheet. High frequencies striking this surface would be almost entirely reflected back, because the plasterboard is reasonably heavy, having a density of about 750 kg/m^3. The weight of a double layer, around 18 kg/m^2, would be sufficient for the mass law, alone, to ensure that the highest frequencies were reflected back. As the frequencies lower, they will be progressively more able to enter the material and set it into vibration. To some degree, these vibrations will set the whole layer in motion, and the opposite surface will then re-radiate the vibrations into the air above. However, to a considerable degree, the particulate/granular nature of the core material will cause frictional losses as the vibrations pass through the material, turning acoustic energy into heat energy. The amount of energy available to re-radiate from the upper surface of the plasterboard will therefore have been reduced, and a degree of isolation will have been achieved. Figure 3.10 shows a typical loss-versus-frequency plot for this type of ceiling. However, the plot can only be taken as approximate, because the isolation provided by such structures is dependent upon their absolute surface area, their rigidity, and numerous other factors.

Not all the isolation is due to absorption, of course, because some is due to the energy being reflected back down towards the floor, but there will be progressively more absorption and transmission via re-radiation, as the frequency lowers and the reflectivity reduces. The overall transmission loss will include the *reflexion* and *absorption*, and the overall coefficient of absorption figure will include *absorption* and *transmission*. Essentially the isolation is what is not transmitted, and the absorption coefficient is what is not reflected. In fact, absorption is traditionally measured in units called sabins, after Wallace Clement Sabine, the pioneer of reverberation analysis, who himself first used 'equivalent open window area' as a unit of absorption. From this concept, it can be clearly understood how absorption can include transmission, because an open window converts almost no acoustic energy into heat. It transmits almost all of it to the other side.

One hundred per cent transmission = 100% absorption = zero reflexion. Zero transmission, achieved either by 100% absorption, 100% reflexion or some combination of the two, still amounts to 100% isolation. The loosely used terminology can be a little confusing, but absorption does get lumped into isolation and transmission figures, depending upon which one is of greatest importance in the context of the problems. Just to help to clarify things, a room of 1 m thick, smooth, solid concrete walls and ceiling would have excellent isolation, but very little absorption. It would tend to have a long reverberation time because the sound energy would be trapped inside of it. Almost nothing would escape. Conversely, a room of tissue paper in the middle of a desert would have very high absorption (and consequently almost no reverberation time), but almost no isolation, because all the energy would be either lost in the sand, or would be free to escape to the outside air.

Anyhow, let us now return to our discussion of Figure 1.1. The vibrational energy that is transmitted through the double plasterboard acoustic shell cap would then proceed across the air gap to the lower surface of the isolation shell ceiling. The acoustic wave would first pass through a layer of 2 cm cotton waste felt, bonded to a 3.5 kg/m^2 layer of an acoustic deadsheet (see Glossary). Here we encounter an additional, loosely coupled mass

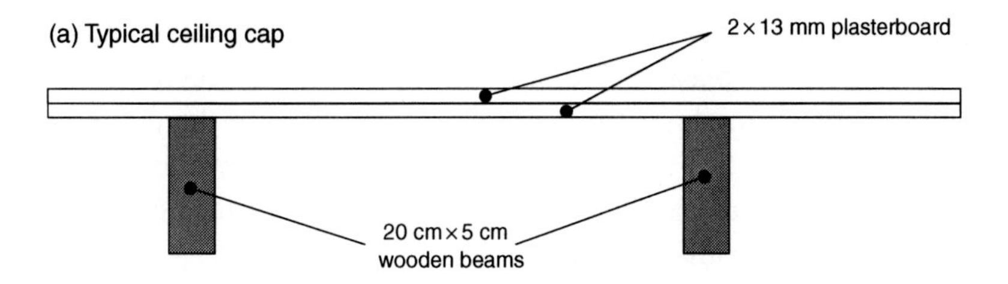

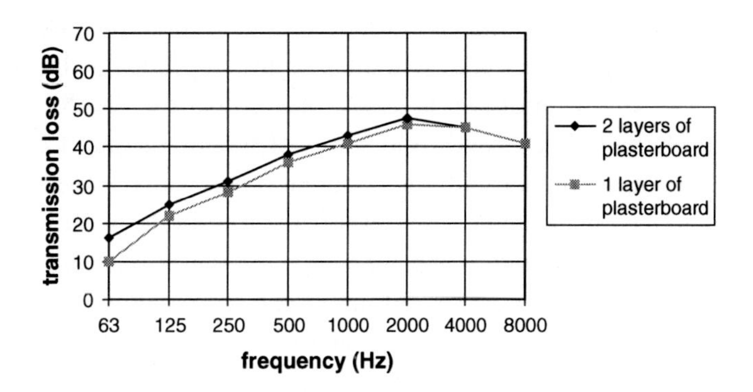

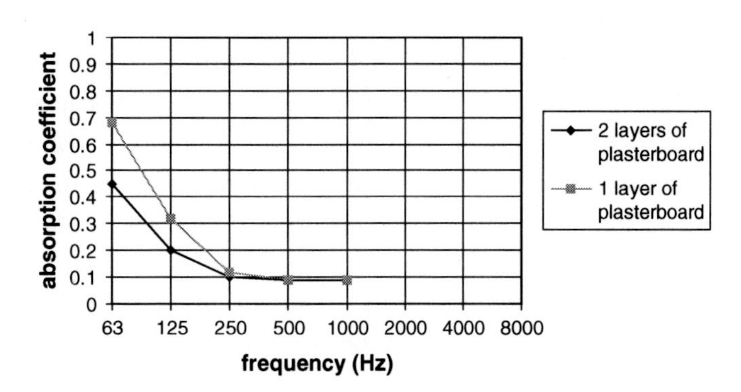

Figure 3.10:

Acoustic performance of a typical ceiling cap: (a) Typical ceiling cap; (b) transmission loss versus frequency; (c) absorption versus frequency. Note how the addition of an extra layer of plasterboard *increases* the transmission loss (isolation), as shown in (b), but *decreases* the absorption, as shown in (c). The addition of a second layer adds more mass to the system and tends to make it more rigid. The effect is that the two layers become more reflective than the single layer. The additional layer therefore improves the isolation by means of reflecting more energy back into the room, but this will of course make the room more reflective or reverberant. If the decay time of the room with only one plasterboard installed was to be maintained after the addition of the second layer, more absorbent materials would need to be introduced into the room. Isolation and absorption are often confused by non-specialists, but this situation clearly demonstrates the difference.

layer with the absorbent felt layer uppermost, in the air gap. The function of these layers is to damp resonances in both the plasterboard and the air. The high viscous losses in the deadsheet tend to suppress any high (see Glossary) resonances in the plasterboard, and the fibrous absorption characteristics of the felt will tend to damp down any cavity resonances, caused by the hard parallel surfaces of the upper surface of the control shell and the lower surface of the isolation shell.

The lower surface of the isolation shell is a classic mass/spring/mass combination, formed by a double layer of 13 mm plasterboard (the lower mass) connected, by contact adhesive only, to a 6 cm layer of reconstituted open cell polyurethane foam of 80 kg/m^3 (the spring). This is in turn connected, by contact adhesive only, to the upper mass. In this case, the upper mass consists of a sandwich of two layers of plasterboard with a 5 kg/m^2 deadsheet in between, nailed to a relatively heavy and rigid series of wooden support beams. The lower plasterboard layers are not in contact with the concrete block walls – there is a gap of a few millimetres all around. This mass layer is therefore free to vibrate, with its vibrations alternately applying compression and expansion forces on the polyurethane foam spring. Such a system might seem a little insecure, but in fact the foam can withstand traction forces of over 3 tonnes per square metre. With a good adhesive, there is little chance of the plasterboard causing problems, as the double layer weighs less than 20 kg/m^2 – less than 1% of the traction limit.

Once again, the lower plasterboard layers will tend to reflect back the incident wave to a degree according to the frequency, and will absorb some of the energy in the internal frictional losses. However, this time its upper surface is not so free to radiate, because it is glued to a spongy foam. The foam will have the action of damping the vibration in the plasterboard, again rather like sticking Plasticine on a bell. The vibrations that do enter the foam layer will be strongly resisted by the mass and rigidity of the upper mass layer. The tendency, therefore, will be for the foam to compress and expand, rather than to move as a whole, and the internal frictional losses in the cellular construction of the foam will be very effective in converting the acoustical energy into heat. This action will be effective down to the resonant frequency of the spring thus loaded.

The upper mass layer acts as a low frequency barrier, and its effect is augmented by the rigidity imparted by the wooden beam structure. In this instance, the two layers of plasterboard sandwich a layer of a 5 kg/m^2 deadsheet, the combination constituting what is known as a constrained layer system. The principle of the damping effect and additional acoustic losses due to the constrained layer are shown in Figure 3.11. Such a system will be reflective down to lower frequencies than the plasterboard alone, but it will also produce more absorption down to lower frequencies. It will certainly transmit less.

Whatever acoustic energy is left to propagate above the constrained layer will then cross the air gap between the wooden beams and proceed on to the layer of 18 mm plasterboard,

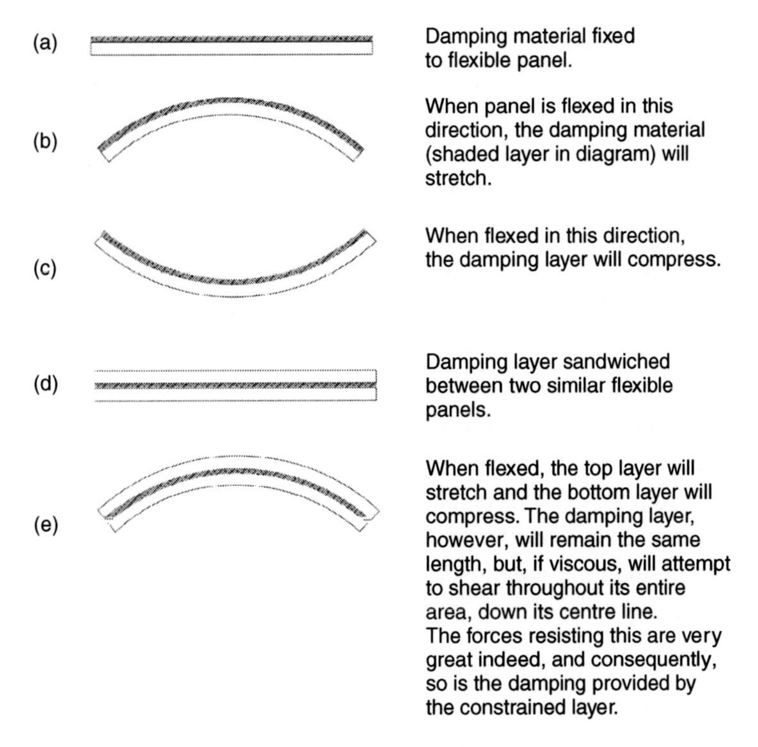

(a)
Damping material fixed to flexible panel.
(b)
When panel is flexed in this direction, the damping material (shaded layer in diagram) will stretch.
(c)
When flexed in this direction, the damping layer will compress.
(d)
Damping layer sandwiched between two similar flexible panels.
(e)
When flexed, the top layer will stretch and the bottom layer will compress. The damping layer, however, will remain the same length, but, if viscous, will attempt to shear throughout its entire area, down its centre line.
The forces resisting this are very great indeed, and consequently, so is the damping provided by the constrained layer.

Figure 3.11:
Constrained-layer damping principle.

attached to a layer of medium density mineral wool by adhesive cement. The mineral wool is itself then attached to the structural ceiling by the same cement-type adhesive. The air gap again contains mineral wool to suppress any cavity resonances, but it also acts as a velocity component absorber, as described in Section 3.1.

The plasterboard/mineral wool/structural ceiling combination, again, is a mass/spring/mass system. As mentioned previously, the reason for using different materials and thicknesses in this upper system was to avoid any weak spots caused by resonances that coincided with the resonances in the lower mass/spring/mass system. Once again, the upper mass of the structural floor acts to resist the bodily movement of the mineral wool spring as it is set in motion by the vibrational energy impinging on the lower mass of 18 mm plasterboard. By this time, the acoustic energy from the source in the recording room will have been reduced to such a degree that what remains to re-radiate into the bedroom is insufficient to create any disturbance to the sleep of the neighbours. By now, also, it will only consist of low frequencies – the highs will be virtually non-existent. If this all seems a little extreme, it must be borne in mind that in order to achieve even 60 dB of isolation, only one one-millionth of the sound power can be allowed to penetrate the barriers.

3.5 Summing the Results

So, from the individual isolation figures of the individual isolation systems we should be able to calculate the total isolation, but there are some practical circumstances which make calculations difficult unless many other factors are known.

If we have a noise source in the recording room, then it is of little help if we only know its output SPL in free-field (anechoic) conditions. If we know that a guitar amplifier can produce 110 dB SPL at 1 m distance in a relatively dead room, it does not mean that it cannot produce more in other circumstances. If we were to take it into a very live room, the level of the direct signal would be augmented by all the reflexions, which could superimpose their energy on the direct sound and easily produce 6 or 8 dB more SPL. Therefore, we need to know the sound pressure levels that will be produced by instruments *in the actual rooms* which we need to isolate, and from this, it will be obvious that for any given sound source, a live room will need more isolation than a dead room. Quite simply if things sound louder in an acoustically live room, then the sound isolation requirements to an adjacent room will be correspondingly greater.

Conversely, if the receiving room is reverberant it will amplify, by means of reflected energy, the noises entering from an adjacent room. In a small bedroom, a double bed provides significant absorption, as do furniture and curtains. In the case that we have been discussing, the bedroom was well furnished, and the noise level measurements were taken at a point 50 cm above the bed. Had the room been stripped of all furnishings it would have become much more reverberant, so the 30 dBA measured after the isolation may then have risen to 35 dBA or so, which could be deemed to be unacceptable for the neighbours to sleep easily. Similarly, if the acoustic control treatment were to be removed from the recording room, rendering it much more reverberant, the maximum SPL produced in the room by typical instruments would also increase, and again the isolation shown in Figure 1.1 may no longer be adequate.

So many variables exist in acoustic control work, which is why it is often so difficult for acoustics engineers to give simple answers to what may appear to be simple questions. What has hopefully become obvious from the previous two sections is that there are often multiple processes at work simultaneously in a typical isolation system, and that they may interact in very complex ways.

3.5.1 Internal Reflexions

There is another reason why the simple summation of the calculated isolation provided by each individual section of the complete structure cannot be relied upon to give the isolation of the whole system. It is that internal reflexions can be set up between the layers, which can add re-reflected energy to the forward-going propagation. This is shown diagrammatically in Figure 3.12. The reduction of this effect requires fibrous layers in the various air spaces. Especially in existing buildings, where the exact nature of the structure is not known, a degree of 'cut and try' can be involved, even for very experienced acoustic engineers.

3.6 Wall Isolation

In general, walls are easier to isolate than floors or ceilings. Figure 3.13 shows a relatively simple solution, the addition of a heavy internal wall with a cavity between the isolation wall and the structural wall. There is usually little problem constructing walls of significantly greater mass per square metre than can be used in ceilings, for example. Floors, of course, will be in some sort of physical contact with the structure, via either springs or mats, but heavy ceilings need to be supported with massive beams or suspension systems, which can often be quite impracticable. A wall, however, as long as it has a solid floor below it, can usually be both very heavy *and* simple to construct. The only problem with the wall system in Figure 3.13(a) is the flanking transmission via the contacts at the floor and ceiling. This is simply avoided by the insertion of spring seals as shown in Figure 3.13(b).

Wall isolation is sometimes also made easier by the fact that the existing structural walls are often more massive than the floors (other than those built directly on the ground) and ceilings. Very effective wall isolation can be achieved by the use of floated concrete block walls filled with sand. These are especially effective if the structural walls are also massive and well damped. In new buildings, it can be advantageous to ask the architects to

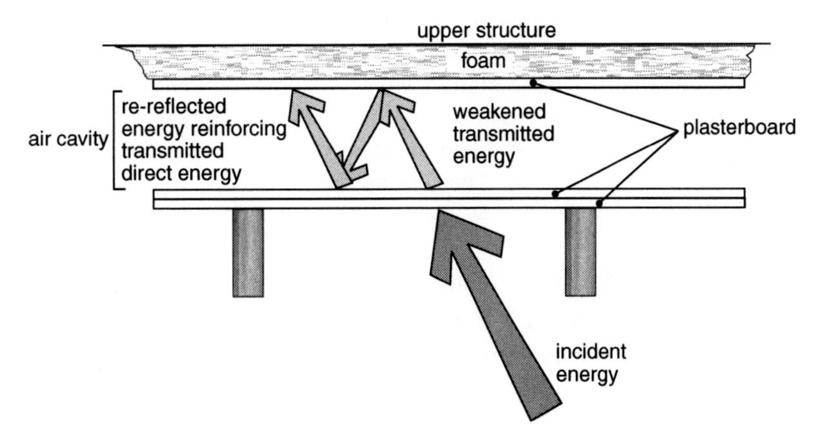

Figure 3.12:
Re-reflexion between layers. Resonant build up in an air cavity can reduce isolation in a similar way to which the build up of reverberant energy in a space can make a source louder than it would be in free-field conditions.

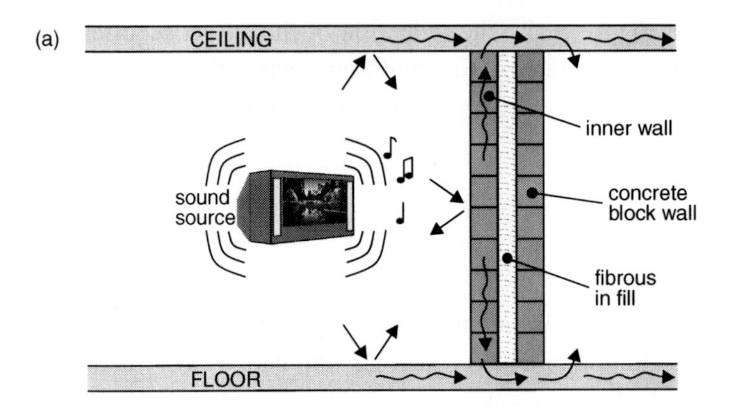

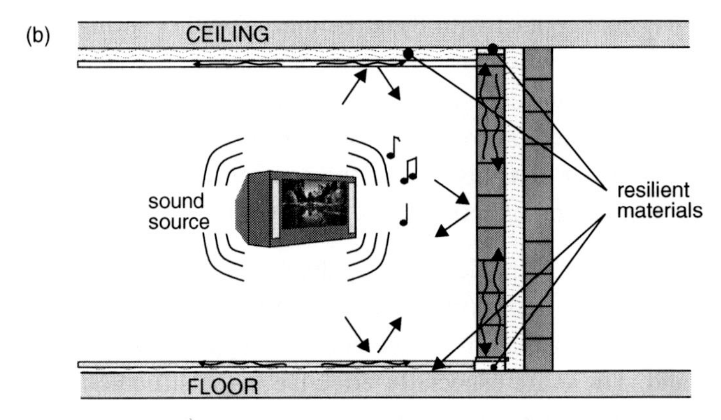

Figure 3.13:
The benefits of floating surfaces: (a) The addition of an inner wall to aid isolation between the rooms can be largely ineffective if flanking transmission via the floor and ceiling pass round it. What is more, resonances within the inner wall structure can themselves pass additional energy into the floor and ceiling, perhaps making matters worse at some frequencies; (b) Protecting the floor and ceiling from direct sound, plus floating the inner wall away from the walls, floor and ceiling, can greatly reduce the amount of sound passed into the structure, and hence to the neighbour.

use sand-filled concrete blocks in the main structure. The combination of structural and floated wall systems so made, with a 5 or 10 cm air space between, can easily produce more than 70 dBA of isolation, which is extremely difficult to achieve in floor or ceiling isolation systems unless these can be used simultaneously, both above and below the structural ceilings and floors.

In shared buildings, therefore, where access to the walls, floors and ceilings of neighbouring premises is not possible, wall isolation is by far the easiest to achieve in one-side-only treatments.

3.7 Lighter Weight Isolation Systems

In some circumstances, either only a limited degree of isolation is necessary, or moderate isolation is needed in buildings with weak floors. In such cases, systems such as those shown in Figure 3.14 can be used. The technique simply involves lining the entire room with open cell polyurethane foam (or mineral wool) of suitable density and thickness, and lining the interior with plasterboard. With low density spring materials (foam or mineral wool) on the walls and ceiling, the system resonances can be kept quite low. On the floor, however, low

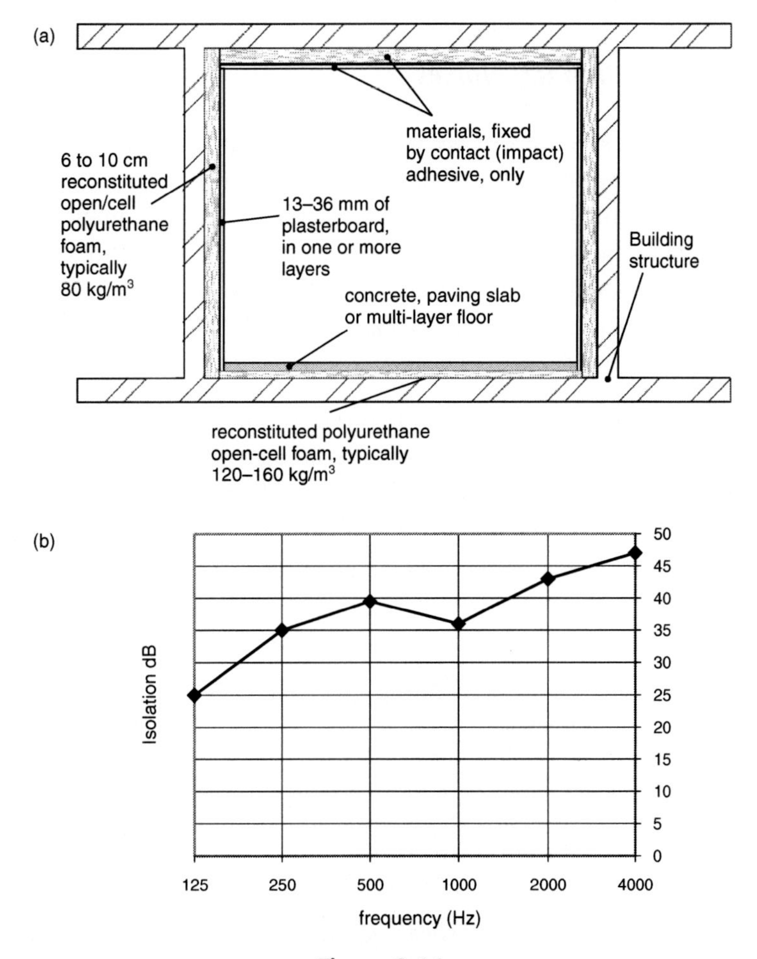

Figure 3.14:
Distributing the isolation load: (a) Typical construction. Isolation of this type can be very effective. The walls and ceiling bear much of the load, which can make this technique useful in premises with weak floors. Alternatively, a similar system can be used with mineral wool and a cement-based adhesive. (See Figure 1.1.) Contact adhesive gives a more instant bond, without the need for support during setting, but the noxious fumes can be a problem during construction; (b) The plot shows the typical extra isolation achieved by the sort of system shown in (a), above that provided by the basic building structure.

density materials would lead to instability, such as shown in Figure 3.6, so to keep the system resonances sufficiently low with the higher density spring, a correspondingly greater weight needs to be added. This can be achieved by laying concrete paving slabs on top of the floor covering, or even casting a reinforced concrete slab if it is feasible.

The results with this type of isolation system can be surprisingly good, though their resistance to the ingress of impact noise from outside the structure can be poor. The other benefit is that the weight is distributed about the room surfaces, and the whole load is not imposed on the floor. Even the walls carry some of the load, which can be very advantageous in weak buildings.

3.8 Reciprocity and Impact Noises

Notwithstanding certain restrictions, if we were to build two adjacent rooms with 60 dB of isolation between them, it would not make any difference which room contained the source of the sound and which was the receiving room. The isolation would remain the same, even though the order of materials in the isolation system may be asymmetrical, as shown in Figure 3.15. However, this only remains true for airborne sound, where the solid materials present an acoustic impedance much greater than that of the air because their densities are enormously different to that of air.

Figure 3.16 shows the vertical separation of two rooms, the lower one having an open cell foam/plasterboard ceiling isolation, which is typical in many rooms with limited ceiling height because it provides a means of, in many cases, adequately protecting the rooms above from airborne noise from below. Once again, 50 dBA of isolation could be achieved, irrespective of the direction, if the sounds reaching the horizontal surfaces were airborne.

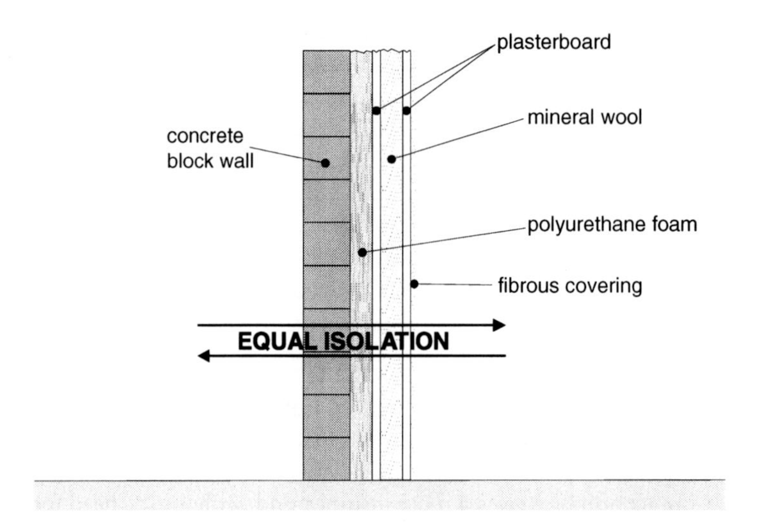

Figure 3.15:
Non-directionality of airborne sound isolation. Despite the asymmetric distribution of the isolation materials, the isolation for airborne sounds will be equal in either direction.

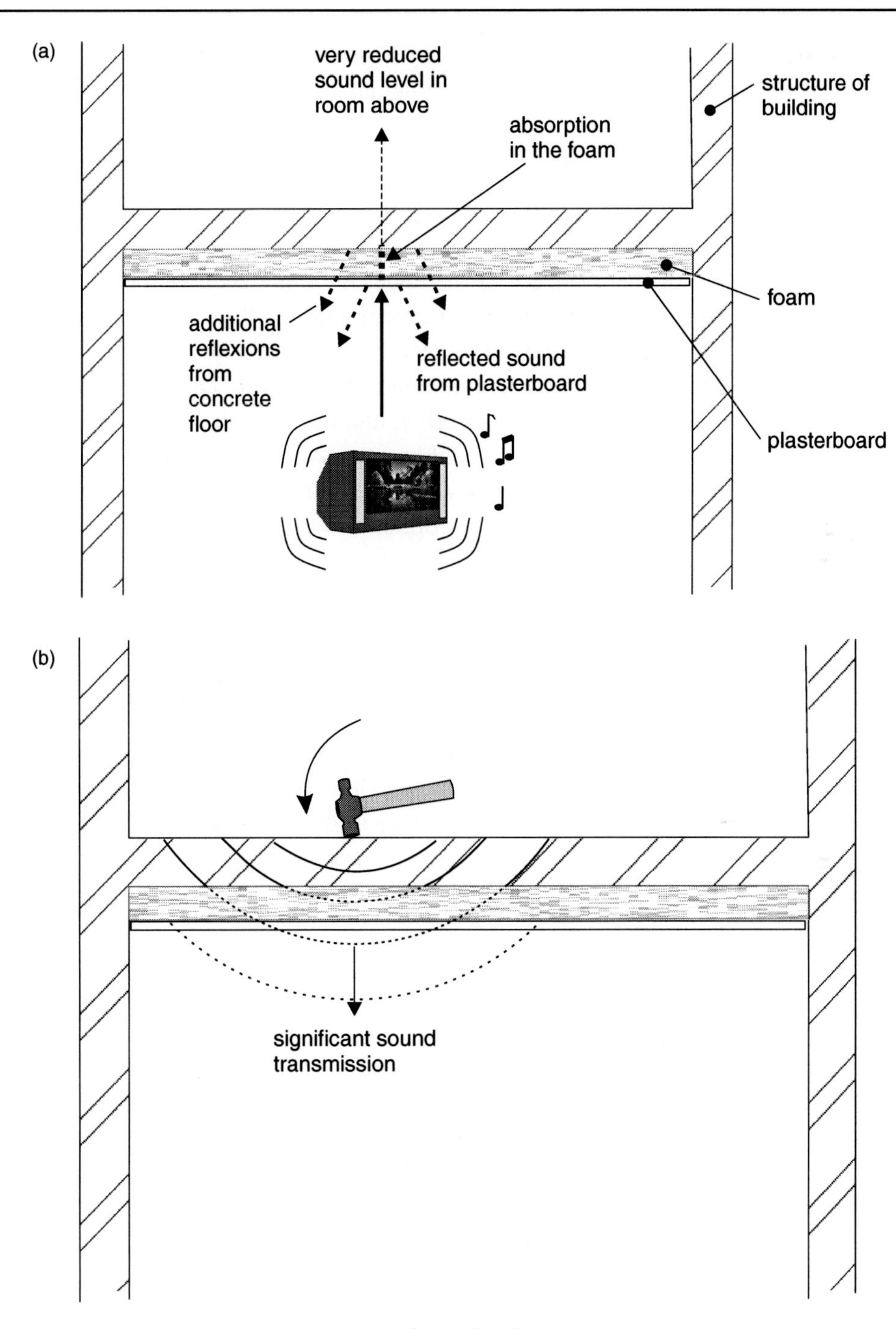

Figure 3.16:
Airborne and impact noise isolation: (a) For airborne sound; (b) for impact noises. Direct impacts on the structure can set the whole thing in motion, including anything that may be attached to it. What is more, a very wide range of frequencies are produced by the impact, which are capable of exciting many resonances.

The problem arises when people, and especially when wearing high, slim heeled shoes, walk on the upper floor. In this case, the weight on a small area when the heel strikes the floor presents a totally different case than if somebody were to strike the lower plasterboard ceiling with a similar shoe heel. The difference between this situation, where the isolation is *not* reciprocal, and the case of airborne noise isolation, where reciprocity exists, lies in the nature of the impedance differences.

When a shoe heel strikes the hard surface of the upper floor, the heel is brought to rest almost instantly. The transmission of the impact energy into the floor is great because the termination is abrupt and resistive (solid). If the heel were to strike the plasterboard surface, below, the surface presents a less abrupt termination to the impact. The plasterboard on the foam will tend to give, acting like a shock absorber, and the springiness will rebound and return some of the energy to the heel. There will be a sensation of a bounce-back; the termination will be more reactive (see Glossary). Such solid object-to-solid surface contacts are *not* typical of the gas/solid/gas contacts of normal airborne sound isolation, and they behave differently.

Without making a complete break in the solid materials it is very difficult to prevent the structural impact noises from penetrating most practical ceiling isolation systems. This is where the problem arises when structural ceilings are too low, and there is too little room to use a separate set of ceiling joists on the floated walls to float a completely separate ceiling structure. One solution to the problem is to use a simple floated floor system, such as that shown in Figure 3.4(d), on the floor above, but this may not be possible when the upper floors are occupied by different owners. In any case it may not be possible to float the floors because of the need to change doors and their frames if the raised floor reduces headroom below legal limits.

In general, adequate isolation may be difficult if not impossible to achieve when chosen premises do not have adequate ceiling height.

3.9 The Distance Option

The degree to which sound isolation is required in any proposed recording studio building is dependent upon several factors: the construction of the building (whether light or heavy); the proximity of noise sensitive neighbours; the proximity of external noise sources, and also the nature of the recording to be undertaken. All affect the noise control calculations. At one extreme, let us imagine a synthesizer-based musical group, who wanted to make recordings in a studio on the ground floor of a massively built farm building in a valley. If this building were sited miles away from any neighbours, and was not subject to any farm machinery or heavy transport noise, then the sum total of isolation needed would probably be to fit double-glazing with 40 cm between the panes of glass, and change the doors to a type with good acoustic isolation.

At the other extreme, let us consider the isolation needs for a recording studio for orchestral and choral use, sited in an office building in a location above an underground railway and surrounded by streets with heavy traffic. In this case, it would be unacceptably expensive for noise-induced delays to affect the recording of a 100-piece orchestra due to underground train rumble during quiet passages. It would also be unacceptable for the fortissimo passages to disturb people who may be concentrating on their work in adjacent offices. The isolation work may require the construction of an entirely floated inner isolation shell, of considerable weight, which may need reinforcement of the floor in order to support it. In turn, this new hermetically sealed box would need to be penetrated by HVAC (heating, ventilation and air-conditioning) systems, which themselves would require considerable isolation. The costs could be very high indeed.

In the first of the cases discussed above (the studio for the synthesiser band), probably the only acoustic recordings would be of close-mic'd vocals and the occasional guest musician. Neither the sound egress nor ingress would be particularly problematical, because there would be nobody to annoy by any external leakage of the sound, and except for the rare coincidences of extreme weather conditions during the recording of an even rarer acoustic guitar, the signal-to-noise ratio of recordings would generally be excellent. The only problem in this type of situation is that it would be restricted in its use. To some degree, though, all recording studios are restricted in their use; none are universal in their applications.

In the second case, after due isolation work and internal acoustic treatment, the studio would be capable of recording just about any type of music at any time of day. The main restriction would be financial. After spending so much money on the studio building and preparation, the hourly rate would need to be prohibitively high for the general overdubbing of vocals, for example. That high hourly rate may be inconsequential though, compared to the cost of an orchestra's expenses when travelling to a studio in a distant out of town location.

It is true, therefore, that an isolated location, by virtue of its distance from noise related problems is an option for reducing isolation costs, because the physical, geographical isolation is acoustic isolation. However, when overall convenience of access is important, this geographical isolation may be completely impractical, so the only solution may lie in the choice of an inner-city site and a considerable amount of acoustic engineering.

3.10 Discussion and Analysis

Clearly, isolation is not a simple subject to grasp, and no simple computer programs can solve the problems. Isolation is a subject for specialists, and where critical situations exist, the cost of calling in a specialist before construction will surely be less than the cost of trying to fix the problems after construction. In fact, the problems may lie deep within the construction, in which case total rebuilding may be required. Intuitively, leaving out a layer of fibrous material between two of many layers in an isolation system may seem insignificant, but if the cavity resonates without the lining, the air at resonance can be a remarkably strong coupling medium.

The addition, or not, of a small component in a complex structure may thus make a disproportionate difference to the results, and even one decibel of improved isolation may be important. Experience has shown that it is better to design for inaudibility for the neighbours, rather than to legal limits, because peace with the neighbours can be crucial to their tolerance of the occasional nuisance. Nevertheless, ultimately the law will decide if the worse comes to the worst, and that is where that 1 dB can mean make or break.

In fact, if the room shown in Figure 1.1 had failed to achieve its low frequency isolation target by 1 dB, then where would one begin to look for the weakness? What would one do to plug the 1 dB leak? If the room had been completed, with all its decorations, and was considered by all concerned to sound good, then how could an extra decibel of isolation be found without perhaps destroying all the finishes, changing the sound within the room, and severely straining the finances of the owners? These are not easy questions to answer; not even for experts, so to fail by 1 dB could be disastrous.

Efficient acoustic engineering implies that the maximum effect is achieved with the minimum use of resources, but isolation in difficult situations does not always lend itself to precise analysis and the presentation of finely tuned solutions. Despite the fact that isolation can be expensive, there needs to be a safety margin in the calculations and assessments of situations, because the science is not exact and the cost of failure can be so grave. It is always better to err on the safe side. The subject of the rather imprecise nature of isolation measurements is discussed in more detail in Section 10.7.1.

Isolation work is something that should not be undertaken without adequate knowledge of the subject, because much of it is very counter-intuitive, and the potential for disappointing results and wasted money are very great.

3.10.1 Fibrous and Cellular Springs – Thicknesses and Densities

To highlight some of the above points, and to answer the queries of some readers of the first edition of this book, it may be informative to discuss the seemingly simple subject of the resilient layers below floated floors. However, despite the fact that it may seemingly be simple, there are many factors which need to be taken into account in order to choose the most appropriate density and thickness for any given application.

As shown in Figure 3.5, if the elasticity of the floor floating springs is either too great or too little, good isolation will not be achieved. Below the resonant frequency of the mass/spring system, such as that formed by a floated concrete slab and a mineral wool layer beneath it, most of the isolation is lost. At and around the resonant frequency, the isolation can actually be worse than with no treatment at all, because of the amplifying effect of the resonance. In general, isolation is considered to begin to become effective above about 1.4 ($\sqrt{2}$) times the resonant frequency; thus if isolation down to 20 Hz is required, the floor resonance should be below about 14 Hz (14 Hz $\times$ 1.4 = 19.6 Hz). Nevertheless, effective isolation does not tend to occur until the resonant frequency is around half the lowest frequency to be isolated. In practice, for full frequency range isolation, 12 Hz is a good target to aim for.

Using a weaker spring will lower the resonant frequency, as will increasing the weight placed upon it, but as shown in Figures 3.6 and 3.7, if the spring material is too weak then floor stability and flatness may be lost. If, for example, 70 kg/m^3 mineral wool were to be used as the base, then doubling the density to 140 kg/m^3 would *increase* the resonant frequency (by virtue of providing a stiffer spring) by $\sqrt{2}$ (1.414). Halving the density to 35 kg/m^3 would reduce the resonant frequency by the same amount. If the higher density was necessary in order to give increased stability, then the only way to reduce the resonance frequency to its original value would be to double the weight resting upon it.

If the resonance frequency were required to be halved, whilst using the same mineral wool spring (or, in fact, any given spring), then the weight would need to be quadrupled. That is, if the frequency of resonance of a 10 cm concrete slab floor resting on a given mineral wool base spring were to be halved, then 40 cm of the same concrete would be necessary. Adjusting the springs is therefore often a more practical (and height saving) option than adjusting the masses. Of course, one could always increase the density of the concrete if the height increase from 10 cm to 40 cm was unacceptable, but even a density change from 2 tonnes per cubic metre to 3 tonnes per cubic metre would still need a thickness of about 27 cm.

Note, here, that the *weight* (mass $\times$ gravity) is what is important. This gives rise to the force acting down upon the spring. In the case of the mineral wool, the density affects the elasticity of the spring, whereas the concrete density affects the weight (and hence force) *applied* to the spring for any given thickness of slab.

However, as we shall see below, we cannot reduce the mineral wool density too far or it may become so compressed that it loses its elasticity and ceases to behave like the required spring. Furthermore, if the mineral wool is compressed to a thickness commensurate with any irregularities in the surface of the floor on which it is laid (or the concrete that may be poured on top of it), then at some places rigid contacts may be made as some points penetrate the material. This problem can be ameliorated by placing a layer of plywood, for example, on top of the mineral wool, and smoothing the floor below it, but such treatments tend to be wasteful of time and materials compared to simply using a thicker layer of mineral wool and/or a higher density.

3.10.2 The General Situation with Masses and Springs

Figure 3.17 shows a graph of the relationship between the mass and the resonance frequency when the mass is loaded on a moulded panel material made from glass fibre, sold under the trade name of Acustilastic. The aforementioned 12 Hz resonance can be seen to be achieved with a load of about 230 kg/m^2, which would signify a 10 cm slab of 2.30 density concrete (2.3 tonnes per cubic metre). Another type of graph which is commonly encountered is shown in Figure 3.18. In this case, it is for a granular, rubber-based matting used typically under brick or concrete block walls, where the loads can be very great. The graph shows the degree of compression under load – up to 300 tonnes per square metre. Although this does not directly show the resulting resonant frequency, it can be calculated from applied force (the weight) and the degree to which the material is compressed under static load as a proportion of its original thickness (see Subsection 3.10.4). In both cases (Figures 3.17 and 3.18), it can be seen that the relationships are described by curves, and not by straight lines, because as the materials are compressed their densities increase. Consequently, a 70 kg/m^3 mineral wool base compressed to 50% of its original thickness would then have a density of 140 kg/m^3, and so would resist a given extra force (weight placed upon it) more than would the uncompressed mineral wool. In other words, if a 50 kg weight over a given area compressed the original 70 kg/m^3 mineral wool by 4 mm, the addition of an extra 50 kg/m^2 would only compress the material by around a further 2 mm.

In the case of Figure 3.17, the graph stops at 2250 kg/m^2. This material (Acustilastic) has an allowed maximum deformation of 30% (which corresponds to a resonant frequency of 4 Hz with the 2250 kg/m^2 load) because above around 2500 kg/m^2 the material goes beyond its elastic limits and begins to permanently deform due to the breaking of individual fibres. The

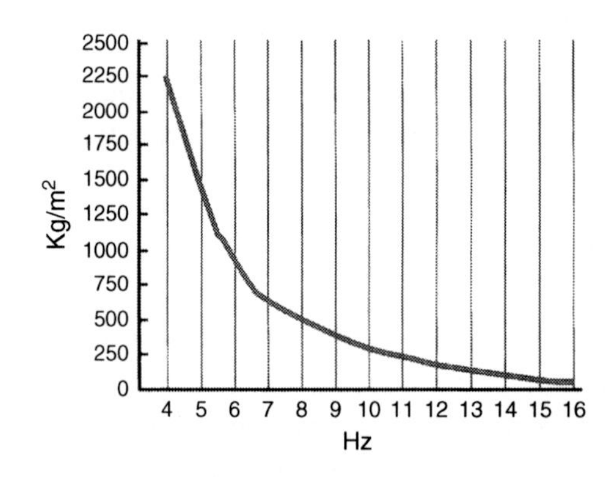

Figure 3.17:
Load versus resonance frequency graph for the material Acustilastic. It can be seen that the weight of the load must be quadrupled in order to halve the resonance frequency; hence to *quarter* the resonance frequency the load must be multiplied by *sixteen*.

(a)

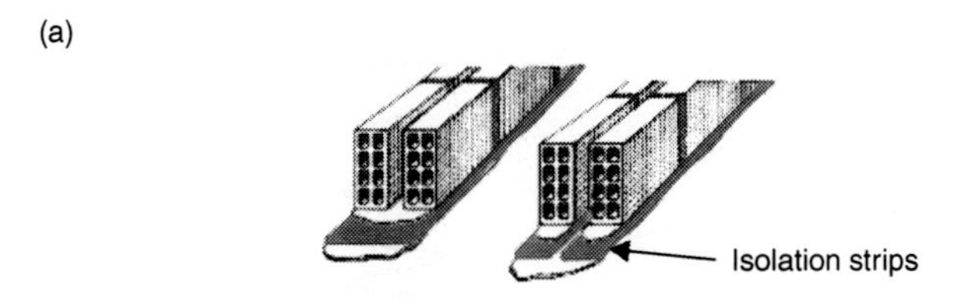

(b)

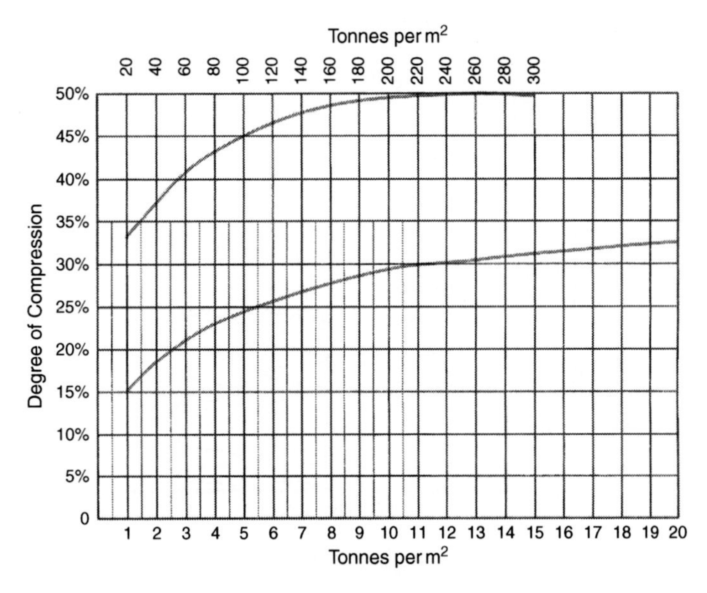

Figure 3.18:
(a) Typical application of the material for floating heavy walls; (b) Plots showing the degree of compression of the material with different loads.

recommended usable load for this material is 2000 kg/m^2. However, if significantly heavy loads are expected to be placed on such a floor, the usable load would need to include the extra load, and cannot be consumed by the floated slab alone. The resonant frequency would be a function of the total load. In the case shown in Figure 3.7, where the walls and ceilings are also loaded on the slab, their extra weight will also contribute to the overall load and so they must also be taken into account when calculating the final resonant frequency of the floor when the whole room is finished and loaded with equipment.

On the subject of the thickness of a 'spring' material, consider the examples shown in Figures 3.19, 20 and 21. In Figure 3.19, a single spring is loaded by a given weight and can be seen to compress by 20% of its original length. In Figure 3.20, two similar springs are placed in parallel and loaded by the same weight, where the deflexion can be seen to be less for each

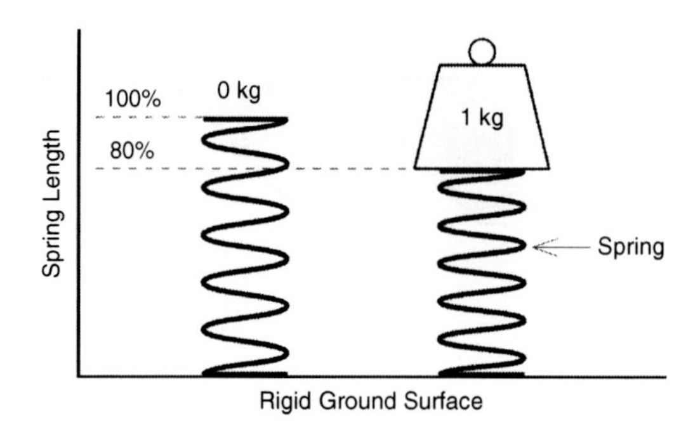

Figure 3.19:
In the example shown, a weight of 1kg has compressed the spring to 80% of its unloaded length.

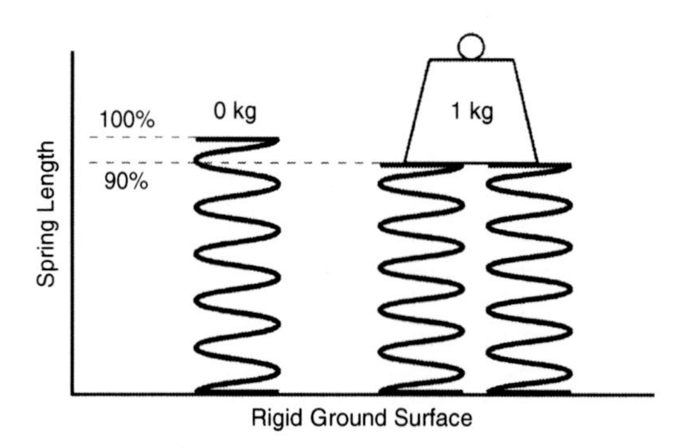

Figure 3.20:
In the case of the two parallel springs, each identical to the single spring shown in Figure 3.19, the compression for each spring is half of the compression of the single spring: the force on each spring is half of the force on the spring shown in the previous figure. The resonance frequency will rise by $\sqrt{2}$ (1.414) compared to the resonance frequency shown in Figure 3.19 because the stiffness has been increased.

spring. In Figure 3.21, however, the two springs are placed one on top of the other – in series – and as the force acting upon them is equal (i.e. they are each carrying the same load – it is *not* divided between them) they each compress to the same degree as the single spring shown in Figure 3.19. The proportional compression of each spring would therefore be the same as for the single spring.

Nevertheless, twice the total deflexion would be experienced, so effectively the spring would be softer. The stiffness would be halved, so the resonant frequency of the mass/spring system would be the resonance of the same weight on a single spring *divided* by $\sqrt{2}$, i.e. 1.414. (The same work cannot compress two springs as many times per second as it can compress only one spring.)

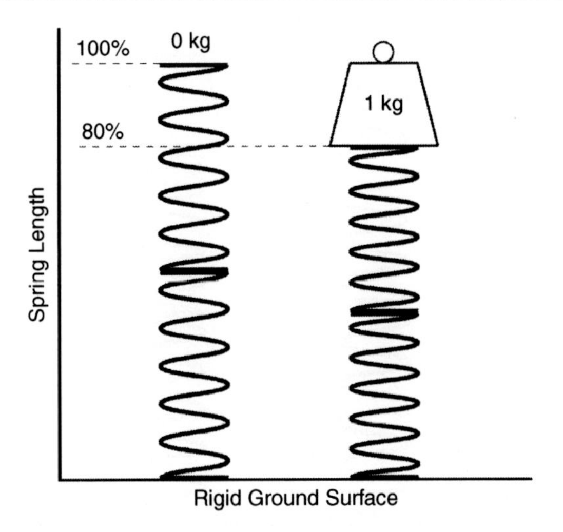

Figure 3.21:
In the above example, two springs, each identical to the spring shown in Figure 3.19, are placed in series. The force acting on each spring is the same, and is also the same as the force acting on the spring in Figure 3.19, so the compression of each spring will also be the same. However, the overall effect is that of a softer spring, as the total compression *distance* is double that of a single spring and the resonant frequency will be that of Figure 3.19 divided by $\sqrt{2}$ (1.414).

Consequently, *doubling* the thickness of a mineral wool base below a concrete slab would reduce the resonant frequency by a factor of about 1.4. *Halving* the thickness would increase the resonant frequency by about 1.4. Doubling the density or halving the density of the mineral wool would also increase or reduce the resonant frequency by a factor of about 1.4, respectively. However, as previously mentioned, if the initial thickness of the material is too little, and/or the density is too low, the material may be compressed to the extent shown in Figure 3.5(c), in which case the spring effect may be lost, and with it much of the isolation if it were the base for a floated floor.

Figure 3.22 shows a somewhat more complicated case. Here, two springs of different stiffness are placed in series, one on top of the other. Despite being in series physically, their stiffnesses behave like two electrical resistors in parallel – the resultant stiffness is always less than the least stiff spring, just as the combined resistance of two parallel resistors is always less than the lower of the two resistances. For resistors in parallel, the combined resistance is calculated by dividing the product by the sum; in other words:

$$\frac{R_1 \times R_2}{R_1 + R_2} \tag{3.1}$$

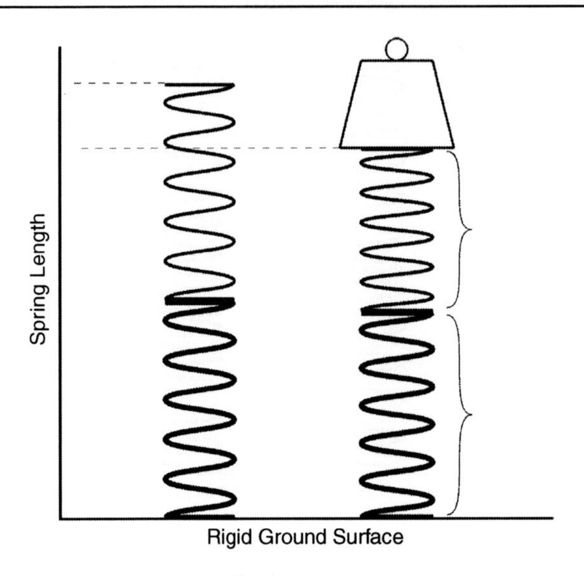

Figure 3.22:
In this example, two dissimilar springs have been placed in *series*. The result is still a single resonance frequency, because the springs combine to yield a total stiffness which can be calculated from the product divided by the sum, as for *parallel* resistors or inductors, or capacitors in series.

For two parallel resistors of 10 ohms and 20 ohms:

$$\frac{10 \times 20}{10 + 20} = \frac{200}{30} = 6.666 \text{ ohms}$$

(In fact, the springs in series behave like two capacitors in series, but the concept of resistors is perhaps more widely understood by people not so conversant with electrical theory.)

The resonance frequency of the system shown in Figure 3.22 with respect to that shown in Figure 3.19 can be calculated from the change in stiffness, but it will always be *below* the resonance frequency of either of the single springs loaded with the same weight. However, if we separate the springs by another mass, as may occur if a floated floor is laid on top of an already resonant floor – which has its own mass and spring components – the situation begins to become rather complicated. The concept is shown diagrammatically in Figure 3.23. If the two resonant systems share the same resonant frequency, individually, then the interactive coupling will move the resonances apart. Prediction of *exactly* what will happen in practical structures is very complicated, as many of the relevant parameters will not be known.

On the other hand, if the resonances of the two systems are well separated, then one resonance may dominate, or the pair may behave chaotically, especially under transient excitation. Parallels can be drawn with the behaviour of double pendulums and their 'strange attractors' described in chaos theory. In the automobile industry, such complicated multiple

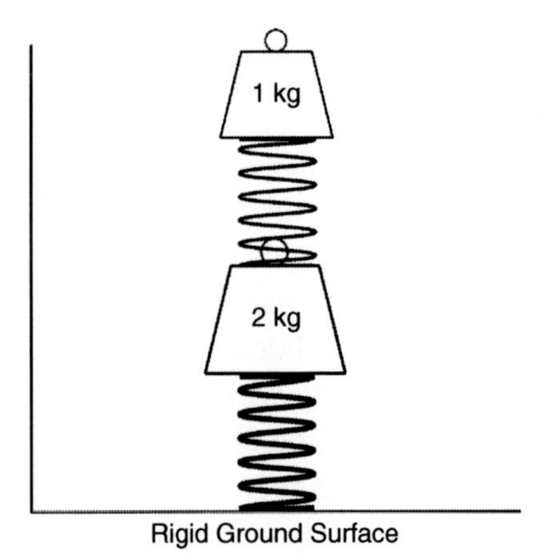

Figure 3.23:
This can be a difficult problem to calculate, because in practice, combined structures are rarely pure springs or pure masses. The springs in this example are separated by a mass, so *two* resonant systems exist. There are two degrees of freedom, and interactions of accelerations. Had the two springs and masses in the above case been equal, they would still not resonate at the same frequency because the resonance would be displaced by the interactions.

spring/mass systems *are* sometimes used, where the huge research and development costs of modelling and measuring can be amortised over a long production run of identical vehicles. Conversely, most acoustic calculations for architectural sound isolation purposes relate to one specific job, so it is usually better to stick to the use of more easily predictable systems, and to avoid double mass/spring systems because the research costs usually cannot be supported.

In multiple mass/spring systems, factors such as the relative masses or stiffnesses will also have to be taken into consideration, as will any damping or loading effects which one system may impose on the other. What is more, if the masses are reasonably similar to each other they may interact in a way which is very different to the case if they were greatly different. The importance of the various rules which govern the behaviour of such systems can therefore change with the relative proportions of each system.

As a practical example of this type of double system, one can imagine a case where a floated isolation floor needs to be laid over a structural floor at ground level, but with a subterranean car park below. It is not uncommon to find such floors with natural resonances in the region of 20 or 30 Hz. In general, if a load-bearing, structural floor exhibits a relatively high resonance frequency, then a floated floor of soft springs and high mass (low stiffness, high inertia) may be needed in order to achieve much isolation below the resonant frequency of the structural floor. If that floor cannot support such a load, then achieving useful low frequency isolation may be out of the question. These are the sorts of situations where the experience of acoustics engineers is invaluable on a case-by-case basis. There tend to be too many variables in this type of equation to expect a simple computer analysis to predict the results.

In practice therefore, combinations of slab thickness and density both affect the *weight* (mass), and changes of the thickness and density of the mineral wool will affect the *stiffness* of the spring. All of these changes will have effects on the resonance frequencies of the systems. When considerations such as floor height or strength must be taken into account the parameters will be chosen for the most practical result. For example, a mineral wool under-layer which is doubled in thickness and halved in density will lower the resonance frequency by half; achieving a reduction of $\sqrt{2}$ from the doubling of the thickness and a further $\sqrt{2}$ from the doubling of the density. The resonance could be returned to its original frequency by quartering the weight of the concrete slab. Such a change may be necessary if a lighter weight or less high floor was required, but the floor would also, obviously, be much weaker with only one quarter of the thickness of concrete. On the other hand, rigidity could perhaps be increased by choosing a thicker layer of a lower density concrete and using more steel reinforcement mesh. Such are the decisions of an acoustics engineer.

3.10.3 Measured Characteristics of Various Suspension Materials

In order to show the effects of the practical application of a selection of commonly used materials for floating floors, a series of measurements was undertaken specifically for the second edition of this book. Materials were cut into samples of 10 cm squares (i.e. 100 cm^2). On each sample, a 1 kg weight would equate to approximately 100 kg/m^3, allowing for small differences due to the different ratios of surface area to edge. The material samples were placed in a device which allowed different weights to be loaded onto them, and to be distributed evenly over their surfaces by means of a light but rigid plate of the same size as the samples. In each case, the height of each material sample was measured unloaded, then measured again at increments of 1 kg between 1 kg and 10 kg, the latter corresponding to approximately 1 tonne per square metre.

Figure 3.24 shows the static loading effect on two samples of a reconstituted polyurethane, open-cell foam, known as Arkobel, having a density of 60 kg/m^3. In three of the measurements the sample is the same, but it has been rotated in each case through 90° to measure each axis separately. There is one axis of recommended use, preferred because of the direction in which the foam is cut after compression during the manufacturing process. The preferred axis can be seen to exhibit a curve which is a little less steep overall. However, what is interesting is that when the sample was *un*loaded, from 10 kg back to 0 kg, the curve did not follow the compression curve. The material was exhibiting hysteresis (see Glossary) under static conditions, but this should not introduce any unwanted effects under dynamic conditions because the vibrational movements would be much too small to exhibit non-linearities.

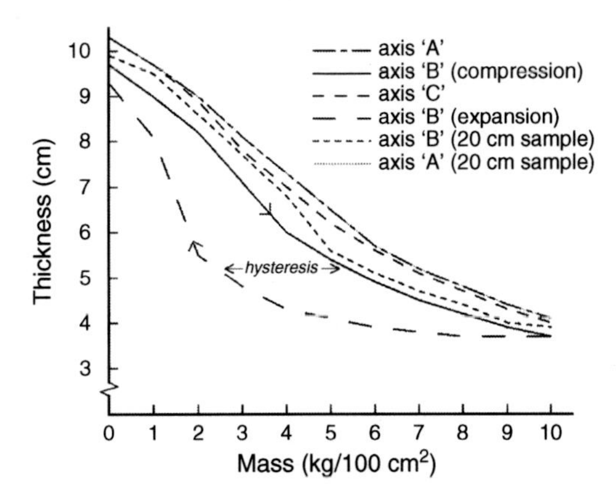

Figure 3.24:
Compression versus load for 10 cm and 20 cm Arkobel (a reconstituted polyurethane foam) of 60 kg/m^3 density. The thicknesses of the 20 cm samples have been divided by two on the plots in order to make easier comparisons with the 10 cm samples. In fact, the 20 cm sample compressed in 'axis A' coincides so well with the 10 cm sample in the same axis that the two lines are almost indistinguishable. Axis B is the normal axis of use. The hysteresis loop shows the different behaviour in compression and expansion.

A 20 cm sample was then tested in the 'A' and 'B' axes, with the figure showing the results divided by two, which overlays the 10 cm and 20 cm curves for easy comparison. In the 'A' axis, except for the section between 0 kg and 1 kg, the plots overlap exactly, showing the practical results of the effect shown in Figure 3.21 relating to springs in series. The 'B' axis of the material, in both the 10 cm and 20 cm samples, exhibits a slightly less regular curve, but the agreement at low and high loadings is relatively consistent.

Figure 3.25 shows the plots for a type of Arkobel with double the density of the first sample – 120 kg/m^3. Samples of 6, 9 and 18 cm were tested. Normally, this higher density is used in smaller thicknesses, such as 3 cm, but for these tests the deflexions would be too small for reliable measurement, so greater thicknesses were chosen. The 18 cm sample measurements have been divided by two for easier comparison of the plots. It can be seen that the proportional compression is much less than for the 60 kg/m^3 sample. Nine point seven centimetres of 60 kg Arkobel compresses to about 3.7 cm under a 10 kg load, whereas 9.1 cm of 120 kg Arkobel compresses to only 6.9 cm under a similar load – a deflexion of 2.2 cm as opposed to 6 cm.

An interesting observation from Figure 3.25 is the comparison of the multiplication of the 6 cm sample measurement by 1.5, overlaid on the 9 cm sample plot. At loadings above 8 kg, the 6 cm sample curve begins to flatten out, showing that the thinner sample is beginning to bottom; that is, compress beyond its elastic limits. This effect is also observable with the thinner samples in Figures 3.26 and 3.27. The bottoming occurs when the material begins to be compressed into a solid mass, and the mechanisms which give rise to the springiness begin to cease to operate. (See Figure 3.5 (c)).

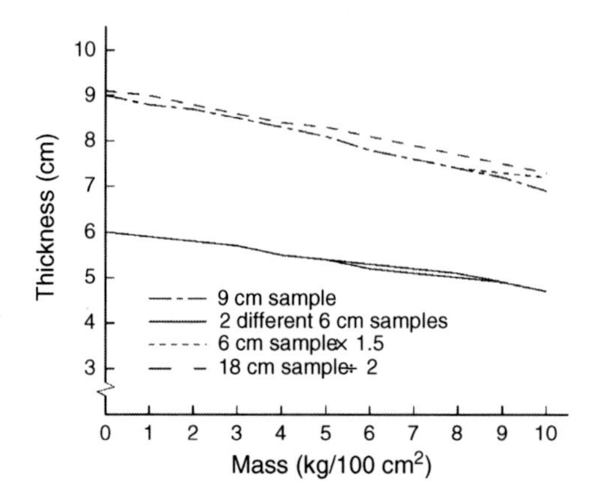

Figure 3.25:
Compression versus load for different thicknesses of Arkobel (reconstituted polyurethane foam) of 120 kg/m^3 density. The 6 cm measurement multiplied by 1.5 is overlaid on the measurement for the 9 cm sample. The two plots correspond very closely up to 8 kg, after which the 6 cm sample begins to compress less, indicating that it is beginning to 'bottom-out'. The 18 cm sample ($\div$ 2) and the 9 cm sample show no such effect.

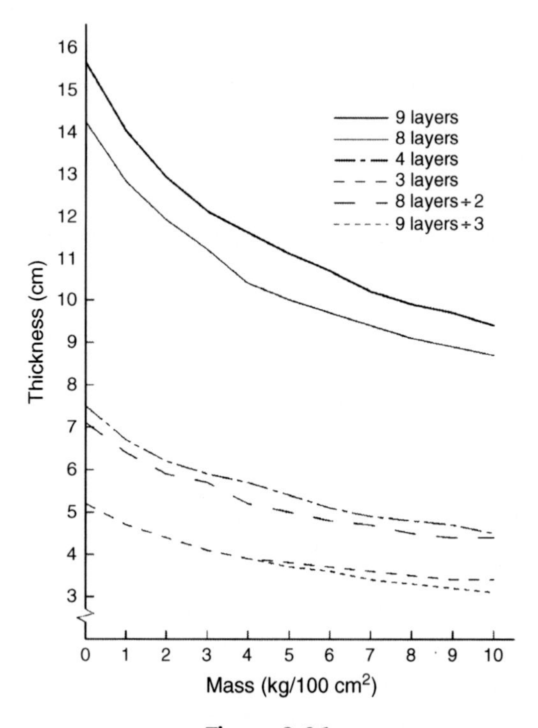

Figure 3.26:
Compression versus load behaviour of various thicknesses of Blancot, a cotton-waste felt material of around 60 kg/m^3 density. The 3 layer plot and the 9 layer plot $\div$ 3 are identical up to 4 kg loading, after which the 3 layer plot shows evidence of bottoming out as the thinner sample begins to compress proportionately much less than the thicker, 9 layer sample.

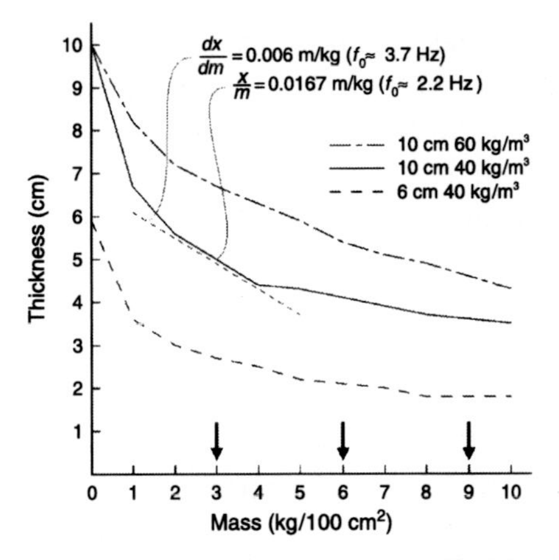

Material	Load (kg/100 cm²)	Resonance (Hz)
10 cm 60 kg/m³	3	4.4
	6	3.4
	9	3.0
10 cm 40 kg/m³	3	3.7
	6	4.5
	9	5.3
6 cm 40 kg/m³	3	5.4
	6	6.0
	9	15.0

The fact that the resonance frequency *rises* as weight is added to the 40 kg/m³ mineral wool samples means that even at 3 kg/100 cm² (or 300 kg/m²) the material is already stiffening to the point where it is losing its spring effect.

Figure 3.27:
Compression versus load for different thicknesses of 40 kg/m^3 and 60 kg/m^3 mineral wool. Again, bottoming-out is evident above 8 kg/100 cm^2 loading for the thinner sample of the lower density material. The resonance calculations relate to the descriptions given in Subsection 3.10.4. Of the two samples which had a thickness of 10 cm when unloaded, the lower density material, as expected, compresses much more as weight is added, up to about 4 kg/100 cm^2, but thereafter compresses less than the higher density sample. This suggests the beginning of a bottoming-out process for the lower density sample under higher loads. In the resonance/load table, note how the resonance frequency begins to rise abruptly once the bottoming-out begins, showing that the spring is becoming very stiff.

The Blancot (a cotton-waste felt material) shown in Figure 3.26 exhibits a different shape of curve to the foam materials in the two previous figures. In this case, the 9 layer measurement divided by 3 and overlaid on the 3 layer measurement shows that the 3 layers begin to bottom at loads above 4 kg (400 kg/m^2), and between 8 and 10 kg there is a horizontal straight line, showing that the material has ceased to act as a spring. The density of this Blancot is the same as that of the first Arkobel sample, 60 kg/m^3, but a comparison of Figures 3.24 and 3.26 shows that the way in which the two materials compress is very different, as shown by the different curve shapes, particularly in the lower half of the weight scale.

The mineral wool plots shown in Figure 3.27 are different from both the Blancot and the Arkobel. Of the lower density samples, the plots show a high initial rate of compression which rapidly begins to flatten out, and indeed bottoms out in the case of the thinner sample. The 10 cm sample of the higher density (60 – 70 kg) is very typical of what is often used in studio isolation floors. Finally, Figure 3.28 shows an interesting comparison of the different slopes and shapes of the different materials as measured for the previous figures. Clearly, there are some very different mechanisms at work which are giving rise to their elasticity. They are most certainly *not* perfect, theoretical springs.

3.10.4 Calculation of Resonance Frequency

From the measured deflexions of the materials, the resonance frequency for each mass/spring combination can be calculated. If the spring is perfect, i.e. 2 × load = 2 × deflexion, the resonance frequency is given by:

$$f_{res} = \frac{1}{2\pi}\sqrt{\frac{g}{x}} \approx \frac{1}{2\sqrt{x}} \qquad (3.2)$$

where x is the deflexion in metres and g is the acceleration due to gravity (9.81 m/s). In general, as shown above, practical materials do not behave as perfect springs and the stiffness of the spring changes with changing load. Under these conditions, the resonance frequency for a particular load may be estimated from the *slope* of the load/deflexion curve, thus:

$$f_{res} = \frac{1}{2\pi}\sqrt{\frac{g}{m(\mathrm{d}x/\mathrm{d}m)}} \approx \frac{1}{2\sqrt{m(\mathrm{d}x/\mathrm{d}m)}} \qquad (3.3)$$

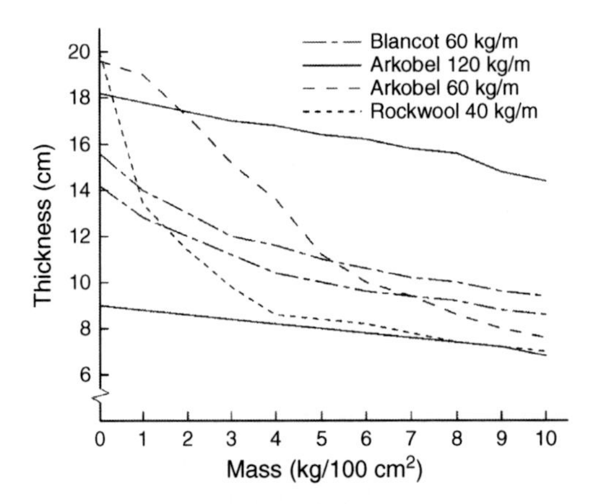

Figure 3.28:
Comparison of the shapes of the compression versus load curves for the different fibrous and cellular materials. The cellular materials can be seen to behave more uniformly, which is evident from the straighter nature of the curves. Clearly, the mechanisms giving rise to the spring-like behaviour are different in each material.

where dx/dm is the slope (gradient) of the load/deflexion curve at the applied mass m (note that for a perfect spring, $dx/dm = x/m$ and the two expressions become identical).

By way of example, marked on Figure 3.27 are two estimates of the resonance frequency of a material (which is clearly not a perfect spring) loaded with a mass of 2 kg/100 cm^2, one based on the static deflexion and the other on the slope of the curve. The static deflexion estimate is shown to be less than half the value of the more reliable estimate based on the slope of the curve, and the predicted resonant frequency is considerably higher.

3.11 Summary

When anything is immersed in a sound field it will, to some degree or other, be set into vibration. Any vibrating structure can, depending upon the degree of isolation that it can achieve, receive the vibrations on one of its surfaces and re-radiate them from another.

When a sound wave strikes any surface, the three possibilities are that the energy in the wave can be reflected, transmitted, or absorbed.

Absorption can be achieved in several ways, such as by tortuosity, particulate friction and internal viscous losses.

Porous absorbers work best in situations of maximum particle velocity, such as at pressure nodes and at high frequencies.

The differences in the speed of sound through the different materials in the structure of a building can give rise to great difficulties in predicting the structural vibration and re-radiation.

Isolation is best achieved by reflexion, because isolation by absorption would generally be too bulky to be practical.

The three basic means of sound isolation are mass, rigidity and distance; but acoustic damping can also play a prominent part.

The use of springs to isolate internal and external structures is usually the best way to achieve high isolation; by *decoupling*.

A good hint to remember is that springs do not like to change their length, and masses do not like to change their velocity. When something tries to vibrate a mass/spring system, both the mass *and* the spring try to resist the vibration, for their own individual reasons. Working together, they can make very efficient isolation systems.

The degree of isolation needed at any given frequency will be dependent upon circumstances and the type of recordings to be made. Low frequencies are generally much more difficult to isolate than high frequencies, though the general reduced sensitivity of the human ear to low frequencies at low levels helps to mitigate the problems incurred.

For floated isolation systems to be effective, their frequency of resonance must be below the lowest frequency to be isolated.

In weak buildings, adhesive mounted isolation systems can be attached to all parts of the internal surfaces.

Complex isolation structures are difficult to analyse, and a degree of empirical experience is often needed to achieve reliable predictions of the outcome of a design.

Impact noises can be much more difficult to deal with than airborne noises.

If possible, careful choice of location can also be instrumental in reducing isolation requirements.

Mass/spring systems are often used for making isolation floors by means of concrete slabs over mineral wool bases.

Doubling or halving the mass, whilst maintaining the same spring, will respectively lower or raise the resonant frequency of the system by $\sqrt{2}$. Quadrupling or quartering the mass will respectively halve or double the resonance frequency.

Doubling or halving the stiffness of the spring, such as by means of doubling or halving the density of mineral wool (the initial thickness remaining the same), will respectively raise or lower the resonance frequency of the system by $\sqrt{2}$. Quadrupling or quartering the stiffness will double or halve the resonance frequency.

Doubling or halving the thickness of a mineral wool spring will lower or raise the resonance frequency by $\sqrt{2}$. Quadrupling or quartering the thickness will respectively halve or double the resonance frequency.

Mass/spring, floated floors begin to become usefully effective in their isolation about one octave above their resonant frequency.

Resonant frequencies can be calculated from the applied force (mass $\times$ gravity, or weight) and the proportional compression of the spring.

Two identical springs in series (such as doubling the thickness of a mineral wool base) will *reduce* the resonance frequency by $\sqrt{2}$.

Two identical springs in parallel (such as achieved by doubling the density of a mineral wool base) will *increase* the resonance frequency by $\sqrt{2}$.

Springs of different stiffness, placed in series, will behave like electrical resistors in parallel, and the combined stiffness will always be below the stiffness of the weaker spring.

Systems with two degrees of freedom, such as mass/spring/mass/spring, can behave in complex ways. If the two masses and springs are identical, the resonances will not be: they will move apart. If the masses and springs are not identical, then their behaviour can depend on their relative proportions. Prediction of the overall behaviour can be very difficult to achieve: one resonance may dominate, or the two resonant systems may even behave somewhat chaotically under shock excitation.

Reference

1 Fahy, Frank, *Foundations of Engineering Acoustics*, Academic Press, London and San Diego, p. 331 (2001)

Bibliography

Gréhant, Bernard, *Acoustics in Buildings*, Thomas Telford Publishing, London, UK (1996). Originally published as *Acoustique en Bâtiment*, Lavoisier Tec & Doc, Paris, France (1994)

Parkin, P. H., Humphreys, H. R. and Cowell, J. R., *Acoustics, Noise and Buildings*, 4th Edn, Faber and Faber, London, UK (1979)

CHAPTER 4

Room Acoustics and Means of Control

The negative aspects of an isolation shell. Basic wave acoustics. Reflexions. Resonant modes and forced modes. The pressure and velocity components of sound waves. Modal patterns. Flutter echoes. Reverberation. Absorption. The speed of sound in gases and the ratio of the specific heats. Porous absorbers, resonant absorbers and membrane absorbers. Q and damping. Diffusion, diffraction and refraction.

The main means by which we achieve the necessary sound isolation in recording rooms is by reflexion, because absorption, as a means of isolation, is rather disastrously inefficient. In typical isolation shells, the sounds produced within the rooms are reflected back from the boundaries, and thus contained within the space. Sounds emanating from without the room are similarly reflected back whence they came, and thus little sound can penetrate the isolation barrier from one side or the other.

Reflective isolation from exterior noises presents no problems, but this method of isolation to the exterior, by the concentration of the acoustic energy inside the room, means that in the extreme we are creating a reverberation chamber. This only serves to make the job of acoustically controlling the room much more difficult than it would have been in its less isolated original state, where much of the sound could leak out. This is especially so at low frequencies. In fact, one reason why so many domestic hi-fi systems sound better in people's homes than the monitor systems in some badly designed control rooms has its roots in the relative degrees of isolation. Very often, the domestic constructions are of a much more leaky nature for low frequencies. Much of the energy escapes not only through the doors and windows but also through the structure of the buildings. This means that it is easier to achieve a more flat, controlled response in most domestic rooms than within the isolation shell of a simple control room, even though musical acoustics never enter the head of most domestic architects. If we start off with a highly reverberant shell, then the control measures must be much more drastic (and proportionally more expensive) than if we begin in a normal domestic room, where the reflected low frequencies are not so concentrated.

There is no doubt that once we provide a studio with a highly effective isolation shell, we are making our own lives much more difficult from the point of view of subsequent acoustic control. Nevertheless, the isolation is usually a prime requirement, without which the studio could not function effectively, so we have little option but to deal with the situation in which we find ourselves.

Figures 4.1 and 4.2 show a reverberation chamber and an anechoic chamber respectively. It is between these two acoustic extremes that all practical studio rooms are constructed. So by understanding the performance of these two extremes we will be able to get some feel for which characteristics of each of them we may require in our desired acoustics. First, however, we need to look at some basic wave acoustics.

There are many text books which wholly or partly dedicate themselves to a rigorous treatment of room acoustics. Some are listed in the Bibliography at the end of this chapter, and they are recommended reading for anybody wishing to take the mathematics and the theory some steps further. The purpose of this book, though, is to try to discuss these principles with people whose lives they affect every day, even if they do not have an academic background in acoustics or mathematics. For this reason the following discussion will be as accurate as can be achieved without the maths, but that does not imply that the mathematics would tell the whole story. In many of the situations that we encounter in room acoustics, there are simply too many variables, so in many cases a good feel for the problem is even more important than knowing the theory without having the practical experience. Anyhow, let us begin with a review of some basic acoustic principles.

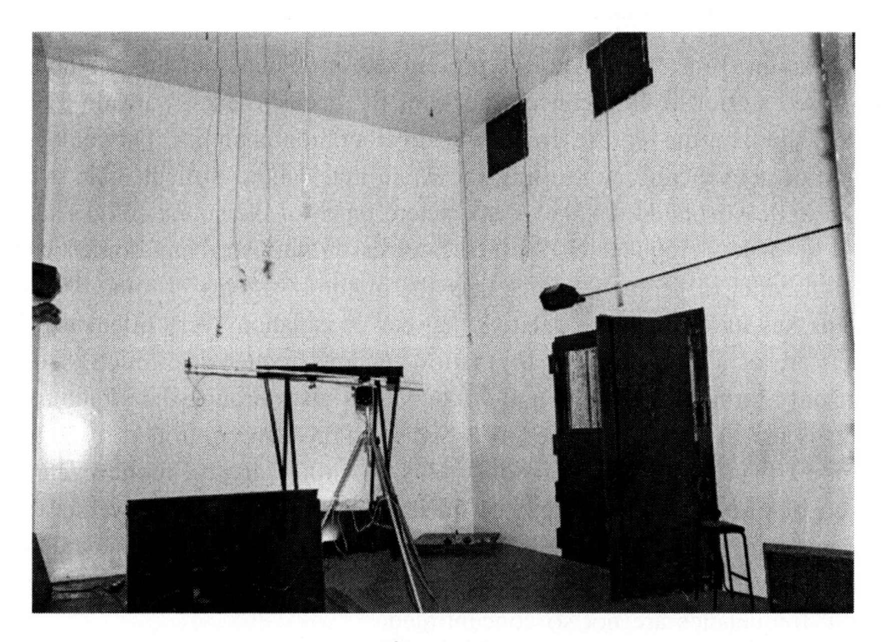

Figure 4.1:
The large reverberation chamber at the ISVR. It has a volume of 348 m^3 with a reverberation time of 10 s at 250 Hz, and 5 s at 200 Hz. All surfaces are non-parallel and are made from painted concrete.

Figure 4.2:
The anechoic chamber previously shown in Figure 2.3, but here with almost all of its floor grids and supports removed, rendering it almost totally anechoic above its cut-off frequency.

4.1 Internal Expansion

When a sound source in a highly reverberant environment emits an acoustic wave, the wave initially expands according to normal free-field expansion.

Before the sound reaches the first boundary, the conditions are, by definition, anechoic. Therefore, let us first take the simple case of a loudspeaker emitting a low frequency sine wave in the region of a hypothetically perfectly reflective wall. The progressive wave expansion is shown in Figure 4.3. The situation for the listener, sat by the loudspeaker, would be absolutely identical if the situation shown in Figure 4.4 existed. In this case the wall has been replaced by a second loudspeaker, placed the same distance behind the position of the wall as that from the wall to the first loudspeaker. The two identical loudspeakers would be fed the same signal at the same level. Were the wall to have an absorption coefficient of 0.5, in which case it would absorb half the acoustic power, then this situation could still be identically mimicked by reducing the signal level to the second loudspeaker by 3 dB – half the power. Even if the wall were to have a frequency dependent absorption characteristic, as all real walls do have, the situation could still be identically replicated by means of an electrical filter having the same response as the wall absorption, in series with the feed to the second loudspeaker.

This leads us to the classic mirrored room analogy, as shown in Figure 4.5, when the behaviour of a sound source in a real room is visualised from the point of view of a listener sitting in

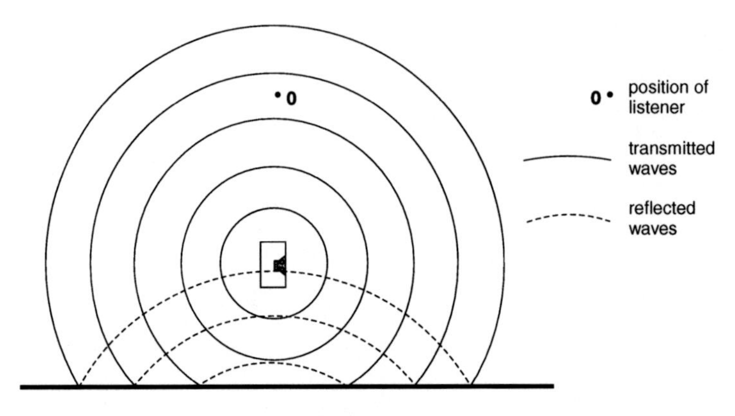

Figure 4.3:
When a loudspeaker is placed near to a solid boundary, some of the waves, which would have continued to expand in free space, will be reflected back from the boundary in the direction of the source and the listener. When the reflected waves arrive at the listener, and if the boundary is perfectly reflective, the situation and perceived sound would be identical to that shown in Figure 4.4.

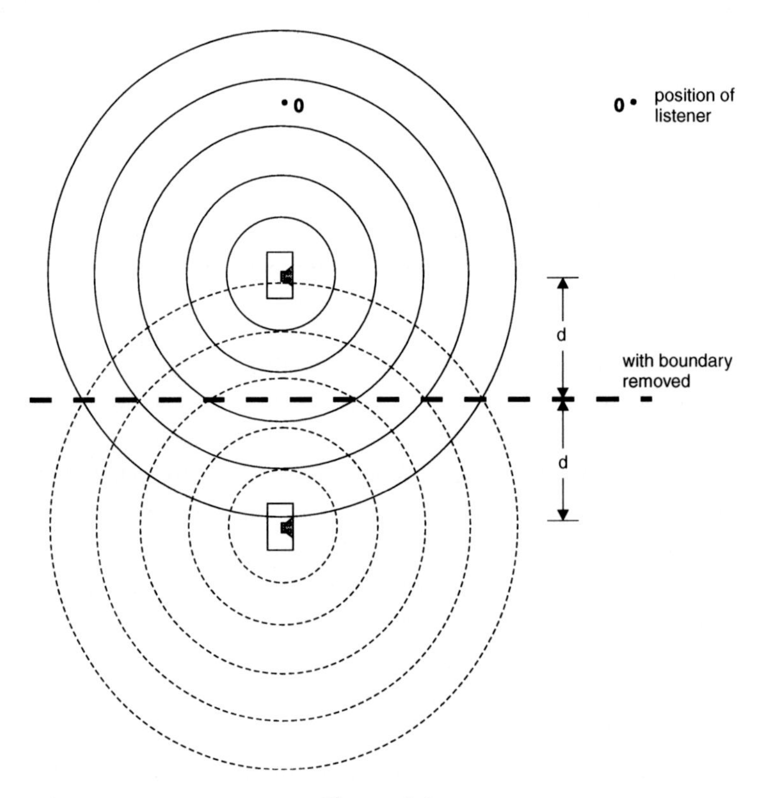

Figure 4.4:
The wave reflected from the boundary in Figure 4.3 can be represented by a loudspeaker placed behind the source loudspeaker with the boundary removed, positioned such that the second loudspeaker is the same distance behind the imaginary boundary as the source loudspeaker is in front of it.

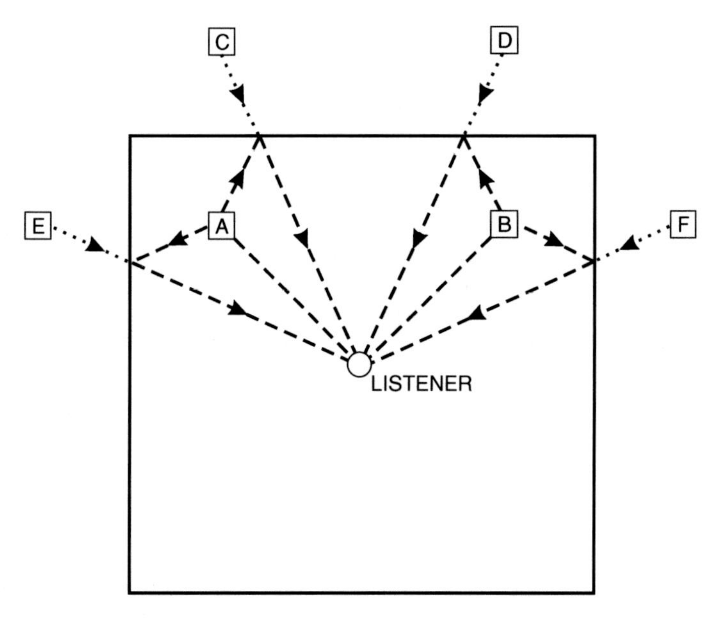

Figure 4.5:
Mirrored room analogy. Reflexions behave as if they were independent sound sources, located at the positions of their images. Floors and ceilings behave similarly: A and B, actual sound sources, C to F, apparent sound sources.

a room with rigid walls and mirrors on all surfaces. Quite simply, from every point where a reflexion or a reflexion of a reflexion existed, the room would behave sonically as though there were no walls and an identical signal were being fed to all the 'loudspeakers' that were visible. It often seems to be easier for many people to grasp the concept of room behaviour in this way, because loudspeakers with volume and tone controls are much more familiar than the invisible concepts of frequency contoured acoustic reflexion. This analogy is useful because it is absolute. There are no differences whatsoever until we begin to introduce complex diffusive and scattering surfaces, but even then we could roughly visualise things by placing obstacles in the way of the reflexions or smearing the mirror with grease at certain points to make the reflected image more fuzzy.

Let us now look at Figure 4.6, which encloses the loudspeaker within the four boundaries of a two dimensional room. In (a) and (b) the walls are anechoic, like the room shown in Figure 4.2, therefore they behave acoustically as though they do not exist and the loudspeaker is in free space. All that arrives at the points X are the direct waves from the loudspeakers. In (c) and (d) the walls are almost perfectly reflective, as in the room shown in Figure 4.1, which is a four-walled representation of the case shown in Figure 4.3. In this case the waves try

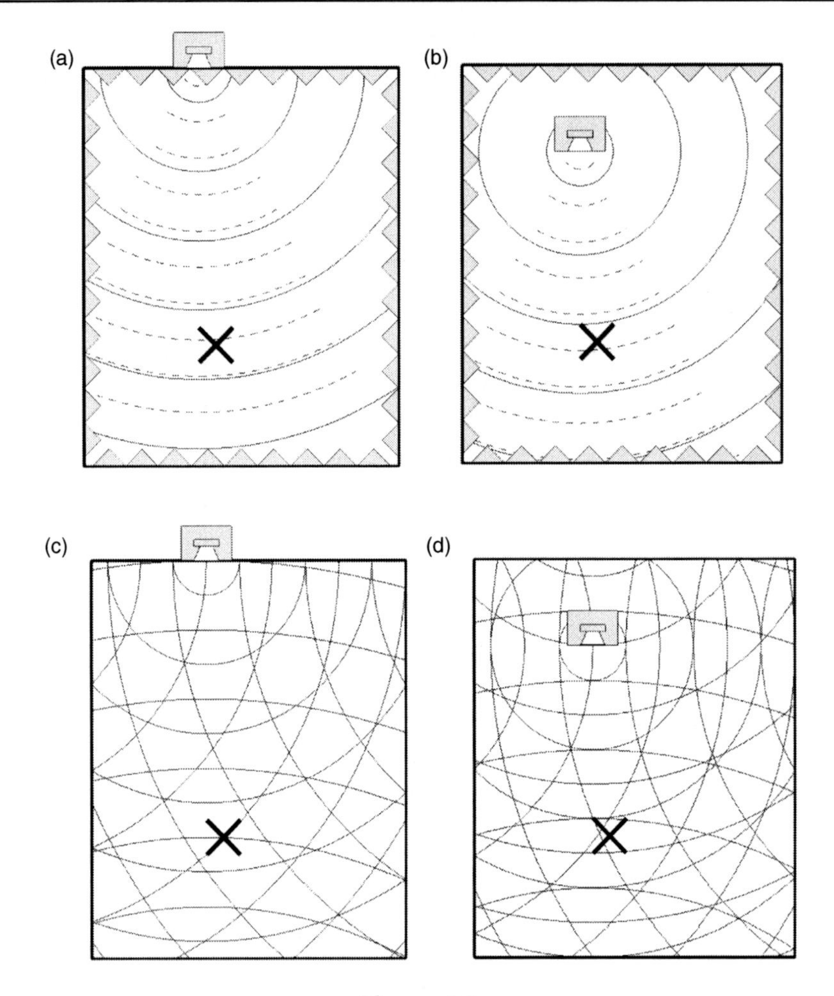

Figure 4.6:
Comparison of loudspeaker sitings in anechoic and reverberant spaces: (a) anechoic, flush; (b) anechoic, free-standing; (c) reverberant, flush and (d) reverberant, free-standing. At point X in the anechoic room, the low and high frequencies are perceived equally for both flush mounted and free-standing loudspeakers. However, in a more reverberant room, the low frequencies rapidly become confused. For the free-standing loudspeaker, the confusion sets in more rapidly than for the flush mounted loudspeaker in any given period. ____ omni-directional low frequencies; ----- directional high frequencies.

to expand, but they are constrained by the perfectly reflective surfaces and so the paths of the waves continuously fold back on themselves. With nowhere for the energy to go, the wave pattern in the room rapidly becomes ultra-complicated, and the acoustic power within a hypothetically perfectly reflective room would build up until the room finally exploded. That is if it were hermetically sealed. In reality, of course, absorption exists, and even in the highly reverberant room shown in Figure 4.1 this acoustic energy drops by 60 dB in about 8 s.

Figure 4.6(c) and (d) also show a further important point which relates very much to control room acoustics. If the loudspeaker is placed *within* the room, as opposed to being mounted flush within the boundary, it also sends low frequencies to the wall behind it, which then add to the reflexions in the room.

It can further be seen from Figure 4.6(c) and (d) that after the same period of time, the wave pattern is much more complex in the room with the free-standing loudspeakers than in the room in which they are flush mounted.

For steady-state signals, such as sine waves, the consequences of the room reflexions are frequency dependent. Figure 4.7 shows the pressure fluctuations in the air between

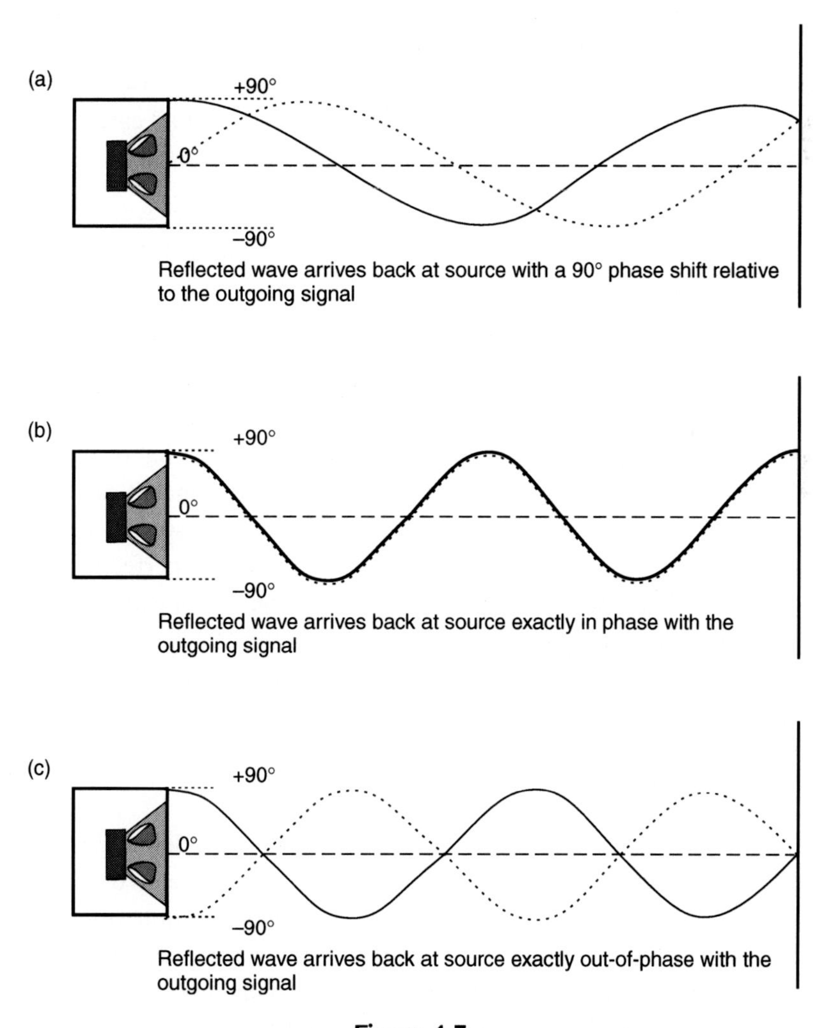
(a)
+90°
0°
−90°
Reflected wave arrives back at source with a 90° phase shift relative to the outgoing signal
(b)
+90°
0°
−90°
Reflected wave arrives back at source exactly in phase with the outgoing signal
(c)
+90°
0°
−90°
Reflected wave arrives back at source exactly out-of-phase with the outgoing signal

Figure 4.7:
Phase relationships of reflected waves at different frequencies.

a loudspeaker and a wall at three different frequencies. In the first case, the reflected wave arrives back at the source loudspeaker with a 90° phase shift relative to the direct wave. In the second case, there is an exact fit for two whole cycles between the source and the wall, so the reflected path corresponds exactly with the outgoing wave in terms of pressure and rarefaction peaks. In the third case, the reflected wave coincides exactly in anti-phase with the outgoing wave in terms of the relative pressure and rarefaction peaks. The sums of the wave pressures are shown in Figure 4.8 at two different positions. It can be seen that the effects at each position are quite different for the three frequencies shown.

This idea is shown in plan view in Chapter 11, in Figure 11.25 where different points in the room exhibit different responses to the combination of direct and reflected signals, dependent upon position and frequency. Whether this complex interaction of direct and reflected signals is a good thing or not is largely dependent upon whether we are listening to live music or to recorded music, and for what purpose. These concepts will be discussed later in their relative contexts. However, what we tend to want to avoid in almost all cases are the stationary waves, or 'standing waves' as they are commonly called. (See Glossary.)

Let us assume that the case of Figure 4.7(b) is modified with another wall behind the loudspeaker, at exactly the same distance from it as the front wall. It is shown diagrammatically in Figure 4.9(a). Because of the exact fit of four cycles across the room, and the position of the loudspeaker on the centre point, the first reflected wave will retrace exactly the path of the outgoing wave. The reflected wave from the rear wall will also retrace the same path, so the pressure changes will linearly superpose themselves, causing a build up in the wave motions.

However, there is another component of an acoustic wave besides that of the pressure. Figure 4.9(b) shows the *velocity* component of the wave motion, which it can be seen is shifted 90° from the pressure component. Thought of logically, this is reasonable enough, because when the wave which is travelling at the speed of sound strikes the perfectly reflective wall, it must obviously stop before returning in the reverse direction. For a stationary wave, the pressure component at a boundary must always be a maximum. This must be so, because with the velocity at zero, all the energy in the acoustic wave has to be carried in the pressure component: the energy simply cannot disappear every time the pressure change passes zero. There is therefore a continuous energy transfer, back and forth, between the pressure and velocity components of the wave.

To help to explain this, Figure 4.10 shows an analogous situation demonstrating the gradual energy transfers during the swinging of a pendulum. At either extreme of travel, the pendulum stops before retracing its path. This is akin to the wave particle velocity stopping as it reaches a reflective boundary, before being reflected in the reverse direction. When stopped, the pendulum has no kinetic energy, but with the height advantage over its vertically down position it has its maximum potential energy for its given distance of swing. At the low point of

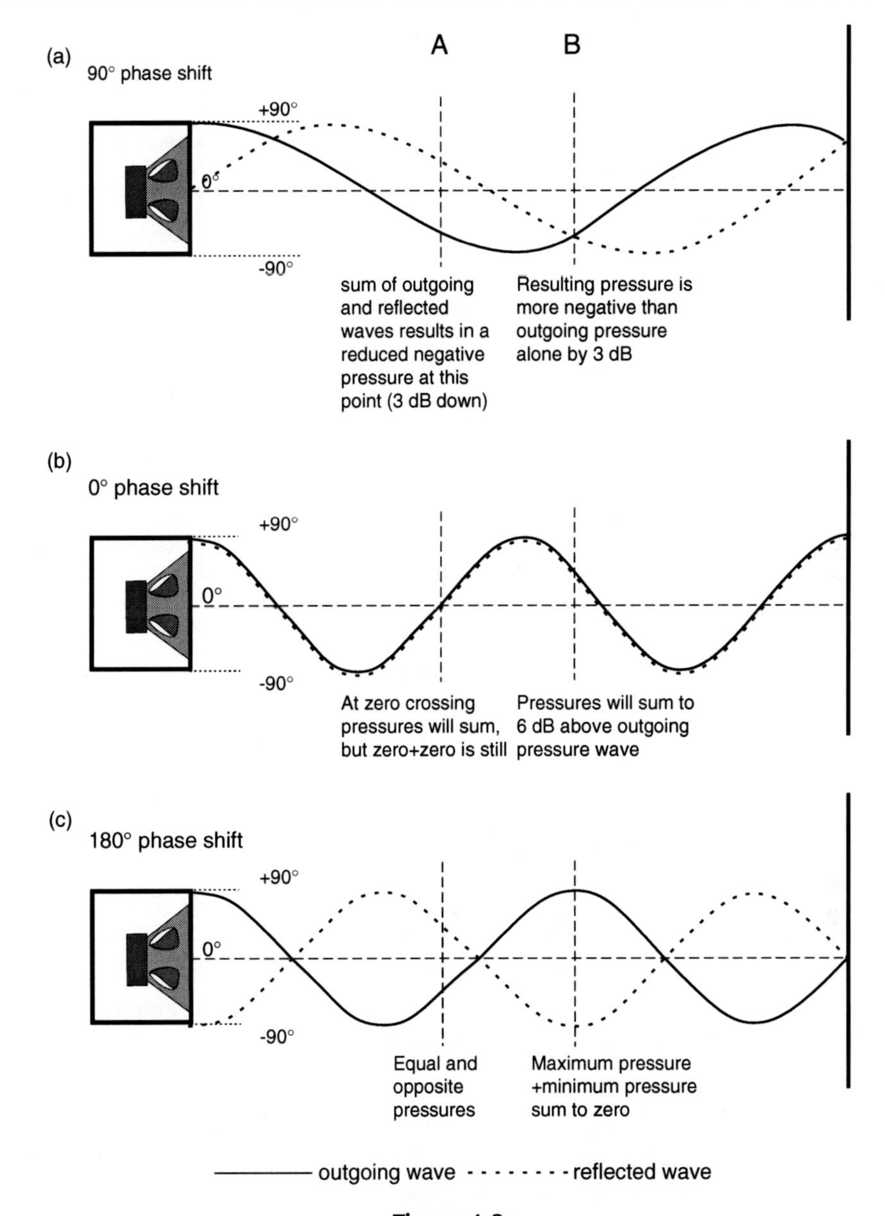

Figure 4.8:

Effects of phase shifts of reflected waves at different positions. Results of wave superposition (assuming that the wall at the right is perfectly reflective): (a) 90° (or 270°) phase shifts give rise to 3 dB variation in summed response. The resulting waveform would be a sine wave displaced to the right; (b) 0° phase shift results in the outgoing and reflected waves *reinforcing* each other at all points. The resulting waveform would be a sine wave of 6 dB greater amplitude; (c) 180° phase shift results in the outgoing and reflected waves cancelling at all points. The resulting waveform would be a flat line.

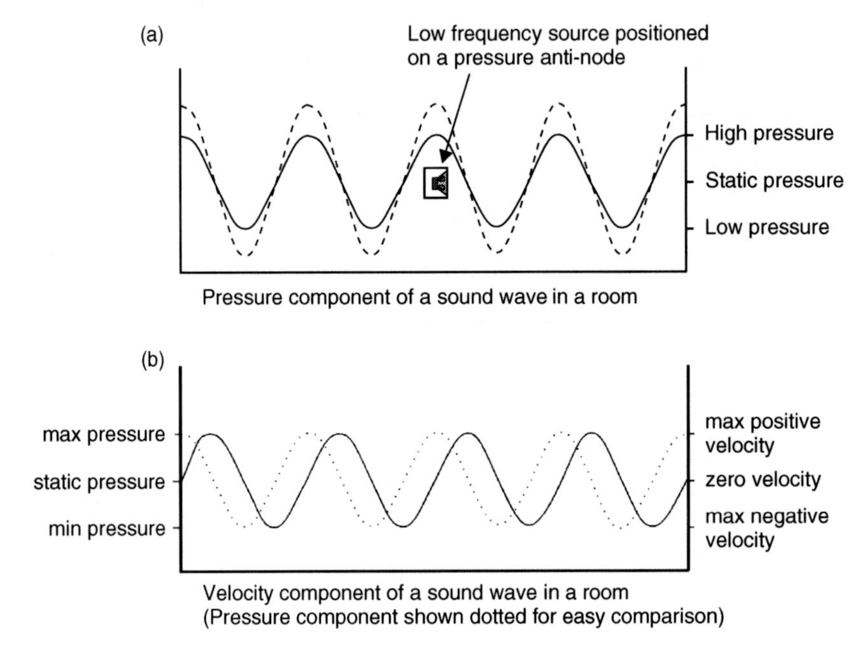

Figure 4.9:
Characteristics of a resonant mode: (a) At resonance, where an exact number of half-cycles can fit between the walls, the direct and reflected waves will superpose constructively to create a resonant build up, shown dashed; (b) The particle velocity component of the wave must always be at a maximum or minimum when the pressure component is at zero, in order to conserve the energy contained in the wave – see Figure 4.10.

its travel, the pendulum passes through its point of maximum velocity, where its kinetic energy is at its maximum and its potential energy is at its minimum. (See Glossary, if necessary, for potential and kinetic energies.) At all points of the swing, other than the three already mentioned, the pendulum has a mixture of both potential and kinetic energy. Similarly, the wave motion has a mixture of potential and kinetic energy at all points other than its nodes and anti-nodes. The nodes are the zero crossing points on either the velocity or pressure plots in Figure 4.9. The anti-nodes are the positive and negative peaks in the plots, and it can be seen from (b) that the velocity nodes correspond with the pressure anti-nodes, and the pressure nodes with the velocity anti-nodes. The velocity and pressure are therefore always 90° out of phase with each other. The importance of this will become evident when, in later chapters, we begin to discuss the acoustic control of rooms.

4.2 Modes

The pathways available for a sound wave to travel around a room are known as modes, and in any room they are infinite in number. When an omni-directional sound source is in a room, it drives the room by radiating sound in all directions. If the sound source is then interrupted,

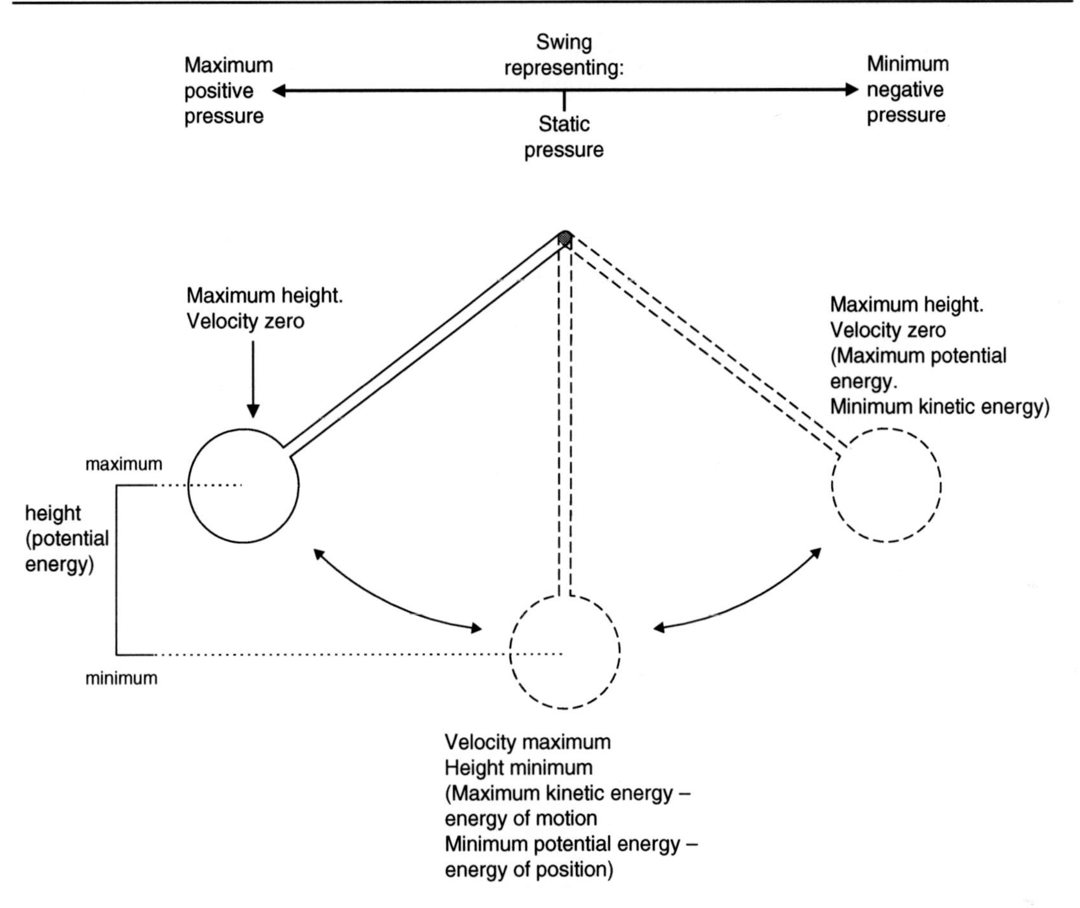

Figure 4.10:
Energy in a pendulum. At the points of *maximum* height there is a *maximum* of potential energy (the energy of position) but a *minimum* of kinetic energy (the energy of motion), [the latter because the pendulum must stop before it retraces its path]. At the place of *minimum* height, the pendulum is passing through its point of *maximum* velocity (and hence *maximum* kinetic energy). At all other places of its swing there is a continuous transfer of energy between potential and kinetic, and vice versa.

some energy will be trapped in the repetitive pathways of the standing waves, indicated by the build up shown in Figure 4.9(a). The resonances associated with these pathways are called eigentones, which is German for the room's 'own frequencies'. These are the natural frequencies of resonance in the room. The non-resonant pathways which are driven by the sound source are known as forced modes. Those pathways which continue to decay for some time after the driving source has ceased are known as resonant modes. Thus, all modes are not standing wave paths, but all standing wave paths are modes. (See Glossary for more on modes.)

There are three types of resonant mode in a rectangular room. Axial modes, which exist between two parallel surfaces; tangential modes, which exist between four surfaces and travel parallel to the other two and oblique modes, whose pathways involve striking all six surfaces before returning to their point of origin to begin retracing the paths. In a room with all surfaces similarly treated, the axial modes of the longest dimension of the room tend to contain the most energy, because they strike fewer surfaces for any given distance travelled, and hence are absorbed less frequently. In real rooms, it is the absorption at the surfaces, on each reflexion, which depletes the energy in the modes. The lowest mode of the room will always be the axial mode between the two most widely spaced surfaces. Despite the longer total pathways for the tangential and oblique modes, their lowest frequencies will always be higher than the lowest axial mode, because it is the shorter side of the pathway which determines the lowest frequency to be supported in any resonant mode. This is because an exact integral number of half-wavelengths must fit into the dimensions of the tangential and oblique modes.

The typical pathways of the modes are shown in Figure 4.11. The combination of these three types of modes can form a very dense set of resonant frequencies in a reflective room. The following equation gives the frequencies of all the possible modes in a *rectangular* room

$$f_{xyz} = \frac{c}{2}\sqrt{\left(\frac{x}{L}\right)^2 + \left(\frac{y}{W}\right)^2 + \left(\frac{z}{H}\right)^2} \qquad (4.1)$$

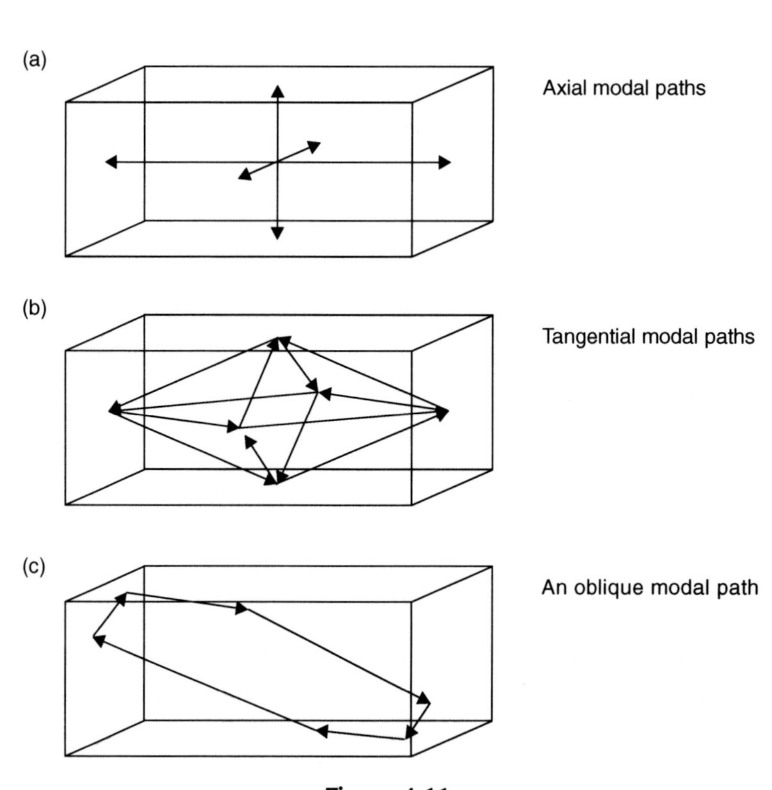

Figure 4.11:
Axial, tangential and oblique modes.

where:

x, y and z represent the number of half-wavelengths between the surfaces (up to the limit of interest – say the first 10 modes)
L = room length (m)
W = room width (m)
H = room height (m)
c = speed of sound (m/s)

For axial modes, only the x term is used; for tangential modes, the x and y terms are used, whilst for oblique modes the x, y and z terms are used.

The lowest axial mode exists at a frequency where half a wavelength fits exactly into the distance between the furthest two parallel surfaces. If the room is 15 m in length, for example, then the lowest resonant frequency would have a wavelength of 30 m, because *two* half-wavelengths would return the energy in phase to the starting point for the cycle to repeat. This would represent a frequency of about 11.5 Hz. The next mode would be at 23 Hz. These frequencies can be derived from the previous formula as:

$$f_x = \frac{c}{2}\sqrt{\left(\frac{x}{\mathrm{L}}\right)^2} \quad (4.2)$$

but for axial modes, only, it can be simplified to:

$$f_x = \frac{c}{2} \times \frac{x}{\mathrm{L}} \quad (4.2a)$$

The following axial mode would be at 34.5 Hz; the next at 46 Hz. On a linear frequency scale, the first 12 modes would appear as shown in Figure 4.12, but on a logarithmic scale, which more closely approximates to the way that we hear, they would be distributed as shown in Figure 4.13. Here, the increasing modal density per octave band can clearly be seen, and this is an important point, as we shall see later.

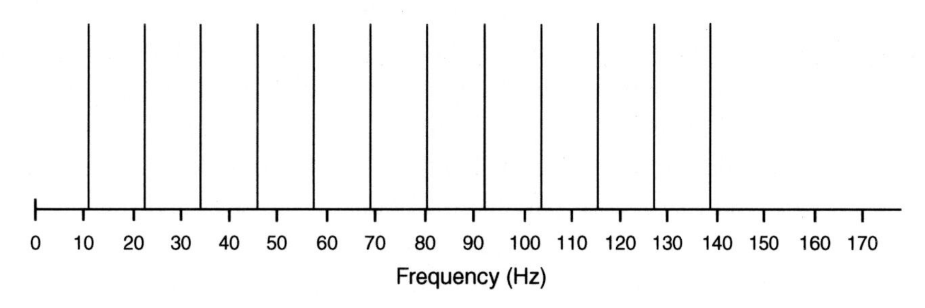

Figure 4.12:
The first axial modes of two room surfaces spaced 15 m apart, plotted on a *linear* frequency scale. Note the absolutely equal spacing.

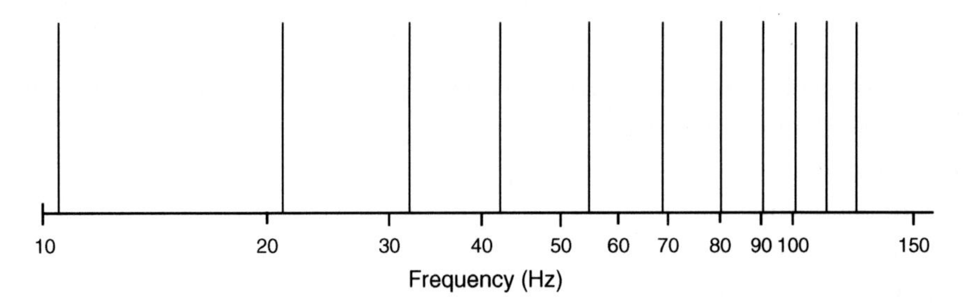

Figure 4.13:
The same modes as shown in Figure 4.12, but plotted here on a *logarithmic* frequency scale. The logarithmic scale relates better to the way in which we hear.

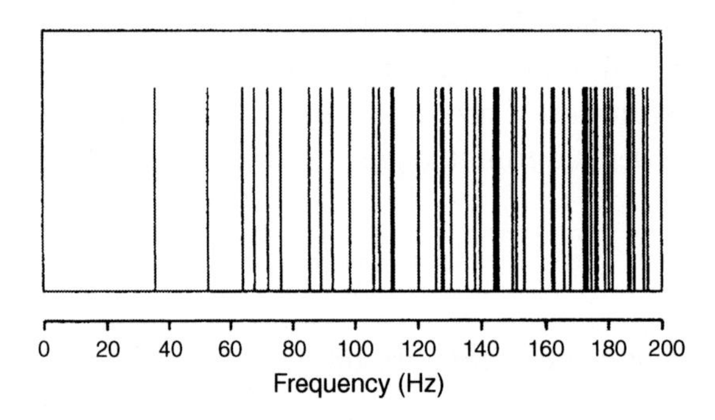

Figure 4.14:
The first 51 modes of a room. The first 51 resonance frequencies of a rectangular room of dimensions length = 4.7 m, width = 3.2 m, height = 2.5 m, plotted on a linear frequency scale. The modes at low frequencies are more widely spaced than those at higher frequencies, which make their perception as a form of colouration more likely. The thicker lines are where different modes are almost coincident in frequency. Not all of the 51 modes are well separated. Thirteen of them merge with other modes (after G. Adams).

Figure 4.14 shows the first 51 modes of a room of dimensions 4.7 m × 3.2 m × 2.5 m, also on a linear frequency scale. It shows how, in three dimensions, the modes definitely increase in number per unit bandwidth as the frequency rises. The problem which faces us is at the lowest frequencies, where the modes begin to separate. Figure 4.15(a) shows the same modes plotted to 100 Hz. The general 'loudness' of the room will be reinforced by the energy in the modes, and the overall perceived frequency response will be somewhat like the broken line which takes a roller-coaster ride across the modes and the spaces. It can clearly be seen that the room is perceived to have a flatter frequency response when the modes are more closely spaced. In Figure 4.15(b) exactly the same pattern exists, but the frequency scale this time represents a room where the dimensions have all been multiplied by three, to 14.1 m × 9.6 m × 7.5 m.

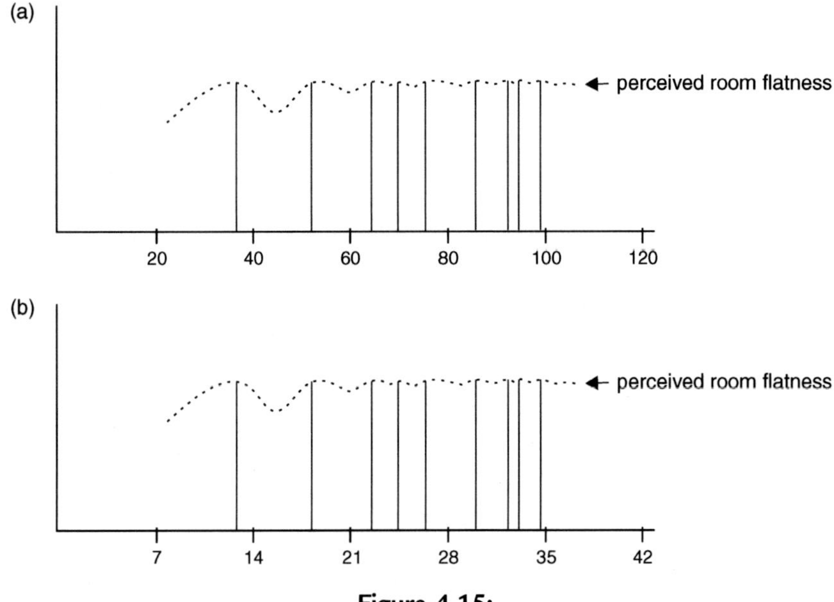

Figure 4.15:
Perceived responses in rooms with significant modal activity: (a) Room similar to that shown in Figure 4.14. A roller-coaster effect begins below 60 Hz; (b) Room of similar proportions, but with dimensions multiplied by three. Roller-coaster on-set is delayed until below 20 Hz. The modal pattern remains similar, and the perceived response shape also remains similar, but in (b) the frequency scale is reduced by a factor of three, yielding a smoother perceived response down to a lower frequency.

Two things are instantly obvious from the comparisons of the two figures. Firstly, that the modal *pattern* is exactly the same. This is because the modal pattern is a function of the *shape* of the room and its *relative* dimensions, not its absolute dimensions. Secondly, in the larger room, the flattest part of the roller-coaster response extends to a lower frequency. The implication of this latter point is that larger rooms with significant modal activity (i.e. they are reasonably acoustically reflective) will have responses which are perceived to be flatter down to a lower frequency than smaller rooms of the same shape. In reality though, the modal energy is not as represented in the previous figures inasmuch as the modes are not at spot frequencies and they do not all have uniform amplitude as represented by the height of the lines. In fact, they are more realistically represented as shown in Figure 4.16.

Figure 4.17 shows the measured power output from a loudspeaker, both in an anechoic chamber and in a room with considerable modal activity. The loading effect of the room which augments the low frequency response will be dealt with in later chapters, but it can be seen how, if the modal density is tightly packed and extended in frequency range, the room can have an overall frequency response which is acceptably flat. The effect of the modal activity is to make the room sound louder in response to whatever sound is being produced within it,

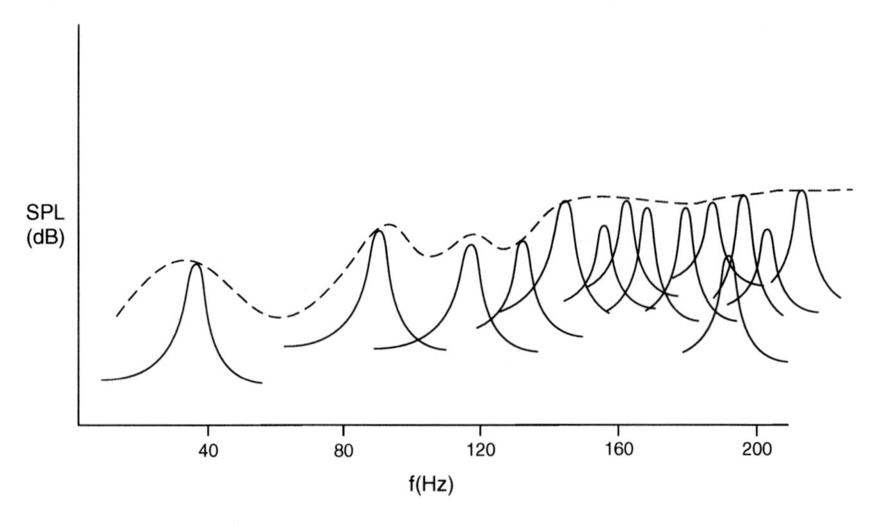

Figure 4.16:
Modal response of a typical room. The individual modes are shown with typical frequency spreading and different amplitudes. This is perhaps more representative of reality than Figure 4.14, which shows the modes as spot frequencies. A typical partially damped mode will be active for around 10 Hz either side of its nominal frequency.

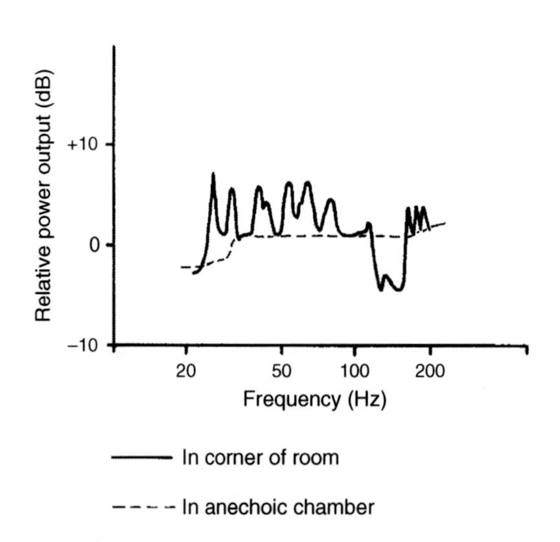

Figure 4.17:
Increase in loudness and extension of low frequency response by modal support. The measured power output of a closed-box loudspeaker system placed in the corner of a room ($x = 0.7$, $y = 0.7$, $z = 0.5$ m). The power output measured in an approximately free space (an anechoic chamber) is shown by the broken curve. The augmentation of the power output due to the room modes is clearly shown; as is a broad dip around 150 Hz due to the early reflexions.

compared with what would be perceived from the same sound source in an anechoic chamber. This is true whether the sound source is a loudspeaker, a cello, a tuba, a pipe organ or any other instrument that can provide a relatively continuous output. Things get a little trickier to explain when we come to impulsive sources and the musical transients from percussive instruments, but we will come to that later.

Nevertheless, in the low frequency range, where modal separation exists, the standing wave activity is usually undesirable, because the response of the room becomes heavily dependent upon the position of both the source and the receiver, be that a microphone or a listener. Figure 4.18 shows the spacial dependence of pressure distribution for two different types of mode. In (a) and (b) the pressure is high at the dots and low at the dashed lines. In (c), the pressure is high where the lines are packed close together, and lower elsewhere. This type of unevenness can be clearly heard when walking around any normal room in the presence of a low frequency tone. Figure 4.19 shows two alternative presentations of the same phenomenon. A conventional loudspeaker or musical instrument placed on a pressure anti-node would be capable of strongly building up a resonance if the output, or a portion of it, coincided with the resonant modal frequency. A listener on an anti-nodal point would also hear a strongly amplified sound. However, a monopole (but not dipole) loudspeaker or instrument placed on a node would not be able to drive the resonance, because the reflexions would constantly be returning to the point of origin out of phase with the drive signal, and so would tend to cancel it. Similarly, a listener on a nodal plane would hear little or nothing at the resonant frequency, even if the source were on an anti-node, because of the cancellation taking place on the nodal planes. Clearly this sort of spacially dependent distribution is not desirable, because it is also frequency dependent, so some notes intended to be equal in level would be perceived to be very different, depending upon where an instrument or listener was positioned and on the note being played. The effect is very easy to demonstrate by passing a low frequency tone through a loudspeaker in a reasonably reverberant room and then walking around the room and listening at different heights. The high and low pressure positions are easily audible, but their location will change as the frequency is changed.

4.2.1 The Summing of Modes and Reflexions

Figure 4.20 shows the gradual way in which a reflective room can upset the frequency balance of a sound source in response to a wideband signal, such as music or pink noise. In this case the source is non-centrally positioned in a rectangular reflective room, with the microphone placed some distance from it. Shown in Figure 4.20(a) is the anechoic output from the source, which is what exists, by definition, within the room in the short time until the effect of the first reflexion is noticed. The affect of the superposition of the first reflexion, from the floor, is shown in (b) and (c) to (f) show the cumulative build up of reflexions from

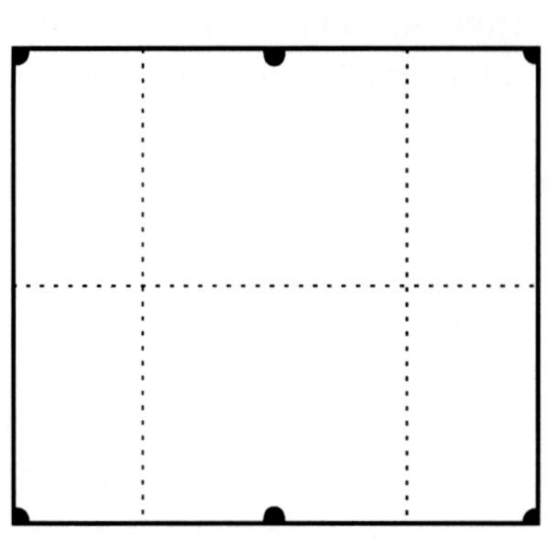

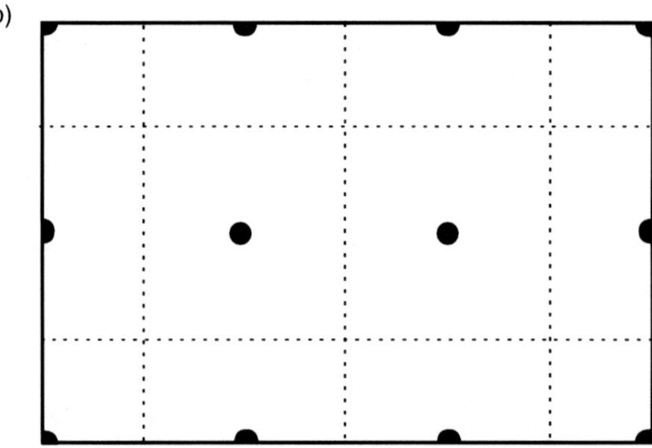

● Antinodes, maximum amplitude

.... Nodal planes, minimum amplitude

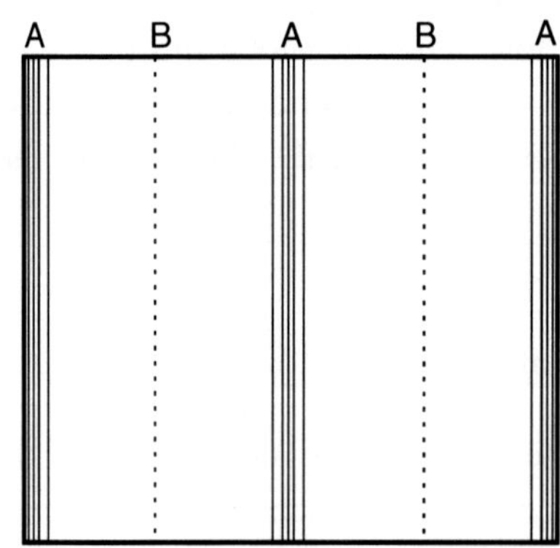

A = pressure maxima

B = pressure minima

each boundary in turn, until the whole room is contributing to the response. Moving the source around the room, whilst leaving the microphone in the same position, would change the pattern of (f), even though (a) would remain the same. In rooms for the playing of music, such variation can have the benefit of adding variety to the sounds, but in quality control listening rooms it is definitely not desirable, as a fixed reference is needed. The frequency and positional dependence of the modes is easily demonstrated by humming in a tiled bathroom with the towels and curtains removed.

If a room has two identical dimensions there will be augmentation of the energy at frequencies where the common dimensions apply. In the case of a cubic room there will be no difference in the axial modes for the three axes, therefore they will superpose themselves on each other and produce very strong resonances due to the common room dimensions. This will result in greater spaces between the modal frequencies, because there will be less frequency distribution than would be the case if the three dimensions were different. Common multiples of the axial distances can also reinforce and separate the higher modes. In a room 15 m × 10 m × 5 m, for example, the 5 m half-wavelength could fit into each dimension, and so could resonate strongly.

In an attempt to avoid such coincidences, many rooms intended for musical performance or reproduction have been designed with only non-parallel walls. In general, such approaches do not reduce the overall number of resonant modal pathways, or the overall number of coincident modes, but they do tend to have the ability to reduce the more energetic axial modes, and this can be an advantage in some circumstances. Unfortunately, the calculation of the modal frequencies in such rooms is too complex to be practicable, and no formulae exist, so such designs tend to rely very greatly on the experience of the designers.

4.3 Flutter Echoes and Transient Phenomena

Whilst on the subject of non-parallel walls, another advantage which they can offer is the avoidance of flutter echoes. These are the rapid repeats which can be heard when, for

Figure 4.18:
(a) Plan of the pressure distribution of a 2.1.0 tangential mode; (b) The distribution of the amplitude of the sound pressure throughout a room of a tangential 3.2.0 mode; (c) Pressure distribution of a single axial mode. Note how in (c) the wave is progressing across the room as a series of compressions and rarefactions. The progress of an acoustic wave (unlike the wave motion of light, which moves in the general fashion of a snake) moves somewhat like an earthworm. The worm moves in a straight line by means of the passage of a series of contractions and elongations through the length of its body. The use of sinusoidal graphs to represent a sound wave merely shows the *level* of the pressure distribution, and not its form. This is why light waves can be polarised. They have a lateral oscillation which can be reorientated, but sound waves cannot.

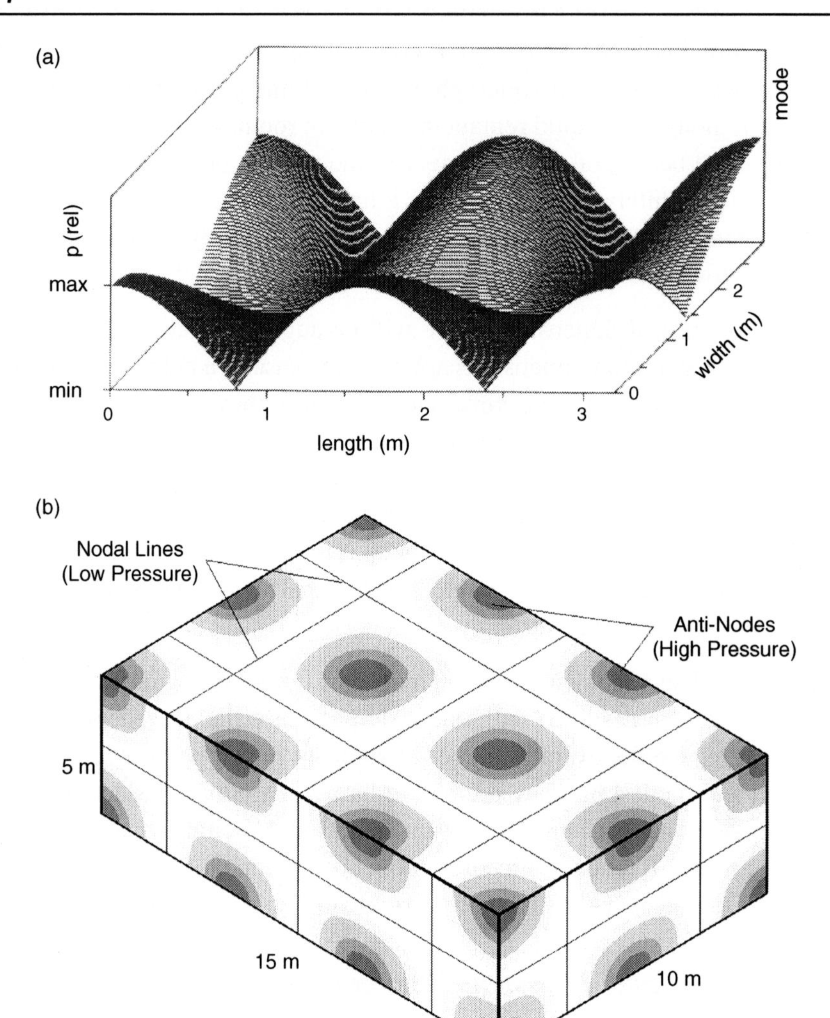

Figure 4.19:
(a) Representation of the sound pressure distribution in a rectangular room of the same tangential mode as depicted in Figure 4.18(a). The vertical axis is pressure. (b) A three-dimensional representation, this time for the 3.2.1 oblique mode at 58.9 Hz.

example, hands are clapped between two hard-surfaced, parallel walls. They usually exist, buried in the general reflective confusion, in all rooms with hard, parallel surfaces, but are usually most noticeable and objectionable when they are left exposed, such as when only two reflective surfaces exist. Such would be the case when a room with curtained walls had a hard surfaced floor and ceiling. Flutter echoes, *per se*, are normally only perceivable if there is a time interval of more than about 30 ms between the discrete repeats. With shorter time intervals, the ear tends to integrate the sound power to produce a continuous sound of

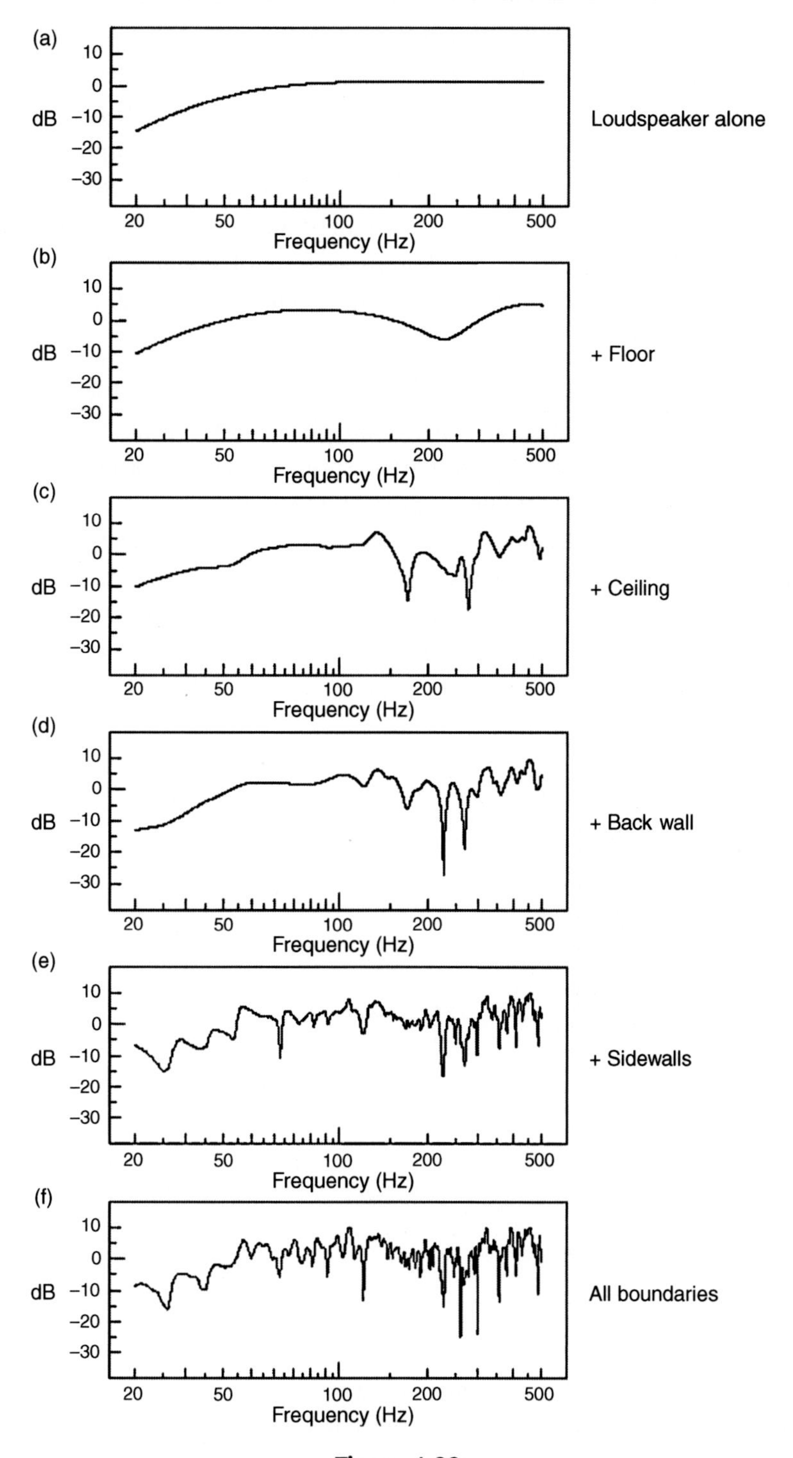

Figure 4.20:
Gradual build up of room reflexions and their effect on the overall response.

greater loudness, but usually with a distinctly nasal or metallic 'ring'. It is the regular, periodic nature of the repeats between a single pair of parallel surfaces which makes the flutter echoes or the ringing colouration so subjectively objectionable. Whenever encountered in a listening room they should be dealt with, either by means of absorption, by geometrical methods, or by diffusion. As a period of around 30 ms is necessary for a true flutter echo to be audible, it follows that a path length of much less than 8 or 10 m between the reflective surfaces will not exhibit the problem as a series of separate, perceivable echoes. What will predominate are the colourations in timbre due to the modal resonances that any impulsive stimulus excites.

Unlike the precise reflected phase relationships which are necessary for resonant standing waves to build up, transient sounds do not exhibit phase shifts. Rather, with temporal displacements, they exhibit phase *slopes*, which represent the rate of change of phase with frequency. As a transient signal contains a great number of frequencies (and some, such as delta functions, contain *all* frequencies), each contributing frequency will have its own individual phase shift. Figure 4.21 shows a phase shift in a sine wave. The waveform itself has in no way been affected; the whole pattern has simply been shifted sideways. On a pure tone, such a shift has no audible effect. However, if these shifts are applied frequency by frequency to a transient signal, the results would be as shown in

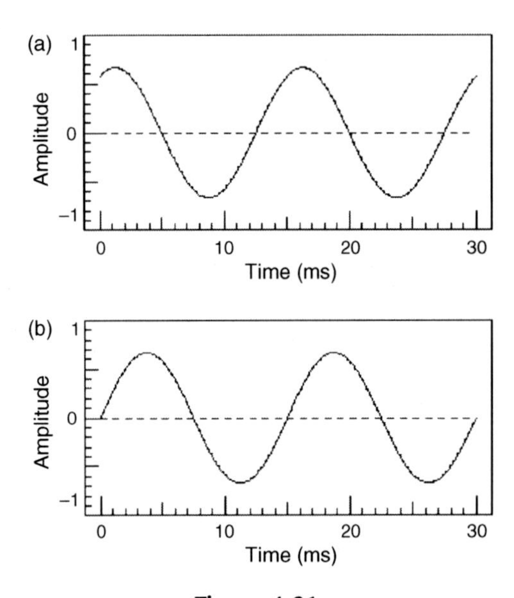

Figure 4.21:
Phase shift in a sine wave. With respect to the sine wave shown in (a), the sine wave in (b) is shifted by 90°. Nothing has changed except its position along the time axis. The quality and quantity of the wave are unaffected (shape, frequency and amplitude).

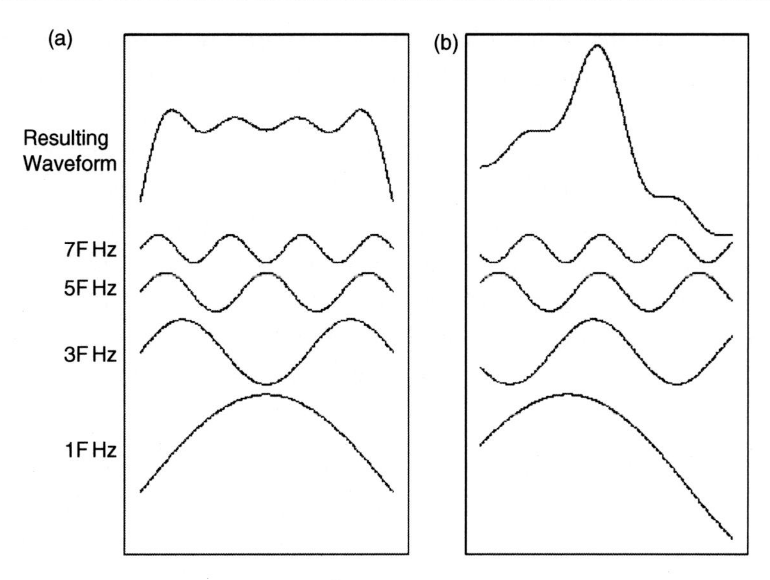

Figure 4.22:
Waveform change due to phase slope difference: (a) Shows a waveform broken down into its sine wave components; (b) Shows a totally different waveform, but consisting of exactly the same quantities of the same sine wave components. The only difference is that a phase slope has been applied, shifting the relative phases of the individual component frequencies.

Figure 4.22 – a great change in the overall waveform. In fact, the behaviour of rooms to steady-state and impulsive (transient) sounds tends to be very different, and later we will address the question of how to simultaneously deal with these aspects of sound which require very different treatments.

4.4 Reverberation

Once we enter the realm of reflexions of reflexions of reflexions, and beyond, we are beginning to develop reverberation. Both the multiple repeats (echoes) of transient events and the modal build up of more steady-state sounds all contribute to the reverberant energy of a room. This consists of a hyper-complex sound field that is said to be statistically diffuse and to contain the same total energy in any given volume in any position in the room. In reality, such a purely diffuse sound field cannot exist. There *must* be a net flow of energy away from the source, because the sound radiates *from* the source. Furthermore, unless the reverberant room is extremely large, there will still be detectable modal activity at low frequencies. In the reverberation chamber shown in Figure 4.1, the field is only statistically diffuse down to around 150 Hz, below which spacial averaging must be performed. This can be done by taking the mean of measurements with the source and receiver in different positions.

Bear in mind, also, that if the chamber has a mean decay time of 8s, then the sound waves will have travelled 344 × 8 m, about 2.7 km, before they decay by 60 dB (344 m/s being the speed of sound in air at room temperature). This helps to give some idea of what is going on in a reverberant field. A further aid to visualisation is the concept of the mirrored room, once again, but with very reflective mirrors. In a room of such long reverberation time (RT_{60} or T_{60}) the room surfaces would appear to be absolutely covered by images of loudspeakers, disappearing into infinity. This concept is also valid for time-varying signals, such as music. The smallest, most distant images would behave like the music originated from them at the same time as it emanated from the source in the room, but the furthest discernible tiny images of loudspeakers would be 2.7 km away, and the sound from them would arrive 60 dB down and 8 s later.

In a very reverberant room, the energy decay is exponential, which means that when it is plotted on a logarithmic scale the decay slope is a straight line.

4.4.1 Measuring Reverberation Time

The length of time that it takes for a sound to decay in a room is a function of the absorption of the surfaces *and* the distance between impacts with the room boundaries. The average distance that a sound will travel between each of the surfaces on its path to decay is known as the 'mean free path', which assumes all possible angles of incidence and position. In a reasonably rectangular room, the mean free path (in metres) is given by:

$$\mathrm{MFP} = \frac{4\mathrm{V}}{\mathrm{S}} \tag{4.3}$$

where:

MFP = mean free path (in metres)
V = room volume in m^3
S = total surface area in m^2

Note: MFP is often written L_{AV} (average length)

The *time* between surface contacts in any given room is given by:

$$\mathrm{T} = \frac{4\mathrm{V}}{\mathrm{S}c} \tag{4.4}$$

which simply divides the former value by the speed of sound. (T = time between reflexions in seconds, and c = the speed of sound in metres per second.) It is worth noting here that there may, in certain rooms, be differences in the above path lengths and times, depending upon whether the reflexions are specular or diffuse, though for the purposes of this chapter it is a minor point.

There are numerous equations for the measurement of reverberation time because there is no universal equation which suits all circumstances. This is because the hypothetical perfectly diffuse field is never achieved in practice, so several 'best fit' approximations are available for different circumstances. (There is always a net flow away from the source.)

The oldest, and still very widely used equation, is the Sabine formula, named after Wallace Clement Sabine who pioneered reverberation time calculation. The Sabine formula is represented by the equation:

$$RT_{60}(\text{or } T_{60}) = \frac{0.161V}{S\alpha} \tag{4.5}$$

where:

$(R)T_{60}$ is the decay time (reverberation time) in seconds
V is the room volume in m^3
S is the total surface in m^2
0.161 is a measurement constant (there is a different constant for imperial measure, 0.049 in feet)
α is the mean absorption coefficient of the room.

This formula is really only valid for rooms of relatively low absorption, where α is less than about 0.3 (see Section 4.5), because it becomes increasingly inaccurate as the absorption increases. In fact, at the extreme, it predicts a reverberation time even when absorption is equal to 1. Perfect absorption ($\alpha = 1$) could only be the case in a room with no walls, which therefore would not exist, so nor could the reverberation. Anyhow, if the total absorption coefficient is less than 0.3, the errors with this equation are less than 6%, which still renders it a valuable equation when used appropriately.

In recording studios, however, we are generally dealing with a much higher absorption coefficient. So high, in fact, that in some very highly absorbent control rooms the sound would decay by 60 dB before even having had time to strike each surface once! Remember that one of the assumptions about reverberation time calculations is that the sound energy visits all surfaces many times with equal probability. In highly absorbent rooms it is better to refer to the T_{60} as the *decay time*, rather than the reverberation time, because the decay really only consists of a short series of reflexions.

Another commonly used means for the calculation of reverberation time is the Norris–Eyring reverberation equation:

$$T_{60} = \frac{-0.161V}{S \ln(1 - \alpha)} \tag{4.6}$$

The minus before the 0.161 V is to restore the result to a positive value after the use of the natural logarithm (ln). The $1-\alpha$ term is derived from the fact that if perfect reflexion is taken as 1, then each time that a sound wave encounters a wall the reflected energy will be equal to 1 minus the energy absorbed by the boundary, which is represented by α, the absorption coefficient. Once again, though, this equation fails in the case of relatively high absorption.

It is perhaps worth noting here, whilst on the subject of reverberation, that as room size increases the reverberation time increases proportionately, even though the average absorption remains constant. This can be deduced from Equation 4.3, which has a term in cubic metres over a term in *square* metres, hence:

$$\frac{\text{Volume}}{\text{Surface area}} \propto \frac{\text{a linear dimension}^3}{\text{a linear dimension}^2} \propto \text{a linear dimension} \tag{4.7}$$

$(\propto \; = \; \text{is proportional to})$

The net result is that as the linear dimensions increase, then so does the reverberation time if other factors remain constant.

It may seem a little strange that we have spoken so much about reverberation, merely to conclude that it is only marginally applicable to most recording studio rooms, and that the equations for calculating it are almost no use at all in any but the very largest of recording studios. Nevertheless, to know this is important, and the greater understanding of what constitutes reverberation is an important aspect of understanding room acoustics in general.

4.5 Absorption

The most important controlling factor in room acoustics is absorption. The coefficient of absorption is the proportion of the acoustic energy which is not reflected when an acoustic wave encounters a boundary. All boundaries absorb to some degree or other, which is why the concept of a perfectly reflective wall is purely hypothetical.

There are numerous means of absorbing acoustic energy, most of which result in converting it into heat. The possible exception is when absorption includes transmission, in which case the energy is lost to the outside world, such as via an open window, but eventually even this acoustic propagation will be absorbed in the air and converted into heat. At 1 kHz, the loss in dry air is about 1 dB per km, but in humid air it can be ten times that figure. Note that the 6 dB SPL reduction for each doubling of distance is *not* energy loss, but merely redistribution over a wider wavefront, as described in Section 2.3.

There are fibrous absorbers, porous absorbers, panel absorbers, membrane absorbers, Helmholtz absorbers, and these days, active absorbers, which will be mentioned in later chapters. They all function in their own ways. Some of them absorb by their action in arresting the particle velocity, in which case they need to be spaced away from the hard boundaries. Others act on the pressure component of an acoustic wave, in which case they need to be placed *at* the boundaries of a room.

The fibrous and porous absorbers can be considered together (fibrous absorbers are a sub-group of porous absorbers in general), as they both work by frictional losses caused by the interaction of the large internal surfaces of the material with the velocity component of the sound wave. Because the rapidity of the pressure changes (the pressure gradient) rises with frequency, the effectiveness of porous/fibrous absorbers also rises with frequency, due to the fact that the frictional losses increase as the particle velocity increases. As a result of this, the effect of these types of absorbers is affected by their density, their porosity, the space between the absorbent material and the wall, and their thickness. There are certain trade-offs, though. Table 4.1 shows the effects of thickness and density on fibrous absorbers. In general, barring any extra production processes that need to be paid for, one tends with most materials to pay for weight. Prices per kilogram are usually what count.

From Table 4.1 it can be seen that if we were to use a material such as 30 kg mineral wool, then at 125 Hz a thickness of one unit (say, 2.5 cm) would have an absorption coefficient of say 0.07, yet a thickness of 4 units (10 cm) of a *similar density* would have an absorption coefficient of 0.38 – over 500% greater for probably four times the price (four times the thickness of the same density = four times the weight). However, if we had our given piece of mineral wool of one *thickness* unit (2.5 cm), and an absorption coefficient of 0.07 at 125 Hz, and another piece of four times the *density* (four times the weight per given volume) but the same thickness, the absorption coefficient would only rise to 0.1. We would therefore be paying about four times the price for an improvement in absorption of less than 50%. Clearly, as we compress the mineral wool from 30 kg/m^3 to 120 kg/m^3, it becomes

Table 4.1: Effect of thickness and density on acoustic absorption

Thickness (Density 30 kg/m^2)	Frequency (Hz)			
	125	500	2000	4000
1.25 cm	0.02	0.12	0.66	0.62
2.5 cm	0.07	0.42	0.73	0.70
5 cm	0.19	0.79	0.82	0.72
10 cm	0.38	0.96	0.91	0.87
Density (Thickness 2.5 cm)	**Frequency (Hz)**			
	125	500	2000	4000
30 kg/m^3	0.07	0.42	0.73	0.70
60 kg/m^3	0.09	0.60	0.75	0.74
120 kg/m^3	0.10	0.70	0.77	0.76

more compact, and there are fewer spaces between the fibres for the air to pass through. In the extreme it would become a solid block, with almost no air inside, and therefore would no longer be porous. In this state it would lose almost all of its ability to absorb.

So, at 125 Hz, weight for weight, and hence in rough terms cost for cost, adding four layers of a lower density material and allowing it to occupy four times the space would produce over five times the absorption increase over a single piece. Compressing those four layers into the space of one would reduce the absorption by a factor of about 4. Note, though, in Table 4.1 how at higher mid frequencies, and above a certain minimum thickness, neither increasing the density nor the thickness has much effect.

4.5.1 Speed of Sound in Gases

In fact, whilst we are dealing with comparative tables, let us look at some of the mechanisms responsible for the effects shown in Table 4.1. Newton originally calculated the speed of sound in air from purely theoretical calculations involving its elasticity and density. He calculated it to be 279 m/s at 0°C, but later physical experiments showed this figure to be low, and that the true speed was around 332 m/s. There was conjecture that sound propagation only took time to pass through the spaces between the air particles, and that only the 'solid' particles transmitted the effect instantly. Newton rejected this, but then proposed that whilst the sound took a finite time to travel through the particles, the particles themselves did not occupy the entire space in which they existed, and that it was this fact which accounted for the difference. This idea also failed to be convincing. The difference continued to mystify people, until, in 1816, Pierre Simon, the Marquis de Laplace, applied what is now known as Laplace's Correction.

It is now common knowledge that air heats when it is compressed, and cools when it is rarefied. As a sound wave travels through air, it does so in a succession of compressions and rarefactions, like an earthworm. As mentioned earlier, Newton's calculations were based on the elasticity and density of the air. Elasticity is the ability to resist a bending force and to 'push back' against it, and the speed of sound through a material is partially dependent upon its elasticity. As the compressive portion of a sound wave compresses the air, it increases its elasticity in two ways; firstly by increasing its density and secondly by the heat which the compression generates. Where Newton had gone wrong was in the omission of the temperature changes from his calculations: he had only taken into account the elasticity increase from the density change. Perhaps this was due to the fact that there is no *average* temperature change as a sound wave passes through a body of air. However, *locally*, there are temperature changes in equal and opposite directions on each compression and rarefaction half-cycle. It is tempting to deduce that the increases and decreases of temperature would cancel each other out, which is perhaps exactly what Newton did, but they do not.

As air is compressed its volume is reduced, and on rarefaction its volume expands. The internal force which *resists* these changes in volume is its elasticity. If a tube containing air is sealed at one end and fitted with an air-tight plunger towards the other end, then as the plunger is pushed and pulled, the air will compress and expand. When the force on the plunger is released, it will spring back to its resting position. If the tube was then filled with a gas of higher elasticity, the force needing to be applied to the plunger for the same volume changes would be greater, as the higher elasticity would be more able to resist the changes.

It is possible to visualise the transmission of sound through air in a manner similar to that shown in Figure 4.23. In this figure, a series of heavy balls are seen contained in a tube, and separated by coil springs. If the springs are quite weak, then an impact applied to the left-hand side of ball A will pass to B, and on to C, and so forth, but there will be a noticeable delay in the transmission of energy from one ball to the next. The wave will clearly be seen to pass along the tube, as shown in Figure 4.24. Now let us assume that we apply heat to the springs, and the effect of the heat makes them much stiffer. Another impact applied to ball A will again pass down the tube, but with stiffer springs the wave will travel more rapidly.

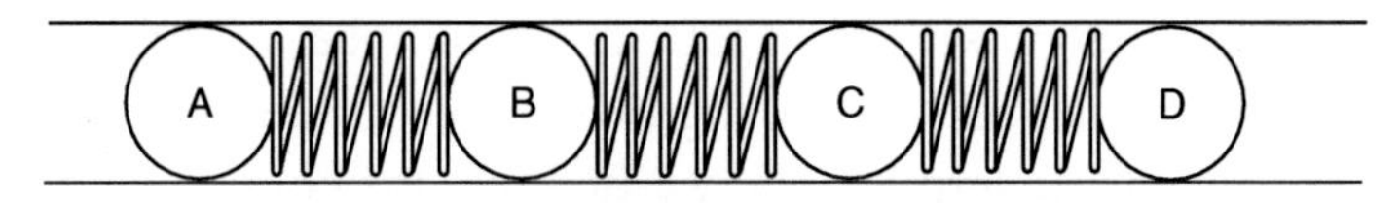

Figure 4.23:
Energy transmission via masses and springs. Four balls A, B, C and D are separated by springs in a glass tube. The system is shown in equilibrium, with no force applied, and the springs in a relaxed state.

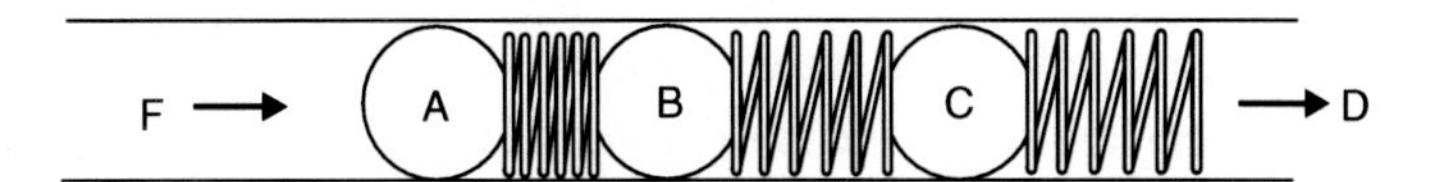

Figure 4.24:
Application of force to the system shown in Figure 4.23. If a force is applied to ball A, it will transmit energy to B via the spring. B will then transmit energy to C via the B–C spring, and so forth, until the whole chain of balls and springs is moved in the direction of the applied force. The speed with which the force will pass along the line is proportional to the stiffness of the springs (the elasticity of the coupling), and the mass of the balls. In the above example, the spring between A and B is seen to compress. The force on B is thus no longer in equilibrium, so B will move towards C, transferring some of the energy in spring A–B and compressing spring B–C until springs A–B and B–C are in equal compression. B will then tend to come to rest. In this state, spring B–C will be partially compressed, so C will have more compression in the B–C spring than in the C–D spring, and so will move in the direction of D. Due to the applied force and the momentum in the moving balls, the system will oscillate at its natural frequency until it finally comes to rest, shifted to the right, when all the applied energy has been expended.

In fact, at the extreme condition of almost rigid springs, the passage of the impact from ball to ball would be more or less instantaneous, as the ball/spring combination would behave similarly to a solid rod. The speed of the transmission of the force through the system can therefore be seen to be proportional to the stiffness of its springs. Effectively, in air, the particles can be thought of as balls being connected by springs, and the elasticity of the air is a function of the strength of these 'springs'. The force on one air particle thus compresses the spring, *heating it up*, and hence increasing the elastic force which acts on the next particle. The heating effect caused by the compression thus serves to augment the elasticity of the gas, and hence the speed with which sound will propagate through it.

In the process of rarefaction, if we consider our tube to again contain coil springs and heavy balls, and if the balls are securely attached to the springs, then a rarefaction will pull on ball A, which will in turn pull on ball B, and on to the other balls in turn. At rest, the elastic force on ball B holds it in position, because the springs A—B and B—C are in equilibrium. By pulling ball A away from ball B, the elastic force on B is reduced on the side nearest to A, thus the greater force from spring B—C will begin to force B towards A, until equilibrium is restored. As B begins to travel towards A, the force B—C will become less, so in turn the extra force D—C will begin to push on C, which will begin to move towards B. The wave of energy will propagate along the tube until all the balls are equally spaced once again, but all shifted slightly further in the direction of the pulling force. This is where, somewhat surprisingly, we find that the cold of rarefaction serves not to cancel the heat of compression, but to work in concert with it by reducing the spring stiffness and aiding the rarefaction.

In rarefaction (in our analogy), the density of the A—B 'spring' is thus reduced, so the force on the C side of B will be greater than on the A side. The cooling produced by the rarefaction will then reduce the elasticity still further (weakening the spring) and thus will act in the *same direction* as the density reduction, making the A—B force even less. This will mean a greater differential between the A—B and B—C forces, so B will be pushed away from C with greater force than that of the rarefaction density change alone.

Hopefully it can be seen from the above discussion that the heat of compression and the cold of rarefaction both work in the same direction as the density changes, and hence both *increase* the effect. The additional help caused by the elasticity changes due to heat is the reason for the speed of sound in air being greater than that first calculated by Newton from the elasticity and density alone. The heat of compression and the cold of rarefaction do *not* cancel, but work together to increase the speed of sound through the air. Once this was realised, experiments were performed to test the ability of air to radiate heat, which was subsequently found to be very poor. This explained why the temperature changes remained in their respective compression and rarefaction cycles, and did not affect each other.

The reason for discussing all this history following the table of absorption coefficients of mineral wool at various densities and thicknesses is that the above fact is one of the mechanisms at work. In loosely packed fibrous material, such as mineral wool, glass fibre, bonded acetate fibre or Dacron (Dralon) fibre, the fibres can act to conduct the heat away from the compression waves, and release it into the rarefaction waves. This removes a great deal of the heat changes available for the augmentation of the elasticity changes, and hence tends to reduce the energy available for sound propagation. It changes the propagation from adiabatic (the alternate heating and cooling within the system) to isothermal, which slows down the speed of sound. It was the isothermal state that Newton had calculated, but it was the adiabatic state that was the normal situation. Due to the above effects, a loudspeaker cabinet which is optimally filled with fibrous absorbent material appears to be acoustically bigger than it is physically, as the speed of sound within the box is slowed down by the more isothermal nature of the propagation through the absorbent fibres. Fibrous absorbents can effectively increase the size of a loudspeaker cabinet by up to about 15%, but new activated carbon materials may be able to increase the effective volume by up to 50%.[1]

Incidentally, Laplace's correction, as referred to earlier, is quite easy to explain. If we place a known volume of air at 0°C, and at a known pressure, into a pressure vessel, then heat it by 1°C, the air cannot expand, so the pressure will rise. If we then place the same volume of air, also at 0°C, in a vessel of the same volume, but fit it with a plunger which can be forced outwards as we heat the air to 1°C, the pressure can be allowed to remain constant by allowing the volume to change. In each case, we heat the same *mass* of air by 1°C, yet the heat necessary to do so is quite different. One is called the specific heat of air at constant volume, and the other, the specific heat of air at constant pressure. In fact, the latter (Cp) divided by the former (Cv) gives an answer of 1.42, and it was the square root of this number by which Laplace found it necessary to multiply Newton's original figure for the speed of sound in order to bring it into agreement with the observed speed. The reason for the difference between the two specific heats is that in the second instance (Cp) extra heat is consumed in the *work* of expanding the gas.

4.5.2 Other Properties of Fibrous Materials

Still on the subject of the absorbent properties of fibrous materials, there are other forces at work besides the ability to convert the sound propagation in air from adiabatic to isothermal. There is a factor known as tortuosity, which describes the obstruction placed in the way of the air particles as they pass through the spaces in the medium. The tortuosity, in increasing the path length for the sound which travels through the fibres, also increases the viscous losses which the air encounters as the sound waves try to find their way through the small passageways available for their propagation. In certain conditions, air can be quite a sticky fluid. There are also internal losses as the vibrations of the air cause the fibres to vibrate, and in order to bend, they must consume energy. Frictional losses are also present as vibrating fibres rub against each other. Energy is required for all of this motion of the fibres, and by these means, acoustic energy becomes transformed into heat energy.

As the above losses are proportional to the speed with which a particle of vibrating air tries to pass through the material, their absorption is greater when the particle velocity is higher. If we consider a sound wave arriving at a wall, the wall will stop its progress and reflect it back. At this point of reversal of direction, the pressure will be great, but the velocity will be zero. The same consideration applies to any ball which bounces from a wall. As it exerts its maximum pressure on the wall, its velocity is zero. The absorbent effect of fibrous materials is thus greatest when they are placed some distance away from a wall, and if any one frequency is of special interest, then a distance of a quarter of its wavelength would be an optimal spacing away from a wall for the most effective absorption by a fibrous material. The quarter and three-quarter wavelength distances are the regions of the maximum particle velocity of a wave. Conversely, membrane absorbers are dependent upon force for their effectiveness, and so should be located near to the point of maximum pressure (i.e. close to a wall) if absorption is to be maximised. The absorption mechanisms are thus very different.

4.5.3 Absorption Coefficients

Acoustic absorption is the property which a material possesses of allowing sound to enter, and *not* to be reflected back. In this sense the absorption coefficient refers not only to the sound internally absorbed, but also to that which is allowed to pass through. A large open window is therefore an excellent absorber, as only minuscule amounts of any sound which reaches it will be reflected back from the impedance change caused by the change in cross-sectional area of the spaces on each side of it. A solid brick wall is a very poor absorber, as it tends to reflect back most of the sound energy which strikes it.

Now let us put some practical figures on some different materials. A 2.5 cm slab of a medium density mineral wool can absorb about 80% of the mid and high frequency sound which strikes it. An open window will absorb in excess of 99% of the sound energy which is allowed to pass through it, and a brick wall, made from 12 cm solid bricks, will allow about 3% of the sound to enter or pass through. Looked at another way, the mineral wool will reflect 20% of the sound energy back into the room, the open window will reflect less than 1%, and the brick wall 97%. If we now look at the same materials in terms of sound *isolation*, the situations are very different. An open window will provide almost no isolation except a small amount at frequencies with wavelengths longer than the largest dimension of the opening. The slab of 2.5 cm mineral wool will provide around 3 dB of isolation (although at low frequencies almost nothing), but our brick wall will provide around 20 dB of isolation. Thus, in these cases, absorption coefficients and sound isolation properties are unrelated. In fact, in the examples quoted, they run in reverse order. Absorption and isolation properties should not be confused.

4.5.4 Porous Absorption

Curtains, carpets and other soft materials are all porous absorbers, but they are also fibrous. There is another group of porous absorbers in which fibrous friction is not present. These include micro-perforated materials with rigid skeletons, such as specially perforated plastics, in which the friction due to tortuosity and air viscosity is the predominant reason for the absorption. Rocks, such as pumice, also come into this category, as do porous plasters and some open-cell plastic foams.

Figure 4.25 shows the typical absorption curves of some porous/fibrous absorbers, and Figure 4.26 shows the effect of spacing such absorbers away from a hard wall. It is clear from the latter figure that such absorbers will absorb more strongly at frequencies whose quarter wavelengths are less than either the distance of the material from the wall, or the thickness of

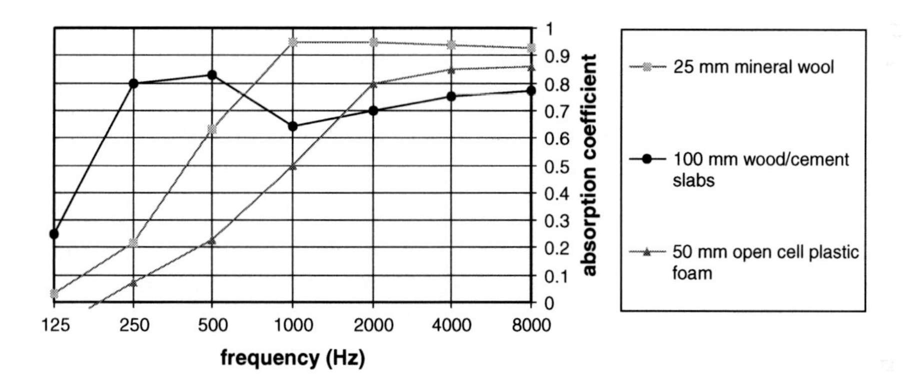

Figure 4.25:
Porous absorber fixed directly to a solid wall.

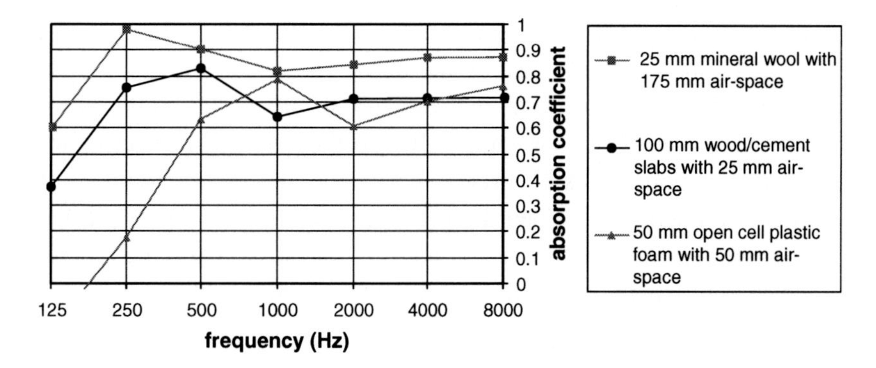

Figure 4.26:
Porous absorbers with air space behind. Note the increased low frequency absorption compared with Figure 4.25.

the material itself if it is bonded to the hard surface. This seems to suggest that as frequencies rise above the quarter wavelength frequency there could be a drop in absorption, because the point of maximum particle velocity may no longer coincide with the position of the absorbent material. However, in practice, due to the spread of the region of the wave over which the particle velocity is quite high, this sensitivity to frequency does not occur unless the absorbent material is very thin.

Before leaving the subject of porous absorbers let us consider the anechoic chamber construction shown in Figure 4.2. It can be seen that the porous/fibrous glass-fibre wool which is used as the absorbent material is in the form of wedges. In Table 4.1 we saw that the effect of thickness and density changes are such that more dense material was less absorbent weight for weight. This is partially because it presents more of a barrier to the arriving sound wave, which can enter into the pores of the less dense material more easily. So, because the surface necessarily presents a change in acoustic impedance from that of the air in the room, some proportion of the acoustic energy will be reflected back. The effect is somewhat like a lot of people all trying to pass through a few doors very quickly, but where they must keep walking at the same speed even if they cannot immediately enter a door. The only way is back! The wedges in the anechoic chamber present less of an abrupt obstacle to the encroaching acoustic wave, and so allow it to enter more gradually into the absorbent material. It also presents a greater surface area of material on each wall than could be provided by a plane surface of the same material. In fact, yet another function of the wedges is that they tend to produce the same amount of absorption independent of the angle of incidence of a sound wave, which in measuring chambers is another important point to be considered, even if it is not too relevant in studio situations.

4.5.5 Resonant Absorbers

In this category we have panel absorbers and Helmholtz absorbers. Figure 4.27 shows the construction of two typical panel absorbers, and Figure 4.28 shows a Helmholtz resonator in the form of a slotted concrete block. Unlike the fibrous/porous absorbers, which act on the velocity component of an acoustic wave, the resonant absorbers act on the pressure component of the wave (see Figures 4.9 and 4.10), so they are most effective when they are placed close to the hard boundaries of a room, where the pressure component of the resonant modes is at a maximum. When the panels are made of wood or plasterboard, for example, it is the internal frictional losses within the cellular or particulate make-up of the materials which converts the acoustic energy into heat, during the panel vibration. In the Helmholtz type of absorber, the frictional losses are due to the high particle velocities that occur in the openings at resonance.

The principle difference in effect between the fibrous/porous absorbers and the resonant absorbers is that the former generally tend to absorb better as the frequency increases, whereas the latter tend to function more effectively at lower frequencies. Figure 4.29 shows the typical sort of absorption curve to be expected from panel absorbers, and comparison with Figures 4.25 and 4.26 should prove interesting. For further comparisons, Figure 4.30 shows typical absorption curves for Helmholtz absorbers.

The frequency of peak absorption of a panel absorber is a function of the vibrating mass of the panel and the depth of the air space behind it. These are mass–spring systems, where the panel is the mass and the air cavity provides the spring. The stiffness of the spring is inversely proportional to the depth of the cavity, so, as the cavity increases in depth, the spring becomes less stiff and the resonant frequency drops. Increasing the mass of the panel also reduces the resonant frequency. All resonant absorbers are mass–spring systems, so in the case of the Helmholtz absorber, the air cavity again provides the spring, but this time the air slug in the neck of the resonator (the slot in the block in the case of Figure 4.28) provides the mass.

A variation on the Helmholtz absorber is the perforated sheet. Although the holes may seem to occupy only a very small proportion of the surface area, they act over a much greater area because of diffraction effects (see Section 4.8) which funnel the sound energy into each mouth from a much larger area. If the resonator mouths are less than a half-wavelength apart, they enhance each other's radiation resistance and are then capable of absorbing a considerable range of frequencies around resonance.

For comparison purposes, Table 4.2 shows the absorption coefficient at various frequencies for a selection of materials. The figures should not be taken as gospel truth, however, because differences inherent in the materials themselves, and in the way that they are applied, can have a great effect on their *in-situ* absorption. Nevertheless, the tendency can clearly be seen from the table that the porous/fibrous absorbers decrease in effect as the frequency lowers, whereas the resonant absorbers tend to increase in effectiveness as the frequency lowers.

4.5.6 Membrane Absorbers

These consist of a flexible, impervious sheet in place of the panel material shown in Figure 4.27. If the sheet is sufficiently 'lossy', and has a low bending stiffness, very significant amounts of acoustic energy can be turned into heat within the material of the sheet membrane. The air behind provides a damping spring. The membranes are typically plasticised bituminous material or mineral loaded flexible plastics. Polymer materials of high internal loss are also used. The early work on these absorbers was done by the BBC using bituminous roofing felt as the membrane. The trials were dogged by the tendency of the characteristics of the materials to change with temperature and time, and they were also inconsistent from batch to batch, even from the same manufacturer.

(a) Absorber wall structure

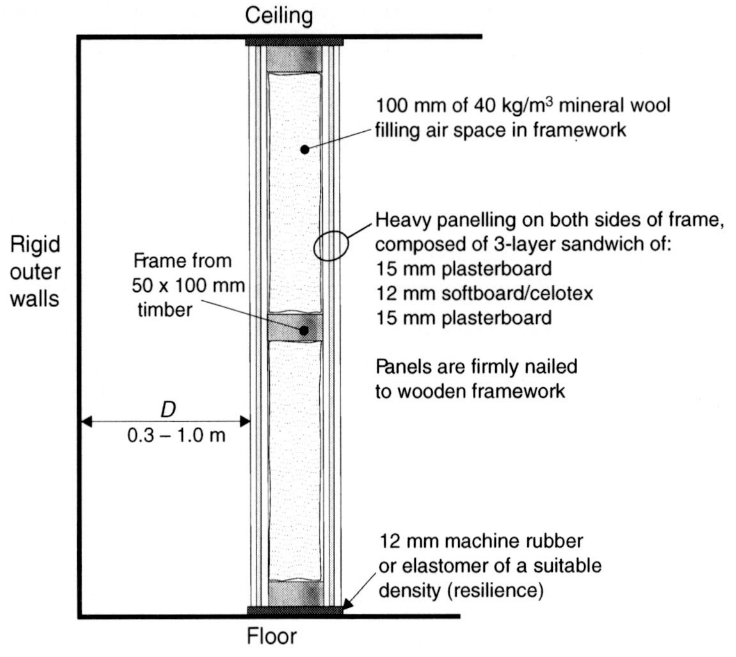

(b) A simple panel absorber

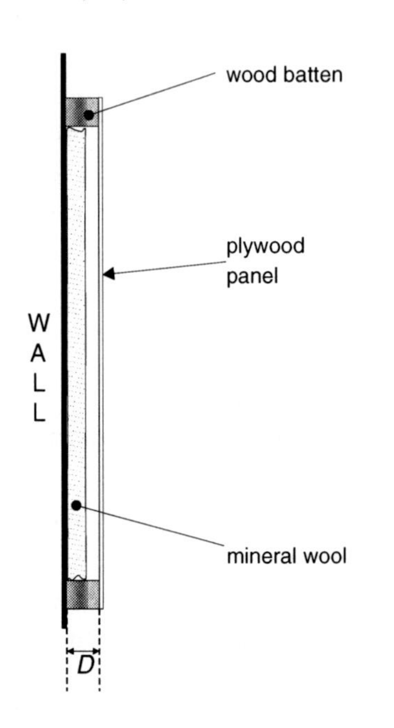

The resonant frequency is given by:

$$f_0 = \left(\frac{c}{2\pi}\right)\left(\frac{p}{MD}\right)^{\frac{1}{2}}$$

where:

c = speed of sound in metres per second (340)
p = air density in kg/m^3 (1.2)
M = panel mass in kg/m^2
D = cavity depth in metres

For 6 mm plywood (about 3 kg/m^2) and a cavity of 5 cm, the resonance would be about 160 Hz.

Specially produced materials are now in widespread use around the world, and many specialist acoustic material manufacturers have their own proprietary types. Dependent on use, such 'deadsheets', as they are commonly known, range in weight from around 3 to 15 kg/m^2. The lower weights are used more for acoustic control, whilst the higher weights are more effective for sound isolation. These materials will be considered further in later chapters. What these deadsheets have in common is that they are all very highly damped.

4.6 Q and Damping

Figure 4.31 shows the typical resonance peak of a standing wave in a highly reflective room. The high Q (quality factor) resonance exhibits a high concentration of energy around the resonant frequency, but very little activity only a few hertz away on either side of it. If any resonant absorber exhibits such a high Q resonance it will be characterised by a tendency to audibly 'ring on' after the acoustic drive signal has ceased. For this reason, acoustic absorber systems are usually heavily damped. The dashed line in Figure 4.31 shows the effect of damping the resonance. The peak of resonant activity is greatly reduced, but the width of the frequency range over which it resonates is considerably broadened. The tendency for the resonance to continue to 'ring on' after the drive signal has ceased is very much reduced. Damping, therefore, tends to be generally beneficial in terms of the desirable properties of acoustic absorbers. They absorb a wider range of frequencies more evenly, and they have fewer tendencies to re-radiate their resonant energy back into the room. However, lowering the Q does reduce the peak absorption, so for any given frequency more area must be used.

Figure 4.27:
(a, b) Panel absorbers: (a) Absorber wall structure. Panel absorber construction for providing absorption at low frequencies. This form of partitioning, often referred to as 'Camden partitioning', can be used to construct a complete self-standing room shell. The resonant frequency of such a structure would typically be about 20 Hz, but depends on the distance 'D', as well as the mass/rigidity/damping characteristics of the outer wall. (b) A simple panel absorber. The resonant frequency is given by:

$$f_0 = \left(\frac{c}{2\pi}\right)\left(\frac{\rho}{MD}\right)^{1/2}$$

where:

c = speed of sound in m/s (340)
ρ = air density in kg/m^3 (1.2)
M = panel mass in kg/m^2
D = cavity depth in m.

For 6 mm plywood (about 3 kg/m^2) and a cavity depth of 5 cm, the resonance would be about 160 Hz.

Figure 4.28:
Helmholtz resonator block.

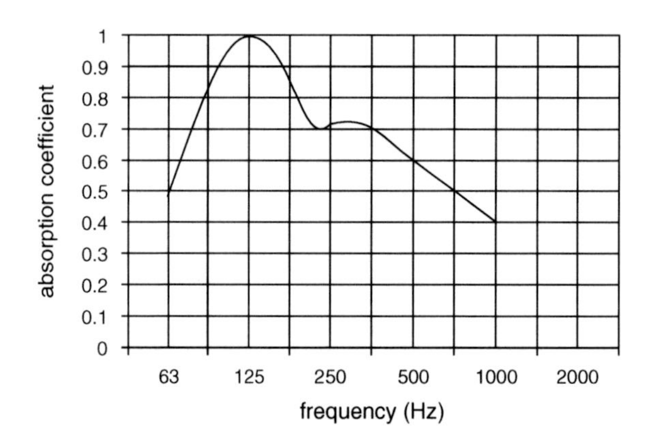

Figure 4.29:
Typical absorption characteristic to be expected from the panel absorber shown in Figure 4.27(b).

Damping refers to *any* mechanism which causes an oscillating system to lose energy. The damping of an acoustic wave can result from the frictional losses associated with the propagation of sound through porous material, the radiation of sound power, or by the internal frictional losses within any material or structure. From this it should be apparent that the damping losses within the materials reduce the Q of resonances within the materials themselves, and the absorption thus produced can, in turn, reduce the Q of the resonances within any room in which they are placed.

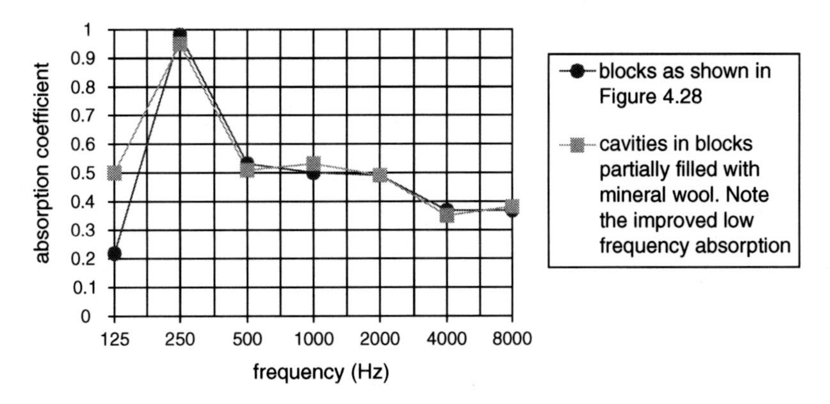

Figure 4.30:
Absorption characteristics of a Helmholtz resonator block.

Table 4.2: Absorption coefficients

	Frequency			
Material (resonant)	**125 Hz**	**500 Hz**	**2000 Hz**	**4000 Hz**
Sonicbloc (See Figure 4.28) (resonant concrete block)	0.47	0.54	0.46	0.36
Plywood panelling	0.28	0.17	0.10	0.11
Glass window	0.35	0.18	0.07	0.04
Plaster on wood laths	0.14	0.06	0.04	0.03
Material (porous)				
Folded curtains	0.07	0.49	0.66	0.54
Carpet on concrete	0.02	0.14	0.60	0.65
Mineral wool (high density)	0.29	0.32	0.50	0.56
Cotton felt	0.13	0.56	0.65	0.70

4.7 Diffusion

The use of acoustic diffusers has been growing steadily since the 1980s. Convex and generally curved surfaces have been used since the design of early concert halls to re-radiate the incident sound over a wide area, and to break up the specular (i.e. behaving like light) reflexions which could tend to concentrate energy in specific areas (see Figure 4.32(a)). Again like absorbers, diffusive surfaces are linked in their size to the frequencies over which they are intended to have an effect. A low frequency diffuser of traditional design would need to be about a quarter of a wavelength deep in order to diffuse effectively.

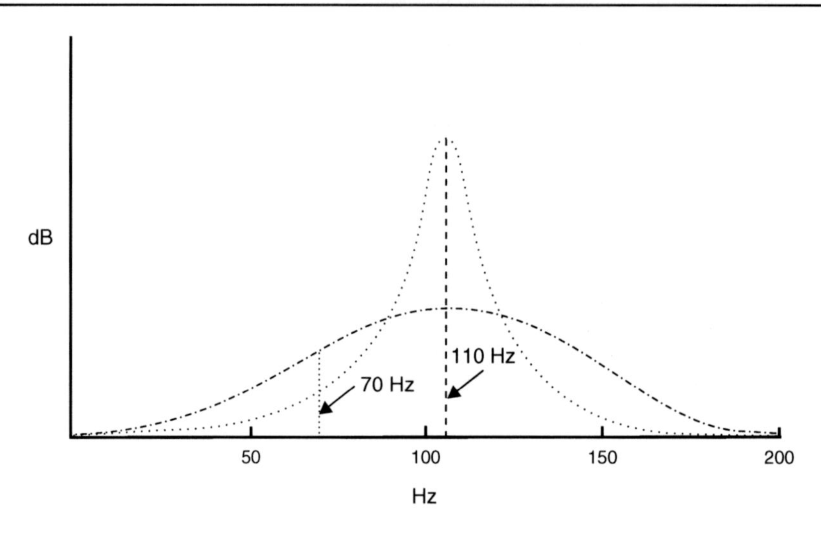

Figure 4.31:
The effect of damping on 'Q'. The dotted line represents a high Q resonance, which will be strongly excited by a stimulus at 110 Hz but will respond only weakly to a stimulus at 70 Hz. The dashed line shows the effect of damping in lowering the Q. In this case, excitation at either 70 Hz or 110 Hz will produce an almost equal degree of resonance. Both the dashed and the dotted lines represent resonances with a 110 Hz centre frequency, and both contain a similar amount of total energy. It can thus be seen that the low Q resonance responds to a broader range of frequencies than the high Q resonance, but the low Q resonance can never achieve the same peak amplitude as the less damped high Q resonance if similarly excited.

Schroeder[2] and D'Antonio[3] proposed designs for smaller diffusers using sequences of wells of different depths which are formally defined by mathematical sequences. To be effective, they still need to have an adequate depth, but the size is generally more manageable than the 1 m that a conventional convex diffuser would need in order to be effective in the 80–100 Hz region. Figure 4.32(b) shows the effect of a traditional tube/cylinder type diffuser, and Figure 4.32(c) shows the plan view of the well structure of a Schroeder-type diffuser.

In both cases, the incident wave is scattered in a wide range of directions other than the specular direction, where, like light, the angle of incidence equals the angle of reflexion, as shown in Figure 4.32(a). Diffusion is a form of scattering, but, where scattering tends to be more random in its effect and is often dependent upon the angle of incidence of the sound upon an irregular surface, effective diffusion aims to scatter the sound over a broader area, and a broader range of frequencies in a more spacially uniform manner. It is also less dependent upon the angle of incidence. In other words they are intended to scatter the energy equally in all directions. Angus[4] has shown, however, that in some cases the mathematically based diffusers *can* create certain 'hot spots', or energy concentrations, that are even more concentrated than would be the case from a plane reflective surface. Nonetheless, this does not detract from the usefulness of such diffusers in many aspects of studio and auditorium acoustics.

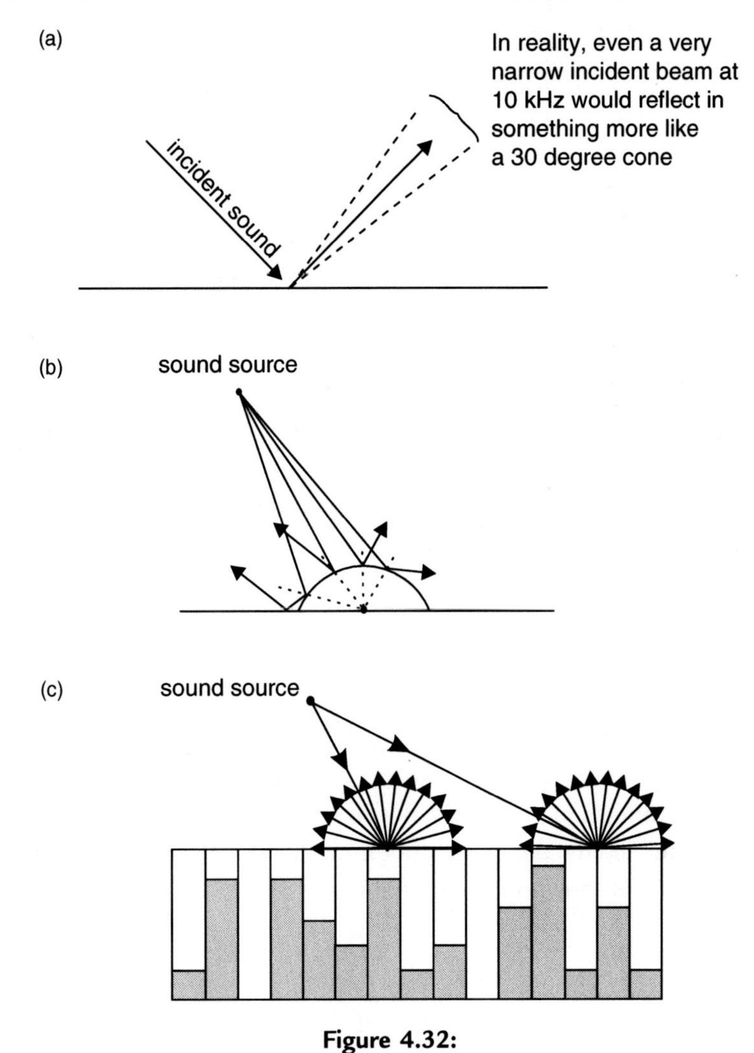

Figure 4.32:
(a) Specular reflexion. The concept of a specular reflexion is that it behaves like the reflexion of a beam of light. In fact this does not happen, because even a narrow beam of 10 kHz will still tend to reflect in a cone of about 30°. Nevertheless, the concept is useful; (b) Reflexion from curved surfaces. Although sound does not travel in narrow rays, the ray concept is useful to describe the diffusive effect of a semi-cylindrical surface. When the wavelength of the sound is longer than the diameter of the curved surface, the diffusive effect begins to reduce. For wavelengths much larger than the diameter of the curved surface, the effect is substantially the same as it would be if the surface were flat; (c) Reflexion from mathematically based diffusers. The concept of the Schroeder/reflective-phase-grating type of diffuser is for sounds of any angle of incidence to be reflected in all directions (two-dimensionally). The depth and distribution of the pits (wells) is determined mathematically, and is usually based on quadratic residues or prime number sequences. The maximum well depth determines the lowest frequency of good diffusion.

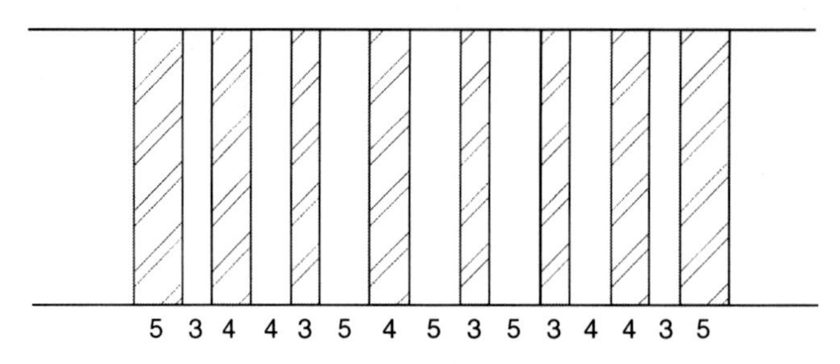

Figure 4.33:
Typical arrangement of wood strips and absorbent openings. The widths of the strips and openings are based on a numerical sequence that provides a more or less equal area of absorbent and reflective surfaces, but without any simple regularity, which could lead to problems at certain frequencies having properties coinciding with the regularity. The numbers in the diagram represent the relative dimensions of the adjacent strips and openings.

Another form of diffuse surface is the 'amplitude reflexion grating', an example of which is shown in Figure 4.33. In fact, any juxtaposition of absorbent and reflective surfaces will cause sound to scatter due to the steering of the wavefront as a result of phase changes at the boundary. There are also mathematical procedures for the optimum relative positioning of absorbent materials to promote diffusion,[5] such as binary and prime number sequences. Amplitude reflexion gratings tend to diffuse not so well as the physically deeper diffusers, but they take up little depth, and the depth does not limit their low frequency performance. A flat, binary amplitude grating is shown in Figure 4.34.

4.8 Diffraction

Diffraction provides another form of acoustic energy scattering by the redirection of the incident energy. It is most commonly encountered at the discontinuities at the edges of partitions and screens, and is evident in control rooms at the edges of mixing consoles and rack furniture. Diffraction also occurs at the boundaries where absorbent and reflective surfaces meet. It is particularly problematical at the sharp edges of small loudspeaker boxes, where the edges can act like independent sound sources. This will be discussed later in Chapter 11. The fact that we are able to hold conversations over high, massive walls is totally thanks to the phenomenon of diffraction. However, the diffraction around complex objects can be hideously difficult to predict. Some of the earliest work on diffraction was done by the French physicist Augustin Fresnel, who invented the lighthouse lens, which is an example of *light* diffraction. Figure 4.35 shows some examples of diffraction, which clearly demonstrate its frequency dependent nature.

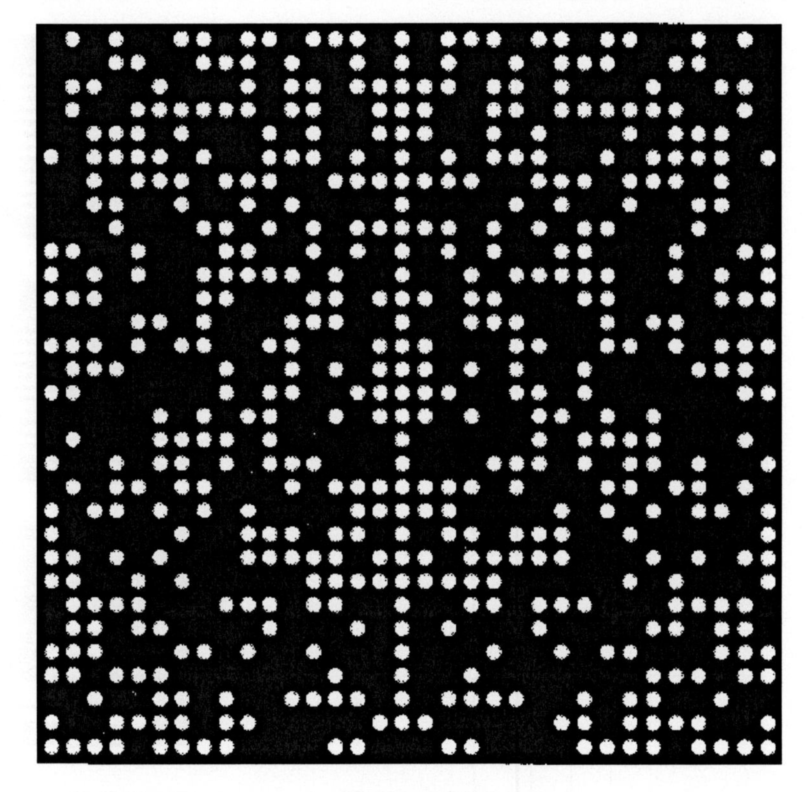

Figure 4.34:
Flat, binary amplitude grating.

Diffraction occurs due to the interaction between adjacent air molecules, because the compressions and rarefactions in a sound wave cannot simply shear at the boundary of an object without exciting the air nearby. If ever a picture told a thousand words, then it is exemplified by the photograph in Figure 4.36. It is also due to diffraction processes that a noise entering a room via a hole can be heard in most parts of the room, even if the room is very non-reflective.

4.9 Refraction

Acoustic refraction exists in fluids as a result of the spacial non-uniformity of the speed of sound. In air, if there is a non-uniformity of temperature in a given volume, the sound will travel at a greater speed in the warmer air, and so the wavefront will bend. It is mentioned here only to clarify the fact that it does not really enter into events in recording studios. It can be apparent at outdoor concerts when the temperature of the air changes with height, and it explains why some sounds can be heard a very great distance away from their source, yet at intermediate distances there is no perception of the sound whatsoever. In recording studios,

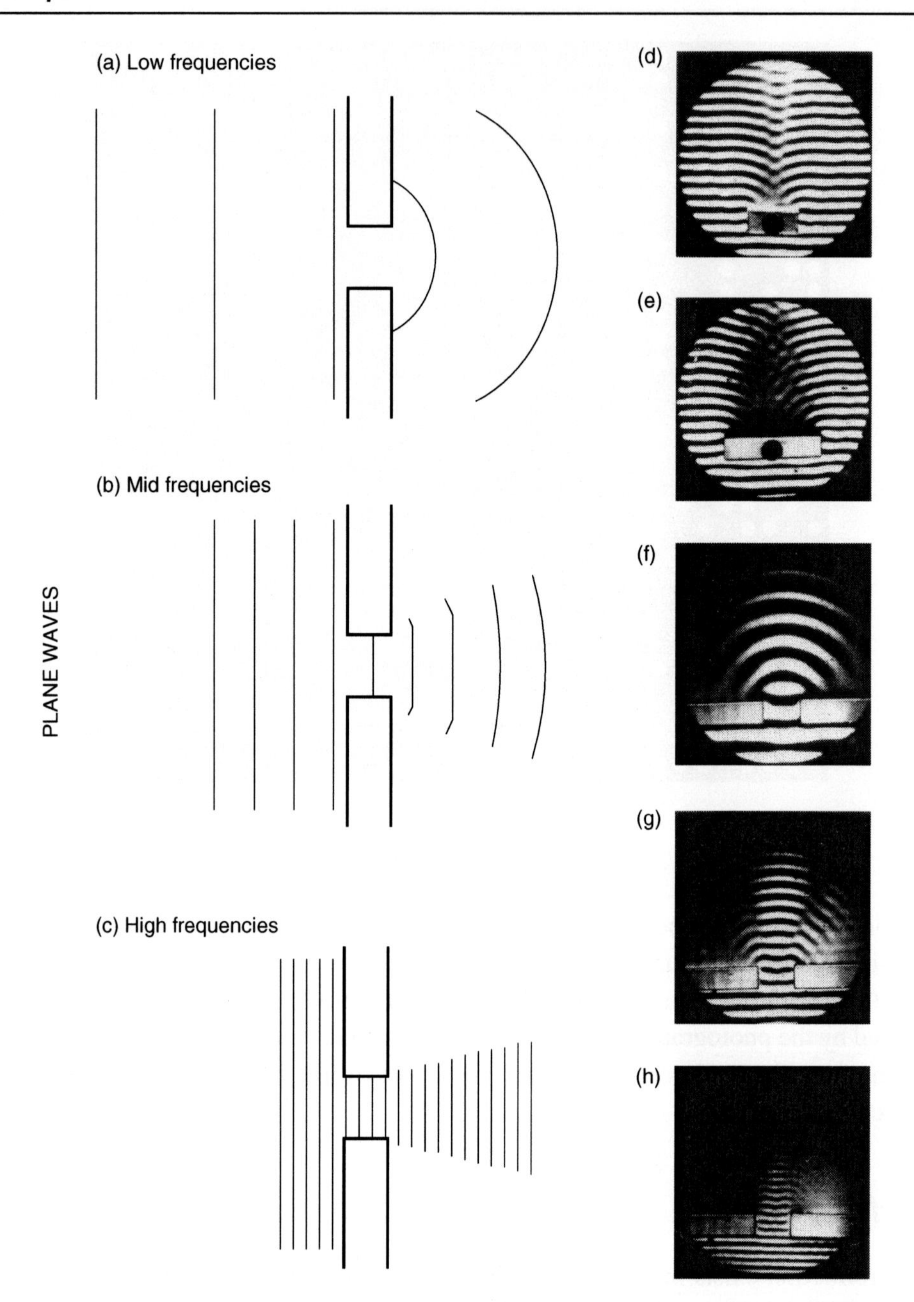

Figure 4.35:
(a–c) Frequency-dependent diffraction through a slot. Alternative presentation, in photographic form, shown in (f), (g) and (h). (d–h) Diffraction of a plane wave by: (d) narrow obstacle; (e) wide obstacle. Diffraction of a plane wave by an aperture in a screen: (f) at low frequency; (g) medium frequency; (h) high frequency. The photographs (d) to (h) were taken from *Foundations of Engineering Acoustics*, by Frank Fahy; Academic Press (2001) (original source unknown).

Figure 4.36:
Diffraction of water waves by a breakwater. Hydraulics Research Station, Wallingford, UK. Reproduced with permission from HMSO.

even if temperature gradients do exist, the distances involved are much too small for refractive effects to be significant.

4.10 Review

Well, one chapter can hardly do justice to the subjects of room acoustics and their means of control, but it has been necessary to deal in a general way with them in order to provide a basis for the discussions which will follow. For more experienced readers, this chapter may have contributed little to their prior knowledge, but even so, a periodic recapitulation of the general concepts does no harm. Unfortunately, though, for a great number of people who are deeply involved in the professional recording industry, such information is not always very readily at hand, so without this discussion of some of the basic principles the remainder of the book would not be so accessible.

In fact, neither the performance of any musical instrument nor any loudspeaker can be realistically defined without a considerable knowledge of the acoustics of the rooms in which they are likely to be used. So, despite the frustrations inherent in the subject of room acoustics, we need to at least grasp the basics in some detail or our knowledge of recording studio design and performance will not be satisfactory. Those wishing greater in-depth knowledge of some of the relevant details are recommended to look at the References and the Bibliography at the end of the chapter.

4.11 Summary

A reflective isolation shell initially worsens the problems of internal acoustic control. A sound control room, well isolated, tends to start from a worse set of acoustics than exists in a normal domestic room.

Walls which are acoustically reflective create sonic images which are akin to the visible images of loudspeakers in a room with mirrors covering all surfaces. The reflexions behave as if they are coming from the distant loudspeakers which would be visible in the mirrors.

Flush mounted loudspeakers create less low frequency confusion in the room than do free-standing loudspeakers.

Sine waves give rise to axial, resonant standing wave patterns in a room with reflective walls when an integral number of half-wavelengths will fit exactly between two acoustically parallel surfaces.

The energy in an acoustic wave is continually changing between kinetic and potential as the velocity and pressure components, respectively, reach their maxima.

Modes are the pathways which a sound wave can take when travelling around a room. There are driven (forced) modes and resonant modes. It is the resonant modes which give rise to the standing/stationary wave patterns.

The lowest frequency of any mode in a room is that of the axial mode where one complete half-wavelength fits exactly between the two walls separated by the longest dimension of the room.

The resonant modal energy increases the loudness of a sound in a room, but unless the modes are closely spaced, they will favour some frequencies more than others, leading to an uneven frequency response.

Modal *patterns* depend on the shape of a room, and not its size. The size will determine the frequency range across which the pattern is distributed.

Flutter echoes result when a transient sound is produced between parallel reflective surfaces which are separated by more than about 8 m. With the surfaces separated by less distance, the result is a metallic ringing sound.

When the reflexion pattern becomes dense and spacially diffuse, reverberation is created. The reverberation time of a room is that which it takes for the reverberant field to decay by 60 dB after the sound source has been stopped.

As room size increases, the reverberation time increases proportionally if the average absorption remains constant.

Absorption is the most important factor in the control of room acoustics. The most common absorbers are porous absorbers, panel absorbers, membrane absorbers and Helmholtz absorbers.

Porous absorbers act on the particle *velocity*, by frictional and viscous losses. They are most effective *away* from wall surfaces and at *higher* frequencies.

Panel, membrane and Helmholtz absorbers are more effective *at the boundaries* of a room, and at *low* frequencies, because they react to the *pressure* component of an acoustic wave.

In air, due to the fact that it is a poor conductor of heat, the heat of compression and the cold of rarefaction *both* act to augment the speed of sound. If these heat changes can be reduced, such as by placing fibrous material in the air volume, then the speed of sound will be reduced. This can effectively enlarge a room (or a loudspeaker cabinet by up to 15%), due to the change from adiabatic to isothermal propagation.

The absorption coefficient of a material represents the proportion of a sound wave that is *not reflected*. Hence an open window is an almost perfect absorber, but it is not a good isolator. Isolation and absorption should not be confused. High *isolation* means low *transmission*. Isolation is usually achieved by reflective containment of the sound waves.

The Q of a resonance can be reduced by damping. Damping refers to any mechanism which causes an oscillating system to lose energy.

Diffusion is the re-radiation of an incident sound wave over a wide area.

Diffraction occurs at discontinuities at the edges of objects, and tends to bend the wavefront around the object. Edges which give rise to diffraction behave as separate sound sources.

Refraction is the process of the change in direction of a wavefront due to the spacial non-uniformity of the speed of sound. It is not usually encountered in recording studios, but may be apparent at outdoor music events when the air temperature changes with height.

References

1 Wright, J. R., 'Increasing the Acoustic Compliance of Loudspeaker Cabinets', *Proceedings of the Institute of Acoustics* [Reproduced Sound 17 Conference], Vol. 23, Part 8, pp. 23–8 (2001)
2 Scroeder, M. R., 'Diffuse Sound Reflection by Maximum Length Sequences', *Journal of the Acoustical Society of America*, Vol. 57, No. 1, pp. 149–50 (January 1975)
3 D'Antonio, P. and Konnert, J. H., 'The Reflection Phase Grating Diffusor: Design, Theory and Application', *Journal of the Audio Engineering Society*, Vol. 32, No. 4, pp. 228–38, (April 1984)
4 Angus, James A. S., 'The Effects of Specular Versus Diffuse Reflections on the Frequency Response at the Listener', *Journal of the Audio Engineering Society*, Vol. 49, No. 3, pp. 125–33, (March 2001)
5 Angus, J. A. S., 'Sound Diffusers Using Absorption Gratings', presented to the 98th Convention of the Audio Engineering Society, Paris, France, Preprint No. 3953 (February 1995)

Bibliography

Angus, James A. S. and McManmon, Claire I., 'Orthogonal Sequence Modulated Phase Reflection Gratings for Wide-Band Diffusion', *Journal of the Audio Engineering Society*, Vol. 46, No. 12, pp. 1109–18 (December 1998)

Borwick, John, *Loudspeaker and Headphone Handbook*, 3rd Edn, Focal Press, Oxford, UK (2001)

D'Antonio, P. and Cox, T., 'Two Decades of Sound Diffusor Design and Development, Part 1: Application and Design', *Journal of the Audio Engineering Society*, Vol. 46, No. 11, pp. 955–76 (November 1998)

D'Antonio, P. and Cox, T., 'Two Decades of Sound Diffusor Design and Development, Part 2: Prediction, Measurement and Characterization', *Journal of the Audio Engineering Society*, Vol. 46, No. 12, pp. 1075–91 (December 1998)

Davis, Don and Davis, Carolyne, *Sound System Engineering*, 2nd Edn, Focal Press, Oxford, UK, Boston, USA (1997)

Fahy, Frank, *Foundation of Engineering Acoustics*, Academic Press, London and San Diego (2001)

Howard, David M. and Angus, Jamie A. S., *Acoustics and Psycoacoustics*, 4th Edn, Focal Press, Oxford, UK (2009)

Parkin, P. H., Humphreys, H. R. and Cowell, J. R., *Acoustics, Noise and Buildings*, 4th Edn, Faber and Faber, London, UK (1979)

CHAPTER 5

Designing Neutral Rooms

The need for neutral rooms. What constitutes 'neutral'. The concept and construction of large neutral rooms. The effect of shape on modal patterns. The effect of a room on a musical performance. Acoustic parallelism. Achieving neutrality in small rooms. Practical construction of a small neutral/dead room. Sound paths within a complex acoustic control structure. The pressure zone. Transfer of sound between high and low density media. Remnant micro-resonances. Dialogue recording rooms.

In the chapters so far we have looked at what a studio needs to be able to achieve; some of the properties of sound and hearing; the means of providing isolation; the internal acoustic behaviour of an isolation shell and the means at our disposal to attempt to control the internal acoustics. What we can now do, therefore, in this chapter, is to apply the information from the previous chapters to try to create a neutral, unobtrusive, yet musically pleasing room. Of course, neutral rooms are not necessarily the most inspiring rooms to design, but here they make a good starting point because the ensuing discussion will take us through a whole range of acoustic control principles. These can then be refined in later chapters when we begin to look at rooms with much more specific acoustics for their own specialised purposes.

5.1 Background

Historically, the recording studios (the word 'studios' here meaning the recording spaces, as opposed to the control rooms) have been relatively neutral environments. This has been partly due to past studios having to cater for a wide range of recordings. Too much bias towards the needs of one specific type of music could lead to a restriction in the amount of work available. Furthermore, it was formerly often considered that the function of the recording was to capture the sounds which the musicians produced as faithfully as possible. The days of a somewhat more creative side to the recording process had still not arrived.

Contrary to what may often be expected, the 'true' sound of a musical instrument is not simply that which it would make in an anechoic chamber. This is because instruments have been developed in the circumstances of more reflective or reverberant surroundings, and it is frequently the combined direct *and* reflected sounds which constitute the 'true' sound of an instrument – the sound as it is *intended* to be heard. Concert halls have long been rated by musicians according to how well they can perform in those halls. Players of acoustic instruments need a feedback from the performing space, as frequently the sound emission from the instruments are inadequate to give directly to the musicians the sensations required for optimum performance. String sections need to hear string *sections*, not a group of individual instruments. When they hear a section, they play as a section; but when they hear separate instruments their playing also often fails to gel into a single, homogeneous performance. Flautists seem almost always to need some reverberant help for their playing, be it natural, or fed electronically into their foldback. Woodwind players also seem to dislike too dry an ambience in their performing space. In fact, an anechoic chamber is a truly awful environment in which to play *any* instrument, and such an environment is not going to inspire a musician to the heights of creativity. Creativity is supremely important, as it would appear to be self-evident that an uninspired performance is hardly worth recording.

So, if we are not meaning a clinically accurate recording space when we speak of 'neutral' spaces, then what *are* we talking about? Essentially, a neutral environment is one which provides sufficient life to allow enough of the character of an instrument to be apparent, but which does not overpower the instrument with the character of the room itself. This means a smoothly sloping reverberation time (or rather, decay time) together with discrete reflexions which add life but do not dominate the natural sound of the instrument. Normally, such rooms will have decay times which rise as the frequency lowers. This is a function of most enclosed spaces other than very small ones, and when one considers the fact that most instruments have been developed for performance *in* such spaces, a recording area with similar characteristics would not be deemed unnatural. Rooms of different sizes, shapes and structures will have their own characteristic acoustics, but as long as those characteristics do not add any significant timbral change to the instrument they can be considered to be neutral.

In general, it is easier to make large neutral rooms than small ones. This is because of two main reasons. Firstly, in large spaces, the resonant modes tend to be more evenly spaced across the frequency spectrum, whereas in small rooms, especially at lower frequencies, they tend to separate, as shown in Figure 4.14. Particularly in the upper bass region, unevenly spaced modes can become strongly audible due to the concentrations of energy beginning to add a strong character to the sound of the instruments in the room. Secondly, in larger rooms there is a greater period of time between the emission of sound from the instruments and the arrival of the reflexions. Floor reflexions are, of course, returned at similar time intervals in all sizes of rooms, but these are usually relatively innocuous, single reflexions, free of resonant characteristics. The more noticeable resonant modal energy must exist between at least two surfaces, and therefore, to reduce undue colouration of the sound, hard, parallel floor/ceiling combinations are usually avoided in studios. In large rooms the greater period of time before the first reflected energy returns to the instrument allows more time for the direct sound of the instrument to stand alone, and thus establish itself clearly in the perception of the listeners, or as captured by the microphone(s).

A further two reasons also lead to the later reflexions having less colouring effect. Reflexions from greater distances have further to travel, so when they do return, they will do so with generally less intensity than those travelling back from shorter distances (given the same surfaces from which to reflect). What is more, when reflexions arrive much more than 30 ms after the initial sound they tend to be perceived by the brain *as* reflexions, whereas those arriving before 30 ms have elapsed they will almost certainly be heard as a timbral colouration of the instrument, and will not be perceived as discrete reflexions. This is, of course, the Haas effect (see Glossary). In a large room, therefore, the resonant modes and reflexions are usually heard as separate entities to the direct sound of the instruments. Unless the direct sound of an instrument is swamped by room sounds which are unduly long in time or high in level, its natural characteristic timbre will be clearly heard.

Generally, a recording room can be considered to be in the acoustically small category if it is impossible to be less than 4 or 5 m from the nearest wall surface. Judicious angling of the ceiling, together with careful use of absorption and diffusion, can permit the use of ceilings of 4 m or less with relative freedom from colouration, so an acoustically small room for recording purposes would tend to be less than about 10 m×10 m×4 m.

5.2 Large Neutral Rooms

To build an acoustically large neutral room is not a particularly difficult exercise as long as a few basic rules are followed. Parallel hard surfaced walls should be avoided, as these can support the strong axial modes which develop, reflecting backwards and forwards between the parallel surfaces. The effects of such reflexions on transient signals are those of 'slap-back' echoes, or series of repeats, usually with pronounced tonal contents which are often quite unmusical and unpleasant in nature. The parallel surface avoidance rule also relates to floors and ceilings. However, we should make a distinction here between what is actually heard in the room and what is heard via the microphones. Although the *ear* is often not unduly troubled by vertical reflexions, because it is much less sensitive in the vertical plane than in the horizontal plane, most microphones tend to be totally ignorant of concepts of horizontal or vertical. Therefore, a floor/ceiling problem or a similar wall/wall problem will be detected by most microphones in exactly the same way. Even though the ear may hear them very differently when listening directly, the perception when listening to a recording will be the effect as detected by the microphones. Consequently, when considering the subjective neutrality of a room, we must consider it from both points of reference: listening directly via human ears, *and* listening via microphones.

Unless two rooms are absolutely identical, not only in shape, size and surface treatments, but also in the structure of their outer shells, they will not sound identical. In reality, there are thousands of 'neutral' recording spaces in the world, but it is unlikely that any two will sound identical to each other. The achievement of acoustic neutrality is all a question of balances and compromises, but unlike the neutrality needed in control rooms, where repeatable and 'standard' reference conditions are needed, such uniformity of neutrality is not required in the studios. The concept of neutrality in a recording room therefore occupies a region within an upper and lower limit of what a room may add. All that is required is that the room sound is evenly distributed in frequency and is subservient to the sound of the instrument(s).

A parallel exists in the realm of amplifiers. Guitar amplifiers and hi-fi amplifiers are quite distinct devices, and normally cannot be interchanged. Guitar amplifiers have relatively high levels of distortions, but those distortions are chosen to be constructive and enhancing in terms of the sound of electric guitars. However, recorded music played through a guitar amplifier and loudspeaker will sound coloured, and will suffer from a lack of definition. The result will certainly not be hi-fi. Conversely, a guitar played through a hi-fi amplifier and loudspeaker (the neutrality of which are more akin to control room neutrality) will be unlikely to sound full-bodied or powerful. In a similar way it is thus entirely justifiable for a neutral recording room to add to the character of an instrument played within that room, as long as that character enhances and supports the instrument and does not in any way become predominant itself. Therefore, anything in the sound *production* side of the record/reproduce chain can be considered to be an extension of the instrument, and hence subjective enhancement is usually desirable. Conversely, things in the *reproduction* or the quality control sides of the chain must be transparently neutral, in order to allow the production to be heard as it is on the recording medium.

5.3 Practical Realisation of a Neutral Room

Neutral rooms are desirable for many types of acoustic recordings, but modern thinking places much more emphasis on the comfort of the musicians than was encountered in a great majority of the coldly neutral rooms which were prevalent in previous years. At each stage of our design we therefore need to consider its effect on the musicians, as well as on the purely acoustical requirements. Anyhow, it seems to be very widely accepted that floor reflexions are in almost all cases desirable, and indeed most live performance spaces have hard floors, so let us begin the design of our large neutral room with the installation of a hard floor.

5.3.1 Floors

Hard floors can be made from many materials, but they generally subdivide into vegetable or mineral origins. On the vegetable side we have a great variety of wood-based choices. There are hardwoods, softwoods, wood composites such as plywood, veneered MDF (medium density fibreboard), cork tiles, parquet and reconstituted boards, to name a few of the most common types. In the mineral domain we have stone in its various forms, ceramic tiles and concrete with resinous overlay, which is often used in television studios where cameras must roll over an extremely smooth, unjointed surface. Usually, in large rooms, wood prevails. It is aesthetically warmer, thermally warmer, less prone to slippage (both by people and instruments), and it is generally richer acoustically. Instruments such as cellos and contrabasses rely on the floor contact to give them a greater area of soundboard as their vibrations travel through the wooden surface. Mineral based floors do not 'speak' in the same way. Wood also represents more closely the flooring which musicians are most likely to encounter during their live performances, and, where possible, studios should seek to make the musicians feel at home. The necessity for this cannot be over-emphasised. The exact nature of the floor structure will usually depend on a multitude of factors concerning the structure and location of the building, but these things will be dealt with in later chapters.

It is very important that floors should not creak or make other noises when people are walking on them, nor should they rattle under the influence of vibrations from instruments or their associated amplifiers. For this reason the floors should be well damped, which means that the upper surface should be securely fixed to the mass layers below, such as those shown in Figures 1.1 and 3.4. Domestic type floated floor systems, laid loosely on a layer of thin polyethylene foam, should not be used unless the foam is omitted and the boards are glued and pinned to the surface below them. They have a tendency to flap about under the influence of high levels of low frequencies if not well fixed down. The reason for not fixing them down in domestic use is that floating them on foam provides more isolation from impact noise, and the non-rigid fixing also allows them to expand and contract at different rates to the floors on which they lie. However, in the (hopefully) well-controlled temperatures and humidity of a professional studio, the expansion problems should not exist. Nevertheless, it is just as well to match the base material of the flooring (on which the veneered surface is fixed) to the upper surface of the underlying floor. From this point of view the floor systems (such as Kahrs, Junckers, etc.) which use plywood or solid wood bases tend to be preferable to the ones with synthetic, resin-based under surfaces.

5.3.2 Shapes, Sizes and Modes

So, we now have a floor, but for the reasons already cited it is of necessity highly reflective, and it has done nothing to neutralise our reverberant isolation shell. We must therefore continue with the rest of the room to find some means of control. From the traditional point of view the worst case starting point would be to have a cubic room, with all dimensions (length, breadth and height) equal. Under such conditions, where all the pairs of parallel surfaces are equally spaced apart, the axial modes will all be similar in path length, and hence will all have similar resonant frequencies. This will lead to a strong resonant build up at the frequencies associated with those modes. Furthermore, the axial modes are the ones which are considered to contain the most energy, and so those frequencies whose wavelengths correspond with the dimensions of the room will predominate, giving the room a highly tuned, strongly resonant character. Such a room could be a 'one note' room, with overpowering resonances destroying the musicality of almost any instrument played in it. At the other extreme, a room of rectangular plan with dimensions of height, breadth and length in the approximate proportions, 1:1.6:2.33 would give rise to a highly varied assortment of modal frequencies, and hence would exhibit the overall least coloured sound. It was long held that this type of room should form the basis of 'standard' listening rooms for the assessment of domestic equipment, but it has been pointed out that the aforementioned modal properties only relate to an empty room with rigid walls. As soon as one installs equipment, people, decorative and absorbent surfaces and so on, the smoothness of the modal distribution may be lost. Nonetheless, a room of such proportions would seem to be a much better starting point than the cubic room, although these proportions only hold good for rooms that are neither extremely small nor the size of large concert halls. Some of these 'golden proportions' (other suggestions being 1:1.4:1.9 and 1:1.28:1.54) are based on calculations which give the most even spread of modes and are intended to yield relatively neutral sounding rooms.

As shown in Figure 4.15, it is the *proportions* of a room, not its absolute size, which dictates the modal distribution pattern. Acousticians such as Louden[1] made some of the earliest in-depth studies of the problem, and as a result of much research came up with proportions of 1:1.4:1.9 after studying the first 216 modes for 126 different room dimension ratios. Gilford[2] adopted a different approach in a later study, in which he concluded that it was the *axial* mode energy which dominated colouration. Consequently, he recommended looking for groupings of axial modes up to about 350 Hz, and if any such clusters were found, then the room dimension should be adjusted to best split up the groupings.

Even later, the Argentinean acoustician Oscar J. Bonello[3] adopted yet another approach by considering the total number of modes which fall in third-octave bands of the frequency spectrum. The principle became known as the 'Bonello Criterion'. He divided the audio spectrum of 'good sounding rooms' into third-octave bands, as an approximation to the critical bands of human hearing (although the relevance of this has now been severely questioned), and then counted the number of modes per band. The resulting criterion states that if the number of modes per band increases monotonically (i.e. increases by varying amounts but never falls) with each higher frequency band, then there is a good chance that the room will be perceived as sounding musically 'good'. It further stated that if there are coincident modes in any band, such as those caused by equal distances or multiples of distances between sets of reflective surfaces (such as shown by the thicker lines in Figure 4.14), then there should be at least three additional, non-coincident modes in the same band, to counterbalance the self-reinforcement of the coincident modes.

Nevertheless, despite the selection of room proportions with the most even modal spreads, if the reverberation time is excessive the room character will still tend to dominate the sound of the instruments, and thus would not meet the conditions necessary to be considered neutral.

Absorption and diffusion will be necessary to bring the acoustic character into the neutral range, and, once such devices have been installed, the tendency is for the modes to weaken in strength and broaden in frequency spread. So, in practice, the 'golden dimensions' are not too relevant to a neutral recording room, which is fortunate because many studios are built in existing premises where ideally proportioned spaces may not be available. It must also be borne in mind that the driving and the reception of all of the modes of a room can only be achieved from three-surface corners. In any other places in the room, only some of the modes would either be driven or heard/picked-up, so under any normal working conditions the calculated modal spread of the room will not be perceived. The concept of the 'special' room dimension ratios therefore tends to be of little practical use.

Irregular shaped rooms tend to significantly eliminate the higher axial mode activity because of the lack of parallel surfaces, but as will be shown in Section 5.4, the lower frequency modes may be very little affected by the angling of the surfaces away from parallel. The other modal resonances will be of the tangential or oblique forms, which generally retain less energy than the axial modes. The non-normal incidence of these sound waves relative to the surfaces tends to make the modes become lower in their 'Q' (or tuning) as the energy becomes spread more broadly, being less tuned to specific notes. The natural reverberance of such rooms is often perceived to be smoother, with fewer predominating frequencies, but in all the above cases, the most persistent problem is how to control the more widely spaced modes in the lowest octaves of the audible range, where wavelengths are long, even when compared to any possible angling of the walls.

5.3.3 From Isolation Shell Towards Neutrality

Perhaps, then, in our quest for neutrality, for investigation purposes it would be wise to look at a relatively difficult, but nonetheless likely, case of a shell of 15 m × 10 m × 5 m high. It is rather awkward because the length and breadth are exact multiples of the height, so resonant modal frequencies of the floor/ceiling dimension (around 34, 69 and 103 Hz) can be supported two and three times over in the length and breadth. Strong irregularities at the resonant frequencies would exist in different locations in the *untreated* room, dependent upon whether the sound sources *or* the microphones were located at nodes or anti-nodes where the pressures of the modes were at a minimum or maximum. Nonetheless, as has just been mentioned, if the intention is to significantly damp the room, in order to make it more acoustically neutral, the modal activity will need to be significantly suppressed, and so the starting dimension ratios will have much less bearing on the outcome than traditional thinking may cause many people to believe.

The lower frequency resonances will be the ones that are most difficult to control. The resonance at approximately 23 Hz, being the second mode of the length dimension (the first is infrasonic), would perhaps not be unduly troublesome; firstly because it is so low, and secondly because it cannot be supported in the other dimensions of the room. As can be seen from Figure 4.19, the pressure distribution around the room would be most uneven and strongly differentiated from one place to another. There is no simple surface treatment which can effectively stop the resonant modes of long wavelength. In fact they are remarkably resilient, especially so in the type of reflective isolation shell which we would probably be faced with here. Greater nuisance would be caused by the modes in the 30 to 100 Hz range.

5.3.4 Lower Frequency Control

An initial approach could be to construct a timber framed internal box structure. A typical structure is shown in Figure 5.1. Considering the size of this room and the need for the walls to support a considerable weight of ceiling, a frame structure of 10 cm×5 cm softwood vertical studs could be constructed, mounted on 60 cm centres. This spacing is sufficient for strength, and is conveniently half of the width of most sheets of plasterboard, which tend to

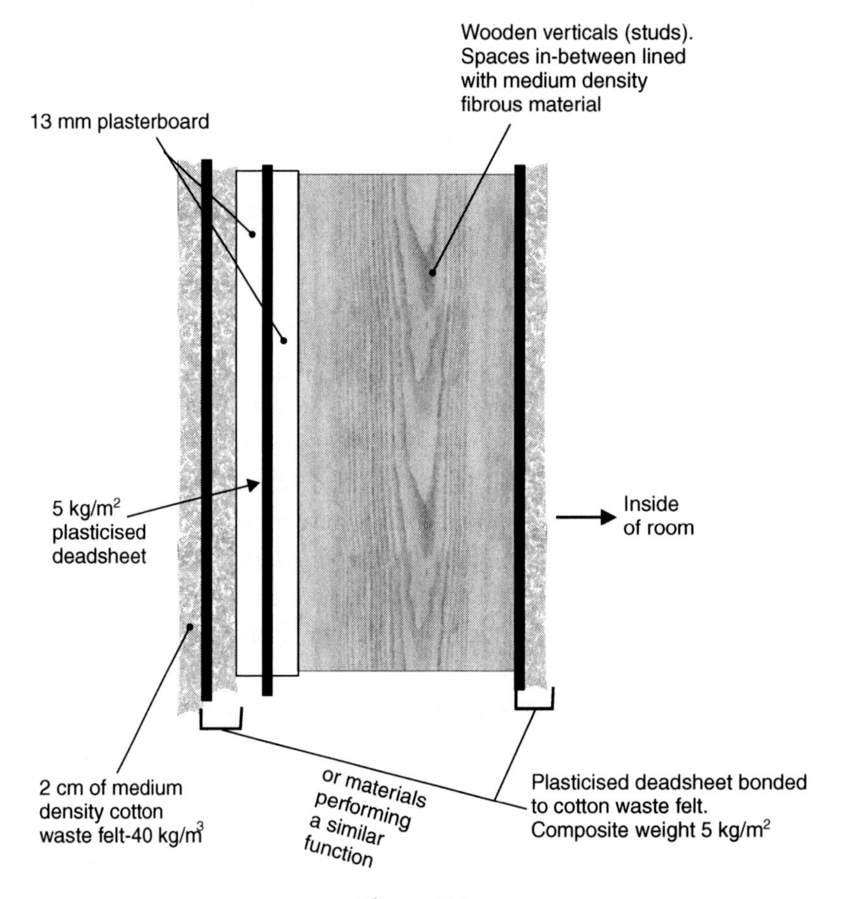

Figure 5.1:
Acoustic control wall.

come in sizes of 120 cm×250 cm, 260 cm or 300 cm. This is important, because to produce our low frequency absorption system the next step is to cover the rear side of the stud wall with the same plasterboard/ deadsheet/plasterboard sandwich as was used in the ceiling shown in Figure 1.1. The wall frames are usually built on the floor, horizontally, where the boards and deadsheet can be laid over the frame and conveniently nailed to the wooden studs with large headed, rough coated, galvanised nails. A further layer of thick cotton waste felt* or other fibrous material is usually fixed to the surface before the wall is raised into a vertical position.

When the four walls are in place, and nailed together at the corners, they are then capable of carrying the weight of several tons of ceiling.

The spacing between this internal wall and the isolation wall is important, as a larger space will usually produce more absorption at the lower frequencies. However, studio owners usually want to *see*, in the finished results, as many as possible of the square metres of floor space that they are paying for. In many cases it seems that no amount of explanation can convince them that by *seeing* a few less metres they will be *hearing* a superior sound, *produced* by the space that they are paying for. It is therefore often necessary to use a further quantity of mineral wool, glass wool, or cotton-waste felt type materials in order to provide at least some augmentation of the absorption.

With the fibrous material applied to the rear of the acoustic control walls, as described above, 5–10 cm of space between the isolation walls and the acoustic walls will usually suffice for the acoustic control of studio rooms. The cotton felt described is about 2 cm thick and of quite high density (60–80 kg/m^3). For safety, it is also treated with a substance to make it self-extinguishing to fire, but in cases where absolute incombustibility is required, more special treatments or mineral based fibrous materials can be used, though the latter are somewhat less comfortable to work with. A further one or two layers of the felt are then inserted in the spaces between the studs (the vertical timbers). These are cut to fit quite well in the spaces, and are fixed by two nails at the top of the frame. The felt not only suppresses resonances in the closed cavity which will be formed when the front surfaces are fitted, but also provides another frictional loss barrier through which the sound must pass twice, once in each direction, because some of what passes through will reflect back from the isolation wall.

If additional isolation and absorption are required, it is possible to add a layer of a material such as PKB2 over the felt on the rear of the wall. PKB2 is a kinetic barrier material. It is a combination of a cotton-waste felt layer, of about 2 cm thickness, bonded by a heat process to a mineral loaded 3.5 kg/m^2 deadsheet; the composite weighing somewhat less than

*Many of the descriptions in this chapter refer to the use of cotton-waste felt. This is not because it is necessarily sonically superior to mineral wool, glass-fibre wool, or other fibrous materials, but it is because it has been found by many people to be much less offensive to work with. The cotton felt is not an irritant, but face-masks should still be used when working with it in large quantities or confined spaces, because the ingestion of the fibres is still unwise.

5 kg/m^2. If this is nailed over the interior surface of the stud wall, with the deadsheet to the studs, it forms a membrane absorber, sandwiched between two layers of felt. Indeed, in the neutral rooms of the type being discussed here, PKB2, or a similar combination, would typically form the first of the layers on the *internal* side of the stud walls. With this composite material covering the 10 cm deep cavity, partially filled with fibrous material and backed by a double sandwich of plasterboard/deadsheet/plasterboard, then felt/deadsheet/felt followed by a further sealed air cavity before the structural or isolation wall, a very effective low frequency absorption system would now be in place. It would also be absorbent in the higher frequency ranges. This can be done without special tools or skills, and all in a space of about 22 cm. Such a combination of panel and membrane absorbers (the plasterboard sandwich and deadsheet composite, respectively) would be effective over the range from 30 Hz to about 250 Hz; the lower octave being controlled predominantly by the heavy plasterboard panels.

The reason for the multiple layers of different materials is because, as described in Chapter 4, different materials and techniques absorb by different means and are effective in different locations and at different frequencies. Large panel absorbers, made out of plywood for example, can produce high degrees of absorption, but they will tend to do so around specific frequencies because they are tuned devices. Obviously, to absorb a wide range of frequencies by such techniques we would need to have many absorbers, but we may have problems finding sufficient space in the room to site them all.

If we lower the Q of the absorber, by adding damping materials, we will reduce the absorption at the central frequency but we will widen the frequency range over which the absorber operates. We can therefore achieve a much better distribution of absorption by filling a room with well-damped absorbers than by filling it with individual, high Q absorbers, in which case the absorption in any given narrow band of frequencies would be localised in different parts of the room. Another advantage of lower Q absorbers is that the resonances within them decay much more rapidly than in high Q absorbers. Resonators of a highly tuned nature, as mentioned in Chapter 4, absorb much energy rapidly, but tend to ring-on after the excitation signal has stopped, and hence may re-radiate sound after an impulsive excitation.

Having now dealt with the floor and the walls in our attempt to 'neutralise' a room, the ceiling can be dealt with in precisely the same manner as the walls, but given the 9–10 m minimum span across this room, joists of either steel or plywood sandwiches would seem appropriate. A typical plywood beam cross-section is shown in Figure 5.2. Prefabricated, laminated beams are also now becoming available, which are capable of supporting considerable weight over long spans without significant bending. The reason for using composite beams for longer spans is that they can be considerably thinner that simple wooden beams. The bending resistance of a beam reduces by a factor of four as the length is doubled. That is to say, if a 3 m beam of a given cross-section is supported at each end, and deflects by *1 cm* if a given weight is put in the middle of its span, a similar beam of 6 m with a similar weight in the centre of its span would deflect by *4 cm*. If it were required that the 6 m beam should also only deflect by 1 cm with the same applied weight, its cross-section would need to be quadrupled. If our first beam was of a section of 20 cm × 5 cm (a cross-sectional area of 100 cm^2) then the 6 m beam would need to be of 40 cm × 10 cm (a cross-sectional area of 400 cm^2). As compared with solid wooden beams, laminated and plywood reinforced beams can achieve the same strength in smaller cross-sections. If the available height is limited, these techniques can be useful, even if slightly more expensive. The laminated beams can also be readily supplied with treatments for resistance to flames, insects and humidity. They may also be certificated for load-bearing constructions.

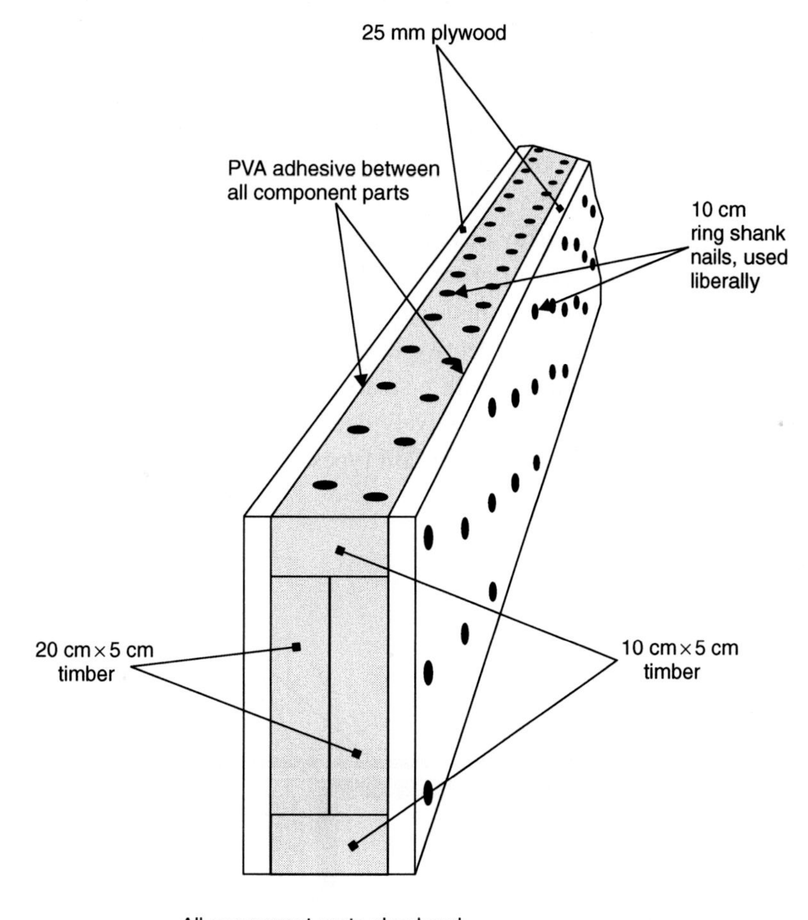

Figure 5.2:
Plywood beam construction. A 30 cm × 15 cm beam of immense strength.

The only significant difference between the wall and the ceiling structures would be on the inside, as shown in Figure 5.3, where the PKB2 or similar material could be placed in arches between the joists. By now, though, we will have taken the acoustics of the room somewhat to the dead side of neutral, so now we must proceed to rebuild a desirable amount of life into our over-controlled space.

A studio room is an instrument in itself, so what we have by now managed to do is to destroy an instrument. But, faced with such an initially troublesome isolation shell, such as a room of these dimensions might be (15 m×10 m×5 m), it is at times prudent to acoustically destroy it, and then rebuild it predictably. With troublesome dimensions the unwanted acoustic characteristics can be difficult to remedy by simple conventional means, and time and experimentation can be needed to assess the remedial work. For this reason the initial acoustical destruction of a problematical room is often a wise choice.

In fact, the principal problem remaining in the room described above would be the incomplete control of the 200–500 Hz region due to the hard floor being in close proximity and parallel to the ceiling. To subdue this problem, panels could be fitted as in Figure 5.4, along with an absorber composed of an array of hanging panels along one wall, as shown in Figure 5.5. More details of this type of absorber, along with various photographs, appear in Appendix 1 at the end of the book. By this stage we would not have a neutral room in the recording sense, but rather, given the reflective floor, we would have something very much approaching a hemi-anechoic chamber, with a small amount of well-damped low frequency modal energy. Such a room would be excessively dead for most music recording purposes, although it could be an excellent basis for certain types of highly damped control rooms. (See Chapter 16 and Appendix 1.)

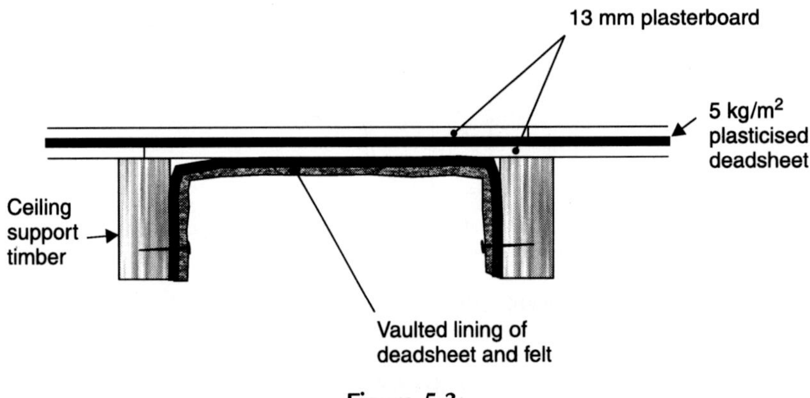

Figure 5.3:
Ceiling construction.

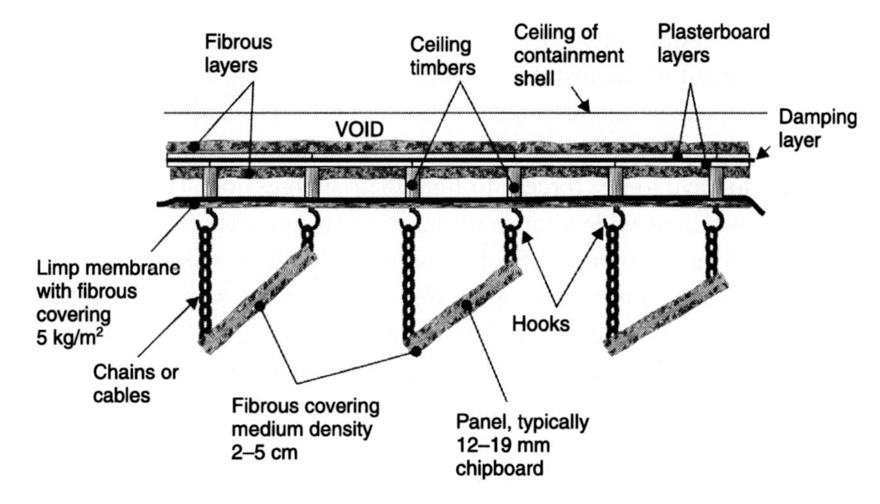

Figure 5.4:
Typical ceiling absorption system.

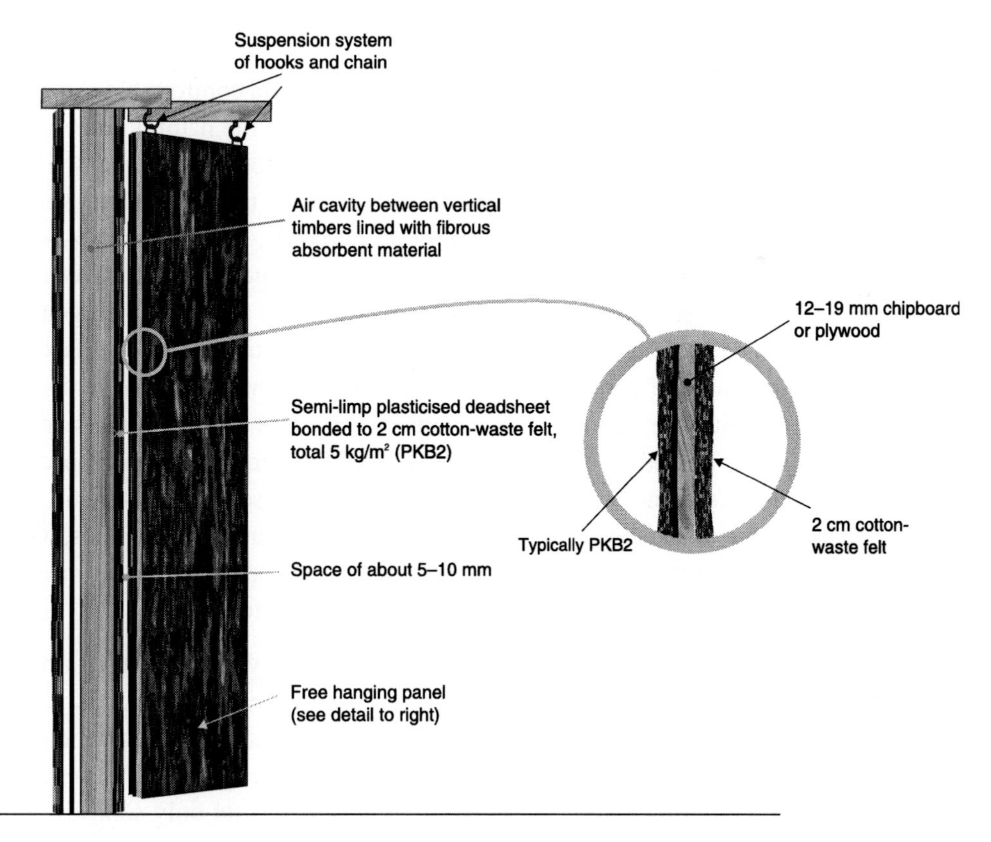

Figure 5.5:
One element of a waveguide absorber, similar to those in Figure 5.4 but hung vertically.

In the case of the studio room which we are seeking to build here, we need to create an acoustic which enhances the sound of the instruments without unduly announcing its own presence. We need a room which, as far as possible, favours all notes reasonably equally, neither producing 'wolf' notes, which stand out due to their coincidence with room resonances, nor causing other notes to have to be 'forced' to fight their suppression. The room should have a sonic ambience in which as wide a range of musicians as possible feel comfortable, both in themselves and with their instruments. Such a room would allow a wide range of choice for the recording engineers in the positioning of microphones. It would also allow a great deal of freedom in the positioning of the different musicians, either for the purposes of improved eye-to-eye contact (which can be very important to them) or for purposes of acoustic separation. However, the wall and ceiling absorbers which we have proposed thus far to overcome the room problems would be rather too absorbent for our requirements of 'neutrality'. So, after controlling the room, we will have to selectively brighten it up, by means that we shall explain shortly. Before doing so, however, perhaps we can briefly digress, in order to look more closely at what we shall be seeking to achieve, and why.

5.3.5 Relative Merits of Neutrality and Idiosyncrasy

Neutral rooms are flexible rooms in which work is usually quick and comfortable. An ensemble placed in a neutral room will tend to be heard and recorded with the natural predominances of that ensemble, with the room favouring neither any instrument nor position to any significant degree. However, if this were the be all and end all of recording, this would be a very short book. Neutral rooms are not the best rooms for all purposes, a point which was perhaps first discovered, at least partially, by accident. There are many studios which have rooms which were no doubt intended to be neutral, but which have failed to realise their goals. From time to time, a resonance in such a room, or a certain characteristic pattern of reflexions, can produce an enhancement of certain types of music and instruments played in them. They can become great favourites for certain types of music. The same is true for the stages of certain concert halls, and indeed of other halls which have not necessarily been specifically designed for musical performance. Unfortunately, in many of these, a characteristic of the room which enhances the music may only do so in certain keys or at certain tempos, where the frequencies of resonance or the timing of the reflexions are appropriate, but this means that their suitability for a wide range of recording becomes more limited.

For example, a symphony played in E major may well be strongly reinforced by a room resonance when certain parts are played with gusto. Perhaps if the coincidence is very fortunate, the main characteristic reflexion patterns will have a natural timing which will produce a powerful effect if they coincide closely with a simple fraction of the beats per minute of the tempo. Such a room may give inspiration to the musicians, not only sonically lifting the music, but also encouraging a more enthusiastic performance. These rooms can have their places in both the recording *and* performing worlds in a way which a neutral room may never achieve, but though these rooms may achieve great results in a case such as that stated above, an orchestra performing a symphony in a different key, and with a different tempo, may have difficulties with the room. If played in the key of F sharp major for example, the resonances around the E may be entirely inappropriate, causing emphasis to notes which should not be emphasised, and masking and weakening the notes which the conductor would prefer to be dominant. In such rooms, for every peak in the response, there will be a dip elsewhere. Furthermore, any series of ill-timed reflexions (echoes) may create confusion and a degree of difficulty with the natural flow of the music. Not all musicians may fully realise what is going on, but many may comment on how they just cannot produce their best in that room with a given piece of music.

This is one of the reasons why much classical music is still recorded outside of studios, either in concert halls (with or without an audience), or in town halls, churches, or similar locations. It gives a choice of ambience for the producer, engineer and conductor to try to achieve the 'ultimate' from selected performances in selected locations. On the other hand, except for some *very* highly specialised recording companies, moving to a different location for each piece of music of less than symphonic length would be financially ruinous. What is more, if recording in one location, choosing an idiosyncratic studio for the main piece would perhaps seriously compromise the remainder of the album. This is one very important reason why neutral rooms are so widely used in the parts of the recording industry where high-quality recordings must be able to be made on a predictable, rapid, and reliable basis. They are especially useful for broadcast studios, where good quality recordings must be made quickly and at reasonable cost, as they are perhaps intended for a once-only transmission.

The next step in the design of our neutral room is therefore how to add into our relatively dead shell as many desirable features as possible, with as few problems as possible. The major pitfalls to be avoided are erratic changes in the reverberation time/frequency characteristic, bunched echoes in terms of their temporal spacing, and strong highly directional echoes (reflexions). Figure 5.6(a) shows the typical sort of reverberation response that we are trying to avoid, and Figure 5.6(b) the response which is more in the order of what we are trying to achieve. The response of (a) shows humps in the curve which are characteristic of unwanted resonances. The peaks are the frequencies which will continue to resonate long after an instrument has stopped playing, and the dips represent notes which may appear weak. The irregularities will therefore favour certain notes, suppress others, and mask many low level details of the sound. Such responses at low frequencies are often the result of large, parallel, reflective surfaces which can support the strong axial resonances. It was stated earlier that the angling of the walls away from parallel would help to redistribute the energy in the axial modes, but at low frequencies the behaviour of sound waves is not always obvious.

In order to reflect at low frequencies, surfaces need to be of a size comparable to a substantial proportion of a wavelength, or the acoustic wave will tend to engulf them and pass around. In our neutral room, we can therefore avoid low frequency resonance

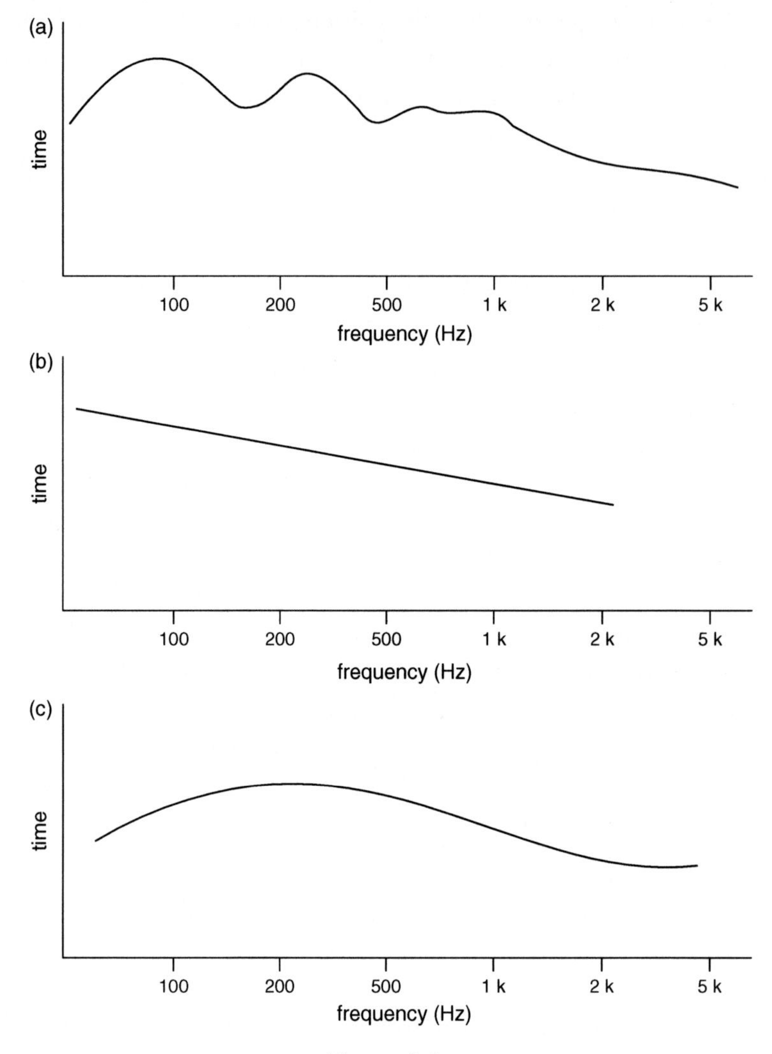

Figure 5.6:
Decay responses (reverberation): (a) Undesirable irregular decay response. This type of curve will cause colouration of the recordings. The humps in the curve are due to room resonances; (b) Smooth, desirable decay responses showing freedom from resonant colouration; (c) This plot would be typical of rooms with less LF energy.

problems by placing any necessarily large, reflective surfaces, such as large glass doors or windows, in positions where they do not directly face each other. Other reflective surfaces, necessary for the addition of life to the middle and high frequencies, can be arranged such that they have gaps between them, the gaps being at intervals of less than half a wavelength of the highest of any troublesome resonances. Alternatively, they can be arranged with suitably random spacing.

5.4 What is Parallel?

The term 'parallel' in its acoustic sense is very frequency dependent. Figure 5.7 shows two reflective walls, each 10 m long and spaced 10 m apart. They are geometrically parallel, and hence are also acoustically parallel at all frequencies. A clap of hands at point X will generate a sound containing very many frequencies, and the sound will propagate in all directions from the source. The waves impinging on points 'Y' and 'Z' will be reflected back through the position of the source, and will continue to 'bounce' backwards and forwards in the form of slap (flutter) echoes along the line 'Y—X—Z'. Frequencies whose wavelengths coincide with whole fractions of the distance between Y and Z will go through positive and negative pressure peaks at positions in the room which coincide on each reflexion. A musical instrument will drive many of these resonant modes, which strongly reinforce each other. They will tend to be audible in some points in the room close to the anti-nodes,

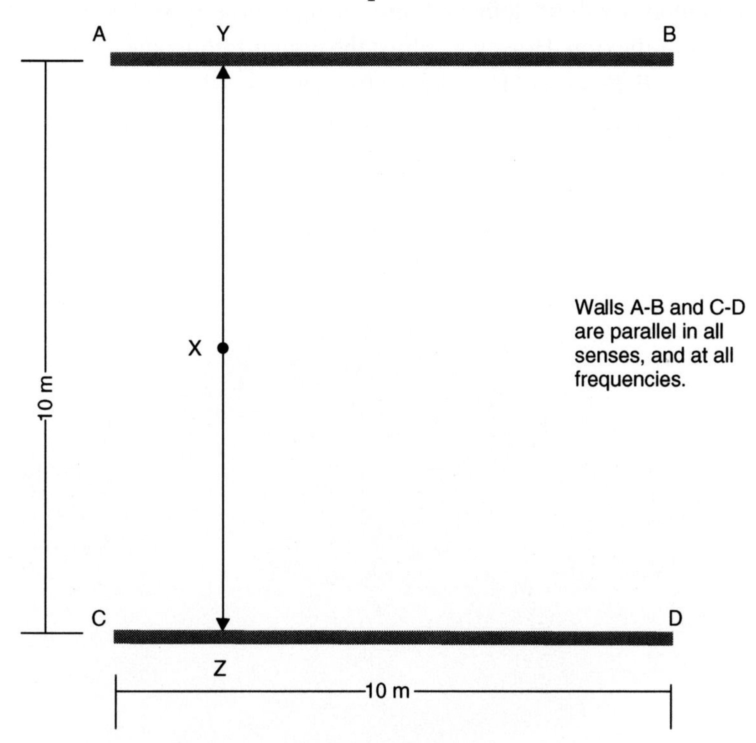

Figure 5.7:
Geometrically parallel walls. A sound, emanating from point 'X' will spread in all directions. However, the sound waves travelling in the directions of points 'Y' and 'Z' will reflect back along the line of their original travel, and will continue to reflect backwards and forwards along the same path, creating flutter echoes, until their energy is finally dissipated by losses in the walls and the air. Such are the paths of axial modes, which, when wavelengths coincide with whole fractions of the distance between the walls, produce modal resonances — see Figure 5.8.

but not in others, closer to the nodes. A 70 Hz standing wave pattern is shown in Figure 5.8. The light areas show regions of low pressure changes, where the waves would be inaudible, and the dark areas show the regions of high pressure changes, where the 70 Hz content of the sound could be clearly heard.

If we now angle one wall in a manner as shown in Figure 5.9, with one end swung in towards the other wall by 1.5 m, we will have two walls with a 15% inclination. Now, a handclap at point X will again send a wave in the direction of Y, which will return to the source point as a reflexion, and will continue on to point Z. A direct wave will also propagate to point Z, and both the direct and reflected waves will reflect from point Z, not back towards point Y, as in the case of the geometrically parallel walls, but towards point F. They will then reflect to point G, and on to point H. Unlike in the case of the geometrically parallel walls in Figure 5.7, a person standing at point X will not hear the chattering echoes, and most of the resonant energy of the room modes will be deflected into the tangential type, taking a much more complicated course of reflexion. However, whilst the higher frequencies will be deflected along the pathways Y–Z, Z–F, F–G, G–H, at lower frequencies, where the wavelengths are long, axial modes may still persist. This suggests that at low frequencies, the walls must still be parallel in an acoustical sense.

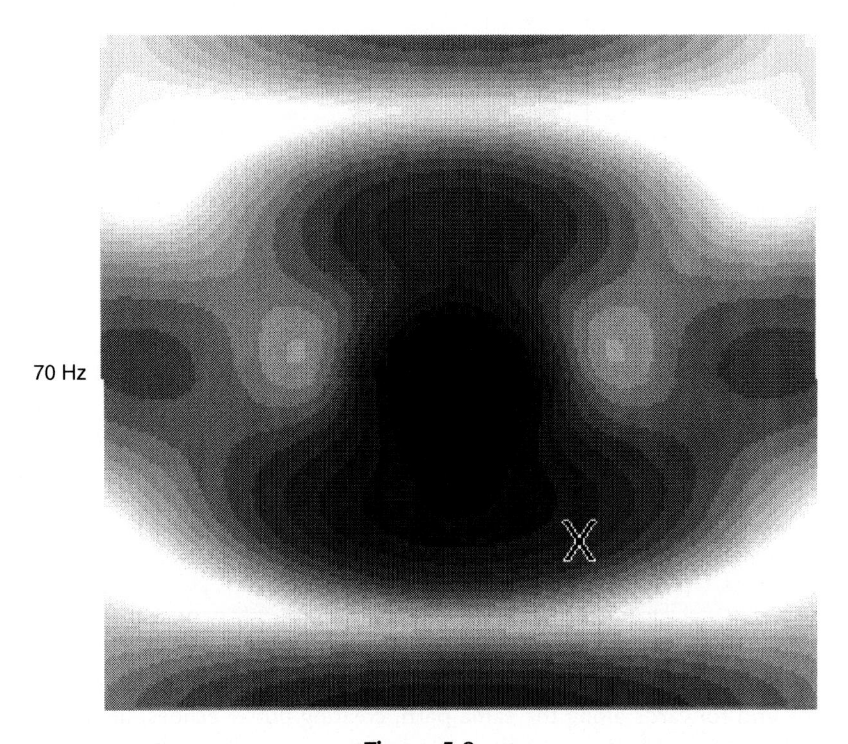

Figure 5.8:
Pressure field between parallel walls, 10 m apart. Magnitude of pressure field due to a point source between the two walls depicted in Figure 5.7.

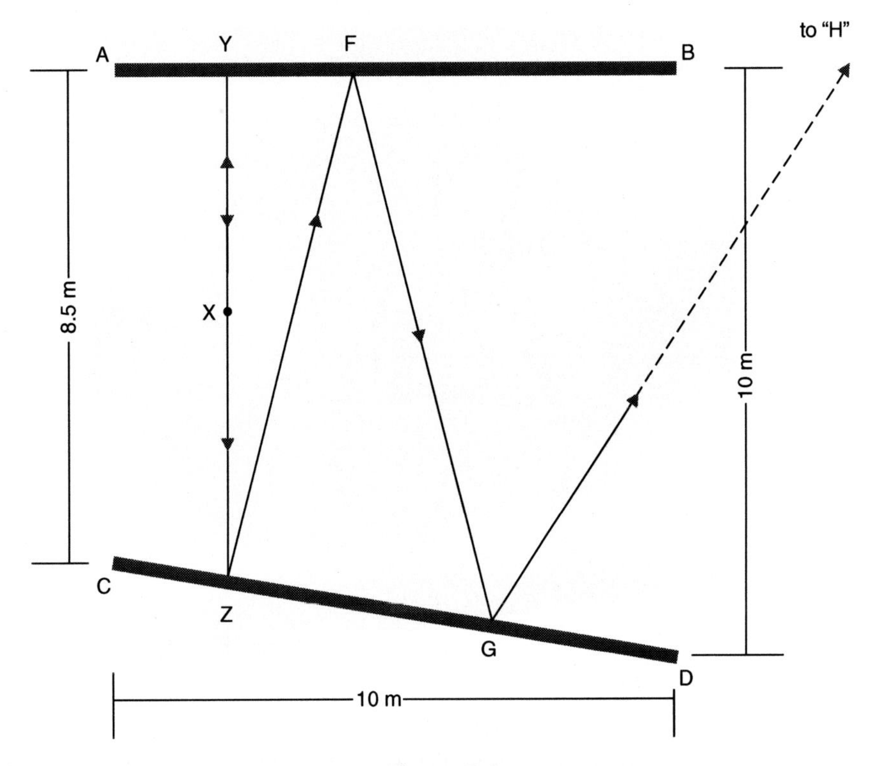

Figure 5.9:
Here, the general situation is that of Figure 5.7, except that one of the reflective surfaces has been moved, to create a geometrically non-parallel arrangement between the surfaces A–B and C–D. Flutter echoes, as produced by the surfaces in Figure 5.7, are not possible, as the reflexions will not follow repeating paths. Sounds emanating in the Y and Z directions, from position X, will subsequently follow the path Z, F, G, H, etc. At low frequencies, however, things may not be too different from the conditions of Figure 5.7 – see Figure 5.10.

Figure 5.10 shows the 70 Hz standing wave pattern with one wall angled to the same degree as that shown in Figure 5.9. The pattern is remarkably similar to the one shown in Figure 5.8. Although Figure 5.9 shows that the angling of the walls has produced a very different path for the handclap echoes, and will be quite dispersive at high frequencies, at low frequencies very little has changed. Essentially, for geometric angling to be acoustically effective, the path length differences for subsequent reflexions must be a significant part of the wavelength. With the wavelength of 50 Hz being about 8 m, the degree of angling required to be acoustically non-parallel would perhaps be possible in buildings the size of concert halls, but would be likely to consume too much potentially usable space in a conventional recording studio.

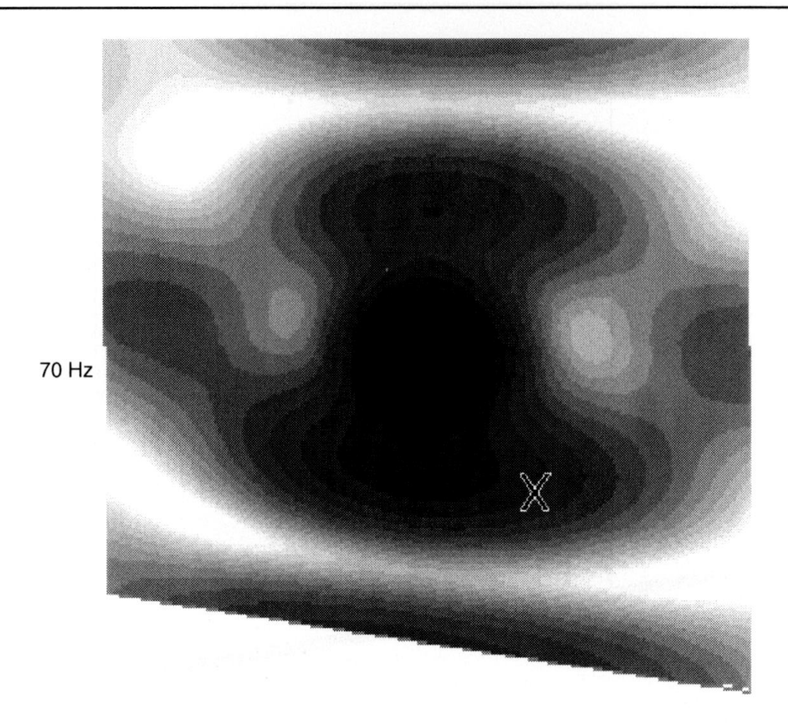

Figure 5.10:
Pressure field between acoustically non-parallel walls. Magnitude of pressure field due to a point source between the walls depicted in Figure 5.9.

The effect of a limited degree of wall angling on the response of a room is shown in Figure 5.11. The two traces show the performance of the walls shown in Figures 5.7 and 5.9. Above 200 Hz, the non-parallel walls show a clear reduction in modal energy when compared with the parallel walls, but below about 80 Hz there is little difference between the two traces, confirming that, in the acoustic sense at least, the walls of Figure 5.9 are still parallel. The reduction in modal energy above 200 Hz is largely due to the fact that the wall angling drives more of the higher frequency modes from the axial to the tangential type. The tangential modes not only have more complicated paths to travel, but also strike the walls at oblique angles, which tends to rob them of more power than is lost in the more perpendicular impacts of the axial modes. It can thus be seen that whilst the angling of walls can have a very worthwhile effect at frequencies above those which possess a wavelength which will be subsequently shifted in position by a half-wavelength or more on their return to the source wall, at frequencies *below* these, the effect will simply be that of a comb filter, as shown in Figure 5.12. Here, as the frequencies are swept downwards from the above half-wavelength frequency, the sweep passes through alternately constructive, neutral and destructive regions. Strong comb filtering at low frequencies is usually musically disruptive and very undesirable in recording studios (and music rooms in general), though it exists to some degree in all reflective spaces. At higher frequencies our ears use it for many beneficial purposes, such as localisation and timbral enrichment.

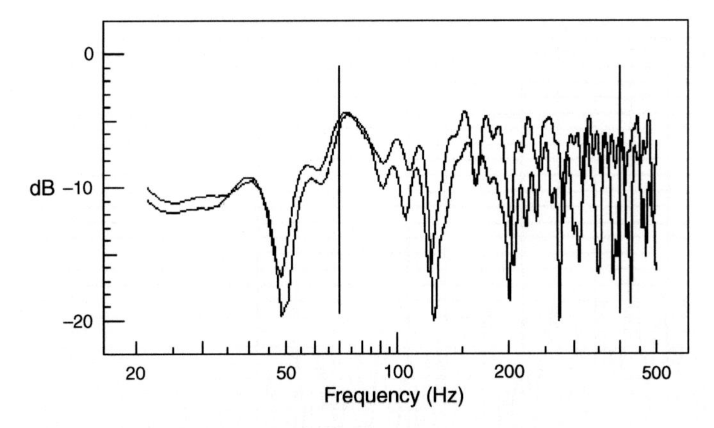

Figure 5.11:
The above plots show the response at point X in Figures 5.8 and 5.10. It can be seen that, at frequencies below around 80 Hz, the effect of the angling of one of the reflective surfaces has had only minimal effect. Above about 200 Hz, however, the effect is quite pronounced.

So, whilst the angling of reflective surfaces is a viable technique for reflexion control at middle and higher frequencies, at low frequencies, geometric solutions are usually not able to produce the desired results, and hence absorption must be resorted to, though diffusive techniques are beginning to become a practical reality. (These will be discussed later in the chapter.) Parallel surfaces also produce the repetitive chatter or 'slap-back' from impact noises, somewhat akin to the ever-decreasing images seen when standing between two parallel mirrors, and this reflective chatter can be equally as undesirable as the resonant modes in their destructive effect on the music. In the following sections, we shall begin looking at practical solutions for the circumvention of these problems, whilst producing a desirably neutral acoustic.

5.5 Reflexions, Reverberation and Diffusion

We are now faced with the problem of how to put the ideas so far discussed into a practical form. Unfortunately, there are so many ways of doing this that a whole book could be filled on this topic alone, so we will have to take an approach which will incorporate a number of solutions in one design. From this, hopefully, it will be possible to gain something of a feel for the range of possibilities open to designers, and the way that these can be put into general recording practice. What is necessary for musical neutrality in rooms of the size under consideration here (750 m^3) is a reverberation time (or, more correctly in these cases, a decay time) in the order of 0.3 to 0.6 s, perhaps rising to 0.5 or 0.8 s at low frequencies, with the rise beginning gradually below 250 Hz. Figure 5.13 shows a typically desirable

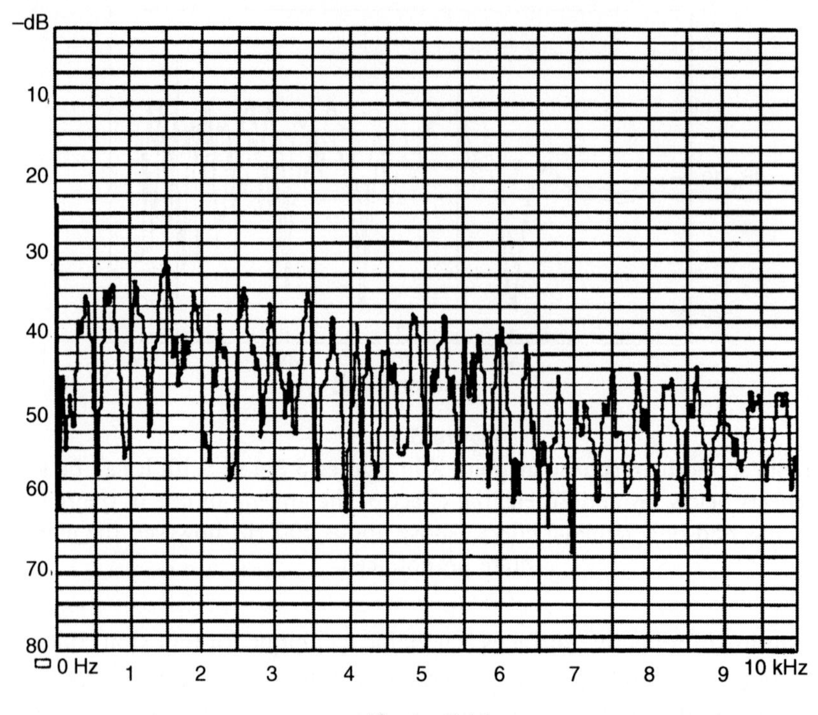

Figure 5.12:
The averaged power spectrum of a signal with one discrete reflexion. Comb filtering is revealed clearly on a linear (as opposed to a logarithmic) frequency scale, where the regular nature of the reflexion-produced disturbances can clearly be seen. In the instance shown above, the additional path length of the reflected signal over the direct signal was just under 1 m, producing comb filtering with dips at a constant frequency spacing of just under 400 Hz. A glance at Figure 11.7 will show the response from a more perfect reflector, from which the origin of the term 'comb filtering' will be readily apparent.

frequency/time decay response of such a room; the time indicated being that taken for a sound to decay to 60 dB below that of its initial steady-state level.

There are two general techniques involved in creating reverberation in such a room; either by reflexions or by diffusion. In recent years, companies such as RPG[4] in the USA have created a range of acoustic diffusers capable of operating over a wide range of frequencies. These are constructed on principles based on sequences of cavities whose depths alternate according to strict mathematical sequences. They can be made from any rigid material, but perhaps wood, concrete and plastics are the ones most frequently encountered. The mathematics were initially proposed by Professor Manfred Schroeder[5–8] in the 1970s, and are based on quadratic residue sequences. The effect of the cavities of different depth is to cause energy to be reflected in a highly random manner, with no distinct reflexions being noticeable. The random energy scatter creates a reverberation of exceptional smoothness. By using such diffusive means, the

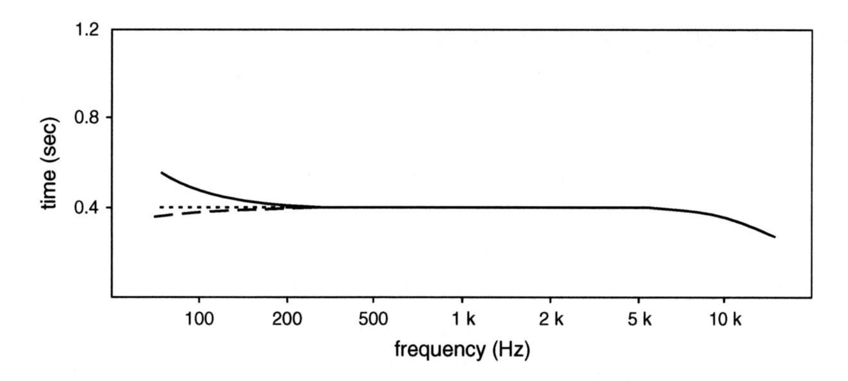

Figure 5.13:
RT_{60} of a good neutral room. Low frequency reverberation times can be allowed to vary with room size.

overall reverberation time can be adjusted by the ratio of diffusive surfaces to absorbent surfaces, though for an even distribution of reverberation in the room a relatively even distribution of diffusive surfaces is required. Except for the floor and windows, all other surfaces are usually available for diffusion.

With the availability of diffusers, achieving the desirable degree of ambient neutrality may well at first seem to be a simple matter of adding diffusers until the desired reverberation time is achieved, but from such a room there would usually be a lack of musicality. What such a room would lack are discrete reflexions. Fortunately, they are easily introduced, as they are of great importance to both the musicians and their audiences. On concert stages, they are needed by the musicians to reinforce the sound of any instrument or ensemble. As mentioned earlier, reflexions divide into two groups, late reflexions and early reflexions. The early reflexions, arriving less than 30 or 40 ms after the direct sound, are heard by the ear as a timbral enrichment of the instruments. The later reflexions, arriving 40 ms or more after the direct sound, add spaciousness to the sound, which for many types of music is essential for its enjoyment.

The dimensions of our example of a large neutral room were chosen such that an instrument in the centre of the room would produce only late reflexions from the wall surfaces, but of course, from the floor and ceiling, the reflexions would necessarily be of the early type. Instrument to floor distances are normally relatively constant; floor reflexions typically being in the 5 to 10 ms region. Any ceiling reflexions in this room would be in a borderline 20 to 40 ms region, dependent not only upon ceiling geometry, but also upon whether diffusive, reflective or absorbent surfaces predominated. In later chapters dealing with more reverberant spaces we shall look at reflexions further, but in the neutral type of room under discussion here, whatever reflexions exist should tend to be reasonably well scattered, otherwise they will develop a character of their own and the room would lose its neutrality.

5.6 Floor and Ceiling Considerations

Let us now look at possible ceiling structures for our neutral room. As we discussed earlier, the nature of the floor has been chosen to be wood. Carpet tends to produce a lifeless acoustic, uninspiring for the musicians and unhelpful for the recordings. Stone was rejected, partly for its 'harder', more strident reflective tendency, but also on the practical ground of slippage. Floors of neutral rooms overwhelmingly tend to be of wood. At this point of the design stage (see Figure 5.3), we have a relatively dead ceiling at a height of about 4.5 m. We cannot come down too much below this because the reflexions which we would introduce would tend to become of the tone colouring, early nature. What is more, too low a ceiling would preclude the siting of microphones above the instruments at a height which may be necessary to cover any given section of musicians.

One solution to such a problem is to construct a ceiling of wooden strips, with spaces between them which would allow a good proportion of the lower frequencies to pass into the absorbers behind. They would therefore provide mid and high frequency reflexions without allowing an unwanted low frequency build up. As mentioned earlier, in order to reflect at low frequencies, surfaces need to be of sizes comparable to the wavelengths to be reflected. We can therefore have some degree of control over the lower limit of our desired reflectivity by providing gaps in our reflective surfaces at appropriately chosen intervals. This alternative juxtaposition of reflective and absorbent surfaces will also produce significant diffusion (see Section 4.7), which in many instances is extremely useful.

Once again with this complex subject there are so many ways of achieving each objective, so here we will only be able to look at some of the possibilities within one technique of specific interest. The strips described could be of hardwood or softwood, and could be plain, varnished, painted, rough or smooth. Each will give its own subtle character to the sound. Hardwoods, untreated, can be quite interesting sounding, but in today's ecological climate many designers refuse to specify any woods of an exotic or not too easily replenishable nature. Sound is of course important, but there are also environmental considerations which we cannot escape.

Obviously, we do not want to encourage the build up of resonant energy in any modes which would unpleasantly colour the sound, so the ceiling surface can be broken into a series of angled sections, set to produce what the designer would consider to be the most appropriate angle for reflexions. A selection of possibilities for ceiling designs are shown in Figure 5.14. Figure 4.33 showed a pattern of wood/space ratios which are based on a numerical sequence, not dissimilar to the one used for the diffuser cavities referred to earlier. This type of arrangement could also be used, though perhaps it would be a little tricky to find a suitably attractive way of mounting the lights. The purpose of all these arrangements, however, is to help to prevent any noticeable patterns forming in the reflected sound-field.

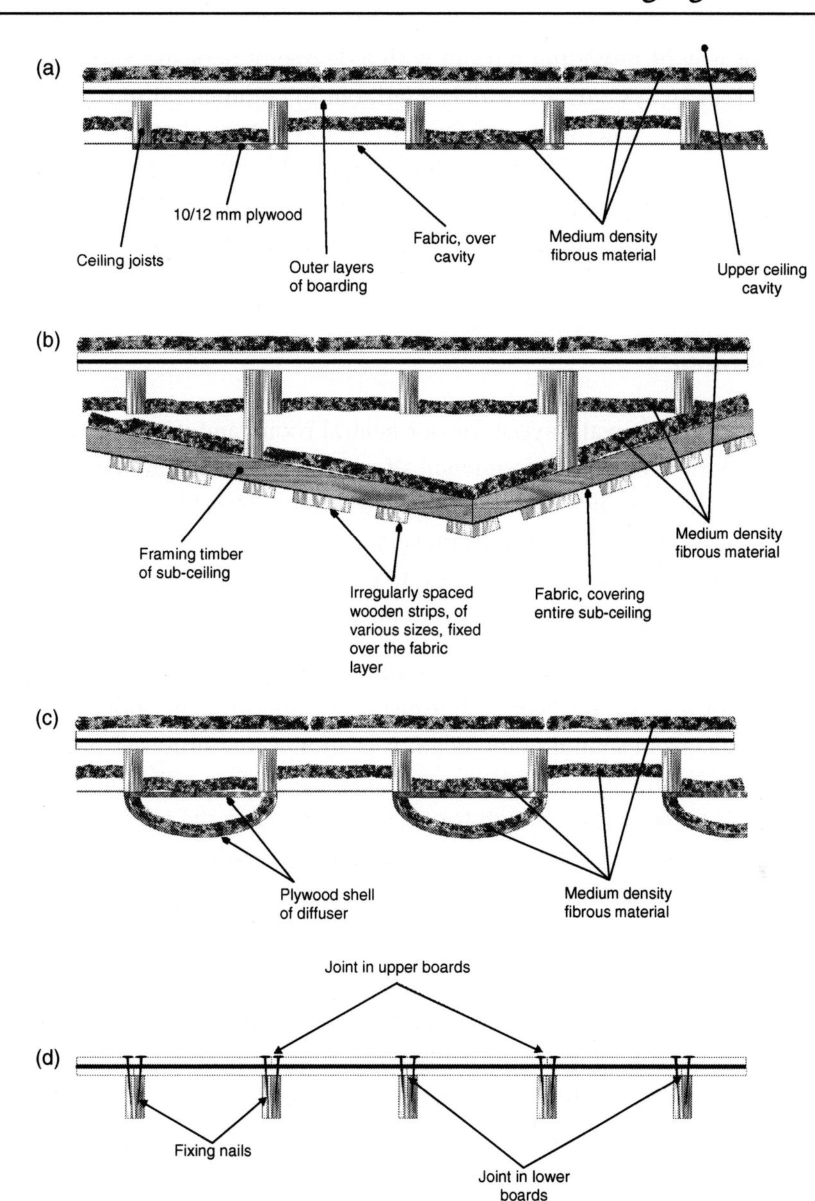

Figure 5.14:
Various ceiling constructions for a neutral room: (a) Alternative hard and soft surfaces; (b) Irregular hard/soft surfaces; (c) Absorbent spaces and curved, diffusive surfaces; (d) Fixing arrangement of plasterboard sheets, with overlapping joints above the beams.

The BBC[9] (British Broadcasting Corporation) have developed a very successful range of ceiling tiles which fit into the typical sort of false ceiling structures used in many offices and broadcast studios. In fact, using a steel-framed grid for a false ceiling allows great flexibility in the introduction of easily replaceable absorbent, reflective or diffusive tiles, which can be an excellent way of providing a room with a degree of acoustic variability. In the broadcasting world, these systems are widely used, but in commercial music recording studios their appearance is generally considered to be too industrial to provide the necessary decorative ambience for musicians to feel comfortable and creative. However, these things are highly personal, and for people who find these prefabricated tiles pleasing to look at they can be safe in the knowledge of their excellent acoustical performance.

5.7 Wall Treatments

Figure 5.15 shows a possible wall layout for our neutral room, and it can be seen that parallel wall surfaces have been avoided. At frequencies for which the walls are still effectively acoustically parallel, the acoustic waves are allowed to pass into absorbers, either directly, or after first reflexion. Care has been taken to make sure that windows and doors do not face directly towards any parallel reflective surfaces. In the room shown, the control room door and window systems are set into a relatively absorbent wall, so one of the walls is now defined.

That now leaves us with three wall surfaces to complete. For musical neutrality, we do not want too much reflective or reverberant energy, but just enough to give the room sufficient life to stop the instruments from sounding too dead. It is similar to a little seasoning bringing out the flavour of the food without overpowering it. The glass surfaces, the entire floor and the ceiling reflexions are more or less enough for this. We also have the problem that if we make the walls reflective to any significant degree, we may create unduly reflective areas for any musicians playing close to the walls. In fact, in the corners of the room, the colouration could become most unnatural if bounded by the floor and two reasonably reflective walls, all providing early reflexions in addition to those from a borderline early/late ceiling.

For reasons of structural integrity, if we have walls covered with fabric for decoration we may need wooden rails, at waist height or a little higher, to prevent people from 'falling' through the fabric. We need skirting boards so that the floors can be cleaned without soiling the fabric, and we may also need to put rails at knee height as a convenient mounting for microphone sockets and electrical outlets. Figure 5.16 shows a small version of a room with these fittings, but if such a room should be considered to be a little too dead, or in cases where only a rather small area of glass is to be used, then an arrangement such as that shown in Figure 5.17 can be employed. The reflective surfaces are based largely on the concepts of Figure 4.33, but the ratio of the areas of the spaces to the facing timber can be adjusted to produce the required degree of acoustic 'life'. The fabric covering is for decorative purposes only, and so should be of a type which is acoustically transparent. Many fabrics can be surprisingly reflective, and if stretched too tight they may also act like drum skins. Lightweight, Lycra-based 'stretch' fabrics are useful here, but specialist 'acoustic' fabrics, which are normally flameproof, are the most common option.

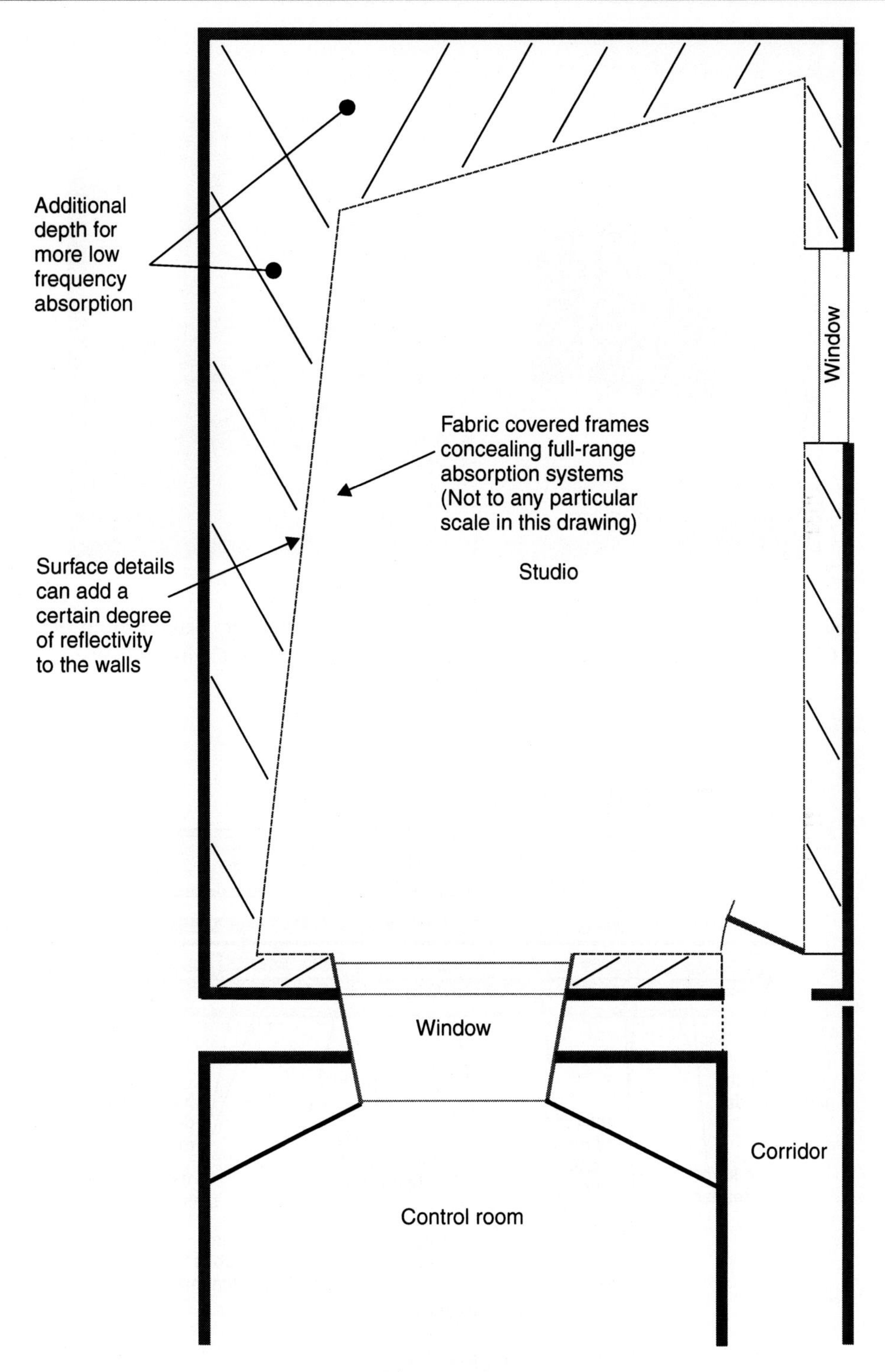

Figure 5.15:
Possible layout of neutral room.

Figure 5.16:
A relatively small room with a very neutral acoustic characteristic. Studio room of the Ukrainian Air Force, Cultural, Educational and Recreational Centre, Vinnitsya, Ukraine (1996).

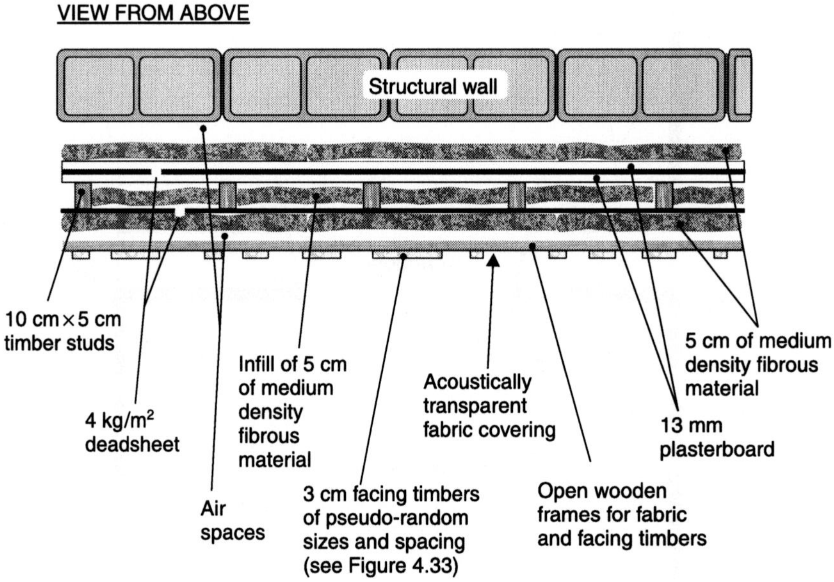

VIEW FROM ABOVE

Structural wall

10 cm × 5 cm timber studs

4 kg/m² deadsheet

Air spaces

Infill of 5 cm of medium density fibrous material

3 cm facing timbers of pseudo-random sizes and spacing (see Figure 4.33)

Acoustically transparent fabric covering

Open wooden frames for fabric and facing timbers

5 cm of medium density fibrous material

13 mm plasterboard

Figure 5.17:
Wall structure and treatment for brightening a 'neutral' room.

So now we have something approximating to a musically neutral room, which tends neither to add character to the sound of an instrument nor to 'suck out' all of its life. Rooms, such as the one being described here, are excellent for recording efficiently, rapidly and predictably, but if they are the *only* spaces available for recording, then the practice still harks back to the philosophies of yesteryear and their more 'technical' correctness. Nonetheless, these rooms can still be very useful, and they are still worthy of consideration in many circumstances. Remember that neutral rooms are very necessary in the broadcast industries and in recording facilities where rapid results are necessary from a wide range of music and instruments. For these purposes, they excel. Figure 5.18 shows a room which may be quite neutral on a macro scale, but which provides local areas of different acoustics.

5.8 Small and Neutral

Until now, we have been discussing the treatment of a room of 650 m^3, which by the standards of the current trend of recording studio building is verging on the huge. A question much more frequently asked today is how to get a degree of neutrality in a room of much more

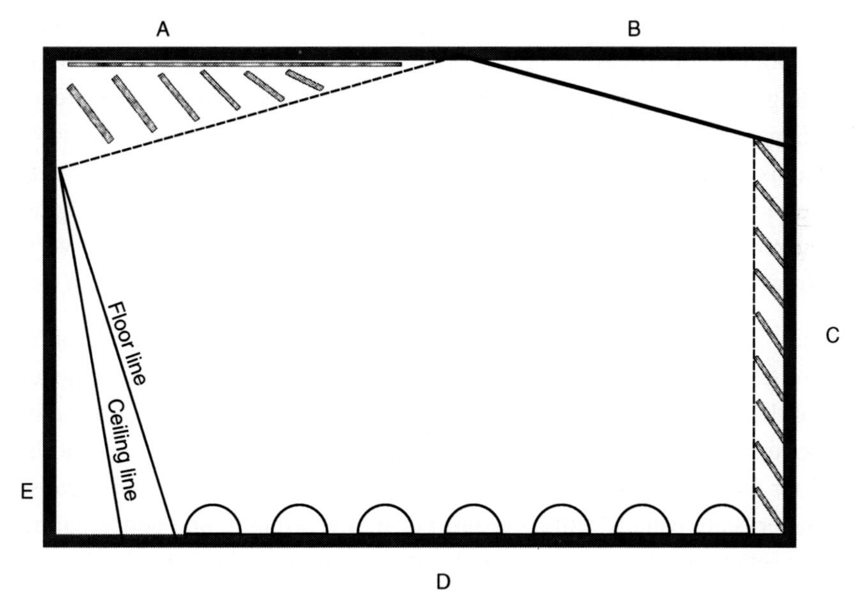

Figure 5.18:
Possible layout of a neutral room with a little flexibility. (Doors and windows omitted for simplicity.) The ceiling could be typically one of those shown in Figure 5.14. The walls: (A) Low frequency absorber with controlled upper mid/high frequency reflectors. Similar surface features to those shown in Figure 5.17; (B) Wood-panelled, reflective walls; (C) Wideband absorber, faced with fabric covered frames; (D) Diffusive wall, with absorbers between diffusing half-cylinders; (E) Reflective, double-sloped wall, faced with wooden panelling.

modest proportions, even to the degree of only beginning in a space of 40 m^3 or less. Now that is a challenge!

When listening to recorded music, and especially on headphones, one frequently becomes aware of the sound characters of the rooms in which many of the recordings were made. In itself this should not be a problem, unless the room sound is 'boxy', or inappropriate to the song or the rest of the instrumentation.

Unfortunately though, this is all too often the case, as the vocals have either been recorded in a small 'vocal booth' to achieve the desired separation from the other instruments, or they were recorded in a small room of insufficient neutrality, perhaps because of convenience or the lack of a better option. The reason could also possibly have been because no large or neutral rooms were available. (Sadly, this is very often the case.)

What is also unfortunate is that far too many control rooms and/or monitor systems are insufficiently neutral in themselves to allow the recording staff to notice the subtleties of the vocal room acoustics. This is especially the case in many multi-media or 'project' studios, where due attention to control room monitoring conditions has often been seriously lacking. Small recording room acoustics often have decay times which are less than those of most control rooms; hence the sound of the recorded room becomes lost in the monitoring acoustics of the control room. In other cases, the staff of a studio can become so used to the sound of a small recording room that they no longer hear it in the background of their recordings. The problem is perhaps now greater than it used to be, as the lower noise floors of digital recording systems have rendered audible (even in the home) sounds which would in earlier years have been lost in the background noise. Ironically, it is also often this digital recording equipment that produces so much mechanical noise in the control rooms that very few of the noises or unwanted sounds on the recordings can be heard, as mentioned in Section 1.2.2.

Where neutrality is required, vocals are usually best performed in the middle of large rooms, where there are few early reflexions. If floor reflexions are a problem, then rugs can always be provided for the vocalists to stand on, and this also helps to reduce the pick-up of the sound of any foot movements. However, most of the vocal energy usually tends not to be directed towards the floor and so very little returns to cause problems. What is more, the selected microphone patterns are usually either cardioid or figure-of-eight, and hence naturally tend to ignore the floor reflexions. Purpose designed vocal rooms usually need to be as neutral as possible unless the acoustic liveliness of a room is being used for effect.

The problem with using *live* rooms for vocals (*live*, in the acoustic sense, that is) is that, usually, the nature of the room ambience that is considered good for instruments is not an ambience that does much good for voices. In large neutral rooms, the space around the microphone is usually conspicuous by the absence of any early reflexions which can colour the sound, but in small rooms, this is not so easily achieved. The problem with small rooms is that *all* reflexions are of the sound-colouring early type, and the spacious sounding late reflexions cannot exist. A general neutrality is difficult to achieve in small rooms, and especially so in very small rooms of the size commonly associated with vocal rooms. In circumstances where no large room is available it is perhaps better for vocal recordings to be made without any room ambience whatsoever, except perhaps for the reflexions from the floor and a window, which, as we have just discussed, most typical microphone techniques would ignore. A further difference between the recording of vocals and the recording of many other instruments is that there is an intelligibility factor to be considered. Many vocal subtleties would be rendered unintelligible in rooms where insufficient 'space' existed around the vocal sound before the reflexions returned to the microphone.

Furthermore, in small rooms there is little space to mount typical low frequency acoustic control devices, and simple treatment of a bare shell will do little to create a neutral sounding environment. Simple attempts at absorption by the placing of acoustically absorbent tiles on the walls and ceiling will not suffice. These will tend to absorb the higher frequencies but leave the lower mid and low frequency modes largely untouched, yielding a room with a heavily coloured ambience which will lack life and add a thickness to the sound, robbing it of much clarity. Rolling out the offending frequencies by equalisation will take the lower frequencies out of the unwanted ambience, but it will also take them from the direct sound of the voice. In turn, this will disturb the natural harmonic structure, and may tend to remove much of the power and body from the sound. Unfortunately, to make a very small room musically neutral is virtually impossible, so, in the vast majority of cases, the only thing to do with a small room is to try to absorb everything, and then provide a few discrete reflexions. The required ambience may then have to be added artificially.

If we make the room too dead, the musicians may find it uncomfortable on first entering the room. In almost all cases they will wear headphones when recording, but nonetheless their initial impressions when entering the room can have lasting effects. They should never be allowed to feel uncomfortable, even if only for the few seconds between entering the room and putting on the headphones, as it is remarkable just how those few seconds can leave many musicians unsure of their sounds. Illogical it may well be, but artistic performances tend to be fragile things, and anything which runs the risk of introducing any extra insecurity into musicians is to be avoided. Fortunately a hard floor and a window, or glass door, can usually provide sufficient reflective life to avoid the musicians experiencing an anechoic chamber effect when they enter the room, and this is easily achievable without creating a boxy character.

5.8.1 Practical Constructions

Figure 5.19 shows a layout for a neutral-to-dead vocal room which only occupies about 9 m^2 of floor area and 3 m of height. It assumes a 'worst case' structural shell of what is more or

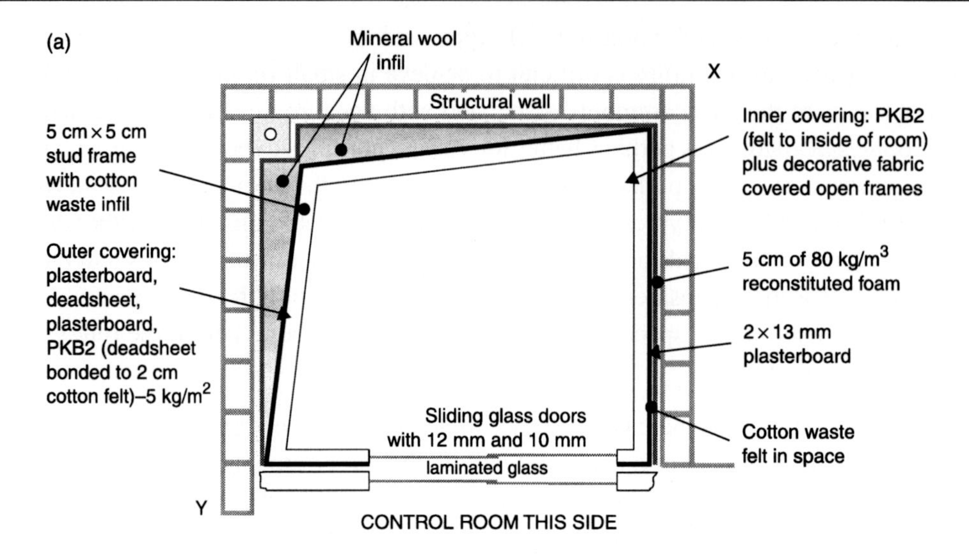
(a)
Mineral wool
infil
X
Structural wall
5 cm × 5 cm
stud frame
with cotton
waste infil
Inner covering: PKB2
(felt to inside of room)
plus decorative fabric
covered open frames
Outer covering:
plasterboard,
deadsheet,
plasterboard,
PKB2 (deadsheet
bonded to 2 cm
cotton felt)–5 kg/m²
5 cm of 80 kg/m³
reconstituted foam
2 × 13 mm
plasterboard
Sliding glass doors
with 12 mm and 10 mm
laminated glass
Cotton waste
felt in space
Y
CONTROL ROOM THIS SIDE

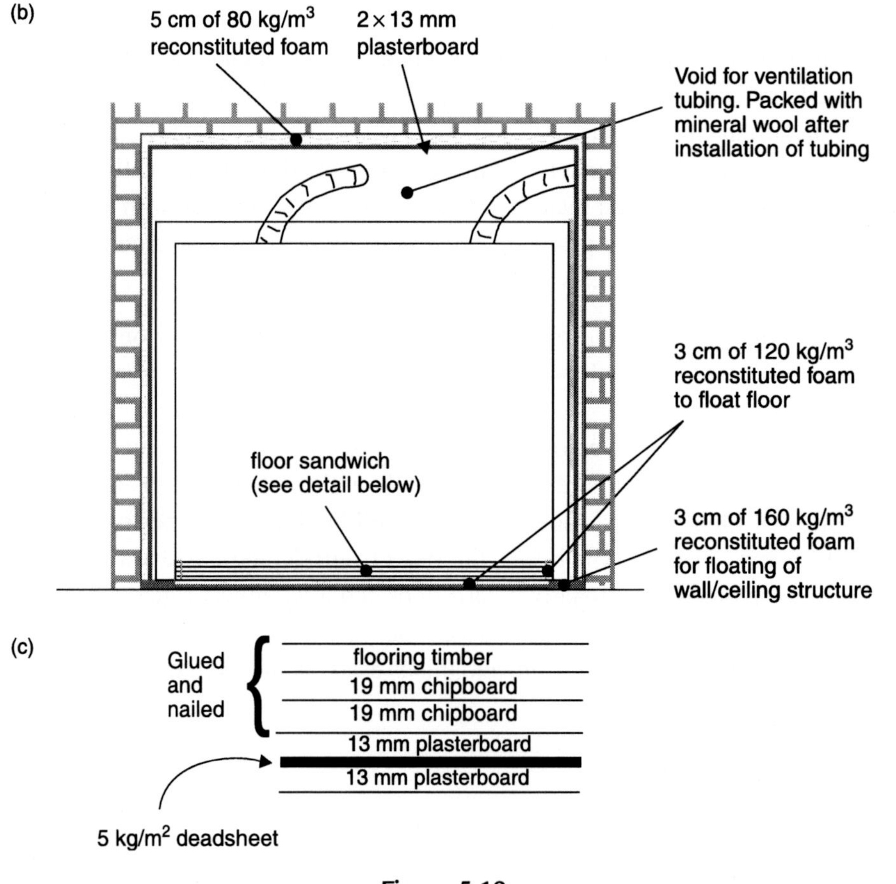
(b)
5 cm of 80 kg/m³
reconstituted foam
2 × 13 mm
plasterboard
Void for ventilation
tubing. Packed with
mineral wool after
installation of tubing
3 cm of 120 kg/m³
reconstituted foam
to float floor
floor sandwich
(see detail below)
3 cm of 160 kg/m³
reconstituted foam
for floating of
wall/ceiling structure
(c)
Glued
and
nailed
flooring timber
19 mm chipboard
19 mm chipboard
13 mm plasterboard
13 mm plasterboard
5 kg/m² deadsheet

Figure 5.19:
Construction of vocal room: (a) Plan; (b) Elevation; (c) Floor sandwich.

less a 3 m cube. In many cases such rooms are not only used for singing but can also be used for voice-overs or dialogue replacement. In these cases there may be no music to mask any extraneous noises, so the room should be well isolated. In the case of Figure 5.19, which is based on the design of a very successful room, one wall adjoins a control room, and the other surfaces are all concrete. The whole structural shell was first lined with a 6 cm layer of reconstituted polyurethane foam of 80 kg/m3 density, except for the floor, where a 3 cm layer of a 120 kg/m^3 foam was used to avoid the problems shown in Figures 3.6 and 3.7. Incidentally, vocal rooms of the type being described here do tend to be excellent for the recording of bass guitar amplifiers, especially if a sound with tight impact is being sought.

The polyurethane foam lining of the walls and ceiling was secured with contact adhesive. In turn, the foam was then lined with two layers of 13 mm plasterboard. This type of combination, mass/spring/mass (plasterboard/foam/wall) is ideal here as it serves two purposes. Firstly, it provides a good degree of broadband sound *isolation*. Secondly, it also provides a good degree of low frequency sound *absorption*. Therefore, it acts like a reasonably good combination of our open window and brick wall from the previous chapter. However, this system largely achieves by internal absorption what the wall and window do by reflexion and transmission, albeit not so effectively. The internal, acoustic control 'box' was constructed on a higher density foam which was first placed on the floor. The walls and ceiling structures were similar in nature to the 'neutralising' acoustic shell described in Section 5.3.4. If the space between the inner, floated room structure and the foam/plasterboard isolation treatment is lined with a fibrous, medium density material, we can then incorporate the properties of fibrous absorption to prevent any resonances from developing in the gap. We can thus combine relatively high absorption, good isolation (low transmission) and relatively low reflexion, all from the same composite lining. This is important because in such a small room we do not have sufficient space for large, conventional, wideband absorber systems. So, perhaps we can now consider the progress of a sound wave as it leaves the mouth of a vocalist and arrives at the room boundaries as shown in Figure 5.19. Remember, we want this room to be sufficiently dead to give no recognisable sound of its own to the recording, but to have just enough life to prevent the room from making the vocalist feel uncomfortable when entering it.

5.8.2 The Journey of the Sound Waves

The sound waves expand from the mouth of the vocalist in a reasonably directional manner. Except at the lowest frequencies, this fact becomes self-evident upon working in these rooms. Another person becomes remarkably quieter if he or she should turn away from the listener, unless they turn to face the reflective glass doors or the floor. In most cases, the musicians *will* be facing glass doors or windows, as visual contact with the control room is usually required for operational purposes. The sound will leave the mouth and strike the glass, which will reflect at least 95% of the energy back into the room with a very wide angle of

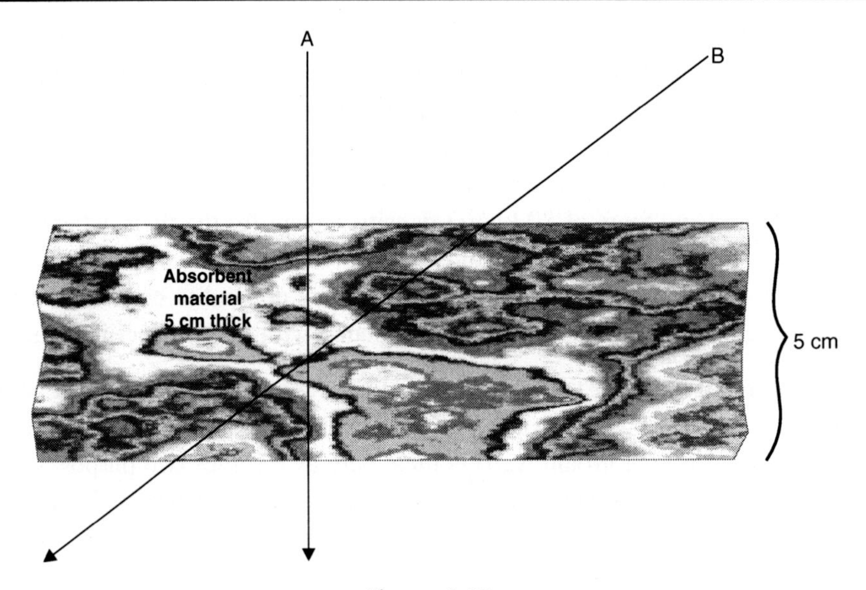

Figure 5.20:
Effect of angle of incidence on absorption. An incident wave 'A' striking the absorbent material at 90° would pass through the 5 cm thickness of the absorbent material. Incident wave 'B', striking at a shallow angle, would pass through approximately 10cm of the absorbent material. For materials of random texture, the absorbent effect is correspondingly increased in proportion to the extra distance travelled through the material. It should be noted, however, that for certain types of material with a pronounced directionality of fibres or pores, the above case may not apply. In general, however, oblique strikes will suffer more absorption than perpendicular strikes.

dispersion. Without headphones, the musicians will hear this, plus any reflexions from the floor, which may either be direct or also via the window. These reflexions will help to alleviate any sense of being in an oppressively dead room.

Energy reflecting back from the glass will strike the two side walls and the ceiling with an oblique angle of incidence. Only the wall opposite the glass will receive any sound at perpendicular incidence. The oblique arrival of sound at the face of an absorber will usually cause a much greater loss of energy, as it must pass through the absorbent material in a diagonal manner, and thus effectively travel through a thicker section of material (see Figure 5.20). Any sound reflecting from the window and the floor will automatically then pass into the walls at oblique angles. If a cardioid microphone is placed between the musician and the window, facing the vocalist, it is only capable of receiving sound directly from the musician's mouth, plus perhaps, a minute amount of the floor reflexions.

The sound which passes into the first layer of felt behind the decorative fabric surface will be partially absorbed, but some will continue through to reach the deadsheet backing. At medium and high frequencies the deadsheet is reasonably reflective, but any high frequencies reflected from it, even at 90° incidences, will still have to pass once again through the felt in order to re-enter the room. For wavelengths in the order of 8 cm (4 kHz) 2 cm of fibrous absorption will be highly effective, as the internal passage of sound through felt at relatively short wavelengths is tortuous. In fact, due to the oblique angles of incidence for many of the reflexion paths, quite high absorption could be expected down to 2 kHz, or below. If only 10% of mid and high frequencies were reflected from the surface of this vocal room on the first contact with a wall, then the second strike would only reflect 10% of that 10%. By the third bounce, which in such a small room may only take 15 or 20 ms, the energy remaining in the reflexions would only be one thousandth (10% of 10% of 10%) of that leaving the musician's mouth. It would thus be 30 dB down within the first 20 ms, and 60 dB down in considerably less than 50 ms. This is a very short decay time (T_{60}).

At low frequencies, such as those produced if a bass guitar amplifier were to be placed within the room, the mechanisms are very different. The low frequency sounds would propagate omni-directionally, and would possess much more penetrative power than the high frequencies because of the wavelengths being very long compared to the wall thicknesses. The first internal lining of the rooms (behind the decorative fabric) is in this instance a kinetic barrier material (deadsheet) covered in a 2 cm layer of cotton-waste felt. The composite weighs just less than 5 kgm^2 and comes in rolls, 5 m$\times$1 m, and this is nailed to the stud framework of the room, felt side to the room. Behind this barrier is an air cavity of 7.5 cm, containing a curtain of the same cotton-waste felt material, hung from the top, and cut carefully to fill completely the cross-section of the gap. On the other side of the stud frame is a double layer of 13 mm plasterboard, with a layer of 5 kg/m^2 deadsheet sandwiched in between. As mentioned in Section 5.3.4, the heavier plasterboard layers tend to be effective absorbers around the lower bass frequencies, whilst the membranous deadsheets absorb slightly higher frequencies. See also Figure 4.27.

All of the above layers are diaphragmatic. They are free to vibrate but they are also all highly damped. The room, to the low frequencies at least, presents itself as a large, limp bag. When the sound waves strike the kinetic barrier, it is somewhat like the effect of a boxer striking a heavy sandbag. The room gives, absorbing much of the energy and converting it into heat. Effectively the inner walls get pushed and pulled around by the compression and rarefaction half-cycles of the sound waves, but their inertia and internal viscous losses are such that they are almost incapable of springing back. Linings of this type have low elasticity, they are more or less inert, which is why such materials are known as deadsheets. In a similar way, one could give an almighty blow to a bell made from lead, but one would not get much ring from it. The lead would yield to the blow, and its high internal damping would absorb the impact energy. Its weight would then ensure that it did not move much, and hence if it could barely move or vibrate, it would have difficulty in radiating any sound.

When the sounds vibrate the deadsheet linings of our room, work is done in moving the heavy, flexible mass, and further acoustic energy is turned into heat energy as a result of the damping which resists the movement. Some sound is inevitably reflected back into the room, but with room dimensions so small, in a matter of a few milliseconds the reflected energy strikes another surface, so suffers losses once again. All in all, the sound decays very rapidly, and the low frequencies, below 150 Hz or so, are effectively gone in less than 100 ms. The *very* low frequencies receive no support at all from modal energy, as the *pressure zone* in a room of such size exists up to quite a high frequency.

5.8.3 The Pressure Zone

Modal support was discussed in Chapter 4, and it was stated that the lowest resonant mode of a room was that for which a half-wavelength would fit exactly into the longest dimension of the room. Once we drop below that frequency, we are entering the pressure zone. If less than half a wavelength can exist within the dimensions of a room, then instead of waves of positive and negative pressure distributing themselves over the room, the whole room will either be rising in pressure, or falling, dependent upon whether it is being subjected to a positive-going or negative-going portion of a long wavelength. The frequency below which the pressure zone will exist is given by the simple equation:

$$f_{pz} = \frac{c}{2Lr} \tag{5.1}$$

where:

f_{pz} = pressure zone upper frequency (also known as *room cut-off frequency*)

c = speed of sound in metres per second (344)

Lr = longest room dimensions, in metres

In the case of the room under discussion here, the maximum room dimension between reasonably parallel surfaces is about 2.5 m, therefore:

$$\begin{aligned} f_{pz} &= \frac{344}{2 \times 2.5} \\ &= \frac{344}{5} \\ \therefore f_{pz} &= 68.8 \text{ Hz} \end{aligned}$$

There could thus be no modal support in this room below 68 Hz, and the heavily damped lowest mode would be the *only* supportable mode within the first octave or so of a bass guitar. The response could therefore be expected to be very uniform. (See Chapter 6, Figure 6.11 for a representation of the pressure zone.) We will return to the subject of the pressure zone in later chapters, because its effect on loudspeaker response can be considerable (Section 13.5).

5.8.4 Wall Losses

When this internal bag inflates and deflates, it does radiate some power outwardly, but the nature of the construction of the room is such that the air cavities created between the vertical studs and the outer and inner linings provide further damping on the inner bag. The air will tend to resist the pressure changes, as its elasticity will apply a restoring force on the inner bag. In turn, it will also exert a force on the outer composite layer of plasterboard and deadsheets. This layer is also very 'lossy' and heavily damped. The internal particles in the plasterboard turn acoustic energy into heat by the friction of the particles rubbing together, and additional work is done by the need to move such a heavy mass. The sandwiched deadsheet forms a constrained layer, tightly trapped between two sheets of plasterboard. The constrained layer concept, discussed previously in Chapter 3 and shown diagrammatically in Figure 3.11, effectively tries to sheer the trapped layer of viscously 'lossy' material over its entire area. The resistance to this sheering force is enormous, so the damping value is high and the acoustic losses are great.

The transmission from the inner layer to the outer layers of the stud wall is partially due to the common studwork (vertical timbers) to which both surfaces are connected. This usually creates no problems as long as the inner surface is of a limp nature. Its lack of rigidity does not provide an effective acoustic coupling to the studs, and its weight, in turn, helps to damp the stud movement. The direct energy impacts on the studwork itself, where the deadsheet is fixed to it, and hence where that deadsheet is effectively more rigid, is only a small proportion of the overall surface area. If the studs are 5 cm in width and spaced on 60 cm fixing centres, they occupy only 5 cm out of every 60 cm, or about 8% of the surface area. In addition, being of narrow section, the lower frequency waves will tend to pass round the studs. However, if every last dB of acoustic isolation is required, the slightly more complex stud arrangement as shown in Figure 5.21 can be used. Here, the studs are double in number, but are interleaved such that the only common coupling between the two surfaces of the wall are at the top and bottom, which are in any case coupled by a common floor and ceiling. Walls of this type take up an extra 2 cm of space, but if the extra performance is needed, the space penalty is low. Thinner studs *could* be used if the space was critical, but they would provide less rigidity, and less damping on the panels, and hence may negate some of the advantage gained by the separation.

The majority of the inter wall-surface coupling is in any case through the air cavity. The cavity should be lined with fibrous material, which helps to increase the losses, but at very low frequencies its effect is only minimal. The losses from the air cavity coupling are great as long as the outer surface is very heavy, as it is difficult for a thing of low mass to move a more

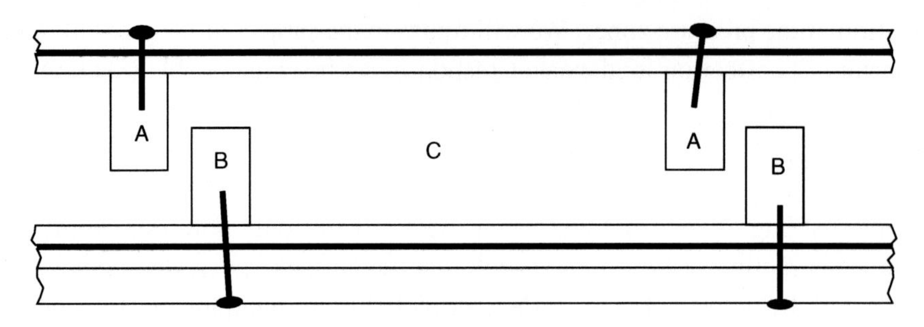

Figure 5.21:
Staggered stud system. The sheet layers on each side of a stud wall can be attached to independent sets of studs, connected only by the top and bottom plates 'C'. By staggering the studs and interleaving them, only a small amount of extra space is consumed compared to that of a conventional stud wall. This system reduces greatly the area of common coupling between the two sets of sheet materials, and usually offers more isolation than the more usual system with common studs. The cavity should be well filled with fibrous material, because resonances in the space can strongly couple the two sheet layers, across the air gap.

massive one. It is therefore relatively difficult for the air which is trapped in the cavity to excite the outer composite layer of heavy sheet materials. Historically, some of the first tests of this principle were carried out by experimenting with cannon in the Alps. So, as this factor of acoustic coupling is quite an important aspect of our isolation systems, perhaps we should now consider it in a little more detail.

5.8.5 Transfer of Sound Between High and Low Densities

In the nineteenth century, two cannon were placed on the side of a mountain, one low down, but not at the foot, and the other high up. The cannon were charged with equal amounts of powder, and observers were placed not only at each cannon position, but also high up and low down on a mountainside at the opposite side of the valley.[10] The arrangement is shown in Figure 5.22. The cannon low down was not placed at the very bottom of the valley so that its sound would not be unfairly reinforced by having the valley floor to push against. When the cannon were fired, the flashes and the smoke were clearly seen from all distant stations, and as the distances to the listening stations were known, the sounds of the cannon were expected to be heard after the appropriate time intervals had elapsed.

The lower cannon was fired first, the flash was seen, and the three listeners at the distant listening stations, A, B and D, awaited its report, which duly arrived at each position after the respectively appropriate time intervals. In each case, when the sound was heard, the listeners signalled its arrival by means of flags. The sound intensity at each station was described as best as could be done in the days before sound level meters. The sound was heard loudly by the listener low down on the opposite side of the valley, and was clearly heard by the two listeners

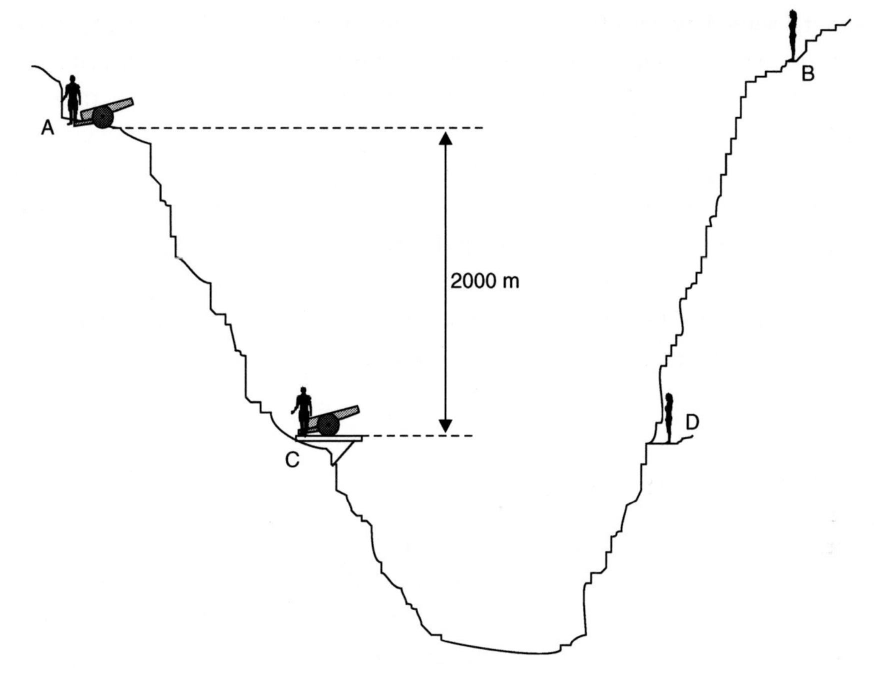

Figure 5.22:
Nineteenth century cannon experiment. Two identical cannons are positioned at 'A' and 'C', and charged with equal amounts of powder. When cannon 'C' is fired, the observers at 'A', 'B' and 'D' will see the flash almost instantaneously. After a length of time corresponding to the local speed of sound in the air and their distance from 'C', the three observers will hear the report from the cannon. When the cannon at 'A' is fired, the observers at 'B', 'C' and 'D' will see the flash. After the appropriate time lapse, the observer at 'B' will hear the sound, but the observers 'C' and 'D' in the denser air, may hear nothing, despite the fact that they are closer to 'A' than observer 'B'.

high up on the two sides of the valley, at positions A and B in Figure 5.22. When the higher cannon was fired from position A, again the flash and the smoke were clearly seen by the distant listeners, this time at positions B, C and D. After the expected time interval had elapsed, the listener at position B signalled that the report had clearly been heard, but after more than enough time had passed for the sound to arrive at positions C and D, no signals were seen, as no sound had been heard.

From this it was deduced that the density of the air in which the sound was produced, relative to the density of the air in which it was heard, was responsible for the efficiency with which the sound would propagate. The explosion in the higher density air, low down in the valley, could easily cause the sound to propagate not only to the listening station D, in the same high density air, but also to the listening stations at A and B, higher up in the lower density air. However, whilst the report from the high cannon could be readily heard at listening position B, which was in the same low density air, the sound could not penetrate effectively into the higher density air at the lower listening stations, C and D. This was despite the fact that position C was closer to A than was position B, where the report was clearly heard. Furthermore, compared to the denser air at the bottom of the valley, the thinner air higher up the mountain provided the explosion of the gunpowder with less air to push against as it left the barrel of the cannon. With less air to push against, less work could be done by the explosion, and hence less work done meant less sound generated. The air pressure reduces by almost 1 millibar for every 8m that one rises above sea level, and if temperatures are equal, the densities will thus be proportionately reduced. In fact, given the 2000 m vertical separation between the cannon shown in Figure 5.22, the air pressure at the upper positions, A and B, would be less than 75% of that at the lower stations, C and D.

5.8.6 Combined Effects of Losses

It can now be appreciated that the acoustic energy in the vocal room will be severely attenuated by the compounding of the low frequency losses in the deadsheet, the damping loaded upon it by the air in the inter-stud cavities, and the mass of the plasterboard to be moved; damped by the sandwiched, heavy deadsheet. As this whole structure sits on a layer of foam of an appropriate density, it thus sits on a foam spring, and is surrounded on the other five sides by air springs. The radiation of what energy is left must now pass through the 'springs' in order to get to the plasterboard layer which covers the foam, which is in turn glued to the structural walls. Springs are reactive; that is they tend to store and release energy, rather than passing it on. The cavity in Figure 5.19 is lined with cotton-waste felt. Any resonances which try to build up in these air cavities, particularly as lateral movement of the waves travelling around the box, must effectively pass through metres of felt, which they patently cannot do. By such means, resonances in the air space are effectively prevented. One acoustic 'short circuit' is therefore avoided, because if the air was allowed to resonate, its ability to acoustically couple the two sides of the gap would increase greatly.

The low density air surrounding the inner, floated, acoustic shell must then transfer its vibrational energy to the much heavier plasterboard wall linings. As described with the cannon experiment, it is difficult for a low mass of low density material to excite a high mass of high density material, so further transmission losses take place as the mass of the plasterboard resists the vibrations of the air in contact with it. The particles in the plasterboard also provide losses due to their friction, so this reduces even more the energy which can be imparted to the foam to which it is stuck, and which in turn is stuck to the wall.

The foam itself, securely bonded to the plasterboard by adhesive, strongly resists the movement of the plasterboard, so yet again, vibrational damping is provided and even more acoustic losses take place. Finally, the foam, glued to the structural wall, flexes under the force being imparted to it from the plasterboard, because its mass and stiffness are insignificant compared to those of the main structural wall which it is trying to move. The exact effect of this final loss depends mainly on the mass of the wall to which it is attached, but typically, from the inside of the finished room to the other side of a structural wall, at least 60 dB of isolation at 40 Hz could be expected from a structure similar to that being discussed here, though obviously not through the side with the glass doors.

Of course, what we have been considering here on our journey through the complex wall structure are transmission losses, but we should not forget that from each layer boundary, and particularly the inner boundaries of the more massive layers, acoustic energy will be reflected back towards the room. We have four heavy layers in this structure, the deadsheet inner lining, the composite plasterboard/deadsheet sandwich on the outside of the inner box, the plasterboard which is glued to the foam and the structural wall. We discussed earlier the progress of the reflected wave from the inner lining, but as we pass to the sandwich layer, the reflexions will be stronger due to the greater rigidity of the material. However, these reflexions cannot pass directly into the room. They must first pass through the inner deadsheet layer, which has already attenuated the incident wave due to its mass, its internal viscous losses and its poor radiating efficiency. The reflexions from the sandwich layer will also suffer similar losses as they seek to pass through the layers on their way back into the room. What is more though, just as some energy of the incident wave is reflected back into the room from the internal lining of deadsheet, the linings will also reflect back into the wall structure, towards the sandwich, a portion of the energy reflected *from* that sandwich, so we lose energy as we pass through each stage. The more internal losses which we can create within our multi-layered wall, by using internal reflexions to trap the sound within the layers until it dissipates as heat or work done, then the more pure absorption (i.e. not counting transmission) we will be able to achieve.

Similarly, the other heavy boundary layers on either side of our 6 cm foam, stuck to the structural wall, will also reflect energy back towards the room, but the deeper we go through our complex wall structures, the more difficult it will be for the acoustic energy to find its way back into the room. Taken as a whole, such a complex wall structure as shown in Figure 5.19 deals with the internal acoustic problems *and* the sound isolation problems by its high overall internal absorption. The structure is progressively absorbent, in that the most reflective layers are those furthest away from the inside of the room. If we remember our open window and brick wall discussion, simple sound absorbing materials tend to be poor isolators, and good isolators tend to be poor absorbers, but in the case of our small neutral-to-dead room we need both good absorption *and* isolation. The option of placing an adequate amount of a sound absorbing material on the inside of a room, then placing a simple isolation wall outside of it, tends to use enormous amounts of space. Using such simplistic techniques in a space of 3 m^3 (the same as the structural space for the room in Figure 5.19) we would end up with a usable space only about the size of a telephone kiosk.

The multi-layered, complex structures described here are much more efficient in terms of the space which they consume. They are also a little like the energy-absorbing front ends of motor cars, which crumple gradually on a frontal impact, absorbing the energy of a crash, stage by stage, and preventing it from doing too much damage to its occupants, or to the occupants of the car which it may have run into. Similar principles are now employed in the new, lightweight armour for battle tanks and warships, in place of the enormously heavy, steel armour-plate which was used almost exclusively up to the late 1960s. The similarities here should not be *too* surprising as, after all, energy absorption is energy absorption, whether it is absorbing the energy of car impacts, shells, or sound waves. However, the impedance differences are great, which is why different materials are necessary in each case.

Well, now that we have dealt with the relatively simple surface treatments which absorb the upper mid and the high frequencies, and we have discussed the limp bag which so effectively controls the low frequencies, all that we are left with is the low mid frequencies. These are the modal frequencies, where resonances can develop between the faces of the room surfaces. Their waves are too long to be absorbed by the superficial layers of absorbent felt, but they are not long enough to be fully affected by the deadsheets and composite layers. In the modally controlled region of the response it is the resonant modes which need to be controlled, as they can continue to ring-on at their natural frequency long after the driving force has ceased. A loudspeaker, when driven by an amplifier, will vibrate at the driving frequencies, but when the drive is disconnected, and if there is little damping, the cone will continue to resonate for a short time at its natural frequencies. The principles in the cases of the room and the loudspeaker are similar.

In the room of Figure 5.19, no parallel surfaces exist, as the walls have been angled in order to use geometrical techniques to help to prevent modal reinforcement. The amount of angling shown would not be enough in a more lively room, as the lower of the lower mid frequencies would still see them as relatively parallel in terms of their wavelength (see also Figures 5.8 and 5.10), but in the case of a room lined with so much felt and deadsheet, the gradual, non-perpendicular impacts tend to suffer much greater absorption and much less effective re-radiation than would be the case when striking similarly angled, but more reflective surfaces.

The ceiling structure of this room is similar to that shown in Figure 5.3. It is vaulted with the PKB2 composite deadsheet/felt material, the felt side facing the floor. The 20 cm depth of the vaulting is sufficient to begin to take the energy out of both the direct mid frequency waves and the ones reflected almost perfectly from the hard floor. As the wooden floor is a better reflecting surface than any of the other surfaces of the room, the design of the ceiling opposing it has to give due consideration to that fact. Above the inner ceiling is a void of about 30 cm in depth. When stuffed with absorbent material, be it fibrous or off-cuts of foam, it provides good absorption of any lower mid frequencies which may penetrate the inner linings. The ceiling void is actually there primarily for another purpose however; the passage of ventilation ducts, as one does tend to have to provide musicians with a good supply of fresh air, delivered as smoothly and silently as possible. The overall response of the room is very similar to that shown in Figure 5.27(a), but without the rise below 50 Hz.

5.8.7 A Micro-Problem

In small vocal rooms, subjectively, the lower mid frequencies tend to be the most difficult ones to suppress, but in a room such as the one which we are discussing here, the remnants of modal energy which may be detectable after making a sudden, loud 'a' sound, as in cat, tend to be no problem in practice. However, these rooms are better assessed through microphones than by listening inside of them. After the completion of an early room of this type, although quite a bit larger than the one being discussed in this chapter, and with an extra window, there was a resonance which was detectable at about 400 Hz, and which lingered for a short time after impulsive excitation of the room. Days were spent looking for ways to control the problem but it simply refused to go away. The studio finally went into use during the time that the investigations were still taking place, and it seemed to be a little strange that there had been no complaints from the recording engineers about the vocal room resonance. In fact, the only comments about vocals recorded in the room were to the effect that they were beautifully clear. The room had a sliding glass door which faced the control room, and a window in an adjacent wall for visibility through to the main studio room. Though this wall was at 90° to the wall with the doors, the window was raked back about 8° from the vertical. There was also a very small window leading to the small vocal booth of the adjacent studio, but this faced directly towards neither of the other windows. The floor was of ceramic flooring tiles. Some thing or things in this room conspired to produce the resonance, but the cause was never found. It was conjectured that the recordings were clean because the directionality of the microphones, when pointing in the usual direction, did not face the source of the problem, which was no doubt partly true. It was finally discovered that the resonance excited by the hard 'a' sound was almost *70 dB* down relative to the signal, but it had been noticed because it was the only resonance left audible in such a highly controlled room. It was never corrected because in practice there was no problem.

In very small neutral rooms, it is usually impossible to achieve the sense of neutrality that is apparent in the larger rooms. It is impossible to achieve the sense of spaciousness simply because this sensation is achieved by means of *late* reflexions, which cannot exist in smaller rooms. The fact that the reflective surfaces are so close to the listener also means that there will be a higher proportion of very early reflexions, and even if these are suppressed to very low levels; it is surprising how sensitive the ear can be to their presence. The problem of the resonance of a transient 'a' sound can be alarming when first heard, but the low level and shortness of duration of the resonance is usually completely masked by the 't' if the whole word 'cat' is pronounced. In these cases, the degree of neutrality must be assessed via microphones, because the purpose of the rooms is recording. They are not specifically designed for walking round in them shouting 'a'.

5.9 Trims

It is customary in such rooms to provide skirting boards to protect the walls when cleaning the floor, and also knee rails, on which to mount sockets for microphone inputs and power outlets. A dado rail is normally also provided at waist height, as can be seen in the room shown in Figure 5.16. If these are only 10 to 15 cm in width, the specular reflexions that they provide add a little extra touch of life without the risk of being able to support any resonances. Frequencies below around 2 kHz just pass round them as a river flows round a projecting rock. Fabrics, though, *can* be reflective if they are too stiff. Fabrics such as calico can be so tight when stretched on their frames that they can ring like drum skins. The best fabrics to use are similar to the ones suggested for use in the large neutral rooms of Section 5.7, soft ones with some elasticity and a relatively open weave.

5.10 The Degree of Neutrality – An Overview

In general, as room size reduces, the maintenance of a neutral sound character tends to be dependent on reducing the decay time proportionally. Decisions on neutrality must be made via microphones, because so many psychoacoustic changes occur as room size shrinks that the ears alone are not the best assessors of what the rooms will behave like when they are performing for their intended use – recording. Although numerically based diffusers may seem to be an obvious solution, at the close distances from the instruments that they would necessarily have to be sited in the smaller rooms, the diffuse re-radiation from them can produce some very unnatural, most un-neutral sounds due to the imperfect nature of their diffusion.

It can be very difficult indeed to accurately measure the characteristics of rooms with very short decay times, especially at low frequencies, where the inherent time responses of the filters in the measuring equipment can be longer than the decay times that they are supposed to be measuring, and hence they can give false and misleading results. The best solution for achieving subjective neutrality in very small rooms tends to be to deaden the rooms to the greatest degree possible, consistent with not allowing the most persistent resonant frequencies to become exposed, and then to add a few discrete reflexions where they can add sufficient life for the room to be comfortable to be in (without headphones). However, these reflexions should return from directions where they will not unduly colour the signal entering the microphones. Of course, if they *are* wanted on a recording, then the microphone positions can be adjusted to capture whatever reflexions may be deemed to be beneficial to the sound.

Such rooms are becoming of growing importance because of the proliferation of small studios, often based around small computerised recording systems, where the diminutive size of the equipment seems to lead to the use of proportionally diminutive rooms. The principle factor conspiring to drive the sizes down is the current financial disproportion between the ever-falling cost of adequate quality (even if not the 'high-end') recording systems, compared to the cost of premises, and especially so in city centres. Nevertheless, without adequate experience, the creation of subjectively good and flexible recording rooms is not an easy task, so unfortunately the increasing number of rather poor (and even very poor) recording spaces is becoming something of an epidemic afflicting a supposedly professional industry.

In fact, when sited in well-constructed rooms with adequate and appropriate acoustic control, such studios can produce (in the right hands) some quite remarkably high-quality results, and they are often orientated towards recordings which do not require many musicians playing simultaneously. Fortunately, many owners of such studios are beginning to realise that they *can* achieve much better results, and get a step up on their competitors, if they have their rooms professionally designed. In almost all cases, unless a very specialised niche market is being sought, the recording rooms need to possess a high degree of subjective neutrality or there will be a tendency for the recording room to stamp its mark on every instrument recorded in it, which after much multi-tracking can become excessive in the final mix. Because of the reflexion time-dependent nature of this characteristic sound, it can be very difficult to remove it from the mix by means of equalisation, or any other process. Such a recognisable sound can soon become tiring, and can damage the reputation of a studio unless that sound is specifically desired. The photograph in Figure 5.23 shows a room, built with similar techniques to those described in this chapter, in the attic of a house whose upper floor was converted into a small commercial studio of some considerable success. Getting such rooms 'right' is not particularly difficult, but in all too many cases it is not done satisfactorily. It is sad that the failures are usually down to ignorance rather than cost.

5.11 Dialogue Recording Rooms

In typical voice-over situations, an actor will watch a picture on a screen, be it a flat panel or a projected image (or possibly still a cathode ray tube), and listen to the original dialogue via a loudspeaker. When the time comes to record, the loudspeaker is muted and the actor performs to the image, synchronising the voice to the lip movements on the screen. The performance by the actor is in every way a performance, and if the room is too absorbent it may cause the voice to be forced and the sound to be unnatural because actors, unlike singers, usually do not wear headphones whilst performing. Conversely, if the room is too live, it will impart its own characteristics on the recorded sound, and these may not correspond well to the environment

Figure 5.23:
The very neutral studio room at Noites Longas studio, in Redondos, Portugal (1993).

depicted in the image. For example, if three people were talking on a sand dune, in a desert, the scene would lose credibility if the voices sounded like they had been recorded in a room. Room sounds can always be added artificially, but it can be very difficult to remove them once recorded.

In dialogue rooms, it can often be useful to provide highly absorptive surfaces behind the actors, at least down to the lower limits of the vocal frequency range, around 90 Hz or so. *Reflective* surfaces (the video or projection screens themselves often help, here) can be placed forward of the actors, and cardioid microphones can be pointed towards them (the actors). Such a directional microphone would not respond to the vocal reflexions, which would be absorbed by the 'trap' towards which the microphone was pointing, yet these reflexions would give sufficient life to the room so that the actors could feel comfortable and natural during their performances. The big difference to remember, here, is that for actors, *the room* is their performance stage, whereas for singers, the foldback *signal in the headphones* is their environmental 'reality'. Much more will be said about this in Chapter 24.

In too many cases, not enough space is allocated to rooms which are 'only' for dialogue recording, and the existence of so many small rooms has led to a general tendency to believe that they are all that is necessary. The reality is somewhat different, because to provide a highly controlled acoustic, with a directional microphone sufficiently far from an actor's mouth to avoid proximity effect problems, requires a room of decent size, say 12 m^2 or so, *before* acoustic treatment, and at least 3 m in height.

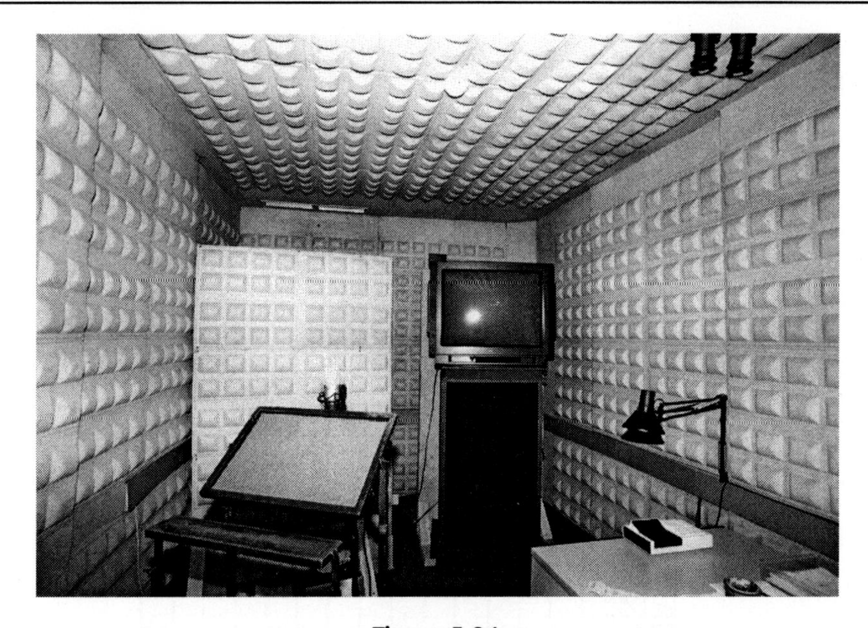

Figure 5.24:
A voice recording room constructed without the required space behind the 'absorbent' panels on the walls and ceiling.

Figure 5.24 shows a room which was built by a supposedly professional acoustic design and construction company, but what used to suffice for talk studios for AM radio will often now not meet the requirements for modern, home theatre presentations. In the room shown, the floor was carpeted and the walls and ceiling were covered in panels of pressed cotton-waste felt, geometrically formed into rounded wedges. The panels were mounted on 5 cm wooden battens, screwed to the concrete walls, and with 4 cm of mineral wool of about 40 kg/m^3 density in the gap. The room sounded dark, and surprisingly dead, but there was always something unnatural at the lower end of the vocal register, and the recording personnel were constantly filtering out these frequencies, which not surprisingly also tended to thin out the voices of the actors because the equalisation was acting on the direct signal as well as on the room sounds.

Figure 5.25 shows the published performance plots for the wall panels, and it is quite clear that, with the distance 'd' equalling almost zero, as was the case in the aforementioned room, the absorption begins to fall off rapidly below around 800 Hz. However, with a spacing of 300 mm between the support frame and the concrete walls, adequate absorption could be achieved at 100 Hz (a coefficient of around 0.4), but many studio owners want to see the space that they are paying for, and many contractors are happy to declare that the customer is always right, despite the laws of acoustics demanding something different from both of them.

One obvious solution to the problem in the room under discussion would have been to mount the panels 'properly' on new frames set 300 mm (1 ft) from the concrete walls, but it was

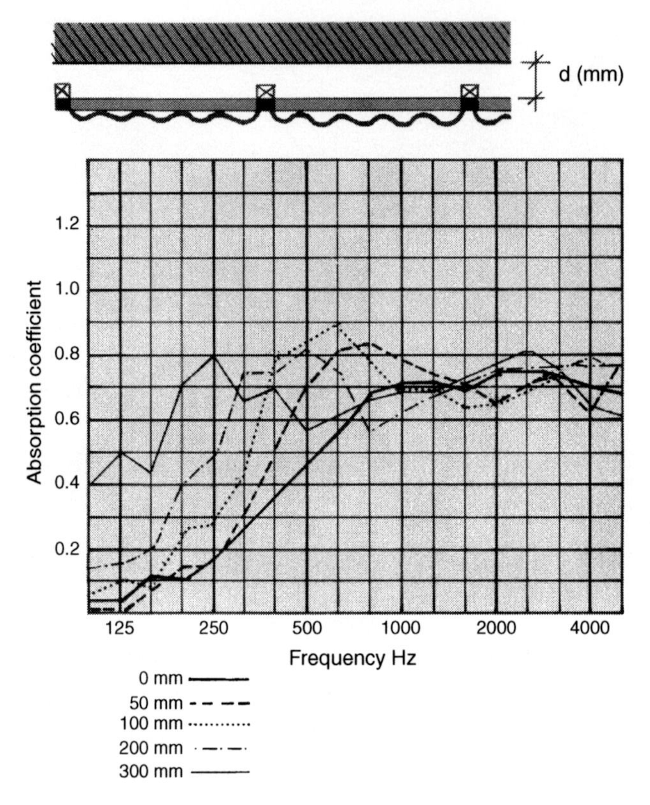

Figure 5.25:
Absorption coefficient versus frequency for the compressed cotton-waste panels with respect to their distance from a solid wall.

generally considered that the remaining area within the room would have been too small for the actors and equipment, so a different solution was sought. (Once again, a room of insufficient size had been chosen.)

Figure 5.26 shows a sister room after reconstruction. In order to only occupy the same space as the original treatment, a totally different approach was devised. To the concrete walls and to the suspended plasterboard ceiling were attached 6 cm panels of an 80 kg/m^3, open cell, reconstituted polyurethane foam, marketed under the name of Arkobel. The foam was attached by contact/impact adhesive, although cement-and-latex adhesives could have been used, which although slow in drying do not give off noxious vapours. To the foam was then attached, by similar adhesive, 3.5 cm panels of a porous nature, made from wood shavings and Portland cement, with ample air spaces in between. This material is quite heavy, being around 18 kg/m^2 for 3.5 cm thickness, so the whole system acts in a combination of ways, as a multi-layer absorber, with absorption being due to the tortuosity of the pores, membrane characteristics and panel absorber characteristics. This type of material was originally manufactured in the United Kingdom as Woodcemair, being made from *wood* shavings,

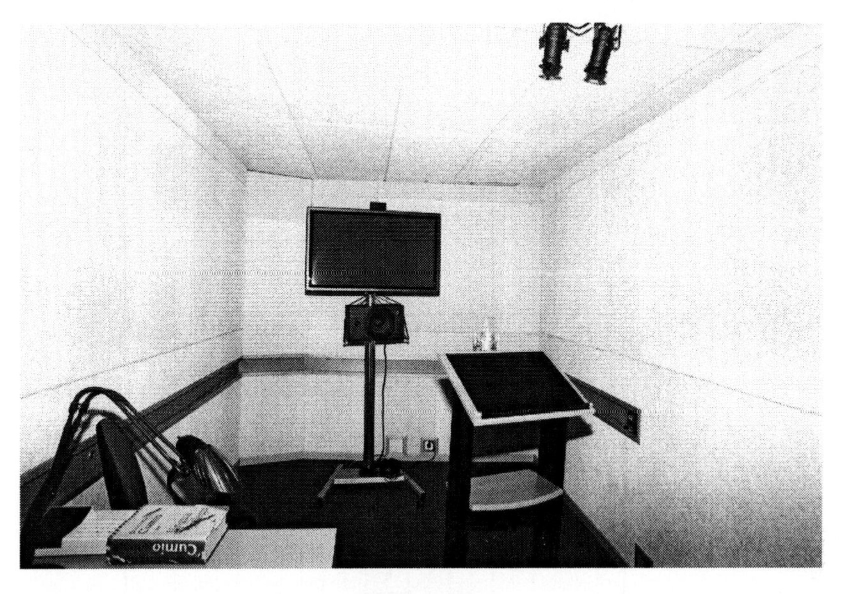

Figure 5.26:
A room similar to the one shown in Figure 5.24 but after treatment with a more appropriate combination of materials, yet still only occupying the same space as the original panels (Sodinor, Vigo, Spain).

cement, and being lightly compressed so that *air* voids still remained. Materials such as Celenit (from Italy), Heraklith (from Austria) and Viruter (from Spain) are similar options based on the same theme, with different characteristics of absorption and aesthetics.

The panels shown in Figure 5.24 were purely porous absorbers, which could only act on the *velocity* component of the acoustic wave. They only become effective absorbers when spaced about a quarter of a wavelength from a solid surface. The wavelength of 800 Hz is 340 (the speed of sound in m/s) divided by 800, or about 42 cm. One quarter of this wavelength would be around 10.5 cm (4 inches) which corresponds well with the average distance of the protrusions in the pressed cotton panels from the concrete walls, but below this frequency the absorption dropped off rapidly. However, the system shown in Figure 5.26 maintained this porous absorption above 800 Hz, but introduced the highly damped, *pressure*-component-dependent absorption of multi-layered membrane and panel absorbers at lower frequencies. The decay time versus frequency plot of the room is shown in Figure 5.27(a), with the corresponding decay time of the original room shown in Figure 5.27(b).

Unlike in music recording rooms, what happens in dialogue rooms below the range of the vocal frequencies is of little relevance, because even if resonances exist there is nothing in the voices to excite them. Even consonant sounds have difficulty in exciting low frequency resonances. The only real low frequency problems occur if the loudspeaker reproduction

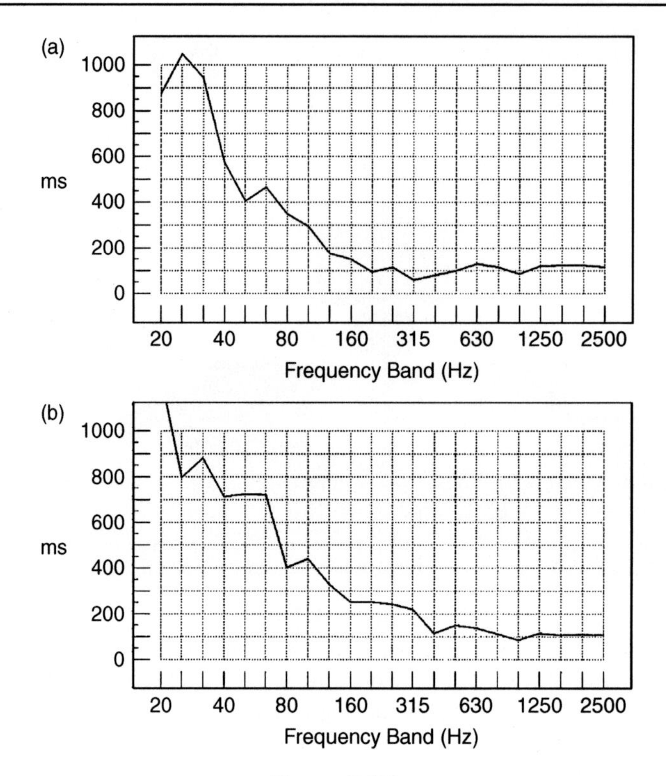

Figure 5.27:
The decay characteristics of the rebuilt room shown in Figure 5.26 are shown in (a), whilst (b) shows the decay of the original room, similar to that shown in Figure 5.24. Note the lumpy, roller-coaster appearance of (b), a sure indication of the existence of resonances in the 63 Hz to 400 Hz region. This contrasts with the smooth curve through the same region in the rebuilt room. The absolute values of the decay times below around 100 Hz are somewhat lower than indicated in the plots. At low decay times, the filters of the measuring system add their own delays. The added delay introduced by typical filters and the loudspeaker used as the acoustic source are shown in Figure A.1.7 in Appendix 1.

becomes less defined during the rehearsal phases before the takes, and even then only if there are low frequency sounds on the original soundtrack.

In these types of dialogue recording rooms the recording technician/engineer is sometimes working in a corner of the same room, monitoring on headphones and using a small mixing console. (Obviously, in such cases, any noisy equipment must be outside the room.) Some people feel that this arrangement can give a closer, more immediate communication between actors and technicians. Alternatively, some rooms have small 'control cubicles' adjacent to them, but some of these are really far too small to allow any form of accurate monitoring via loudspeakers. A somewhat ludicrous control cubicle is shown in Figure 5.28, ludicrous at least if serious attempts at monitoring via loudspeakers are to be made. In fact, if the only realistic

Figure 5.28:
An absurdly small 'control booth'. Rooms of such small sizes are effectively too small to allow a flat monitoring acoustic, and colouration is unavoidable.

means of monitoring in such a small room is via headphones, then the technician/engineer may as well be in the same room as the actors, remote controlling the machinery, and monitoring via headphones. The room shown in Figure 5.28 is simply too small for the introduction of any worthwhile acoustic treatment, and so a coloured, boxy sound is virtually inevitable. Playback of a take via a good loudspeaker in the well-controlled room in which the actors perform would be a much more realistic way of judging the quality of the recording, so there is good justification for working that way if no decent-sized and acoustically treated control room is available.

In general, the greatest error in the design of most rooms for dialogue recording, be they recording rooms or control 'cubicles', is that they are usually too small to allow good acoustic characteristics to be enjoyed, and the control of their deficiencies by equalisation is almost certain to degrade the natural qualities of the actors' voices. Better recordings are usually made in adequate sized rooms, with space available for the appropriate acoustic treatment down to at least 100 Hz.

5.12 Summary

Neutral rooms are good, general-purpose recording spaces. They lightly enrich the character of musical instruments without imposing their own presence on the sounds.

Some life in a room is usually necessary to make the musicians feel comfortable. Uncomfortable musicians rarely play at their best.

Neutrality is easier to achieve in larger spaces than in small ones.

To achieve neutrality, parallel reflective surfaces should be avoided, strong resonances should be suppressed, and sufficient specular reflexions should be present to give a sense of spaciousness.

The neutrality of a recording room should not only be judged directly by the ear, but also via microphones.

Floors should normally be of wood, but should neither creak nor rattle.

Certain room proportions are known to give rise to a more even spread of modes. In some circumstances these can give useful starting dimensions.

Plasterboard-based acoustic control shells can usefully control the very low frequencies, and deadsheet membranes can be used to control the mid-bass frequencies. Fibrous layers can be used to control cavity resonances and frequencies from 2 kHz upwards. The upper bass/low mid frequency region is best controlled by a combination of geometry, membranes and fibrous materials.

The proximity of the ceiling to the floor, and to the microphones and instruments, can be rendered benign by combinations of diffusion, reflexion and absorption.

To neutralise an existing room, or a sound isolation shell, it may be best to suppress the reflexions as much as possible then to rebuild the desired life into the rooms.

Neutral rooms are very flexible in use, but they may not be able to match the character of more idiosyncratic rooms in some circumstances, even with the aid of artificial (electronic) processing. (See Chapter 6.)

What is acoustically parallel at low frequencies may be far from geometrically parallel.

Neutral rooms of 500 to 1000 m^3 would tend to have decay times of around 0.3 to 0.6 s in the mid frequencies, but reflexions should also be present.

Large surfaces of diffusers tend to lead to decay times which are too long for many neutral rooms. Rooms with too much diffusion can lack the necessary specular reflexions which give rise to a sense of spaciousness.

Fabrics used for wall coverings should be non-resonant when stretched and mounted.

Small neutral rooms are difficult to achieve, because the reflexions are rapid and dense, and late reflexions are unobtainable. For general neutrality, small rooms may have to be rather dead, with a few surfaces reflective at high frequencies to achieve a little ‘air’.

Perversely, it is the very same digital recording equipment that exposes low level room sounds in the recordings which also, due to its mechanical noise and bad positioning, frequently masks these room sounds in the control room monitoring. This leaves them exposed to many domestic listeners, whilst going unnoticed by the recording personnel.

It is difficult to find any simple surface treatments to make a small room neutral-to-dead. Complex structures of low depth may have to cover almost the entire surface area, other than the floor, to be able to adequately control the room resonances.

Post-processing of the recording will not usually be able to remove the sound of a bad room.

Multiple-layer treatments can be used to trap the sound energy within the layers.

Very small rooms can have very flat low frequency responses because they are operating within their pressure zones, where no modes exist. The pressure zone is in the region below the frequency of the lowest room mode.

Very small rooms can be controlled at low frequencies by a 'limp bag' concept and resilient walls.

Sound vibrations have difficulties passing from a lighter medium to a heavy, dense one.

Small, strange resonances in highly damped rooms are often only audible because there are no other room sounds to mask them. What may sound strange *in* a room may be inaudible via microphones in normal use.

Rooms for dialogue recording may differ in their needs from rooms for recording musical performances.

Highly damped rooms can be beneficial when dry voices must be artificially processed to match a picture.

Rooms used for dialogue recording are often too small to allow adequate acoustic control down to the 100 Hz region, and so suffer from colouration.

Room problems below about 80 Hz are usually of no consequence in dialogue recording because voices contain no frequencies sufficiently low to excite such low resonances.

References

1 Louden, M. M., 'Dimension Ratios of Rectangular Rooms with Good Distribution of Eigentones', *Acoustica*, Vol. 24, pp. 101–4 (1971)

2 Gilford, C. L. S., 'The Acoustic Design of Talks Studios and Listening Rooms', *Journal of the Audio Engineering Society,* Vol. 27, No. 1/2, pp. 17–29 (Jan/Feb 1979)

3 Bonello, O. J., 'A New Criterion for the Distribution of Normal Room Modes', *Journal of the Audio Engineering Society,* Vol. 29, No. 9, pp. 597–605 (September 1981)

4 D'Antonio, Peter, 'Two Decades of Diffusor Design and Development', AES Pre-print, 99th Convention, New York (1995). (See also Bibliography below.)
5 Schroeder, M. R., 'Diffusive Sound Reflection by Maximum Length Sequences', *Journal of the Acoustical Society of America*, Vol. 57, No. 1, pp. 149–50, January (1975)
6 Schroeder, M. R., 'Comparative Study of European Concert Halls: Correlation of Subjective Performance with Geometric and Acoustic Parameters', *Journal of the Acoustical Society of America*, Vol. 56, No. 4, pp. 1195–201 (October 1974)
7 Schroeder, M. R., 'Progress in Architectural Acoustics and Artificial Reverberation in Concert Hall Acoustics and Number Theory', *Journal of the Audio Engineering Society*, Vol. 32, pp. 194–203 (1984)
8 Schroeder, M. R. and Gerlack, R., Response to 'Comments on Diffuse Sound Reflection by Maximum Length Sequences', *Journal of the Acoustical Society of America*, Vol. 60, No. 4, p. 954 (October 1976)
9 Walker, R., The Design and Application of Modular, Acoustic Diffusing Elements, *Proceedings of the Institute of Acoustics*, Vol. 12, Part 8, pp. 209–18 (1990)
10 Tyndall, J., DCL, LCD, FRS, *On Sound*, 6th Edn, Longmans Green and Co., London (1895)

Bibliography

D'Antonio, P. and Cox, T., 'Two Decades of Sound Diffusor Design and Development, Part 1: Application and Design', *Journal of the Audio Engineering Society*, Vol. 46, No. 11, pp. 955–76 (November 1998)

D'Antonio, P. and Cox, T., 'Two Decades of Sound Diffusor Design and Development, Part 2: Prediction, Measurement and Characterization', *Journal of the Audio Engineering Society*, Vol. 46, No. 12, pp. 1075–91 (December 1998)

Cox, T. J. and D'Antonio, P., 'Acoustic Absorbers and Diffusers', Spon Press, Abingdon, UK and New York, USA (2004)

CHAPTER 6

Rooms with Characteristic Acoustics

Unusual recording environments. The influence of the environment on the musicians and the music. The search for special sounds. The negative effects of a containment shell. The sounds of different construction materials. The effects of the directional characteristics of loudspeakers and microphones. Stone rooms. Acoustic and electronic reverberation characteristics. The 20% rule. Reverberant verses bright rooms. Low frequency considerations. Orchestral rooms. Psychoacoustics and spacial awareness. Dead rooms. Foley rooms.

6.1 Definitions

From Morfey's *Dictionary of Acoustics*[1]:

> *Dead room*: A room in which the total absorbing area approaches the actual room surface area. Note that such a room cannot qualify as 'live'; compare *live room.*
>
> *Live room*: A room in which the amount of sound absorption is small enough that the mean free path L_{AV} is greater than $\sqrt{A}$, where A is the *room absorption.* (For 'mean free path' see Section 4.4.1.)
>
> *Room absorption*: A measure of the equivalent absorption area of a room – units, m^2.

The above definitions are quoted because there is a growing tendency, due to the popularisation of recording, to refer to any room in which musicians play together as a 'live room' because they play live, as opposed to recorded. Throughout this book, the term live room is used in its acoustical sense – a room with considerable acoustic life; neither neutral nor dead.

6.2 A Brief History of Idiosyncrasy

In the early days of recording, when studios were expected to be able to handle any type of recording which their size would allow, live rooms were more or less unknown. To this day, if the *only* recording space in a studio is a live room, then it is either in a studio which specialises in a certain type of recording, or the room is used as an adjunct to a studio which is mainly concerned with electronic music. Live rooms are distinctly individual in their sound character, and tend to impose themselves quite noticeably upon the recordings made in them. However, when that specific sound character is *wanted*, then effectively there is no substitute for live rooms. Electronic or other artificial reverberation simply cannot achieve the same results, a point which will be discussed further in Section 6.7.

When, in the late 1960s, many British rock bands began to drift away from the more 'sterile' neutral studios, gravitating towards the ones in which they felt comfortable and in which they could play 'live' in a more familiar sonic environment, a momentum had begun to grow which would radically change the course of studio design. In 1970 the Rolling Stones put the first European 16-track mobile recording truck into action. This was not only intended for the use of live recordings, but also for the recording of bands in their homes, or anywhere else that they felt at ease. It was soon to record the Rolling Stones *Exile on Main Street* album in Keith Richard's rented house, Villa Nelcote, between Villefranch-sur-Mer and Cap Ferrat in the south of France, but another of its earliest uses was the recording of much of Led Zeppelin's fourth album. This album was a landmark, containing such rock classics as *Stairway to Heaven* and *When the Levee Breaks*; the latter perhaps inspiring a whole generation of recorded drum sounds. Previously, Led Zeppelin had made many recordings in the famous, old, large room at Olympic, in London. For the fourth album, they rented Headley Grange in Hampshire, England, and took along the Rolling Stones' mobile, and engineer Andy Johns. The house had some large rooms, but none had been especially treated for recording. Nevertheless, as the house existed in a quiet, country location, the ingress and egress of noise was not too problematical.

6.2.1 From a Room to a Classic

It seems that *When the Levee Breaks*, with its stunning drum sound for the time, was never planned to be on the album, or indeed to be recorded at all. In the room in which they were recording, John Bonham was unhappy with the sound of his drum kit, so he asked the road crew to bring another one. When it arrived, they duly set it up in the large hallway, so as not to disturb the recording, and waited for John to try it. At the next available opportunity, he took a break from recording and went out into the hallway to see if he preferred the feel and sound of the new kit. The other members of the band remained in their positions, relaxing, when suddenly a huge sound was heard through their headphones. The sound was from the drum kit in the hallway.

John had failed to close the door when he went out, and the sound from the kit that he was playing was picking up on all the open microphones in the recording room. The hall itself had wood panelled walls, with a large staircase, high ceiling, and a balcony above. It was thus diffusive, absorbent to low frequencies, reverberant, well supplied with both late and early reflexions, and very much of a sonic character which matched perfectly the style and power of the drumming. He went into a now famous drum pattern, over which Jimmy Page and John Paul Jones began playing some guitar and bass riffs which they had been working on. Robert Plant picked up on the whole thing, and sung along with some words of an old Memphis

Minnie/Kansas City Joe McCoy song. Subsequently, Andy Johns, who had been recording the sounds out of pure interest, reported from the mobile recording truck that they should consider this carefully, as he was hearing a great sound on his monitors.

Such was the birth of this classic rock recording. The story was related to the author by Jimmy Page, a dozen years or so after the event, during a telephone conversation relating to the production of some recordings for Tom Newman (co-producer of Mike Oldfield's *Tubular Bells I & II*), as one of the songs being recorded was the one referred to above. The problem related to the fact that no matter how many times Newman listened to the Led Zeppelin version, he could make absolutely no sense of the words in the bridge section. 'Probably that is because they *don't* make sense.' replied Jimmy. Evidently, during Robert's sing-along, there was no bridge lyric which he knew or remembered, so he sung what came into his head. Why this story is relevant to this chapter is because it shows, most forcefully, how a *room* inspired an all-time rock classic. It is almost certainly true to say that without the sound of the Headley Grange hallway, Zeppelin's *When the Levee Breaks* would never have existed.

Had Led Zeppelin been recording in a conventional studio of that time, they could perhaps now have been considered to be one rock classic short of a repertoire. *But*, it must also be remembered that had John Bonham been playing a different drum pattern in Headley Grange, or had Jimmy Page and John Paul Jones opted for a different response, then the sound of the drums in the hallway may have been totally inappropriate. This highlights the limitation with live rooms; they can be an inspiration and a unique asset in the creation of sounds, or they can be a totally intrusive nuisance. Furthermore, there is no one live room which will serve all live room purposes.

6.2.2 Limited, or Priceless?

Let us suppose that we had a reverberation chamber, similar to the ones used for acoustic power measurements in universities, institutes and research departments. These are constructed to give the broadest achievable spread of modal resonances, in order not to favour any one frequency over another. They *can* produce excellent results as reverberation chambers for musical mixes, but they lack the specific idiosyncratic characteristics that make *special* live rooms special. Let us now consider an analogous situation. Imagine that guitar manufacturers' were to opt for a totally even spread of resonances in their instruments; to have 'life', but as uniformly spread as possible. All of a sudden, almost all guitars would begin to sound much more alike than they currently do. In general, it is impossible to make a Fender Stratocaster sound like a Gretch Anniversary, or *vice versa*, because the time domain responses of the resonances and internal reflexions cannot be controlled by the frequency domain effects of an equaliser. Something halfway in between a Stratocaster and an Anniversary would be neither one thing nor the other, and whilst it may well be a valid instrument in its own right, it could not replace either of the others when their own special sounds are wanted.

Another analogy exists here in that most professional guitarists have a range of instruments for different purposes: different music, different arrangements, different styles, and so forth. Such is the case with live rooms, where a 'universal' room could perhaps be used as an acoustic reverberation chamber for the recording and mixing process, but it could never be used to substitute for the special sounds of the special rooms. What is more, no large, self-respecting recording studio has banks of identical 'best' electronic reverberation devices, but rather a range of different ones. It is generally recognised that a range of different artificial reverberation is greatly preferable to more of 'the best'. In so many cases, neutrality is quite definitely *not* what is called for.

Dave Purple, who was formerly at dbx in Boston, USA, related a story about the reverberation plates in Chess Studios. In the 1960s, he used to be an engineer at Chess Records in Chicago, when they had the big EMT 140 electro-mechanical reverberation plates. These consisted of spring-mounted steel sheets, about 2 m by 1 m, with an electromagnetic drive unit and two contact pick-ups. Dave described how they used to heat the room where the plates were kept to about 30°C, then tension the plates. When the room cooled down on removal of the heaters, and if they were lucky and the springs did not snap as the steel contracted, the plates produced a unique sound. More often than not, however, the springs did snap, and the whole process would have to be tediously repeated until it succeeded in its aims. Some of the recordings which inspired many people in this industry were the Chess recordings of Bo Diddley and Chuck Berry. This was partly due to their powerful and distinctive music, but partly also for their unusual sounds. Somewhat unusually, for the time, the Rolling Stones went to Chicago specifically to record in the old Chess studios at 2120 South Michigan Avenue, and its unique sounds can still clearly be heard on a number of their older recordings.

The old EMT plates were very inconsistent, and though their German manufacturers guaranteed their performance within a quite respectably tight specification, the sonic differences achievable within that specification were enormous. What Chess were doing would have had the EMT engineers tearing their hair out, but it got Chess the sound which was a part of their fame, so in such circumstances specifications were meaningless. The irony here is that whilst EMT were trying to make widely usable, relatively neutral plates, Chess were doing all they could to give *their* plates a most definitely un-neutral sound. This situation was a perfect parallel to what was happening in the development of live rooms for studios, where the musicians and engineers were looking for something other than what the then current room designers were offering them.

Perhaps it would be appropriate to relate a few more personal experiences at this juncture. I recall in 1978, when building The Townhouse studios in London, desperately trying to buy a specific EMT 140 plate from Manfred Mann's Workhouse studios. I had done some recordings there for Virgin Records, and had been impressed by the sonic character of one of their plates in particular. I offered them a really ludicrously large amount of money for the plate, but short of buying the whole studio, towards the success of which the plate had no doubt contributed, there was no way that Manfred Mann and Mike Hugg would sell it. In fact at one stage, I actually did come close to acquiring on the whole studio, which also had a wonderful sounding API mixing console.

In the Silo studios in West London, in the early 1980s, they had intended to build a large studio room with a small drum room, connected to the studio via a window. The proposed drum room was about 3.5 m × 2 m, and about 2 m high, which in fact was far too small for the purpose. However, this was happening right at the time when the backlash against small, dead, isolated drum rooms was reaching a peak, and once the owners realised that their ideas were out of date they stopped work on the room. When the studio opened, the proposed drum room remained just a concrete shell, with a window and a heavy door. I recorded some of the best electric rock guitar sounds that I have ever recorded with amplifiers turned up loud in that room. This pleased the owners, as their 'white elephant' had become a great guitar room. The density of reflexions when the room was saturated with an overdriven valve (tube) amplifier combo was stunning. The 'power' in the sounds, even when heard at low level in the final mixes, was very impressive, yet the problem with this room was that for almost any other purpose it was a waste of space. Neither this room, nor Manfred Mann's plate, nor the Chess plates, nor the Headley Grange entrance hall would have had any significant place in a large recording studio complex outside of their uses with certain very specific types of music and instrumentation.

In another instance, I was fortunate enough to be one of the engineers on a huge recording of Mahler's Second Symphony for film, television, and what was later to be one of the CBS 'Masterworks' CD series. It was performed by the London Symphony Orchestra, the 300 piece Edinburgh Festival Choir, two female soloists, an organ, and an 'off stage' brass band. It all took place over four days in Ely cathedral, England, which the conductor, Leonard Bernstein, had specifically chosen for its acoustics and general ambience. This was a masterful choice by a genius of a conductor, but despite the wonderful acoustics of the cathedral for orchestral/choral purposes, it would have been entirely unsuitable for any of the other recordings discussed so far in this chapter. They would simply have been a mess if recorded there. Yet, conversely, perhaps there was no studio that could have hoped to achieve a recording with the sound which we captured in the cathedral.

The Kingsway Hall in London was an example of another great room for the recording of many classics, despite its sloping floor and problems from the noises of underground trains, aeroplanes, and occasional heavy traffic. It was an assembly hall, never specifically designed for musical acoustics, yet it far outstripped the performance of any large orchestral studio in London at the time.

Many specifically designed concert halls have been criticised for their failure to live up to their expected performance, though such criticism has not always been fair. Concert halls must cater for a wide range of musical performances, so they tend to some degree to have to be 'jacks of all trades' rather than the 'masters of one', the latter of which is perhaps more the case with Ely Cathedral. To experience Mahler's Second Symphony in Ely, or to have heard some Stockhausen in Walthamstow town hall, would show just how appropriate those recording venues were; but Stockhausen in Ely Cathedral? … Perhaps not!

Unfortunately, though, purpose-built concert halls would be expected to reasonably support both, at least adequately, if not optimally. As yet, one cannot design a general purpose concert hall or recording studio to equal the *specific* performances of some accidentally discovered recording locations for certain individual pieces of music. Neither can one build one live room to equal the performances of Headley Grange, The Silo's concrete room, Manfred Mann's plate, or the one in the old Chess Studios. Brass bands, chamber orchestras, symphony orchestras, choirs, organs, folk music, pan pipes, and a whole range of other instrumentation or musical genre all have their own requirements for optimum acoustic life. The bane about the 'best' live rooms is that they gain their fame by doing one thing exceptionally well, but for much of the rest of the time they lie idle. If one is not careful, they can be something of an under-used investment, which is why many studio owners opt to use whatever limited space they have for rooms of more general usability.

The complexity of the acoustic character of live rooms is almost incomprehensible, and they are often designed on intuition backed up by a great deal of experience, as opposed to any rules of thumb or computers. Just as no computer has yet analysed the sound of a Stradivarius violin and shown us how to build them, full computer analysis of what is relevant in good live rooms is beyond present capabilities: the modelling of surface contours is not possible unless the position and shape of every surface irregularity is known in advance, and in great detail. Designs are thus usually 'trusted' to designers with experience in such things, but in almost every case, some engineer or other will declare the result 'rubbish' because it fails to do what he or she wants it to do, and there is no knob to change the setting. The lack of fixed specification means that the value of such rooms is largely dependent upon taste.

There is also another aspect of recording in live rooms which, whilst unrelated to acoustics, is so fundamental in their use that it should be addressed. Reproducing the exact acoustic of Headley Grange would not guarantee a Led Zeppelin drum sound. There were five other very important factors in that equation. John Bonham, his drums, the other members of the band, the song, and the engineer Andy Johns. Designers are often asked to produce rooms which sound like some given example, but in the case in point, John Bonham was a drummer of legendary power, and he also had the money to afford to buy good drum kits. The engineer, Andy Johns, the brother of another legend, Glyn, had been well taught in recording techniques. He also had the excellent equipment of the Rolling Stones mobile truck at his disposal, and considering his brother's fame, perhaps also had a natural aptitude and a pair of musical ears. He knew where to put microphones, and where *not* to put microphones. He trusted his own ears and decisions, and he had excellent musicians and instruments to record.

Hopefully this point is not being unduly stressed, but a great live room is not a magic potion in its own right. It can enhance the performance, or even, as we have seen, *inspire* the performance of musicians, but *nothing* can substitute for starting off with good musicians, good instruments, good recording engineers and good music. Good recording equipment is also important, but is generally secondary to the previous items. Good and experienced musicians will react to a good room, and play to its strengths. They will not simply sit there playing like robots. The whole recording process is an interactive process, hence the care which must be taken to avoid anything which could make the musicians uncomfortable.

6.3 Drawbacks of the Containment Shells

The question is frequently asked as to why so many great acoustics are found in places not specifically designed to have them, and why so many places which *are* specifically designed are frequently not so special. For example, why is the fluke sound of the entrance hall at Headley Grange so difficult to repeat in a studio design? Well, part of the problem was discussed in Chapter 4. Studios tend necessarily to be built in isolation shells which reflect a lot of low frequency energy back into the room. This reflected energy must then be dealt with by absorption, which in turn takes up a lot of space and generally changes the acoustic character of the rooms to a very great degree. The old Kingsway Hall, in London, was used for so many classical recordings because of its very special acoustic character, but its drawbacks were tolerated in a way which they never would be in a recording studio. Many good takes were ruined by noise and had to be rerecorded, but because it was *not* a recording studio, this was a hazard of the job. Conversely, a purpose-built recording studio with similar problems would be open to so much argument and litigation that it would not be a commercial proposition. Even if it *had* the magic sounds, people would not tolerate such problems in a building actually being marketed as a professional studio; their expectations would be different.

The Kingsway Hall had windows through which many noises entered; as they also did through the floor and general structure. However, these were the selfsame escape routes which allowed much of the unwanted sounds to leave, such as the sounds which would cause unwanted low frequency build ups. Structural resonances could, in turn, change the sound character, and remember, once again, that most instruments were developed for sounding at their best in the normal spaces of their day. Once we build a sound containment shell and a structurally damped room, we have a new set of starting conditions. However, if we are to operate a studio commercially, without disturbance to or from our neighbours, and offer a controlled and reliable set of conditions for the musicians and recording staff, such a containment shell is an absolute necessity.

If we consider again the entrance hall at Headley Grange, it is a relatively lightweight structure which is acoustically coupled via hallways, corridors and lightweight doors into a rabbit warren of other rooms. Effectively, the sound of the hallway is the sound of the whole Grange, and if we were to contain the hallway in an isolation shell and seal all the doors and windows, then we would end up with a room having a very different sound to the one which it actually has. It would be rare, indeed, to find a client for a studio design who would provide a building the size of a mansion house only to end up with a recording space the size of its hallway, yet this is what it would possibly take to create a sonic replica.

6.4 Design Considerations

When designing live rooms it is often necessary to recreate the acoustic of a more usual space within a very unusual shell, and this may in itself require some unusual architecture. The materials which are used to create these internal acoustics are very important in terms of the overall sound character of the rooms. Wood, plaster, concrete, soft stone, hard stone, metal, glass, ceramics and other materials all have their own characteristic sound qualities. Within the range of current response specification, it may be almost impossible to differentiate between the response plots of rooms of different materials in terms of determining from which materials they were made. Yet the ear will almost certainly detect instantly a woody, metallic, or 'stony' sound. In general, all the above materials are suitable for the construction of live rooms, and it is down to the careful choice of the designer to decide which ones are most appropriate for any specific design. The overall sound of the rooms, however, will tend to have the self-evident sound quality associated with each material. Wood is generally warmer sounding than stone, and hard stone is generally brighter sounding than soft stone. Geometry and surface textures also play great parts in the subjective acoustic quality.

6.4.1 Room Character Differences

For this discussion, live rooms should now be split into two groups; reflective rooms and reverberant rooms. The former tend to have short reverberation times, but are characterised by a large number of reflexions which die away quickly. The reverberant rooms tend to have a more diffusive character, with a smooth reverberant tail-off. The reflective 'bright' rooms also often employ relatively flat surfaces, though rarely parallel, and they often contain a considerable amount of absorption to prevent excessive reverberant build up. The reverberant rooms, on the other hand, tend to employ more irregular surfaces and relatively little absorption. It is possible to combine the two techniques, but the tendency here is usually towards rooms which have very strong sonic signatures, and consequently their use becomes more restricted.

The question often seems to be asked as to why flat reflective surfaces in studios usually sound less musical than they do in the rooms of many houses or halls. Notwithstanding the isolation shell problem, studios rarely have space-consuming chimney breasts, staircases, furniture, and other typical domestic characteristics. These things are all very effective in breaking up the regularity of room reflexions, but studio owners usually press designers for every available centimetre of space. As mentioned in Section 5.3.4, it would seem that far too many of them are more interested in selling the studio to their clients on the basis of floor space rather than acoustic performance. (Never mind the sound, ... look at the size!) Perhaps this is to a large degree the fault of the ignorance of the clients as much as that of the studio owners. There is possibly too much belief today in what can be achieved electronically, and the importance of good acoustics is still not appreciated by a very large proportion of studio clients. Of course, those who do know tend to produce better recordings, by virtue of having had the luxury of better starting conditions, and leave the mass market wasting huge amounts of time trying to work out exactly which effects processor program they used in order to get that sound. The answer of course is that good recorded sound usually needs little or no post-processing, unless, that is, the processed sound is the object of the recording.

There are almost no absolute rights or wrongs in terms of live room design, and it seems that whatever a designer provides, there will always be people coming along who have heard 'better' elsewhere. Nevertheless it is also surprising how one famous recording made in a room will suddenly reverse the opinions of many of the previous critics, who will then flock to the studio for the 'magic' sound. Creating good live rooms is like creating instruments, certainly to the extent that the skill, intuition and experience of the designer and constructors tend to mean more than any text-book rules. It is also difficult to give a list of taboos, because almost every time that something could be considered to be absolutely out of the question for wise room design, an example of a well-liked room can be found which apparently flouts the rules. Large live rooms are unusual in that, beyond their existence in a somewhat acoustically controlled form, as, for example, the rooms used for selected orchestral recordings, they are normally to be found outside of purpose-designed recording studios. Their use tends to be too sporadic to allocate such a large amount of space to occasional use. It is usually via mobile recording set-ups that the benefits of these rooms are enjoyed.

6.5 Driving and Collecting the Rooms

Smaller live rooms do at least appear to have one thing in common; recording staff must learn how to get the best out of each one individually. The modal nature of such rooms defies any reasonable analysis: the complexity is incredible. The positions of sound sources within the rooms can have a dramatic effect on determining which modes are driven and which receive less energy. An amplifier facing directly towards a wall will, in all probability, drive some axial modes very strongly. However, precisely to what degree they will be driven will depend upon such things as the distance between the amplifier and the facing wall, and whether the loudspeaker is in a closed box, or is open-backed.

Positioning a closed-back loudspeaker on an anti-node, where the modal pressure is at a peak, will add energy to the mode, and the room will then resonate strongly at the natural frequency of the mode. Placed on a node, where the pressure is minimal, that mode will not be driven, and its normally strong character in the overall room sound will be overpowered by other modes. Positions in between will produce sounds in between. If the cabinet is open-backed, it will act as a doublet (or figure-of-eight) source, radiating backwards and forwards, but very little towards the sides. A closed-back cabinet at lower frequencies, say between 300 or 400 Hz, will radiate omni-directionally and will drive the low frequency axial modes of all three room axes, that is, floor to ceiling, side wall to side wall, and front wall to back wall. With an open back, however, only one set of axial modes will be driven, in the axis along which the loudspeaker is facing. What is more, because the output from the rear of the cabinet is in anti-phase to the frontal radiation, it acts as a pressure-gradient source, and not a volume-velocity source, so it couples to the *velocity* anti-nodes, which are the pressure *nodes*. Open and closed-back cabinets therefore drive the room modes very differently. Figure 6.1 shows the typical radiating pattern of open and closed-back cabinets.

Microphone positioning in such rooms is also critical. An omni-directional microphone placed at a nodal point of a mode will not respond to that mode because it is at the point of minimal pressure variation. Conversely, at an anti-node (a point of peak pressure change), the response may be overpowering. Microphone position can be used not only to balance the relative quantities of direct and reverberant sound, but also to minimise or maximise the effect of some of the room modes. Furthermore, more than one microphone can be used if the desired room sound is only achievable in a position where there is too little direct sound. Therefore, a parallel to the ability of amplifiers to drive a room is the ability of the microphones to collect it. Variable pattern microphones can produce greatly different results when switched between cardioid, figure-of-eight, omni-directional, hyper-cardioid, or whatever other patterns are available. What is more, certain desirable characteristics of a room may *only* be heard to their full effect via certain specific types of microphones. Exactly as with the open and closed-back loudspeakers, omni-directional microphones will respond to pressure anti-nodes in three axes, whereas figure-of-eight (pressure-gradient/velocity) microphones will respond best at pressure *nodes* and in one axis only. Cardioid microphones behave somewhere in between.

If open-backed amplifiers are set at an angle to the walls, they will tend to drive more of the tangential modes, which travel around four of the surfaces of the room. If the amplifiers are then also angled away from the vertical, they will tend to drive the more numerous, but weaker, oblique modes. At least this will tend to be the case for the amplifiers which radiate directionally. Precisely the same principle applies to the directions which the microphones face in respect of their ability to *collect* the characteristic modes of the room. If more than one microphone is used, switching their phase can also have some very interesting effects. Given their positional differences, the exact distance which they are apart will determine

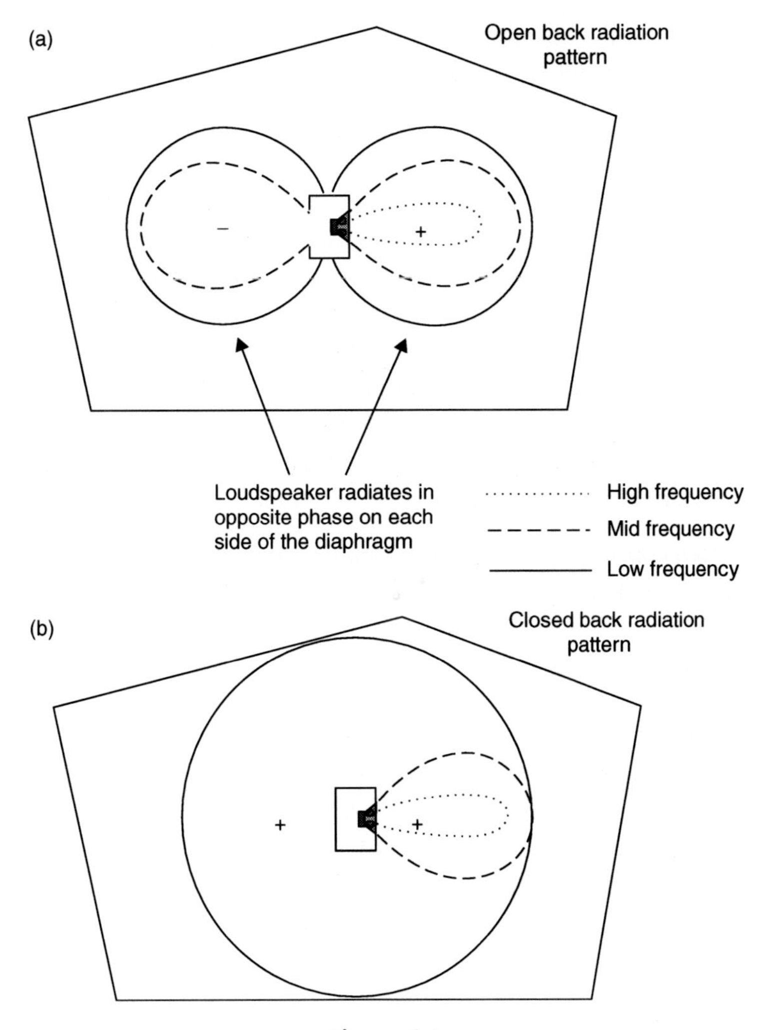

Figure 6.1:
Comparison of open and closed-back loudspeakers in the way that they drive the room modes: (a) High frequencies propagate in a forward direction, but low and mid frequencies largely radiate in a figure of eight pattern; (b) As in the above case, the high frequencies still show a forward directivity. However, in this case, the mid frequencies also radiate only in a largely forward direction, but the low frequency radiation pattern becomes omni-directional.

which frequencies arrive *in* phase and which arrive *out* of phase, but there is no absolute in or out of phase condition here, except at very low frequencies. A pair of congas which sound great in one live room may not respond so well in another. Conversely another set of congas in the first room may also fail to respond as well. But, there again, in another position in the room, perhaps they will. It could all depend on the resonances within the congas and if they matched the room dimensions. The whole thing depends on the distribution of energy within the modes, the harmonic structure of the instruments, and the way the wavelengths relate to the coupling to the room modes.

Live rooms are great things to have as adjuncts to other recording facilities, but they are a dangerous proposal if they form the only available recording space; unless, that is, the studio is specialising in the provision of a specific facility and the recording staff know the characteristics of the room very well indeed. These rooms have grown in popularity as it has become apparent that electronic simulation of many of their desirable characteristics is way beyond present capabilities. They have also been able to provide individual studios with something unique to each one, and this point is of growing importance in an industry where the same electronic effects with the same computer programs are becoming standardised the world over. If a certain live room sound is wanted, it may well bring work to the studio which possesses it, because the option to go elsewhere does not exist. But be warned, they can bite! The design *and* use of these rooms are art forms in their own right, and very specialised ones at that.

Live rooms have been constructed from wood, metal, glass, ceramics, brick, concrete and many other reflective materials. Undoubtedly, the materials of construction affect the timbre of the resulting sound and it is very difficult, if not impossible, to make one material sound like another. Wood is frequently used in live areas, but largely due to the success of certain early rooms, stone rooms have grown in popularity over the past two decades. Stone is unique amongst the other live room materials in that its surface, in a readily available state, is far less regular and hence more diffusive than the other materials. Hard stones sound different to soft stones, just as hard woods sound different to soft woods. No one live room can be all things to all people, but where a specific live room is required, as opposed to a live area *within* a room, stone does seem to be a particular favourite for many. The widespread use of stone rooms is a relatively recent phenomenon and, as with so many other aspects of the recording industry, its origin and acceptance can be traced back through some quite unpredictable chains of events.

6.6 Evolution of Stone Rooms

Twenty-four track tape recorders became generally available in early 1973, bringing what was an unprecedented luxury of being able to record drum kits by using multiple microphones on a one microphone to one-track basis. Previously, anything more than two or three microphones and one track of the tape recorder for a drum kit was seen as either wanton extravagance or the actions of an ostentatious 'prima donna' recording engineer. Almost inevitably, the one-microphone-to-one-track recording technique led to experimentation in the recording of drum kits.

Equalisation of the individual drums in the kit was a novelty that was much pursued, but the desired signal processing was seen to be being made less simple or effective by overspill from adjacent drums in the kit picking up on the other microphones. The answer seemed to lie in increased acoustic separation, which led in turn to studio designers being asked to produce isolation booths which not only separated the drum kit from the rest of the instrumentation, but also enabled a new degree of separation between the individual drums within a kit. The mid-1970s witnessed the industry experiencing the aberration of the use of very dead drum booths, which were later to be largely relegated to use for the storing of flight cases or microphone stands.

The highly damped drum booths had enabled the period of experimentation with separation to run its course. Achieving separation was undoubtedly facilitated by the use of these booths, but unfortunately they posed two major problems. Firstly, once the novelty of individually equalised and processed kits had begun to wear off, the essential ambient 'glue' which held a kit together was conspicuous by its absence. Secondly, and probably even more importantly, most drummers did not enjoy playing in these booths. The drums, themselves, seemed unresponsive to the sticks. An uncomfortable drummer can rarely produce an inspired performance, and it soon became apparent that the human requirements of the drummers needed far more attention paying to them than had previously been allocated. Drummers soon began moving back into the main studio areas; often reverting to the older practice of being shielded behind acoustic screens, but the re-recognition of the importance of the ambient 'glue' was beginning to signal the end of the dead isolation booths for the drums. The problems of separation causing the integrity of the kit to disintegrate became even more apparent with the advent of digital recording.

The arrival of digital reverberators and room simulators, especially the programmable variety, seemed initially to many people to sound the death knell for the live rooms, but as the strengths and weaknesses of the acoustic *and* electronic approaches began to become more widely appreciated, it was soon understood that each had their place. By the mid-1980s, the live rooms had returned with a vengeance. The stone room in Townhouse Studio Two is shown in Figure 6.2, where Phil Collins recorded the drums on his classic *In the Air Tonight.* This one recording spawned a whole generation of stone rooms, though after 22 years of use the Townhouse room was finally demolished to make room for a five-channel surround mixing room. Ultimately, more money could be made from mixing in the space than from recording in it. This again highlights the commercial pressures on such specialised spaces. In fact, not long after the conversion to a mixing room was completed, the whole studio complex closed because the building became worth more money as a residential development.

It is worth noting here that what directly led to the construction of the stone room in The Townhouse was that the first two recording engineers scheduled to work there, *and* its design co-ordinator (the author), had all had a lot of experience with mobile recordings. Many were made in old castles and country houses, especially for television programmes, and the crews were bringing the recordings back to the studios for mixing. 'How do you get these sounds?' was

Figure 6.2:
Townhouse Two, London (1978).

a common question from the staff of the studios, recognising that they were hearing things which were rarely encountered in studio recordings. Many people were beginning to recognise that a lot of interesting sounds could be recorded in spaces which were not purpose-designed studios.

It was thus not difficult, with the two senior engineers and the technical director all in favour of the idea, to persuade Richard Branson that Virgin should (as was quite normal for the company, anyway) take a bold step away from traditional studio design and put part of an old castle into The Townhouse. In fact, Virgin's Manor Studio had a considerable amount of exposed stone in its main studio area after its rebuild in 1975, so Virgin already had a half-precedent. However, the stone room in The Townhouse was a bit of a shock to some recording staff when it was first built. For the first few months it was seen as a white elephant by many who tried to use it, because circumstances had led to the three main supporters of its construction working principally in Studio One, and not Studio Two where the stone room was sited. Finally, though, it only took one single highly successful recording to turn the room from a white elephant into a legend. It is interesting to contemplate what may have been the future of the room had Phil Collins *not* chosen to record his *Face Value* album in that particular studio. The life of the room could have been only a matter of months, instead of decades. Survival can hang on delicate threads.

6.6.1 Construction Options

One thing which most certainly *can* be said of stone rooms is that they are all different. In these days of preprogrammed instruments and factory set effects programs, stone rooms add an extra degree of variation. In effect, each studio owner has a genuine 'first edition', something unique, which, as experience of its performance is gained, can supply a sound unattainable elsewhere. As well as sizes and shapes, the type of stone can also be varied. The early rooms designed by the author were Oxfordshire sandstone, which was a little on the soft side and slightly crumbly. Consequently, a thin coat of polyurethane varnish was applied to reduce the dust problem. Subsequently, Purbeck and York stones were used, and later still, Spanish and Portuguese granites.

Granites allow a greater degree of variation in acoustics, and, being much harder, they have fewer tendencies to shed dust. Once a PVA adhesive has been added to bind the cement, the choice of varnishing the stone or leaving it bare introduces an interestingly different variable. Varnish noticeably softens the sound when compared to the natural granite, whereas with the softer stones, the varnish treatment is less readily noticeable. The high concentration of PVA adhesive in the cement, together with expanded metal backings to hold everything securely, allows the cement between the stones to be cut back, and thus exposes in a great deal of relief the outline of each individual stone. This technique was developed for the large drum room in Blackwing, London, initially as a cosmetic measure to produce a castle-like atmosphere, but its acoustic advantages also soon become clearly understood. The deep crevices between each stone gave a much more diffuse sound field, especially as the stones in that particular studio were laid in a highly random manner; again initially for cosmetic effect. The room is shown in Figure 6.3.

One major problem which was solved in the design of the first Blackwing room was how to stop the low frequency build up, which had previously been problematical in rooms of that size and over. Blackwing was a large room by normal standards, 24 ft × 16 ft × 10 ft (8 m × 5 m × 3 m) after the internal finishes. The shell of the room was much higher, as the space had previously been used as a rehearsal room, with a mineral wool suspended ceiling. There was also a large amount of mineral wool, a metre deep, over this ceiling, and the aversion of everybody concerned towards being rained-on by such unpleasant substances during their removal concentrated minds wonderfully in the pursuit of alternative solutions. It was eventually elected to leave the ceiling in position, especially as, unusually for such ceilings, it sloped. A coat of bonding plaster, roughly applied, took advantage of that slope to produce a non-parallel, reflecting surface, opposing the concrete floor, with the roughness of the plaster helping the high frequency scattering.

The intention was to reflect mid and high frequencies back into the room whilst allowing the low frequencies to penetrate. For the low frequencies to be reflected back into the room they would have to penetrate the ceiling, suffer absorption by the metre or so of mineral wool

Figure 6.3:
Splendid (Blackwing), London (1988).

overlay, reflect from the structural and isolation ceilings and return through the same obstacle course. Obviously, such a path would introduce severe attenuation of the low frequencies, so the ceiling would provide an escape path for them, and act as a high pass filter (from the point of view of the room response) in one of the three main axes of the room (the vertical axis). During the construction of this room there occurred a situation which is worth relating, because it sheds light on a rather important aspect of live room design in general.

On the day that the room was completed in all respects other than the plastering of the ceiling, it brought general disapproval from a number of people who had great expectations of its reverberation. The room with its concrete floor and granite walls was certainly bright, but no obvious reverberation existed. The studio owner had asked for a reverberant room that could double as an 'echo chamber'. After the plasterer had done his work there was little change, nevertheless as the plaster slowly dried, little by little, the room came to life. The next day, with the plaster fully hardened and dry, the room delivered all that had been promised. This point will be returned to in Section 6.8, containing discussions about an empirically derived '20% rule'.

6.7 Live versus Electronic Reverberation

Each stone room is unique, so it is difficult to say if one is better than another, or not. The knack of using them is to play to their individual strengths and to avoid their individual weaknesses. They add a degree of uniqueness to a studio which is simply not yet available with electronic devices; programmable, or not. There are some subliminal reasons for this, but there are also some very hard engineering reasons. Although low level effects in the tails of digital reverberator responses are generally very low indeed, we do often seem to detect them by their absence in natural reverberation tails. Some aspects of these decay differences have been difficult to measure, as we do not have analytical equipment even approaching the discriminative ability of the human ear/brain combination. What is more, many of the arguments about just what is, or is not, audible has been based on research into hearing thresholds relating to language and intelligibility. There have been many cases of people who have received accidental injuries resulting in severe impairment of their ability to communicate verbally, yet their appreciation of music has been unimpaired, implying that the areas of the brain responsible for the perception of speech and music are quite separate. Much more research is still required into these differences; however, there is no integrated signal reaching the brain which resembles an analogue of the eardrum motion. The ear presents the brain with many component factors of the 'sound', and it is only by way of a massive signal processing exercise by the brain that we hear what we hear.

It is the degree of these subtleties which still confound the manufacturers of digital reverberators. The late Michael Gerzon had studied in detail the then current state of electronic reverberators in the early 1990s. Michael was co-inventor of the SoundField microphone and the main developer of the Ambisonics surround sound system, and had much experience in the world of sound-field perception. He was also the first to propose dither noise shaping for digital audio, and was co-inventor of the Meridian Lossless Packing Audio data compression system, so his experience in the realms of both digital audio and spacial recording and perception were considerable. He said that the then state-of-the-art digital reverberation units, in electrical terms, represented something in the order of a *ten thousand* pole filter. The complexity of the inter-reaction of the sound field within a moderately sized live room, and the ability to simulate the directional and positional aspects of the microphones and loudspeakers, would need to be simulated by something in the order of a *one hundred thousand million* pole filter. Even if the current rate of acceleration of electronic development were to be maintained, it would be 40 years or so before a room could truly be simulated electronically; and even then, at what cost, and with what further restrictions?

A room simulator may well go a long way towards reconstructing the reflexion patterns for a sound emanating from *one point* in that phantom 'room'. However, the nub of the issue is that in real life, a band, or even a drum kit, does not inhabit one point in space in the room. Different instruments, or different parts of *one* instrument, occupy different spaces in the room. Sounds are generated from very many different positions in the room, some at nodes, some at anti-nodes, and others at many points in between. All excite different room resonances to differing degrees, and all produce reflexions in different directions. This subject was discussed in a previous section, and depicted in Figure 6.1. For example, the room behaves differently towards the snare drum than it does to a floor tom in the same kit. With current digital reverberation, all the instruments, and indeed all the individual *parts* of all the instruments, are injected into the phantom 'room' from, at most, only a few points in the theoretical space. All injections into the same space are driving a similar series of resonances; all are equally distanced from the rooms' nodes and anti-nodes. Such occurrences do not exist in nature. What is more, in real rooms the microphones are also in different positions, so they all 'hear' the room in a different way.

Acoustic reverberation chambers, when driven by a large stereo pair of loudspeakers, can overcome this problem to some degree, but what any 'after-the-event' processing system lacks is the interaction between say, a drum kit, a live room in which it may be played, *and* a drummer. The drums excite the room, and the room resonances, in turn, interact with and modify the resonances of the kit. These processes undertake reiteration until energy levels fall below perceivable thresholds. The instrument, the room, and the musician are inextricably linked; they behave as one complex instrument. Physical separation of the playing and the addition of the reverberation break this very necessary unification. The room resonances modify the *feel* of a drum kit to the drummer, and the drummer will also perceive the room effects via any headphones, bone conduction, and general tactile sensations. The room will modify the musicians' performance; and this *cannot* be accomplished after the event. Performances are unique events in time. It was on these grounds that the drummers rebelled against the mid-1970s, dead, high-separation drum booths which were then in vogue. It seems probable that electronic simulation will never have an answer for the human performance interaction problem, as no subsequently applied artificial reverberation can acoustically feed its effects back into the feel of the instruments themselves. Only artificial reverberation in the room itself, at the time of playing, could achieve that, which is one direction that the future may exploit.

Building stone rooms, or live rooms in general, is a very long way from the acoustic discipline of control room design. Control room design usually seeks neutrality in which the sound of the room is perceived to as small an extent as possible. For a control room to add any characteristic sound of its own is greatly frowned upon. Conversely, if a live room *sounds* good, then it *is* good. It is possible to walk away from a completed stone room with a degree of satisfaction, pleasure, excitement, and a sense of achievement in having created something different. Their only drawbacks would appear to be that they take up more space than a digital room simulation unit, they cannot be taken from studio to studio, and they cannot readily be traded-in or sold-off. People seem to expect to have all of their equipment encapsulated in boxes these days. Nevertheless, from this point of view, the stone rooms *can* comply … other than for the fact that they are not rack-mountable, they are somewhat large, and they tend to weigh in the order of twenty tons!

An important point to be made here is that natural reverberation and artificial reverberation are two distinct things. They each have their uses, and sometimes either will suffice, but neither one totally replaces the other. Whilst sounds can be obtained from stone rooms that no artificial reverberator can match, it is equally true that electronic reverberators can produce some wonderful sounds that no live room could give.

6.8 The 20% Rule

The story of the ceiling at Blackwing, in London, in Section 6.6.1 highlights a point of general significance in terms of the percentages of room surfaces which are needed to create any significant effect. With the mineral wool ceiling it was difficult for the room to achieve any reverberation, as almost all of the energy in the oblique modes, which pass in a chain around all the surfaces of the room, would be absorbed upon coming into contact with the ceiling, and could thus never become resonant. In fact, even the energy in the tangential and axial modes would gradually expand into the ceiling. In the 8 m×5 m×3 m room, the total surface area is about 160 m^2. The ceiling (8 m × 5 m) has an area of 40 m^2, which is about 25% of the total.

A reverberant room generally needs to have reflective material on all of its surfaces, and usually, only about 20% of the total surface area needs to be made absorbent to effectively kill the reverberation. On the other hand, in a very *dead* room, introducing about 20% of *reflective* surfaces will usually begin to bring the room to life. Equally, a room with troublesome modal problems will usually experience a significant reduction of those problems by the covering of about 20% of its total surface area with diffusers. If one wall creates a problem, the covering of about 20% of that wall with diffusers will usually render the wall more neutral, but the diffusion should be reasonably evenly distributed over the surface to be treated.

At Blackwing it was absolutely fascinating to listen to the plaster dry, or rather to listen to the effect of its hardening on the room acoustics. The wet plaster was not very reflective, so upon initial application it did not significantly change the room, but once the hardening process began, after a few hours it was possible to witness an empty room changing very noticeably in its character, in a way which was almost unique. Without any physical disturbance or any abrupt changes, the room was 'morphing' from a bright, reflective room, to a highly reverberant one. The luxury of such experiences can provide much insight into the acoustical characteristics of rooms of this nature.

6.9 Reverberant Rooms and Bright Rooms – Reflexion and Diffusion

The terms brightness and reverberation often tend to get confused by the loose use of the term 'live room' when studio acoustics are discussed with many recording personnel. Stone rooms can be produced to either bright or reverberant specifications, and the flexibility is such that rooms looking very similar may sound very different. Figures 6.4, 6.5, 6.6 and 6.7 show rooms of very different acoustic characters. All four are built with Iberian granite, and an explanation of the different construction techniques used in each case will help to give an understanding of the respective processes at work.

Figure 6.4:
Planta Sonica, Vigo, Spain (1987).

Figure 6.5:
Discossete 3, Lisbon, Portugal (1991).

Figure 6.6:
Regiestudio, Amadora, Portugal (1992).

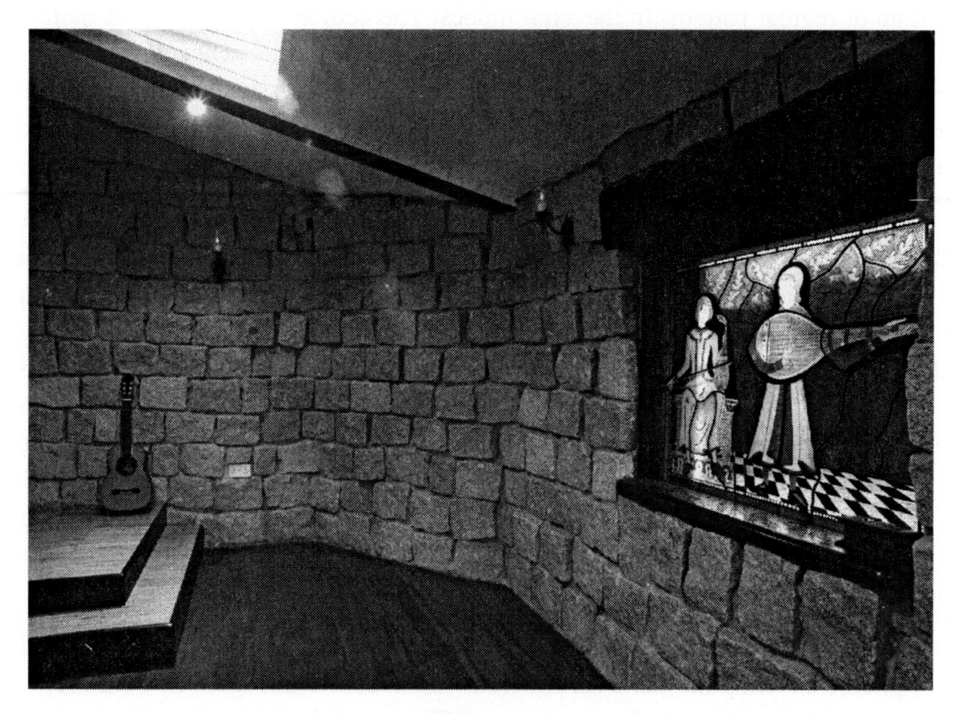

Figure 6.7:
Shambles, Marlow, England (1989).

Clearly, the reverberation which we are referring to in these rooms is not reverberation in its true acoustic sense. Reverberation refers to a totally diffuse sound-field, where its intensity and character is the same throughout the room. This cannot occur in small rooms as the existence of modal resonances and discrete reflexions will always ensure that places of different character can be found within the room. Even the absolute decay time of all reflective and resonant energy can be position dependent, but in general, the term reverberation as understood amongst most recording engineers is perhaps the most widely understood term which can be used in these descriptions. Certainly in the names of many programs in digital effects processors the term 'reverb' is now so universal that, except in academic acoustic circles, it would be like trying to swim up a waterfall to be too pedantic about its accurate use at this stage of the development of the recording industry. Bearing the terminological inexactitude in mind, let us look at some different 'reverberant' room designs.

In Figure 6.4, the room is about 5 m×4 m×3 m high, and is built using a facing of granite blocks, about 10 cm in thickness. The stones are bonded by cement to a studwork wall, which is covered in various sheet materials such as plasterboard, chipboard and insulation board. The blocks have reasonably irregular surfaces, but are all laid flat against the wall. After the cement behind and between the block had dried, the gaps were pointed with a trowel, to smooth over the cement and bring it more or less level with the face of the blocks. The resulting wall is hard and relatively flat, but the irregularities are sufficiently large to be somewhat diffusive at frequencies above about 3 kHz. None of the walls are parallel, which helps to avoid the build up of regular patterns in the axial modes. The ceiling is heavy, and of quite a solid structure behind the plaster. A fabric panel at one end of the ceiling covers the entrance to a low frequency absorber system, which helps to reduce excessive low frequency build up. The room has two sliding glass doors, each about 1.8 m wide and 2 m high, one leading to the control room, and the other to the main recording space.

The above room, at the former Planta Sonica studio in Vigo, Spain, produced some excellent recordings of drums, electric guitars, acoustic instruments, and especially the traditional Celtic bagpipes which are very popular in Galicia (the Celtic influenced province of Spain where Vigo is situated). The room decay was smooth, without excessive low frequencies and spacially very rich sounding. The empty reverberation time was about 3 s, but of course the empty state is not really relevant in such cases, because it would not be used empty, except perhaps as a reverberation chamber during mixdown. In such cases, a loudspeaker/amplifier would be fed from the mixing console, and a stereo (usually) pair of microphones would pick up the room sound for addition to the mix. However, in normal use, the influx of people and equipment can have a great impact on the empty performance, as they tend to be absorbent and can occupy a significant portion of the total room volume.

The rooms in Figures 6.5 and 6.6 are built in a very similar manner to each other. Both have layers of granite covering the same sort of stud wall structure of Figure 6.4, but they are built more on the lines of Blackwing, as shown in Figure 6.3. Both also have the granite blocks

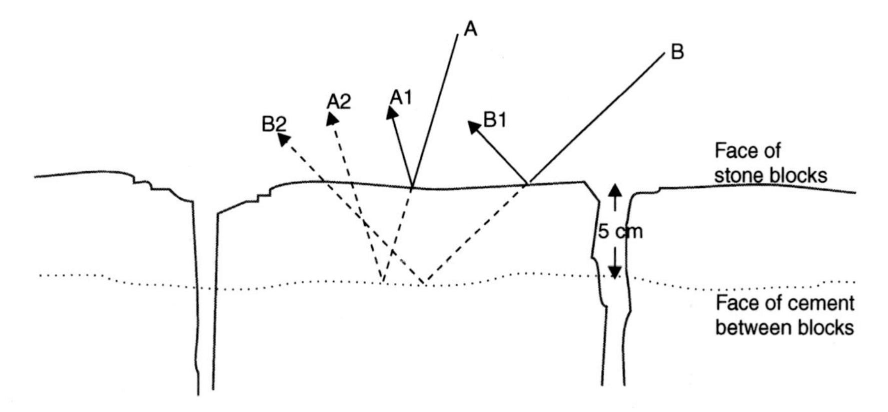

Figure 6.8:
Effect of incidence angle on the reflexions from the pits and blocks. From a nominal depth of cavity of 5 cm, incident wave 'A' will travel about 10 cm further, before returning to the room, when it reflects from the back of the pits, as compared to when it reflects from the face of the blocks. Incident wave 'B', striking at a shallower angle, will show an even greater path length difference between the reflexions from the back of the pits and the reflexions from the faces of the blocks.

laid in a more three-dimensional configuration, with many of the blocks protruding from the walls. They have somewhat similar types of ceiling structures, but are much smaller in size. The room shown in Figure 6.5 has a total surface area of about 90 m^2, and the Figure 6.6 room about 60 m^2. The room shown in Figure 6.6 also differs in being made from granite blocks having only a quarter of the surface area of those in the other rooms.

The room in Figure 6.5 is a strongly reverberant room, but lacks the powerful reflexion patterns of the old Planta Sonica room shown in Figure 6.4. The deep cutting back of the cement between the blocks creates a series of randomly sized pits which render the surface much more diffusive, a property also enhanced by the protrusion of many of the blocks. Due to the non-parallel nature of the surfaces of the room, most of the energy will be concentrated in the tangential and oblique modes, thus the depth of the pits and protrusions is effectively increased because the modal incidence will be at varying angles to the wall surfaces, as shown diagramatically in Figure 6.8. The effect is that the walls are diffusive down to much lower frequencies than the walls of the room in Figure 6.4. There is therefore more diffusive energy in the decay tail than in the room of Figure 6.4. The more diffusive room does not produce the same haunting wail of bagpipes as the other rooms, but it can produce some very powerful sounds from congas, and rich enhancement of saxophones or woodwind. These two types of rooms are not really interchangeable; they are sonically very different.

6.9.1 Bright Rooms

Let us now move on to consider the room in Figure 6.6. This is a room with a total surface area of about 60 m^2, about 4 m^2 of which are non-parallel glass surfaces. There is also 2 m^2 of flat, wood panelled door, and around 8 to 10 m^2 each of wooden floor and sloping, highly irregular, plastered ceiling. The remaining surfaces of the walls, which form the majority of the surface of the room, are of small granite blocks, each having a face area of 80 to 100 cm^2. The cement has been deeply cut back in the gaps, producing a series of irregular cavities, but unlike the rooms in Figures 6.2 to 6.5, the ratio of the surface area of the granite to the surface area of the pits is much less; in fact only around 20% of that in the other rooms.

The effect of this is to produce far fewer specular reflexions because there are fewer flat surfaces, at least not acoustically so, until much lower frequencies. Any wave striking the wall surface will be reflected differently from the stones and pits, but let us consider the case of what happens at 500 Hz, where the wavelength is around 60 cm. If the pits average 8cm in depth, then a reflexion travelling from the face of the stones will travel about 16 cm less than when entering and being reflected from the back of a pit. This will create about a quarter-wavelength phase shift on a single bounce, and the irregular shape of the stones and pits will tend to scatter the wavefront as it reflects from the wall. At least down to around 500 Hz, such a room surface becomes very diffusive until specular reflexions begin again when the dimensions of the stone faces become proportionate to wavelength, say above 5 kHz. However, at these frequencies and above, the irregularity of the stone surfaces themselves begin to become diffusive, so the walls do begin to scatter effectively from around 500 Hz upwards. There will also be a considerable degree of diffraction from the edges of the stone blocks, which will also add to the diffusion.

The room in Figure 6.6 is very bright, emphasising well the harmonics of plucked string instruments, and adding richness to flutes and woodwind. Somewhat surprisingly perhaps, the reverberation time is much shorter than one would tend to expect from looking at the photograph. In this type of room, which is also very small, the energy passes rapidly from surface to surface. As the surfaces are so diffusive, the scattering of the modal energy is very wide. Thus, in such a room, the number of times that a sound wave will strike a wall surface as it travels around is many times greater than would be the case in a room such as the one shown in Figure 6.3. Each impact with a wall surface, especially at a near grazing angle as opposed to a 90° impact, will take energy from the reflected wave, either by absorption (including low frequency transmission), or by the energy losses due to the interaction of the diffusive elements. Consequently, in two rooms of any given surface material and construction, the small room will have the lower reverberation time because more surface contacts will take place in any given period of time. The smaller room will also have a higher initial reflexion density. The room in Figure 6.6 produces a brightness and thickness to the recorded sound, but it falls off within about one second.

Figure 6.7 shows a further variation on the theme, and is a room with a character somewhere between the ones shown in Figures 6.4 to 6.6. It is the only recording space of a small studio, so it has to be slightly more flexible than the 'specific' rooms shown in the other photographs. When empty, it has a reverberation time of just over 2 s, but this can readily be reduced by the insertion of lightweight absorbent 'pillars' containing a fibrous filling, especially when they are positioned some way out from the corners. As was discussed in Chapter 4, fibrous absorbers are velocity dependent, so they should not be placed too near to a reflective surface or their effect will be reduced.

From the descriptions of the rooms shown in Figures 6.2 to 6.7 it can be appreciated that the cutting back of the cement to form pits between the stone blocks both increases the diffusion and lowers the decay time. Figures 6.8 and 6.9 will help to show the mechanisms which create these effects. Figure 6.8 shows the way in which the reflexion paths differ from the front of the blocks and the back of the pits. It can also be seen that for angles of incidence other than 90°,

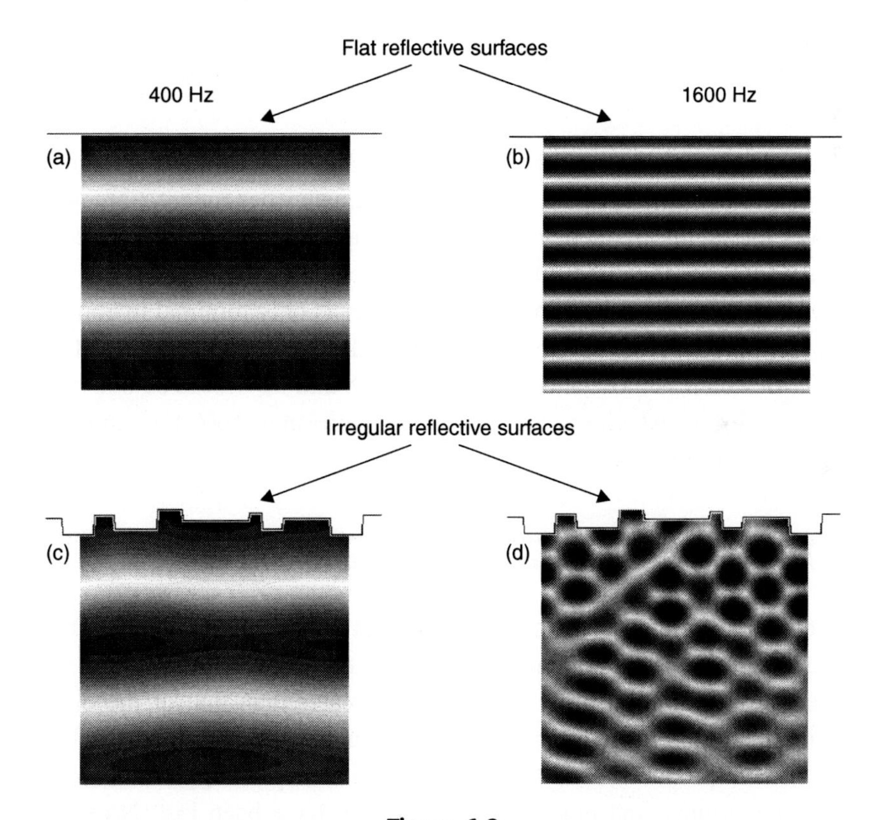

Figure 6.9:
Effect of surface irregularities on reflexion patterns. Figures (a) and (b) show the interference patterns of 400 Hz and 1600 Hz plane waves when reflecting from a flat surface. Figures (c) and (d) are again 400 Hz and 1600 Hz plane waves, but this time the interference fields are those produced by reflexions from an irregular surface, such as a rough stone wall. In (c) the 400 Hz pattern is little changed from (a), but at higher frequencies, as shown in (d), where the size of the irregularities becomes significant in proportion to the wavelength, the reflexion pattern becomes fragmented, and loses energy more rapidly after reflexion.

the disruptive effect will be greater as the path length differences increase. The frequency dependence of the effect of the irregularities is shown in Figure 6.9. At 400 Hz there is little difference in the interference patterns created when a plane wave strikes either the flat surface or the irregular surface at a 90° incidence angle. At 1600 Hz, however, the reflected wave is well broken by the 10 cm depth of surface irregularity. It can also be seen that the energy in the pattern reduces significantly more rapidly further away from the wall, which is a result of the diffusive effect.

In normal situations the effect is even more pronounced than that shown in Figure 6.9, because the rooms which employ such techniques usually have non-parallel surfaces. This tends to cause more of the sound waves to strike the wall surfaces at angles of incidence other than 90°, where the path length differences caused by the irregularities will be greater. The two primary effects of this are that the disturbance of the interference field will extend to lower frequencies and the energy losses after reflexion and diffraction will be greater.

Section 5.4 looked at the question 'What is parallel?' and showed that the degree to which a pair of surfaces was *acoustically* parallel was very frequency dependent. Somewhat similarly, Figure 6.9 shows the highly frequency-dependent nature of the effect of surface irregularities. At 1600 Hz, the effect of the surface irregularities can clearly be seen (and heard). At 400 Hz, the effect of the surface irregularities is only minimal in comparison to the interference pattern produced by an absolutely flat wall. Down at 50 Hz the effect of the surface irregularities such as those shown in Figures 6.8 and 6.9 would be non-existent. So, in acoustical terms, a surface which can be highly non-uniform at 1600 Hz can be seen to be absolutely regular at 50 Hz.

6.10 Low Frequency Considerations in Live Rooms

As discussed earlier in reference to the large room at Blackwing (Figure 6.3), all 24 m^2 of the ceiling was used as a low frequency absorber. Yet even with this amount of absorption the low frequency reverberation time is still much greater than that of the other rooms mentioned, principally due to its larger size. (See Equation 4.7.) Without such an absorber, the room would have produced a build up of low frequency energy which would have muddied all the recordings, and much definition would have been lost. No such absorption was needed in the room shown in Figure 6.6 because a similar low frequency build up cannot develop in rooms of such small dimensions. The modal path-lengths are too short to support long wavelength resonances. We shall consider this point further in the following section.

Figure 6.10 shows the way in which the characters of rooms are controlled by the different acoustical properties of their dimensions and surfaces. At the high frequencies, the room

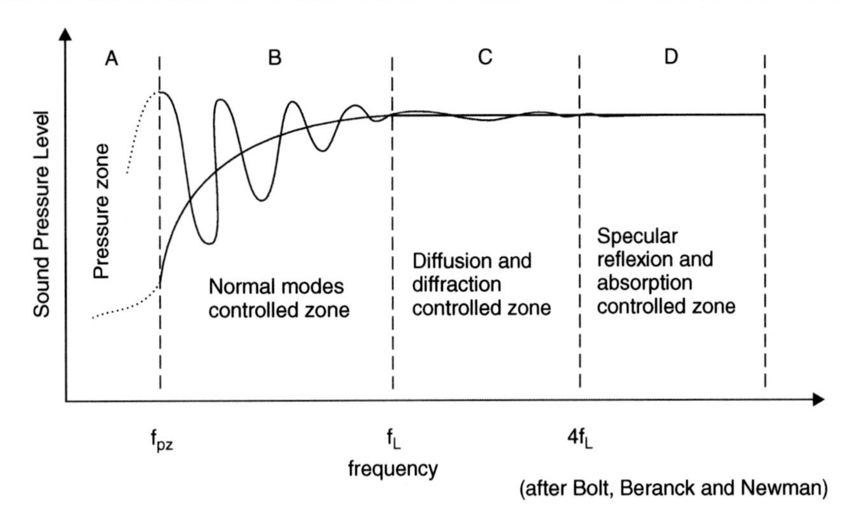

Figure 6.10:
Controllers for steady state room acoustics. The above diagram shows the frequency ranges over which different aspects of room acoustics are the predominant controllers. Explanations of f_{pz} and f_L are given in the text. In the pressure zone, the response of a loudspeaker is not supported by room effects, so loudspeakers driving the pressure zone may need to be capable of considerable low frequency output.

response is generally controlled by the relationship of specular reflexions to absorption. Effectively, here, the sound can be considered to travel in rays, like beams of light. In the mid frequency band, control is mainly down to the diffusion and diffraction created by the irregularities and edges of the surfaces. At the low and lower mid frequencies, the room response is mainly that which can be reinforced by modal energy. This region is usually dealt with in terms of normal wave acoustics, but as the room size reduces, it tends to produce a greater spacing of modal frequencies, hence the energy is concentrated in more clearly defined frequency bands, which leads to a more 'coloured' or resonant sound characteristic. As the room dimensions are reduced, the lowest modal frequencies which can be supported are driven upwards, so excessive low frequency build up becomes less likely.

Small rooms tend to sound 'boxy' because the modal energy is in a higher frequency range, reminiscent of the sound character of a large box, hence the 'boxy' sound. The modally controlled frequency region is bounded in its upper range by a limit known as the large-room frequency, and at its lower range by the pressure zone, as previously discussed in Section 5.8.3. (The pressure zone is also discussed further in Section 13.5.) In Figure 6.11(a), a sound wave can be represented by a line crossing the room. This is a snapshot in time, and shows the positions of high and low pressure for that instant. For a travelling wave (i.e. *not* a resonant modal path) another snapshot taken a few moments

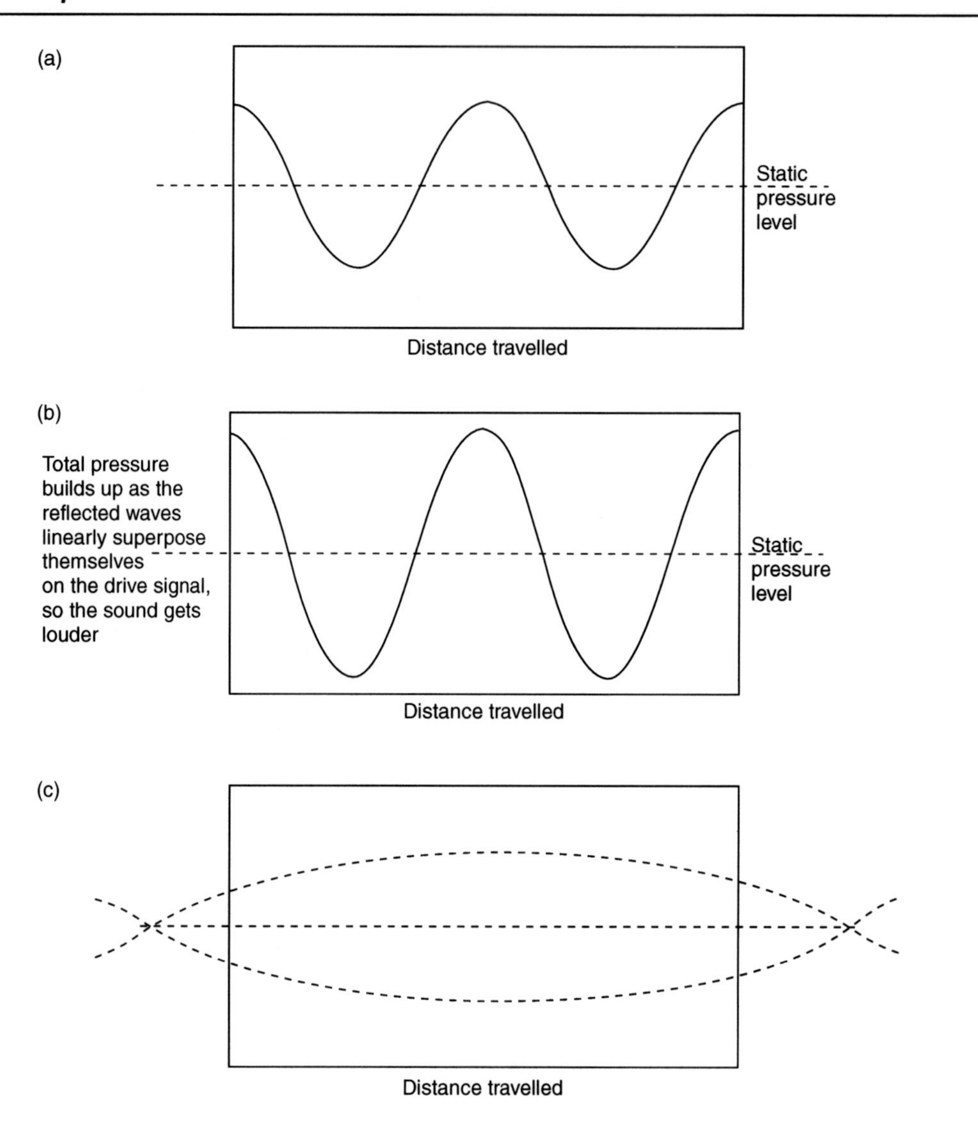
(a)
Static
pressure
level
Distance travelled
(b)
Total pressure
builds up as the
reflected waves
linearly superpose
themselves
on the drive signal,
so the sound gets
louder
Static
pressure
level
Distance travelled
(c)
Distance travelled

Figure 6.11:
The pressure zone: (a) A wave traversing a room will produce areas of high and low pressure that correspond to the compression and rarefaction half-cycles of the wave. Upon striking the opposing wall, it will reflect back, and, if the path length is an integral multiple of the wavelength, the returning reflexions will be exactly in phase with the drive signal; (b) As the drive signal continues, it will add in-phase energy to the resonant energy, which will build up the total energy in the resonant mode. This is similar to the effect of adding energy at the right moment to a child on a swing, when the arc of swing can be made to increase with little effort; (c) Where the wavelength is longer than twice the size of the room, the whole room will be rising or falling in pressure, more or less simultaneously. This is the pressure zone condition, where alternate regions of high and low pressure are not evident. No resonances can therefore occur in this region.

later would show the peaks and troughs uniformly shifted in the direction of travel. At resonance, however, the interference pattern of the direct and reflected waves becomes fixed in space. A resonance occurs when the distance between two opposing, reflective surfaces bears an exact relationship to the wavelength of the resonant frequency, and a stationary wave results.

The resonant modal pathway tends to trap the energy, with the peaks and troughs of the direct and reflected waves exactly coinciding with each other. The energy build up in such circumstances can be very great. The effect is analogous to a child on a swing. If the start of the energy input to the swing (a push) is timed to coincide with the peak of its travel, the swing can build up great oscillations with what appear to be only minimal inputs of applied force. (See Figure 6.11(b).) However, Figure 6.11(c) shows that when wavelengths become sufficiently long, there are no noticeable alternate regions of the room with above and below average pressure, but rather the whole volume of the room is simultaneously either rising or falling in pressure. The room response is thus unable to boost the initial sound wave by adding any resonant energy, because it cannot provide any resonant pathways, and the sound thus decays rapidly. This is one reason why the low frequency reverberation response of rooms falls as room size reduces, as well as, of course, the energy losses due to the increased number of wall impacts per second. Conversely, in large halls, where the pressure zone frequency is in the infrasonic region, even the lowest audible frequencies can be reinforced by modal energy. These frequencies are notoriously difficult to absorb, and are not significantly lost through doors or windows unless these are proportional to their wavelengths. They also largely refuse to be thrown from their paths by obstructions, unless they are of sizes in the region of a quarter of a wavelength of the frequencies in question. In the smaller rooms, however, when the pressure zone frequency rises into the audible spectrum, added to the increased losses due to the greater number of wall strikes per unit of time, it is therefore to be expected that the low frequency reverberation time will tend to fall off more rapidly than in larger rooms.

The large-room frequency, bounding the frequency region dominated by normal modal energy and an upper frequency range which is mainly diffusion/diffraction controlled, can be estimated from the simple equation given by Schroeder:

$$F_L = K\sqrt{RT_{60}}\,/V \qquad (6.1)$$

where:

F_L = large-room frequency (Hz)
K = constant: 2000 (SI)
V = room volume (m^3)
RT_{60} = apparent reverberation time for 60 dB decay (seconds)

6.11 General Comments on Live Rooms

The rooms depicted in Figures 6.3–6.7 all have very different sonic characteristics which are derived from their physical sizes, shapes, and the nature of the arrangement of their surface materials. This is despite the fact that, except for their floor surfaces, they are basically all constructed from the same materials. In these cases, the floor material was of relatively little overall significance, although stone floors, perhaps, would add just a little extra brightness. Rooms such as the ones described here, as with all live rooms, take some time to get to know, but once known they can be very acoustically productive. They also take a great deal of experience to design and construct. Bad ones can be really useless, or of such restricted use that they are more of a liability than an asset. There is no current hope of computer modelling these rooms, partly because the complexities of the interactions are enormous, and also because the influence of complex room shapes on the acoustical virtues of the subjective perception are not yet sufficiently understood to program them into a computer. The influence of the great variety of equipment and people who may 'invade' the acoustic space is also beyond current modelling capability. However, for engineers wishing to record in such rooms, hopefully this discussion will have provided enough insight to make the process more productive.

6.12 Orchestral Rooms

Orchestral music was designed to be performed live in front of an audience. When many of the great classical works were written there was no such thing as recording, so the instrumentation and structure of the music was aimed only at its performance in spaces with audiences. Transferring the performance into a studio, perhaps of only just sufficient size to fit the whole orchestra, imposes a completely different set of conditions. As was discussed in Section 6.3, a big constraint on the achievement of a natural orchestral sound in studios is the fact that the whole process is usually entombed in a massive acoustic containment shell. This even further removes the ambience of the studios from that of the concert halls.

Today, much orchestral recording is for the soundtracks of films, and in such circumstances, when the conductors need to see the films as well as to be in close contact with the musicians, the facilities of a studio are more or less mandatory. Nevertheless, (under less technically demanding circumstances) ever since the earliest days of recording, a large proportion of orchestral recordings have been made outside of purpose-built studios. When one begins to consider the deeper elements of achieving good orchestral recordings, the above fact comes as no surprise.

6.12.1 Choice of Venues, and Musicians' Needs

Around the world there are some very famous and widely-used locations for orchestral recording. Not too surprisingly, concert halls are one member of this group, but assembly halls, churches and cathedrals are also popular locations. One requirement of such a space is that it needs to be big enough to house the orchestra, but usually, the apparent *acoustic* space needs to be even larger.

If we first consider the obvious, recording in a concert hall, there are two conditions likely to be encountered; recording *with*, or recording *without* an audience. Performances intended for recording tend to be assessed in greater detail than live performances. A small defect which may pass in a live performance may be irritating on repeated listening to a recording. The straightening out of these small points can become very time consuming. Frank and open discussion of sensitive issues would rarely be possible in front of an audience, nor would such occurrences be likely to constitute an interesting spectacle for the paying public, so almost out of necessity, many such recordings take place out of public view. Unfortunately, if such recordings are made in a concert hall, then that hall will probably have been designed to have an appropriate reverberation time when full. Its empty acoustic may not have been a prime object of its design; although in an attempt to make the acoustics independent of audience numbers, some halls have seats with absorption coefficients which, even when empty, try to match that of an occupied seat.

With recording techniques using close microphones this may or may not be of great importance, but for recordings with more distant microphones the ambience of some empty halls may be undesirable. Surprising as it may seem, even when full, not all concert halls are highly rated for orchestral recording purposes. A concert hall is an expensive thing to build, and few can be dedicated solely to orchestral performances. Consequently, ideal orchestral acoustics may have to be compromised to accommodate other uses of the halls, such as for conferences, operas, ballets, electrified music concerts or jazz performances. Each use has its optimum set of acoustic conditions, both in terms of hall acoustics *and* stage acoustics. Furthermore, in addition to the optimum reverberation requirements, lateral reflexion characteristics may also form another subdivision of the ideal acoustics for each use.

Some of the old halls, built before recording existed, are still very well liked. In fact, this is not too surprising, for they frequently did not have to make quite so many performance compromises as for halls of more recent construction. Moreover, when much of the classical music repertoire was written, it was written with many such concert halls in mind. The fact that much classical music can be expected to sound good in those halls is something of a self-fulfilling prophesy. Acoustic live performances were the *only* performances to be heard by the public when those halls were built, and no compromises needed to be made to allow for electric amplification. Many orchestral pieces were even written with specific concert halls already considered for their first public performances. However, this is somewhat akin to writing and recording a piece of music in a studio with idiosyncratic monitoring: it may not easily transfer to more acoustically neutral surroundings.

The great composers usually also had a comprehensive understanding of the needs of the musicians: because *they were* musicians. Orchestral musicians need to hear themselves in a way that is both clear and sonorously inspiring. They also need to hear, clearly, many of the other musicians in order to develop the feel of the performance. Composers often bore these facts in mind when arranging the instrumentation, so the music, the instruments themselves and the halls in which they frequently performed developed not in isolation, but in concert with each other. It is thus not surprising that many of the shoe-box shaped halls, which have been commonplace for centuries, are still well liked by musicians, recording personnel and audiences alike.

In recent years it has become apparent that, for a rich sense of spaciousness, lateral reflexions are of great importance. The shoebox shaped designs provide plenty of these, but they can be very problematical for many other purposes for which the hall may be used, often wreaking havoc, for example, with the intelligibility of the spoken word. Conferences or speeches in such halls can be difficult. In fact, the reason why so many religious masses are chanted, and not spoken, is because the longer sounds of a chant are less easily confused by the reflexions and reverberations of most churches than would be the more impulsive consonants of normal speech. Hard consonants would tend to excite more modal resonances, as they contain more frequencies than the softer chanted consonants. Ecclesiastical chanting therefore appears to have been of acoustic origins rather than religious ones.[2]

Nowadays, however, we must perform much of the old music in the multi-functional halls, or in churches, town halls and the like. What all of these locations have in common though, which sets them aside from most recording studios, is the large amount of space which they have for the audiences or congregations. They also tend to have many windows and doors, which allow an acoustic coupling of the internal spaces to the outside world. They are *not* constrained within the bunker structure of a sound isolation shell, and all, therefore, have room to 'breathe' at low frequencies; even if this is not immediately visually apparent from the interior dimensions of the rooms themselves. Given this acoustic coupling through the structure, they tend to appear to be acoustically larger than they are physically.

6.13 RT Considerations

The optimum reverberation time (RT) for orchestral recordings tends to vary, according to the music and instrumentation, from about 1.6 to 2.5 s in the mid-band, with the 5 kHz RT usually being around 1.5 s. The low frequency RT is a continuing controversy, with differences of opinion as to whether the 100 Hz RT should rise or not, and if so, by how much. In general, there are compromises with different reverberation times between definition, warmth and majesty. Inevitably the optimum requirements for each piece of music, or each orchestral arrangement, will call for somewhat different conditions, as will the presence or absence of an organ, chorus, solo piano, or number of contrabasses in the orchestra. There is a general tendency to believe that a moderate rise in RT at low frequencies is usually desirable, but that introduces yet another variable; personal preference.

The contours of equal loudness could have something to do with the LF debate. Figure 2.1 shows the classic Fletcher–Munson curves alongside the more modern equivalent, the Robinson–Dadson curves. The latter are now generally accepted as being more accurate, but the differences are too small to have eclipsed the former in general usage, so both sets of curves are shown for comparison. What they both show is how the sensitivity of the ear falls at the extremes of the frequency range. If the 0 phon threshold line at 3 kHz is followed to 30 Hz, it strikes the sound pressure level line at 60 dB. If the line passing through the 25 phon point at 3 kHz is followed down to 30 Hz, it will be seen to pass through the SPL line at around 65 dB. These are curves of equal loudness, and the above observations mean two things. Firstly that 60 dB more (or one million times the acoustic power) is needed at 30 Hz to reach the threshold of audibility than is needed at 3 kHz. An extra 40 dB (or 10,000 times the power) is needed at 30 Hz to sound as loud as a tone of 25 dB SPL at 3 kHz. So, it can be seen that the ear is vastly more sensitive to mid frequencies than to low frequencies at low sound pressure levels. Secondly, the rise of 25 dB in loudness from 0 dB SPL to 25 dB SPL at 3 kHz needs only a 10 dB increase at 30 Hz to produce the same subjective loudness increase. Looking at these figures again, 25 dB above the threshold of hearing at 3 kHz (the 30 phon curve) will sound as loud as only 9 or 10 dB above the threshold of hearing at 30 Hz. The perceived dynamics are considerably expanded at the low frequencies. At high SPLs, above 100 dB, the responses are flatter.

Let us consider a symphony orchestra in a hall, playing at 100 dB SPL. The direct sound will, according to the 100 phon lines on the equal loudness curves, be perceived in a reasonably even frequency balance. However, when the reverberant tail has reduced by 50 dB, and is still very clearly audible at 50 dB SPL in the mid-range, the lower octaves will have fallen below the threshold of audibility. Considering the fact that much of the reverberation that we perceive in concert halls is that of the tails, after the effect of the direct and reflected sounds has ceased, then much of our perception of that reverberation will be in the area where the low frequencies would be passing below audibility if a uniform reverberation time/frequency response existed. This suggests that halls with a rising low frequency reverberation time may be preferable for music which is performed at lower levels, in order to maintain a more evenly *perceived* frequency balance in the reverberation tails. Rock music, on the other hand, played at 120 dB SPL, may well degenerate into an incomprehensible confusion of low frequency wash in halls in which the RT rises at low frequencies. In fact, electrified rock music usually requires a much lower overall RT than orchestral music, and, quite definitely for rock music, a flatter RT/frequency ‘curve’ is desirable, or even one where the low frequency RT falls.

6.14 Fixed Studio Environments

As has probably already been gathered from all of this, to produce a large single recording studio to mimic all of these possibilities would be a mighty task indeed. It is one very significant reason why so many orchestral recordings are done in the aforementioned wide range of out-of-studio locations.

Live performances inevitably must be recorded in public halls, though the reasons for recording live are not always solely for the hall acoustics. It can be very difficult to lift definitive performances out of orchestras in studios. The adrenaline of a live performance will, in most cases, give the recording an 'edge' which the studio recordings frequently fail to achieve. Nevertheless it is still heard from musicians that far too many studios also fail to inspire a performance from an ambient point of view. All too often, the 'technical' environment of many studios is not conducive to the 'artistic' feel of a hall. Comfort and familiarity with the surroundings can be a very important influence on musicians, as can their own perception of their sound(s). In most concert halls, assembly halls, and other such places used for recording, the performing area is usually surrounded on at least three sides by reflective surfaces. As mentioned before, these lateral reflexions are of great importance to the sense of space, and by association, with the sense of occasion. If the acoustic conditions in a studio can help to evoke the sense of the occasion of a concert stage, then that studio may well be off to a good start in terms of the comfort factor for the musicians.

The one big obstacle in recreating realistic performance space acoustics in a studio is the normal lack of the open space in front of the orchestra. During concerts, the musicians face from the platform into a hall, from which there will probably be no significant reflexions, but much reverberation. The musicians thus receive the bulk of their reflexions from the stage (platform) area, and the subsequent reverberation from the hall in front of them. Human hearing is very directional in the horizontal plane, so the perception of the necessary reflexions, which help to reinforce and localise the instruments, can be easily discriminated from the equally necessary reverberation. The reverberation usually does not swamp the reflexions because of the spacial separation of the two.

In a studio of practical size we have a great problem in trying to recreate this situation. If the orchestra faces a reflective wall, the reverberation will be confined within the area of the performance, and additional reflexions will also arrive from the front wall. If we make the wall absorbent, we will tend to kill the overall reverberation. Either result is unnatural for the orchestra. It would seem that perhaps the only way to overcome this would be with a relatively absorbent wall containing an array of loudspeakers, coupled to a programmable artificial reverberation system. This way a reasonably sized studio could be created, say about the size of a concert platform, but with the effect of a reverberant hall in front of the musicians. The directional characteristics of the sound could thus be preserved whilst keeping the studio to a reasonable size.

It is doubtless the directional sensitivity of our hearing which is at the root of the phenomenon that when using conventional microphones (Sound-Field microphones *can* be an exception to this) the necessary microphone position for a subjectively most similar direct/reflective/reverberant balance is usually considerably closer to the orchestra than would be the corresponding seating position for a listener. However, it must be said that the microphone positions are usually *above* the audience, so they perhaps suffer less from local absorption by the seating, but this does not account for all of the effect.

What tends to set orchestral studios apart from most other studios is that a generally rectangular acoustic shape is rather more common, a typical example of which is shown in Figure 6.12. Walls are normally faced with a significant amount of diffusive objects to break up unwanted cross-modes, but the more or less parallel-sided nature can help greatly in the reproduction of the environment of the platform of a hall. Ceilings, though, as in halls, are almost never parallel to the floors. In some multi-functional rooms, they *may* be found to be parallel, but they are most likely to be relatively absorbent. Many concert halls have acoustic reflectors above the platform, usually angled to project more of the sound towards the audience, though frequently the sound also rises into scenery and curtain raising mechanisms which can be quite absorbent in their overall effect.

Figure 6.13 shows a studio designed for a wide range of recordings, including medium-sized orchestras. It is designed with a shape reminiscent of many concert platforms, and is surrounded on three sides by walls of wood and glass panelling. The wall adjacent to the control room, which would normally be facing the orchestra, is very absorbent, although some energy, quite well diffused, will return from the control room window. The lateral walls are fitted with rotating triangular panels of a type shown in Chapter 7, Figure 7.9(a). These panels are mounted in front of wood panelled walls, and have absorptive, diffusive and reflective (flat) sides. The ceiling has a mix of diffusive, absorptive and reflective properties. The design brief was to construct an orchestral concert platform but with acoustics sufficiently variable to allow a wide range of recordings. A live room (stone), a dead room and a well-controlled vocal room are sited behind the various glass doors.

However, in no cases can studio acoustics be separated from the controversy on the subject of the use of close versus distant microphones. Many people, when listening to recorded music, enjoy the dynamism of the close mic'd recordings, yet often they also can equally enjoy the more integrated grandeur when listening to many of the 'stereo pair', distant microphone, or SoundField recordings. The problem is, of course, that a listener at a live concert cannot be both near and far from the orchestra at the same time. With microphones, via a recording, it *is* possible, but to the ears of many people the result is perceptually confusing unless either one is greatly subservient to the other, and present only for added detail or richness. The problems of which option to take usually revolve around the fundamental aspects of psychoacoustic perception, some of which are mutually exclusive.

(a)

(b)

Figure 6.12:
Namouche studios, Lisbon, Portugal (1994).

(a)

(b)

Figure 6.13:
Estudios MB, Canelas, Portugal (2001).

Essentially, close and distant microphone techniques are two entirely different things, and the use of either is a matter of choice. They are not in competition for supremacy, but are merely options for the producers, who will choose the techniques which are considered to be appropriate for any given occasion. However, insufficiently flexible studio acoustics may force the decision one way or another, irrespective of which one would have been the more ideal artistic choice.

6.15 Psychoacoustic Considerations and Spacial Awareness

Not all of the problems of the design of recording studios for orchestral use are of direct concern to the recording engineers. Many other things which affect only the comfort and sense of ease of the musicians must be taken into account if the best overall performances are to be recorded. Sometimes these things can be better explained by looking at extreme situations, which help to separate any individual characteristics which may only be subconsciously or subliminally perceived in the general confusion of the recording spaces. So, let us now take a look at some of the spacial effects which can play such great parts in the fields of awareness and comfort.

Let us consider the Anglican Cathedral in Liverpool, UK. Its pipe organ was originally commissioned in 1912 as the largest and most complete organ in the world. It has 145 speaking stops, over 30 couplers, around 9700 pipes, and on Full Organ can produce a 120 dB SPL in the 9 s reverberation time of the cathedral. It is blown by almost 50 horsepower of motor, and it needs it all because the cathedral is one of the ten largest enclosed spaces in the world. Its size is truly awe-inspiring.

When inside the cathedral, one is aware of a reverberant confusion from a multitude of noise sources, but occasionally, on windless days and when not open to the public, there are occasions of eerie silence. When speaking in a low-to-normal voice to a person close by in the centre of the great nave, it is almost as though one is speaking outside in a quiet car park. It is not anechoic, as there is a floor reflexion to liven things up, but the distances to the other reflective surfaces are so great that by the time they have returned to the speaker they are below the threshold of audibility. However, a sharp tap on the floor with the heel of one's shoe is followed almost half a second later by an explosion of reverberation, lasting eight seconds or more. The clarity of the initial sound is absolute, as coming from the floor it has no floor reflexion component; yet a subsequent tap, originating before the reverberation of the first one has died down, is almost lost in the ambient sound. This highlights well the temporal separation discrimination which plays so much part in orchestral acoustics.

In the early 1980s, Hugo Zuccarelli, the Argentinian/Italian 'Holophonics' creator, visited many recording studios in the UK and elsewhere, demonstrating his new spacial recording system. On headphones at least, he could pan things around the head with stunning realism;

and realism not only of position, but of clarity as well. He also had a demonstration cassette of rather poor quality, yet the sounds of jangling keys remained crystal clear in front or behind the head, whilst the tape hiss remained fixed between the ears. It was impressive because the wanted sounds became absolutely separated from the noise, existing totally naturally and noise-free in their own spaces. This demonstrated most clearly the powerful ability of the ear to discriminate spacially, even in what amounts to an entirely artificial environment. The Holophonics signal supplied equalised, phase/time-adjusted signals to the ears, not present in mono signals, or necessarily in multiple microphone stereo recordings. The extra information provides a powerful means for the brain to discriminate between different signals, including between the wanted and unwanted sounds.

The old chestnut of 'The Cocktail Party Effect' has been around for years, and is no doubt already familiar to most readers. To recapitulate, if a stereo pair of microphones is placed above a cocktail party and auditioned on a pair of headphones, but panned into mono, then the general murmur of the party would be heard. However, it would be difficult to concentrate on any individual conversation, unless one of them happened to be occurring very close to the microphones. When auditioned in stereo, on the other hand, at exactly the same level, suddenly it would become clear that many separate conversations could be easily recognised and understood. The precise audiological/psychoacoustical mechanisms behind this have still not been fully explained, but nonetheless the effect itself is very well established as fact.

If we lower the level of the cocktail party recording into the domain of the stereo background noise of the tape hiss, then the conversations will begin to get lost in it, masked by the hiss even whilst still relatively well above the absolute hiss level. But this is not the case with the effect of Holophonics, which can lift a recognisable pattern out of the hiss by spacial differentiation beyond the in-head localisation of conventional headphone stereo. The phenomenon, once recognised, has a lasting effect. The significant difference between the effects of Holophonics and conventional stereo is that conventional stereo only provides information in a single plane, the Holophonics effect is three-dimensional. The single plane of conventional stereo contains all the information, both wanted and unwanted, so the tape hiss shares the same sound stage as the wanted signals: they are all superimposed spacially. In Holophonics, however, the wanted signals can be positioned three-dimensionally, yet the tape hiss is still restricted to its single plane distribution. Once sounds are positioned away from the hiss plane they exist with a clarity which is quite startling, and it is remarkable how the hiss can be ignored, even at relatively high levels, once it is *spacially* separated from the desired signals.

Thus, we have several distinct mechanisms working in our perception of the spaces that we are in, and all of the available mechanisms are available to the musicians in their perceptions of the spaces in which they are performing. Clearly, all of these mechanisms cannot be detected by microphones, nor can they be conveyed to the recording medium, and consequently many of them will never be heard by the recording engineers in the control rooms. In fact, although the recording engineers may well, consciously *or* subconsciously, perceive these things, they often fail to fully realise their importance to the musicians. To the musicians, however, who work with and live off these things daily, their importance cannot be over-stressed. In the recording room, therefore, all of these aspects of the traditional performance spaces need to be considered. Quite simply, if the musicians are not at ease with their environment, then an inspired performance cannot be expected.

From time-to-time, there are situations which require even orchestral musicians to wear headphones when recording; either full stereo headphones or single headphones on one ear. However, perhaps the major reason why this is usually avoided is not the enormous number of headphones required, but that inside headphones, certainly if they are of the closed type, it is not possible to reproduce the sort of complex sound-field that the orchestral musicians are accustomed to playing within. Human aural pattern recognition ability is very strong, often even when sounds are hidden deep within other sounds, and even deep in noise. If headphones isolate the musicians from the usual patterns, the effect can be very off-putting.

For many of the reasons discussed, orchestral studios should not be designed primarily for what is good from the point of view of the recording personnel. First and foremost a performing environment should be created to aid the *whole* recording process. There are some hi-fi fanatics who would disagree, but most people would rather hear a compromised recording of an excellent performance, than an excellent recording of a compromised performance. Obviously, though, the real aim of the exercise is to make excellent recordings of excellent performances, and this is why so many things must be considered.

Many musicians of all kinds play off their own sound; it is like a feedback mechanism that both reassures and inspires them. This is true of more or less every musician and of whatever type of music that they perform. If their sounds are not given to them in their foldback as they need, then it can be very disconcerting. In the case of orchestral musicians, their foldback is not usually via headphones, but via the reflective surfaces of the room. Somewhat confusingly the orchestral musicians often refer to it as feedback. If playing to a backing track, it will possibly be only the conductor who will listen to it on headphones. In these cases, the acoustic of the performing space provides the foldback (feedback) to the musicians, so it should be given its due consideration as such.

Another frequent conflict between the needs of musicians and those of the recording engineers is the use of acoustic screens. In the close microphone type of recording, separation is a factor which is often considered desirable by the engineers. This can be improved by the use of acoustic screens between different sections of instruments, which normally as a concession to the musicians have windows to allow eye-to-eye contact. Unfortunately, this can disrupt the perception of the desired acoustic sound-field by the musicians, and in most cases they would prefer the screens not to be there.

It has been quite amazing how, over the course of recording history, musicians needs have so often been neglected in the recording process. Time and time again, if the recording engineer has had a problem with overspill, screens have been imposed upon the musicians, without due consideration of the effects of their insertion on other aspects of the recording process. In many cases recording staff have totally failed to appreciate the artistic damage which can be done by delays and disturbances caused by technical adjustments. Awareness of such things has been one thing which has set apart the specialist recording engineers and producers, who by understanding things in a more holistic way have gained the cooperation and respect of the musicians, and have thus produced more inspired recordings from which they have deservedly built their reputations.

This discussion on some rather peripheral aspects of audiology and psychoacoustics has not been a digression, but is a fundamental requirement in understanding what is necessary for the design of good recording spaces for orchestral performances (or perhaps rather, for the design of good *performance* spaces in which orchestras may be recorded).

The room variability needs to be in terms of overall reverberation time, and if possible, the relative balance of low and mid/high frequency reverberation. Reflexions need to be controllable in terms of time, direction and density, with the availability of some reasonably diffusive surfaces to add richness without undue colour. Always, however, the consideration of what the *musicians* need should be given at least equal weight to the needs of the recording staff when acoustic adjustments are being made. It is very necessary to strike a balance between these priorities and to achieve the close cooperation between all parties involved. Hopefully, what this chapter has shown so far is that whilst rooms of strong characteristic acoustics can be very valuable and even insuperable when their specific strengths are precisely what are required, they can also be very inflexible. This can lead to their use being limited, and when space is at a premium, something capable of a wider range of acoustic options is more desirable. The following chapter will therefore look at means of creating rooms with highly variable acoustics. However, before we leave the subject of specialised rooms, we must consider those with almost *no* acoustic characteristics.

6.16 Dead Rooms

Before leaving the subject of specialised recording rooms, perhaps we should look at the design of a very dead room. Indeed, one such room exists in each of the studios shown in Figures 6.13, 7.1 and 8.2. They can produce interesting sounds when used as isolation rooms for bass guitar amplifiers, and quite a number of vocalists have been noted as liking the 'different' sounds which recording in such rooms can achieve. Figure 6.14 shows two

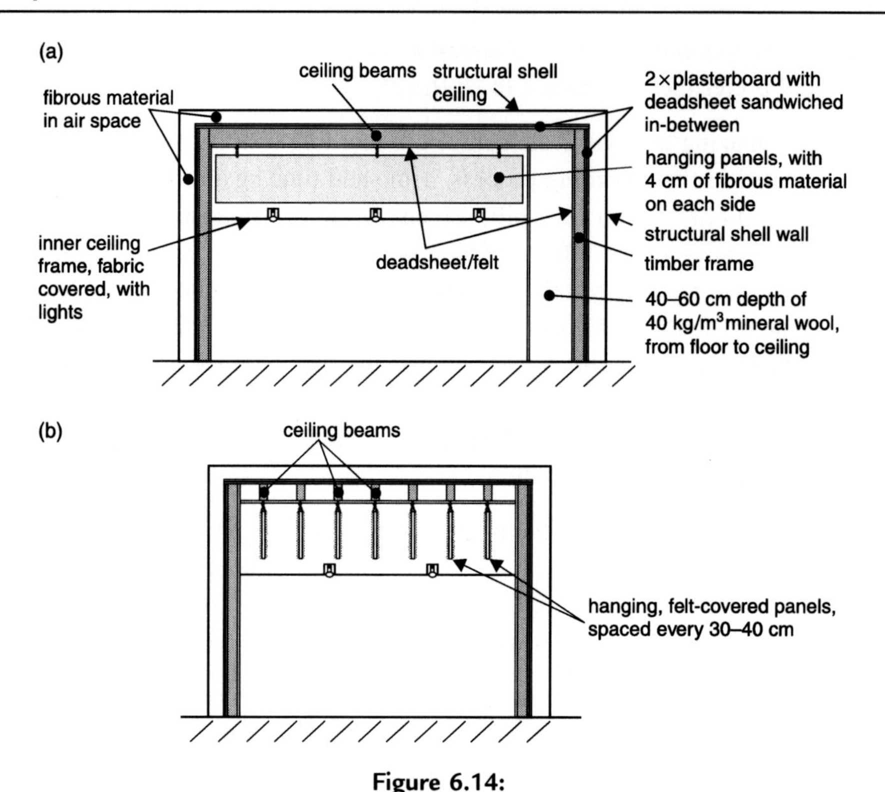

Figure 6.14:
The construction of a dead room: (a) Side elevation; (b) end elevation.

cross-sections of such a room. The ceiling, above 2.5 m, consists of a series of hanging panels of plywood, spaced about 30cm apart and lined on each side by 4 or 5 cm of 40 kg/m^3 fibrous material. One wall is treated with a layer of 60 cm thickness of 40 kg/m^3 mineral wool. The other walls are constructed as shown in Figure 5.1.

The floor is hard and the glass doors are useful for good visibility if the musician is in the room, and not playing via a remote amplifier head. Nevertheless, the same conditions apply as were discussed about the room shown in Figure 5.19, whereby the directionality of the microphones can be used to avoid picking up any undesirable reflexions. In the cases of bass amplifiers facing the super-absorbent wall, the microphone would normally be between the amplifier and the absorber, and so would be shielded from any room effects. When money and space allow, these rooms can prove to be very useful adjuncts to a studio. They can also be highly useful where voice-over or dialogue replacement requires a totally dry voice, such as for a scene in a snow-drift, where conditions would tend to be anechoic, and any trace of room ambience on the recording would spoil the effect. The principal difference between the room shown in Figure 6.14 and that of Figure 5.19 is the much greater low frequency absorption of the former room. Obviously, this can only be achieved at the expense of requiring a larger space for the construction. A specialised version of such rooms is the Foley Room.

6.17 Foley Rooms

Another type of studio room which requires a relatively dead acoustic, and which is important in the film industry, is the Foley room. Named after Jack Foley at Universal Pictures, who was Hollywood's first specialist in this artform, they are used for the recording of minor sound effects such as footsteps, opening and closing doors, jangling keys, hands knocking on windows, and other such sounds which may have been badly captured by the original recording microphones or which may have been lost when original dialogue soundtracks have been replaced, such as in foreign language versions. Obviously, when such recordings are re-mixed into the soundtrack they must acoustically match the image on the screen. It can be detrimental to the realism to hear the voices of two men rowing a boat on a lake whilst the splashing of the oars clearly sounds like it has been recorded in a room. Such things may be acceptable in a comedy film but they would greatly affect the credibility of a serious drama.

In general, it is better to record the Foley effects in an acoustically dead room, and then use artificial means to match the ambience of the sound to the context of the picture. Figure 6.15 shows a specially designed Foley room which can also be used for the mixing of stems (pre-mixes of effects, music etc.) in 5.1 surround. On the floor can be seen different

Figure 6.15:
The Foley room at Cinemar Films, Milladoiro, Spain, with the perforated screen dismounted and the different walking surfaces, for the recording of footsteps, clearly visible. This room, with a mixing console installed, can also be used for pre-mixing 'stems' of dialogue, music or effects, hence the surround loudspeakers which are visible at each side of the photograph. For this reason, the commonly used false doors and windows, for recording opening and closing sounds, are removable so as not to interfere with the room's mixing capability.

surfaces – wood, stone, carpet, water, metal and gravel – for the Foley walkers to synchronise their footsteps to the picture. There are spaces under some of the surfaces to introduce different quantities of damping materials, and the water depth can be varied. The room is usable either with projector and screen or with a video monitor.

The construction of the room shown in the figure is generally similar to the principles shown in Figures 5.1, 5.3, 5.4 and 5.5. The front wall needs to be very rigid in order to provide an extended baffle for the loudspeakers, and so cannot be absorbent at lower frequencies. At higher frequencies, the perforated screen is also fairly reflective. However, in use, the actor will be facing the screen, and the microphone(s) will be pointing away from the reflective front surfaces of the room. As the rest of the room is highly absorbent, and the microphones which are used tend to be cardioid in their response pattern (in order to minimise the room sounds in general), the reflective front wall tends not to be a problem.

Figure 6.16 shows a different concept of Foley room design. In this case, the room is 'constructed' from 10 cm of polyurethane, open-cell foam of 80 kg/m^3 density, glued to the structural wall surfaces. Next, a 3.5 cm layer of panels of 'Celenit' was glued to the foam. In each case, the glue was a contact/impact adhesive. (This type of construction was described previously in Section 5.11.) On the right-hand wall of the room shown in the figure, a door and a window can be seen. These are for recording the sounds of doors and windows being opened and closed, and also the sounds of people knocking on them. They do not actually lead to anywhere, but are simply mounted on the surface of the wall.

Figure 6.16:
A Foley room in the Tobis complex in Lisbon, Portugal.

The ceiling of the room is made from the same foam and Celenit, but glued to plasterboard fixed to wooden beams at 60 cm centres. A 30 cm cavity, lined with cotton-waste felt, gives rise to good low frequency absorption. The overall effect is a room with a very low decay time over a wide frequency range, enabling the recording of sound effects with a very low level of room colouration.

6.18 Summary

Strictly speaking, a live room is one where $L_{AV} > A$.

Some rooms with appropriate acoustics can be inspirational to the musicians.

Idiosyncratic rooms tend to add variety to a studio's facilities, but are much less flexible than more acoustically controlled rooms.

Non-purpose-built recording rooms have often been found to be excellent for recording, partly due to the fact that they have not been built inside acoustic isolation shells. However, this can also limit their usability in some circumstances.

All materials of construction tend to impart some of their sonic characters to the rooms from which they are built.

Live rooms tend to come in two basic types: reverberant or reflective (bright).

In live rooms, the directional characteristics of loudspeakers and microphones can greatly influence the recorded sound.

Live rooms feed energy back into acoustic instruments, producing a sound and a tactile feel that cannot be achieved by electronic post-processing. They can therefore affect a performance in their own, unique ways.

Spacially distributed instruments, such as a drum kit, drive the room differently according to the position of each part of the instrument.

Usually about 20% of the surface area of a room needs to be acoustically changed before it has a significant influence on the overall sonic characteristics of the room.

Adjusting the degree of irregularity on the surface of a stone wall can greatly affect the acoustics of a room. More relief leads to more diffusion. Greater depth of irregularity leads to diffusion down to lower frequencies.

The modally controlled range of a room is bounded by the large-room frequency and the pressure zone.

Much orchestral music has been composed with live performance as the principal objective. Modern studios do not always conform to the expectations of the composers. Much orchestral recording is therefore carried out outside of conventional studios.

Lateral reflexions can be important, both for the recordings *and* the musical performance.

The optimum RT_{60} for orchestral recording tends to be from 1.6 to 2.5 s at mid frequencies.

Many psychoacoustic factors need to be taken into account when designing spaces for orchestral recording.

Recording personnel should be attentive to the needs of musicians if the best overall results are to be achieved. Lack of thought can lead them to break up the acoustic character of a room solely from the point of view of the recording techniques, which can make the job of the musicians much more difficult.

Foley rooms, for the recording of incidental sound effects for cinema and video sound tracks, need to be acoustically dead so that the recordings can be artificially matched to the sound required by the picture.

References

1 Morfey, Christopher L., *Dictionary of Acoustics*, Academic Press, London and San Diego (2001)
2 Lubman, D. and Kiser, B. H., 'This History of Western Civilization through the Acoustics of its Worship Spaces', *Proceedings of the Seventeenth International Congress on Acoustics*, Rome (September 2000)

Bibliography

Barron, M., *Auditorium Acoustics and Architecture*, E. & F. N. Spon, London, UK (1993)
Beranek, L., *Music, Acoustics and Architecture*, John Wiley & Sons, Chichester, UK (1979)
Kutruff, H., *Room Acoustics*, 4th Edn, E. & F. N. Spon, London, UK (2000)
Beranek, L., *Concert Halls and Opera Houses*, Springer-Verlag, New York, USA (2004)

CHAPTER 7

Variable Acoustics

Rooms with acoustically different zones. Rooms with directional acoustics. Changeable surfaces. Acoustic change by rotating panels. Considerations concerning different room sizes.

7.1 The Geometry of Change

A degree of acoustic variability in a recording studio is a useful asset. It allows areas of the recording spaces to be better suited to the 'natural' acoustics for which many instruments were made. The variability allows a studio to accommodate different types of recording, as opposed to just specialising in the recordings that suit the fixed acoustics. As discussed in Chapter 5, acoustic neutrality, although very useful in many circumstances, is not the answer when a more naturally ambient type of recording is required. Let us now, therefore, look at some ways in which acoustic variability can be brought to rooms, whether they are large, medium or small.

Figure 5.18 shows a room of fixed acoustics, although a good degree of acoustic variability can still be achieved in such a room by choosing different positions for individual instruments and microphones. On a macro scale, say by using the whole room for one instrumentalist in the centre, the room would be remarkably neutral. Subtle but significant changes in the acoustics could be obtained by rotating the musician and/or microphone(s) in order to face either the reflective, diffusive or absorbent surfaces. However, some quite marked changes in recording acoustics could be effected if the instruments were moved close to the different boundaries. The principal problem with this type of room arises when one needs to record an ensemble. This is not only because the room is not balanced about any axis, but also because the room would have a fixed and very short decay time. Nevertheless, both problems would tend to be ameliorated as the room size increased. An actual room using this type of control is shown in Figure 7.1, although the diffusers in this case are not of the cylindrical type.

Figures 7.2 to 7.4 show three possible ways of achieving acoustic variability in larger spaces. Although the type shown in Figure 7.2 is relatively cheap to construct, and highly effective, it again has the drawback of being awkward to use if a large ensemble is in the room, as

Figure 7.1:
Dobrolyot Studio, St Petersburg, Russia (1996). Note the ex-Soviet era perforated diffusers over the glass doors and on the adjacent left-hand wall.

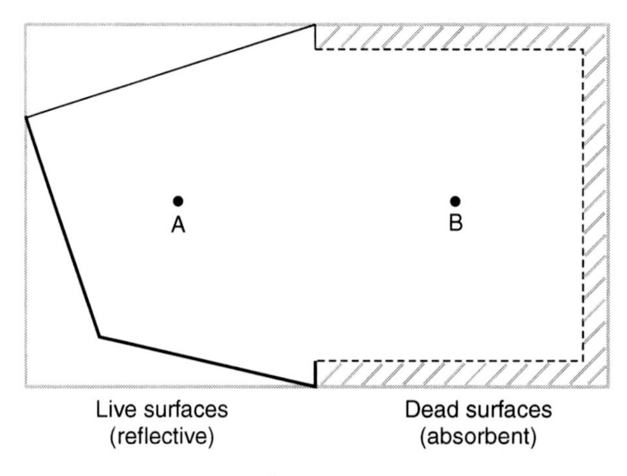

Figure 7.2:
Recording room with fixed live and dead areas.

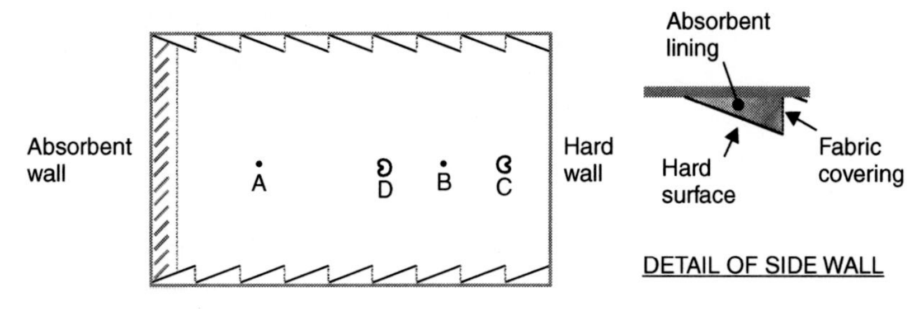

Figure 7.3:
Recording room with graded acoustics.

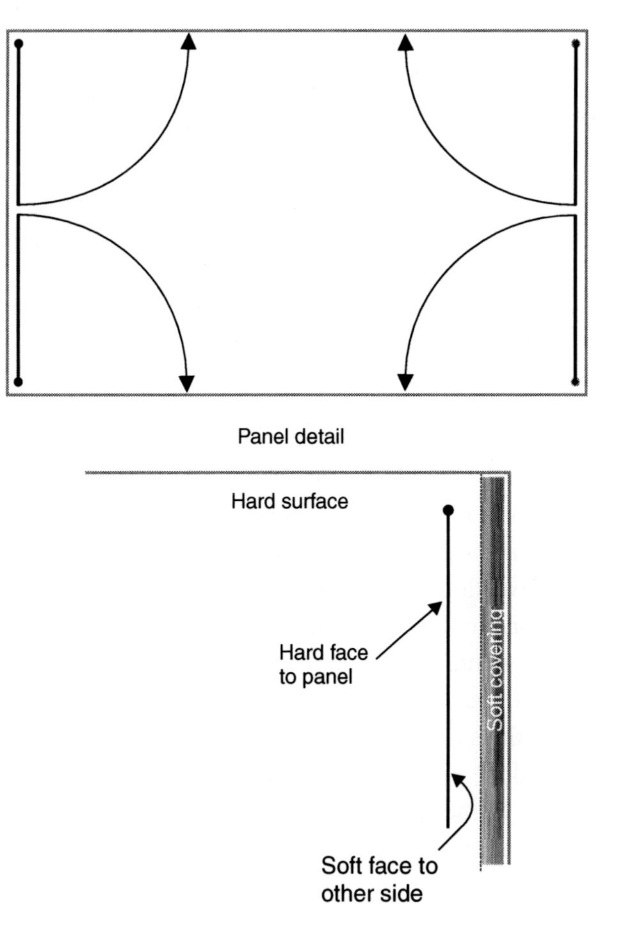

Figure 7.4:
Recording room with hinged acoustic panels.

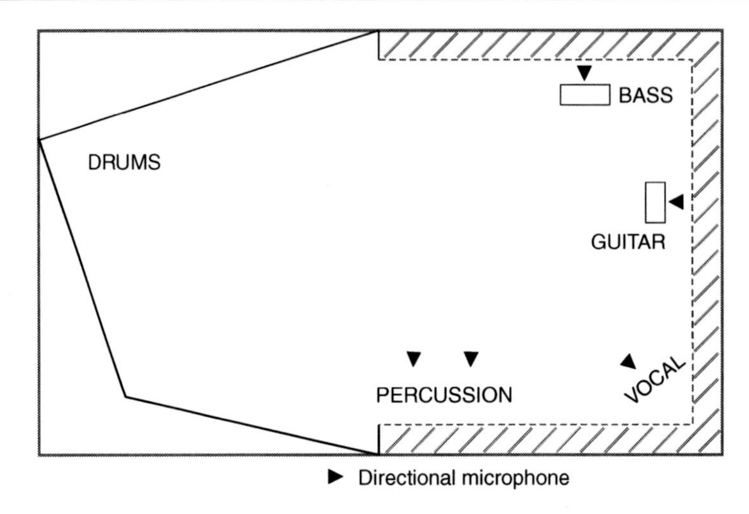

Figure 7.5:
Typical usage of room shown in Figure 7.2 with rock group. In the example above, a five-piece group could record 'live' using the characteristics of the room to good effect. The drums could be set up in the live area, to produce a full sound and a good feel for the drummer. Bass and guitar amplifiers face the absorbent walls, thus reducing the overspill back to the other instruments. The amplifiers also help to shield the microphones from the drums and percussion. The percussionist faces the other musicians, and the percussion microphones do not point in a direction where they are likely to collect excessive overspill, even though the percusion is in a relatively live part of the room. The vocalist, in a dead corner, faces the other musicians, but the directional vocal microphone faces the absorbent surface.

a uniform acoustic could not be shared by the whole group. Figure 7.5 shows how the room could be used effectively for a typical rock group, allowing the musicians to play as one unit in close contact with each other, yet with each instrument in its own, desired, acoustic space. The results can be excellent, but unfortunately, if a larger ensemble is placed in such a room, filling more than one half of the space, either the front/back or left/right balance of direct to reverberant sound would not be uniform. It would simply not be possible to achieve a balanced, overall sound. In the area between the two halves, an interesting zone can exist where very different recordings can be achieved by the varied positioning of directional microphones. This can be done either in the vicinity of the instrument, or more distantly with ambient microphones. In the live end of the studio, walls can be made as reflective as judged to be appropriate for the work most typical for the studio.

Referring again to Figure 7.2, position A would enjoy a rather live acoustic whilst position B would be much more dead. Position A is surrounded on three sides by reflective walls, whilst position B is surrounded on three sides by very absorbent walls. The only live surface directly facing position B is the far wall of the live area, which is quite some distance away. Figure 7.3 again has positions A and B, live and dead respectively, though this time the method of achieving the effect is rather different. The sawtooth arrangement of the reflector/absorber surfaces creates a situation whereby as one moves from the absorbent wall to the reflective one, there is a progressive reduction in the reflective surfaces facing the sound source. If an instrument is at position B, with a cardioid microphone at position C, facing B, then an almost dead room response would be recorded, because all the reflexions would be passing away from the front of the microphone diaphragm. Conversely, by positioning the microphone at D, many reflexions would be captured by the microphone, both from the hard wall and from at least the first three 'sawtooth' reflectors of each of the side walls.

The rooms shown in Figures 7.2 and 7.3 are acoustically variable in terms of both the positions of the sources of sound *and* the microphones, but Figure 7.4 shows a move toward a more truly variable room. In this instance, hinged panels can be moved through 90°, exposing either hard or soft surfaces to the room. By this means, either the whole room or different sections of it can be radically altered between live and dead. A variant of this theme exists where the hinged panels are in the centre of the long walls, allowing the rooms to be subdivided. They have generally been very well received, but in the case of larger rooms this type of system is somewhat impracticable. An actual example is shown in Figure 7.6.

Figure 7.7 shows a concept for a large room with quite comprehensively variable acoustics. It can probably be guessed from this that a large room with seriously variable acoustics does not come cheap. Space is consumed by the variable elements, which means that considerably more floor area is needed in the building shell than will be realised in the final recording area. Highly variable rooms tend to be both structurally complicated and expensive, but they can also be very effective recording tools when a truly multi-functional room is needed. Essentially, though, it will be seen from Figure 7.7 that almost every surface of the room needs to be capable of being changed from hard to soft if a very high degree of variability is to be achieved. In practical examples, such as shown in Figure 7.8, the ceiling panels are motorised, and their positions can be changed remotely from the control room whilst listening to the effect as it happens.

Figure 7.9 shows the details of the rotating panels that would be typically used in rooms of the type outlined in Figure 7.7 and shown in Figure 6.13. Four different designs are shown, though many other variations on this theme are possible. The diffusive sides show the options of either curved surfaces of different radii, or the quadratic residue types in both the pit and relief forms. This type of room variability technique is now widely used in concert halls. It was largely the Japanese designer, Sam Toyashima, who brought it to high profile in recording studios around the world, though the principle dates back to German broadcast studios of the 1930s. The wall panels of Figure 7.9 have three surfaces and can be rotated into

Figure 7.6:
Acoustic variability by hinged panels – Sonobox studios, Madrid, Spain (2001).

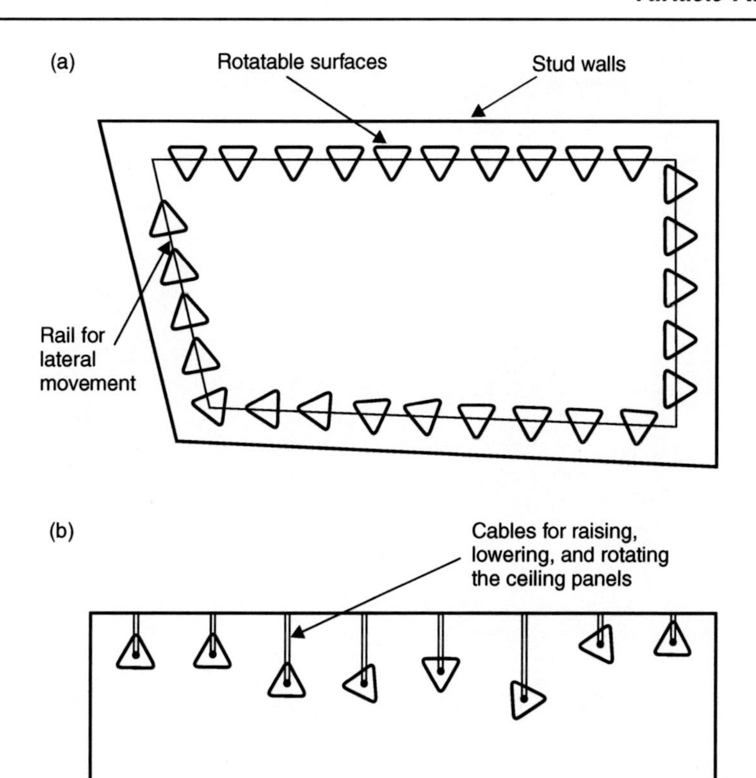

Figure 7.7:
Room with high degree of acoustic variability (but at a cost of much floor area): (a) Plan; (b) Side elevation.

any position desired. Having three surfaces gives the option of mixing any of the diffusive, reflective or absorbent characteristics by intermediate positioning. The room can be divided into live or dead areas, and, by the rotation of the panels, the reflexions are to a large degree steerable, which can create interesting effects for ambience microphones. Similar ceiling panels can be raised, lowered or rotated, and therefore can be absorbent, diffusive or reflective with varying degrees of timing for first reflexions.

With rooms such as this it is often surprising to the uninitiated just how much area needs to be changed to achieve any significantly noticeable change in the general acoustics. Except for microphones and instruments in close proximity to the variable panels, rotating only three or four of them would usually be virtually unnoticeable in the context of

(a)

(b)

a change in the overall acoustic of the room. It really takes a change of 15 to 20% of the overall surface area of the room to have any readily noticeable general effect. In a 15 m × 10 m × 5 m room, accepting for now a hard floor, it means changing something around 80 to 90 m^2 of surface area for a worthwhile change in acoustics. Once over that threshold, however, the effect, in whichever direction it is acting on the room ambience, begins to develop rapidly.

The rotating wall-panel technique takes up more permanent spacc than the method of varying the wall surfaces by movable panels, either hinged, as in Figure 7.4, or attached by some sort of hook system. However, with the hinged panel system it is more difficult to achieve the intermediate situations and it may also be less rapidly adjustable. Obviously, the hinged panels need a free space to swing out, which is perhaps not too much of a problem in a very large room, but in smaller rooms it could mean dismantling a whole drum kit just to make an experimental change. On the other hand, if the changes of acoustics can be pre-planned, the hinged/hung panel system does cause less permanent loss of space inside the isolation shell than does the rotating panel system of Figure 7.7. As usual, compromises exist in each case.

A variation on the rotating panel technique is to use multiple, small, hinged panels, as shown in Figure 7.10. The panels are made from 12 mm plywood, and about 15 mm is left between them when they are fully closed so that they can be operated individually without fouling each other. In the case illustrated, the panels are mounted over a wideband (full frequency range) absorber of about 1 m depth. The choice of 12 mm plywood was made so that the very low frequencies would still enter the absorber, as the opposing wall was stone, and there was the risk of an undesirable build up of bass frequencies if the plywood was too thick. The arrangement is very flexible, and a great number of permutations of panel combinations are available which can either deal with small zones in close proximity to the panels, or which can act on the whole room in a more general way. In the case shown in Figure 7.10, when the panels are all fully open, their effect on both the absorber and the rest of the room is negligible. This system of providing acoustic variability has proved to be useful in rooms of many different sizes. However, the absorber still needs to be quite deep if the variability is required over a wide frequency range.

Figure 7.8:
(a) The large, variable acoustics room at Olympic Studios, London, UK, designed by Sam Toyashima. The ceiling panels can be raised and lowered electrically, from the control room, enabling the acoustic changes to be monitored as they occur.
(b) Another Toyashima variable room at the Townhouse, London, UK, with semi-cylindrical, rotatable wall panels (photograph by Bob Stewart).

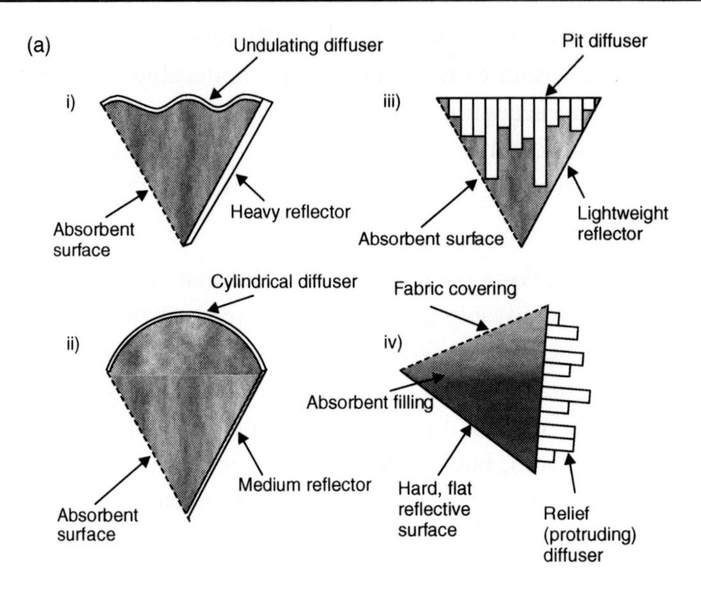

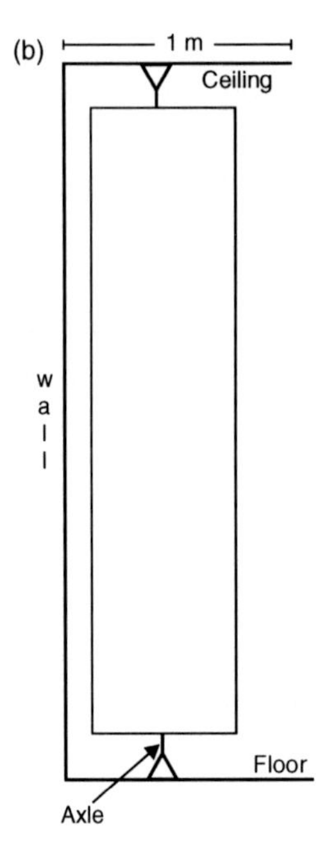

Figure 7.9:
Rotating acoustic control devices: (a) Detail of rotating units – 4 variants – end views. Rotating contrivances, such as shown above, can provide reflective, diffusive or absorptive surfaces, either wholly, or in combinations by being rotated to intermediate positions. They can be applied to walls or ceilings, and can be motor driven, operated from the control room, where their effect on the recording acoustics can be judged whilst listening via the microphones.
(b) Mounting arrangement for rotating acoustical devices shown in (a).

Figure 7.10:
Multiple, hinged panels covering a large, full-range absorber. These provide considerable variability to the acoustics of the room. The different combinations available are almost infinite – Rockaway Studios, Castellón de la Plana, Spain (2008).

7.2 Small Room Considerations

Small recording rooms, of less than around 50 m^3, tend to be more difficult in terms of variability. They tend to flip-flop from one state to another, passing through some strange sound characters on the verge of the main changes. What is more, in a small room, the number and positions of people and equipment occupy a much greater proportion of the room volume, and hence may themselves have a great bearing on the acoustics. The variations in the surfaces of small rooms usually have to be judged once the room is ready to record, with all personnel and equipment in place. As mentioned earlier, with wheeled or hinged panels, changing things after the instruments have been positioned can become a very disruptive and slow process in a congested room. Nevertheless, once the recording staff become accustomed to how the room sounds in their typical recording situations, they can pre-set the room to a state which their experience suggests will be most appropriate for the music which they expect to be recording. A technique such as that shown in Figure 7.7 would be unlikely to be a good choice in a small room, because the overall loss of space due to the movable sections would be too great a proportion of the total space of the room.

The important point to remember is that the movable panels of Figures 7.7 and 7.9 will not scale. They cannot be built one half of the size in a room of one half of the volume. The effects of the panels are related not to room sizes but to wavelengths, and as the frequency ranges of the instruments in a small or large room remain the same, some of the dimensions of the variable wall sections must remain the same. If an absorber needs to be 1 m deep for any given frequency range, then in a room of half the total area, despite the fact that only half the surface area of absorbers may be needed, they still need to be 1 m deep. There have been many situations in which small-scale versions of these devices have been tried, but they have only been effective over much narrower frequency bands, and their sonic effect has frequently been very unnatural. For so many reasons, recording spaces tend to be like boxers; a good big one will almost always beat a good little one.

In the smaller rooms it is almost impossible to achieve situations whereby the first reflexions back to the instrument are of the late type (echoes). There are many more early reflexions returning to the musicians and the recording microphones, so the reflexions of this type produce tonal colouration rather than a sense of spaciousness. This can be off-putting for the musicians, whether they hear the sound directly or via the microphones and foldback systems, both of which can affect the way that they play. All in all, the tendency is that smaller rooms are better suited to either fixed design concepts, or to being provided with the means of gross state changes such as shown in Figure 7.6. Subtle variability rarely seems to achieve its aims in small spaces. Diffusers, also, suffer from wavelength limitations, and for good low frequency diffusion they currently need to be relatively large in their front-to-back dimension. Even the shallower amplitude gratings (see Figures 4.33 and 4.34) have a tendency to sound coloured at close distances. However, some interesting research is afoot whereby actively driven end walls at the back of the pits may be able to simulate in a shallow pit the effect of a deep one. Time will tell whether these techniques eventually deliver what they are promising. Nonetheless, in experienced hands, the idiosyncratic nature of some small, variable rooms can be put to good use, especially in the more 'creative' environment of modern music. In a situation where unusual sounds are being sought there can be some interesting possibilities, but one has to be careful that similar room-sound characteristics do not build up track by track to become overbearing in the final mix.

Most of the characteristic sound of the small rooms are not frequency aberrations *per se*, but are characteristics of the time response. It is not just the frequency content of the room sounds which stand out, but the time-performance of the resonant modes and reflexions which give rise to the idiosyncrasies of the overall decay. As such, with the problems being in the time domain, there is little hope of using the equalisation controls of the mixing console to correct the problem. Any attempts at such 'corrective' equalisation can seriously detract from the frequency balance of the direct sounds, and in many cases the effect of the medicine will be worse than the illness. Such room problems can usually only be solved by acoustic changes in the rooms themselves. If small rooms cannot be significantly changed acoustically, then at least the positions of subsequent recordings should be changed. If we are to use another medical analogy, although the room-related problems may not be *terminal* to a recording, once in the recording system they do tend to be incurable.

In smaller rooms, where the room sound is noticeable, it is of paramount importance to monitor it carefully at the recording stage, and, if any sign of a characteristic acoustic build up becomes evident, steps must be taken to ensure that further recordings on any given song are done with a varied acoustic. This may mean moving the musician(s) and/or the instrument(s); moving the microphone(s), or making use of any acoustic variability. Although changing a microphone *type* may change the direct/reflective pick-up characteristic, it is not likely to be as effective as the former measures because the offending time characteristics are quite likely to still find their way into the recordings. Nevertheless, idiosyncratic acoustics do have their place when they are available for effect, but to have a degree of acoustic variability as an option in the recording space is often as useful as a whole rack of effects processors in the control room.

7.3 Summary

Rooms with variable acoustics can be very useful recording tools.

Variability can be achieved in a room by constructing different zones with different characteristics, or by the use of changeable surfaces.

Hinged panels, dismountable panels and rotatable panels can all be employed to good effect.

Rooms with different acoustic zones may not be very usable when recording an ensemble, because the overall acoustic will not be spacially balanced.

Hinged panels tend to take up less space than rotating panels, but changing their position in a small room full of instruments and microphones can be very inconvenient.

Rotating, triangular-section panels can be very effective in changing the acoustics of a room, by offering reflective, diffusive or absorptive surfaces.

It is much more difficult to make subtle changes in small rooms than in large ones.

Absorber and diffuser depths are usually dependent upon the wavelength of the lowest frequencies to which they are expected to work, and they will *not* scale with room size.

Actively driven diffuser pits may offer the hope of shallower diffusers in the future.

Many characteristic room sounds are time domain effects, and hence cannot be adjusted by conventional electronic equalisation.

CHAPTER 8

Room Combinations and Operational Considerations

Juxtaposition of different working areas. Choice of appropriate acoustic spaces. Door and window isolation. Multiple glazing considerations. Modal control of small rooms. Special recording techniques.

As has no doubt become clear by now, no one room can perform all the functions which may be required for the recording of the whole range of probable musical performances. The large, variable room is perhaps the best option if money and space allow such a room to be built, but even the best such room can never mimic the performance of a good, small live room, or a good stone room; at least not without absurd complexity of structure and mechanisms. We should now, therefore, consider how we could put together the pieces so far discussed in order to make a viable, practical and effectively flexible studio package.

8.1 Options and Influences

Let us imagine that we were presented with a building in which to construct a studio, and that we had the luxury of 500 m^2 of floor space and 6 m of ceiling height. This would give us enough space to construct a large, variable, general recording area, plus a stone room, a moderately live room, a vocal room and a dead room. However, the layout of such rooms is full of compromise, forced upon us by many conflicting priorities. Ideally, everybody needs to be able to see the control room, but also, everybody needs to be able to see each other. At the same time, the general influence of the positioning of the smaller rooms should not detract from the optimum shape of the main studio area. This may all seem rather obvious, yet it is surprising just how many studios are built without the most basic of these requirements being given their due consideration; this even includes some studios which have ostensibly been professionally designed.

One practical option for our 500 m^2 studio could be that which is shown in Figure 8.1. This first appeared in the book *Recording Spaces*[1] in 1998, which directly led to the construction of the two studios shown in Figures 6.13 and 8.2, though on slightly more modest scales than the 500 m^2 proposed. They were built in 320 m^2 and 220 m^2 respectively, and as can be seen they are practical realisations of many of the points discussed in previous chapters.

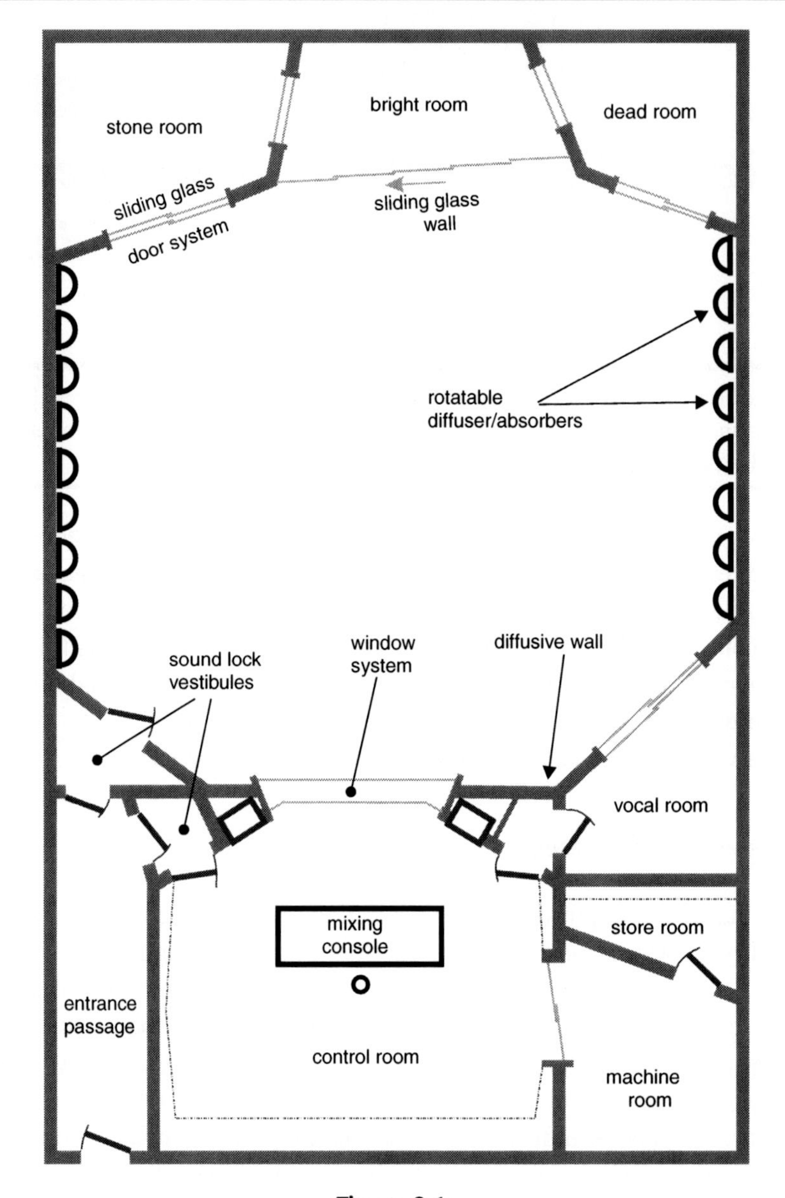

Figure 8.1:
Layout of hypothetical 500m^2 studio.

As a contrast, Figure 8.3 shows an actual situation that was in a 90% completed form before it was realised that it was not going to perform as expected, and over €60,000 had already been spent on the rooms. The designs had been done by a theoretical acoustician with some theatrical acoustics experience, but who was not a specialised studio designer, and all the rooms had been designed around the use of large semi-cylindrical diffusers, Helmholtz resonators and absorbent/diffusive tiles made from gypsum and mineral wool. All the rooms

Figure 8.2:
The main room in Tio Pete studios, Urduliz, Bilbao, Spain, loosely based on the drawing of Figure 8.1 (1999).

had been designed to classic RT_{60} figures, and somewhat surprisingly all had been aimed at a somewhat similarly neutral acoustic performance, except for the slightly longer RT in the larger room. The windows and doors had been situated in accordance with structural simplicity rather than acoustical, operational, or musicians' needs. Perhaps this fact, more than any other, made the task of modifying the studio all but impossible.

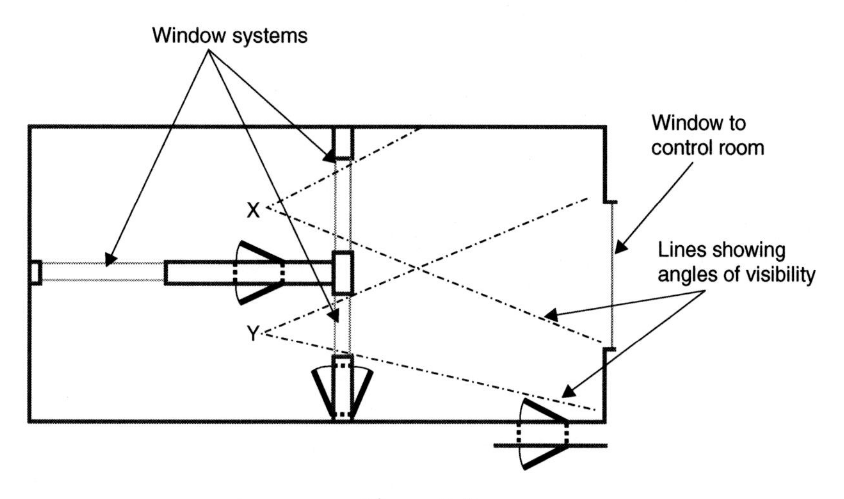

Figure 8.3:
A studio designed by a general acoustician. Musicians in positions X and Y (the natural positions for good visibility into the larger recording room and control room) are unable to see each other. Insufficient thought had been given to practicalities.

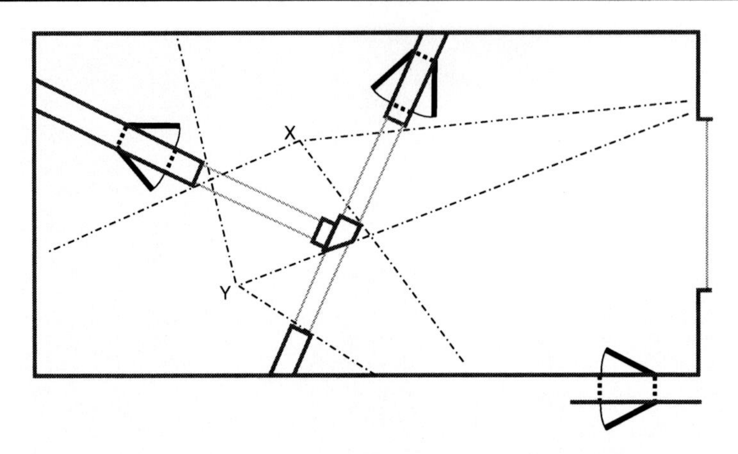

Figure 8.4:
Layout that could have been workable, using similar space and materials as Figure 8.3. Note also the improved angles of visual contact between the rooms.

It was ultimately the owners themselves who suggested a total rebuild. What was so distressing was that if it had been built with the internal walls, doors and windows, such as shown in Figure 8.4, it would have been much more rescuable. What is more, it would have cost no more to build the Figure 8.4 version than to build what had actually been built. It seems however that the only design considerations had been separation and acoustic decay time control. The designer had failed, almost totally, to consider the needs of the musicians or the recording engineers. It had been built for a partly grant-funded arts organisation, and the bureaucracy had dominated events.

The sad thing about the above situation is that with a little more forethought and experience, the use of the layout of Figure 8.4 would have allowed the construction of an excellent live room in section 'Y', a good vocal room in section 'X' and the shape of the larger room would have allowed much better acoustic control without having to resort to the 'battleship' methods which lost so much usable space in the original layout. Unfortunately, such things are all too common in community and educational establishments, where specifications and tenders are dealt with by administrators, and not by artistes or engineers. Even in many universities!

8.1.1 Demands from Control Rooms

Even 'external' things, such as the control rooms, will make their demands on the design of the studio rooms. As will be discussed in later chapters on control room and monitoring design, whether you have a wonderful set of studio acoustics or not, recording is like driving a car with a greasy windscreen if the control room monitoring and acoustics are not sufficiently sonically neutral and free from idiosyncrasies of their own. To make the best of what the studio acoustics, microphones, and musicians have to offer, it is imperative that the control room should not be compromised. The most likely compromise in this area is to visibility, which is indeed very important, but there are optimal locations in control rooms for loudspeaker positions, and it is generally unwise to push the monitor loudspeakers into high locations, where the ear perceives things very differently than from a horizontal direction, all for the 'luxury' of an enormous front window.

Large windows are problems in themselves. The bigger they get, the more they tend to allow sound through, or to resonate in unwanted frequency bands. That is, unless they are of a very great weight, in which case they may need to be custom-made by a specialist glass factory, the cost of which can be alarming. Two-way closed-circuit television (CCTV) is one possible solution for visibility between the control rooms and any out of the way places. With the older, cathode-ray tube (CRT) based systems, there is often the problem of the 10 or 15 kHz time-base whistles from the TV monitors. Some people find these to be extremely irritating, but perhaps more dangerously from a recording point of view, many people, particularly over the age of 35 or 40 years, cannot hear them at all. In such cases, if the whistles do pick up on the recording microphones, and especially if any top boost is applied on the console, then they can pass unnoticed on to the master recording, only to subsequently irritate the purchasers of the final product. More recently, the advent of cameras and monitors which are not based on CRT technology has made this problem less common, and so has opened up more possibilities for the application of CCTV, but there is no real substitute for direct eye-contact.

Access to and from the control room is another important point to consider, but the position of doors may be dependent upon control room design philosophies. In situations where the rear wall of a control room is sacrosanct, and solely to be used for absorber systems, the need for front or side entrances is fixed. The rear wall is perhaps the most critical surface, as it takes the full impact of the incident wave from the monitor system, and anything coming back from that wall will be likely to colour the perception in the room. The low frequency absorber or diffuser surfaces need to be as large as possible, so penetrating them with doors is generally undesirable. Sliding glass doors in the centre of the front wall is another option, but such relatively lightweight structures so close to the monitors can be problematical. Unless the glass is *very* heavy, so much so that perhaps opening and closing the doors becomes a difficult exercise, any LF resonance which exists in the door system will not only cause a perceivable resonant overhang, but will also absorb some of the impact from the bass end of the monitor system.

If the control room is sufficiently wide, then a heavy window can be placed centrally, between the monitors, and heavy doors of composite nature can be located beyond the outer sides of the monitors. These can provide access to the studio rooms and to the reception areas or access corridors if necessary. If side doors must be provided to a control room, it is often wise for them to be as far to the back of the room as possible, to avoid disturbing the absorption system on either side of the mixing position. However, if they must be at the sides, they can often be angled so as to tend to reflect any sound which strikes them into the rear absorber or diffuser. Every situation is different, with its own set of specific demands. If the control room is to be of a type such as the Live-End, Dead-End, with a live rear half to the room, then the compromises could be somewhat different. The whole design must be integrated if the best results are to be turned from expectation into reality.

8.2 Layout of Rooms

In Figure 8.1, the control room is flanked on one side by an access corridor, and on the other side by a machine room and a vocal room. If any one of the studio rooms needs to be alongside the control room, it is better to be one of the most acoustically dead of the rooms, because the SPL which can develop in the live rooms can present much greater sound isolation problems if they are adjacent to the control room. Also, usually, the dead rooms will be used for much of the vocal overdubbing work. This process can consume a great deal of time, so a location close to the control room is convenient and expeditious. The arrangement shown allows access and visibility directly from the vocal room to and from both the control room and the studio, so whether doing live vocals or overdubs the vocalist will not feel too isolated.

Already, having discussed so little about the studio layout itself, we have been diverted by considerations from some seemingly secondary sources, but their pull is great, and any studio designed without due consideration of the widest spheres of influence will likely fail to achieve its full potential. Different designers will have their own very special priorities which they will try to avoid compromising, and each of these points is likely to be based not only on knowledge of facts, but also on insight gained through years of practical experience. This is largely what separates studio designers from other acousticians. Even amongst studio designers their experiences will have been different, and the individual tastes of the designer will be different, so this will lead to some very different designs. This is actually quite fortunate, as without variety things would be a little dull. There is therefore no 'bible' which governs this subject absolutely.

8.2.1 Priorities and Practice

Figure 8.1 shows a very attractive layout, but such a flexible option is expensive to build. The prospective owners of such studios need to be sure of their market. Studios are usually built on budgets that must be earned back over a matter of a few years, and so must be constructed with the appropriate economic considerations in mind. The studio shown in Figure 6.13 was built in 2000 at a cost of about €800,000 for the specially constructed building and all the internal acoustic work and decoration. When we cut our budgets we usually also cut our options, so a designer must consider carefully the most likely uses to which a studio will most frequently be put. If it is proposed to record many voices, or chamber orchestras, then obviously a stone room and a dead room would not be an appropriate choice of combination. There are a few studios which earn their livings from a single live room, but they tend to cater for niche markets. It is likely that if any of them were to build a second studio room, then it would either be a neutral room, a dead room, or a room with some degree of variability, but not another live room.

So let us now discuss some of the important points which went into the design of a small pair of studios which were squeezed into a space of only 100 m^2. They were designed for a wide range of music recording, plus dialogue replacement for cartoons and short films. The discussion can be used to highlight many constructional details and pertinent specifications. In the design shown in Figure 8.5, Control Room One was the main control room for music recording. Previous experience had led the owner to believe that they needed a good live room and a vocal room which would not colour the sound. The second control room was not intended to be used for serious mixing purposes, but it should be neutral enough to make a good editing room and to be able to function as a control room for overdubs or dialogue replacement. In Figure 8.5, it can be seen that Control Room Two has an adjoining Studio Two, which is only large enough for one musician and instrument at a time, or perhaps a trio of backing vocalists, but this room also had to serve as a third recording space for Control Room One. The problem was that Studio Two would best serve Control Room Two if it were a room which was as dead as the ones shown in Figure 5.19, but Control Room One already had such a room adjacent to it, so doubling up on such rooms would seem to serve little purpose. It was eventually decided to make Studio Two a bi-directional room, which could serve as a relatively dead room when being used with Control Room Two, but would have more life when used with Studio One.

It was presumed that when being used as an extension to Studio One, the musicians would be facing Control Room One, and when working with Control Room Two, the musicians could face the opposite direction without losing visual contact with Control Room Two. Figure 8.6 shows the idea in more detail. The main problem with this option was how to make a small room with the required brightness, but without the boxiness which usually accompanies such designs.

8.3 Isolation Considerations – Doors and Windows

Isolation was a great priority, not only because the two studios had to be able to work separately, but also because the poor reputation from bad separation between the previous studios which had occupied the building (and which had partly led to the need to rebuild) had to be ended once and for all. The wall between Studios One and Two was a massive,

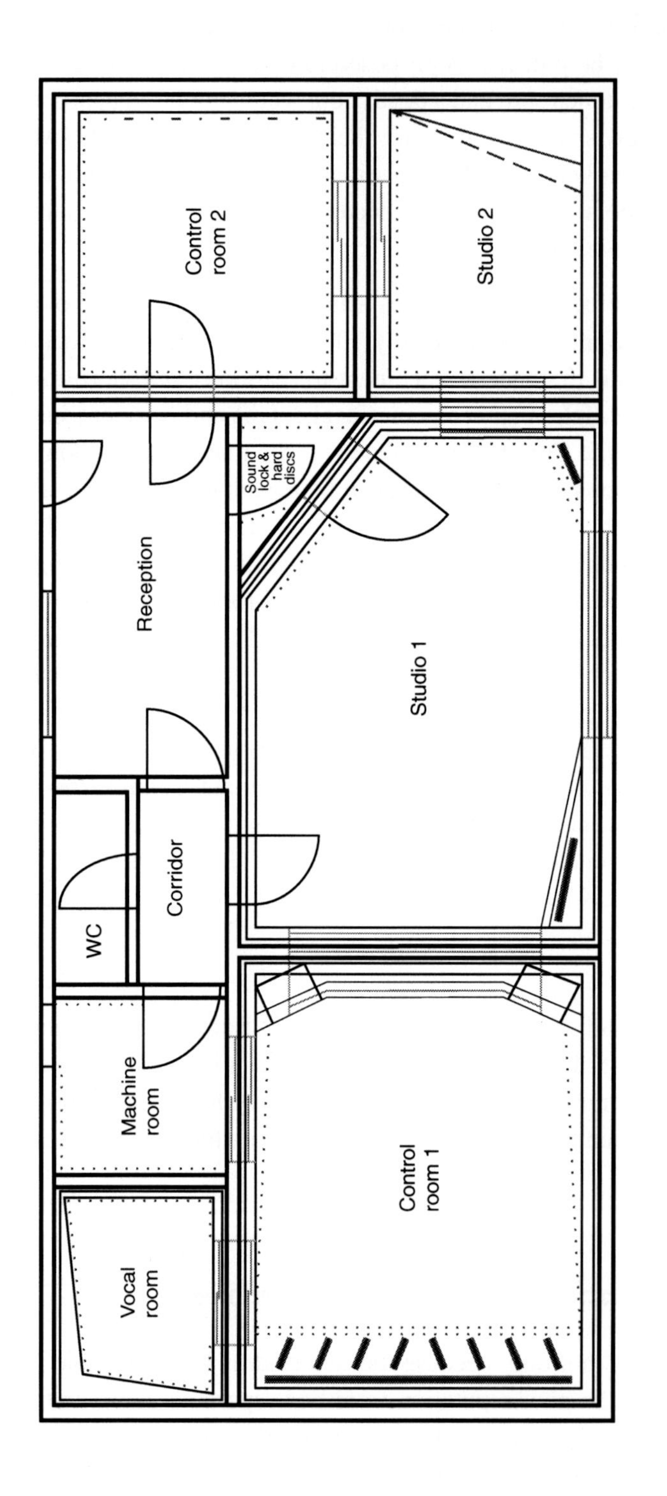

Figure 8.5:
General layout of a small studio complex – 'Tcha Tcha Tcha' Lisbon, Portugal, (1996).

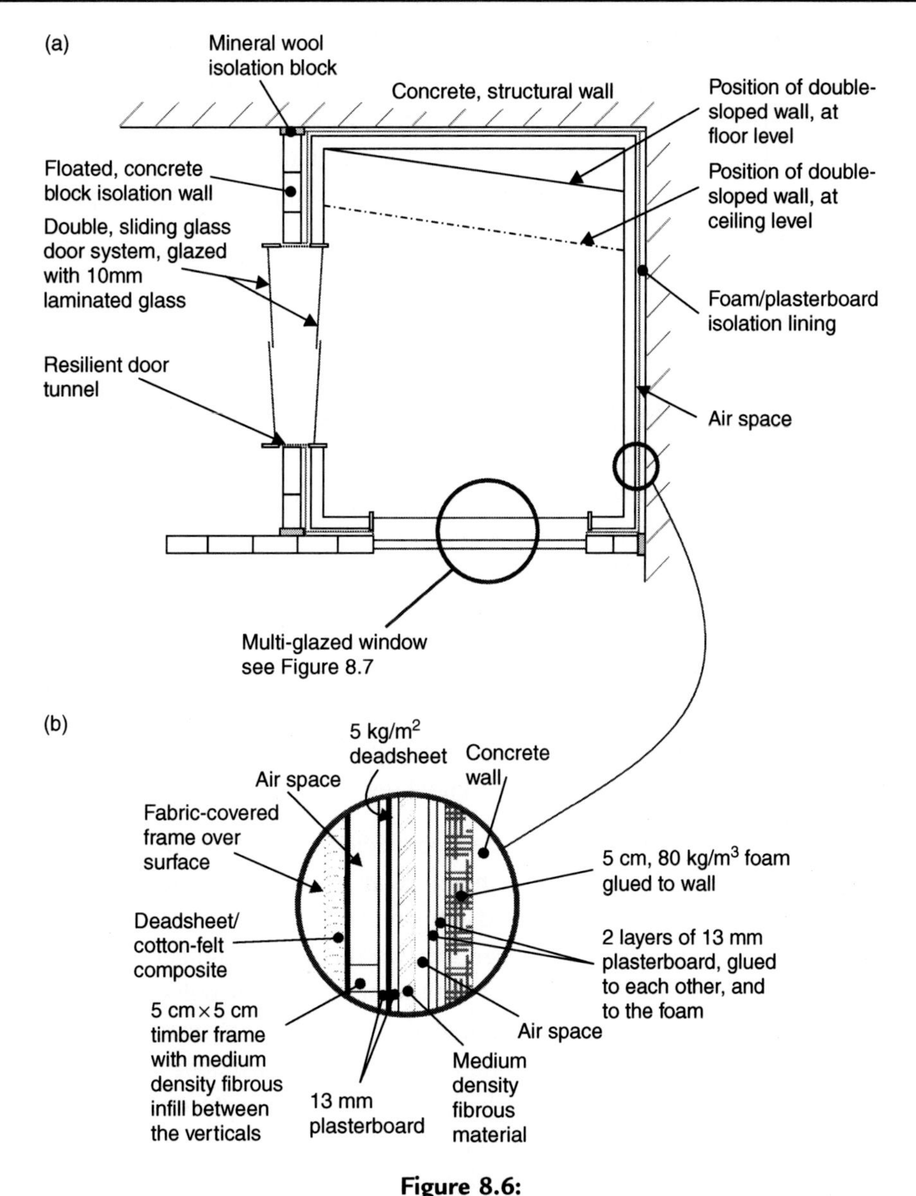

Figure 8.6:
Small room using the double-sloped 'Geddes' wall principle: (a) Plan view; (b) Detail of wall structure.

sand-filled concrete block construction, floated away from the structural walls, floor and ceiling with high-density mineral wool. Each side of the wall was treated with 6 cm of 80 kg/m^3 open cell polyurethane foam, then two sheets of plasterboard, all layers bonded only with contact adhesive. All three studio rooms and both control rooms were separately floated on 120 kg/m^3 polyurethane foam (Arkobel) but Studio Two had a second floated floor, different in density and thickness to the other floated floors. This was to avoid any common resonances in similar construction methods which could have reduced the isolation at any common resonant frequencies. The shell treatment of Studio Two was essentially the same as that described in Chapter 5. The room began as a dead room, but with the additional life of the double glass door system to the control room, plus a quadruple-glazed window to Studio One. However, before looking in detail at the system of livening Studio Two, perhaps we should first look here at the door and window systems.

8.3.1 Sliding Doors

The two door systems were each made out of two panels, one fixed and one opening (sliding). The doors were mounted in the frames of each of the adjacent, floated boxes, whose walls were separated by a sand-filled concrete block wall. The non-parallel mounting of the doors, and the soft walls of the tunnel between them, helped to reduce the resonant energy in any modes in the resulting cavity, and also allowed them to avoid forming parallel surfaces with the opposing walls inside their respective rooms. The glass chosen in this instance was 10 mm laminated glass, which is quite heavy, and also, by virtue of the laminating, very acoustically dead. The laminating layer acts as a constrained layer in the same way that the deadsheet layer operates when sandwiched between the two sheets of plasterboard in the wall structures. The laminated glass also adds an extra element of safety, as it is *very* strong, easily resisting the impact of wooden blocks thrown at it quite forcibly. There is little risk of people breaking it by accidentally walking into it or knocking a guitar against it.

Between the two doors a tunnel was constructed, connected to one door frame rigidly and to the other only by flexible bonds. The tunnel is therefore an extension of one room, but connecting resiliently with the other room. By this means, no significant solid-borne transmission of sound can take place between the two rooms. The tunnel passes through the concrete isolation wall, but it was spaced off with mineral wool or polyurethane foam, again to avoid direct contact with the isolation wall, and hence to prevent the transmission of sound into the main structure. Ideally, if conditions allow, the two doors should be of different widths, and the glass of different thickness, in order to reduce any transmission of sound via any common resonances, but one has to be careful here in applying rules too generally. A 12 mm and an 8 mm glass would probably be *marginally* better than two 10 mm glasses, but the difference is rarely worth the extra cost of different door frames, and there is a tendency for glass prices to rise disproportionately with thickness. Twenty-five per cent extra thickness can sometimes mean 100% more price. Varying the *area* of the panels is often just as effective as varying their thicknesses, and is often a less expensive solution.

The precise isolation figures of sliding glass door systems is difficult to generalise about, but they are affected by the thickness of the glass, the spacing between the door pairs, and the completeness of the sealing of gaps, although the latter only tends to affect higher frequencies. Sixty decibels at mid frequencies and 40 dB at low frequencies (60 Hz) are perfectly achievable, and such degrees of isolation suffice in many cases, but one must also beware when using large glass door systems in control rooms that any resonances may affect the monitoring. Where systems are used with differing thicknesses of glass for each set of doors (in order to stagger resonances and avoid weak spots in the isolation), the heavier glass should usually be used on the control room side to reduce the likelihood of problematical resonant absorption and overhang.

In general, double glass panes which are vacuum sealed are for thermal isolation and a sensible reduction of traffic and general street noise. They do not usually provide much low frequency isolation, and also tend to be rather fragile in inter-room locations. Heavy, toughened (heat-treated) glass is much more robust, but single panes still tend to resonate like ordinary plate glass, and may also exhibit frequency dependent weak points in the isolation. Laminated glass is normally the wisest choice, and incorporates robustness (it is a technique which can be used to make so-called 'bullet-proof' glass) with high acoustic damping and good low frequency isolation when used in adequate thickness −24 mm (12 m + 12 m) in large control room windows, for example − and with a suitable laminate in the centre of the sandwich. One must always be careful not to confuse *thermal* doubling-glazing with *acoustic* double-glazing when dealing with low frequencies.

8.3.2 Window Systems

The window system between Studios One and Two raises a number of interesting topics regarding window design. In this case, quadruple-glazing was used, with two panes mounted as widely apart as possible on the foam/plasterboard linings of the concrete centre wall and one further pane on each of the floated boxes. The sound isolation between these rooms was more important than usual because there would be times when the two studios were working independently, perhaps with a rock band in Studio One and a voice-over in Studio Two.

As explained earlier, the 20 cm, sand-filled concrete block wall was covered on each side with 6 cm of 80 kg/m^3 reconstituted foam and 2.6 cm of plasterboard, which gave a total wall thickness of just over 40 cm. The hole in this wall was sufficiently large to allow as much visibility as necessary, but no larger, as the bigger the hole in the wall, the less the resulting isolation. The lines of sight were carefully assessed and the minimum practical hole size was used for the windows. The central panes were different; one was a vacuum sealed, double unit and the other a pane of 8 mm plate. The windows in the walls of the floated boxes were correspondingly larger to allow a wider angle of visibility. The whole system is shown in Figure 8.7.

Again, as with the doors, tunnels run from the outer windows to the inner ones, fixed to the frames on the floating rooms and lightly attached by silicone rubber to the central windows. These are usually lined with carpet to avoid chattering in the gap, and are normally made of a relatively acoustically dead material of not too great a thickness, which will allow sound to

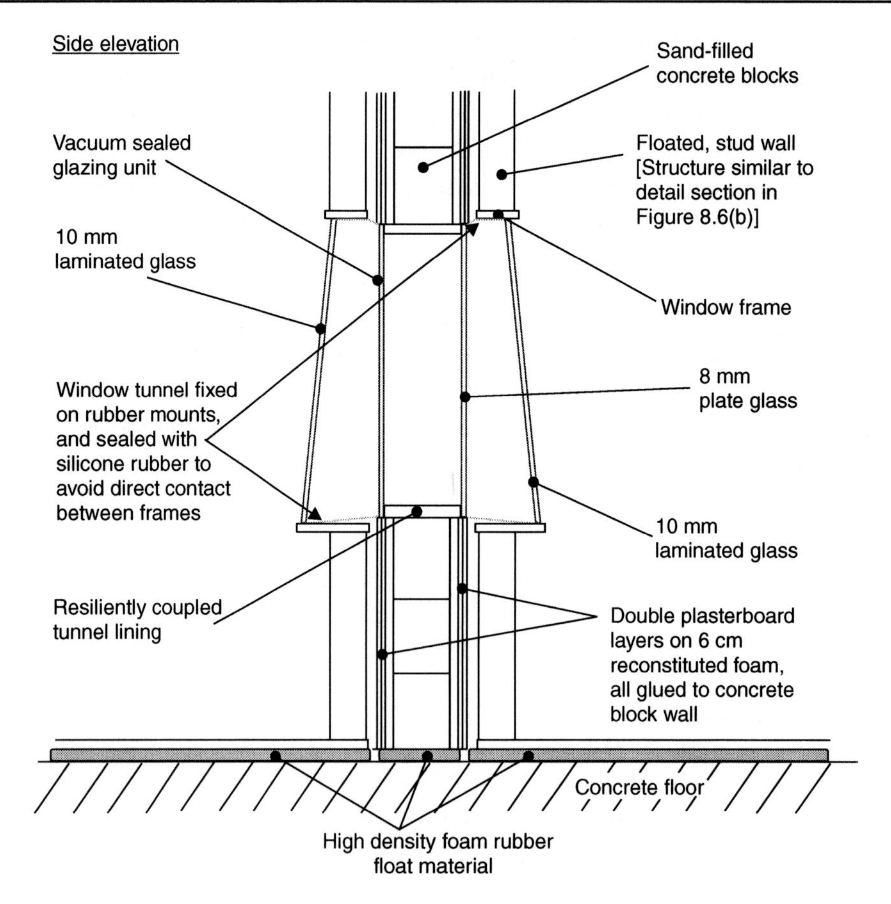

Figure 8.7:
Quadruple-glazed window system.

be absorbed not only by the tunnel lining itself, but also by the soft materials packed around the tunnel. In this specific case, the outer panes were of 10 and 12 mm laminated glass; different thicknesses were used because each was the same width and height. They were also angled such that they would tend to reflect any sound striking them towards the absorbent ceilings of each of the rooms.

It is perhaps appropriate, here, to discuss the subject of reflexion from panes of glass, or any other similar material for that matter. There is a tendency for people to think of the way that sound reflects off a hard surface in the way that light reflects off a mirror. Indeed this is the basic concept of many computerised ray-tracing programs, in a crude approximation to real life. Figure 4.3 shows the case which actually applies for low frequency reflexion, and whilst it is true that the reflective pattern narrows as the frequency rises, even a narrow beam of sound projected at a window at 10 kHz will tend towards a reflected cone of energy of around 30° apex angle, as shown in Figure 8.8.

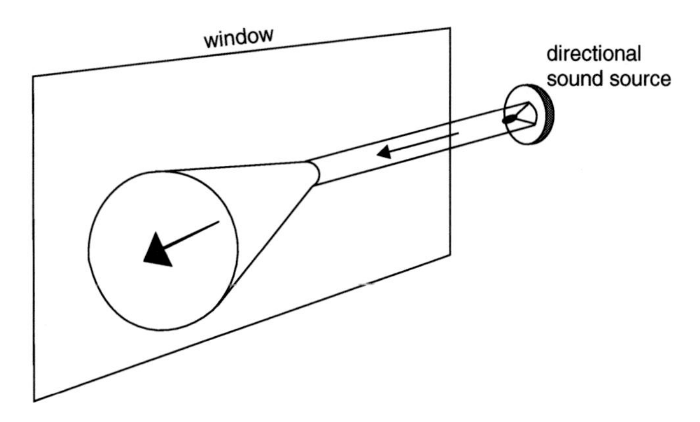

Figure 8.8:
Even a very narrow, controlled beam of high frequencies will tend to reflect from a flat surface, such as a window, with a conical pattern which will not be likely to be less than 30°, even at the highest audio frequencies.

Consequently, an angle of only 15° from vertical will still allow some energy emanating from the same height of the window to return to its source, but the general tendency will be for most of the higher frequencies to be reflected in the direction of the inclination.

8.3.3 Multiple Glazing Considerations

In the case shown in Figure 8.7, quadruple-glazing was used in order to reduce the sound transmission through each independent wall structure, and in particular to prevent the sound from reaching the concrete wall. Effectively, the inner pair of windows was on the plasterboard layers, which in turn were isolated from the concrete wall by polyurethane foam. However, in many normal circumstances, say between two simple walls, a greater space between the windows can often be more effective than more window panes. In other words, if the space between two walls is 80 cm, then it *can* be more effective to use two panes of glass of, say, 10 mm and 12 mm, than to split the tunnel into spaces of 20, 35 and 25 cm by the use of four panes of mixed 10 and 12 mm glass. It is probable that the mid and high frequency isolation of the quadruple-glazing would be better, but that the low frequency isolation would be worse, as the *space between* the panes can be a more important factor in low frequency isolation. There are few absolute, hard and fast rules to this, as the wall structures, the degree of possible angling of the glass panes, and the type of work in which the studio is usually engaged will all add their own special conditions. For example, a voice-over studio, or a talk studio, is unlikely to be placing great importance on very low frequency isolation, because low frequency sounds are not likely to exist at high level on either side of the window. In this case, the quadruple-glazing option may be preferred, whereas for a studio mainly producing dance music, with lots of low frequencies, then quite possibly the double-glazing/larger space option would be preferable. Each case must be assessed individually.

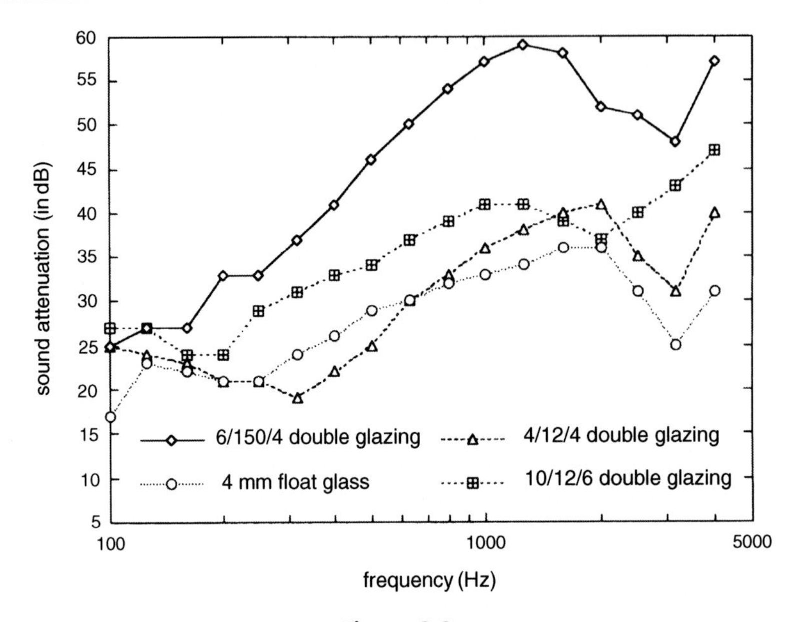

Figure 8.9:
Sound transmissions versus frequency for single and double-glazing (after Inman).

Figure 8.9 shows some interesting results first published by Inman in 1994.[2,3] It shows the measured results from examples of single and double-glazing made with similar and different thicknesses of glass and spacings. Now 4 mm glass is not something that one often encountered in studios, but in principle it is worth noting how the double-glazed unit with 4mm glass is actually worse in isolation between about 200 Hz and 700 Hz than a single pane of the same 4 mm glass. The results also show clearly how the increased spacing of the 4 and 6 mm glass (150 mm) gives far superior results to the 10 and 6 mm glass with only a 12 mm spacing.

In fact, if absorption can be placed above, below, or at the sides of the tunnels which flexibly join the glasses, a significant improvement in isolation can be achieved, but this needs to be done without compromising the isolation by causing any contact between the floated walls. One must also be careful not to leave any spaces which may allow insects to enter, or a miniature zoo may be the result after a few months. Desiccants, of course, are totally unnecessary because no condensation can exist in double-glazed windows between the internal walls of adjacent rooms. Desiccants are only relevant in window systems in exterior walls, though usually vacuum-sealed panes on the outside wall are a better solution.

8.3.4 High Degrees of Isolation

There is obviously no need to produce isolation in a window system that is any greater than the isolation provided by the walls which they penetrate. A 70 dB window system in a 55 dB wall system only succeeds in wasting money. However, if *very* high isolation window systems are required, they may become absurdly expensive. Eastlake Audio[4] installed a window system in Belgium, where an isolation of 80 dB was required between the control rooms and the studio. It was required that the various control rooms and performing rooms could be used in different permutations; hence a loud group may be recording alongside a control room which was mixing another piece of music, for example.

The glass panes were 11 cm thick (yes, centimetres!), and weighed almost 1 tonne each. Eleven centimetres was the maximum that the local glass making machines could stand the weight of, and the transport was very difficult. The original specification called for 14 cm glass, but when it was found that it could not be acquired in Europe, panes of 11 cm were used with a greater space between them than had originally been planned. The cost of all of this is better not to even contemplate. In fact, the total weight of the various rooms at Galaxy, floated on steel springs, is almost 2000 tonnes, with 9 tonnes of rubber in the walls. It is certainly impressive, but such would hardly be practical solutions in most cases. In fact, the following quotation from David Hawkins, the owner of Eastlake Audio, just about sums up the reality: 'Normally, when people ask for isolation figures of this type (80−85 dB), I generally suggest that they build the rooms in different streets.'

Conventional hinged doors also pose their problems. However, doors are usually able to be positioned in areas of the structure which are less critical than the normal window locations. What is more, right-angled bends and isolation vestibules can often be incorporated into the designs of door systems. In general, doors need to be heavy, acoustically lossy and well sealed into their frames, but perhaps an extreme case can once again be outlined by describing the doors at the above mentioned Belgian studio. They used industrial doors with a nominal isolation value of 55 dB, which were employed either in pairs, or in triple sets between critical areas. Each door weighed 300 kg and was suspended on ball bearing hinges. When the closing handles were turned, they applied a pressure of 400 kg on the door seals. A discussion of more typical studio isolation doors can be found in Section 8.8.

It is often difficult to get the fact across to people that 70 dB or 80 dB of isolation demands the same degree of treatment, irrespective of the purpose for which it is needed. 'But I only want it for practising my drums' is a cry so frequently heard. It is as though people expect that the cost of isolation somehow reflects the cost of what they are doing. Because they are only spending the cost of a drum kit, they expect that the isolation to the bedroom of a neighbour will somehow be cheaper than if they were spending a hundred times as much on a complete set of studio equipment. Obviously, (or at least it should be obvious) the two situations will have identical isolation requirement, whether the drums are used for practice only, or for a very serious recording process. Basically there are only two factors involved in extreme cases of sound isolation; great weight or great distance, and these concepts also apply to door and window systems. Section 8.8 will take a more detailed look at a rather more practical acoustic door design. Not surprisingly, though, such doors still tend to be heavy.

8.3.5 The Optimum Amount of Isolation

The question is often asked as to what is the optimum amount of isolation between the various rooms of a recording studio. Not surprisingly, the answer depends on the circumstances, but extreme isolation is certainly not always necessary. In the case of the layout shown in Figure 8.5, very high isolation was required between Studios One and Two because they could be simultaneously in use for entirely different recordings. On the other hand, it was not expected that the studio shown in Figure 8.2, which was based on the layout of Figure 8.1, would ever be working on more than one recording at a time. Consequently, whilst the external isolation is sufficient to render passing vehicles inaudible inside the studio and not to be captured as rumble by the microphones, the isolation *between* the rooms is more modest. A rock group playing at 115 dB in the main room would be measured in the control room at around 60 dB, and in the stone room at about 85 dB. In the dead room and the vocal room the level would be a little lower, around 75 dB. Even though the isolation systems from the main room to the stone room and the dead room is the same, the reverberation of the stone room tends to amplify whatever sound exists within it, whether that be from an internal source or leakage from an adjacent room.

These levels of isolation have proved to be perfectly workable in almost all circumstances. If a recording was being made of a rock group, the probable monitoring level at the listening position in the control room would be likely to be around 90 dB. The acoustic leakage of the same music at a level 25 or 30 dB higher than the monitoring level would not influence the decisions made about the sound in the control room in any significant way. At the more critical mixing phase, there would be nobody playing in the studio, and hence the leakage would not exist. Obviously, as the isolation is constant with level, quieter instruments would produce correspondingly less leakage level, so even if the monitor level were lower, the isolation would still be adequate.

In the case of the isolation between the main studio and its peripheral 'isolation' rooms, they are used as often for their different acoustic characteristics as for the isolation *per se*. Of the isolation figures quoted, the majority of the leakage will be at low frequencies. The mid and high frequency isolation will be considerably greater. If necessary, the low frequency situation can be improved by the careful choice of the location of the bass guitar amplifier. In fact, this is the only instrument which would be likely to cause problems.

As described in subsection 8.3.4, as the required isolation increases beyond the normal levels, the cost and weight can become somewhat unwieldy. Special building techniques and reinforced foundations may be needed. There is no real advantage in having more isolation than necessary, and going beyond this is only likely to waste space and construction materials.

The weight of very high isolation systems may simply render them impossible to realise in any normal building; even an industrial building.

What is more, as wall or window sizes increase, if all other aspects remain the same, the isolation will reduce. A 12 m^2 wall with a 1 m^2 window system may provide X dB of isolation. Increasing the wall size to 40 m^2 and the window size to 6 m^2 may reduce the isolation by 5 or 10 dB, even though the construction materials and techniques remain the same. This is due to the reduction in rigidity as the sizes increase. It may be very difficult to bend a 1 m^2 of 12 mm laminated glass, but a panel of 1 m × 6 m may be relatively easy to bend by pushing on it in the centre. Therefore, in order to maintain the same rigidity, it would be necessary to considerably increase the thickness of the glass. The same situation would also apply to the wall. It is therefore not possible to give a list of methods for achieving precise amounts of isolation between rooms, because it can all depend on size. Studio designers have to deal with many variables at the same time. Deciding on the required isolation and the means of achieving it is not a trivial task.

8.4 The Geddes Approach

Returning to the subject of the studio under discussion, in Studio Two of the layout shown in Figure 8.5, one great problem was how to introduce life into the room without creating any boxiness in the recorded sound. Figure 8.6 shows a floor plan, with the dotted line showing the line where the angled wall strikes the ceiling. Figure 8.10 shows a side elevation of the room. From these figures it can be seen that the wall opposite to Studio One is double-sloped. Dr Earl Geddes,[5] proposed this double-sloped wall concept in an otherwise rectangular room, initially for the more even distribution of modal energy in control room shells. Nevertheless, the technique seems to provide a useful solution to some of the more intransigent small recording room problems.

Essentially, the double slope allows a relatively steeply angled wall surface without undue loss of floor space, although this is all in relation to the room size. The steep angling tends rapidly to steer the modal energy into the oblique form, which passes round all six room surfaces. On each contact with a wall, a sound wave will lose energy by absorption, and that absorption tends in many cases to be greater for waves striking the surfaces at oblique angles than for perpendicular impacts. The oblique modes also strike more surfaces more often, which also contributes to the increased absorption. By making all surfaces absorbent, other than the glass, the wooden floor and the double-angled wall, the modal energy can be suppressed very quickly.

Figure 8.11 shows the room in typical use with Control Room Two. In this situation let us assume that voice replacement work is being done. The person speaking, or singing, is facing the double-sloped wall. The wall is panelled with wood strips, to add life to the room,

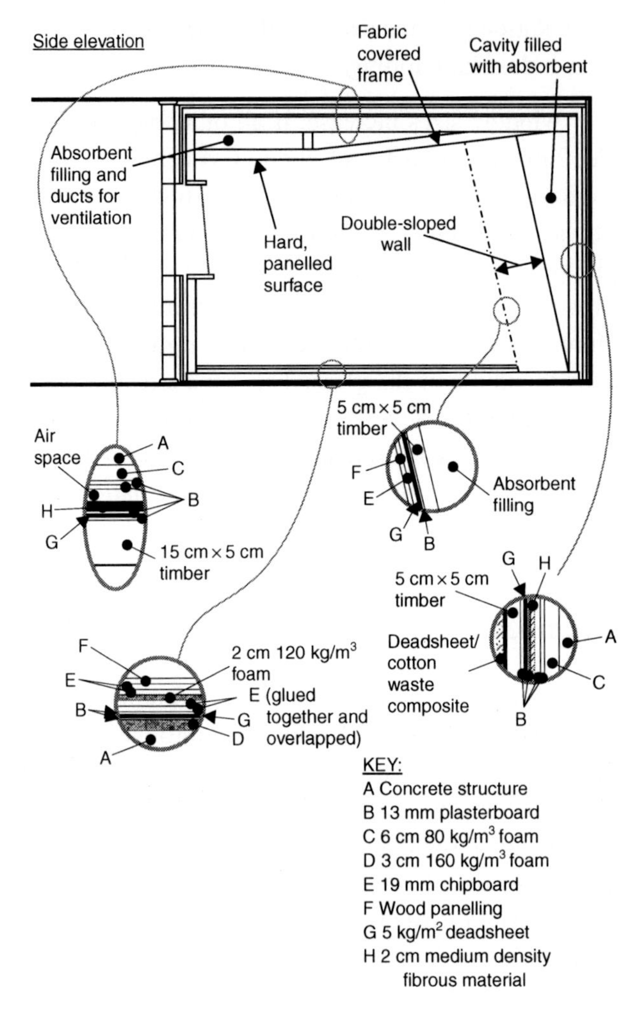

Figure 8.10:
Isolation and acoustic treatment details of a small 'Geddes' room.

and into it has been recessed a monitor television for voice synchronisation. The whole wall is hard, but only little reflected energy reaches the vocal microphone because the window behind the vocalist is angled steeply upwards, and deflects much of the energy into the absorbent ceiling. What is more, when the room is being used in this way, it is often desirable to draw curtains across the window. This gives more privacy between the two studios, and the curtains, heavy and deeply folded, help to suppress reflexions. The double-angled wall is of a relatively lightweight construction, and the whole cavity behind it is filled with mineral wool and scraps of felt and deadsheet, which press against the rear of the wall to provide considerable damping. The cavity itself is also heavily damped by the filling.

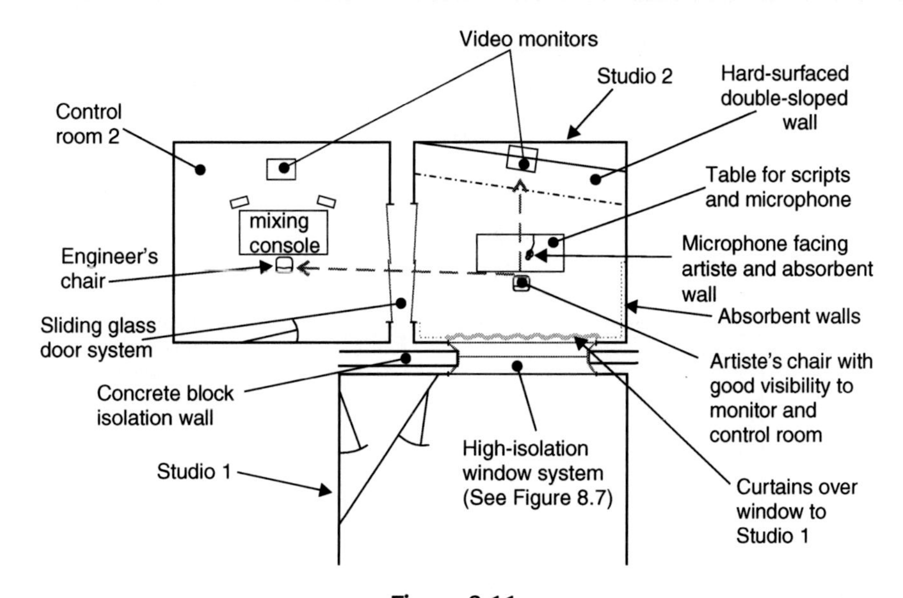

Figure 8.11:
Studio 2 in use for voice-overs.

If we now consider this room one frequency band at a time, we will see how the overall design is effective for wideband control. Due to the geometry, the mid and high frequencies can find no simple, effective, reflective path back to the microphone. The upper range of the low frequencies, for which the angling of the window may be insufficient, are not reflected much by the sloping wall, and either pass through into the great depth of absorbent filling or are absorbed by the highly damped panels of the wall structure itself. The very low frequencies are in the pressure zone of the room, so are in any case rapidly lost. The pressure zone frequency for this room is about 65 Hz. Visibility between the engineer and vocalist is good. Each can face their own video monitors, and with just a small turn of the head can see each other clearly.

In the case of the room being used in conjunction with Studio One, the musician, in this case let us presume that it is a drummer, will face away from the double-sloped wall. This will give direct visibility to Studio One and on through to Control Room One. There is also excellent visibility to Control Room Two in the event that it should be being used as an additional isolation room. This situation is shown in Figure 8.12, and it can be seen that the microphones around the drum kit are generally pointing backwards and downwards. They are thus pointing not only at the drums, but also towards the reflective floor and double-sloped wall, and hence will pick up a great deal of early reflected energy which will 'fatten' the sound of the drums. The drum kit will not sound the same as a kit recorded in a live room, nor one recorded in a large room, but it will certainly, for many types of music, be preferable to the recording of a kit in either a more conventional small room or in a dead room. There will be a more potent drum sound and more 'feel' for the drummer than would be the case in

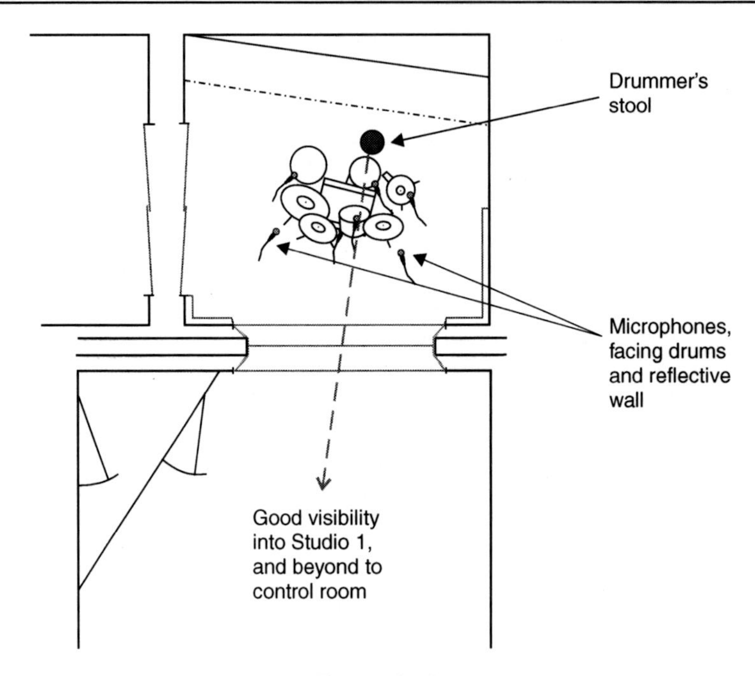

Figure 8.12:
Studio 2 in use in conjunction with Studio 1.

a small dead room, and yet only very low levels of unnatural room colouration result. The room does not possess the small-room 'boxiness' which spoils so many recordings, and has been very well received by musicians and recording engineers.

8.5 Recording Techniques for Limited Acoustics

In the small studio complex depicted in Figure 8.5, the main recording area was relatively live, having windows to Control Room One and to Studio Two on two of its sides (its two ends), and a large window to the outside on another side. This wall was of angled brickwork, installed in a sawtooth arrangement, and faced a dead wall on the other long side of the room. The floor was of wood, over which rugs could be laid if required, and the ceiling was 'V' shaped, being hard at the side nearest to the control room, progressing to soft at the Studio Two side. The room contained a grand piano, the open lid of which could face the hard surfaces if a rich tone was called for, or it could face the absorbent 'trap' wall if more separation was needed, such as when recording a jazz quartet or similar.

In this room, other than with the laying down of rugs or the use of movable acoustic screens, there was little variability provided, as the owners deemed the *range* of rooms that they had to be sufficient for all of their normal needs. Because they had insisted on such high degrees of sound isolation, and considerable space had therefore already been consumed by the massive structures, they did not want to lose any further space by the installation of systems of variable acoustics. Nevertheless, despite the lack of variability, they have a variety of ways to circumvent the restrictions of limited acoustics.

8.5.1 Moving Musicians and Changing Microphones

Figure 8.13 shows a stereo pair of microphones in a room with an appropriate ambience. If it is desired that two vocalists build up the sound of ten by means of five unison or harmonised recordings, then instead of grouping them around a single microphone and panning the different tracks into different locations on the stereo mix, a single stereo pair of microphones can be positioned at the chosen spot in the room. The vocalists then move into different locations for each take. If ten recording tracks are available for this, each stereo pair would be panned left and right in the mix, or to whatever other desired position, and the spread of panning would be built up automatically. Usually, the sound of such recordings is much more spacious, powerful and natural than the 'five, panned mono tracks' approach. If ten tracks are not available, and especially if digital recorders are used and noise build up and generation loss are not problems, each subsequent recording can be mixed with the playback of the previous recordings, carefully balanced of course, and recorded onto another pair of tracks. The previous pair of recorded tracks can then be used for the next 'bounce', and so forth. This way, only four tracks are used, and anything which may be needed to 'repair' the final

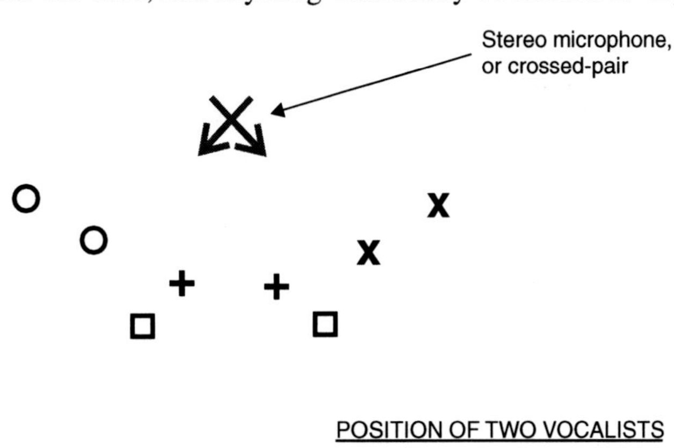

Figure 8.13:
Method of creating a more natural ambience when multi-tracking backing vocals. This technique can create a better choir effect than merely panning voices that have been recorded in a fixed position.

build up, such as earlier takes which are subsequently considered to need reinforcing, can be recorded separately on the last but one pair, and left separate till the final mixdown. By the use of this technique there is no extra ambient build up, because no extra energy is driven into the room by recording five pairs of vocalists separately than by ten vocalists singing simultaneously. The same ambient spread is also achieved in each case. It is a technique worth trying, because it reduces the build up of the same ambience on each take.

Techniques such as this can greatly improve the flexibility of rooms in which the quantity of acoustic variability is not great. Another possibility is to change microphones for different overdubs. If other microphones can be used which are still considered to be acceptable sounding for whatever is to be recorded, then as each microphone type has a distinctive and different polar pattern/frequency characteristic, the build up of the room artefacts by multiple recordings in the same place can be significantly reduced. There are limits, however, because any strong room responses may still predominate.

8.6 A Compact Studio

The studio shown in Figure 8.14 was built in London, and required not only a considerable degree of acoustic variability, but also good separation between the instruments. Its owner was a drummer, and, not too surprisingly, drum sounds had a strong input into the design considerations. Nevertheless, there was also great importance put on vocal recordings and electric guitars and basses. The main studio area subdivides by the employment of hinged panels, similar to those shown in Figure 7.6, having hard surfaces on one side, and soft, absorbent surfaces on the other. The whole studio is situated in a warehouse, with relatively few problems caused by a small amount of escaping sound. It is not on a main road, so the low frequency rumble of fast, heavy traffic is no problem either. A controlled leak of low frequencies was allowed to pass into the warehouse, thus preventing excessive build up in the smallish rooms.

Drums could be recorded in the main area, in an acoustic ambience of whatever 'liveness' or deadness that the room could provide with its variable states. Separation for guide vocals or other instruments could be achieved by creating booths out of the hinged partitions. Outside of the main recording area, and separated from it by a corridor, is an independently floated box containing three small booths. Into these, guitar or bass amplifiers can be placed, fed from instruments in the other rooms via line amplifiers. The musicians themselves can be located in the main studio, or the control room, which can also house keyboards and their attendant musicians. Once the rhythm tracks have been laid, the main room can be reconfigured for the recording of acoustic guitars, lead vocals, backing vocals, or anything which may be required for completion of the recordings. This is yet another approach to the concept of a multi-room studio with a combination of acoustics. However, it should be stressed that the

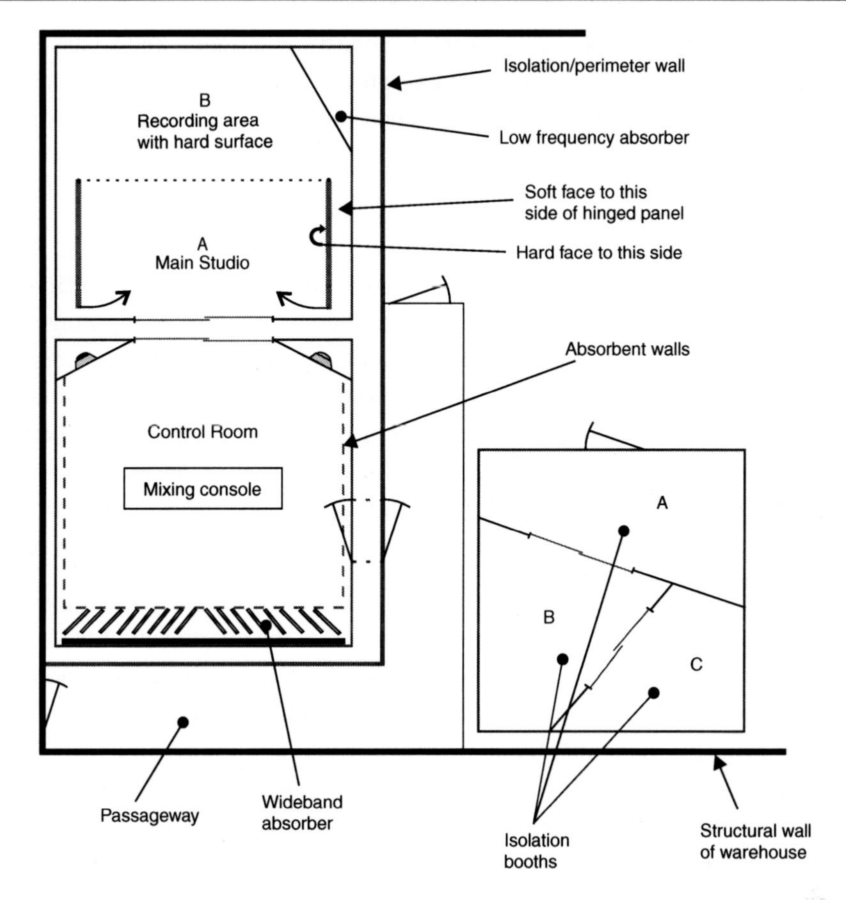

Figure 8.14:
Studio in a warehouse, with external booths. With the hinged panels in the positions shown, the whole of area A/B becomes one large, live room. When the panels are swung out to the dotted line, section B remains live, but in section A, soft, absorbent surfaces become exposed, rendering the acoustic in that area much more dead. Panels can be moved individually, or partially, to change the sound or to use them as screens between instruments. The panels are quite heavy, with wheels supporting their outer ends, so isolation between sections A and B is quite reasonable when they are swung out to the dotted line. The ceiling is an inverted V-shape, with its apex above the dotted line. Over area A, it is absorbent, and over area B, finished with wooden panelling. The isolation booths, intended for high-level recording of bass or guitars, are well isolated from the main studio and control room. Booths A and C are acoustically rather dead, whilst booth B is relatively live. Sliding glass doors are used to maximise the usable floor space. With windows in the control room entrance doors, and in the side wall of booth B, visual contact between the engineer and booths B and C could be provided. If such a warehouse were not in a noise sensitive location, then the walls could be relatively lightweight, allowing a controlled leak of low frequencies, and thus giving the rooms better LF performance without the use of too much internal LF absorption. With a lighter weight construction, costs can also be greatly reduced.

three isolation booths do not have space for much acoustic treatment, and so have a somewhat coloured sound. Their purpose was for recording guitar amplifiers via very close microphones. The isolation booths were not intended for recording acoustic instruments.

8.7 Review

Figures 8.1, 8.5 and 8.14 show three actual, workable, successful, but very different methods of tackling the problem of how to combine different acoustic options. The range of possibilities is almost endless, but these three examples give a feel for three very dissimilar approaches which could be variously adapted and modified to fit a wide range of likely starting conditions. Again, however, there are few absolute rules, which is why the experienced studio designers are so frequently relied upon for their expertise. Their function is not only to offer design suggestions, but also to point out things which may lead to trouble in the future. Such points are not always obvious to the clients, and the problems are often only realised after past experience has pointed them out.

As a further point of possible interest, the rebuilt version of the aborted attempt to construct the studio as depicted in Figure 8.3 is similar to that shown in Figure 7.1. The studio now incorporates a wooden live room, an acoustically dead room and a moderately large general studio area with reflective, absorbent, and diffusive walls, not too dissimilar to the ideas outlined in Figure 5.18. It works well, and it also ended up costing far less to build than the €60,000 which had already been spent on the partial completion of the originally proposed design.

The idea of yet another whole studio concept is illustrated in Figure 8.15. As shown in Figure 8.15(a), the studio takes the form of a 'village' within a building. The hexagonal room in the centre of the photograph is a vocal room, which leads off a granite live room. The main neutral recording room is beyond, with skylights rising above it. At the back, with an external entrance on the right of the photograph, is the control room. The balcony from where the photograph was taken has facilities for the preparation of food and drink, and also incorporates a lounge area with satellite television projected onto a large screen. The idea behind the design was to take the musicians into a self-contained space, isolated from the outside world, in which they could relax in absolute privacy. Despite the sense of spaciousness and tranquillity, it is less than 30 m from a busy commercial street, with shops and many restaurants of different ethnic origins.

8.8 Typical Isolation Door Construction

In many cases, people without sufficient experience opt for 'seemingly' high isolation industrial doors, but their high degrees of isolation may not extend to low frequencies. It should

(a)

(b)

Figure 8.15:
The Pink Museum (now The Motor Museum), Liverpool, UK (1988): (a) The studio takes the form of a village within a building; (b) View from the stone room into the vocal room, which can be fitted with curtains when necessary. This room can be seen in the centre of photograph (a).

be remembered that few industrial processes produce 115 or 120 dB SPL in the 40–80 Hz band, but such levels in a rock music studio are quite normal. The isolation curve of a 37 dB 'acoustic' door is shown in Figure 8.16. Indeed, in the mid frequencies the isolation *is* over 37 dB, but by around 100 Hz it is only 20 dB, and falling with frequency. Thus, at 60 or 80 Hz with 120 dB SPL on one side of the door one could expect 100 dB, or so, on the other side – hardly high isolation! Obviously, with two such doors and a decent space in between them,

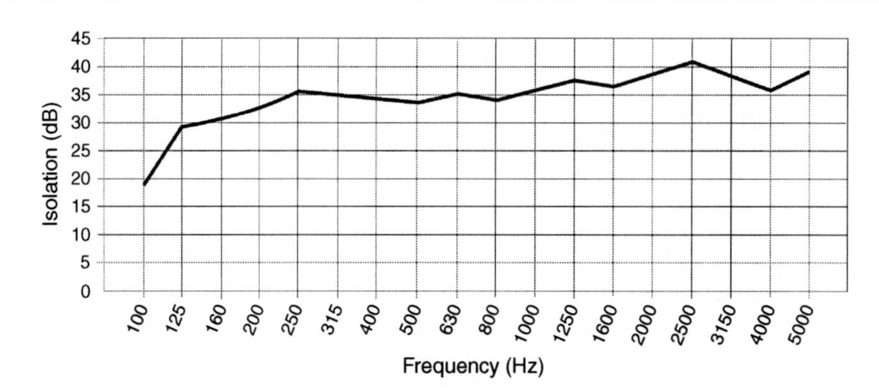

Figure 8.16:
Door isolation versus frequency. Isolation plot of a nominally '37 dB' acoustic isolation door. Note how below 250 Hz the isolation begins to fall. Standard isolation graphs such as this one rarely quote isolation below 125 Hz, and often the isolation around 50–63 Hz can be *much* lower than the nominal isolation figure, which approximates to a dBA curve.

perhaps 50 or 60 dB of isolation could be achieved, but that may still allow 60 dB SPL on the other side of the door system.

If extreme isolation is needed then the 300 kg doors described in Section 8.3.4 may be needed, but much success has been achieved with double door systems made from plywood sandwiches as shown in Figure 8.17. The doors also look a lot more 'creative' than industrial doors; a point which is worth considering in an artistic environment. (However, in radio stations or voice studios, ready-made doors may be an easy solution because low frequency isolation is not usually an issue in these recording environments.) Another advantage of the doors shown in Figure 8.17 is that they can be made to measure, and can usually be made by a carpenter for considerably less than the cost of buying the industrial style doors.

Figure 8.18 shows the hydraulic closing systems which are suitable for such doors. When used in conjunction with simple grab handles, to pull the doors open and closed, the hydraulic retainers also act as shock absorbers, making it impossible to slam the doors against the frames. The retainers should be positioned in the top, centre of the doors unless structural restrictions necessitate alternative locations.

8.9 Rectangular Room Shells

It became evident after the publication of the earlier editions of this book that many readers were surprised to find that almost all of the rooms described were in rectangular, and not irregularly-shaped, isolation shells. As discussed in Section 5.4, what is meant by parallel depends very much on the wavelengths of the frequencies involved, Figures 5.8 and 5.10 clearly show the futility of the angling of the walls, even with frequencies as high

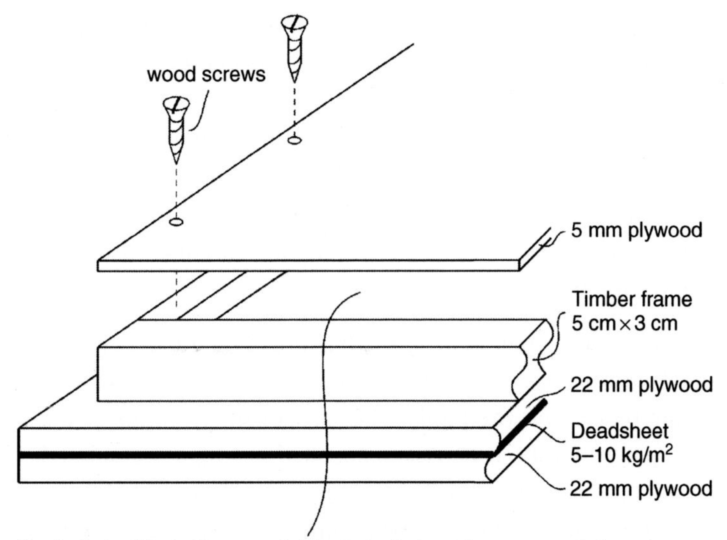
wood screws
5 mm plywood
Timber frame
5 cm × 3 cm
22 mm plywood
Deadsheet
5–10 kg/m2
22 mm plywood
Cavity to be filled with a sandwich of plasterboard, sponge or felt, and deadsheet to just above the level of the wooden frame. When the 5mm plywood lid is screwed down, it will belly and put the internal materials under compression, significantly damping the door. The upper and lower layers, in contact with the plywood, should be the soft materials.

Figure 8.17:
Acoustic doors.

Figure 8.18:
The hydraulic brake on the door engages with the hook on the frame when the door is closed. A hydraulic damper controls the rate at which the door can close the last few centimetres (as the roller on the brake is forced vertically upwards by the shape of the hook). Once 'over centre' the hydraulic ram works in the opposite direction, forcing the roller upwards and maintaining the door tightly closed.

as 70 Hz; Section 5.3.2 discussed the concept of 'ideal' room dimension ratios for rectangular rooms and the formulae for calculating the modal frequencies were given in Section 4.2. In reality, these dimensions and formulae have little bearing on recording studio design. To dispense with the formulae first, they relate to sealed rooms with perfectly flat, infinitely rigid walls. However, nobody records in such conditions, and so once the rooms or their internal linings become less than rigid, due to the installation of diaphragmatic acoustic control shells, or significant amounts of absorption or diffusion, the formulae become rather useless.

The 'special' room ratios only give the true modal spread of frequencies for conditions where the source of sound *and* the receiver are placed in three-surface corners of the rooms, where they excite and receive *all* the modes. For any other positions of the source and/or the receiver, the modal spread will be less dense, and will not be as calculated. Concepts of listening rooms based on these dimension ratios are very flawed, and have largely been discredited. Unfortunately, there is an enormous amount of nonsense on the Internet about the values of such room ratios, but it should be utterly ignored.

The effect of angling the walls of a room is only really beneficial in the reduction of flutter-echo related problems between hard surfaces at higher frequencies. Once a room is *not* perfectly rectangular, and does not have perfectly rigid walls, there are no formulae for accurately calculating the modes. In general, the angling of the walls of the outer shell of a studio only serves to waste space. In all the studios discussed in this chapter, the degree of internal acoustic control which has been introduced into the rooms damps the modes to such a degree that the shapes of the isolation shells are largely unimportant. Yes; the rooms still have low frequency modal activity because they are not large anechoic chambers. Yes; it is still wise to avoid a perfectly cubic room if it is only going to be lightly treated, acoustically, but once something such as is shown in Figures 6.14 and 8.10 is introduced into a room, the shape of the isolation shell becomes completely insignificant.

8.10 Summary

For the greatest flexibility in recording, a range of different acoustic spaces is desirable.

Many things need consideration, such as the control room requirements, inter-room visibility, and ease of access to frequently used areas, for example vocal rooms.

Specialist studio designers tend to have much more knowledge about such needs than general acousticians.

Building a studio correctly usually costs no more than building it incorrectly.

Inter-room visibility is important, but it should normally not be allowed to compromise the acoustic needs of the rooms.

Closed-circuit television systems based on cathode-ray tube technology can create problems with their time-base whistles.

Triple-glazing is not always preferable to double-glazing. Double-glazing with a large space between the panes can be better at low frequencies than triple-glazing in the same overall space.

Window panes should be of different sizes or thicknesses, to avoid weak spots in the isolation at the frequencies of coincident resonances.

Door and window systems generally do not need to achieve isolation any greater than that of the walls which they penetrate.

The need for very high degrees of isolation can lead to greatly increased costs.

Double-sloped walls can be used in small rooms to drive modes into the oblique, more 'lossy' form, without using too much floor space.

When rooms have only limited acoustic flexibility, microphone and recording techniques can introduce an extra degree of variability into the recording process.

Special room ratios are of very limited value in the acoustic design of rooms. They tend not to relate to the normal working layouts of studio rooms, nor to practical constructions with doors and windows.

When adequate acoustic treatment is installed, rooms of rectangular shape are often at no disadvantage compared with non-rectangular rooms. In fact, they tend to offer a more efficient use of space.

References

1 Newell, Philip, *Recording Spaces*, Focal Press, Oxford, UK (1998)
2 Inman, C., 'A Practical Guide to the Selection of Glazing for Acoustic Performance in Buildings', *Acoustics Bulletin*, Vol. 19, No. 5, pp. 19–24 (September/October 1994)
3 Howard, David M. and Angus, James A. S., *Acoustics and Psychoacoustics*, 2nd Edn, Focal Press, Oxford, UK, p. 319 (2001)
4 Schoepe, Zenon, 'Galaxy', *Studio Sound*, Vol. 36, No. 10, pp. 42–4 (October 1994)
5 Geddes, Earl, 'An Analysis of the Low Frequency Sound Field in Non-Rectangular Enclosures using the Finite Element Method', Ph.D. dissertation, Pennsylvania State University, USA (1982)

CHAPTER 9

The Studio Environment

Lighting. Ventilation and air-conditioning. Foldback systems. Studio decoration. AC mains supplies. Earthing.

9.1 Some Human Needs

Much has been written in the preceding chapters about the studio rooms being, first and foremost, rooms for musicians to play in. Obviously, all the individual tastes of every musician cannot be accommodated by any one studio design, so the 'ultimate' environment cannot be produced. Nonetheless there are a few general points which are worthy of consideration. Light colours, for example, make spaces feel larger than they would do if finished in dark colours. In general, they tend to create less oppressive atmospheres in which to spend long periods of time. Daylight in studios is also widely recognised as being very popular.

9.1.1 Daylight

Traditionally, it has been the out-of-town studios which have been more inclined towards allowing daylight into the recording areas. Initially it was probably easier to do this in the relative tranquillity of rural life, as the sound isolation problems of inner city locations were not so great. The knowledge of the changes from day to night, and more unpleasantly, after very late sessions from night to day, along with an awareness of the changing seasons, are all instrumental in helping a person's well-being. This fact is perhaps more medically recognised now than it was only a few years ago. Once it had been established that the sound isolation in studios with outside windows was adequate, daylight soon became widely accepted as a desirable asset for studios.

9.1.2 Artificial Light

The use of imaginative lighting to create moods can be highly beneficial to the general ambience of a studio. Fluorescent lighting, apart from the 'hardness' of its light, is generally taboo in recording studios because of the problems of mechanically and electrically radiated noises. The mechanical noise problem can sometimes be overcome by the remote mounting of the ballast chokes (inductors), which prevent the tubes from drawing excess current once they have 'struck' (lit), but the problem of the radiated electrical noise can be a curse upon electric guitar players. In some large studios, fluorescent lights can be used without apparent noise problems in very high ceilings (6 m), well above the instruments, but in general they are likely to be more trouble than they are worth.

Lighting is a personal thing, of that there is no doubt, and the design of studio lighting systems is quite an art form in itself. Many people do not like the small, halogen lights, as they find their light too stark. Argon filled, tungsten filament bulbs are still a very practical solution, though interior designers seem to eschew them. Figure 6.2 shows a room with floor-level, concealed, tungsten strip-lights, illuminating the stonework and casting shadows of the surface irregularities up the wall. The down-lighting is provided by old theatre floods. Figure 6.3 shows the use of wall lights and an 'old' candelabrum, to create the effect of an old castle. The room in Figure 6.7 was expected to be used by many session musicians, so in its ceiling it had both tungsten 60° down-light reflectors and a skylight to allow in daylight. These forms of lighting would facilitate the reading of sheet music. If desired, for more mood, the skylight could be covered, and daylight would then only enter via the stained glass window, which could also be externally lit at night to great effect. 'Mood' lighting was also provided by electric 'candles' on the walls.

In all of the above cases the lighting was controllable not only by the switches of individual groups of lights (or even single lights in some cases) but also by 'Variacs'. The Variacs are continuously variable auto-transformers, which despite their bulk and expense (roughly €80 each for the 500 VA [watts] variety, and 15 cm in diameter by 15 cm deep) are the ideal choice of controller in many instances. They cannot produce any of the problems of electrical interference noise, which almost all electronic systems are likely to create from time to time. Variacs simply reduce the voltage to the bulbs, without the power wastage and heat generation of rheostats, which dim by resistive loss. It *is* possible by more complex means to produce a 'clean', variable AC, but the methods of doing so tend to be even larger and more costly than Variacs, whose pure simplicity is a blessing in itself.

Electronic dimming systems using semiconductor switching can be a great noise-inducing nuisance, both via the mains power (AC) and by direct radiation. It is simply not possible to switch an AC voltage without causing voltage spikes. So-called 'zero voltage switching' may still not be zero current switching, as the loads which they control will be unlikely to be purely resistive. A theoretically perfect sine wave can only exist from the dawn of time to eternity, and cannot be switched on or off without transient spikes. Even if you find this hard to understand; believe it! (A positive-going half-cycle of a sine wave starts as a gradual continuation of the previous negative-going half-cycle. On a plot it can be seen as a smooth transition. On the other hand, if the positive half-cycle begins from zero, it has to begin from a horizontal line on the display. There will therefore be an abrupt change in the direction of the plot, and this can only be achieved by the inclusion of multiple higher frequencies in the overall signal.)

From the early 1970s to the mid-1980s many studios installed coloured lighting as well as the white light for music reading and general maintenance. Suddenly, they became *passé* for a decade or so; but recently they have again begun to occasionally emerge. Such are the cycles of fashion. Anyhow, whatever lighting system is used, the general rule should be 'better too much than too little'. One can always reduce the level of lighting by circuit switching or by voltage reduction; but if there is insufficient total lighting for music reading or maintenance, work can become very tiresome.

9.1.2.1 Low-consumption lighting

Since the beginning of the twenty-first century, the pressure has increased on designers to try, where possible, to use lighting of low consumption. The most common types of such bulbs tend to be of the mini-fluorescent type, or LEDs (Light-Emitting Diodes). Whether these are suitable in any given studio environment may be a question of trial and error, and may depend on the ceiling height and the types of instruments to be recorded. Many old, classic, electric guitars and amplifiers are very high impedance devices which do not reject interference very well. A useful test of the suitability of various types of lighting can be to make a cable with a power (mains) plug on one end and a bulb-holder on the other end. The bulb under test can be mounted in the holder, the plug can be connected to an AC power socket and then brought close to the guitar and the amplifier, with the volume set to a realistic level. Some of the low-consumption lights will give rise to buzzing sounds, either by injecting interference into the guitar pick-ups or into the amplifier circuitry. Clearly, an array of 20 or more of such lights in the studio ceiling would be a very likely source of problems.

In many countries, modern electrical regulations seem to be demanding the use of low-consumption lighting, but in almost all cases there are 'get out' clauses for special circumstances. Unfortunately many professional electricians are unaware of these exemptions, and so insist on using the low-consumption bulbs and telling the clients that the exemptions do not exist. This seems to be a growing problem with the ever-changing and expanding regulations, and the general 'dumbing down' of training programmes. In 2011, tungsten filament bulbs such as R63s are still being routinely fitted into new studios under the full view of the inspectors where alternatives can be shown to be unsuitable. In fact, it can also be shown that the use of appropriate lighting, even with filament bulbs, can often use no more power than the indiscriminate use of unsuitable low-consumption bulbs. What is more, there is a question of health. Inappropriate lighting can lead to headaches and eye fatigue. The colour of some of the LED lighting can also be rather harsh and unpleasant.

As with so many 'green' issues, the reality of whether any ostensible energy-saving measure will *really* save energy is often very difficult to calculate.

9.1.3 Ease and Comfort

General comfort is also an important issue, as again, comfortable musicians are inclined to play better than uncomfortable ones, and few things seem to kill the sense of comfort and relaxation more than an untidy mess of cables and other 'technical' equipment. Musicians should never be made to feel in any way subservient to the technology of the recording process. The studio is there to record what *they* do; they are not there to make sounds for the studio to record. Sufficient sensitivity about this subject is all too frequently lacking in studio staff, especially in many of those having more of a technical background than a musical one.

Easy access to the studio, for the musicians, is another point worth considering well when choosing the location for a studio. Studios have been sited on the fourth floor of buildings with no lifts, but their existences have usually been rather short lived. Especially for musicians with busy schedules, easy access and loading, together with easy local parking, can be the difference between a session being booked in a certain studio or not. It is also useful to have some facility for the storage of flight cases and instrument cases, *outside* of the recording room. Apart from the problems of unwanted rattles and vibrations which the cases may exhibit in response to the music, when a studio looks like a cross between a warehouse and a junk room it is hardly conducive to creating an appropriately 'artistic' environment in which to play.

The points made in this section are not trivial, though they are not given their due attention in all too many designs. Experienced studio owners, operators, engineers, users and designers, appreciate these points well, but in the current state of the industry, a large proportion of studios are built for first-time owners. As so many of them fail to realise the true importance of these things, they are frequently the first things to be trimmed from the budget, especially when some new expensive electronic processor appears on the market midway through the studio construction. Most long-established studios do provide most of these things though, which is probably one of the reasons why they have stayed in business long enough to *become* long-established.

9.2 Ventilation and Air-Conditioning

It is certainly not only in studios that people complain about the unnatural, and at times uncomfortable, sensation of air-conditioned rooms. Unfortunately though, with sound isolation usually also being good thermal isolation, and given that people, lights and electrical equipment produce considerable amounts of heat, air-conditioning of some sort is more or less a mandatory requirement in all studios. In smaller rooms, such as vocal rooms, where few people are likely to be playing at one time, sometimes a simple ventilation system is all that is required. In fact, it can be advantageous in many instances to provide a separate, well-filtered, ventilation-only system in addition to any air-conditioning. It is remarkable how many times these systems have refreshed the musicians without having them complain of dry throats.

9.2.1 Ventilation

In order to get the best out of a ventilation system there are a few points worth bearing in mind which are of great importance. One of the greatest rules is never to only extract air from the room. If an extraction-only system is used, the room will be in a state of under-pressure; a partial vacuum. Whenever a door is opened (or for that matter, via any other available route), air will be drawn into the room. This air will be dirty, because it cannot be filtered. Dirt and general pollution will also enter the room along with the air, perhaps affecting the throats of singers and leaving dust everywhere.

In rooms which only require a low volume of air flow, such as vocal rooms, it is often possible to simply use an input-only fan, with the air being allowed to find its own way out via an outlet duct of suitable dimensions. When high volumes of air flow are necessary, extraction fans *are* usually fitted, but there are some basic rules which should be followed. For the reasons mentioned in the previous paragraph, the extraction fans should never be operating without the inlet fans, as they would constitute an extract-only system and would suffer all the drawbacks. What is more, the flow rate of the extraction system should never exceed that of the inlet system, as this would still produce an under-pressure in the room. It is prudent to restrict the flow rate of the extraction system to between 60 and 80% of the inlet flow, dependent largely upon the degree to which air can find its way out of the rooms by routes other than the intended outlet system. In purpose-designed studios, the extraction flow rate can usually be almost as high as the inlet flow rate, because the rooms are normally more or less hermetically sealed at all points other than via the air-change system(s). However, multi-functional rooms, which may serve as studios from time to time, are often somewhat more leaky through their doors, windows and roofs, in particular. In these cases a proportionally lower extract flow rate would be desirable (or even none at all), as over-pressures may be more difficult to achieve.

The normal way to ventilate is to draw air in from outside through a filtration system. This would usually consist of either a single filter or a series of filters, with removable elements which can easily be cleaned and/or replaced. Air will then pass to an in-line fan and on to a silencer, or series of silencers, before entering the room. This way, the room is kept in over-pressure, so, any time a door is opened, the clean, filtered air in the room will escape through the door, keeping the dirty, outside air from entering through the doorway. Such systems keep the rooms cleaner, and ensure that the air which enters the rooms is always clean. Outlets will normally pass first through silencers (both to prevent street noise from entering and studio noise from leaving) and then on out to the outside air, perhaps via a back-draught damper. The damper is a one-way flap arrangement which only allows air to flow out. If the ventilation system is switched off, and the wind is in an unfavourable direction, then should air attempt to enter via the outlet, the flaps close off the ducts. This also prevents dirt from entering via the unfiltered outlets. Fire dampers are sometimes also installed, which close the air ducts completely should the room temperature rise above a predetermined limit. By this means, if the temperature rise is due to a fire, the oxygen supply to the fire will be cut off, thus slowing the spread of the fire even if not extinguishing it completely. Another point to remember is to turn off the ventilation system when the studio is unoccupied, so that if a fire should begin it will not be supported by a constant supply of oxygen. A typical ventilation system outside of a studio with three rooms is shown in Figure 9.1.

The ventilation ducts will often be of the 'acoustic' type, such as Decflex Sonodec, having a thin, perforated aluminium foil liner, then a wrapping of 5 cm of fibrous absorbent material, and finally an outer metalised foil covering. In this type of ducting the air is allowed to pass through a relatively smooth inner tube, which presents only minimal friction, and hence loss of air flow. This smooth inner lining must be as acoustically transparent as possible to allow the sound to be absorbed by the fibrous layers around it. If the sound cannot easily penetrate the inner lining of the duct, it will travel very effectively along the tube. Sound can travel along kilometres of smooth bore tubing with remarkably little loss, as no expansion

Figure 9.1:
The ventilation system of a studio with three rooms. On the shelves are, from left to right, the filter boxes, the axial inline fans, and the 90 cm long silencers. All the flexible ducting is of the acoustic type, which itself acts as an in-line silencer, except for the extensions on the outlet ducts to the right, which are simple, flexible aluminium tubes. The air inlet is where the cable tubes can be seen, at the bottom left of the picture. Normally, to comply with building regulations, the inlets and outlets should be at least 5 m apart, to prevent recirculation of exhausted air. Audio Record Studios (now 'El Estudio de Domi), Morón de la Frontera, Spain (2001).

of the wave can take place. (The double distance rule [see Section 2.3] does not apply.) This is the principle by which the speaking tubes of the old ocean liners were able to provide excellent communications, often over long distances and in noisy surroundings; bridge to engine-room, for example. Figure 9.2 shows an installation of acoustic ducts running through the back of an absorber system.

Duct *size* is also very important. For any given quantity of air to pass, the air flow down a duct of large diameter will be much slower than down a narrower duct. A 20 cm diameter duct has an approximate cross-sectional area of 315 cm^2, and a 30 cm diameter duct an area of 709 cm^2, which is more than double. Therefore, for any given rate of air flow, the speed of the air down the 30 cm duct need only be just under half of that down the 20 cm duct. As the noise caused by turbulent air flow follows something like a sixth-power law, then for any given flow rate, ventilation system noise will rise rapidly with falling duct diameters. For adequate air flow, even in the smaller rooms, 20 cm seems to be about the minimum usable duct diameter. Appropriate fans with a flow rate of about 700 m^3 per hour on full speed will usually suffice for this diameter of tube. Speed control should again be provided by variable or switchable transformers, and not by electronic means, to avoid interference.

Some local authorities may demand that ventilation systems meet certain building regulations, but common sense can usually be applied. In many cases, the incorporation of some of the regulations intended to save energy can often *consume* more energy than necessary. Ten litres per second per person is absolutely adequate for ventilation systems in recording studios, especially now that in many countries people no longer smoke in studios. This equates to about

Figure 9.2:
'Acoustic' ventilation tubing passing through the absorbent rear 'trap' of a control room, during construction, before the mounting of the principal absorption materials. Sonobox, Madrid, Spain (2001).

35 m^3 per hour. A room for four persons would therefore suffice with an air change of around 150 m^3 per hour, yet the building regulations may require 10 times this figure. This can be enormously wasteful, especially if the ventilation system is of the constant loss variety, which means that outside air is simply pumped through the room, from inlet to outlet. With an unnecessarily high flow rate, any heating or cooling in the room will be severely reduced in efficiency, but such is the blind stupidity of many regulations. They are meant for more normal circumstances, which recording studios are *not*.

Heat recovery systems are also sometimes required, which pass the incoming air and the outgoing air across heat exchangers, but once again they are not always as 'green' as they seem. In principle, if the air outside is cold, the outgoing air at room temperature can warm the incoming air to some degree (and cool it if the air outside is hot) if the two are passed across different surfaces of the exchanger. However, when flow rates are low, the power used in the motors of the recovery systems, plus the additional flow resistance to the air which can lead to increased power being required from the ventilation fans (which may then produce more noise and require more silencing) can consume more energy than they save. Studios can frequently be much greener when some of the 'sledgehammer' green regulations can be avoided (legally, of course).

9.2.2 Air-Conditioning Systems and General Mechanical Noises

Traditionally, professional recording studios have used conventional, ducted air-conditioning systems, and noise floors of NC20 or less have been achievable, even with quite high rates of air flow. Such systems are still the only way to properly air-condition a studio, but as has always been the case, they are relatively expensive.

Since the early 1980s, the real cost of studio equipment has been falling. Partly as a consequence of this, and somewhat unrealistically, the charges per hour for the studios have plummeted. At a time when €100,500,000 are spent on the recording equipment for a studio, to have to pay €150,000 for an air-conditioning system may not seem too disproportionate. On the other hand, when manufacturing technology has made it possible to buy for €150,000 an entire set of equipment that will not perform far short of the equipment costing €1,500,000, people seem to baulk at similar charges for good air-conditioning. Competition between studios has forced real prices ever downwards, and a state of affairs has now developed where the cost of ducted air-conditioning systems, for most of the mid-priced studios, has become insupportable. It must be said, though, that this is a situation driven by commercial realities, and that the need for good, conventional air-conditioning systems is as much a professional requirement today as it ever was.

It is a strange situation, exactly analogous to that which exists with sound isolation costs. People seem to expect sound isolation to be cheaper if they 'only' want to use a room to practice drums, as opposed to the more 'serious' purpose of recording. Similarly, people seem to expect to have to pay less for the air-conditioning systems if they are paying less for the recording equipment. There is simply no logic in any of this.

Largely for economic reasons, therefore, there has been a great increase in the number of 'split' type air-conditioning systems coming into studio use. Although these are by no means ideal for this purpose, they are many times cheaper than the ducted systems, but as the heat exchangers and their fans are in the studio, with only the compressors remaining outside, there is an attendant noise problem. In control rooms the units can usually be left running in 'quiet' mode, as the noise which they produce is often less than the disc drives and machine fans which may also be in the room (but which really should not be there), but in the studio rooms they usually must be turned off during quiet recordings. Unfortunately this intermittent use can lead to temperature fluctuations which may not be too good for the consistency of the tuning of the instruments. Nevertheless, despite their problems, there are now many split systems in studio use. The use of multiple, small, quiet heat exchangers is normally preferable, where practicable, to the use of larger, single units. Some small units now produce less than 30 dBA of background noise at a distance of 1 m when used on low speed. Such a unit is shown in Figure 9.3.

Studio customers have become used to inexpensive recording, and all too few of them now want to pay for facilities which can provide all the right conditions for optimal recordings. Unfortunately, as this has become widespread, it has also apparently become *acceptable* in these 'market forces' days. Air-conditioning systems have in very many cases been tailored to a 'reasonable' proportion of the recording equipment budget, which has led to unsatisfactory air-conditioning, but this reality exists. On the other hand, market forces have led to much competition between manufacturers of split-type systems on the subject of noise, which has been gradually reduced in level.

Some of the Daikin units have now become just about acceptable for control room use when running on low speed, even for critical listening, but it often requires that they should be mounted in front of absorbent surfaces. Nevertheless, the marketing wars have led to some rather confusing publicity battles regarding the quoted noise figures. Customers should take great care when choosing systems because some that are quoted as having extremely low noise

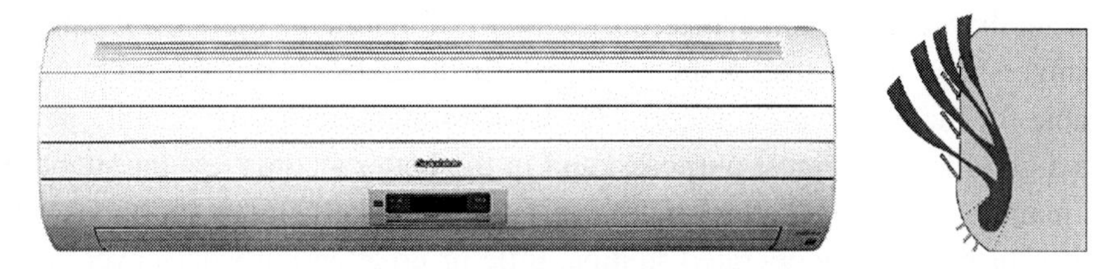

Figure 9.3:
Towards quieter, split air-conditioning. If split air-conditioning systems *must* be used, then it is wise to choose units with aerodynamically streamlined inlet and outlet vents, to avoid the higher levels of turbulence noise associated with the more common grills.

figures on their 'quiet' settings have air flow rates that are so low as to be almost useless in a studio. There seems to be missing any means of readily correlating noise with air flow when looking at the publicity material. For example, one machine with three speeds may have a marginal, but acceptable performance on low speed, with a quoted noise of say 23 dBA at 1m. Another machine, with five speeds, may quote the same maximum heating and cooling but with a noise figure on low speed of only 20 dBA. However, the low speed of the second machine may be so low that it is not adequate for the room, and on higher, more practical speed, it may make more noise than the first machine. Therefore, in reality, the first machine would be the quieter option, despite the publicity for the second machine claiming a lower noise performance.

All this can be very confusing for inexperienced people to sort out. It can be very difficult to make choices based on publicity material alone unless a more in-depth study is made of the specifications (which also may not be very clear). What is more, subjectively, not all dBs of noise are equal. In fact, two machines can produce equal noise figures yet one may be subjectively noisier than the other. It may all depend on the type of noise and in which frequency bands it is concentrated. Studio designers may often have their own preferred choices of split air-conditioning systems, based on experience, but it can sometimes be difficult to convince clients of their superiority when faced with a small army of sales representatives from air-conditioning suppliers who are all trying to sell their own products. It is lamentable how so many studies operate with totally inadequate ventilation and air-conditioning. The owners are such easy prey for the sales representatives.

There are limits, however, below which the lack of suitable environmental control will pose serious problems, not only for the musicians but also for instruments such as pianos and drums. Draughts of air are generally disliked, most musicians would agree about that, but there may be some considerable differences in their preferences of optimal temperature for their own comfort. If a studio is block-booked for a week or so, then the chosen temperature can be selected at the start of the session and maintained, but pianos should not be tuned until the temperature has had time to stabilise. With shorter bookings, the temperature should be kept at a suitable compromise, as frequent changes in air temperature will cause great problems with the tuning of any permanently situated pianos, and many other instruments for that matter.

Humidity is another factor which needs consideration. If maintained too low it can dry out the throats of singers, and may cause piano sound-boards to crack. If it is too high it can be uncomfortable for the musicians and corrosive to instruments. Sixty or seventy per cent humidity is a good level for most purposes, and in the better studios regular attention is paid to the maintenance of the appropriate levels. Unfortunately, in an enormous number of smaller or less professionally operated studios, little or no attention whatsoever is paid to humidity control. Once again, the relentless driving down of studio prices has rendered it impossible for many 'professional' studios to provide the sort of environmental controls that are musts for *truly* professional studio operations.

There is now a huge 'consumer' recording market, which, although providing standards below what the 'professional' market was accustomed to, has blurred with the professional market to such an extent that the lower sets of standards have begun to influence the professional world. Some of this has no doubt been due to the enormous influence of electronically based music, where studio acoustics and air-conditioning noise have not been problems. But it is difficult to understand how some people can hear any detail at all in control rooms whose noise floors are made ridiculously high by the presence of hard disc drives and numerous equipment fans. Fortunately though, there seems to be a swing back to the use of acoustic and electrified (as opposed to electronic) instruments, which is just as well if much of the passed down experience that exists in the recording industry is not to be forgotten through the loss of a whole generation of recording staff to the computer world. The provision of acoustically isolated machine rooms, with good temperature control, is now an important feature of many studios in order to keep the control room background noise level below acceptable limits. In these cases the 'split' air-conditioning systems are ideal, and the background noise levels which they may add are of no practical consequence. In fact, given their efficiency they are probably the *preferred* option.

When very low noise levels are required in studios, a very viable option is to employ batteries of water coils in the ventilation ductwork. Figure 9.4 shows such a system above a film dubbing theatre, and Figure 9.5 shows the noise inside with all systems running. The rise towards 1 kHz is due to – somewhat perversely – the Dolby processors; when the requirement for low noise in the room was made by Dolby, themselves. The low frequency noise from the HVAC (heating, ventilation and air-conditioning) system can be seen to be around NC10 (approximately 10 dBA). Normal, 'split' air-conditioning systems, and other systems using ammonia or similar halogen-free refrigerants, tend to operate in the cooling mode with very low temperatures, often below 0°C in their heat exchangers. These low temperatures are efficient coolers, but they tend to condense too much of the humidity out of the air. This is what often gives rise to the well-known discomfort which many people feel when working long hours in air-conditioned rooms, and it can also badly affect the voices of singers and actors. When water is used to cool the heat exchangers, and high flow rates can be allowed, temperatures as high as 14°C can be employed in order to cool rooms down to 20°, even in relatively hot weather. High volumes of relatively cool water dry the air much less than low volumes of refrigerants operating below zero. Where possible, when water systems are used, it is best to try to site the heat exchangers in places where leaks will not cause disasters (i.e. not directly over the studio rooms) and where there is no danger of them freezing up in winter conditions, which will probably split the pipes and spring leaks once they thaw out. Turning them off during the winter, weekends or holidays, is not an option if outside temperatures dip below 0°C, unless the systems are first drained of their water. Nevertheless, using water as a temperature controlling medium can be beneficial. In winter, the systems can be reversed to provide heat for the rooms, and normally

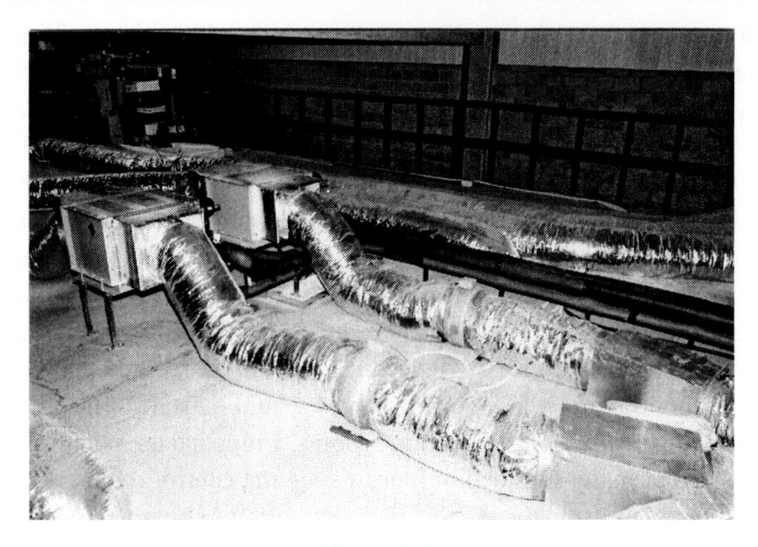

Figure 9.4:
In the system shown, the air enters through the rectangular filter boxes, on the extreme right of the photograph, before passing through the fans. The 'water battery' heat exchangers are in the large rectangular boxes, which also act as expansion chambers and quite effective silencers. The air then loops round through the 'acoustic' ducting before entering the room. An external heat pump controls the water temperature between about 28°C in winter and 12°C in summer (depending upon the outside air temperature) in order to maintain a constant temperature of around 20°C in the studio. Stepped transformers control the fan speed, thus regulating the air flow through the room.

the sensation is of a very pleasant, natural heat. In general, these systems are very effective, but they tend to be more expensive than some of the other options, especially as all the water pipe work usually needs to be made of stainless steel. All the ductwork shown in Figure 9.4 is Sonodec, which is both flexible and also acts as a silencer. The inner tube must be stretched as much as possible to open the 'accordion' folds and avoid the introduction of turbulence noise into the system at higher air flow rates.

9.3 Headphone Foldback

The topic of foldback will be discussed in detail in Chapter 24, but it is also relevant in this chapter because it is an extremely important part of the studio environment. It can be the entire acoustic reality for the musicians during their performances, because in by far the majority of cases musicians will record whilst wearing headphones. In these instances, the acoustics of the studios can only be heard via the microphones, mixing console and headphones, so the musicians can find themselves in a totally alien world if due consideration is not given by the recording engineers to the creation of the right foldback ambience.

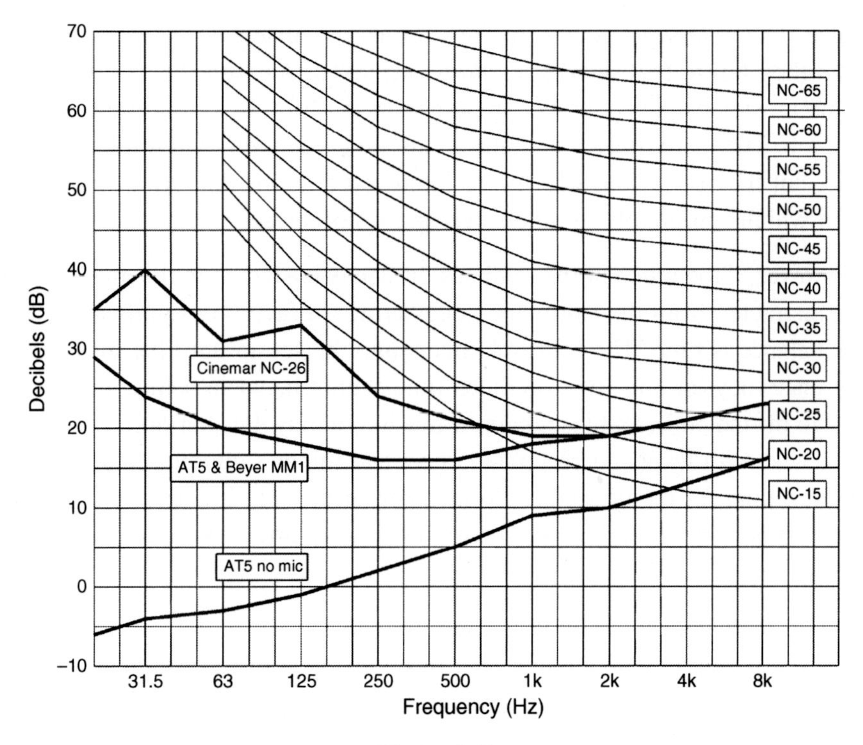

Figure 9.5:
The above plot shows the noise level inside the dubbing theatre whose temperature control system is shown in Figure 9.4. The upper trace shows the noise level with all systems running. The middle trace shows the noise floor of the measuring microphone. Somewhat ironically, the noise in the region around 500 Hz was due to the ventilation fan in the Dolby processor, mounted close to the mixing console. Without the processor the noise floor, below 1 kHz, remains always less than NC 15.

If musicians need to hear what they need to hear in the studio, then they need to hear it in their headphones as well. Many musicians play off their own tone, so if they cannot hear themselves with their usual sound they may 'force their tone', or perhaps hold back, and neither is satisfactory for optimal recording. If musicians need to hear richness in the direct sound, then they ideally would need to hear it in their headphones; if they need the reinforcement of lateral reflexions, then they need to hear those in the headphones also. At times, it can be considered beneficial to put up ambience microphones which will be used in the foldback only. If this helps to improve the sense of 'being there' for the musicians then it is surely worthwhile, though it is rarely done.

There are two big problems with the optimisation of foldback. Firstly, few recording engineers have spent enough time as recording *musicians* to be sufficiently aware of the complexities and importance of the requirements of different musicians in terms of foldback. They cannot be blamed for this, as they cannot spend their lifetimes doing two things at once. The reverse of course applies. So many musicians are used to poor foldback that they fail to realise what *could* be achieved. As with the case of the recording engineers, the musicians have also not spent a working lifetime doing the other end of the foldback job. There are times when musicians are given their own foldback mixer for use in the studio, but they rarely have access to the reverberations that are available from the control room which can greatly aid their perceptions of space.

Secondly, however, if too much of the foldback burden is loaded on to the musicians, then they can become distracted from their primary job: playing. In fact, one significant restriction on foldback balance is time. The foldback mix cannot be set up until the musicians are playing, but once they *are* playing, too much time cannot be spent setting up the foldback before they 'go off the boil', and lose their motivation to play. Furthermore, too much fiddling with the foldback, with levels going up and down and things switching in and out, is absolutely infuriating for the musicians. Remember, effectively it is their whole audible environment that is being disturbed. For them, it is like an artist trying to paint a picture with the lights going on and off randomly.

Where possible, foldback should be very carefully considered. A stereo foldback system is much easier to hear clearly than a mono one. In stereo, even things which are perhaps a little too low for the ideal balance for a mono mix can be perceived much more easily due to the spacial separation which stereo provides. Whenever possible, it seems beneficial to provide the facility of systems where the recording engineer can monitor exactly what the musicians are hearing, which includes listening at the same level. Obviously, on systems where all, or at least many, of the musicians are in a position to make their own balances, this is hardly practicable, but in cases where many headphones are driven from a common power amplifier, it is. In these cases, it can be beneficial for the engineer to have a line in the control room which is connected to the same power amplifier output as the studio headphones, and, where possible, to monitor on the same type of headphones that are being used by the musicians. There can in this way be little doubt that the engineer is monitoring the selfsame space that the musicians are immersed in, and there is therefore much less chance of misunderstandings.

In many cases where the control room foldback monitoring system is only via headphones plugged into the mixing console, or where it is heard on loudspeakers, the recording engineers can never truly know what the musicians are hearing. This has often led to time wasting, or, if the problems have not been pursued, to the musicians having to try to play with the problems still unresolved. In either case, the musical performance will probably suffer. The musicians also seem to feel an added sense of being understood and appreciated when they feel that their environment is being shared by its controller, and this helps to allay any insecurities which they all too frequently feel.

It is imperative that the design of any studio extends to ensuring that sufficiently flexible foldback systems are available, because the 'virtual' spaces in which the musicians may have to perform can be just as important to the recording process as the very real spaces in which they physically play. A full understanding of this is a fundamental requirement of being a good studio designer or a good recording engineer, and good foldback systems are an equally fundamental part of a good studio. In effect there are therefore three aspects to the acoustics of a recording space. Firstly, the acoustic as heard in the room; secondly, the acoustic as collected by the microphones and thirdly, the combination of the two as perceived by the musicians if they must use headphones. All should be considered very carefully in both the design and the use of the rooms.

Choice of open or closed headphones can also modify this environmental balance. Closed (sealed) headphones tend to add to the feeling of isolation, and take the musicians one step further away from their 'real' acoustic space. Nonetheless, there are times when closed headphones are very necessary. Drummers, for example, may need closed headphones in order to avoid having to use the painfully high foldback levels which may be necessary so that the other instruments can be heard over the acoustic sound of the drums. In this case, closed headphones are used to keep unwanted sounds *out.* Conversely, vocalists, or the players of quiet, acoustic instruments and, in particular, those instruments which require the positioning of a microphone close to the head of the musician, may also need to use closed headphones. This is especially the case during overdubbing, when the 'tizz tizz' reminiscent of sitting next to somebody using a personal stereo system may be picked up by the microphone. It may subsequently be difficult to remove these sounds (especially timing 'clicks') from an exposed track. In this instance the closed headphones are used to keep the unwanted sound *in.*

It is a pity that foldback is so often subject to so much compromise, but in a way such is its nature. As so much modern music has developed out of domestic recording technology it is not surprising that aspects of the limitations of the prior technology should be carried along. The simple headphone systems often used in small studios get used in bigger studios, for which they are often inappropriate, when studio companies expand without the necessary experience for a larger operation. It is quite amazing how many people in relatively large studios are totally ignorant of many excellent practices that were well-established 40 years ago. So many people are now self-taught, and they can lack so much knowledge of many simple things which make recording practices so much more effective. But it is imperative to remember that whatever wonders may be created in the acoustic design of the studio, often for the benefit of the musicians, as soon as they put on a pair of headphones those musicians can be in a different world, and it is best not to leave them feeling lost in it. This is one very important difference between designing live performance spaces and designing recording spaces. In the latter case, it is not only how the musicians hear the space which predominates in the design considerations, but how *microphones* react to the space. This point should never be forgotten.

9.3.1 Loudspeaker Foldback

Many studios *and* musicians find the facility of being able to provide foldback via a loudspeaker a useful asset. The concept of the 'tracking loudspeaker' goes back almost to the dawn of electrical recording. Many vocalists found that it was easier for them to perform without headphones, so that they could clearly hear their own, natural voices. A loudspeaker would be positioned facing the vocalist, and the backing track would be played back through the loudspeaker at the minimum level needed to allow the singer to perform well, but not so loud that the overspill into the microphone would cause problems. The directional characteristics of a cardioid microphone were employed in order to reject as much of the loudspeaker output as possible. This was further aided by placing non-reflective screens behind the vocalists, to prevent the playback signal from bouncing off a wall behind and subsequently entering the microphone.

These loudspeakers also became useful as a means of playing a recording back to the musicians without them having to leave their positions and go into the control room, which would in any case perhaps be too small to house them all. This was never a particularly good means of assessing the *sound* of the recording, but was a useful tool for discussing performance quality or mistakes with the producer, conductor, or musical director. They were also useful in orchestral recordings, when the conductor needed to highlight some point in the performance to the musicians, as taking a whole symphony orchestra into the control room would clearly be out of the question. Once again, ease and comfort go a very long way towards getting the best performances out of musicians, so they should be given great attention when designing, *or* operating, a studio. The recording process begins with the musicians, and they are the foundation on which the rest of the process is built. If these foundations are weak, then all the subsequent proceedings may never achieve their potential strength or quality.

9.4 Colours and General Decoration

These are the areas over which a studio designer often has least control. They are very subjective; very personal. Each studio owner or manager seems to want to make their own input to the design, and this is an area where they feel free to do so. Nevertheless, some guidance can be given by the designers, based on previous experiences, which may help the studio chiefs to avoid making any great errors.

Dark colours tend to make spaces feel smaller. In large studios this may help to create a more intimate atmosphere, but in small studios the effect can be claustrophobic. Very often, of necessity, the ceilings of control rooms and small overdubbing booths are not very high, and light colours can go a long way towards lifting any sense of oppression. Light colours also reflect more light back into the room, so they tend to diffuse the light, making it less hard on the eyes. Also, the extra reflected light means that less overall power of lighting is needed, therefore less heat is generated and less air-conditioning is necessary.

In a studio in Watford, England, the owner began seriously running out of money towards the end of the construction. Wisely, and it was a decision he never regretted, he continued to invest more in the acoustics than the decorations, which he felt could be easily improved at a later date. He went to a local outdoor market, looking for inexpensive fabric, and returned with several short rolls of leftover material which he had bought cheaply. The comments of the building crew when he showed them the fabric that they were to work with would not be printable in any decent publication. He had bought white fabric with yellow stripes, blue flowers on a white background, red dots on a white background, and various other designs. Nevertheless, once the different fabrics had been juxtaposed carefully, the result was a small studio with a spacious feel and a remarkably happy and pleasant atmosphere to work in. Even the most severe of the original critics admitted that the result was, surprisingly, very agreeable.

The other extreme of this situation is when a studio owner employs the services of an interior designer who is allowed to severely compromise the acoustics. This happens in a great number of cases, especially when marketing and financial people, in their customary absolute ignorance of recording needs, believe that if a studio looks absolutely great, but sounds only mediocre, it is a better business option than sounding fantastic but perhaps lacking the 'designer' touch. It must also be said that some studio designers *will* bend to such compromises if it ensures more work for their business, and hence more profit.

There is, of course, no law against any of this, and some adherents to market forces philosophies may even applaud the existence of this state of affairs, but it really does nothing for the advancement of studio functionality. This is being mentioned here to stress the fact that a studio which looks great in publicity photographs, and has the name of a reputable studio designer attached to it, does not necessarily mean that it is either a good studio, *or* that the named designer had as much control over the acoustics in the way that the publicity is implying. One of the problems of the 'low profile' of acoustic work is that good acoustics can be easily sacrificed to other, more visible trivia. Cases abound such as where money has been short for good anti-vibration mounts for an air-conditioning system because the owner decided to spend €60 each on 24 door handles. Or there is an inadequate electrical installation, because the owner spent €100/m^2 on a hardwood floor that would largely be covered by a mixing console and effects racks. These are examples of common occurrences.

There is perhaps a philosophically different outlook between studio designers who make money from building good studios, and studio designers who are in business to make money. The latter will probably be more easily swayed into being subjugated by interior designers. For a proposed studio owner looking for a designer, it is perhaps a case of *caveat emptor* – let the buyer beware – because he or she alone is responsible if the results are disappointing. It also must be said that some owners *do* want something which looks great, even if the acoustic and operational results are compromised, but how to achieve that would be a totally inappropriate subject for discussion in a book such as this.

The aesthetics of recording studios are a very important aspect of studio design, but good design *can* incorporate aesthetically pleasing ideas. However, the aesthetic demands of some interior designers are frequently acoustically insupportable. An air of respectful cooperation between the specialists is the obvious solution.

9.5 AC Mains Supplies

Appropriate AC mains supplies are a fundamental part of the infrastructure of any good recording studio, but general electrical contractors are often totally unfamiliar with the needs of high quality recording studios, and they are often greatly resistant to suggestions which they consider to be out of the ordinary. Sometimes there is good reason for this, because standardised systems of installation and power distribution have been developed as a safety measure. Standardised systems also mean that any qualified electrician can make modifications or tests on a system which will not upset the rest of the installation, or show up as fault conditions.

However, standardised industrial or domestic power installations may be inadequate for highly sensitive recording equipment. Hospital power supplies which feed delicate equipment upon which lives depend are another example of installations where normal power distribution techniques can be inadequate. The life or death circumstances which exist in medical facilities ensured that hospitals and the like have had exemptions in many countries, for many years, which have allowed special AC supply techniques to be used. Electricians who work on medical installations will be trained in the use of these special techniques. However, wiring systems in recording studios are generally not so well regulated or supervised. The recording industry is not usually perceived as being a particularly responsible or necessary profession, so such exemptions have been slow in arriving.

Indeed, there have been many cases where special AC power installations *have* been incorporated into new studio designs, and where much care and attention has been paid to detail. Nevertheless, within a few months of the opening, perhaps where a studio manager or head of maintenance has been replaced, other electricians have arrived, working to 'the book'. Under such circumstances, chaotic systems have begun to develop, sometimes with dangerous results. Studios have not done themselves any favours by this sort of lax behaviour, so it is little wonder that they have not been granted their technologically necessary exemptions. Even in a technically advanced country like the USA, electrical codes were not modified until the 1990s to allow balanced power installation, and even these were restricted to studios with a competent, qualified electrical supervisor. It was recognised that despite its advanced technology, the USA was also the home of the very market forces which would seek to cut costs by *not* employing a qualified electrical supervisor if at all possible. Technological advancement does not always go hand in hand with responsible business attitudes.

It is therefore quite dangerous and irresponsible in a book such as this to try to explain how to make AC power installations. It is also not practicable, because the rules and regulations in not only each country, but often in each state or region, can be very different. What *can* be done, however, is to discuss a few of the ideal requirements, and these will be discussed in further detail in Chapter 25. Nevertheless, it would still be up to any individual wishing to pursue these to find a knowledgeable and flexible electrical engineer or contractor who could then interpret them in such a way that would not break local installation rules. Whilst it is true that many *qualified electricians* are neither knowledgeable of electrical engineering theory nor are they willing to change their habits, it is usually possible to find specialised electrical *engineers* who *can* make the appropriate arrangements. Remember also that a non-standard installation can be dangerous if it is worked on during later visits by electricians who, quite justifiably, expect the installation to be standard. Adequate and permanently accessible documentation that will *not* be lost when staff change is not something which one can expect to exist in 99% of the recording industry. Technologically advanced countries such as the UK, France and the USA are no exceptions, either, but Chapter 26 will look in detail at audio wiring systems which are generally tolerant of use with more standard power wiring installations.

9.5.1 Phase

The ideal number of phases to which all the audio equipment should be connected in *any* studio, or complex of studios which can be interconnected, is ONE! All audio equipment, together with any other equipment to which it may be connected, including computers, video, film or radio equipment, should all be connected to the same power phase. The only exceptions are where equipment, such as radio transmitters, are fed via galvanically isolated audio connections; i.e. through transformers with adequate insulation. Electricians often do not like to put all the equipment on one phase, because they prefer for good power engineering reasons to balance the load. Nevertheless, the reasons to connect to one phase only are first, safety, and second, noise.

In countries where the nominal mains supply voltage is in the 230 volt region, the voltage between the live connections of any two phases will be in the region of 380–400 volts. Under certain fault conditions, connecting the audio cables between two pieces of equipment on different phases can have instantaneously lethal results. Under normal circumstances this should not happen, but with weak insulation on two devices, even if the insulation holds on monophase operation, the extra potential when two pieces of such equipment are connected together can initiate a fatal breakdown.

What is more, when audio signals (analogue or digital) pass repeatedly through equipment with different ground plane potentials, as is almost inevitably the case if interconnected pieces of equipment are connected to different mains supply phases, signal contamination is likely to result. It is true that extremely well designed equipment can be immune to these effects, but unfortunately even some very *expensive* audio equipment does not come under the heading of 'extremely well designed'. Capacitive coupling in power supply transformers, for example, can easily lead to fluctuating ground plane voltages. In general, the higher the gain of any equipment, the more sensitive it will be to supply-phase differences.

If three phases are available it is best to dedicate the cleanest, most stable phase to the audio equipment. Lighting, mono-phase air-conditioning systems and ventilation equipment can use a second phase; and refrigerators, coffee machines and general office equipment can be connected to a third phase. General-purpose wall outlets outside of the recording area can be connected to the least heavily loaded of the second and third phases, but some local regulations may modify this. Although this may not seem to balance the phases very well in terms of equal current drain on each phase, electricians frequently fail to realise just how variable is the current drawn by a recording studio. Equipment such as power amplifiers often draw current in pulses, dependent on the musical drive signal. Musicians connect and disconnect their instruments and amplifiers, lighting gets turned up and down, air-conditioning currents are temperature dependent, tape recorder motors stop and start (although there are less of them, these days); many things change during the course of a session. There is a tendency for non-specialist electricians to look at the maximum power consumption of each piece of equipment, and then try to distribute it presuming that the current drain is constant. The reality is nothing like this, and phases sensibly balanced for technical reasons can easily be made to give a distributed average load that is better balanced than many electricians would achieve by distributing things according to their 'standard' calculations. The problem is often how to convince them about this when they perceive *themselves* as being the experts on such matters. There is more on this subject in Section 25.6.1.

9.5.2 Power Cabling

Guitar amplifiers and audio power amplifiers have a tendency to draw current in pulses. For this reason oversized cable sections are often specified. General electricians will size cable according to the heat produced by the cable resistance in response to the power being consumed by the equipment. To many electricians, anything much beyond what such power requirements would need for reasons of safety is usually seen as a waste of money. However, pulse-drawing equipment *needs* oversized cables. There are two reasons for this. Excess resistance in the power supply cable may restrict the peak level of the required current pulse, which can limit the high-level transient response of audio power amplifiers. Furthermore, cables lacking sufficient cross-section for the current pulses can create problems in other equipment, and can even crash computers. If a current pulse causes a voltage drop on a power cable, then even if this cable is not shared by any other piece of equipment, the voltage drop can cause harmonic distortion on the supply line. This can then bypass the power supply filtering of some sensitive equipment and interfere with the proper operation.

In fact, the problem of harmonic induced interference can also be caused by uninterruptible power supplies (UPSs) which do not have waveform feedback. Not all audio computers are happy working on all UPSs. It would seem a pity for a UPS first to save a recording when the mains power fails, only to subsequently crash the computer due to its bad voltage waveform. Many computer crashes that are blamed on software problems can be traced to bad power installations or inappropriate UPSs.

9.5.3 Balanced Power

In some hostile power line environments, one of the best solutions to avoid interference problems coming in via the electricity supply can be to balance the power. This involves the use of power transformers with centre-tapped secondary windings, producing, for example, 115 V–0–115 V in place of a single 230 V supply. Electricians should be consulted if this sort of measure is needed, in order to avoid making any illegal installations, but the technique can reject interference in exactly the same way that balanced audio lines are less sensitive to interference than unbalanced lines. Signal-to-noise ratio improvements of 15–20 dB have been reported as a result of power balancing.

In many countries, it may be illegal to supply balanced power to normal wall-socket power outlets. Special installation work may be required, by expert engineers. Many electricians may simply refuse to do it, as they cannot justifiably put their signature to something with which they are not familiar. Nevertheless, specialist engineers can usually find legal ways of installing such systems if there appears to be no other option for interference suppression. It should also be stressed that with such a specialist job as studio electrics, it is *all* really a job for electricians with some experience in this field if the highest standards of performance are required.

9.5.4 Mains Feeds

It is important to try to ensure that the main cables feeding the studio in-coming fuse-board are connected as close as possible to the main distribution board for the building. This ensures that the large-section cables will continue to the street, and onwards to the principal supply. If this is not done, then any common cable between the point of connection of the studio to the mains supply of the building, and the main feed connecting the building to the street, will be a common impedance. Any currents and noises generated elsewhere in any part of the building which shares the same feed cables will superimpose themselves on the supply to the studio, and the supply will not be clean. In addition, if these cables are only rated in terms of the power consumption, they may not have the cross-section sufficient for the non-distortion of surge currents, as described in Section 9.5.2. A low impedance supply is highly desirable.

Once again, independent supply cables to the main feed to the building may be inconvenient for the electricians, and they may resist it on the grounds of being unnecessary because they are still thinking in terms of power consumption and standard cable sizes. Nonetheless, it needs to be impressed upon them that a recording studio is no ordinary installation in terms of its power cable requirements. Again, an experienced specialist electrical engineer may need to be called in, if only to convince the electrical contractors that the necessity is a very real one.

9.5.5 Earthing

A good earth is essential as a safety measure, and it is normally also needed for a good signal-to-noise ratio, but if very elaborate earthing systems seem to be the only way to reduce noises the implication is that there are problems in the system wiring; either the power wiring or the audio wiring. Normally, in buildings with steel frames or steel reinforced concrete in the foundations, connection of the earthing (grounding) system to the steel provides an excellent earth (ground).

In the book by Giddings[1], on wiring in general, the first 113 pages are dedicated to power and grounding systems, so it should be evident that a few paragraphs, here, cannot deal with the subject. However, what *can* be made clear is that similar rules apply to the earthing cables as to the power feed cables. The studio power system, and its technical earth, should be connected to the best earth point via the shortest cable run and with the largest practical cross-section of cable.

Another point worth mentioning, which is often overlooked, is the degree to which the earthing system of a studio may be being polluted from within. Obviously, sharing the long tail of an earth cable with other users of a building may result in a lot of electrical noise on the common (shared) section of the earth cable. The studio earthing system therefore needs to be connected *directly* to the building safety earth, and *not* via any shared lengths of cable. However, even from *within* the studio, the earth can be polluted by bad choices of electrical filters. Many filters of inappropriate design, which frequently still find use in studios, do little other than remove the interference from the live and neutral cables and dump it all on the earth, which in some cases is *more* sensitive to the noise than the live and neutral are. Power balancing can be a big help in these cases, because the noises on the live and neutral *cancel* on their way to earth. This will be discussed at greater length in Chapter 25.

Very well designed systems using well-designed equipment can often work well with the most simple of earthing systems. Unfortunately, all recording equipment is not so cleverly designed, so in buildings with earth noise problems, and on ground which is geologically bad for earthing, expert advice may be needed. Standard electrical contractors may only know of safety earths, and perhaps a little, learned by hearsay, about technical earths. It is not always wise to rely on their advice. In cases of persistent problems, a specialist engineer should be sought.

9.6 Summary

In the majority of cases, the presence of daylight in a studio is a desirable asset.

Choice of artificial lighting should be made carefully, so as not to risk compromising thc electroacoustic performance of a studio.

Variable transformers tend to be the best form of lighting control.

Good ventilation, air-conditioning and humidity control is usually essential. Extraction-only ventilation systems should *not* be used.

There has been a great increase in the number of split air-conditioning units being used in studios. These are not ideal, but if they *must* be used, the quietest, aerodynamically profiled ones should be chosen.

Foldback systems are very much part of the studio environment for the musicians, and the necessity of seeking the most appropriate systems for the musicians' needs should not be neglected.

Colours and general decoration are a very important part of a studio environment, but interior design should never be allowed to degrade or limit the work of the acoustic designer.

Studio power wiring needs are often beyond the experience of normal electrical contractors. Specialist advice may be needed.

When any non-standard installation is completed, documentation should be thorough, and available for consultation by any electrical contractors who may do work in the studio in the future.

Electrical regulations can change from country to country, and even from state to state or region to region, so local advice should always be sought.

It is strongly advisable to connect all audio and associated equipment to the same electrical power phase. This is for reasons of both safety and noise.

The main studio breaker-board/fuse-board should be connected to the incoming supply of the building, without sharing any cable runs with other tenants. Oversized cable sections should also be used, to help to provide the lowest possible source impedance.

In very problematical installations, from the point of view of electrical supply noise, balancing the power can be a very effective cure, but it should only be done under the supervision of an electrical engineer experienced in such techniques.

Good earthing systems should be installed, again avoiding common cabling with other tenants. Direct connection to the steelwork in building foundations can make a good technical earth. Despite concrete being perceived to be a good insulator, it still grounds well the steel buried inside concrete foundations.

Many mains filters can often be a *source* of noise on the earthing system.

Reference

1 Giddings, Philip, *Audio Systems – Design and Installation*, Focal Press, Boston, USA, Oxford, UK (1990)

CHAPTER 10

Limitations to Design Predictions

Measuring techniques for the evaluation of room performance. Modelling techniques for the prediction of room performance. The strengths and weaknesses of different approaches.

10.1 Room Responses

Much of the space in this book could have been filled up with corresponding plots of the reverberation times (RT_{60}) and the energy/time curves for each of the rooms discussed and for each of the photographs shown. The 'waterfall' plots for each room could also have been discussed, and indeed, many people would perhaps have expected such, but specifications often tell us very little about the perceived sound characteristics that we have been discussing. What is more, in the wrong hands, they can be misleading, so perhaps we had better examine some of the different representations and their different uses. This may be an appropriate time to discuss these points because they are also very relevant to control rooms and monitoring systems which will be the principle subjects of the following chapters.

The classic RT_{60}, which is often now written T_{60}, is the time taken for the sound of an event in a room to decay to one millionth of its initial power, which is -60 dB. The typical RT_{60} (reverberation time) presentation is shown in Figures 10.1(a) and (b). The two plots shown are actually the reverberation time responses for two famous concert halls within 20 km of central London. In these plots, reverberation time is plotted against frequency, and the curves are useful to see the 'frequency response' of the halls. It is not surprising that the subjective quality of (a) is warmer and richer but less detailed than (b), as (a) has a much longer low frequency reverberation time, which not only serves to enrich the bass, but can also mask much low level transient and high frequency detail. Unfortunately, such plots only tell us what is happening at the -60 dB level, and not what happens during the decay process itself. Depending on *how* the reverberation decays, it is possible that seemingly obvious deductions of the subjective quality could have been erroneous if based only on $(R)T_{60}$ information.

Beside the reverberation representations such as those in Figure 10.1, that plot time against frequency, there are other ways of looking at the reverberation decay, such as methods that plot reverberant *energy* against time. One such representation is the Schroeder plot. Figure 10.2(a) shows the decay 'curve', or Schroeder plot, that would be expected from a good

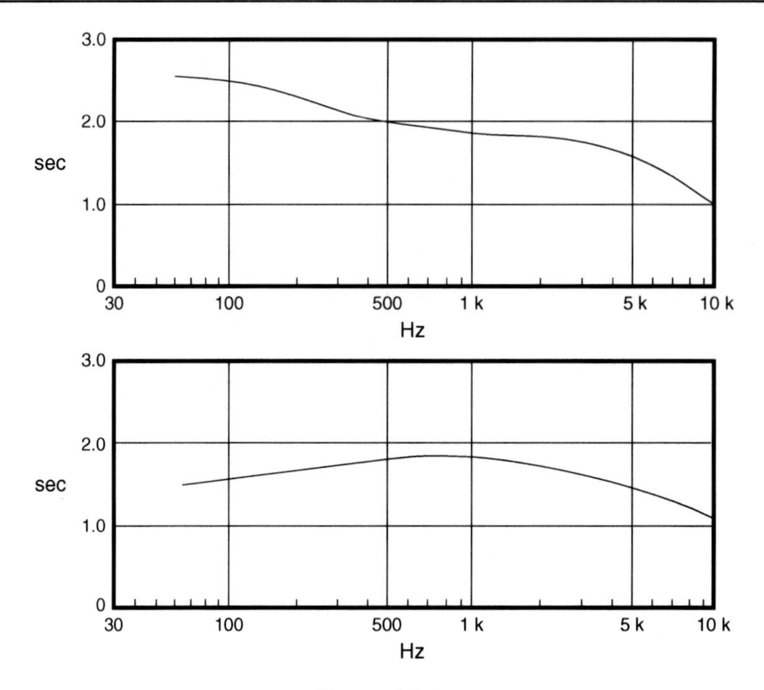

Figure 10.1:
Reverberation times of two famous concert halls in England.

reverberation chamber. However, in studio rooms there is usually absorption, diffusion, and a whole series of reflexions, both of the early and late variety in the larger rooms, which all serve to modify the smooth decay of a reverberation response. Plots more typical of studio rooms are shown in Figure 10.2(b). Figure 10.3 shows a series of 'reverberation' responses from different rooms. They have nominally the same T_{60}, and all could conceivably produce very similar plots to each other of the type shown in Figure 10.1, although clearly, the solid curve contains less overall energy than the other curves. In a room showing this type of characteristic response, with an *initial* decay that is much more rapid, there will be less of a tendency for the reverberation to mask the mid-level detail in any sounds occurring almost immediately after the occurrence of a loud sound. The broken curves, on the other hand, would represent spaces which sounded richer than the one represented by the solid line. In many ways, it is the onset of the decay (the time taken for the sound to decay to 10 dB below its initial level), which tells us much more about a room than its T_{60}. However, there can be many uncertainties about exactly where a decay begins, so the concept of the *early decay time* has evolved.

The early decay time (EDT) is usually expressed as the time taken to decay by 60 dB, but extrapolated from a straight line drawn through the −5 dB and −15 dB points. Alternatively, it can be calculated by taking the time from the 0 to −10 dB levels and multiplying the figure by six. If the resultant figure is greater than the actual T_{60}, then the early decay time is slower than the

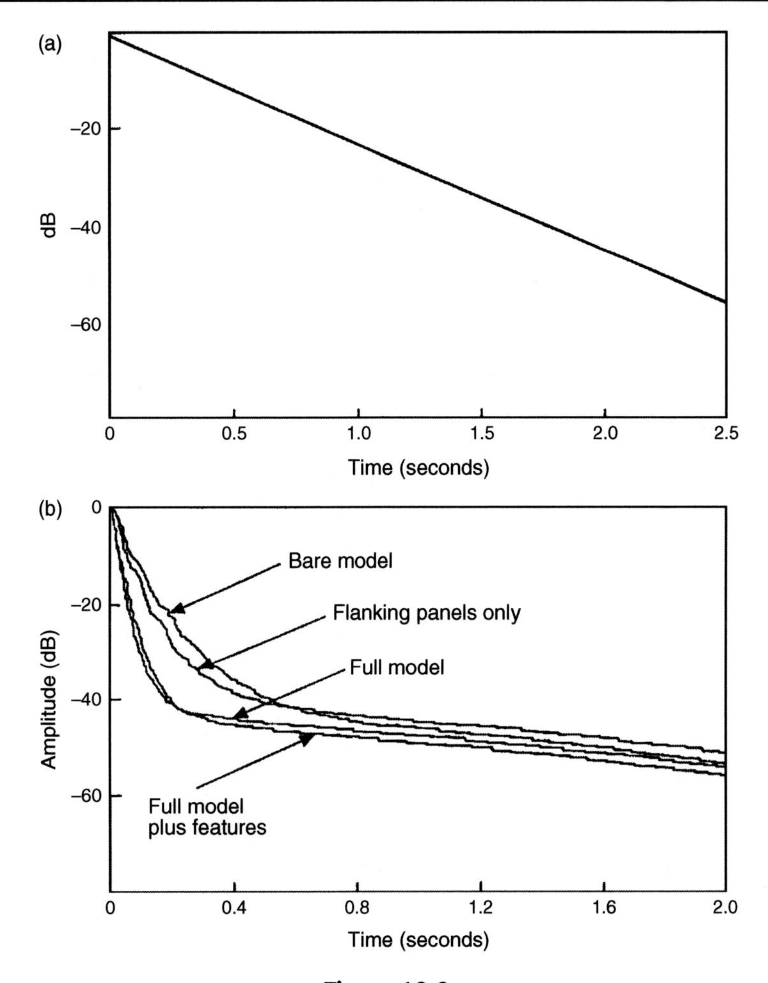

Figure 10.2:
Schroeder plots: (a) In a perfectly reverberant room, the Schroeder plot would show a straight-line decay. In the case depicted here, the RT_{60} is slightly over 2.5 s; (b) This Schroeder plot of a test room shows how the installation of acoustic control items removes the energy from the early part of the decay curve, 'cleaning up' the room.

average decay time, and *vice versa*. Measurements made in this way are often referred to as RT_{10}, even though the resultant figure is that of the time for the sound to decay to −60 dB. In a reverberant room with a uniform decay rate, the two figures would coincide. The measurement is conventionally performed in octave bands. The concept is shown graphically in Figure 10.4.

In small rooms, the truly diffuse sound-field required to produce reverberation can never develop, but the relative energy in the modes, discrete reflexions and diffused sound all contribute to the overall perceived sound. The Schroeder plots of Figures 10.2 and 10.3 show the overall envelope of the decaying energy, which is a very useful guide to the general behaviour of a room, but when problems exist it is sometimes necessary to see more detail.

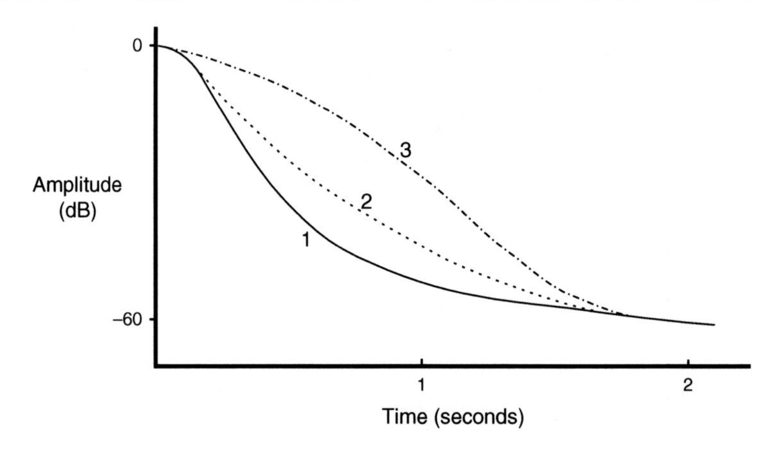

Figure 10.3:
Different decay characteristics.

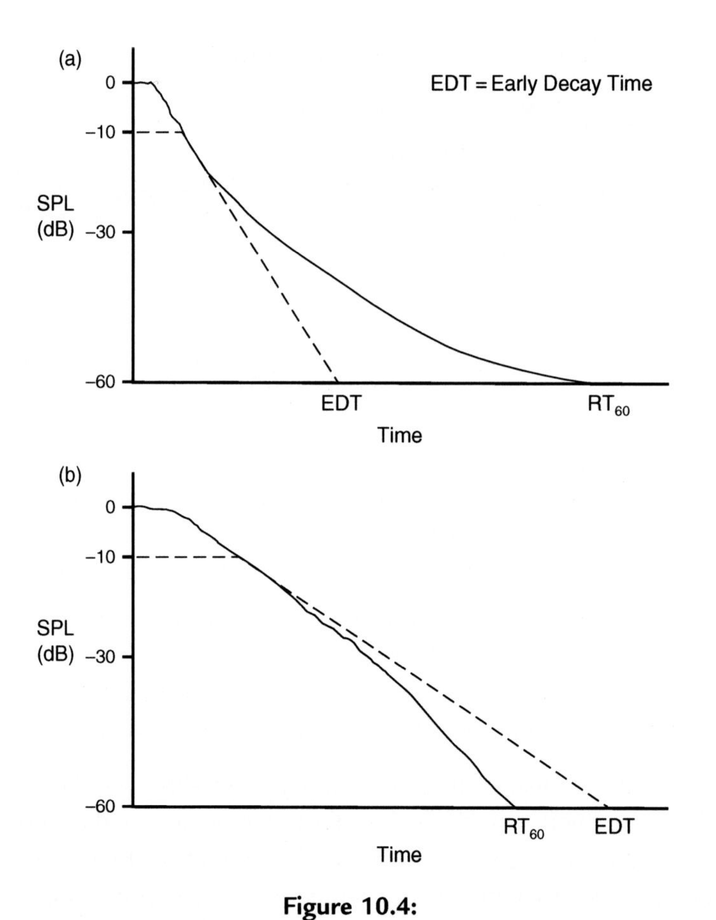

Figure 10.4:
(a) Early decay time shorter than RT_{60}; (b) Early decay time longer than RT_{60}.

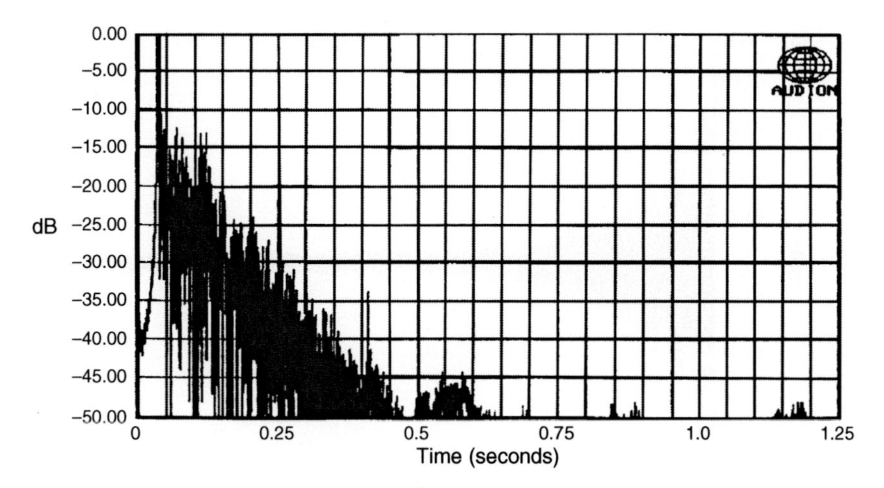

Figure 10.5:
Energy/time curve (ETC). This ETC is a representation of a room not too dissimilar from the bare model plot in Figure 10.2(b). The bumps at 0.55, 0.85 and 1.15 s were due to traffic noise. The plot was from an unfinished room.

Figure 10.5 shows an energy/time curve (ETC) in which the individual predominant reflexions can be seen as spikes, above the general curve of the overall envelope. These plots are, like the Schroeder plots, showing level against time, so the *time* between the initial event and any problem reflexions can be determined. When the time of travel is known, the path from the sound source to the microphone can be calculated. Any offending surface can then be located, and if necessary dealt with.

10.1.1 The Envelope of the Impulse Response and Reverberation Time

Despite the Schroeder plots and the ETCs both being representations of energy versus time, they each have their distinct uses. They are also generated in different ways in order to highlight best the aspects of the responses of which they are intended to facilitate the assessment. Figure 10.6 shows the impulse response of a room, which is a simple representation of sound pressure against time. It may at first seem plausible that half of this plot could be 'smoothed' to give a representation of the decay of the room, but it will be noticed that the plot is not symmetrical about its horizontal axis. It will also be noticed that there are many crossings of the horizontal axis, where the representation of the sound pressure level is zero. If we simply summed the two halves, or folded up the lower half of the plot on top of the upper half, then the zero-crossings, such as are indicated at points X, Y and Z on Figure 10.6(b), would retain their zero values. However, on listening to the decay of a room, it soon becomes intuitively obvious that the energy decay does not alternate between rapid bursts of energy separated by points of zero energy, but that energy is present in the room from the inception of the sound until its decay to below audibility.

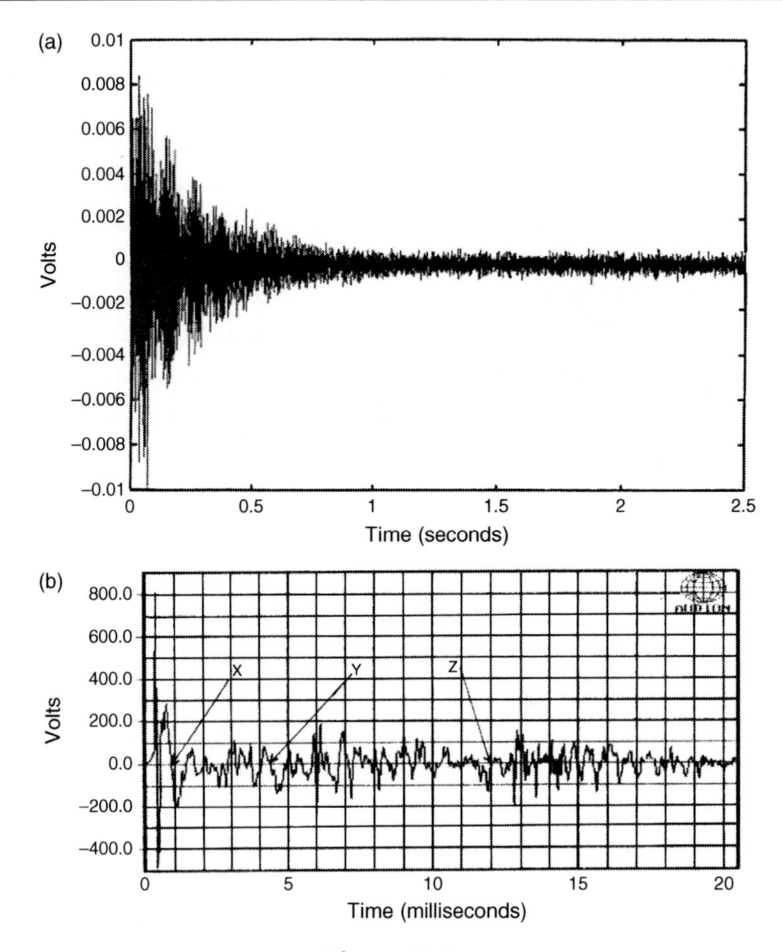

Figure 10.6:
(a) and (b). Impulse response of room (bare model) shown in Figure 10.2(b). (a) The impulse response shown here does not appear on first glance to relate directly to the more intuitively apparent presentation of the Schroeder plot of Figure 10.2(b), though both were generated from the selfsame measurement. The asymmetry of the upper and lower halves of the plot can clearly be seen, but due to the long time-window, the zero crossings are not obvious. (b) When the time scale of an impulse response is stretched (in this case the time scale is only 20 ms) the multiple crossings of the zero amplitude line can clearly be seen, such as at points X, Y and Z, and many other points.

For this reason, some integration is necessary if the simple instantaneous sound pressure representation of the impulse response is to be made into a better representation of energy decay. The impulse response shows only *pressure* against time, but, as shown in Figure 4.10, in reality the acoustic energy is constantly changing between the pressure and the velocity components.

A measure of the *power*, or energy per unit time, in such a signal can be calculated by squaring the signal and then averaging it, over a suitable length of time, to yield the mean-square value of the signal (the familiar rms [root-mean-square] value is the square root of this). The mean-square is a continuous positive value, which is only zero if the signal is zero for longer than the averaging time, and thus it does not contain the multiple zero-crossings present in the original signal. Provided the averaging time is long enough, *continuous* signals, such as a sine wave, have a mean-square value which is independent of time and independent of the length of the averaging time. However, estimates of the changing mean-square value of a *transient* signal, such as an impulse response, can be very dependent upon the length of the averaging time.

10.1.2 Schroeder Plots

In the early 1960s, Manfred Schroeder[1] had been frustrated in his attempts to accurately measure the reverberation time of some large concert halls. He was having trouble achieving repeatability in his results because, dependent upon the precise phase relationship of the elements within the random excitation noise, and the beating between the different modal frequencies thus excited, he could get some very different readings which could even halve, or double, on subsequent measurements. Confidence in such measurements was therefore not good. The method that he developed to overcome these problems involved the excitation of the room by a filtered tone burst, and the recording of the signal onto a magnetic tape recorder. The tape was then replayed backwards, and the output was squared and integrated by means of a resistor/capacitor integrating network. The voltage on the capacitor would then represent (in reversed time) the averaged energy decay of the room. Because the squared tone burst response is a positive function of time, its integral is a monotonically decreasing function of time. The resultant plot is therefore a gradually falling line, free of the up and down irregularities of many other decay time representations, and is thus easier to use for accurate decay time *rate* assessments.

Nowadays, the generation of Schroeder plots is usually by means of digital computers, which have made redundant the need for the cumbersome use of tape recorders in the making of these measurements. The use of Schroeder plots is widespread, because they show, perhaps more clearly than any other decay time representation, the presence of multiple slopes within a complex decay tail. It is now widely appreciated that the initial slope, for the first 10 dB of decay, is perhaps one of the most important characteristics in the subjective assessment of the performance of a room.

The decay curve for a Schroeder plot is derived by squaring the decaying signal and integrating backwards in time from the point when the response exceeds the noise floor (t_1), back to the start of the response (t_0). The value of the decay curve at any point in time (t) is therefore the integral of the response over the interval from t to t_1. The resultant decay curve is a smoothly varying, and (necessarily) monotonically decreasing function of time. This characteristic is a necessary condition of the total sound energy in a system (such as a room) after the source of energy has ceased. The technique is thus valuable in determining the decay slopes, particularly in the early part of the response, which are necessary for reverberation time estimates in conditions where background noise does not allow measurement down to −60 dB. However, any fine detail in the impulse response is lost, due to the time integration, so the Schroeder plots cannot show the individual reflexions which can be seen on an ETC.

For anybody unfamiliar with the term 'monotonic', it means either forever rising, or forever falling, although fluctuations in the *rate* of rise or fall can be allowed. The monotonically falling characteristic of the Schroeder plot decay curve is easy to appreciate, as once the energy source has been turned off, the absorption in the room can only lead to a gradually reducing net energy level in the decay tail (because energy cannot reappear once it has been absorbed).

10.1.3 Energy/Time Curves

The energy/time curve (ETC) is the result of applying a technique to a transient signal which does *not* depend upon time averaging, but which yields an estimate of a mean-square-like value, which varies with time. The seminal paper on this technique was written by the late Richard Heyser and was published in 1971.[2] As previously mentioned, the multiple zero-crossings of the impulse response do not represent points of zero energy. Before we look in more detail at the generation of an ETC, perhaps we had first better look again at the pendulum analogy, in order to understand why an instantaneously zero SPL is not necessarily representative of zero energy.

When the pendulum is at either end of its swing, it is at its maximum height but has zero velocity. When the pendulum is at the bottom of its swing, it has minimum height but maximum velocity. The two forms of energy present in the pendulum are *potential* energy, which is a function of the *height* of the pendulum above equilibrium, and *kinetic* energy, which is a function of the *velocity* of the pendulum. As the pendulum swings, there is an alternating transfer of energy from potential to kinetic to potential, with one form of energy being a maximum when the other is a minimum. The total energy in the pendulum at any time is the sum of the potential and kinetic energies, and is independent of time unless the pendulum slows down. A graph of either the height of the pendulum or its velocity would show multiple zero-crossings, and would *not* therefore be a measure of the total energy in the pendulum. One could, however, obtain a good estimate of the total energy by comparing one graph with the other, or even estimating one from the other, then calculating the two energies and summing them. The latter method is the basis for the ETC calculations.

A signal, such as the impulse response of a room, is treated as though it represents one form of energy in an oscillatory system, such as the pendulum. Using powerful signal processing techniques, such as the Hilbert Transform, it is possible to derive a second signal which represents the other form of energy. The two signals are similar, but 'appear' to be 90° out of phase with each other, as shown in Figure 4.9(b). One has a maximum or minimum when the other has a zero-crossing, and *vice versa*. The ETC results from squaring these two signals and adding them together.

The ETC does not rely upon time averaging (or integration) and therefore, to such an extent that the time response of the filters allows, does not mask any fine detail in the instantaneous energy of the signal. In general, the ETC of an impulse response has a characteristic which can increase as well as decrease with increasing time (it is *not* monotonic), and is thus useful for identifying early reflexions and echoes.

To summarise, the Schroeder plot is most suitable for reverberation time estimation where the main point of interest is the rate of decay of the energy in a room. The ETC is better for identifying distinct reflexions, or other time-related details of a specific impulse response, such as those generated by a loudspeaker system.

10.1.4 Waterfall Plots

Some of the characteristics of the ETCs *and* the more conventional T_{60} plots can be combined into what is commonly known as the 'waterfall plot'. Normally computer generated, these plots show a perspective view in a three-axis form, as shown in Figure 10.7. The vertical axis represents the amplitude of the sound, and the two horizontal axes represent time and frequency. Such plots are very useful, but care should be taken when assessing them. On first

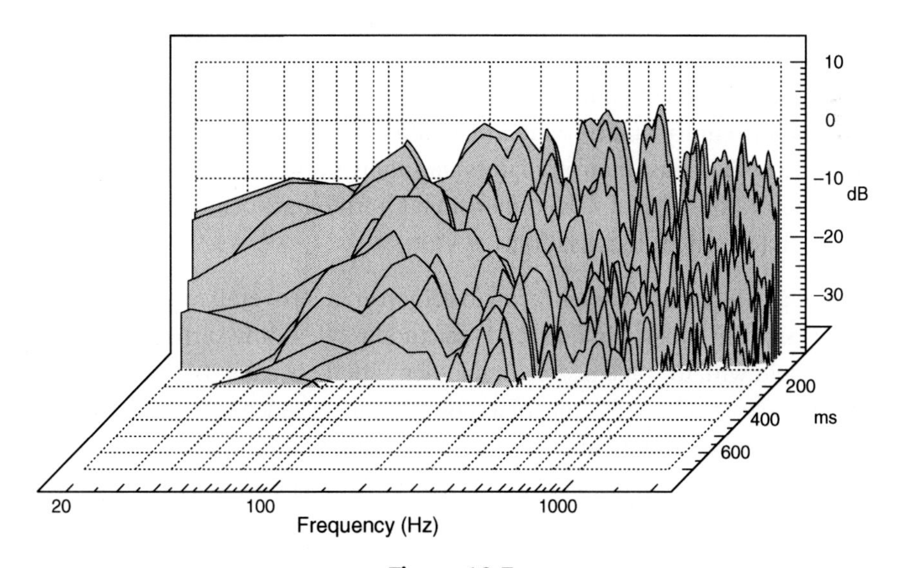

Figure 10.7:
This waterfall plot is of the finished control room described in Appendix 1. The cumulative spectral decay is represented in a three-axis format, the vertical axis representing amplitude, the front/back axis representing time, and the left/right axis representing frequency. In the case shown, only the low frequency range was being studied.

impressions it often seems that they contain all the values represented by the three axes, but the nature of their perspective representation can hide some details in 'valleys' which lie behind some of the 'hills'. Nevertheless, waterfall plots are very useful forms of analysis, as in a single plot they allow an 'at a glance' assessment of reverberation time against frequency, and decay rates at different levels *and* frequencies. They not only have the ability to show discrete resonances, but also the ability to show the predominant frequency bands inherent within any individual reflexions. Despite all of this information though, one must always remember that what a room *sounds* like can still not accurately be assessed from *any* piece of paper, because measurement microphones, no matter how sophisticated, are not ears, and neither are they connected to human brains.

10.1.5 Directional Effects

Unfortunately, even when supplied with the RT_{60} information about a room, *and* the Schroeder plots, *and* the ETC details, *and* the waterfall plots, this still tells us nothing about the diffusion or diffraction effects which may be taking place within the curves. We would still know little about any directional effects, which audibly may make one seemingly objectively innocuous reflexion more subjectively objectionable than an obviously higher-level reflexion approaching the microphone position (or listening position) from a direction which causes only little offence to the ear. Floor reflexions, for example, can be collected by a measuring microphone at a high level, yet our sensitivity to vertical reflexions, in general, is far less than to horizontal ones of the same intensity. Human beings have poor vertical sound discrimination, as our vertical position is usually known, and evolution has caused the development of horizontal discrimination at a more urgent pace. The knowledge of the proximity and direction of predators or prey have been great survival advantages. Because humans, their predators *and* their prey have historically largely been confined to living on the ground, two-dimensional horizontal localisation has been much more important than vertical localisation.

Another directional effect that is *very* difficult to demonstrate in overall response plots is the effect of any coupling between acoustic modes and structural modes. Any surfaces immersed in an acoustic energy field will, to some degree or another, vibrate because of the coupling to the acoustic energy. Sections of the structure of a room, or panels attached to the surface of the room, may vibrate *and* re-radiate energy into the room. The re-radiation will not only modify the modal interference patterns of the room, but will also act as a secondary sound source. To a person in the room its effect may well be audible, and detectable from a definite direction. This is another subjectively undesirable possibility that would be difficult to locate in most visual representations of room responses. It will no doubt by now be realised that if we wish to find out how a proposed room may sound, in advance of its construction, simply building to some preconceived specifications will obviously not suffice. If we need to know how a room may sound, we need to resort to other techniques.

10.2 Scale Models

Short of building full-size prototype rooms, the next best possibility would seem to be to build and test scale models. At concert hall level this is a proven method, at least as far as the major characteristics of the room are concerned. In a one-tenth scale model of a studio room, music can be played through miniature loudspeakers in the model room, speeded up by ten times, and recorded via miniature microphones placed in the ears of a one-tenth scale head. The sounds can be recorded, and then slowed down by ten times, and the result heard on headphones should be a reasonable representation of what the full-scale room would sound like. The above description of the technique is something of a simplification of the whole process, but the technique is useful and is used on some large-scale projects. One-fiftieth scale modelling is also used for larger halls, but because the frequencies must be scaled up by the same amount, air absorption losses are such that dry nitrogen is frequently used to fill the models, and in many cases, predictions are restricted to the low frequency range of the final, full-scale room. However, with rooms of studio size, the scale-modelling method would be expensive to put into operation. Furthermore, as the characteristics of acoustically small rooms are so dominated by surface features, which will not easily scale, the scale-modelling technique is likely to fail. Whilst frequencies, sizes and general modal responses *can* be scaled, the effects of absorbent treatments, the irregular surfaces of stonework, or the effects of floor resonances cannot be scaled. Mineral wool is almost impossible to scale, as are carpets or curtains. Scale models of small rooms can only be used to determine gross effects, and if effects are gross, they can normally be deduced from pure theory.

10.3 Computer Models

The power of computer modelling seems to increase day by day, but computers still cannot *design* rooms. They can be an aid to the experienced designer, who knows what he or she is looking for, but there is always a danger that in less experienced hands the graphics and apparent simplicity of the operation of the programs can lead to a belief that they give all the answers. They do not; the underlying rules and calculations behind the programs must be understood if gross over-simplification and over-confidence in the results are to be avoided. In many cases, exact solutions are only possible for simple shapes, whereas in practice, the same things that cause problems in the scaling of physical models will cause similar problems with computer modelling. Lack of ability to deal with the surface irregularities of different types of stone is one example of this general failing.

Computer models can be very useful as an interface between acousticians and any clients who do not have a good grasp of theoretical acoustics. The graphics can be used to explain to a client what a proposed room could look like, or to highlight the problems, which may be caused by an inappropriate choice of shape, size, or room features. There is always a danger, though, that in this world where computers are seen by many as essential tools for just about everything, the over-reliance on computer graphics and predictions will lead to a belief that it will achieve something that it will not. There are just so many parameters which contribute to the subjective character of a room; and not enough is known about their interrelationships to be able to program fully a computer sufficiently to give accurate predictions.

What follows are some words of the late Ted Uzzle, from a presentation which he made to the 72nd Convention of the Audio Engineering Society in Anaheim, California, in 1982. They are worth repeating here, as they do help to put the design process in perspective; 'No sound system, no sound product, no acoustic environment can be designed by a calculator. Nor a computer, nor a cardboard slide-rule, nor a Ouija board. There are no step-by-step instructions a technician can follow; that is like Isaac Newton going to the library and asking for a book on gravity. Design work can only be done by designers, each with his own hierarchy of priorities and criteria. His three most important tools are knowledge, experience and good judgement.' Ted Uzzle, incidentally, was not anti-computerisation; in fact, his presentation was on the *progress* of computerised design. In the almost 30 years separating then and 2011, there have been huge developments in the programs for computer-aided design, but still his words hold true. We still do not know enough about subjective acoustics to effectively program computers for design, and that is why many 'design' software packages are referred to as CAD – computer *aided* design. In many ways, it is what we do *not* know which can make acoustic design work so fascinating. There is still so much to be learned.

As stated in earlier chapters, the music recording rooms are extensions of the musical instruments; just as a guitar amplifier is an extension of the electric guitar. Studio rooms are not like control rooms. Not only are they usually *allowed* to have a sound of their own, but also they may be *desired* to have a sound of their own. In such cases, the design of a room is very much like the design of a musical instrument, and just as no computer program has yet, from within itself, come up with the design for a violin to equal the sound of a Stradivarius, nor can computers yet design musical rooms. Stradivari was a structural engineer as well as a musical instrument builder. His in-depth knowledge of the stresses and loading which his designs would demand allowed him to choose the appropriate thicknesses and cuts of the wood that were very close to their loading limits. It is such in-depth knowledge of the small details of processes which lead to refined designs.

Another drawback with computers is that one cannot have a conversation with them over dinner. So much of the work of a designer is involved with getting a feel from the client about the required characteristics of a proposed studio. Very frequently, prospective studio owners will only have a limited knowledge of the options available, and may not be able to articulate fully their wishes. In fact, many of their wishes may be based more on misguided beliefs or intuition rather than fact. Part of the job of a studio designer is to hold detailed discussions with the studio owners, offering options which may not have been considered, or relating anecdotes which may help them to understand points more clearly. In turn, the designer hears of the needs, the problems, the frustrations and the successes of the studio owners, and these shared experiences help to build up the designer's own store of practical experience, which may serve in the future to forewarn of problems or to reinforce other experiences.

This appears to be one great failing of the typical Internet services offered, such as 'send us the plans of your building, and for $X we will send to you an advanced studio design, the product of the very latest acoustic design technology', and so forth. There is no attempt to discuss the client's wants or needs. There is no personalised service, no meeting of minds, and no 'heart' in the process. It could be that people providing many such services do a great disservice to the whole acoustic design industry, and actually mislead many clients who fail to realise the limitations of the computer 'designs' that are being offered to them. In this technical age, however, the claims for computer designs can be deceptively seductive; but beware the pitfalls.

In a somewhat back-to-front way, though, the advancement of computer modelling has very much increased the total body of acoustic understanding. Rather ironically, this has not all come from the *results* of the computer analyses, but from the great stimulation which the need for programming information has given to basic acoustic research. In order to make a computer program, a great number of facts and figures are necessary. Without a rigorously factual input, a computer cannot produce a factual output. In recent years, the need for more programming information has provided both the need and the funding for further basic scientific work, and this has been a great boon to the whole science of acoustics and its applied technology. In turn, due to their great analytical capability and the speed of their calculations, computers have fed back a great deal of additional information, and have brought new people into the world of acoustics who otherwise would not have taken up this obscure science. With acoustics, there is always so much more to know. In fact, the application of computers to acoustic design is rather like their application to the production of this book. Except for the photographs, all the figures in the book were produced by computers, some were even *generated* by computers, but they were all *conceived* by human brains.

10.4 Sound Pulse Modelling

Some of the earliest attempts to 'see' what sound was doing involved shining light beams on to mirrors fitted to resonating tuning forks, and projecting the reflexions on to a screen as the tuning fork was rotated. In the days before oscilloscopes, this was one of the only ways to 'see' sine waves. To see what was happening inside models of buildings, light was also used in the technique of sound pulse photography, as used by Sabine in 1912. The principle goes back to 1864 when Teopler showed that when parallel light rays cross a sound-field at 90° to that of the 'sound rays', the part of the sound wave-front that is met tangentially by the light rays produces two visible lines, one light, the other dark, on a projection screen behind the sound-field. In the case of 'sound pulse photography', a spark is used as an impulsive sound source, and a second spark illuminates the model room, which is exposed to a photographic screen. The screen is shielded from the direct flash from the spark in order to prevent it from washing out the faint, refracted image. The images travel along the photographic screen at the speed of sound, and they can be sufficiently sharp to allow 1 mm wavefronts to be clearly perceived. Accurate representations of sound diffraction and diffuse reflexions can be observed, even in very small models. The progress of sound waves at different instants can be studied by altering the time intervals between the two sparks.

10.5 Light Ray Modelling

Light ray models have also been used in room analysis. In these cases, mirrors were used in the models in the places where acoustically reflective surfaces were to be situated in real life. Absorbent surfaces were represented by matt black painted surfaces. This technique was mainly used when there was interest in the effects of how particular reflective surfaces in the rooms could affect any parts of the modelled room where the principle reflexions would reach. By moving the light source(s) around to all the locations of interest, the distribution of complex reflexions from an entire surface could be studied. The main limitation to this type of technique as a design tool is that the wavelengths involved bear no resemblance to the actual wavelengths of the sound in the room under study. The diffuse reflexion and diffraction expected with low frequency sound waves therefore cannot be evaluated by this process. The technique only makes possible a reasonable estimate of the behaviour of sound waves at high frequencies.

10.6 Ripple Tank Modelling

Another modelling technique is the use of water trays, into which are inserted profiles of the room under investigation. A simulated 'sound' can be generated by allowing a drop of water to fall into the tank in an appropriate position, and the progress of the wave can be seen by watching the ripples cross the surface of the water. The model has the advantage that the slow progress of water ripples can make the wave propagation clearly visible, but there is the drawback that the model only operates in a two-dimensional plane. Ripple tank models are usually carried out in a glass-bottomed tank, illuminated from below by plane-parallel light rays. This method is excellent for demonstrations and is easily photographed. By selecting a depth of water of around 8 mm, the effects of gravity and surface tension can be effectively nullified such that long and short wavelengths travel at more or less equal speeds, just as they do in acoustic waves. Water wave models were used by Scott Russell for acoustical investigations as far back as 1843.

10.7 Measurement of Absorption Coefficients

Of course, design predictions cannot be well-achieved if the input data is erroneous. Care must be taken when using published absorption coefficients in calculations because the figures are generally only approximations measured under standard test conditions, whereas real applications will depend very much on things such as batch to batch similarity and mounting conditions. What is more, the very measurement systems, themselves, are fallible, so any coefficients measured by such systems cannot be definitive.

The traditional method of using a reverberation chamber places a sample of at least 10 m^2 in a room of long reverberation time and with nearly uniform distribution of very low wall absorption (usually hard, shiny walls). The empty room is first calibrated by measuring the spacially averaged reverberation time in one-third octave bands. The measurement is then repeated with the sample of absorbent material suitably placed in the room, and the absorption coefficient is calculated from the reduction in the reverberation time. However, despite the time-honoured nature of the technique, it is known to be unreliable because the conditions assumed in the reverberation time calculation depend on a well-distributed absorption, so by putting all the absorbent material in one place the statistically random incidence of all the reflexion paths is seriously disturbed. Unfortunately, the wide distribution of smaller samples would give rise to many more absorbent edges of the material, which would change the amount of surface area exposed to random incidence waves, and the diffraction produced by the edges of the material could also augment the measured absorption. Also, the effect of small patches of distributed material would be negligible at low frequencies, where their sizes were small compared to the wavelengths.

Perhaps, a more accurate measurement can be made in a plane-wave tube, but small measuring tubes cannot accommodate samples where the material is not relatively uniform on a small scale, or where a small sample may not be representative of the normal conditions of application. Such would be the case with a wall panel, where the overall absorption when mounted could not be assessed from a small sample. There are *in situ* measurement techniques for large, constructed surfaces, but they are complex, require much care and precision, and are also prone to being disturbed by other room effects such as spurious reflexions. Figure 10.8, taken from *Acoustic Absorbers and Diffusers*, by Cox and D'Antonio,[3] shows a comparison of absorption coefficients for a single sample as measured in 24 different laboratories. In some cases, the differences are almost by a margin of 0.4, which on a 0 to 1 scale can hardly be considered to be trivial! The 'impossible' values above 1 (perfect absorption) show the effects of the diffraction and edge problems mentioned in the previous paragraph.

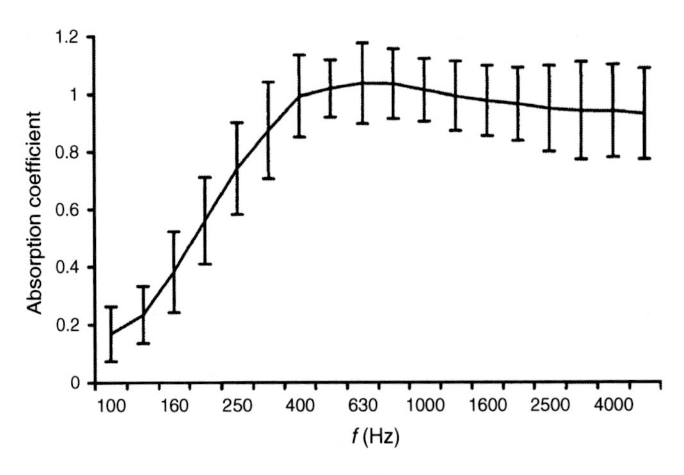

Figure 10.8:
Comparison of measured absorption coefficients for a single sample in 24 laboratories. The mean absorption coefficient across all laboratories is shown, along with error bars indicating the 95% confidence in any one laboratory measurement.

The thing to be most wary of is that absorbent materials measured under standard conditions may not perform in a similar way under different mounting conditions, and measurement under the precise mounting conditions used may not be either simple or reliable. Many acoustic design services offered on the Internet, for example, are based on calculations using standard absorption and diffusion figures, very little modified (if at all) with regard to their actual conditions of use. One cannot therefore consider such services as anything much more than generalisations.

In fact, even the reverberation times, themselves, are not always consistent when measured by different techniques. Schroeder's frustration with reverberation time measurements was mentioned in Section 10.1.2. Fifty years later, the two plots shown in Figure 10.9 show the decay time of a film dubbing theatre, one measurement made by an engineer from Dolby Laboratories in England, and the other by a scientist at the prestigious Institute of Sound and Vibration Research (ISVR), one of the world's leading acoustics research institutes; again in England. One measurement was made from a dual-channel FFT analysis and the other with a dedicated RT analyser. The starting conditions were good, with well over 50 dB of signal-to-noise ratio.

Especially with very low decay times, the options as to exactly how to go about the analysis are considerable. In these days, where almost every enthusiastic amateur has a program which they 'know' is absolutely accurate, how come we have the differences shown in Figure 10.9? The people who made the analyses were experienced professionals in very respected laboratories. It really cannot be the case that the enthusiastic amateurs know better, so in fact it must be concluded that it *is* difficult to have 100% confidence in decay time measurements, just as Schroeder noted in the early 1960s. In Figure 10.9, the difference between the two plots in the 600 Hz to 1200 Hz range is significant.

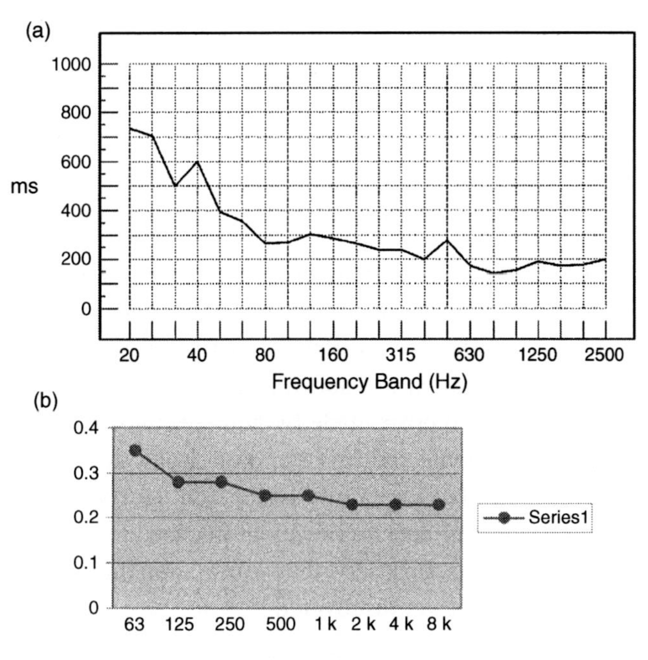

Figure 10.9:
Plot (a) shows the decay time characteristics of the dubbing (mixing) theatre shown in Figure 21.8, as analysed at the ISVR from a dual channel pink noise recording using an inverse Fourier transform and a reverse Schroeder integration; Plot (b) shows the decay time as measured by Dolby Laboratories using an AT5 reverberation-time analyser.

The differences between the plots shown in Figure 10.9 are partially due to Dolby working with ANSI (American National Standards Institute) recommended filters, and the measurements at the ISVR working to 'best academic principles'. However, differences may also be due to the different systems making calculations which had been based on different concepts of where the slope should be measured from (there is no absolute rule for this). There are obvious advantages if people use standard filters for this process, such as the ANSI recommended filters, but standard filters need to be widely available and realisable in a wide range of measurement systems if everybody is going to make measurements which can easily compare with each other. On the other hand, in academic institutions, less easily derived filters and processing may be more definitively accurate, but their less simple realisation in a wide range of portable measurement systems may make room-to-room comparisons more difficult if different people have used different analysis techniques.

10.7.1 General Limitations to Precision

It is important to understand that there *is* no universally accepted technique for measuring the low frequency decay time in rooms with very short decays. It is perhaps worth quoting here something from the conclusion of an article by Philip J. Dickinson, first published in

Acoustics Australia in 2005, then re-published in *Acoustics Bulletin* in 2007.[4] 'We can measure the light from a star millions of kilometres away, we can measure the time for light to travel a distance less than a tenth of a millimetre, we can measure the heat output of a candle more than a kilometre away — all to an accuracy of 3% or better — but we cannot measure a sound, even under *strict laboratory control* conditions, to better than ±26% (±1 dB in terms of percentage of power), or in the *open air* (it would seem) to much better than ±300% (±5 dB)'.

Much of the skill in many acoustic measurements is therefore in the *interpretation* of the results, not simply in the making of the measurements. This goes not only for reverberation (decay) time measurements, but also for absorption coefficient measurement and many others. Furthermore, even things which measure substantially similar may still sound very different to the ear. A very common error made by non-specialists in acoustics is to put too great a faith in absorption coefficients and reverberation/decay time figures. These are not things on which it is advisable to bet the family fortune on their veracity. It seems totally incredible when some mastering engineers write about the need to set their loudspeaker levels to be within 0.1 dB of each other, but perhaps they know something that the world's greatest acoustic scientists have not yet discovered.

The problem also extends into sound isolation measurements between rooms. Let us imagine that we have two adjacent rooms in a studio: a live room and a dead room. How do we measure the realistic isolation? Where do we put the noise source? Where do we put the measuring microphone? What do we use as the noise source? All these variables can significantly affect the measured results. Let us imagine that we have a source in the live room and a microphone in the dead room. If we then swap them round, reciprocity demands that the isolation would be the same in either case. However, if we change the position of the source and/or the receiver, in either room, we will measure different results. The position of either the source or the receiver in the dead room will influence the results mainly in terms of distance *from* the live room, but little else. Conversely, the changing of the position of either the source or the receiver in the live room will change things more dramatically, depending on how they couple to the resonant modes. It is therefore difficult to give a client a precise figure for the inter-room isolation because if measured by standard techniques it may not represent the normal way in which the rooms are used.

In many countries there are regulations for the minimum isolation between classrooms in schools. In these cases, the classrooms are usually neither very live nor very dead, as the former would lead to intelligibility problems and the latter to having to use raised voices, which would be very fatiguing for the teachers. In these more 'normal' rooms there are standard ways to make the isolation measurements, based on experience of how to make them in a representative way. It is usually expected that a certain minimum of isolation when measured in the standard way will ensure that problems will not occur due to the noise from one classroom disturbing the concentration of the pupils in the adjacent classroom. However, such figures are not necessarily representative of the *definitive* isolation from one room to another. They are figures to meet legal requirements. In fact, if we changed the furnishings in either of the rooms, say from hard seats to soft sofas, we would probably increase the measured isolation between the rooms by reducing the ability of the room resonances to amplify the sound.

As mentioned in Section 3.10, *legally*, a problem could arise because a neighbour was receiving noise from an adjacent studio which was 1 dB over the permitted limit. Whether this figure was acted upon or not could depend on whether the person responsible for making the measurement considered that the studio or the neighbour had the greater credibility. It has been recognised that in the UK, around half of all complaints made about noise nuisance are actually motivated by reasons *other* than the noise, hoping that the noise nuisance, if proved, would lead to the closure of the place producing it, which the person(s) complaining primarily wanted closed for some other reason. Environmental noise officers often take this into account when making measurements. The imprecision of the measurement techniques allows some leeway, and gives the officials a margin for using their own judgement in dubious cases, although if a *gross* noise nuisance exists, the case would be clear.

The nonsense which exists on the Internet about many measurement concepts is monumental, and can be very misleading for the non-specialists. Acoustics deals with three-dimensional wave expansion, which cannot be fully described by reference to a couple of points in space – the measurement positions. As shown in Figure 10.8, not even the properties of the raw materials of construction can be relied upon to be accurately documented. It all depends on how the measurements are done.

10.8 Review

In computer models, the effects of all of the various physical modelling techniques can be incorporated into the programs, but nevertheless they are only analytical tools; just like the physical modelling methods. For achieving the best results from computer models it is essential that the users understand not only the limitations of the expressions, but also those of the mathematical models used by the programs. Unfortunately, this is rarely the case. There is some very heavy mathematics involved in these models, and there are a *very* limited number of individuals who understand fully the mathematical *and* the acoustical implications. These few people tend to reside in the academic world, and consequently may not be devoting their working lives to studio design. Neither will they necessarily have the practical design experience of professional designers. The whole team of specialists that it would take for a full understanding of many of the computer-derived implications is the sort of thing which could only be afforded by governments, and not by any normal prospective studio owners.

All modelling techniques are useful methods of scientific and technical investigation, and in computer form are very marketable items, with a ready multitude of customers wanting to buy them. However, one should never expect to glean more from the computer program than one knows how to interpret. The graphic displays are not gospel. They are perhaps more useful in locating problems than they are in suggesting what should be done about them. But that is generally true of the whole science/art of studio design, or, for that matter, the design of any room for musical performances. The previously quoted words from Ted Uzzle still hold very true, and are likely to do so for a very long time to come.

10.9 Summary

A great range of measurement techniques and their displays is needed in order to see many audible characteristics of a room. Each measurement technique has its strengths and weaknesses.

Schroeder plots show a good representation of the average energy decay, but lack detail within the curve.

ETCs clearly show individual reflexions, but only poorly demonstrate the curve of the decay.

RT_{60} plots show time against frequency at the −60 dB level, but give no indication of what happens *during* the decay.

Waterfall plots show level against time and frequency, but cannot give any information about directional effects.

For the *prediction* of room responses, various forms of modelling can be used. Again, they all have their strengths and weaknesses, but they can be useful tools for investigating some specific areas of the performance of a proposed room design.

Beware of using published figures for absorbent materials too rigidly. The measurement techniques are not perfect, and even respected laboratories are not always consistent in their findings. Mounting and application conditions can also greatly affect the coefficients.

Reverberation/decay-time and isolation figures are also subject to considerable measurement error.

References

1 Schroeder, Manfred R., 'New Method of Measuring Reverberation Time', *Journal of the Acoustical Society of America*, Vol. 37, pp. 409–12 (1965)
2 Heyser, Richard C., 'Acoustical Measurements by Time Delay Spectrometry', *Journal of the Audio Engineering Society*, Vol. 15, p. 370 (1967)
3 Cox, Trevor, J., D'Antonio, Peter, *Acoustic Absorbers and Diffusers – Theory, Design and Application*, Spon Press, London and New York (2004)
4 Dickinson, Philip, J., 'Changes and Challenges in Environmental Noise Measurements', *Acoustics Bulletin*, Vol. 32, No. 2, pp. 32–36 (March/April 2007)

CHAPTER 11

Loudspeakers in Rooms

Room influences on a loudspeaker response. Radiation pattern differences and their effect on a room response. Boundary induced loading of loudspeakers. Critical distance. Room equalisation pitfalls. Minimum and non-minimum phase. Electronic room-correction considerations. Modulation transfer functions. Electronic bass-traps.

11.1 From The Studio to the Control Room

The first ten chapters of this book have considered sound, some characteristics of the ear (and hearing in general), principles of sound isolation and acoustic control measures for performing rooms. The fundamental principles of room acoustics have also been discussed. All of these subjects come under the heading of 'Acoustics and Psychoacoustics' but now we must extend our discussion into the realms of *electroacoustics* before continuing to look at the very controversial issue of the design of sound control rooms and monitoring systems. Moreover, where we have discussed the performance of relatively empty rooms, we also need to look at the sonic performance of rooms that contain considerable amounts of equipment. Much of this equipment, sadly, is not designed with the monitoring sound-fields in mind, even though it is specifically built for use in professional sound control rooms. Such factors do not help matters.

11.2 Room Influences

In an anechoic chamber or in a true free-field situation, excellent loudspeakers may produce a frequency response which is flat within ±2 dB over their designed performance range. In general, the wider the frequency range to be covered, the more design skill is necessary to maintain these limits, which is usually reflected in the cost. To people more accustomed to working in the electronics side of audio, ±2 dB would seem an absurdly wide range of tolerance for professional use, but in the realm of electroacoustic transducers it is a tight specification. Even many microphones have difficulty in maintaining ±1 dB over just *nine* octaves of the audio bandwidth, let alone ten. Very flat measuring microphones which can achieve this sort of performance over ten octaves are precision instruments, made with great care, and this *also* usually reflects in the cost.

However, the reality is that the ±2 dB tolerance on the loudspeaker specification seems to pale into insignificance when one considers that in the process of getting the sound out of a loudspeaker and across the room to the ears of the listener, ±10 dB would not necessarily be considered to be too bad. This does not apply to the smoothed, one-third octave responses, but to the *unsmoothed* responses that loudspeaker manufacturers rarely show in their literature. Figure 11.1 shows a pair of response plots for a well-known loudspeaker in an anechoic chamber. In (a) the response is unsmoothed, and in (b) it is as presented in the manufacturer's literature. Compare these plots with Figures 11.2(a) and (b), which show the loudspeaker in a normal room. Here it can be seen that the difference between the smoothed and unsmoothed plots is gross. To be fair to loudspeaker manufacturers, the in-room responses are not entirely their responsibility, because the rooms are so variable that there is no typical response which would be representative. However, even the anechoic responses are sometimes smoothed for marketing purposes to an extent that is questionable in terms of providing accurate and honest information.

Room reflexions and resonances all contribute to the overall response perceived by the listeners, but the additional acoustic loading provided by the room boundaries can cause major response changes at low frequencies. These can be in the order of up to 18 dB in the relative balance of the low and high frequencies between loudspeakers in the centre of a reasonably large room or placed on the floor in a corner of the same room. The variability in the frequency balance in different places in the room can be very great indeed. The situation is not only complicated by the nature of the radiation pattern of the loudspeaker with respect to frequency, but also the reflexion density and decay time within the room. Room resonances, and whether the source is monopole or dipole, also add their complications. Before going into detail about the individual effects, we can perhaps look briefly at them in order to get an impression of how they manifest themselves. They all form characteristic parts of a loudspeaker/room *system*, as it should be clear by now that the loudspeaker response, alone, does not describe what we will hear in practical circumstances.

11.2.1 Radiation Patterns

Figure 11.3 shows three polar plots of low frequency loudspeaker radiation. The first pattern (a) is that of a dipole source. This is the typical radiation pattern of a conventional woofer on a simple open baffle board. The effect is also typical of the flat electrostatic loudspeakers with open backs. When the diaphragm is driven forwards, a positive pressure is created in front of the diaphragm and a negative pressure is created behind it. The lack of enclosure behind the loudspeaker allows the pressure to equalise at the sides, simply by travelling around the edges of the baffle. The size of the baffle board on which the loudspeaker is mounted determines the frequency below which the cancellation will be effective. For a listener directly in front of the loudspeaker the perceived frequency response will be flat (assuming a perfect source) down to the frequency where the cancellation begins due to the finite size of the baffle, below which a roll-off will begin, which will finally reach 18 dB/octave.

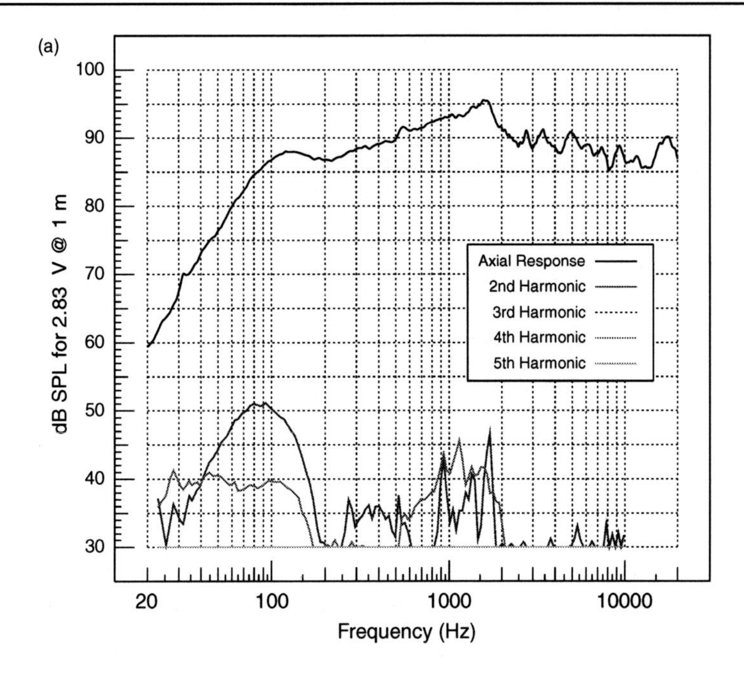

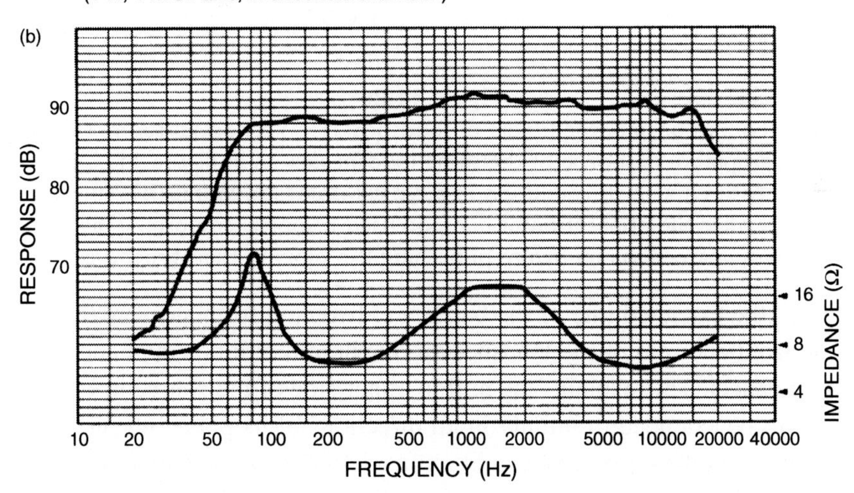

Figure 11.1:
Frequency response presentations: (a) Unsmoothed anechoic response as measured in a large, reputable, acoustics research institute. Response ±5 dB, 85 Hz to 20 kHz; (b) Response of the same loudspeaker from the manufacturers' literature, perhaps after the application of octave band smoothing. Frequency range quoted as 60 Hz–20 kHz, without specifying deviation limits. The plots suggest ±2 dB 85 Hz to 20 kHz, as opposed to the ±5 dB of the unsmoothed plot, (a). The version that relates best to what we actually hear is still a moot point.

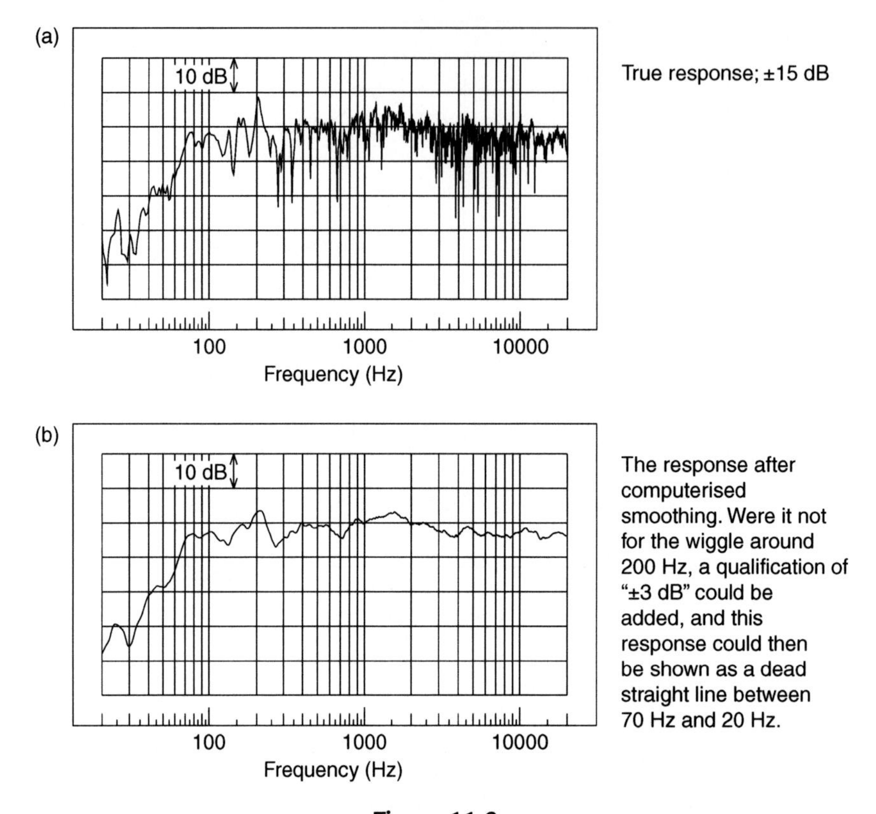
(a)
10 dB
True response; ±15 dB
100
1000
10000
Frequency (Hz)
(b)
10 dB
The response after computerised smoothing. Were it not for the wiggle around 200 Hz, a qualification of "±3 dB" could be added, and this response could then be shown as a dead straight line between 70 Hz and 20 Hz.
100
1000
10000
Frequency (Hz)

Figure 11.2:
Smoothed and unsmoothed plots of loudspeakers in a normal room: (a) The pressure amplitude response of the same loudspeaker as in Figure 11.1, but this time measured at high resolution in a typical domestic room; (b) As in (a), above, but after the application of computerised response smoothing, as used by many manufacturers before publishing the plots in their brochures.

For a listener on the extended line of the baffle, i.e. listening from side-on, the pressures radiated at the front and rear of the loudspeaker diaphragm will be equal and opposite, and hence they will cancel. This explains the null to either side of the source in Figure 11.3(a). If the baffle were to be extended to infinity in all directions of its plane, no such cancellation would exist. Hence, for the same drive conditions as shown in (a) the radiation pattern would be as shown in Figure 11.3(b). Here the areas of positive and negative pressure still exist at the front and rear of the driver, but the truly infinite baffle has prevented the cancellation from occurring around the sides. The radiation pattern becomes a sphere, with the baffle dividing the region of positive and negative pressure. (Incidentally, for non-native English-speaking readers, 'to baffle' means 'to stop the progress of'. The phrase 'I'm baffled', is often used by people whose thought trains are not progressing, in other words they have no idea. In the case of a loudspeaker, the baffle impedes the progress of the equalisation of the pressures between the front and rear of the diaphragm.)

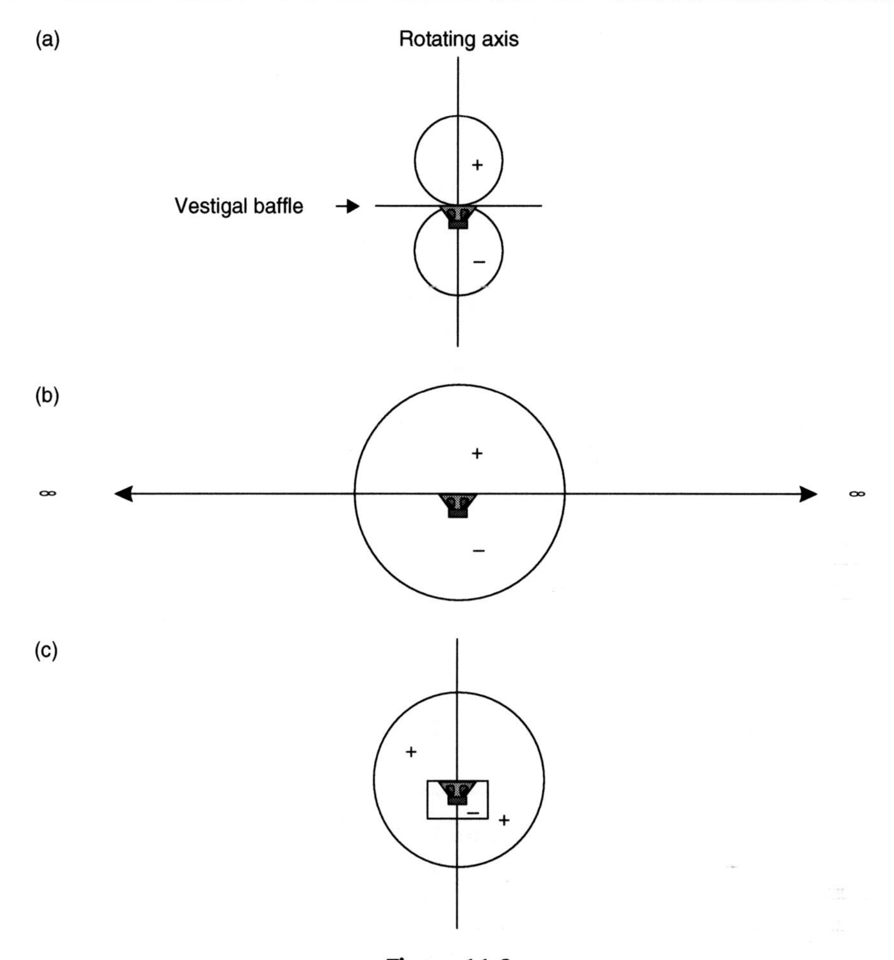

Figure 11.3:
Loudspeaker radiation patterns at low frequencies: (a) Dipole source; (b) True infinite baffle; (c) Monopole source (such as a sealed box).

The term infinite baffle is often wrongly used to describe a sealed box loudspeaker. Figure 11.3(c) shows how the sealed box in fact resembles more a monopole source. A *monopole* source describes the radiation pattern of a small (compared with the wavelength involved) pulsating sphere. As the sphere expands, a positive pressure would be radiated in all directions. Indeed, when a loudspeaker cabinet is completely enclosed and in free space, a positive pressure in front of the diaphragm *would* also radiate in all directions at low frequencies, with the negative pressure being trapped inside the box as the diaphragm moved outwards.

As the frequencies rise to a point where the wavelength begins to equal the circumference of the radiating diaphragm, the radiation angle begins to narrow. This is shown diagrammatically in Figure 11.4. Here it can be seen that, at positions off-axis, the distance to the nearest and farthest points on the radiating diaphragm are such that the path length difference to the listening

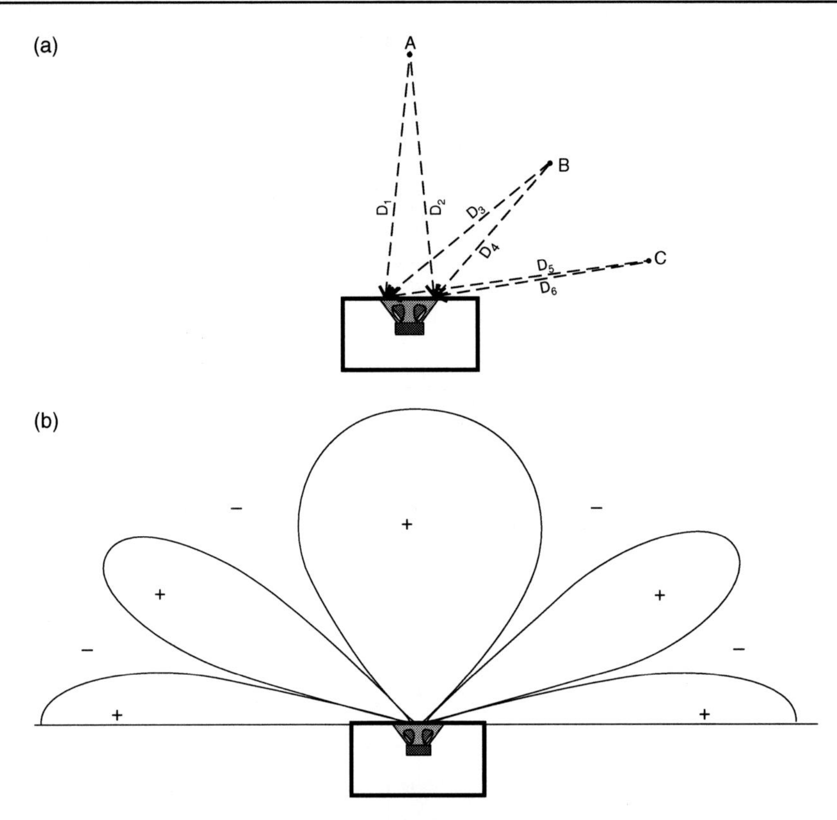

Figure 11.4:
Radiation from a sealed box loudspeaker: (a) At point 'A', distances D1 and D2 are equal, and both are very similar to the distance from 'A' to the centre of the cone. The radiation from all points of the cone will therefore arrive at 'A' substantially in phase, and so will sum constructively. However, the distances D3 and D4, from each side of the cone to point 'B', are *not* equal, and have, in the example shown, relative lengths of 5 to 4. If we assume D4 to be 1 m, then a frequency of 2000 Hz, which has a wavelength of about 16 cm, would travel six full wavelengths from the right-hand side of the cone to position 'B'. The path length from the left-hand side of the cone to point 'B' is 125 cm, so a 2000 Hz sound wave would travel seven and a half wavelengths, arriving at point 'B'; half a wavelength (180°) out of phase with the wave from the right-hand side of the cone, and hence would cancel. Further off-axis, at point 'C', the respective distances to the left and right sides of the cone are 133 cm and 1 m. The distances represent about 8 and 6 full wavelengths at 2000 Hz, which means that the waves would once again be in-phase by the time that they reached position 'C'. This continuously varying phase relationship as one moves around the front of a loudspeaker gives rise to the typical polar pattern shown in (b); (b) Response lobing. At frequencies with wavelengths less than the diameter of the radiating surface, the + and − signs show, at a given frequency, the regions where the radiation from opposite sides of the diaphragm superpose constructively or destructively. Each frequency will have its own particular pattern. (See Figures 11.22 and 14.3.) For this reason, it is customary to cross over to a smaller drive unit before the frequency reaches the point where the wavelength approximates to the radiating diameter of the diaphragm.

position becomes out of step by half a wavelength, and hence the radiation from these points cancel. Traditionally, when this point is approached, where the uniform forward radiating angle becomes too narrow, it has been customary to 'cross over' to a smaller loudspeaker, in order to avoid the forward 'beaming' of high frequencies. In most full range loudspeaker designs there is a gradual narrowing of the response towards the mid frequencies, and then, by means of selecting drivers of the appropriate source sizes, that response angle would be maintained until the highest frequencies. The typical radiation pattern of a conventional box loudspeaker is shown in Figure 11.5. The response pattern shown in the above figure is typical of by far the majority of loudspeakers designed for use in recording studios.

11.2.2 Loading by Boundaries

If we were to take a loudspeaker such as the one represented in Figure 11.5, and take it towards a room boundary, the effect at low frequencies would be as shown in Figure 11.6. The radiation pattern shown dotted is the reflected wave from the boundary. If the boundary were of a rigid, massive, highly reflective material, the reflected wave would be the same in all respects, except direction, as the section of the wave would have been had it been able to expand without the boundary (shown dashed). The main difference in the response at the loudspeaker is due to the fact that the reflected energy will superimpose itself on the direct energy instead of being 'lost' by expansion. At low frequencies, where the distance from the loudspeaker to the wall is small compared with the wavelength, the reflected pressure will sum with the direct pressure. As the frequencies rise, however, frequencies will be reached where the reflected pressure arrives back at the radiating diaphragm out of phase with the source, and the pressures will then cancel. At even higher frequencies, the direct and reflective waves will go through regions of summation and cancellation as the wavelengths vary. The effect will be comb filtering of the response, as shown in Figure 11.7, so called because on a linear frequency scale the effect looks like a comb with many teeth.

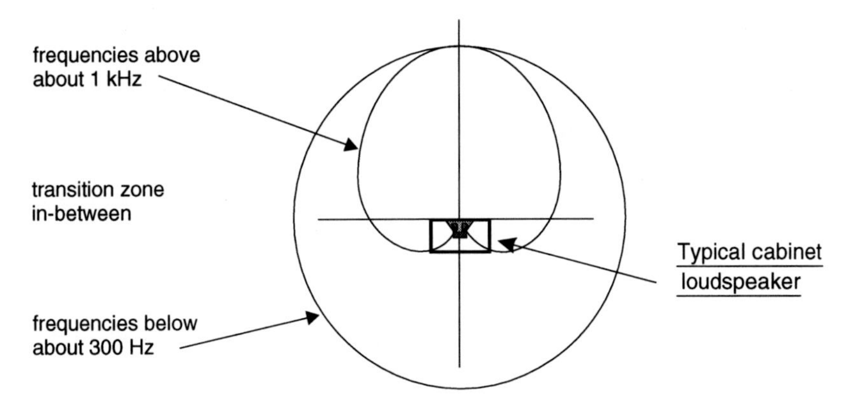

Figure 11.5:
Directivity versus frequency.

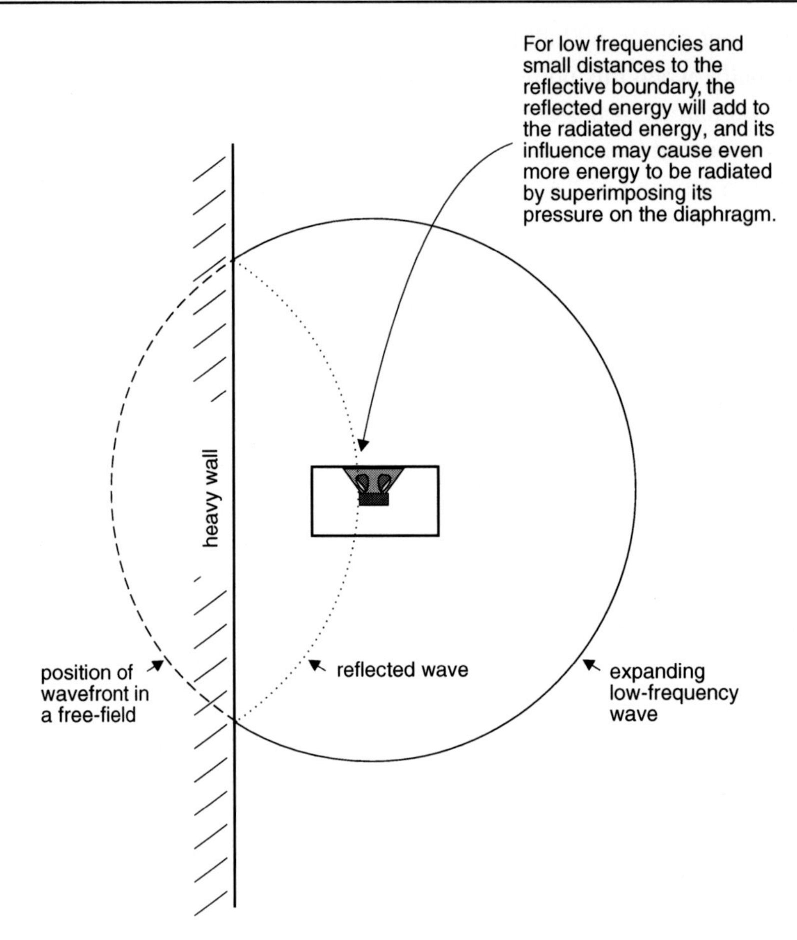
For low frequencies and small distances to the reflective boundary, the reflected energy will add to the radiated energy, and its influence may cause even more energy to be radiated by superimposing its pressure on the diaphragm.
heavy wall
position of wavefront in a free-field
reflected wave
expanding low-frequency wave

Figure 11.6:
Reflective loading on a diaphragm.

When we place a monopole loudspeaker as shown in Figure 11.6, the logarithmically scaled response will tend to look like the plot in Figure 11.8. At higher frequencies, the response is dependent upon the receiving position as well as the source position. At low frequencies, where the response is *not* dependent upon the receiving position, the effect would be a general boost in the pressure response. Every position for the source and reception points thus has its own individual response. In order to give some idea of the magnitude of the effects, Figures 11.9 and 11.10 show the way in which the received response of a loudspeaker can vary, solely according to where it is positioned within a room.

There is also another mechanism at work, besides the simple reflected superposition of the pressure waves. There is mutual coupling between the reflected surfaces and the radiating surfaces. When the reflected wave returns *in-phase* to the radiating diaphragm, it imposes a greater pressure on the diaphragm than would exist due to the ambient air pressure alone.

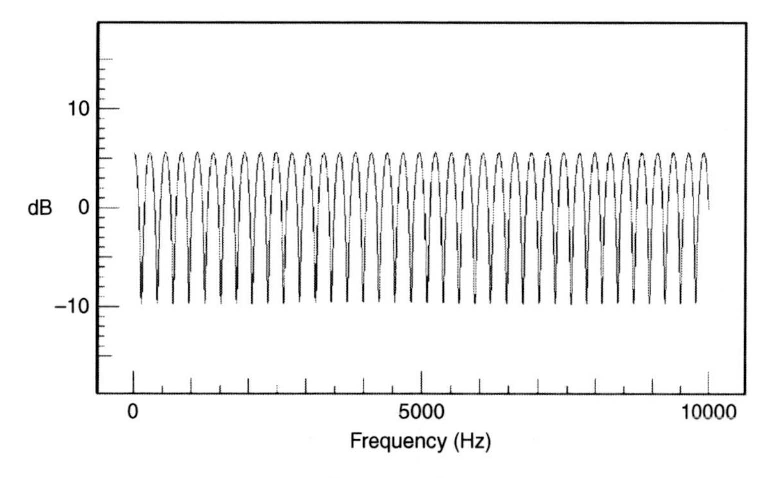

Figure 11.7:
Comb filtering. The result of the superposition of the reflexion on the direct signal, as shown in Figure 11.6, will give rise to a frequency response as shown here, which, plotted on a linear frequency scale, clearly shows the origin of the term comb filtering.

The increased pressure gives the diaphragm more to push against, and so more work is done, which results in more acoustic power being radiated by the diaphragm. A conventional electromagnetic, cone loudspeaker is a very inefficient device, largely because it is a volume-velocity source whose mass is great in respect to that of the local air upon which it is acting. Its velocity tends to be controlled by its own mass, and hence the volume of air that it moves is largely independent of the air pressure resisting its movement. Therefore, if

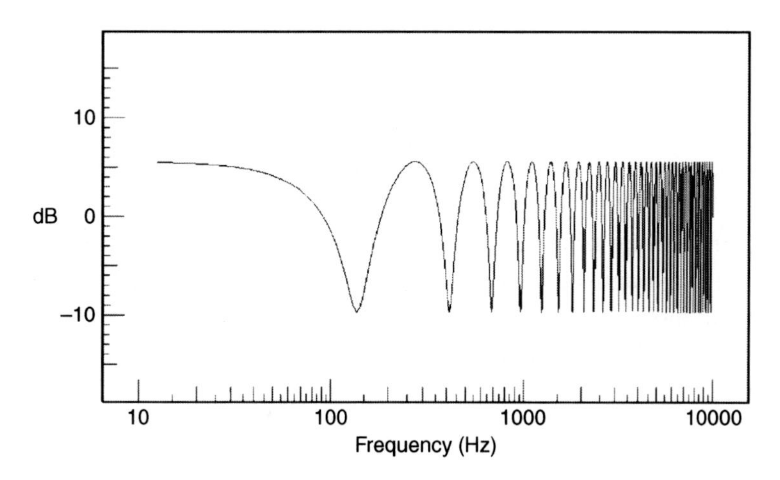

Figure 11.8:
The same response as in Figure 11.7, but here plotted on a more conventional logarithmic frequency scale. This relates better to how we perceive the effect, but for analytical purposes it does not show the periodicity of the effect as clearly as the linear plot in Figure 11.7.

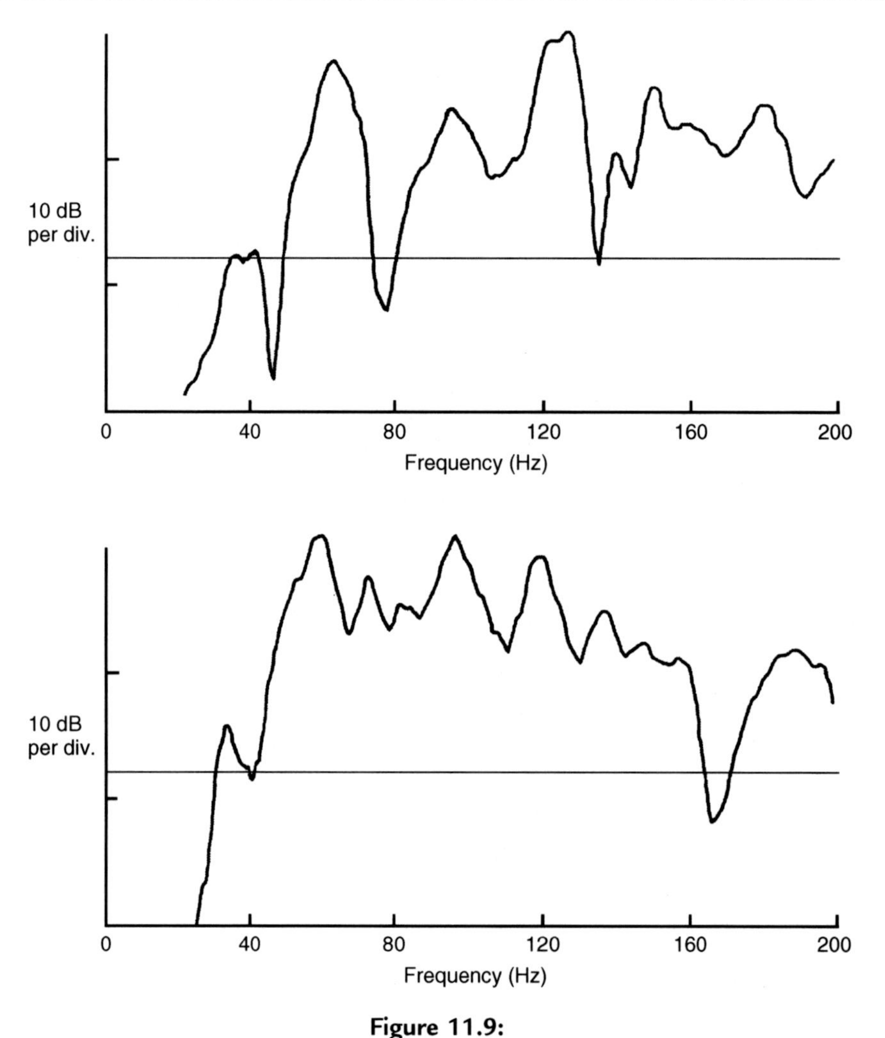

Figure 11.9:
Low frequency response plots of one loudspeaker in one control room but in two different positions.

the air pressure on the face of the diaphragm is increased, the diaphragm speed will be unaffected. As the diaphragm becomes more heavily loaded it can exert more *force* on the air, consequently doing more work and radiating more power. The mutual coupling and the in-phase superposition of reflected pressures act together in the positive *and* negative senses. Pressure increases and decreases will both be augmented.

Figures 11.11–11.14 show the effect of the reflexions on the radiated sound power of a non-directional monopole source when placed near three mutually perpendicular boundaries in a room, i.e. close to a corner. In each case, the output in free space would be the 0 dB level. From these figures, it can be seen that the reflective loading effects of the room boundaries are enormous. Figures 11.15–11.18 show the effects on a dipole source, and it can be seen that the

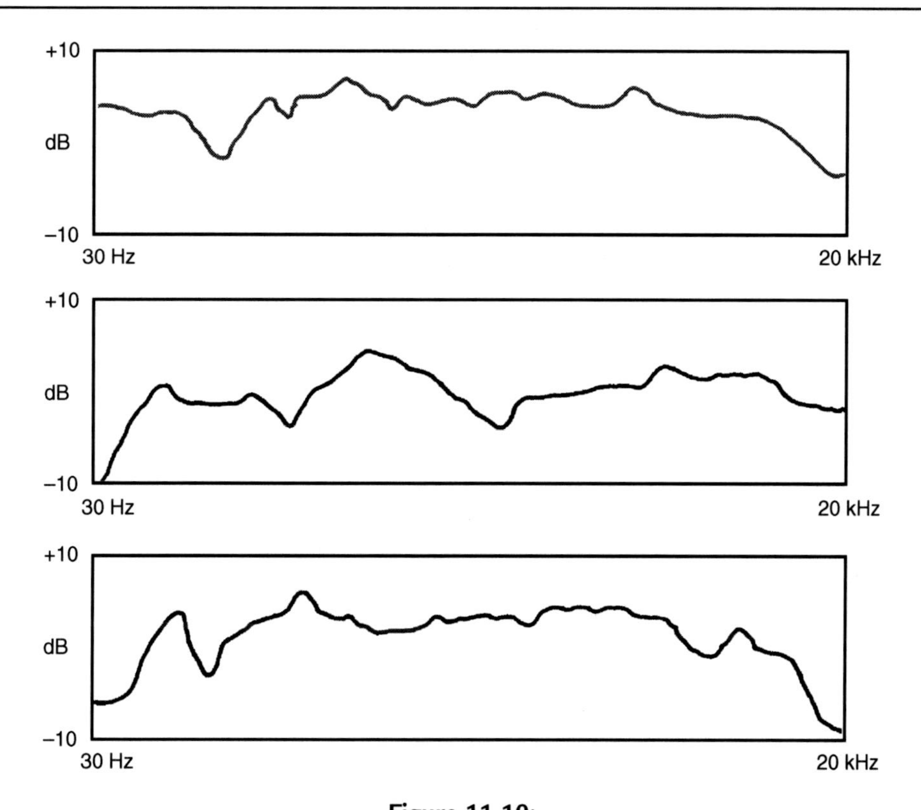

Figure 11.10:
Response of one loudspeaker at three different locations in the same room. Differences are entirely due to room position.

effects on the two types of sources are very different. The figures highlight clearly the consequences of the positioning of loudspeakers within rooms. This is the principal reason why loudspeaker manufacturers tend to publish only the free-field response, because what rooms can do to modify it is simply too variable for any general representation of an in-room response to be meaningful.

The mirrored room analogy from Figure 4.5 also still holds true, because loudspeaker diaphragms mutually interfere in the same way that reflexions interact with diaphragms, and *vice versa.* We could place real loudspeakers at the image positions and the same response variations shown in Figures 11.11−11.18 would be exhibited. We could reduce the effect of the interference by turning down the drive signal to the image loudspeakers. If we turned them off, the loading effect would disappear. Acoustically we can simulate this by making the boundaries absorbent. In the 99.99% absorption of an anechoic chamber, the 'image loudspeakers' (the reflexions) would be turned down by 40 dB relative to the principle loudspeaker (99.99% absorption means only 0.01% reflected energy; 0.01% = 1 part in 10,000 = −40 dB). The single loudspeaker response would then return to what it would be in free-field conditions.

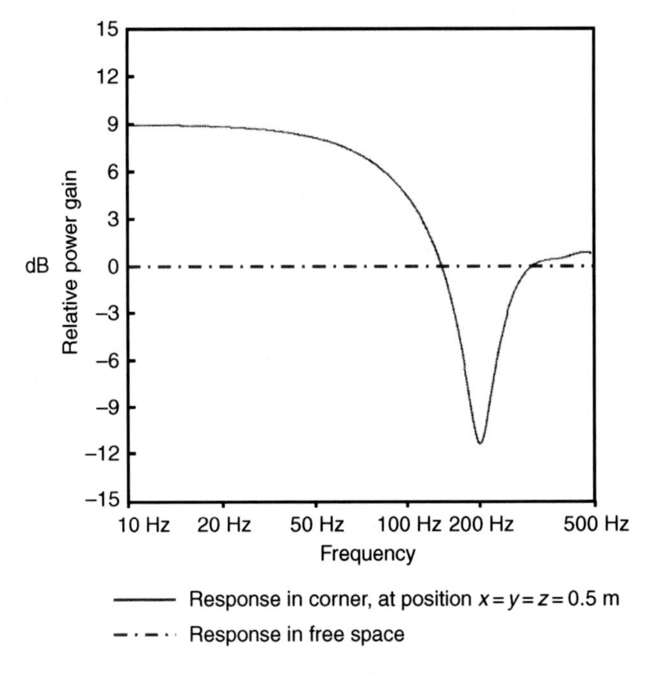

Figure 11.11:
The sound power output of a non-directional sound source placed near three mutually perpendicular room boundaries, relative to the power output in free space, shown dashed. The source is positioned 0.5 m away from each boundary ($x = y = z = 0.5$ m).

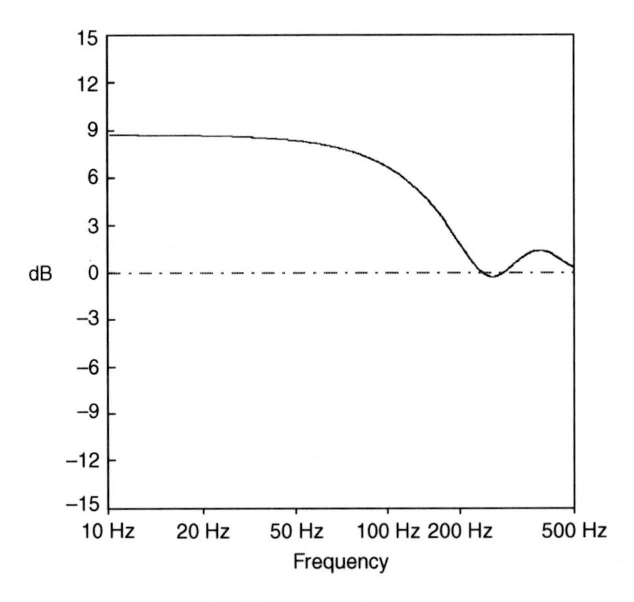

Figure 11.12:
The sound power output relative to free-space loading, as for Figure 11.11, but for asymmetric placement ($x = 0.5$, $y = 0.35$, $z = 0.15$ m).

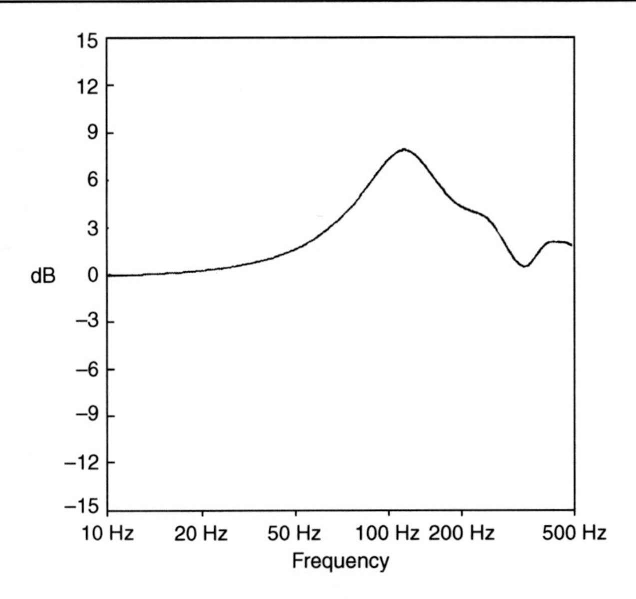

Figure 11.13:
The power gain (the difference) obtained by moving a non-directional source from one position ($x = 1$, $y = 0.7$, $z = 0.3$ m) to a second position nearer the corner ($x = 0.33$, $y = 0.23$, $z = 0.1$ m). The increase in radiated power around 100 Hz is almost 9 dB.

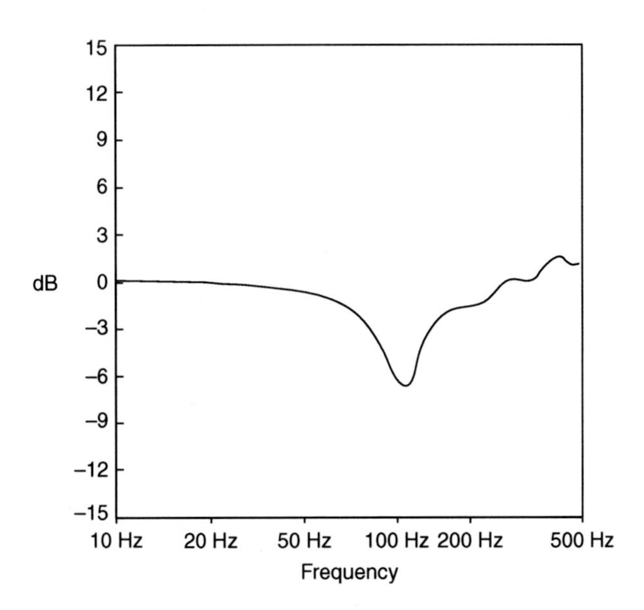

Figure 11.14:
The power gain (the difference) obtained by raising a non-directional source from one position ($x = 1, y = 0.7, z = 0.3$ m) to a second position 0.5 m higher ($x = 1, y = 0.7, z = 0.8$ m). Here there is a clear reduction in the radiated power around 100 Hz, although there is a slight increase around 400 Hz.

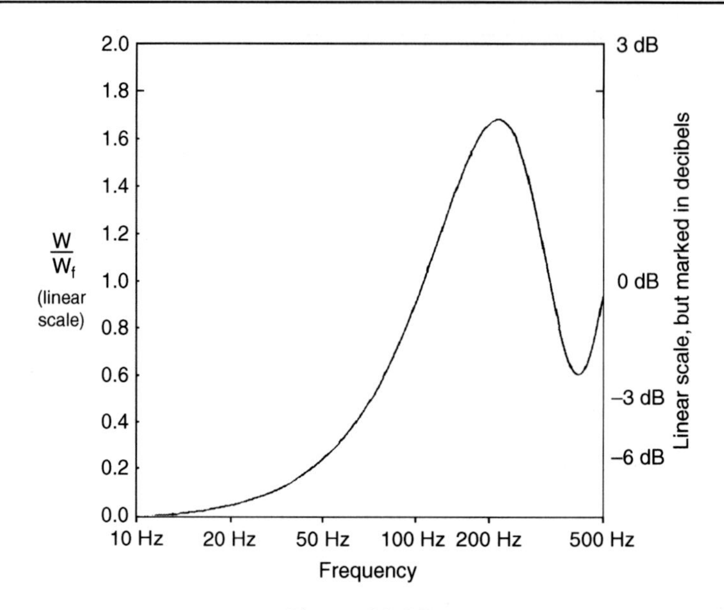

Figure 11.15:
Sound power output relative to free-space loading of a dipole sound source with baffle parallel to a single solid boundary. Distance from boundary 0.5 m.

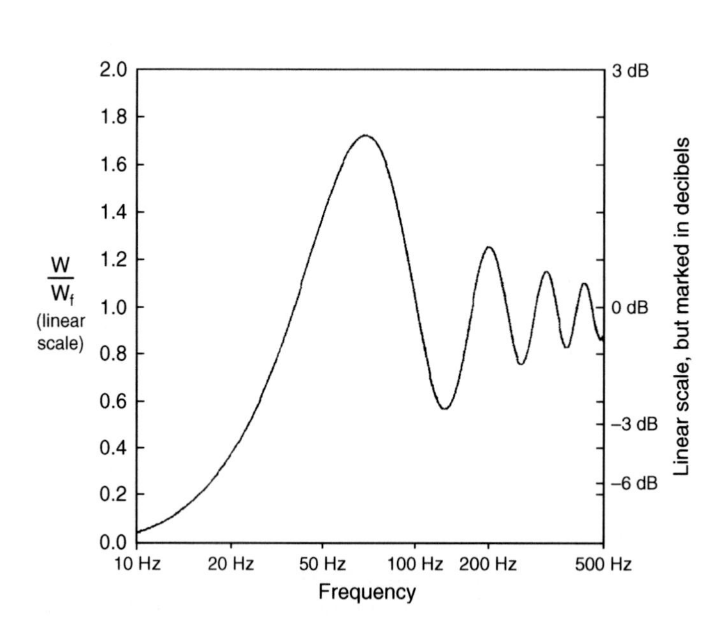

Figure 11.16:
Sound power output relative to free-space loading of a dipole sound source with baffle parallel to a single solid boundary. Distance from boundary 1.5 m.

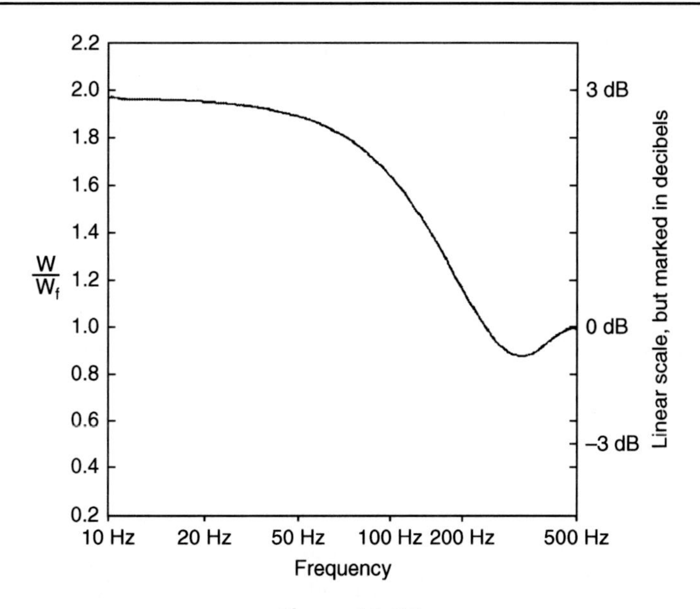

Figure 11.17:
Sound power output relative to free-space loading of a dipole sound source with baffle placed at right angles to a single solid boundary. Distance from boundary 0.5 m.

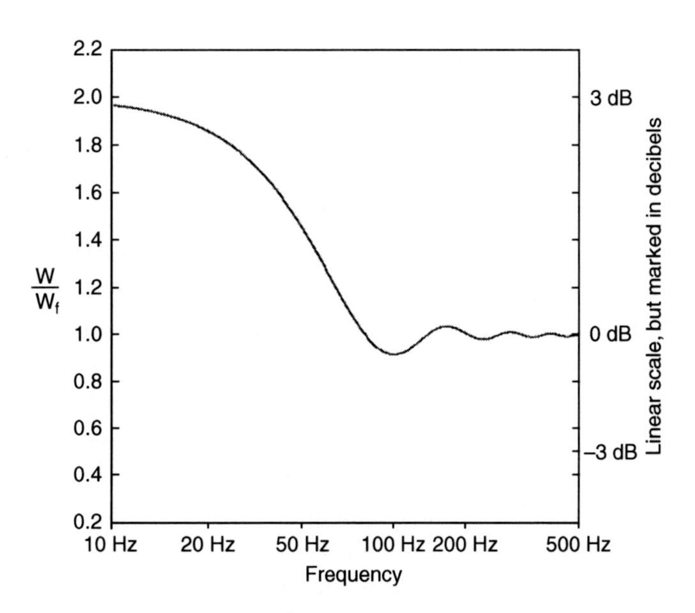

Figure 11.18:
Sound power output relative to free-space loading of a dipole sound source with baffle placed at right angles to a single solid boundary. Distance from boundary 1.5 m (for further study, see chapter by Adams in *Loudspeaker and Headphone Handbook* – details given in the Bibliography at the end of this chapter).

If the boundary has an absorption coefficient of 0.5, then half the energy would be reflected, which would equate to turning down all the image loudspeakers by 3 dB. Unfortunately, as can be seen from Table 4.1, the absorbent materials in normal rooms do not absorb uniformly with frequency. Therefore, all our ideal response modifications shown in Figures 11.11–11.18, which presumed a set of reflexions from perfectly reflective walls (absorption coefficient 0 at all frequencies), would be further modified in real life due to the frequency dependent absorption by the boundaries. This could still be represented in the mirrored room analogy by not only providing the image loudspeakers with level controls, but also with equalisers. This concept is useful in that it helps many people visualise the complexity of the reflective coupling.

11.2.3 Dipole Considerations

A comparison of the monopole response plots shown in Figures 11.11–11.14 with the dipole source plots in Figures 11.15–11.18 should make it evident that the two types of radiating sources behave very differently in rooms. Even though the on-axis anechoic responses may be virtually identical, the presence of boundaries affects their responses in very different ways, and the fact becomes even more pronounced when the loudspeakers are used in stereo pairs.

Most stereo programme material has the majority of its low frequency content distributed within the mixes as central phantom images. Two monopole sources (such as most cabinet loudspeakers) radiate the low frequencies in all available directions. For relatively steady-state signals, a standing wave pattern is produced by the interference field (see Glossary) and the two separate sources are very effective in exciting many, if not most, of the room modes to varying degrees.

By contrast, the bi-directional, figure-of-eight radiation patterns of the dipole loudspeakers principally tend to excite modes in the front/back axis only. They also excite the modes most strongly when they are on the velocity anti-nodes (the pressure nodes), which is the exact opposite of monopole source behaviour, because they are pressure-gradient sources, not volume-velocity sources. The modal excitation pattern is thus very different for the two types of radiation, not least because the electrostatic dipole loudspeakers, with very light diaphragms, also tend towards being true pressure sources, whose output is independent of the pressure load on the diaphragm. Any increased work done by the extra loading (pressure on the diaphragm) will be offset by the resulting reduction in diaphragm displacement. Note how the boost due to the reflective loading in Figures 11.17 and 11.18 are much less than those shown for the cabinet loudspeakers (volume/velocity sources) shown in Figures 11.11 and 11.12.

A further aspect of dipole radiation is that because in almost all practical cases the loudspeakers present their side-nulls towards the other loudspeaker in a pair, virtually no radiation from one loudspeaker can impinge directly upon the diaphragm of the other. Mutual coupling, the mechanism by which multiple drivers mutually boost each other's output (see Chapter 14 and Glossary), can therefore only take place via the less energetic reflective pathways. The far-field response of dipoles, as compared to monopoles (in any given, typical room), therefore tends to be characterised by less modal density, but stronger excitement of some of the modes that *are* energised. Furthermore, except in central positions some distance in front of the loudspeakers, the mutual coupling boost at the lower frequencies, which is characteristic of monopole sources when radiating coherent signals, will be absent from the combined output of the dipole pair. This can lead to less low frequency content in the ambient response of the rooms.

These factors together, plus the fact that the dipole sources cannot drive a pressure zone (see Section 13.5) no doubt contribute greatly to the oft-held belief that many electrostatic loudspeakers, most of which are dipoles, have a tendency to be bass light, even when the individual on-axis anechoic responses compare quite closely with many monopole sources. It can therefore be concluded that great care must be taken in the siting of dipole loudspeakers with regard to any room boundaries, *and* the modal distribution within a room if a relatively uniform response is to be produced at the listening position. In fact, these considerations may well dictate the listening position.

11.2.4 Diffraction Sources

Figure 4.35 showed the diffraction effects due to obstructions, and spaces *in* obstructions, on sound propagation. Figure 11.19 shows the diffraction caused by the edges of loudspeaker cabinets. Effectively what happens is that the expanding radiated wave travels along the front surface of the loudspeaker cabinet until it arrives at a discontinuity, such as the cabinet edge. From this point on it suddenly has to expand more rapidly around the corner, but the same pressure cannot be sustained in the new, larger volume into which it expands. This sudden increase in expansion rate is due to a change in acoustic impedance, and any change in impedance will send back a reflected wave. (If there is a sudden pressure change at any point, then sound will radiate from that point. After all, a point source, or a tiny loudspeaker, is radiating power solely by virtue of the pressure changes that it creates.) The radiation that turns the corner is in-phase with the source, but the reflected wave, due to the sudden *drop* in pressure (as the wave reaches the corner it is free to expand more easily as it is less restrained), radiates back to the source *out* of phase. This gives the effect of a spherical type of wave expansion that appears to emanate from the corner discontinuity, giving the effect to the listener as shown in Figure 11.20. The total effect may be considered as though the loudspeaker were in an infinite baffle, but with additional sources mounted at the positions of the cabinet edges.

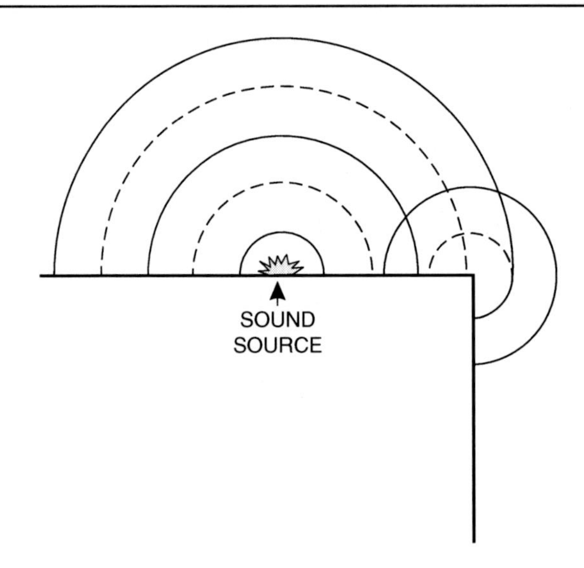

Figure 11.19:
Diffraction sources. Representation of the sudden increase in the rate of expansion of a wavefront at a sharp edge. The diffracted wave in the region round the corner from the sound source has the same phase as the incident wave on the edge, but the diffracted wave, travelling back from the edge *towards* the source, is phase-reversed.

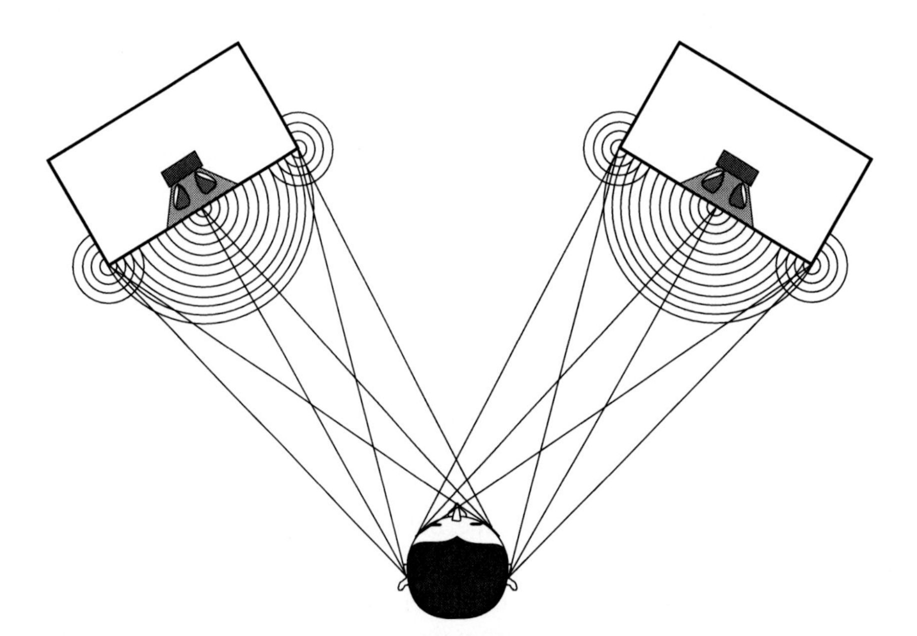

Figure 11.20:
Cabinet diffraction. The sharp edges of loudspeaker boxes, mixing console meter bridges, and any other abrupt cross-sectional changes between the loudspeakers and the listener will form secondary diffraction sources that can smear the stereo imaging. It is to help to avoid this problem that many loudspeaker cabinets have rounded or bevelled edges.

Objects in any room in which a loudspeaker may be positioned, and especially objects close to a loudspeaker, such as the mixing consoles on which they may be mounted, can all act as diffraction sources, disturbing in yet another way the free-field response of a loudspeaker. (See also Section 4.8.)

11.3 Room Reverberation and the Critical Distance

True reverberation exists as a diffuse sound field, with equal energy being present at any one time in any equal volume anywhere in the room. However, modal resonances, the juxtaposition of different surfaces, and the net energy flow away from the source ensure that in recording studio control rooms such a spacially diffuse sound-field never exists. Nevertheless, especially in *some* control room designs, a reasonably diffuse decay field does normally exist, which means that after a sound is emitted by the loudspeakers, a quasi-reverberant sound-field will develop in the room. There will therefore, at any time except immediately after a silence, be a mixture of the direct sound from the loudspeaker and the reflected sound-field from the boundaries. In a large reverberant space, with a loudspeaker at one end and a listener at the other, the direct sound field will reduce by 6 dB for every doubling of distance from the source, whereas the reverberant field will be relatively uniform in all parts. Where the listener stands, the reverberant sound field may be 20 dB or more *above* the direct sound field, and the room response would clearly be the dominant source. This is shown diagrammatically in Figure 11.21, which shows that as one

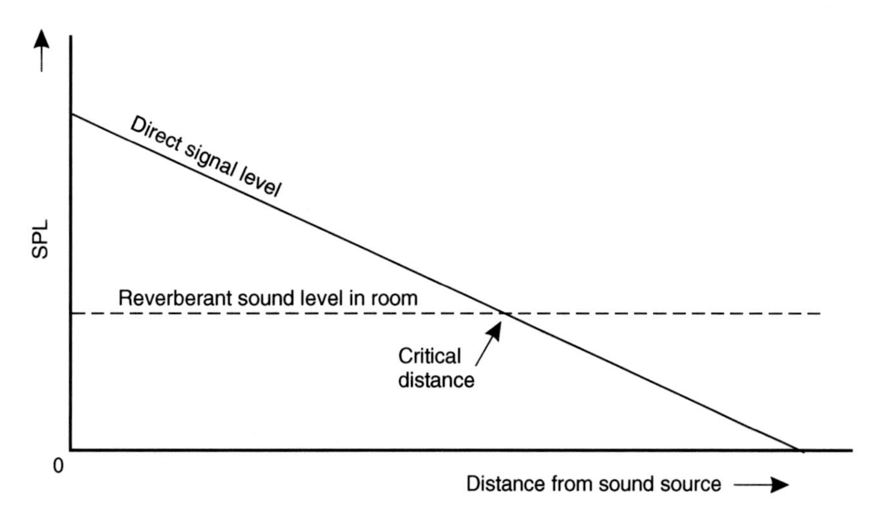

Figure 11.21:
The critical distance concept. In a room, the direct sound decays with the distance from the source, but the reverberation level in the room is, by definition, evenly distributed. The critical distance is that at which the falling direct signal level (as one moves away from the source) reduces to such a level that it is equal to the reverberant level. At less than the critical distance, a listener is predominantly in the direct field, whereas beyond the critical distance, the reverberant field predominates.

moves away from a sound source, in non-anechoic conditions, one will gradually leave the domain of the direct field and enter that of the reverberant field. The point where the two sound-fields are equal is known as the critical distance, beyond which the level of the sound will soon tend not to reduce any further as one moves away from the source.

Figure 11.5 shows that the low frequencies tend to radiate omni-directionally, whereas the high frequencies are more directional, and principally travel forwards from the source. The significance of this on the critical distance is that the low frequencies will strike more boundaries more rapidly, and hence they will drive the reverberant field more powerfully than the more directional high frequencies. The result is that the critical distance will be frequency dependent and so the effect of moving away from the source will not be perceived equally at all frequencies. In fact, the critical distance is dependent only upon the room constant (R) and the directivity of the loudspeaker (Q). The room constant is given by:

$$R = \frac{S\alpha}{1-\alpha} \tag{11.1}$$

where S is the surface area of the room (m^2) and α is the average absorption coefficient of the surfaces.

The Q is effectively given by dividing the area of a sphere by that portion of the area that is illuminated (insonified) in the direct field of the loudspeaker. The directivity index (DI) is the ratio in decibels of the sound pressure radiated along the forward axis of the loudspeaker, compared to the sound pressure that would exist at the same position if the equivalent sound energy were to be radiated omni-directionally. Consequently, for omni-directional low frequency radiation, the directivity index and the Q are both 1.

In a typical domestic room with a surface area of 85 m^2, plaster walls and a moderate degree of furnishing, the average absorption coefficient could be about 0.2. In such a room, and with an omni-directional source, the critical distance can be given by the formula:

$$\text{Critical distance } (r) = 0.141\sqrt{RQ} \tag{11.2}$$

The omni-directional loudspeaker has a directivity of 1, and the term R can be found from Equation 11.1:

$$R = \frac{85 \times 0.2}{1-0.2}$$

$$R = \frac{17}{0.8}$$

$$R = \underline{21.25}$$

Therefore, the critical distance for the low frequencies in the room under discussion would be:

$$= 0.141\sqrt{RQ} \qquad \text{(given in equation 11.2)}$$

$$= 0.141\sqrt{21.25 \times 1}$$

$$= 0.141 \times 4.61$$

$$= \underline{0.65 \text{ metres}}$$

If the directivity of the high frequencies was around 10, as would be the case for a pattern typically as shown in Figure 11.5, the critical distance (r) would be:

$$r = 0.141\sqrt{21.25 \times 10}$$

$$= 0.141\sqrt{212.5}$$

$$= 0.141 \times 14.6$$

$$= \underline{2.06 \text{ metres}}$$

Figure 11.22 shows the increasing directivity-with-frequency from a typical piston radiator (such as a conventional cone loudspeaker). If these plots were spun round in the axis which runs through the page from top to bottom, one can imagine the spherical space and the proportion of the whole sphere taken by the reducing areas (or volumes if three-dimensional) within the directivity fields. One can imagine how the insonified volume in Figure 11.22(f) would occupy less than one-tenth of the solid sphere. Hence, less than one-tenth of the power would be needed to fill this space with a given quantity of sound, as opposed to that which would be needed to fill the entire sphere. It requires twice the power either to double the sound intensity over a given area, or to supply the same sound intensity over twice the area. It thus follows that it requires ten times the power (+10 dB) to maintain the same sound intensity over ten times the area, as is more or less the case under discussion here.

A room of 5 m × 4 m × 2.5 m would have a surface area of 85 m^2, which was the figure used in our previous calculations. As counter-intuitive as it may seem, the fact is that in such a room, with an absorption coefficient of 0.2, the room response will begin to dominate the overall low frequency response at listening distances greater than only 65cm from the loudspeaker. At high frequencies, the distance would be 2 m, or thereabouts, with the loudspeakers described, but such is the case in many typical domestic listening rooms (and bad control rooms).

11.4 Sound Power Radiation

Figure 11.23 shows the on-axis pressure amplitude response (frequency response) of a loudspeaker in an anechoic chamber. Figure 11.24 shows the same loudspeaker response measured in a reverberation chamber, which integrates the total power output. The difference

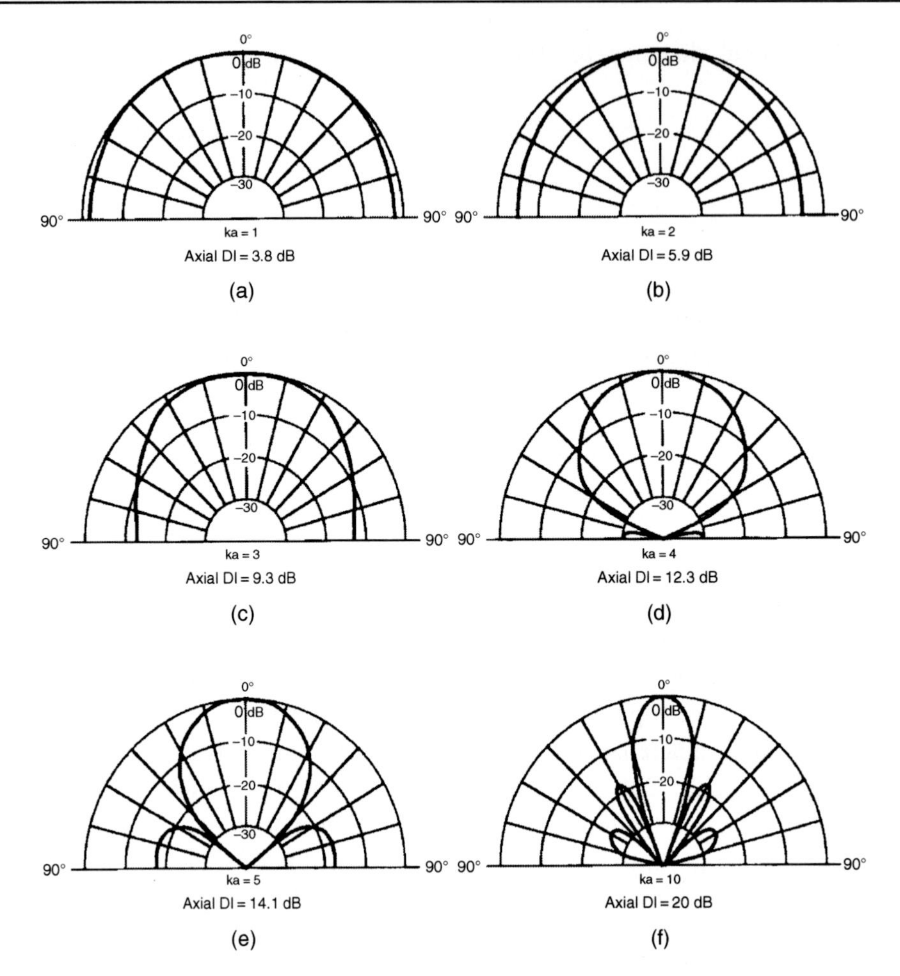

Figure 11.22:
Directional response of a piston mounted in a large baffle. It can be seen that the directivity increases (the radiation angle decreases) with frequency. The effect of specific diaphragm sizes and frequencies can be calculated from the ka figure, where

$$k = \text{the wave number} = \frac{2\pi}{\text{wavelength in metres}}$$
$$a = \text{the radius of the piston (diaphragm) in metres}$$

between the plots shows how more power is required in order to maintain a flat on-axis response with reducing directivity as the frequency lowers. In Figure 11.24, it is clear that about 10 times the power (10 dB more) is being radiated around 100 Hz compared to around 2 kHz, despite the fact that the on-axis levels are equal in Figure 11.23. This also means that more power is available to excite a room at low frequencies from this typical loudspeaker source, which gives rise to the shorter critical distance at low frequencies, as described in the previous paragraph.

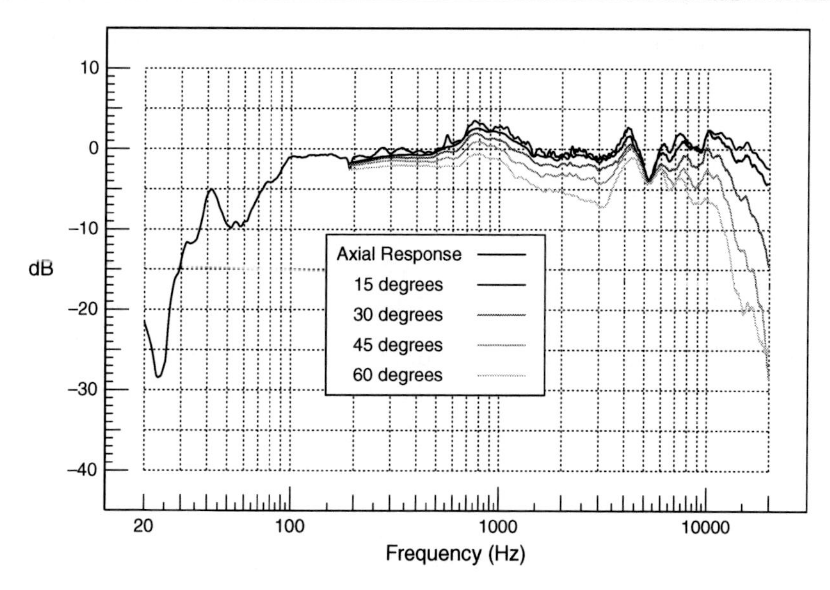

Figure 11.23:
Axial and off-axis responses of a loudspeaker in an anechoic chamber.

In domestic circumstances, we often cannot (or at least we do not find it socially desirable to) exercise very much control over the acoustics of domestic living/listening rooms. We must therefore juggle with the things that we can easily change, and loudspeaker directivity is one of those things. In circumstances where reflective side walls are unavoidable, narrower high frequency directivity can help to avoid too many high frequency reflexions. This can help to

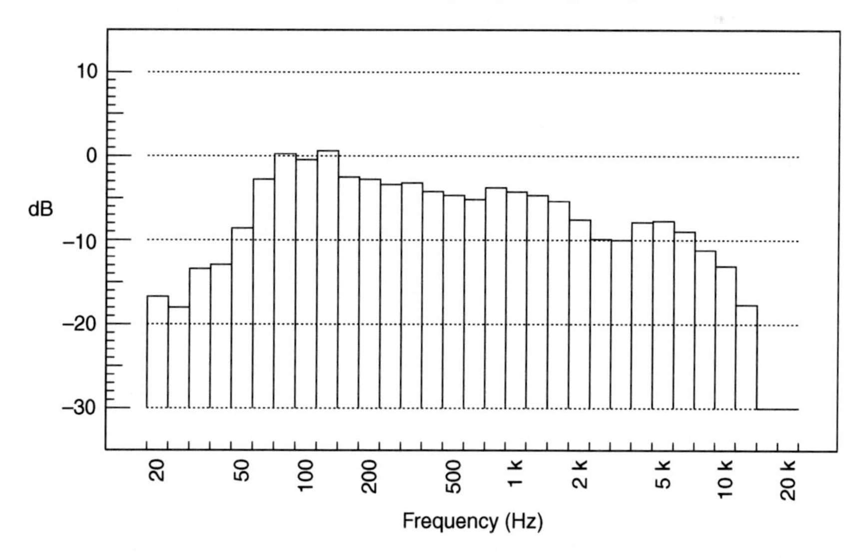

Figure 11.24:
Total power response of the same loudspeaker as in Figure 11.23, but made in the reverberation chamber at the ISVR.

maintain more detail in the sound by beaming more high frequencies directly at the listener. However, there would be a tendency for the reverberation to have a 'dark' character, because it would predominantly be driven by the greater amount of omni-directional low frequency energy. A loudspeaker that was omni-directional at *all* frequencies, such as some models by Canon and Bose, in a room with a relatively uniform reverberation time with frequency, would give a more 'correctly' balanced direct *and* reverberant response. However, the critical distance would be very short at all frequencies, and hence the more confused room sound could tend to mask much detail in the direct sound, which it would tend to swamp.

In an anechoic chamber there would be little difference on-axis between a flat response from a varying directivity loudspeaker or an omni-directional loudspeaker, except that the latter would need to radiate much more power at high frequencies because much of it would be uselessly radiating in unnecessary directions to no effect.

Many domestic loudspeakers have what would in professional use be considered an excessive amount of on-axis high frequencies. They are often designed this way specifically because many domestic situations have a large amount of soft furnishings that absorb the high frequencies much more than the low frequencies. The bright treble response is intended to help to counterbalance the darker, low-frequency-dominant reverberation in the room, but for serious listening such systems can be tiring. Nevertheless, the enormously wide ranges of loudspeakers that are available for domestic use are not only the result of marketing battles. They are necessarily available because of the even greater range of domestic acoustic conditions. When wishing to listen purely for pleasure, many people tend to choose a loudspeaker in a hi-fi showroom, which is often very different to their home listening conditions. This is the reason for the common complaint that the loudspeakers were disappointing when the customers took them home. Clearly, if the room response is so dominant, the 'best fit' to the acoustic conditions in the showroom may well not be the best fit in another room. The positioning within the room also plays a big part in this; for example, when the loudspeaker sounded well balanced in its position in the centre of a wall in the showroom, but was bass heavy when placed in a corner at home. (See Figures 11.11–11.14.)

The variation in overall frequency responses and radiation patterns does allow some control over the domestic listening experience, and it is good that such choice exists. By judicious choice of both the loudspeakers *and* their positioning in a room, a balanced response can often be found. A bass light loudspeaker can be chosen if the only practical places for them in the home are the corners. However, if well-liked loudspeakers were moved into a new house, but were considered to be bass light in their initially chosen positions, moving them towards the corners may be able to augment the response sufficiently to restore an even overall balance. These are examples of electroacoustic balancing. The objective, of course, is to deliver the desirable transfer function from the recording to the ear, and whether the bass is supplied more by the loudspeaker or the room loading is not too relevant unless other aspects of the response are upset.

11.5 Corrective Measures

Let us imagine the cellos and percussion sections of an orchestra, playing together and being auditioned via loudspeakers in a poorly controlled room. Let us also imagine that the listening position had a peak at 120 Hz due to a room resonance problem. If we were to use a 'room' equaliser to reduce the loudspeaker response boost at 120 Hz, at the given listening position, the effect of the resonance could be considerably reduced, and an unpleasant 'boom' each time the cellos were playing notes with a strong 120 Hz content would be removed. However, by reducing the 120 Hz region from the loudspeakers, we would also be robbing the percussion of the same frequency region.

Cellos and percussion are very different instruments. A cello tends to 'speak' as its response develops, so the steady-state 'continuous' response of a listening room, as dictated by its modal activity, can be crucial for its accurate perception. By contrast, the character of a percussion instrument is largely in its attack, much of which will have died away before the room resonances fully take effect, so for percussion the *direct signal* from the loudspeaker will be of greatest importance. If the direct signal from the loudspeaker is altered, the whole character of a percussive instrument may be degraded. We thus, in this one instance alone, have conflicting interests with respect to the 'correction' of the room/loudspeaker combination, dependent upon the nature of the instruments.

The problem does not stop there, though. As air at room temperature has a speed of sound of around 344 m/s, 120 Hz will have a wavelength of 344 divided by 120, or just under three metres. The half wavelength would therefore be about 1.5 m. So if the listeners in our above room moved 1.5 m, in either direction, along the axis of the mode which caused the cellos to boom (away from the area producing the 120 Hz resonance peak at the anti-node), they would move from an anti-nodal region into a nodal region. The nodal region would exhibit a 120 Hz *dip*, which would be exaggerated by the reduction of 120 Hz in the equaliser feeding the signal to the loudspeakers. The cellos and percussion would both have had their 120 Hz content reduced in their loudspeaker drive signals, and from the new listening position, neither would receive any support from the room resonance. Whilst it could be argued that the equalisation had improved the response of the cellos at the previous position, albeit to the detriment of the percussion, neither of the instruments would benefit from the equalisation at the new position. In fact, both would have been degraded by the equalisation. From this it can be understood how room response 'correction' by equalisation of the drive signal to the monitor loudspeakers is both very position dependent *and* instrument dependent. Transient and steady-state signals do not respond similarly.

Let us now consider the low frequency response boost created by placing the loudspeakers close to a wall, where we will observe a very different set of circumstances. There will be a boost which is a function of the room positioning creating a greater load on the diaphragm, and hence producing more power output from the loudspeaker. This response alteration is virtually instantaneous, and is equal for all places in the room. That is, if 100 Hz is 3 dB above 200 Hz at the loudspeaker, then for the direct signal it will also be 3dB above at all places in the room. The loading effect would boost the cellos and percussion in identical manners, in their transient *and* steady-state components, and would have the same relative effect in the drive to any reflexions and resonances in the room. Correction by equalisation of the drive signal will therefore tend to correct simultaneously *all* the response changes caused by the proximity of the wall. In this case, equalisation of the drive signal would be perfectly appropriate, though such equalisation in professional use is usually carried out by means of fixed filters, and not by any sonically questionable graphic equalisers. The above is thus a very different state of affairs from a situation where the response errors are created by complex reflexions and room resonances.

The general rule for the use of loudspeaker/room equalisation is that when a room affects the response of a loudspeaker directly, such as by loading the diaphragm, then corrective equalisation *can* be used. When the effect of a room is indirect, such as by the superposition of reflexions and resonances on to the direct signal, then equalisation will *not* tend to be corrective in any overall manner, and will be as likely to create as many response disturbances as it solves. The two situations are minimum phase, and non-minimum phase, respectively.

11.5.1 Minimum and Non-Minimum Phase

A response modification such as the low frequency boost experienced when mounting a loudspeaker close to a wall is a modification which takes place more or less simultaneously with the propagation of the sound waves from the source. In the case of flush mounted loudspeakers, the wall provides a block to the low frequency radiation that otherwise tends to be omni-directional, and reflects the sound pressure back towards the source. If the pressure wave cannot travel behind the loudspeaker, all the pressure is concentrated in the forward direction, so the normal forward propagation is augmented by the would-be rear radiation. What is more, the constrained radiating space restricts the ability of the air to move out of the way of the diaphragm movements. This increases the load on the diaphragm, which in turn increases the work done, and thus increases the radiated power as compared to free-space radiation. In either case, the effect is virtually instantaneous, it is equal throughout the space into which it radiates, and it can be considered a part of the actual loudspeaker response in those given conditions. A response modification of this nature is minimum-phase.

Alternatively, a non-minimum phase response modification would be produced by the multiple reflexions from room boundaries superimposing themselves upon the direct sound from a loudspeaker. In such cases, every different listening position would also receive a different balance of direct sound and reflexions from each boundary, and no causal filter (see Glossary if necessary) could correct the complex, spacially dependent response irregularities by equalising the loudspeaker response. Each frequency at each position would also exhibit a different phase relationship to the direct signal, as shown in Figure 11.25, and so they could not be generally equalised.

An important point to get across is that the term 'minimum phase' is a mathematical one, relating to how the amplitude and phase responses track each other. It has no relevance to the absolute quantity of a phase change. Essentially, a minimum phase response is one where every change in the amplitude response has a corresponding change in the phase response, and *vice versa*. When the restoration to flatness of either response does *not* restore the other, the response is said to be 'non-minimum phase', and cannot be corrected by a causal inverse filter. The degree of non-minimum phase deviation is known as 'excess phase', and tends to build up with the summation of many types of time-shifted signals, which even occur in the recombination of most crossover filters. We will meet this again in Section 20.6.

Figure 11.26 shows a typical response modification from a minimum phase low frequency boost caused by the flush mounting of a loudspeaker in a wall. The relative response of frequencies x and $2x$ will be the same throughout the room, and so can be corrected by equalisation of the loudspeaker drive signal. Figure 11.27, on the other hand, shows the response of a non-minimum phase effect, where the relative responses of frequencies x and $2x$ are different at each position. Clearly, if they are different at each listening position, they cannot be universally corrected by equalisation of the loudspeaker drive signal. In fact, the disturbance caused by a *single* reflexion can be of a minimum phase nature, as the effect is somewhat dependent on the relative levels of the direct and reflected waves. However, if the reflected level exceeds the direct level, as can occur in some frequency bands with front wall reflexions from a rear surround loudspeaker, or when multiple reflexions are involved, the 'excess phase' builds up, and the overall response rapidly tends towards being non-minimum phase, and hence not correctable by conventional equalisation.

Two other observations can be made from Figures 11.26 and 11.27. The minimum-phase effect in Figure 11.26 can be modelled by a relatively simple analogue equalisation circuit, the inverse of which can produce a mirror image of the amplitude *and* phase characteristics of the disturbed response. Such circuits can readily be used to return the transfer function to its original response (i.e. in this case, as it was before the wall-mounting boost was superimposed upon the free-field response). However, the response of Figure 11.27 is very difficult to mirror with analogue filters, and anyhow, as the problem is non-minimum phase (the result of time-shifted superposition) even perfect amplitude correction could not restore the original phase response by any known practical analogue means. Only by digital signal processing could an *almost* perfect response be restored. Attempts to 'correct' such responses by the use of one-third octave filters would perhaps create a slightly better overall *amplitude* response, but such quasi-restoration would inevitably produce phase absurdities. Any such phase anomalies contrive to distort the time response, which can lead to conflicts in the steady-state and transient responses of a loudspeaker/room combination. It is in this area where typical

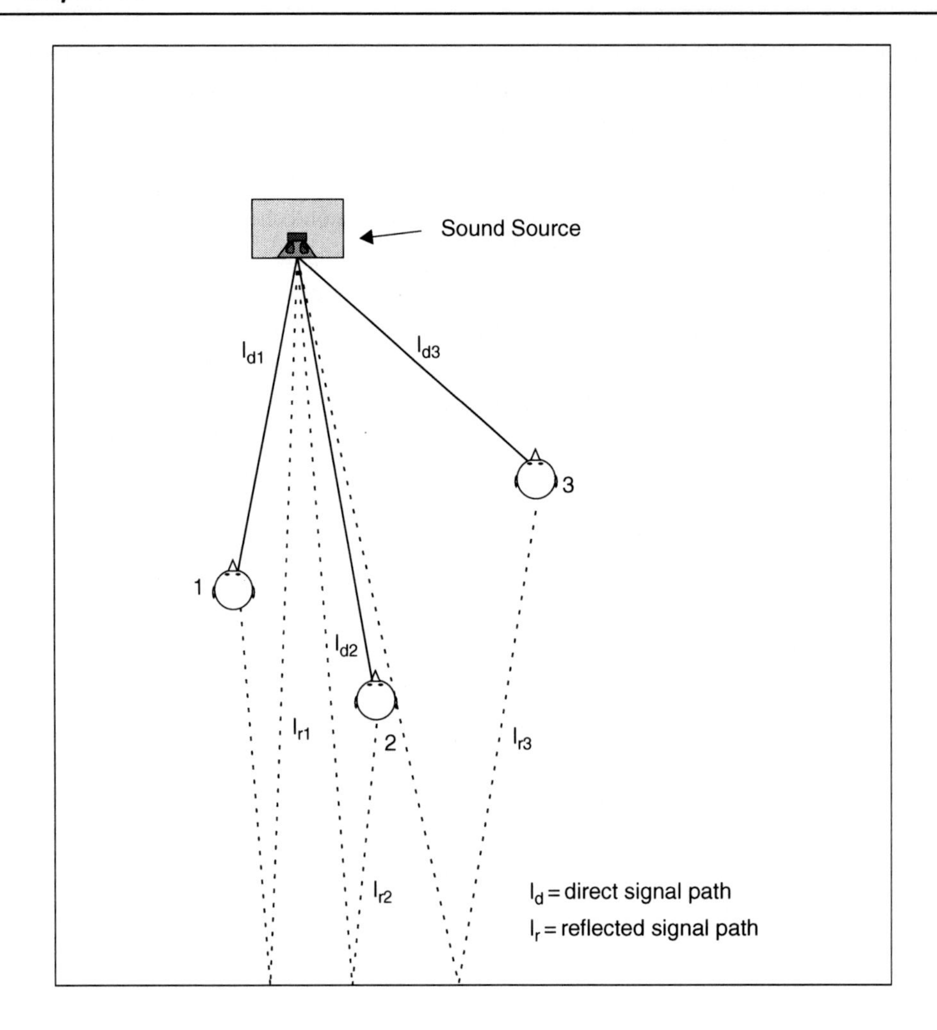

Examples for 344 Hz and 100 Hz

Position 1	$l_d = 5.7$ m	path difference = 10.2 m
	$l_r = 15.9$ m	
	relative phase	@344 Hz = 172°
		@100 Hz = 267°
Position 2	$l_d = 7.5$ m	path difference = 6.7 m
	$l_r = 14.2$ m	
	relative phase	@344 Hz = 252°
		@100 Hz = 158°
Position 3	$l_d = 5.7$ m	path difference = 12.5 m
	$l_r = 18.2$ m	
	relative phase	@344 Hz = 180°
		@100 Hz = 228°

Figure 11.25:
Position dependent phase relationship of direct and reflected waves.

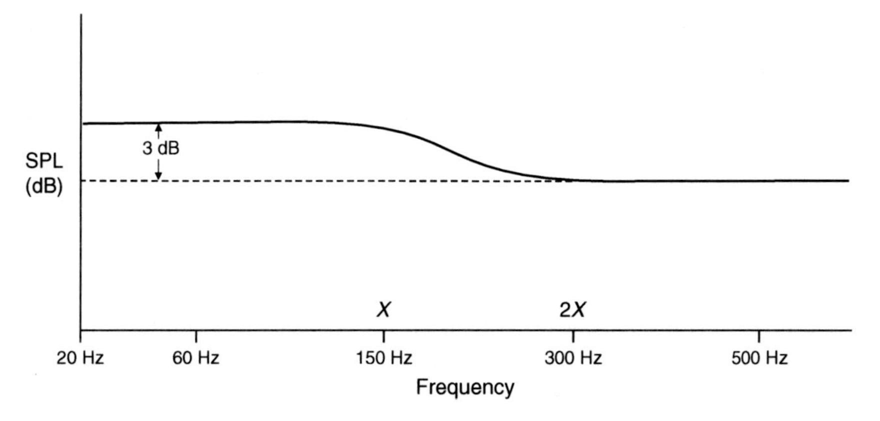

Figure 11.26:
Example of a minimum phase effect, such as flush mounting a loudspeaker. Typical boost in output to be expected at low frequencies when a loudspeaker designed for free-standing is flush mounted. The dotted line indicates the low frequency output when free-standing. For the direct output from the flush-mounted loudspeaker, at any position, frequency 2*x* is 3 dB down on frequency *x*. Equalising them to the same level at the loudspeaker will therefore correct the direct response at all positions in the room.

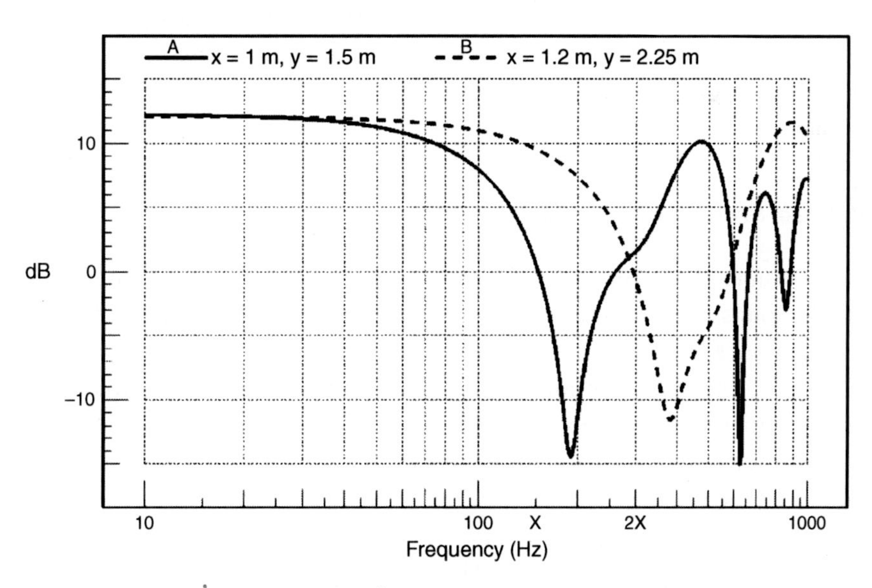

Figure 11.27:
Example of a *non-minimum* phase effect, such as created by boundary reflexions. Measured response at two different microphone positions in a reflective room. The drive signal from the loudspeaker had a flat response. Frequencies *x* and 2*x* vary in relative level at each position in the room. At positions 'A' and 'B', the resultant total response is different; therefore loudspeaker equalisation cannot correct the response at anything other than one point. It cannot correct the response at both points simultaneously.

third-octave equalisation of monitors fails badly. It is the lack of understanding of these general concepts that leads people to misapply 'room' equalisers, then wonder why the results are not satisfactory.

11.5.2 Digital Correction Techniques

We have, until now, been referring to correction by analogue means; but in the digital domain we can have more control over events. Adaptive digital filtering *can* model, very accurately, the inverse responses necessary to correct either the minimum or non-minimum phase components of a transient *or* a steady-state response anomaly, although these systems, too, have limits to their abilities. By means of measuring-microphones, modelling delays (to allow the implementation of acausal [i.e. effect before cause] correction filters) and adaptive filtering processes, the digital systems can be made to 'learn' what a given room will do to a loudspeaker response, and apply acausal corrections to cancel the disturbances. The digital filters can be adjusted to give an almost absolute correction in the amplitude, phase and time domains at one point in the room, or they can produce a less accurate correction over a wider area. However, they absolutely cannot correct to a high degree of accuracy over a large area, and in any case, all corrections in one area are gained at the expense of response deteriorations elsewhere. The confusion of the reflective field of a room is ultra-complicated, and can only be dealt with at the boundaries themselves. To attempt to correct completely these complex problems by equalisation of the loudspeaker drive signal is futile. This is why acoustic engineering is such a fundamental part of good listening room design. An example of correction by adaptive digital filtering is shown in Figure 11.28.

11.5.3 Related Problems in Loudspeakers

It should also be noted that there are frequently both minimum and non-minimum phase problems in the loudspeakers themselves, which are related to the room problems. The loudspeakers can have internal reflexion and resonance problems within both the cabinets and the drive units, and these non-minimum phase disturbances can lead to narrow-band response irregularities which cannot be equalised out for the same reasons that some of the room boundary effects cannot be equalised. The summation of the outputs of multiple drive units in multi-way loudspeaker systems is a particularly common source of non-minimum phase problems, because the minimum phase summation of crossover outputs tends to be difficult due to group delays in the filters. These effects can be observed in the various responses which are described in Chapter 19. Nevertheless, as with the room response problems, the *gradual* boosts or roll-offs caused by the electro-mechanical characteristics of the drivers or their enclosures *are* of a minimum-phase nature, and *can* be equalised to benefit the overall response.

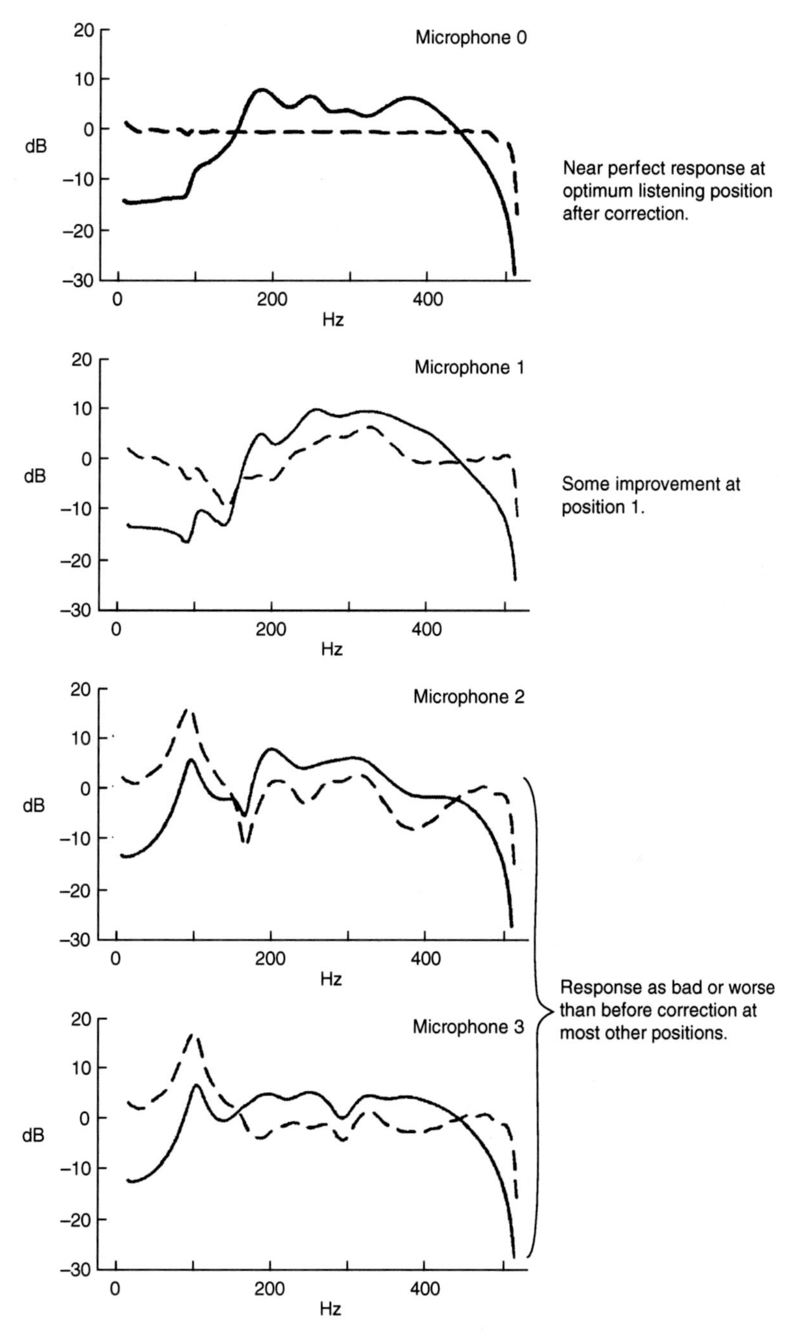
Microphone 0
dB
Hz
Near perfect response at optimum listening position after correction.
Microphone 1
Some improvement at position 1.
Microphone 2
Microphone 3
Response as bad or worse than before correction at most other positions.
Single point equalisation of steady state response. Solid line, original response; dashed line, after correction for position 0.

Figure 11.28:
Room correction by means of digital active signal processing. At microphone position 0, the designated listening position, the correction can be made to be almost perfect. However, for other positions in the room, the response tends to be worsened as position 0 is corrected. (after Clarkson, Mourjopoulos and Hammond, JAES 33, 127–132 (1985))

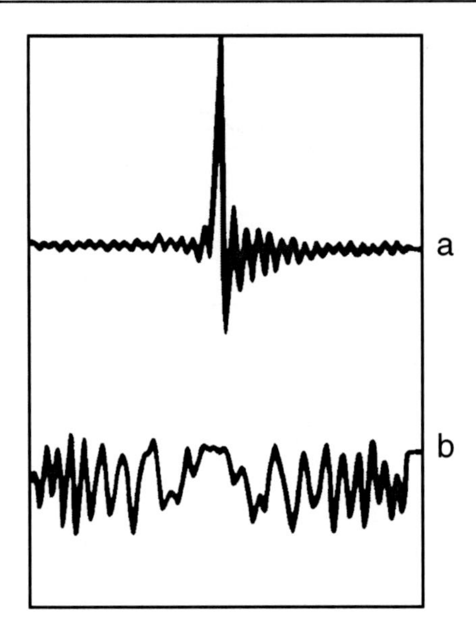

Figure 11.29:
Effect of phase relationship on waveform. In the above plots, the relative levels of the component harmonics are exactly equal, only their relative phases have been changed. The effect on the waveform is gross. (after Schroeder, Models of Hearing (1975) – see Bibliography)

Any time (waveform) response can be represented by a unique combination of amplitude and phase responses (as in the Fourier Transform) so any disturbance in either of the latter responses will inevitably affect the time response, which also represents the transient response. Figure 11.29 shows an example of how a transient spike can be absolutely destroyed, solely by the manipulation of the relative phases of the component frequencies. This very clearly shows why attempts to correct responses in the amplitude domain only, whilst allowing phase responses to be further disturbed, can be disastrous to the transient response of any system to which it may be applied.

A gradual roll-off in the high or low frequency response of a loudspeaker does *not* involve reflexions or response delays, but *will* still involve phase shifts and will consequently affect the transient response. As these are minimum phase problems, the correction of the amplitude response by electronic means *will* apply a phase shift in the opposite sense, and thus will tend to restore the phase and time responses to their correct values. However, to boost by 6 dB requires four times the power, so one can soon run into overload problems with equalised systems, because the overall response headroom will be lower as the flat response is extended.

11.5.4 Summary of Correct Applications of Equalisation

Corrective equalisation of a conventional form *can* be applied to room/loudspeaker response irregularities when those effects are of a minimum phase nature and are time and space invariant, such as loudspeaker roll-offs and rises, or room loading effects. It *cannot* be applied to irregularities that are the result of effects which are separated temporally or spacially from the initial event, such as reflexions or resonances. As the latter are the principal culprits in room response problems, they can only be dealt with by treatment of the rooms themselves. In such cases, the use of electrical filtering, other than of the digitally adaptive type, is a *total misapplication of equalisation.* The failure to appreciate this point has led to the ruination of many natural sounding loudspeaker systems in an alarming number of rooms. Equalisers cannot fix rooms; only acoustic control measures can do that.

11.5.5 The Modulation Transfer Function and its Implications for Electronic Room Correction

In recent years, some manufacturers have begun to introduce monitor loudspeaker systems with built-in, electronic signal processing which is designed to improve the response in less-than-perfect rooms. Concurrently with these developments, a series of investigations has taken place using the concept of the modulation transfer function (MTF)[1–6] in an attempt to assess the limits of this type of processing. The basic concept is relatively simple to understand. A band-limited pink noise signal is modulated by a sine wave, and then passed through a loudspeaker system, or a loudspeaker system and a room, before being captured by a measurement microphone. The received signal is then compared with the transmitted signal and the accuracy of the modulation is graded on a scale from one to zero, one being identical to the transmitted signal and zero showing no resemblance whatsoever to the transmitted signal.

Figure 11.30 shows the MTF of a high definition, full-range monitor system in a control room with very heavily damped modal activity. Its MTF at distances of 1 m and 4 m is shown in (a) and (b). Two things are immediately noticeable: firstly, that the MTF score is generally high, even down to the lowest frequencies, and secondly that the scores at 1m and 4 m are largely similar. Figure 11.31 shows the MTF scores of a small monitor loudspeaker in a recording room, much less acoustically damped than the control room of Figure 11.30. Again, two things are immediately obvious: firstly, the low frequency MTF scores fall at frequencies where the loudspeaker output cannot be maintained, and secondly that the MTF at 4 m is much worse than at 1 m. However, it should be noted that the MTF at low frequencies is not just a function of the pressure amplitude response ('frequency response') of the loudspeakers. Figure 11.32 shows the pressure amplitude and MTF responses of two loudspeakers, one a sealed box and the other a reflex enclosure with electrical protection filtering. The more accurate transient response of the sealed box enables it to exhibit a better MTF due to its more accurate signal waveform output at low frequencies. The time-smeared response of the filtered reflex enclosure, whilst maintaining the pressure amplitude response down to a lower frequency range, does so by time-distorting the signal (and thus the modulation) and therefore cannot maintain the modulation accuracy. The MTF can therefore be considered as a plot of *quality against frequency*, as opposed to the more normal plots of *level against frequency.*

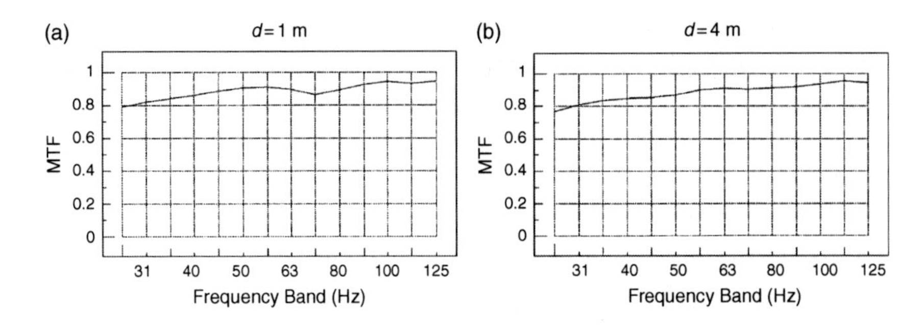

Table 1 Frequency-averaged MTF values corresponding to the above figures. Note the absence of MTF-change with distance

	$d = 1$ m	$d = 4$ m	change
Average MTF	0.88	0.88	0%

Figure 11.30:
MTFs of a wide range monitor system in a very highly damped studio control room at different distances (d) from the loudspeaker.

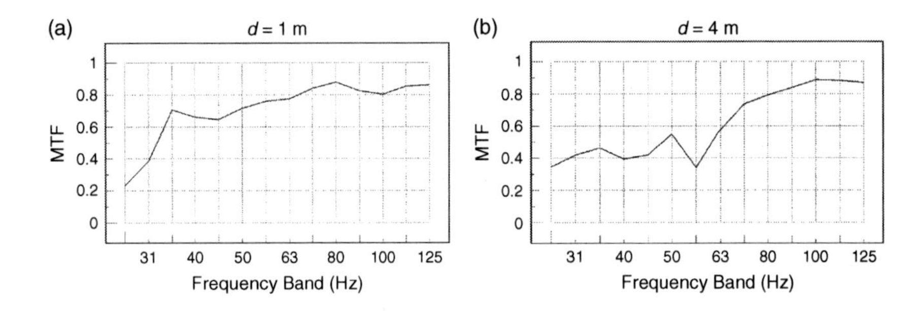

Table 2 Frequency-averaged MTF values corresponding to the above figures.

	$d = 1$ m	$d = 4$ m	change
Average MTF	0.71	0.60	15.5%

Figure 11.31:
MTFs of a loudspeaker in a reasonably well damped studio recording room at different distances (d) from the loudspeaker.

Figure 11.33 shows the MTFs of the same loudspeaker in the same room as shown in Figure 11.30. However, this time the responses have been subjected to simulated, perfect, real-time equalisation. Probably, the most noticeable aspect of these plots is that not a lot has changed when compared to Figure 11.30, other than that a few squiggles and bumps have been smoothed out. On the other hand, Figure 11.34 shows the MTF of the same loudspeaker and room as Figure 11.31, but again after simulated, perfect, real-time equalisation. In this

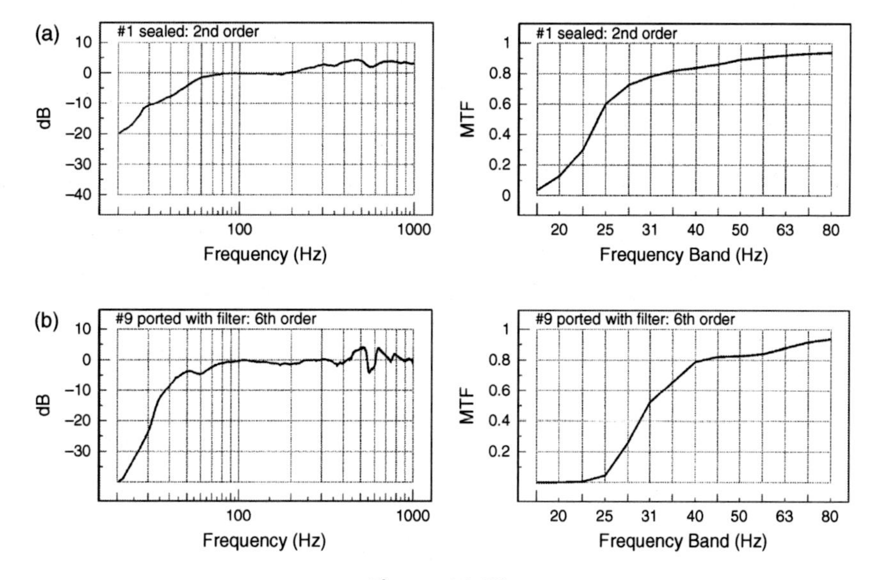

Figure 11.32:
MTF Comparisons. The two frequency response plots (the left hand plots), are generally very similar from 40 Hz to 800 Hz, but the MTF of the sealed box (a) remains high to a much lower frequency than the MTF of the filtered reflex enclosure (b), suggesting that at low frequencies the acoustic output from the sealed box is more faithful to the input signal than are the time-smeared low frequencies from the reflex enclosure.

case, *two* things are noticeable. Firstly, the MTF at 1m has been improved to a degree which makes it almost on par with the full-range system in the highly damped control room. The digital equalisation has worked very well indeed, although in practice the low frequency headroom would have been severely compromised by such high degrees of boost at low frequencies. By contrast, at 4 m distance, the equalisation has not been able to improve the response accuracy very much because of the severity and non-minimum-phase nature of the room effects in the far field.

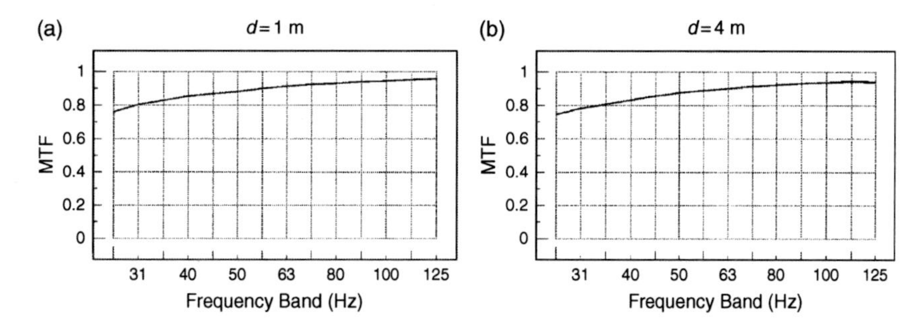

Figure 11.33:
MTFs of a wide range monitor system in a very highly damped studio control room at different distances (*d*) from the loudspeaker, after equalisation. There is little difference from the plots shown in Figure 11.29.

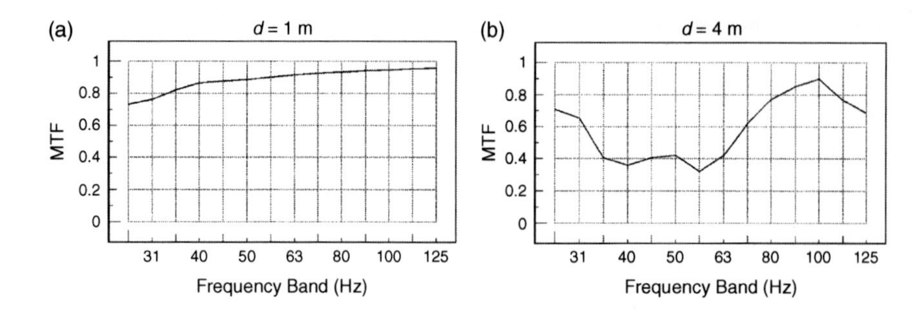

Table 3 Frequency-averaged MTF values corresponding to the above figures.

	d = 1 m	*d* = 4 m	change
Average MTF	0.88	0.59	32.9%

Figure 11.34:
MTFs of a loudspeaker in a reasonably well damped studio recording room at different distances (*d*) from the loudspeaker after equalisation.

The important implication here is that whilst room response correction may be possible in terms *of flattening* the response when measured in decibels, the flat response in the pressure domain may not necessarily mean that the response is accurately maintaining the *information* content of the signal. If the time response is smeared, then the information will *not* be accurately maintained, even despite the pressure amplitude being flat. *Quantity* is not the be-all and end-all.

Think of it this way: take the words 'information content' as an example, and think of the quantity of each letter in that term. These are, alphabetically, in what we could call the 'alphabet response':

a = 1

c = 1

e = 1

f = 1

i = 2

m = 1

n = 4

o = 3

r = 1

t = 3

By changing the 'time response' of the order of the letters we could write:

information content
infromation content
infrotmaion ontcent
infrtomaion ontecnt
frintomaoin tecontc

and so forth. The 'alphabet response' is the same in each case, as exactly the same numbers of each letter are used in each example, but as the time at which each letter arrives is more out of the original sequence, the information carried in the sequence of letters is gradually reduced. Whilst this is not a direct analogy, it does give some idea of the concept that the quantity of pressure that arrives at each frequency, which is what is measured by a typical 'frequency response' curve, may only very poorly relate to the information content which is being carried by the signal. The *information* in the musical signal will be degraded if the times of arrival of the relative frequencies are mixed up, even though the pressure response may be ruler-flat.

The MTF looks not only at how much pressure there is at each frequency, but by means of the modulation signal, measures the arrival times of the various components of the signal. The implication from what has been shown in the previous figures is that if equalisation is used to flatten a loudspeaker/room pressure amplitude response in the far-field, the influence of the delays and subsequent time jumbling of the room reflexions will not allow the information content to be maintained, despite the apparent 'perfection' of the pressure response. As can be seen from Figure 11.30, in a highly damped control room, in which reflexions have been kept to a minimum even in the far-field, the MTF is substantially maintained. Indeed, in the extreme, in an anechoic chamber, only the level would fall with distance. Whatever the MTF was at 1m would be maintained with distance.

As a consequence of the above, it would appear that electronic 'room correction' may well be able to flatten a frequency response curve, and at short distances, where the direct response from the loudspeaker dominates, make a general improvement to the fidelity even in a room with relatively poor acoustics. Nevertheless, in the far field, where the room response begins to dominate the direct response, quality (fidelity) losses are still to be expected. It would seem that, despite some of the claims in the publicity, room correction programs have not dispensed with the need for good room acoustic control where the highest fidelity is required over a wide and deep listening area. Nonetheless they have certainly made possible more accurate close-range monitoring in rooms of a more conventional nature or in rooms such as television or video post-production rooms which are often sparsely treated and filled with racks of equipment which are not exactly 'acoustic-friendly'.

11.5.6 Electronic Bass-Traps

Despite the fact that electronic room correction tends to be less than perfect, there do exist some means by which low frequency response problems in small rooms can be ameliorated by the use of tuneable electroacoustic devices. Such systems can at least deal with the most troublesome modal resonances, even if they cannot achieve the overall response flatness of an acoustically controlled room. Devices such as the Bag End E-trap are essentially loudspeaker cabinets with integral amplification, microphone and processing. In the case of the E-trap, two individually tuneable frequencies can be simultaneously processed. These resonant room frequencies are then damped by the device, which is much smaller than any purely acoustic control structure which could be used for the same purpose. Where a room is not required for optimal monitoring, but work within the room is being hindered by one or two very offending resonant modes, devices such as the E-trap can be very useful indeed, especially when one considers that the size of this example is only about 45 cm × 33 cm × 25 cm. Places used as temporary recording or control rooms would be the sort of spaces which could benefit from such systems. Another means of electroacoustically controlling low-frequency room modes is discussed in subsection 21.8.2, dealing with the application of processed multiple sub-woofers, but whereas the E-trap will work with an *acoustic* source, the latter will only work with electrical sources.

11.6 Phase and Time

The precise effect that reflexions have on a direct signal depends upon their amplitude and phase. The reflexion amplitude is dependent upon the distance travelled and the coefficient of absorption of the boundaries. The phase is dependent upon the distance travelled, the wavelength, the absorbent or diffusive nature of the reflective surfaces *and* on the angle of incidence. Hence, there are few simple ways to calculate the precise result of reflective superposition.

The time (t) taken for any reflexion to arrive at the listening position, after the direct sound, is given by the equation:

$$t = \frac{d_2 - d_1}{\mathrm{c}} \tag{11.3}$$

where:

d_1 = direct path (in metres)

d_2 = reflected path (in metres)

c = speed of sound (in metres per second)

For any given frequency, the relative phase of the direct and reflected waves can be found from the equation:

$$\text{Phase shift}, L = 360 \times \frac{\text{L}d}{\lambda} \tag{11.4}$$

where:

Ld = path length difference (in metres)

λ = wavelength (in metres)

For example, if a wave with a frequency of 172 Hz (having a 2 m wavelength) had a direct and reflected path length difference of 1 m, then this formula would give:

$$\begin{aligned} L &= 360 \times \frac{1}{2} \\ &= 180^\circ \text{ phase shift} \end{aligned}$$

which means that the reflected wave would tend to cancel the direct wave. For a path length difference of 6 metres, the equation would give:

$$\begin{aligned} L &= 360 \times \frac{6}{2} \\ L &= 1080 \text{ degrees} \end{aligned}$$

(For answers above 360, the result is simply divided by 360, and any remainder is the difference in phase angle.)

Now 1080° is 3 × 360°. In this case, the reflected wave would show a 0° phase difference, and thus would be in phase with the direct signal, meaning that it would add its pressure to that of the direct wave, but the fact that it had travelled through three full revolutions of phase is a function of the arrival delay caused by the path length difference. Phase shifts due to distance travelled are frequency dependent, because they are wavelength related, and as sine waves are continuous (or at least quasi-continuous), 360°, 720°, or 1080° phase shifts all yield the same result: in-phase summation. However, the same case does not hold true for transient signals, which have phase slopes; i.e. a rate of change of phase with *frequency*, whereas a single frequency only has a rate of change of phase with *distance.*

The main difference between a phase shift of 360° and 3600° on a 172 Hz reflected signal would be that the latter reflexions would have travelled nine full cycles further. As each cycle of 172 Hz has a wavelength of 2 m, 9 cycles represent a distance of 18 m. This may be a significant factor in the degree to which a 360° or a 3600° reflexion sums with the pressure of the direct wave, because the 18 m extra distance which the 3600° reflected wave travels will also reduce its sound pressure level compared to the 360° reflexion (if reflected from similar surfaces). The strength of the returning reflected wave will also be dependent upon the absorption coefficient of the wall. Just for a single reflexion of a single frequency, therefore, we must consider its relative phase, the distance that it has travelled, and the reflexion coefficient of the reflecting surface before we can even begin to calculate its effect on the direct signal. To add a few complications more, remember that the absorption coefficient will be frequency dependent, angle of incidence dependent, and may itself affect the phase of the reflected wave. It is also worth noting here that whilst a reflected wave which is 6 dB down on a direct wave with which it superimposes itself will not affect the measured SPL, it may well still be very audible due to the spacial discrimination of the ear.

11.7 The Black Art

From this chapter it can be seen that the combination of a loudspeaker and a room is a complex interactive system that defies simple explanation. Every situation requires individual assessment if the best results are to be achieved.

Most rooms consist principally of four walls, a floor and a ceiling, but the capacity for variability, in this simple combination alone, is infinite. There are also rooms of many different forms, such as with circular walls or domed ceilings, and almost all rooms contain furniture. In general, the behaviour of sound waves in such spaces is not intuitively obvious, and the number, location and sizes of doors and windows further complicate matters. It is perhaps not surprising, then, that the room environment, which is the interface between the loudspeaker and the listener, can impose a great degree of its own character on the performance of any sound system within it. No doubt as a result of this, together with the limited degrees of response predictability and the few clear 'rules of thumb' which can be followed, the subject of room acoustics has come to be widely perceived as a 'black art'. This situation is perhaps not helped by the fact that a good set of room acoustics for one type of music may be bad for another, and a room that is good for speech may be awful for an orchestra. Large rooms, small rooms, live rooms, dead rooms, conference rooms, theatres, sports halls and many others all have their unique design criteria, and rarely do those criteria coincide. The subject of the effects of rooms on loudspeakers is therefore truly huge, and whole books exist dealing with specific areas alone.

One chapter can hardly do the subject justice, but in the general context of this book, the most relevant area of study is the part played by relatively small rooms in the music production/reproduction chain. In fact, the performance of neither a musical instrument nor a loudspeaker can be wholly defined without knowledge of the room in which it will be used. This is because in any signal chain relating to music recording or its reproduction via loudspeakers, the room will be the final and the most variable link, possibly greatly affecting the overall transfer function from the recorded source to the ear. So, despite the frustrations inherent in the subject, we need to address it in some detail or our understanding of recording studio requirements will not be complete. The following chapter will begin to look at how to deal with the problems.

11.8 Summary

Rooms tend to change significantly the responses of loudspeakers located within them.

The radiation pattern of a loudspeaker greatly influences the degree to which various room effects will be perceived.

Rooms can actually increase the radiated output from loudspeakers, but the effects are frequency dependent.

The critical distance from a loudspeaker is that at which the direct and reverberant sound fields are equal in level. At low frequencies in a typical domestic room the critical distance can be as small as half a metre.

Some loudspeakers for domestic use have on-axis responses that rise with frequency. This is to try to achieve a more frequency-balanced overall effect in the common conditions of greater low frequency reverberation.

Attempts to correct room response problems by means of loudspeaker equalisation will tend to affect transient and steady-state signals differently. No overall improvement may be possible by such means.

When a correction to an amplitude response irregularity also tends to correct the phase response (and hence also the time response) the effect is said to be *minimum phase.*

When a correction to an amplitude response cannot also correct the phase response, the effect is said to be *non-minimum phase.*

Digital correction techniques can be used to correct a non-minimum phase problem in both amplitude *and* phase, but the correction can only be made accurately for one point in space.

The modulation transfer function may prove to be a useful tool in determining the limits up to which digitally signal-processed loudspeakers are able to compensate for room acoustic problems, and also to indicate where the subjective losses can become greater than the gains.

Loudspeaker crossover filters and the displacement of the drivers can also produce non-minimum phase responses.

Sine waves have phase *shifts*: a rate of change of phase with *distance.* Transient signals have phase *slopes*: a rate of change of phase with *frequency.*

The effect of reflective superposition on a direct sound wave is dependent upon many variables, such as distance (relative phase and level), degree of reflectivity of boundaries (relative phase and level), and boundary absorption with respect to frequency (frequency content, level *and* phase of reflected wave). The angle of incidence of the direct wave on the reflective boundary may also influence the relative frequency, phase and level proportions of the reflexions.

With all of these influences to consider, it is wise to think of a loudspeaker/room combination as a single system, and to treat it as such.

Some electronic solutions can now deal with the more gross irregularities of room responses.

References

1 Holland, Keith, Newell, Philip, Mapp, Peter, 'Steady State and Transient Loudspeaker Frequency Responses', *Proceedings of the Institute of Acoustics*, Vol. 25, Part 8, Reproduced Sound 19 conference, Oxford, UK (2003)
2 Holland, Keith, Newell, Philip, Mapp, Peter, 'Modulation Depth as a Measure of Loudspeaker Low-Frequency Performance', *Proceedings of the Institute of Acoustics*, Vol. 26, Part 8, Reproduced Sound 20 conference, Oxford, UK (2004)
3 Holland, Keith, Newell, Philip, Castro, Sergio, Fazenda, Bruno, 'Excess Phase Effects and Modulation Transfer Function Degradation in Relation to Loudspeakers and Rooms Intended for the Quality Control Monitoring of Music', *Proceedings of the Institute of Acoustics*, Vol. 27, Part 8, Reproduced Sound 21 conference, Oxford, UK (2005)
4 Fazenda, B. M., Holland, K. R., Newell, P. R., Castro, S. V., 'Modulation Transfer Function as a Measure of Room Low Frequency Performance', *Proceedings of the Institute of Acoustics*, Vol. 28, Part 8, pp. 187–194, Reproduced Sound 22 conference, Oxford, UK (2006)
5 Harris, L. E., Holland, K. R., Newell, P. R., 'Subjective Assessment of the Modulation Transfer Function as a Means for Quantifying Low-Frequency Sound Quality', *Proceedings of the Institute of Acoustics*, Vol. 28, Part 8, pp. 195–203 Reproduced Sound 22 conference, Oxford, UK (2006)
6 Newell, Philip, Holland, Keith, *Loudspeakers for the Recording and Reproduction of Music*, Focal Press, Oxford, UK (2007)

Bibliography

Borwick, John, *Loudspeaker and Headphone Handbook*, 3rd Edn, Focal Press, Oxford, UK, Boston, USA (2001)
Charalampos, Ferekidis and Kempe, Uwe, 'Room Mode Excitation of Dipolar and Monopolar Low Frequency Sources', presented at the One Hundredth Convention of the Audio Engineering Society, Preprint No. 4193, Copenhagen (1996)
Colloms, Martin, *High Performance Loudspeakers*, 5th Edn, John Wiley & Sons, Chichester, UK (1997)
Eargle, John M., *Loudspeaker Handbook*, Chapman and Hall, New York, USA, London, UK (1997)
Newell, Philip, *Studio Monitoring Design*, Focal Press, Oxford, UK (1995)
Schroeder, Manfred R., 'Models of Hearing', *The Proceedings of the Institute of Electrical and Electronic Engineers (IEEE)*, Vol. 63, No. 9, pp. 1332–50 (September 1975)

Essential reading

Toole, Floyd E., *Sound Reproduction: The Acoustics and Psychoacoustics of Loudspeakers and Rooms*, Focal Press, Oxford, UK (2008)

CHAPTER 12

Flattening the Room Response

Subduing the room effects. Further discussion on electronic room equalisation. Standard listening rooms. The concept of a listening room. Close-field monitoring. The two near-fields.

From the discussion in Chapter 11, it can be seen that the influence of a room on the response of a loudspeaker is not trivial. In domestic circumstances, the degree of response variation is enormous, and ± 10 or 12 dB is no exaggeration, even before the public get their hands on the 'tone' controls. Clearly, for critical listening, such response variability is absurd, so for professional recording studios an entirely different approach is needed if there is to be any hope of tonal consistency between recordings.

In the days of mono, it was usually not too difficult to find a combination of loudspeaker position and listening position which could avoid the worst modal disturbances in a room. The listening distance could also be adjusted to allow some control over the direct to reverberant sound balance. Indeed, when listening in mono, the ambience of the listening room is an important part of the experience, because the mono signals emanating from a point source in a relatively dead room can be rather uninspiring. Books on loudspeakers and hi-fi from the 1950s, such as those by Briggs,[1,2] deal at great length with means of adding diffusion and 'life' to loudspeakers in order to help to spread the sound stage. Such ideas almost became taboo in the stereo era because the general tendency would be for the stereo definition to suffer if the sources were diffuse, although the advent of distributed mode loudspeakers (DMLs) is currently challenging that thinking to some degree.

The old mono sound control rooms in music recording studios were generally rather similar to the broadcast control rooms of the same era. Surfaces were usually parallel, and treated with whatever degree of acoustic treatment that was deemed necessary in order to achieve a decay time of 0.5 s, or thereabouts. Many of these rooms were awful when stereo systems were installed in them. It was no longer so easy with stereo to make the positional changes that facilitated the 'optimisation' of mono rooms. A triangle had to be maintained between the two loudspeakers and the listener(s). Many options of placement were no longer workable, because when one of the loudspeakers was moved into a good position, the triangulation could force the other loudspeaker into a totally impractical place. Even by 1970, in a major recording centre such as London (UK), the number of stereo control rooms that a broad consensus of recording personnel considered excellent was pitifully few.

Many people in the industry knew a good deal about what the problems were, but a lack of knowledge about how to solve them, together with established practices such as the presence of tall metal racks and the use of rather 'industrial' furniture in general, delayed significant progress. What is more, domestic hi-fi systems were not in widespread use, so even many of the bad control rooms were still capable of quality-control monitoring way ahead of what could be heard in the vast majority of homes. The pressure to improve control-room acoustics was still largely driven by the people in the industry who knew that they could do better. What a recording sounded like on the average radio was already an important factor, at least for popular music recording, so the small loudspeakers provided on or in the mixing consoles were used to make a crude comparison. Nevertheless, the general trend in the industry was to try to advance, and to make better recordings.

12.1 Electronic Correction Concerns

By the mid-1970s, the majority of top studios around the world were using monitor equalisation to try to achieve a more standardised frequency response at the listening position. This was almost invariably done by means of putting pink noise into each loudspeaker system in turn, and adjusting graphic equalisers to give the desired response at the listening position (usually flat, with a slight top-end roll-off) on one of the newly available, third-octave, real-time analysers. The development of solid-state electronics had enabled the production of portable analysers, spanning 20 Hz to 20 kHz in third-octave bands, which displayed the output from a microphone or an electronic input in real-time (i.e. as it happened). Previous methods of room measurement had often involved the use of bulky equipment, which measured each band sequentially. The overall response usually had to be analysed some time after the event (hence *not* in real-time). In some ways though, the apparent simplicity of use of the portable equipment led to its use by many non-acousticians, who often grossly misused it in totally inappropriate circumstances.

By the late 1970s, the warning bells began to sound when it started to be realised on a widespread basis that rooms which were supposedly equalised to within very tight limits were often still sounding very different to each other. Over the next few years the whole concept of monitor equalisation came into serious question, but by that time, the idea that professional studios used room equalisation had become so deep in the folklore of audio that the misuse continues to this day. Although only a small proportion of top-line studios now use monitor equalisation for response corrections, lower down the scale of professionalism equalisers can still be found in many recording studios staffed by people of lesser experience. Somewhat perversely, the studios that do still use monitor equalisation are often the ones with the less acoustically controlled rooms, and it is in these rooms where the use of equalisers is least appropriate.

It was well understood that the critical bands of human hearing were in one-third octaves, and that the ear 'sampled' loudness in packets every 50 ms or so. It thus seemed to follow that if the one-third octave bands were adjusted by means of equalisers to provide more or less equal energy in each band (in response to a pink noise input, which contains equal energy in each one-third octave band), then a natural frequency balance would result. The poor waveform discrimination by the ear was thought to render inconsequential the phase anomalies brought about the application of the equalisation. Unfortunately, the reality was not so straightforward. Once again, results from hearing tests done on speech and noises had been taken as absolute, and applied incorrectly to musical perception.

The response plots shown in Figure 11.9 depict the pressure amplitude curves for one loudspeaker at two different places in a room. It ought to be intuitively obvious that if these plots represented the left and right loudspeakers of a stereo pair, then no simple application of such a crude device as a one-third octave equaliser could make the two responses identical. It would be like trying to repair a watch with a crowbar. Figure 12.1(a) shows an example of such a crude attempt at correction. The solid line shows a typical response dip caused by a floor reflexion, which is centred on the 160 Hz one-third octave band. The display on a one-third octave analyser would look something like the pattern shown in (b), which also clearly shows the dip at around 160 Hz.

The typical 'fix' by means of equalisation would be to raise the 160 Hz band until the analyser showed it to be up to the reference level, but this would also raise the adjacent bands, as shown in (c). Subsequent downward adjustment of the 125 and 200 Hz equaliser controls would tend to flatten the display, but the 100 and 250 Hz bands may then need raising a little. A process of reiteration could probably reach a situation whereby the one-third octave analyser showed a flat display, but only after the manipulation of controls up to one octave away from the narrow dip. When looked at in finer detail, it would be revealed that the true response after 'correction' would be that shown by the dotted line in Figure 12.1(a), which in fact looks no more correct than the response with the original dip, despite now having equal energy per one-third octave band.

So, what would have been achieved by all this equalisation? A dip at 160 Hz would still exist, but it would be flanked at either side by small peaks. This is *not* a correction, and neither is it necessarily an improvement. The direct signal, which *was* flat, would now be exhibiting a response similar to that delineated by the knobs on the equaliser – that familiar pattern of one band up, the next band down, the next up, the next down, and so on. This is only *detrimental* to the direct signal, and is in no way an improvement. The phase response, and hence the transient response of the direct signal, would also have been altered, and this again could only be detrimental to the natural character of the sound. All in all, the damage done would have been much greater than any derived benefit, but such has been, and still is, the fate of thousands of studio monitoring systems in inexperienced hands.

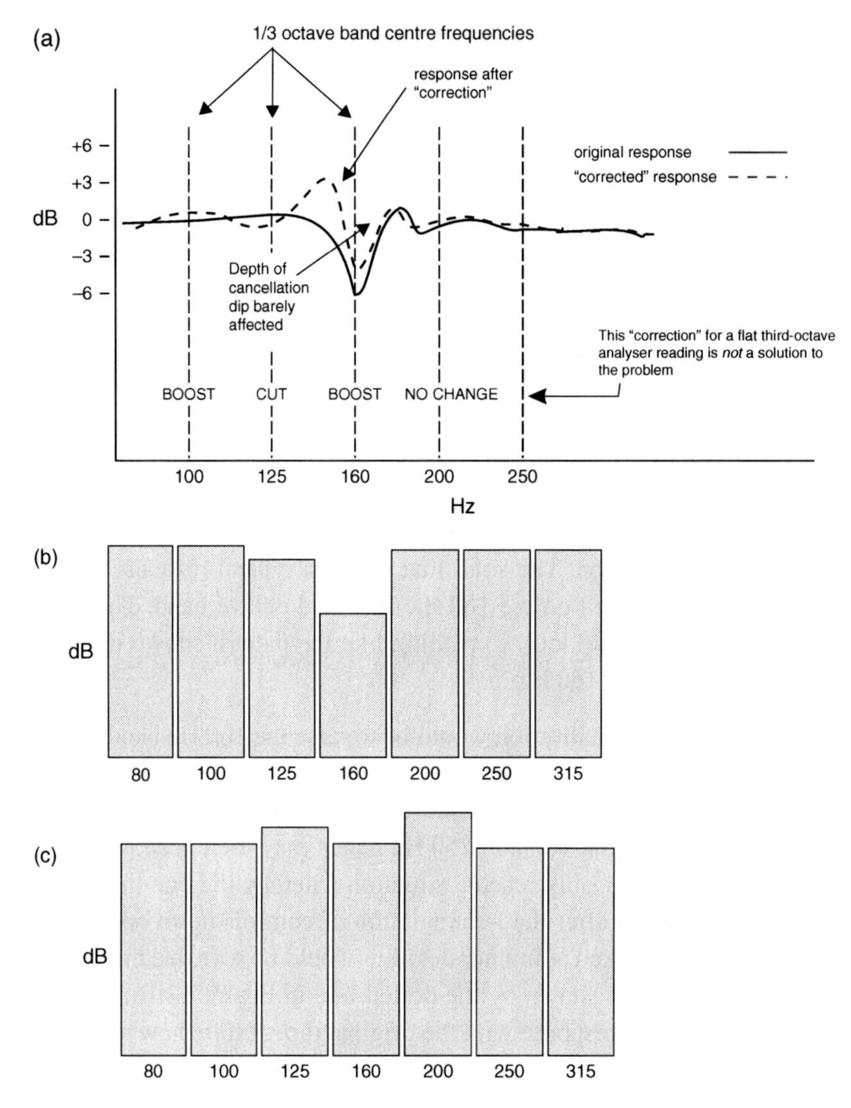

Figure 12.1:
Attempted one-third octave correction of a floor dip: (a) In the above figure, the solid line represents the response of a loudspeaker in a room, affected by a floor reflexion. Below the curves are shown some typical boosts and cuts that could be applied by graphic equalisers to try to flatten the response. The result after equalisation may well read flat on a one-third octave spectrum analyser, but the true response would be more likely to be that shown by the dotted line, which subjectively may be no better than before the supposed 'correction' was applied; (b) The original response (the solid line in (a)) may well look like this on a one-third octave analyser, but attempts at correction would probably, by raising the 160 Hz band on a graphic equaliser, give rise to a response as shown in (c); (c) This response, resulting from a boosting of the 160 Hz band, could be further corrected to overall flatness by manipulating adjacent frequency bands, but any resultant overall flat reading would, in fact, still have a response like the dotted curve in (a).

An *acoustic* correction of the response dip could be achieved by putting mattresses on the floor between the loudspeakers and the listener. This is hardly a practical solution, but some recording studio designers have installed 'pit traps', or other in-floor absorption systems, to address this common problem. An electroacoustic solution is shown in Figure 12.2, but this is more a function of a loudspeaker being designed to reduce floor dip effects rather than being a fix for a floor dip problem on an existing loudspeaker. The fact is that there is no effective electronic fix to the floor dip problem, because even digitally adaptive flattening can introduce unwanted artefacts. The saving grace is that such response dips are usually of quite narrow bandwidth, and that the ear's sensitivity to them is low. In many ways, however, the ear's sensitivity to the effects of attempts to correct the problem is such that many solutions are subjectively worse sounding than the problem itself.

The thing to remember about the floor dips is that they are *cancellations*. Therefore, the more energy that is fed into the drive signal the greater will be the energy in the cancelling reflexion. For this reason, a disproportionate amount of extra drive may be needed to raise the measured response because the power in the reflexion is increasing as the drive is increasing. More, or less, it still cancels the same proportion of the direct signal. What is more, as can be seen from Figure 12.2(b), the different distances from the loudspeakers to the ears give rise to dips at different frequencies at different places. Every different distance from the loudspeaker will have a different frequency of cancellation. As a listener walks backwards and forwards in the room, the l_d to l_r ratio would change, and so therefore would the *frequency* of the dip. From this it can be seen that the frequency of the floor dip is entirely position dependent, so any 'correction' at any given place would give rise to a *worsening* situation elsewhere. Any reflexions from other surfaces would then return to the listening position with frequency-distorted balances. Nevertheless, despite all of this evidence, in many studios people do still try to correct the uncorrectable, but in these circumstances it must be stressed that what one sees on a spectrum analyser is *not* necessarily what one hears.

Another problem with electronic equalisation in general is that any boosts in level eat into the headroom of the monitor system. A 6 dB boost in one frequency band would require the amplifiers to supply four times the power at those frequencies. In cases where the sonic improvement due to the equalisation is marginal at best, the loss of headroom and possible resultant increase in distortion are further good reasons to avoid the use of monitor equalisation. In reality, there are only three *bona fide* uses for monitor equalisers:

1. To correct for minor discrepancies in the loudspeaker drive units when these are of a minimum phase nature.
2. To compensate for any acoustic gain caused by the proximity of boundaries.
3. To apply a desired curve the monitor response, such as the gentle high frequency roll-off which many studios employ, or cinema standard equalisation. (But see also Section 21.2.4.)

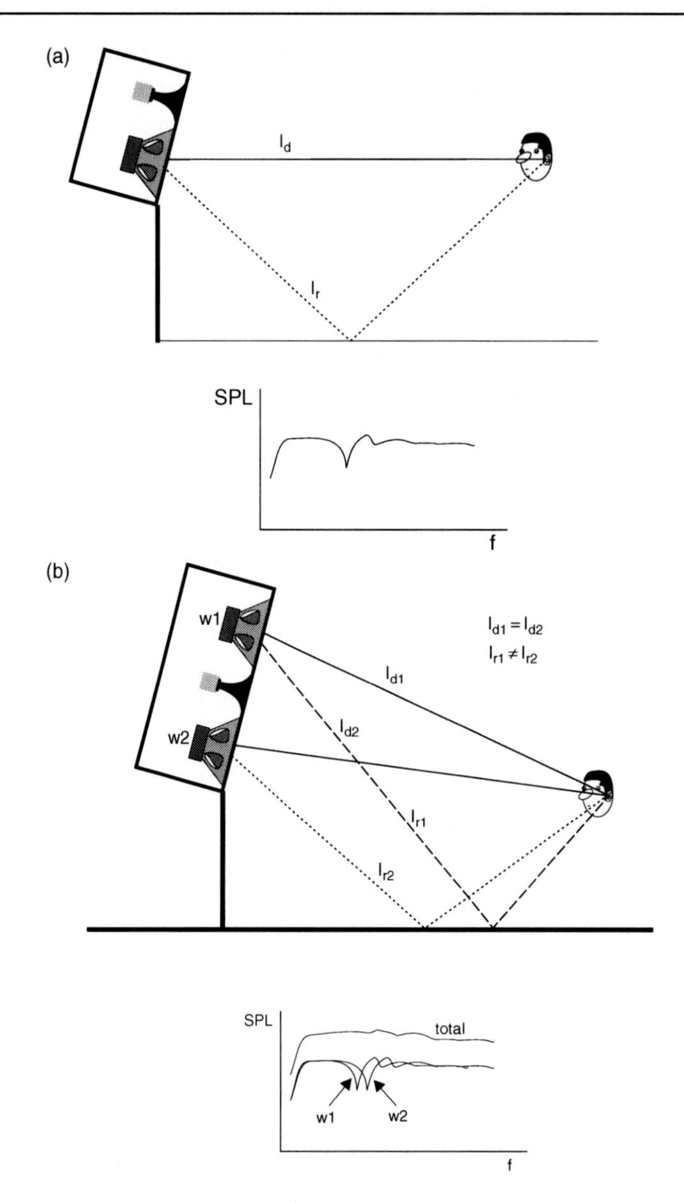

Figure 12.2:

(a, b) Typical floor reflexion disturbance: (a) The superimposition of the reflected wave on the direct wave tends to produce a combined response similar to that shown above. The extra length of reflected pathway (l_r) over the direct pathway (l_d) causes summations and cancellations at different frequencies, and a deep cancellation around one frequency, where the distance $l_r - l_d$ equals half a wavelength.

(b) Non-minimum phase response-dip correction by the addition of a second low frequency driver. By mounting a second low frequency driver vertically above the other one, the floor dips are displaced due to the different reflected path lengths. The total response, as shown above, shows an almost complete absence of response irregularity (after Kinoshita, Rey Audio, Japan).

All of the above are amplitude/frequency corrections. It is entirely appropriate to use amplitude/frequency corrections as long as the filters being used are of an appropriate quality and do not introduce any extra distortion or loss of response transparency. What cannot be treated by frequency domain corrections are time domain aberrations, such as those caused by room reflexions or resonances. Strictly speaking, a resonance *can* be equalised, but if the equalisation, as it must be, is applied to all signals in a loudspeaker system, it will be inappropriate for the non-resonant content of the music in the same frequency ranges, as explained in Section 11.5. Crossover filter induced dips are also not correctable by traditional equalisers, because of the delays inherent in the filter circuits. These are akin to the dips produced by electronic or tape-delay phasers, and almost anybody who has used studio equipment will be aware of the fact that although a phased sound sounds equalised, no equaliser in the effects rack can be used to de-phase the sound.

Clearly, therefore, from the evidence presented in this section, electronic correction was never going to be the answer to the standardisation or the sonic neutralisation of control rooms.

12.2 The Standard Room

In Section 5.3.2, reference was made to the preferred ratios of room dimensions for the achievement of the most even spread of modal density within a room. The Bonello Criterion was also discussed. The preferred ratios led to the concept of the 'standard listening room'. The primary use of such rooms was so that broadcasting companies and equipment manufacturers could assess their work in a set of reference conditions that were reasonably repeatable in other, not identical but similar, rooms in other places. The idea was that frequency balance, distortions, tonal colouration, stereo imaging and other aspects of sound could be reasonably assessed in meaningful ways. Accurate comparisons obviously could not be achieved in *ad hoc* conditions. For European homes, an International Electrotechnical Commission (IEC) standard[3] for listening tests on loudspeakers recommended a reverberation time of 0.4 ($\pm$0.05) seconds between 250 Hz and 4 kHz, with a maximum of 0.8 s at low frequencies. A room volume of 80 m^3 was also recommended, which could typically be a room of approximately 6.5 m × 4.5 m × 2.8 m.

The recommended specifications in these standards are intended to represent typical domestic listening conditions, and it would seem reasonably logical to record and mix music in conditions typical of those in which the results were intended to be heard. The problem with this concept is that what is statistically typical may not be representative of the majority of cases. The number 5.5 may be the statistical average of the integral numbers from 1 to 10, but although it may be reasonably representative of the numbers 4, 5, 6 and 7, it is *not* a reasonable representation of the *majority* of the numbers in the range, such as 1, 2, 3, 8, 9 and 10.

Another drawback to the use of such rooms for recording control rooms is that the colourations inherent in such designs are too great and too different from room to room for the critical assessment of the timbre of a recorded instrument. Furthermore, the reverberation time, in particular at low frequencies, is far too long for the recording and mixing of 'dry' sounds, such as those produced by synthesisers. Much electronic music simply fuses into a blur in 0.8 s of low frequency reverberation, which the IEC specification allows. Whilst it is perhaps true to say that the colouration and room-to-room differences could be reduced by means of adequate diffusion, this was never a part of the specifications, which called for specific acoustic treatments such as those shown in Figure 12.3.

It should also be noted that the IEC rooms were based on a European standard, which is perhaps relevant for European broadcasting companies and national record companies. However, the music recording industry is very *international*, and as record buyers pay the same money for a recording, whether they live in an Indonesian wood hut or a Scottish castle,

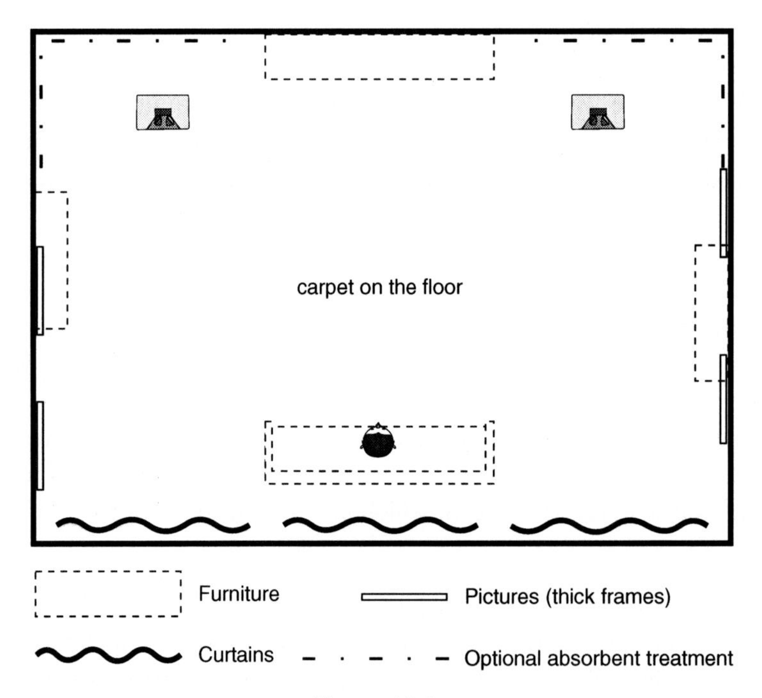

Figure 12.3:
An IEC listening room. Recommended layout of a typical domestic listening room. By placing the loudspeaker systems along one of the longer walls they can be positioned away from the room corners at different distances from the nearest boundaries. The listener is then also able to sit near the boundary opposite the loudspeaker systems, away from the central part of the room where nodes occur at low frequencies.

it would seem that they deserve equal consideration. Perhaps one of the things that most discredits the concept of the standard room is that there is no set position for the loudspeakers. All the rooms of different shape that lie within the IEC room specification will have different modal patterns. Whilst these modes are all intended to be evenly distributed, they will not all be driven. The loudspeakers will, due to their position, couple to some modes more strongly than to others. Therefore, with two different pairs of loudspeakers in two different IEC rooms, even with less than a 1 m difference in any room dimension and with the loudspeakers placed more or less in corresponding positions, the response differences, and hence the perceived colouration, may be significant. All the modes could only be driven by the siting of the loudspeakers at the intersections of three room boundaries, such as in the corners of the floor or ceiling, but for many reasons this is a truly awful location for most loudspeakers.

Clearly, such 'standard' rooms, although intuitively appealing, are no answer to the problem of achieving the desired acoustics for control room monitoring. They tend to exist principally for loudspeaker manufacturers and broadcasters, for the verification of compliance with various other standards and specifications. In fact, the specification of international 'standard' rooms has often been done more to satisfy the most *politically acceptable* compromises, rather than the most *psychoacoustically desirable* compromises. This is not unreasonable, as the latter tend to be too subjective for worldwide agreement on standardisation. Overall, the use of IEC rooms has not been particularly widespread, and when conditions of super-critical listening are required, many professionals consider them unacceptable because of the degree of different colourations that they can produce. They still contain too many characteristics of an idiosyncratic nature.

12.2.1 Beyond the Standard Room

In November 2006, Fazenda et al. published a paper on the subject of the use of modulation transfer functions (MTFs) (as introduced in Subsection 11.5.5) to assess the difference in fidelity of reproduction as firstly the room sizes and secondly the dimension ratios were changed.[4] The third change made to the rooms was the degree of modal damping, and it was shown that only this aspect of room control could significantly improve the accuracy of the perceived signal received by the ears from a loudspeaker across a listening room. A related paper by Harris et al. investigated the psychoacoustic link between the MTF and perceived quality of the low frequency reproduction.[5] Walking around a listening room will take the listener through regions of different modal effects, and will give rise to a perception of a changing frequency response, but in general there will not be a noticeable change in the *quality* of the music. However, fine detail may be lost as the listener goes further away from the loudspeakers in a relatively lively room, or listens on the anti-node of a strongly excited mode.

The above effect was shown in terms of MTF in Figure 11.31, and Fazenda has shown that the only viable way to reduce this effect of distance is by restricting the modal Q by damping the room.[4,6] In terms of the use of standardised room ratios, and with implications of 0.4 s mid-band decay time being suitable for high definition listening, the case has now been effectively closed. Such rooms do *not* meet the requirements for detailed quality-control listening (although the definition of 'detailed' may vary from person to person).

Frustrated by years of attempting to use their 'reference' rooms to assess all aspects of loudspeaker responses, the KEF loudspeaker company approached the author in 2006 about the design of *three* listening rooms to be constructed in their new factory in southern England. One of these rooms was to be a quality control room, for listening to the responses of the loudspeakers themselves. The second room was to be constructed in the manner of a typical British domestic lounge, and the third room was to be constructed in the typical manner of a Californian domestic lounge. The reason behind this decision was that after 40 years of trying to use single rooms for multiple purposes it finally became apparent that no single room could be used to extrapolate the complex performance of other rooms of significantly different construction. However, in this case, KEF were very concerned about the British and American markets, and how loudspeakers aligned for typical British homes could sound bass light in many larger, wood-framed American homes and *vice versa*. Listening rooms for the precise assessment of specific characteristics therefore tend to need to be *built* with those characteristics. Extrapolation from 'standard' conditions can be rather imprecise.

Nevertheless, work carried out by Genelec and JBL on a very wide range of listening conditions, from hotel alarm-clock radios to ghetto-blasters, from motor vehicles to discothèques, from Scottish castles to tropical wooden huts, from home theatre systems to headphones; the average of all the responses was FLAT! This, perhaps surprising result, suggests that control rooms in studios which are recording music for all reproduction circumstances should have flat responses if they are to be fair to *all* reproduction conditions. In practice, listeners then receive the responses according to the circumstances under which they are listening, and not a response which has been biased towards any specific listening conditions which may prejudice the enjoyment of music under different circumstances. A flat response is the fairest and the most representative response for a control room to exhibit.

Therefore, the work by Fazenda et al. is trying to arrive at a threshold of specification for a room decay which would allow flat mixing conditions and neutral timbral responses. It is expected that these listening conditions would, even at low frequencies, not be far different from what would be perceived when listening to the same loudspeakers in an anechoic chamber, but without the sensation of a dead environment. Some of the control room concepts described in later chapters also seek to achieve a similar goal, but control rooms are highly specialised rooms, often with flush mounted monitor systems; they are not typical domestic rooms.

The aim of the new type of listening room is to be more tolerant of loudspeaker placement and listening position, yet to have acoustic characteristics with absolute minimal colouration of the sounds reproduced within them. Fazenda is therefore working towards a room, not with the most even modal *distribution*, but with all modes damped to a point where they are not subjectively considered to give rise to misleading colouration.

It could, of course, also be argued that such listening rooms as this would also not be typically domestic, but nevertheless they would allow quality control listening in rooms which were not acoustically bi-directional. The majority of top class stereo control rooms *are* acoustically bi-directional, so reversing the direction of the loudspeakers and listening position could be subjectively disastrous. Obviously, in these typical control rooms, arbitrary loudspeaker placement is not an option.

12.3 The Anechoic Chamber

Figure 4.2 shows an anechoic chamber used for acoustic measurement at Southampton University in the UK. The chamber has also been used for subjective/objective comparisons on loudspeakers. There is absolutely no doubt that anechoic chambers provide the optimum conditions for achieving the flattest transfer from the loudspeakers to the ears, but the problems with their use as control rooms are two-fold. Firstly, many people are nervous when entering such rooms because they are deprived of the aural sensation from the surroundings to which they are accustomed, and which they need psychologically for their sense of normality. One *can* become accustomed to working in such circumstances, but during listening tests involving many inexperienced people it has been usual for some of them to be on the verge of a state of panic. The second reason why anechoic chambers are not suitable for control room monitoring is that at least some lateral reflexions are needed in order to develop a sense of spaciousness. If these are missing, the sound field collapses totally into the plane between the loudspeakers. This is highly untypical of the listening conditions in which the end-result is likely to be heard. Anechoic listening is too far away from normality.

There are also two further reasons why anechoic conditions are not the answer to the control room problem. A large increase in loudspeaker power would be needed due to so much energy being absorbed by the boundaries, with the room giving no help whatsoever to the low frequency loading of the loudspeakers. What is more, most loudspeakers are not designed for truly free-field operation, so it would be unlikely that very many commercially available loudspeakers would deliver their intended tonal balance in an anechoic chamber. Although the last two points are technically solvable, the first two (from the previous paragraph) are not, and so they are the prime reasons for the rejection of anechoic chambers as ideal control rooms. The cost of suitable wedges, at around −1,000/m^2, is also a consideration. Finally of course, in order to be effective down to low frequencies, the rooms also have to be huge.

12.4 The Hybrid Room

In a paper presented to the 1982 Audio Engineering Society convention in Anaheim, USA,[7] a proposal was put forward for a room such as is shown in Figure 12.4 as a solution to the neutral listening-room dilemma. The equation that the authors proposed was:

$$\text{Anechoic chamber} + \text{Reverberation chamber} = \text{New standard room}$$

Their idea was to use large areas of full-frequency-band absorption, alongside large areas of full-frequency-band reflexion. This was achieved by the use of recesses almost 2 m deep, filled with glass fibre wool, adjacent to slabs of heavy reflective materials. The juxtaposition would also lead to considerable diffusion. The system maintained a reverberation time of about 0.4 s with no more than 0.1 s total deviation between 20 Hz and 20 kHz. The rigid walls were non-parallel in order to prevent slap echoes (flutter echoes), and were massive so as not to vibrate and produce any resonant colouration.

The idea was well conceived. Wideband absorption and wideband reflexion, spacially well distributed, automatically provides good diffusion and some specular reflexions without flutter or resonance. The drawbacks, however, are numerous. The loss of 2 m depth on each of three wall surfaces, plus the need for sufficient distance from the reflective surfaces to avoid the reflexion returning to the listening position too early, means that all such rooms would necessarily be large; too large, in fact, for the spaces available for control rooms in most studios. As discussed in Section 7.2, all such control measures are time and wavelength related, and so to scale them down in a smaller room is not possible. Despite the paper referring to its use in studios, this room concept (and the T_{60} of 0.4 s) still seemed to relate more to listening room standards rather than the somewhat more critical conditions necessary for control room monitoring.

Nevertheless, the Ishii-Mizutoni paper proposed a room that would exhibit generally acoustically neutral listening conditions, where the direct to reverberant sound fields could be balanced by adjustment of the distance between the loudspeakers and the listeners. Interestingly, though, in their AES paper, they stated 'The back of the front stage should be made of all rigid wall construction, because as a listening room, the difference in position of the test loudspeaker should be minimised'. This is interesting inasmuch as it recognises the need for a relatively fixed loudspeaker position for repeatable critical assessment; in this case very close to a massive, rigid wall.

(a)

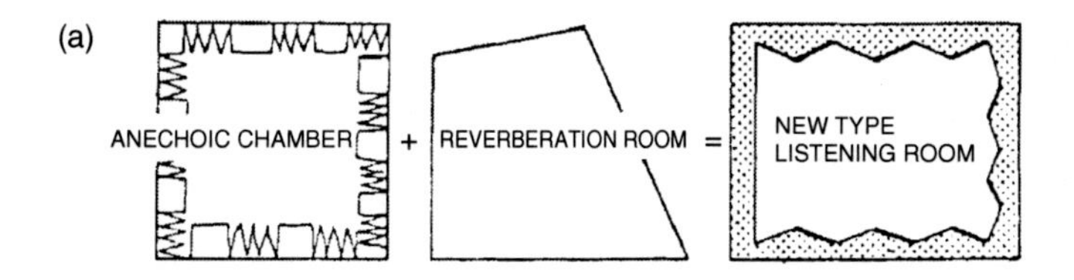

(b)

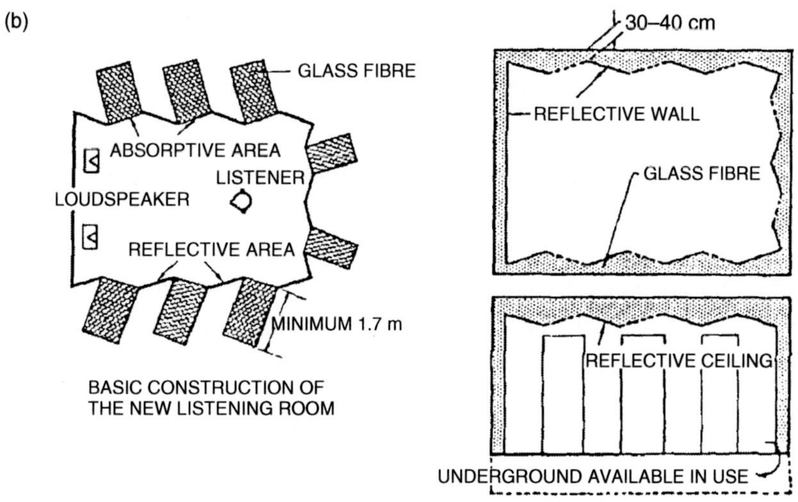

(c)

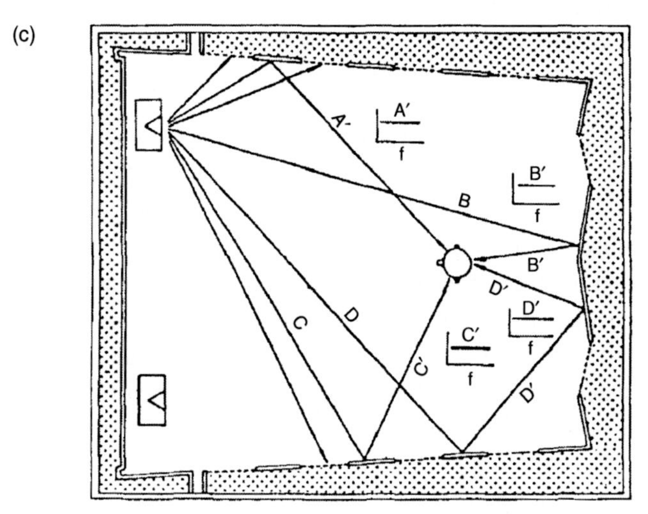

Figure 12.4:
The Ishii and Mizutoni proposal.

12.5 A BBC Solution

Large broadcasting organisations such as the British Broadcasting Corporation (BBC) may need to transfer programme material between many studios in many different locations. Staff may also frequently move between studios, and to work efficiently they need to be able to settle quickly into an unfamiliar room with a minimum of time for acclimatisation. They need to work rapidly and reliably, and to feel confident that the ensuing broadcasts are well balanced in people's homes. There is also a tendency for the broadcast industry to rely on free-standing loudspeakers, with all the attendant problems of edge diffractions, the confusion caused by rear radiation, and the response variability with respect to position.

Broadcasting staff often have their own ways of working, which necessarily differ from the music recording industry, and any new rooms that are introduced into a studio complex usually need to be closely compatible with existing rooms. The deader acoustics of many music industry control rooms are generally not well liked in the broadcast world. The tendency is still to favour rooms with characteristics not unlike the IEC rooms, but this type of construction is not always appropriate because of limitations on size.

In 1994, Bob Walker presented a paper[8] on a newly developed type of room for the BBC, which was to exhibit a relatively uniform decay time with frequency. It had the added specification that no reflexions should arrive at the ear within 15 ms of the direct signal, and that no reflexions arriving after that time should be more than −15 dB relative to the direct signal. The concept is shown in Figure 12.5. The ceiling, as can be seen, is of generally similar shape to the walls, which seems logical considering the three-dimensional expansion of the sound waves. The floor dip problem, discussed in Chapter 11, is partially blocked by the siting of a mixing console between the loudspeakers and the listener.

The design concept relies on the positioning of reflective surfaces in such locations that they will not reflect energy from the loudspeakers directly towards the listener, which is not as easy as it may seem. In reality, unlike the images shown on computer ray-tracing programs, the sound waves do not travel in straight lines according to Euclidian geometry. They travel as expanding waves that are greatly influenced by diffraction effects at most audio frequencies. The diffraction ensures that non-specular (i.e. unlike light) reflexions will travel in unpredicted directions, only to be rediffracted from other boundary discontinuities. The design of such rooms is therefore not a simple matter. Nevertheless, the above concept has much to offer in some difficult circumstances, and its concepts are valuable when free-standing loudspeakers *must* be used.

12.6 On Listening Rooms in General

When rooms are to be used for listening to music for serious enjoyment, almost all of the concepts discussed so far in this chapter must be taken into account. Simply placing loudspeakers in any convenient location will not suffice. It must be understood that in a highly

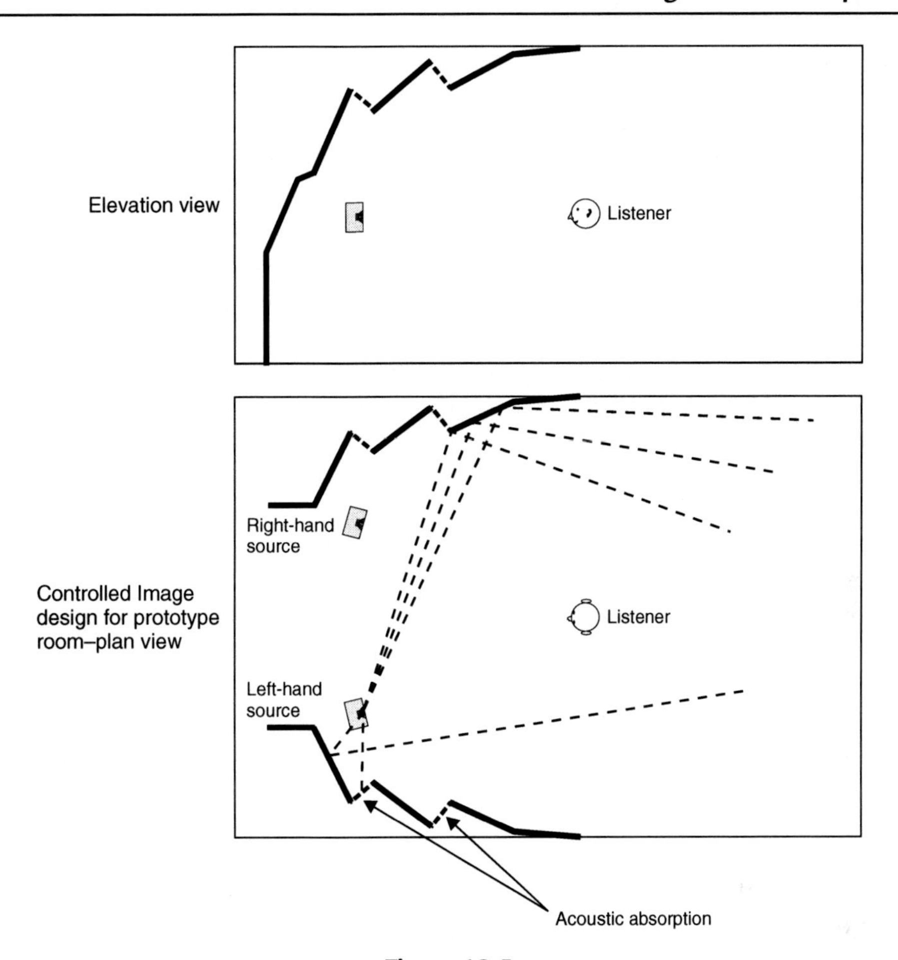

Figure 12.5:
Controlled reflexion room, in the style of Bob Walker, for free-standing loudspeakers.

reverberant room, with little furniture and plastered walls and ceiling, it is unlikely that pleasant listening conditions could be achieved no matter what type of loudspeakers were used, nor however carefully they were positioned. What is needed, when listening to music for pleasure, is a relatively flat amplitude response from the room/loudspeaker combination, and a quantity of reverberation that is appropriate for the music. Except for the purposes of the quality control of recordings, it is probably true to say that all recorded music benefits from a little reverberation in the listening environment. Surround sound systems, which can provide different ambiences for different types of recording, probably benefit from being reproduced in rooms with reverberation times slightly lower than the shortest reverberation time in any recordings to be auditioned in them. However, for stereo reproduction in relatively small rooms, reverberation times (or more correctly in these cases, decay times) should be between about 0.3 and 0.5 s from 100 Hz to 1 kHz, with allowable increases below 100 Hz and reductions above 1 kHz. In general, rock music and electronic music with high transient contents tend to favour the shorter decay times, whilst much acoustic music tends to favour the upper limits.

Control of the decay time can be accomplished by furnishings or by specific acoustic treatments, but the situation is greatly affected by the structure of the room. Diaphragmatic walls and ceilings, such as those made from timber and plasterboard (variously known as sheetrock or Pladur) tend to produce better listening rooms than those made from granite blocks, in which large low frequency build ups are commonly encountered. This is the reason why some loudspeaker manufacturers, with their principal markets being in known countries, specifically tailor the low frequency responses of their products to take into account the absorption characteristics of the typical structures and furnishings of the most common homes in which their products will be used. Flat anechoic chamber responses are most definitely *not* the objective in such cases. A flat response in the end-use environment is what is being sought.

One of the main requirements of any length of reverberation time is that it should not vary significantly between adjacent third-octave bands, and that it should show a slight monotonic roll-off above about 100 Hz. In smaller rooms, the decay time below 100 Hz will usually fall, but in larger rooms it can be allowed to rise, though the 50 Hz reverberation time should not exceed twice that at 1 kHz. If the reverberation/decay time is *not* smooth, it will colour the reproduced sound, because certain bands of frequencies hang on in isolation when the rest of the music has decayed. When such cases exist, they are usually the result of the existence of poorly damped resonant modes.

In purpose designed listening rooms, internal room structures can be built if the main structure is unsuitable. Figure 12.6 shows a typical wall structure that could be used in a serious listening room. This is a useful means of damping the room modes and reducing the reflected low frequency energy. Bookcases on rear walls (i.e. facing the loudspeakers) can produce excellent diffusion, and heavy upholstery and deeply folded curtains are effective wideband absorbers. Wooden floors tend to absorb much more bass than do concrete or stone floors. However, the absorption of deeply folded curtains is affected by their distance from the walls. As with the partition wall shown in Figure 12.6, the greater the space between the absorber and the structural wall, the greater will be the very low frequency absorption.

12.7 Close-Field Monitoring

From the concepts discussed in this chapter, it can be seen that many apparently obvious (and other less obvious) solutions are not necessarily solutions at all to the problem of creating ideal critical listening conditions for music in stereo. It is true that chaos effectively rules in the domestic listening world, because the range of domestic listening conditions must include loudspeakers of all qualities, response variations of at least ± 10 dB, and decay times of

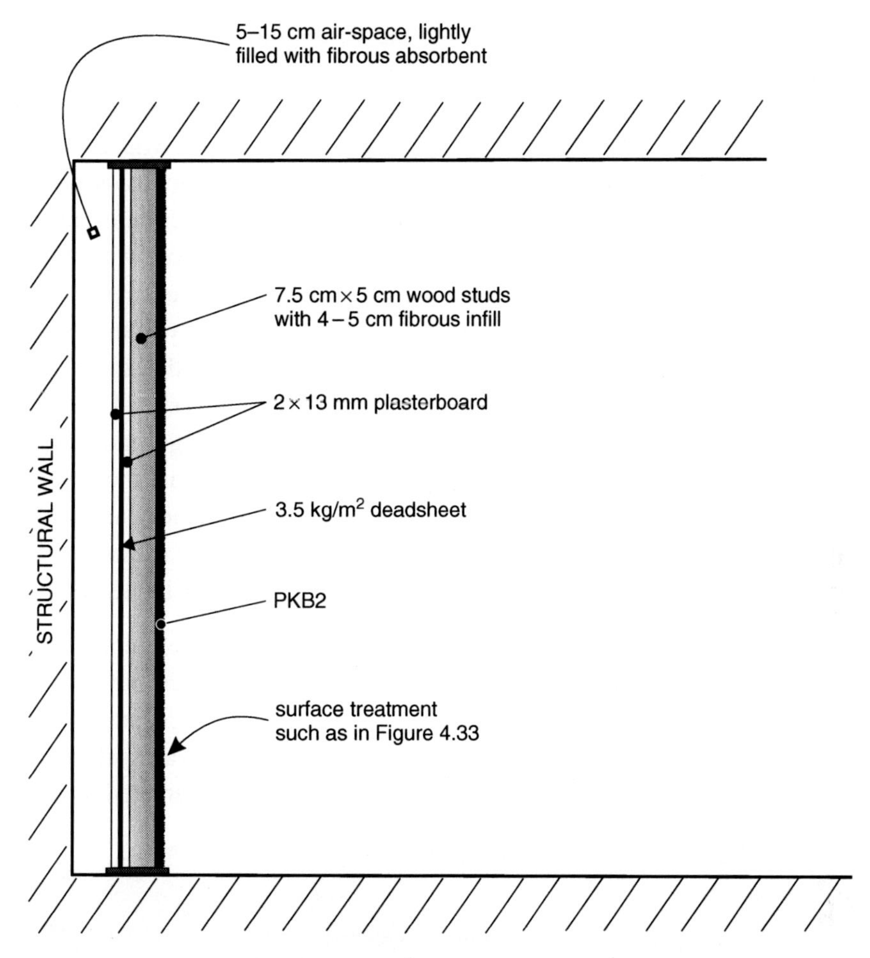

Figure 12.6:
Composite acoustic control wall. The wall shown in this figure is a combination of membrane absorber, panel absorber and amplitude reflexion grating. By such means, a room can be well-controlled down to very low frequencies, and the internal 'life' in the room acoustic can be adjusted by varying the ratio of hard to soft surfaces on the surface, either strip-by-strip or section-by-section.

between (usually) 0.3 and 1.2 s; so what *is* the reference point? How do we find a set of monitoring conditions which will allow recordings to be produced in the studios which will 'travel' to the wide range of domestic listening conditions? An answer is needed so that the people making the recordings can be reasonably sure of the balance of instruments that will be heard in the diverse range of home situations.

In the early 1970s, with no apparent answer to the above problem, a fashion swept through the western recording world. It was the use of the Auratone 5C Sound Cubes. These consisted of boxes about sixteen centimeters cubed usually placed on top of the mixing consoles, which were used to check mixes for compatibility in domestic circumstances (note to non-native English speakers – *sixteen centimetres cubed* – 16 cm × 16 cm × 16 cm – 4096 cm^3). The Auratones consisted of a 4.5inch loudspeaker (110 mm) in a sealed box which, for their time, had an unexpectedly good low frequency response. They led the way to what became known as 'near-field' or, more correctly, 'close-field' monitoring. These small loudspeakers, in stereo pairs at close range, gave surprisingly reliable reproduction. In most circumstances, the listening distance of around 1m ensured that monitoring was carried out within the critical distance, except at the lowest frequencies (of which there was very little, anyway). This helped to introduce a form of standardised monitoring conditions – listening within the close-field of Auratones on top of a mixing console. Of course, the limitation was that the loudspeakers could not be considered full range, so what was going on at the lowest and the very highest frequencies could not be heard. Nevertheless, in conjunction with larger monitors, recording personnel began to learn how to use the technique, and often found that the more direct sound which the close-field provided allowed them to work more rapidly and with better compatibility with the bulk of domestic listening circumstances. This subject will be raised numerous times in later chapters, and will be discussed at length in Chapter 19.

By the early 1980s a new contender arrived for the role of standard reference; the Yamaha NS10M. This was originally conceived as a domestic hi-fi loudspeaker, but it was a commercial failure as such. However, it found rapid international acceptance as the new close-field monitor. Louder and deeper than the Auratone, this two-way system, shown in Figure 12.7, provided something on which the heavily synthesiser orientated music of the day

Figure 12.7:
The Yamaha 'NS10M Studio'. Shown, incidentally, as it should *not* be used, see Figure 18.5.

could be heard more realistically. The high driver failure rate of the original NS10 s led Yamaha to create the 'NS10M Studio', specifically designed for studio use with greater power handling and similar sensitivity. Although there were some early complaints that it did not sound quite the same as the original, to which so many people had become accustomed, the studio version nevertheless soon became a *de facto* reference for the next 15 years or more.

Clearly, the use of the newly available, relatively wide response, compact loudspeakers, which had enough SPL and sufficient damage tolerance for adequate monitoring, opened up new possibilities for close-field monitoring. The need for this, and the various ways in which it was taken up, was a clear statement by the recording industry that control rooms and their monitor systems were, in general, not performing as well as they should. In many ways, it was a sad reflection on the state of control room design. It seemed like almost a whole industry had abandoned the control rooms and withdrawn into the close-field.

Incidentally, the term 'close-field' is preferred because 'near-field' already has two, well established and clearly defined acoustical senses, neither of which relate to the way that small monitors are used in studios. The geometric and hydrodynamic near-fields are described in the glossary. Briefly, however, the geometric near-field exists where the distance from a measuring point to all parts of the radiating surface cannot be considered to be equal. This depends on wavelength, so the extent of the near-field is frequency *and* source-size dependent. This is why, when calculating sound pressure levels by counting back to the source, and increasing the value by 6 dB for each halving of distance, one does not get infinite power at the source. For example, in a free-field, 100 dB at 1 m would suggest 106 dB at 50 cm; 112 dB at 25 cm; 118 dB at 12.5 cm; and 124 dB at 6.25 cm; until infinite pressure at the diaphragm. In reality, at some distance, the SPL will reach a level which will be the limit at any given frequency. Going closer to the source diaphragm would not produce any higher SPLs because one would be entering the geometric near-field.

Another type of pseudo near-field exists with multi-way loudspeakers, when the listening distances are so short that, in effect, the highs can be heard coming from the tweeters and the lows from the woofers; in other words, where the multiple drive elements fail to blend into one common, apparent source. In this region, small movements of the head can lead to large changes in the sound character, and in such conditions of inconsistency, monitoring would be unwise, if not ridiculous. The frequency balance would not be consistent within this near-field, so it is wiser to listen in the close-field.

In general, the popularity of close-field monitoring suggests that listening within the critical distance is desirable not only for the perception of detail, but also because the frequency response (pressure amplitude response) of the loudspeakers are more consistent, and less variability thus exists between what is heard in different rooms. There is no reason why this cannot also be achieved using large monitors in well controlled rooms, but the main obstacle in the development of such control rooms was a very deeply entrenched reluctance to accept that what was required of the IEC listening rooms, in terms of modal help, decay time and reflexions, could be dispensed with. Somewhat ironically, many of the people demanding loyalty to the old ideas were themselves totally rejecting them by their reliance on close-field working.

When people are listening at normal distances from loudspeakers, they are usually listening at or beyond the critical distance, as explained in Chapter 11. They are therefore listening to the room sound at a level that is equal to or exceeds the direct sound. If the music is cut, the natural decay down to 60 dB below the previous level would be what is perceived from the room. However, when listening in the close-field, the direct sound may be up to 10 dB higher than the room ambience. In such cases, when the music is cut, the sound level immediately drops by the difference. Figure 12.8 shows the two cases. In the latter, it is when the room sound has decayed only by 50 dB that it will be perceived to be at the −60 dB level relative to the music. Therefore, when wishing to work in the close-field, a person is effectively requesting to work in a room with a decay time that is shorter than the nominal decay time in the room in which they *are* working. The practice of requesting a given room decay time (RT_{60}, T_{60}), rejecting the main monitors as being too affected by the room, and then electing to work in the close-field, is a triumph of ignorance over acoustics. The answer could seem to lie in reducing the room decay time for the large monitoring systems; then, if the room were deemed oppressive, increasing the reflected energy from the speech and actions of the people working within the room by means of surfaces out of the line of sight of the loudspeakers. Now we are finally beginning to touch upon the subject of control room design.

12.8 Summary

The control of a room, to flatten the transfer function from loudspeaker to ear, is best achieved acoustically.

In the case of mono, the loudspeaker and listening positions can be adjusted to find a best compromise response, but this is more difficult in stereo, where a triangular geometry between *two* loudspeakers and the listener must be maintained.

Monitor equalisation is not an answer, but perversely it is usually the studios with the least acoustic treatment – the ones *least able* to be corrected by equalisation – that are the ones where it is currently most likely to be found in use. The tendency is for monitor equalisers to shift problems around rather than to cure them. Equalisers can only legitimately be used for the correction of amplitude/frequency problems that are not caused by time domain effects.

IEC standard listening rooms are generally too coloured for critical listening purposes.

Anechoic chambers are too far removed from real listening conditions to be used for music mixing.

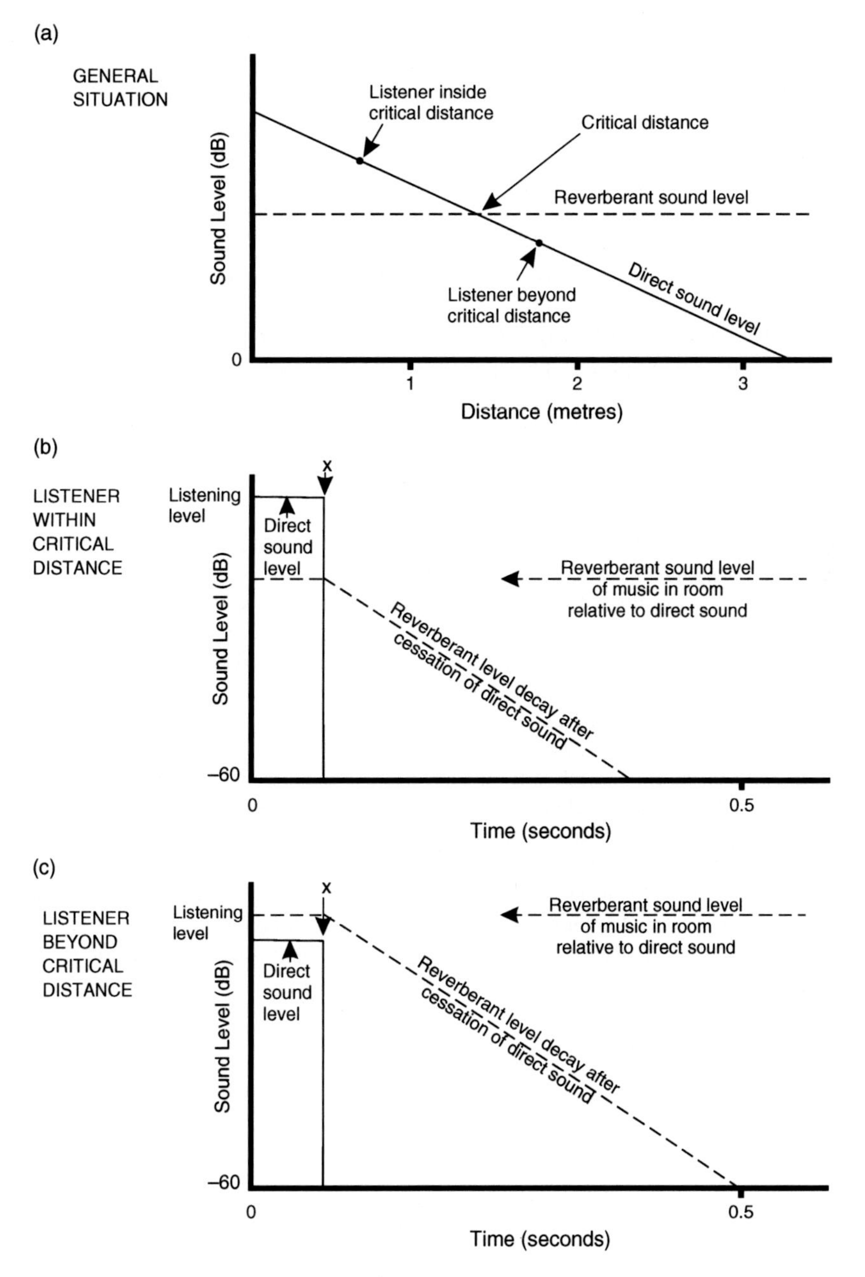

Figure 12.8:
Perception from within and beyond the critical distance: (a) General situation; (b) When the signal is cut at time *x*, because the *direct* sound is higher than the reverberant sound, a sudden drop in level is perceived, then the reverberation is heard to decay in accordance with the nature of the room; (c) If the listener is in the reverberant field, which *already* dominates the direct sound beyond the critical distance, no sudden drop in level is noticed when the signal is cut at time *x*, only the steady decay of the reverberant field is perceived.

The Ishii-Mizutoni concept is too bulky, and will not scale because of wavelength dependency.

The Walker/BBC rooms are a reasonable solution when free-standing loudspeakers must be used.

The rooms used for serious listening for musical *enjoyment* are not necessarily the ones best suited for critical quality control listening.

Heavily damped rooms, with low modal Qs, offer better room-to-room compatibility.

The concept of close field monitoring has, in effect, been a request for working inside, or quite close to, the critical distance. This can be reinterpreted as being a request for control rooms with a shorter decay time.

References

1 Briggs, G. A., *Loudspeakers*, 5th Edn, Wharfedale Wireless Works, Bradford, UK (1958). [Later re-prints published by Rank Wharfedale Limited, into the 1970s.]
2 Briggs, G. A., *Sound Reproduction*, 3rd Edn, Wharfedale Wireless Works, Bradford, UK (1953)
3 IEC, *Sound System Equipment, Part 13*: *Listening Tests on Loudspeakers*, IEC Publications 268–13 (1983)
4 Ishii, S., Mizutoni, T., 'A New Type of Listening Room and its Characteristics – A Proposal for a Standard Listening Room.' Presented at the 72nd AES Convention, Anaheim, USA, Preprint No. 1887 (1982)
5 Walker, R., 'The Control of Early Reflections in Studio Control Rooms', *Proceedings of the Institute of Acoustics*, Vol. 16, Part 4, pp. 299–311 (1994)
6 Fazenda, B. M., Holland, K. R., Newell, P. R., Castro, S. V., 'Modulation Transfer Function as a Measure of Room Low Frequency Performance', *Proceedings of the Institute of Acoustics*, Vol. 28, Part 8, pp. 187–194, Reproduced Sound 22 conference, Oxford, UK (2006)
7 Harris, L. E., Holland, K. R., Newell, P. R., 'Subjective Assessment of the Modulation Transfer Function as a Means for Quantifying Low-Frequency Sound Quality', *Proceedings of the Institute of Acoustics*, Vol. 28, Part 8, pp. 195–203, Reproduced Sound 22 conference, Oxford, UK (2006)
8 Fazenda, B. M., Avis, M. R., 'Perception of Low Frequencies in Small Rooms', Proceedings of the European Acoustics Symposeum (Acústica 2004), Guimarães, Portugal (2004)

CHAPTER 13

Control Rooms

Basic needs in control rooms. Special requirements of studio monitor systems. Differing control room philosophies. Flush mounted loudspeakers. Loudspeaker responses in the pressure zone. Problems in small rooms.

Although the function of almost all sound studio control rooms is to produce recordings (or live broadcasts) which will ultimately be heard in domestic rooms, there are other demands made of the monitoring systems and acoustics that lead to some very different solutions than the use of hi-fi loudspeakers in 'standard' listening rooms. Control room monitoring usually needs to be capable of a degree of resolution that will show up any errors in the proceedings to such an extent that the recording personnel will always be one step ahead of the consumers. Control rooms must also be practical working environments. Several people may need to be able to work in different places in the rooms whilst listening to substantially the same musical and tonal balances. This implies that position-dependent differences within the working area should be minimised.

Music recording also frequently tends to take place at higher levels than will be used for most domestic reproduction or standard listening tests, which can further alter the balance of compromises. The reasons for this are numerous, and many are discussed in Chapter 15. However, in general, it can be said that most recording studio control rooms should be more acoustically dead than most domestic rooms. As can be seen from Figure 4.17, the reverberant/reflected energy in domestic rooms can substantially enhance the subjective loudness of a loudspeaker. An increase of 5 or 6 dB is shown, which is not untypical, so loudspeakers for use in rather dead control rooms need to be able to supply the extra power.

In fact, this 5 or 6 dB, plus the few dB s extra that are needed for other reasons, such as greater listening distances and the need to inspire musicians, explains why loudspeakers for recording use may need to be able to supply up to 100 times the output power of domestic loudspeakers. However, one hundred times the power is only four times as loud, and much of this can be soaked up in highly controlled rooms. In engineering terms, the 100 times (20 dB) power increase poses many more electro-mechanical design problems than 'only' four times the loudness would suggest. (An increase of 10 dB [at least at mid frequencies] is generally considered twice as loud as the original.) Increased power usually means increased size, both to move more air and to lose the additional waste heat. Many loudspeakers in professional use are typically only 0.5 to 5% efficient, therefore 500 W into such a loudspeaker would produce between 475 and 497.5 W of heat in the voice coil. This must be dissipated in the metal chassis and by means of ventilation. Such problems can also lead to the use of different construction materials, and this can also move studio loudspeakers one step further from their domestic counterparts.

13.1 The Advent of Specialised Control Rooms

As mentioned in Chapter 12, back in the days when most recording was done in mono, judicious movement of the loudspeakers and/or listening positions could often achieve a more or less desirable sound, both in terms of critical distance considerations and the avoidance of troublesome nodes or anti-nodes in a room. When stereo arrived, a new set of restrictions arrived with it. The listening position became a function of the loudspeaker distances *and* the subtended angles between them. Thus, it was no longer a relatively simple matter of adjusting the positions of a single loudspeaker and/or listening position, but the moving of a whole triangle, formed by the two loudspeakers and the listener. In other words, all three items may need to be moved not only in relation to their individual responses in the room, but whilst maintaining relative angles and distances between them. They also introduced the necessity to maintain their stereo imaging, which can be very fragile. Moving loudspeakers away from troublesome modes may disrupt the stereo balance or drive them into impractical locations. Improving the siting of one loudspeaker may, through the need to maintain the triangulation, move the other loudspeaker into an acoustically worse position, or into an inconvenient place, where it may need to be moved every time the door needs to be opened. This would clearly not be practical in professional use. These requirements suggested a need for the suppression, to a very high degree, of the temporally and spacially dependent characteristics of the room, but the better solutions took a long time to develop.

13.1.1 Geometrically Controlled Rooms

Commercial recording studios put up with a lot of bad rooms until the early 1970s, when serious efforts were made, on an international scale, to try to find control room designs which could be relied upon to produce recordings which generally travelled well; both to the outside world *and* between studios. This was an era when work really began to travel from studio-to-studio, and even country-to-country, during its production. One of the first big commercial efforts to produce acoustically standardised 'interchangeable' rooms, was by Tom Hidley at Westlake Audio in California, USA. Soon, rooms of this type were in use worldwide, and were designed to have reverberation times of less than about 0.3 s. They incorporated large volumes of 'bass traps' in an effort to bring the low frequency reverberation time into relative uniformity with the mid-band times, and also to try to avoid the build up of low frequency standing waves, or resonant modes.

An attempt to reduce further the resonant mode problem was effected by the use of entirely non-parallel construction, to deter the formation of the more energetic axial modes. Monitor equalisation was *de rigueur* as the rooms were 'sold' on their flat frequency response at the listening position, and third-octave equalisation was employed to achieve that goal. The rooms were generally quite well received at the time, and were a significant improvement on much of what was then current. Nevertheless, their responses were by no means as subjectively similar as their pink noise, real-time, third-octave spectrum analysis led so many people to expect. It soon became widely apparent that the control of reverberation time and a time average of the combined direct/reflected (loudspeaker/room) response was not sufficient to describe the sonic character of a room. Some people already knew this, but they were a small minority and few of them held any significant sway in the recording world. A typical design of an early Westlake-style room is shown in Figure 13.1.

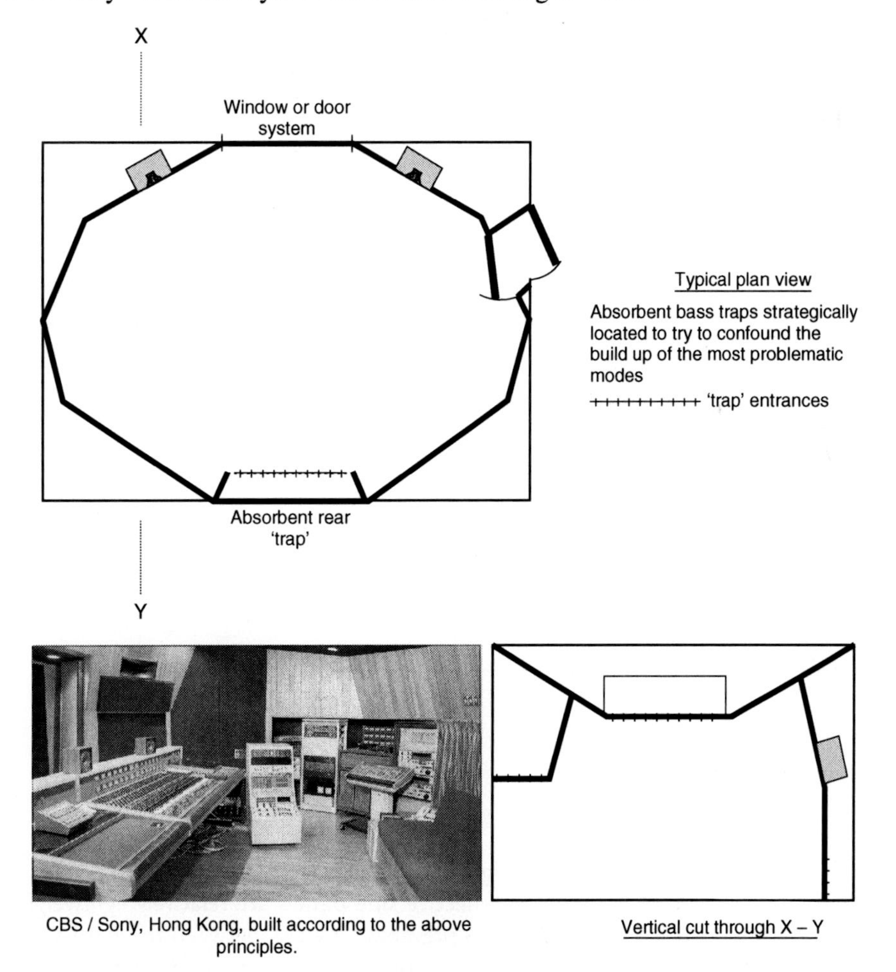

Figure 13.1:
Typical room design of Westlake/Hidley style of the late 1970s.

13.1.2 Directional Dual Acoustics

Also during the mid-1970s, Wolfgang W. Jensen, in Europe, was producing rooms as depicted in Figure 13.2. These rooms used 'sawtooth' absorbers, which tended to absorb much of the incident wave from the loudspeakers. They had reflective surfaces at angles where they could reflect back sounds created within the room, such as by the speech and actions of the personnel, but they would *not* cause reflexions directly from the monitors. Total absorption of the incident wave was not intended, because the Jensen rooms still sought to maintain a room decay time on the low side of the 'standard' domestic range, in accordance with well-established recommendations. Reflective rear walls were quite common in these rooms, though most examples had a rather absorbent rear wall.

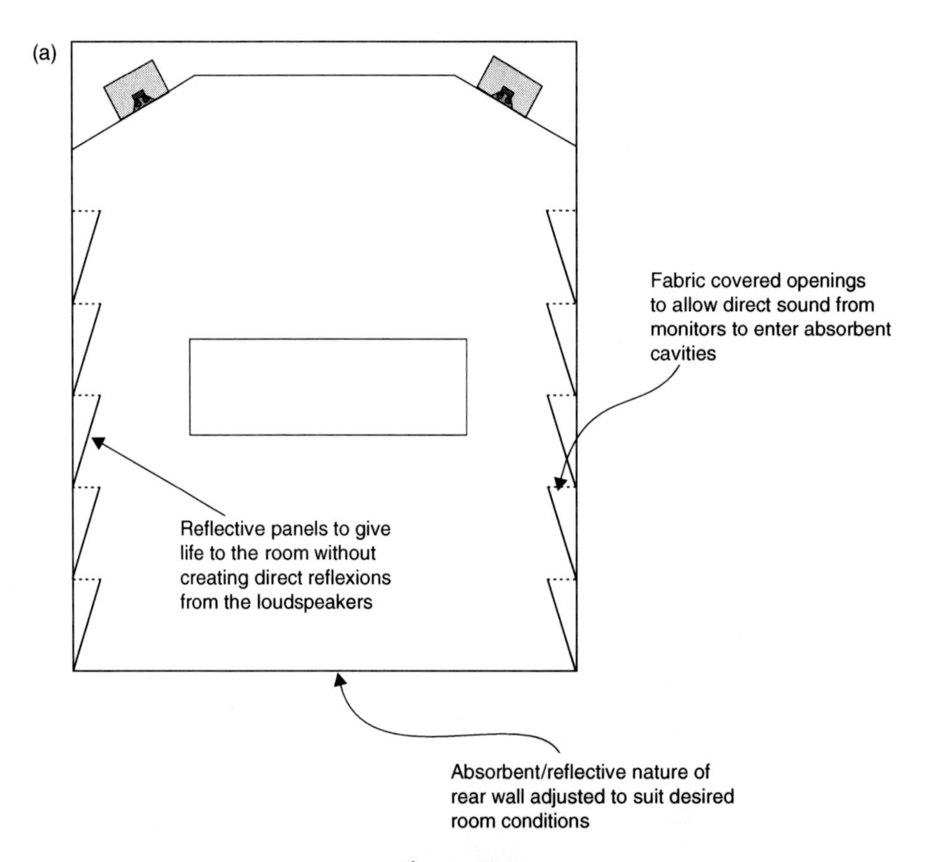

Figure 13.2:
(a) Typical room in the style of Jensen. The reflective side panels are relatively lightweight, and act as low frequency absorbers. Their angling also prevents 'chatter' between the hard-surfaced side walls. (b) View towards the rear of the Jensen designed control room at Sintonia, Madrid, Spain, showing the exposed entrances to the sawtooth absorbers on the side wall. (c) View towards the front of the Sintonia control room, showing only the reflective surfaces on the side wall.

(b)

(c)

Figure 13.2: (continued)

The rooms of Jensen were interesting because they made a clear distinction between the acoustic conditions required for the monitoring (relatively dead) and the perceived acoustic needs for a sensation of comfort within the room (relatively live). The absorber openings were facing the loudspeakers only, although the wooden panelling, used to provide reflective energy to speech within the room, would also act as a general low frequency absorber. The rooms were clearly bi-directional in the same way as the recording room shown in Figure 7.3, and the much later BBC rooms shown in Figure 12.5. The Jensen rooms, like the Westlake rooms, usually featured flush mounted (built-in) monitors as standard, which were normally of their own designs and custom tailored to the rooms. This was an important advance.

13.1.3 The LEDE

In the late 1970s, Don and Carolyn Davis were keenly investigating many acoustic and psychoacoustic phenomena with the then newly developed Time Energy Frequency/Time Delay Spectrometry (TEF/TDS) measurement systems. TDS measurements made at Wally Heider Studios, Los Angeles, USA, and at RCA and Capitol Records, in Hollywood, had given them a lot to think about, leading them to the concept of the 'Reflection Free Zone' and the 'Live-End, Dead-End' (LEDE) control rooms.

Concurrently, Carolyn (Puddie) Rodgers was presenting new ideas about how certain room reflexion characteristics could confuse the ear by giving rise to response filtering which closely mimicked the pinnae (outer ear) transformations used by the brain to facilitate spacial localisation.[1] This work gave some very explicit explanations of the psychoacoustic relevance of the Energy/Time Curve (ETC) responses in the above-mentioned room measurements, and reinforced the concepts of the Davis' LEDE principles. Don Davis and Chips Davis (no relation) then wrote their seminal paper on the 'LEDE' concept of control room design,[2] and these rooms came into very widespread use in the subsequent years. The concepts were further developed by Jack Wrightson, Russell Berger and other notable designers. Sadly, however, Puddie Rodgers died of cancer before ever seeing the fruits of her labours fully mature.

The LEDE rooms rely on some psychoacoustic criteria such as the Haas effect and the directional aspects of human hearing. The Haas effect,[3] otherwise known as the precedence effect, manifests itself when two short sounds are heard in rapid succession. The human auditory system appears to suppress the separate identity of the second sound and give precedence to the first. The pair of sounds is perceived as a single sound coming from the direction of the first sound, but with greater loudness than the first sound alone. For this effect to be operative there must be a minimum separation of about 1 ms between the two sounds, or the ear will tend to confuse the source position if the two are not co-located. Beyond about 30 or 40 ms, the two sounds are heard as separate events. The effect can be overridden if the second sound is more than 10 or 15 dB higher in level than the first. Haas stated that within the 1–40 ms, or thereabouts, time zone, where the effect is in operation, the second sound to arrive would need to be at least 10 dB higher in level than the first sound if it were to be heard to be separate and equal in level. The 'just detectable' threshold for the second sound to arrive is about 4 to 6 dB above the level of the first sound, depending upon conditions.

Figure 13.3 shows the LEDE concept, where the front half of the room is largely absorbent. Its geometry is designed to produce a zone free of early reflexions around the principal listening position. The idea is to allow a clean first pass of the sound, directly from the loudspeakers, and then to allow a suitable time interval before the first room reflexions return to the listeners' ears. The rear half of the rooms are made diffusively reverberant, allowing the perception of a room 'life', which should then not unduly colour the perception

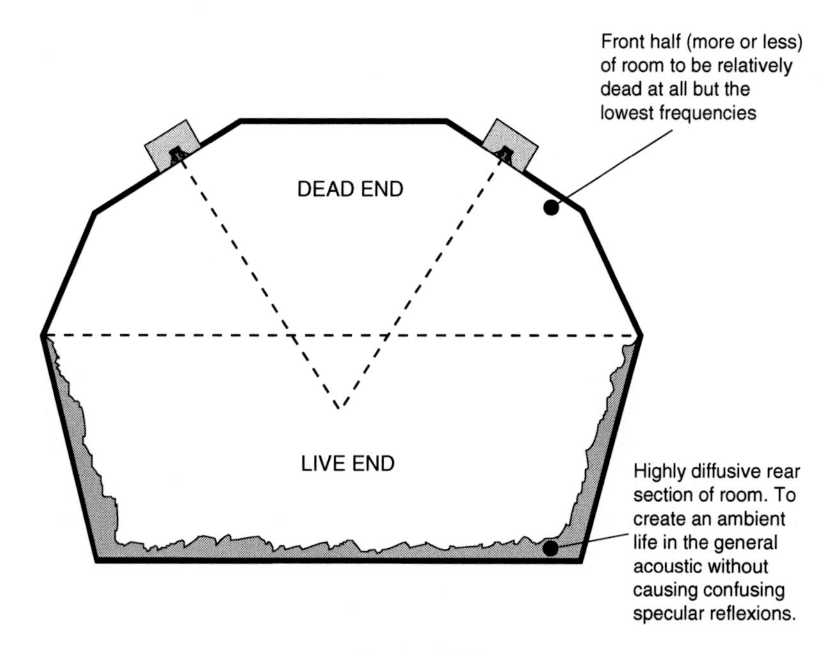

Figure 13.3:
Concept of a Live-End, Dead-End room. This figure shows the concept of the Live-End, Dead-End stereo control room, in which a reflexion-free zone is created around the listeners. The rear half of the room is made to be highly acoustically diffusive, usually without specular reflexions, to simulate a sense of ambience within the room that does not obscure the general clarity of the monitoring. The front half of the room is acoustically dead, to prevent any room effect from returning from the frontal direction which could spacially superimpose itself on the direct monitor sound. It should be noted, however, that some of the early proponents of the concept stated that the room is defined more by the characteristics of its ETC response rather than by the physical distribution of hard and soft surfaces. These, strictly speaking, need not go hand-in-hand, although very often they do.

of the directly propagated information. The rooms require a diffuse reverberation/decay characteristic, and proprietary diffusers such as those developed by Dr Peter D'Antonio, and marketed by his company RPG, are widely used in such rooms.[4] Strong, discrete, specular reflexions are to be avoided, as they produce position dependent colouration and general response flatness irregularities.

There was a period of time when specularly reflective panels *were* used at the rear/side corners of many LEDE control rooms. These were known as 'Haas kickers', and were intended to prolong the Haas effect, but their subjective effects failed to live up to expectations and they were generally soon abandoned. Despite the Haas effect, a reflexion coming from the opposite rear corner to the source did not help the stereo imaging. LEDE rooms will be discussed in Chapter 17 at much greater length.

13.1.4 The Non-Environment

After growing criticism of his 1970s designs, Tom Hidley took a break between 1980 and 1983. During this lay-off, he thought over the old problems and came up with a new concept that he called 'the Non-Environment' room principle. In these rooms, he made the front wall maximally reflective, and, other than for a hard floor, made all the other room surfaces as absorbent as possible.[5] The principle sought to drive the loudspeakers into something approximating an anechoic termination. With the monitors set flush into the front wall it would act as a baffle extension, but it could not reflect sound from the loudspeakers because they were radiating away from it. The wall would, however, in concert with the floor and equipment surfaces, provide life to the speech and actions of people within the room, thus relieving any tendency for the rooms to feel uncomfortably dead. The principles were not unlike those of the aforementioned Jensen rooms, but were taken to a greater extreme, approaching a hemi-anechoic chamber (not to be confused with a semi-anechoic chamber – see Glossary). It is also not unlike the Ishii-Mizutoni room, as described in Section 12.4, but with the loudspeakers flush mounted and the wideband reflectors removed.

By introducing more absorption into the room and reducing the quantities of reflexions, the ratio of the direct sound to the reflected sound is increased, and hence the levels of colouration are reduced. This is, however, achieved at the expense of any consideration whatsoever for mimicking domestic listening acoustic conditions. The consistency between Non-Environment type rooms is, perhaps, greater than that between most other types of rooms. One very famous recording engineer/producer who loves these rooms for mixing (George Massenburg) freely admits that for certain types of music recording, especially when the musicians are creating music in the control rooms (as is often the case these days), he must use rooms with a more inspiring life. Such things are very subjective, though, and he also produces people who love the rooms for the whole recording process. A typical 'Non-Environment' construction is shown in Figure 13.4, and the concept is discussed at length in Chapter 16.

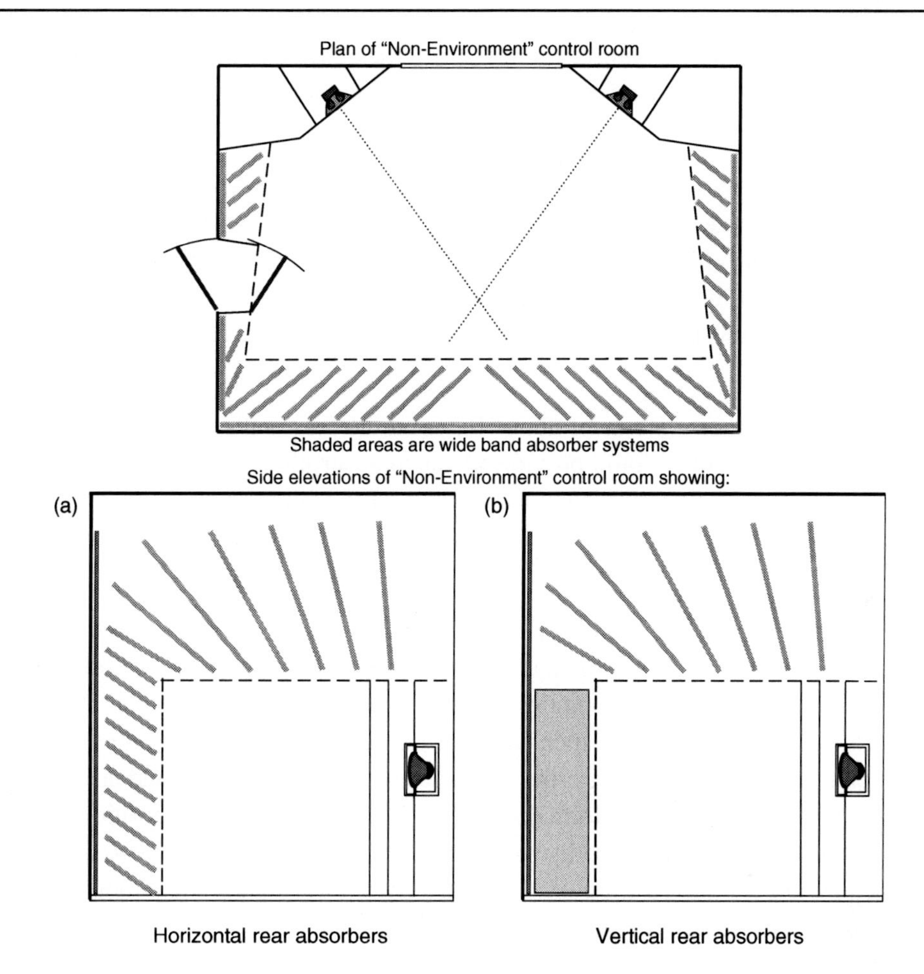

Figure 13.4:
The Non-Environment concept.

13.1.5 Toyoshima Rooms

In 1981, Hirata et al, presented a paper on optimum reverberation times for control rooms and listening rooms.[6] They made some interesting statements for the day, such as 'It is easier for a recording engineer to assess the clarity of phantom images of reproduced sound in a dead room than in a live room, though this is not the purpose of listening to music'. They concluded that perhaps, for super-critical monitoring, a decay time of around 0.2 s would be optimum, with perhaps about 0.4 s being the optimum decay time for a listening room. The clear statement of different optimum decay times for quality control and listening for pleasure was important.

Around the same time, in Tokyo, Sam Toyoshima was designing studios for some of the major recording companies, which were also beginning to be noticed in Europe and the USA. In

1986, he presented a paper on control room acoustic design.[7] The first of Tim Hidley's Non-Environment rooms had also been built in Tokyo, so a momentum was beginning to build up from the east in the direction of dead rear walls.

Toyoshima stated unequivocally in his 1986 paper, 'A control room should be designed for the live front wall and the dead rear end'. 'To suppress … standing waves at low frequencies, the rear wall must be fully absorptive'. 'If the rear wall is reflective, the wall must be designed to provide a high degree of diffusion. However, for 85 Hz, for example, the dimensions of the diffusing members must be as large as 4m, comparable to the wavelength of 85 Hz, which is practically impossible'.

Hidley was using absorber systems of about 1.2 m on the rear walls, and Toyoshima was claiming to have designs that would provide adequate absorption in a depth of approximately 60 cm. The decay time versus frequency for a typical Toyoshima room is shown in Figure 13.5. A typical wall construction is shown in Figure 13.6. By way of contrast, the innards of a large Hidley Non-Environment room are shown in Figure 13.7.

Thus, a clear division had arisen between people such as the Davises, who favoured *live* rear ends in control rooms, and Hidley, Toyoshima and others who opted for maximally *absorptive* rear walls. Effectively the option was becoming between 'Live-End, Dead-End' and 'Dead-End, Live-End'. But what they all agreed upon was that, for best results, control rooms should have directional acoustics whose properties depended on source positions, and not a generally diffuse sound field of uniform decay time, *irrespective* of source position. Differences in opinion about which end should be live and which should be dead continue to the present day, and each philosophy has its followers. The underlying psychoacoustic philosophies will be discussed in detail in Chapters 16 and 17, when the concepts and constructions will be outlined by experienced practitioners specialising in one or other of the two points of view.

It must be appreciated, though, that the room concepts must be well understood if they are to be applied to their greatest effect, and that the principles involved are not necessarily interchangeable. Each one is a system in itself. It is the misapplication of many of the components of these philosophies which has led to some rather poor monitoring conditions. Nonetheless, some aspects of the designs are more or less common to all of the rooms described above, and it is worthwhile looking at some of these points.

13.2 Built-in Monitors

In the top-level control rooms it is general practice to build the monitor loudspeakers into the front walls. The flush mounting of the loudspeakers ensures that all the sound radiates in a forward direction. This avoids the low frequency problems associated with free-standing loudspeakers, where the rear-radiation strikes the wall behind the loudspeaker and returns to the listener with

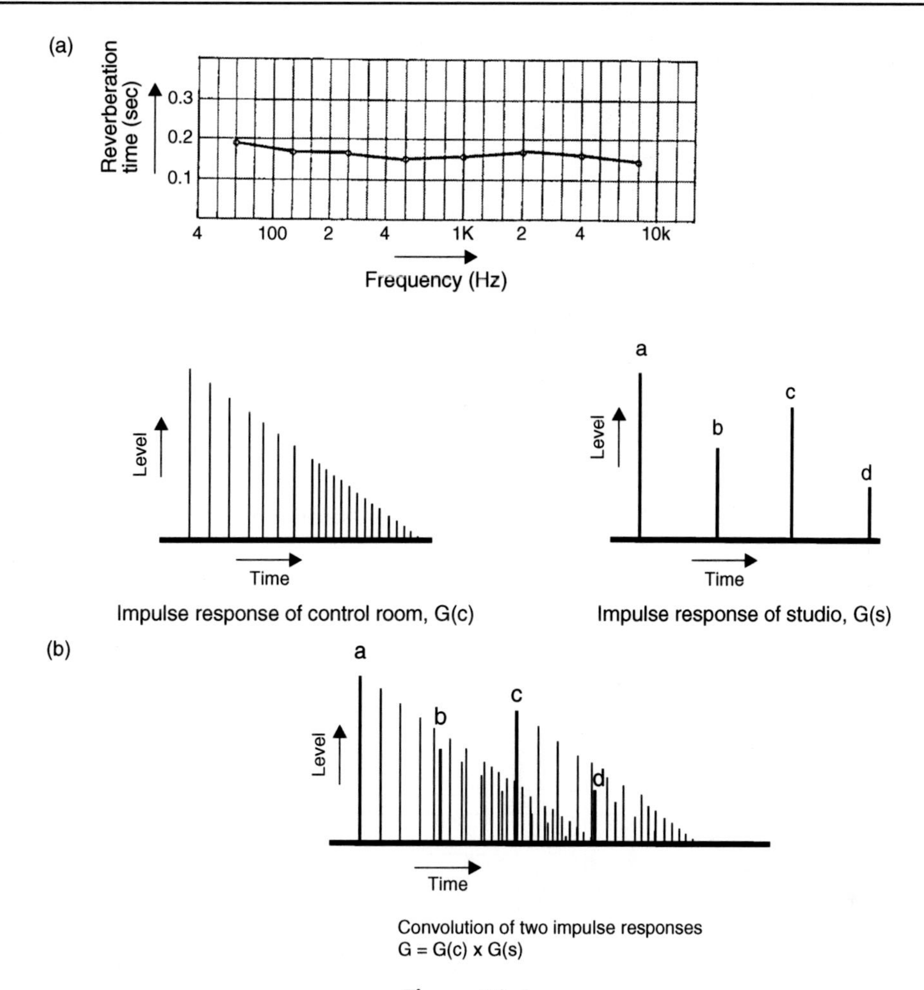

Figure 13.5:
(a) Example of reverberation characteristic of a Toyashima control room; (b) Toyashima's concept of the need for short decay times in control rooms. Relationship between impulse responses of control room and studio. The control room decay should be more rapid than that of the studio, and should generally be lower in level than the ambience of the recorded music to avoid masking caused by the superimposition of the two responses.

a delay, giving rise to response irregularities due to the relative phase differences of the direct and reflected waves. None of the current control room concepts has front walls that are sufficiently absorbent at low frequencies to avoid such effects from free-standing loudspeakers, although the Walker rooms of Figure 12.5 do go some considerable way towards limiting the damage.

Incidentally, flush mounting is sometimes, wrongly, described as 'soffit' mounting. In many 1970s designs, the loudspeakers were flush mounted above soffits – 'soffit' being the architectural term for the underside of an overhang, or the ceiling of a recess. (The Italian word

(a)

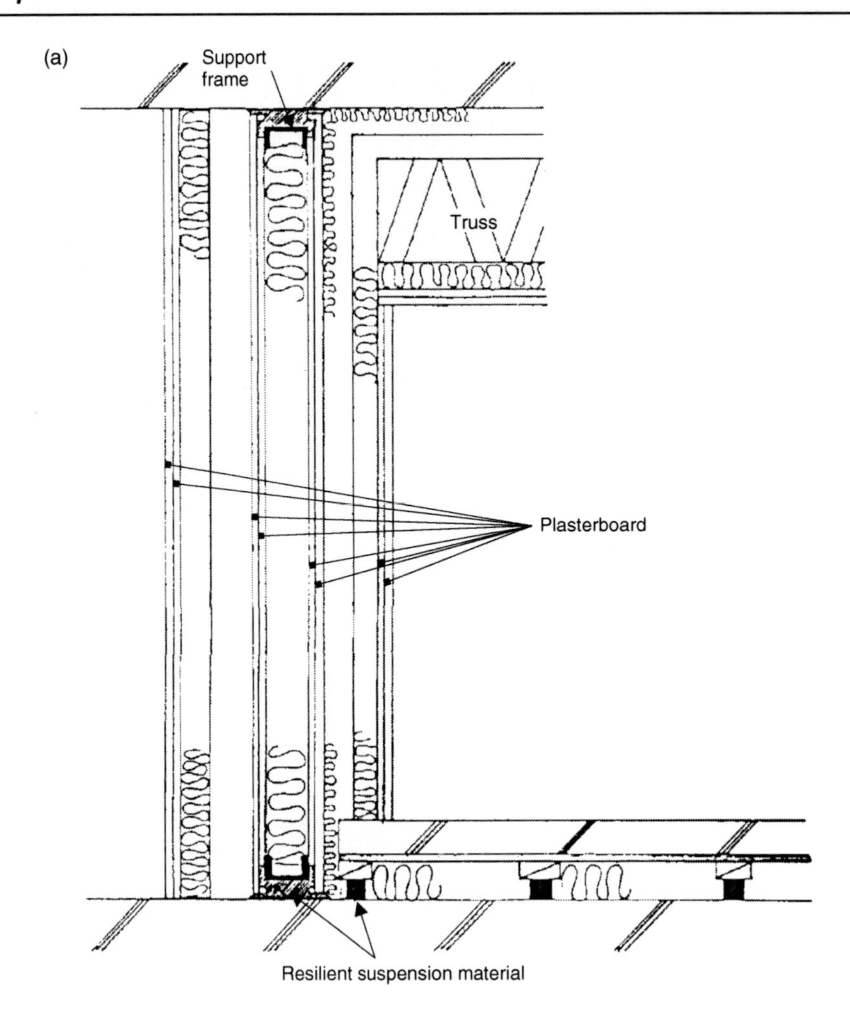

(b)

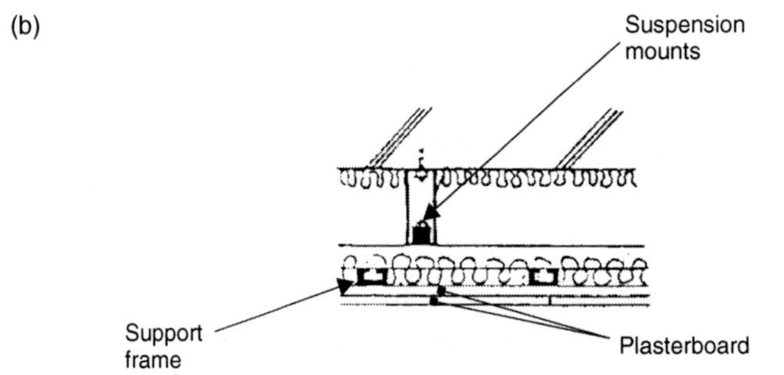

Figure 13.6:
(a) Construction of ceiling, floor and wall in the style of Sam Toyashima; (b) Alternative ceiling construction; (c) The Sam Toyashima designed control room in Studio 4 at the Townhouse, London, UK.

(c)

Figure 13.6: (continued)

for ceiling is 'soffitto'.) These overhangs were often built above windows, or to create recesses in which to house tape machines. This was never a good idea from an acoustic point of view because the recesses could create resonant cavities. They also allowed a certain amount of rearwards sound propagation below them, leading to the problems described in the previous paragraph. Nonetheless, the concept was widespread, and the loudspeakers were mounted above soffits. Somehow, by erroneous extension, the term 'soffit mounting' has come into rather common usage, especially in the USA, referring to flush mounting in general. Figure 13.8 shows a genuinely soffit mounted loudspeaker, whereas the photographs in this chapter all show flush mounted loudspeakers. Anyhow, loudspeaker positioning is a source of variability of room response, so it is a good idea for serious control rooms to standardise on a monitoring practice which has more in common from room to room, and flush mounting fulfils this purpose. The benefits of flush mounting are many.

Flush mounted loudspeakers drive a room from a boundary, as shown in Figure 4.6(c) so there is no rear radiation to add to the general confusion in the room as shown in Figure 4.6(d). Flush mounted loudspeakers, with well-designed, smooth, unobstructed front baffles, also significantly reduce diffraction problems because there are no edges to act as secondary diffraction sources. Flush mounted loudspeakers enjoy an extended baffle plane, which means that even when driven from the boundary of a hemi-anechoic chamber, the radiation impedance on the loudspeakers is increased because of the constrained angle of radiation, as explained in Chapters 11 and 14. This can mean a greater low

(a)

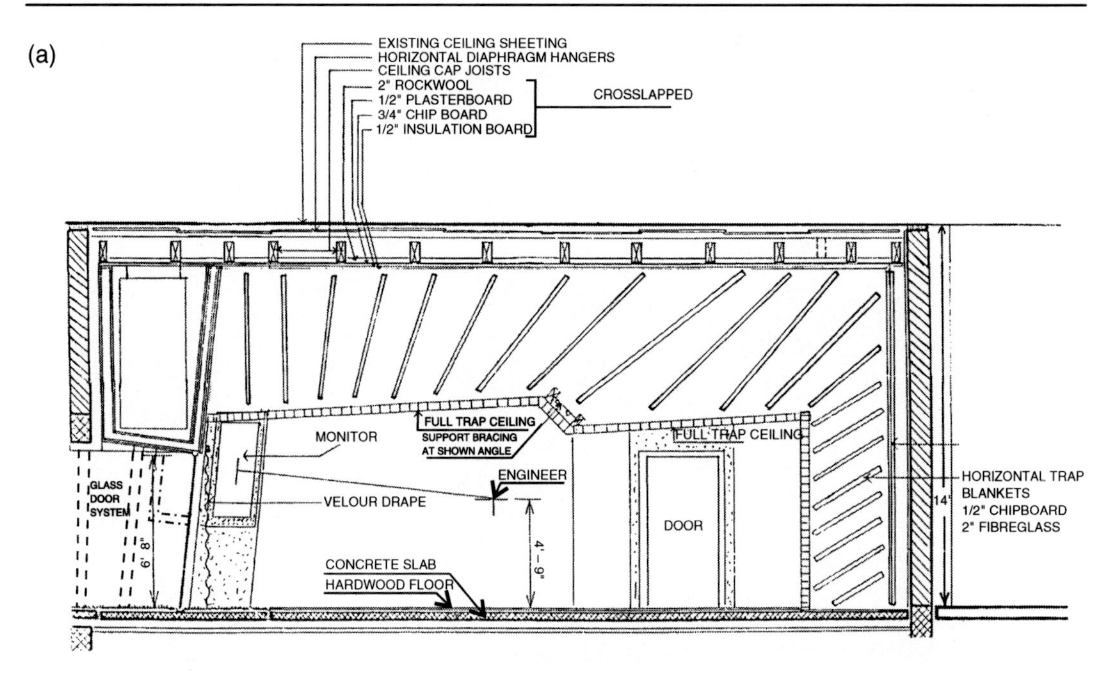

(b)

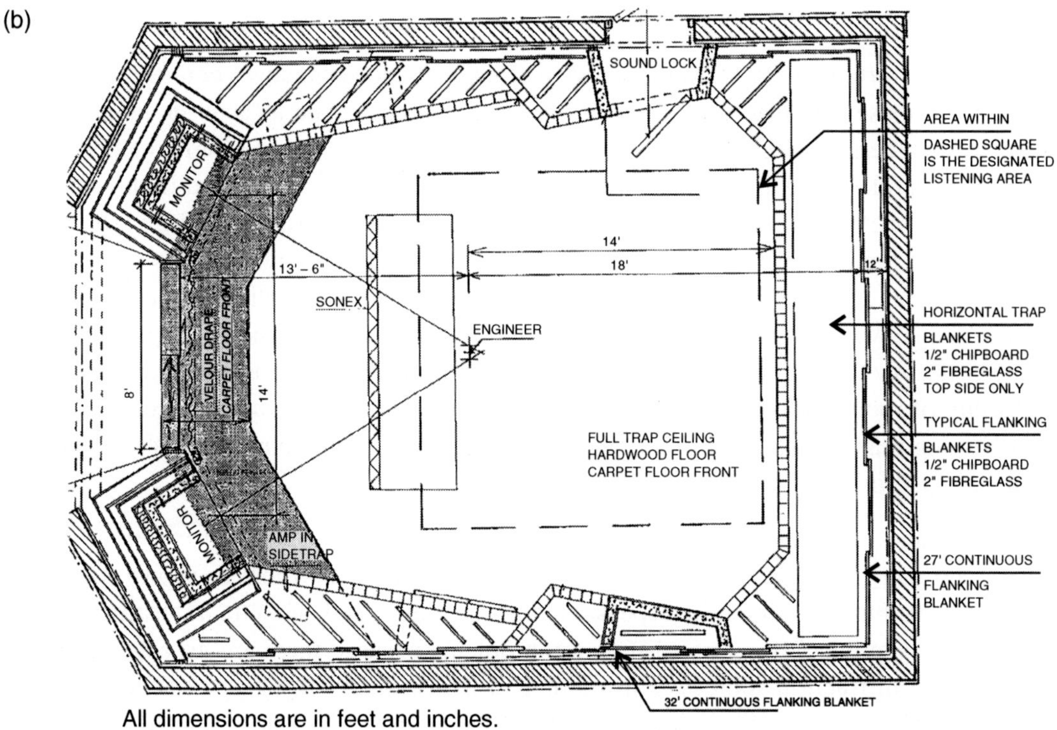

All dimensions are in feet and inches.

Figure 13.7:
Extreme control: (a) Elevation; and (b) plan of a typical 'Non-Environment' absorbent control room (courtesy of Tom Hidley).

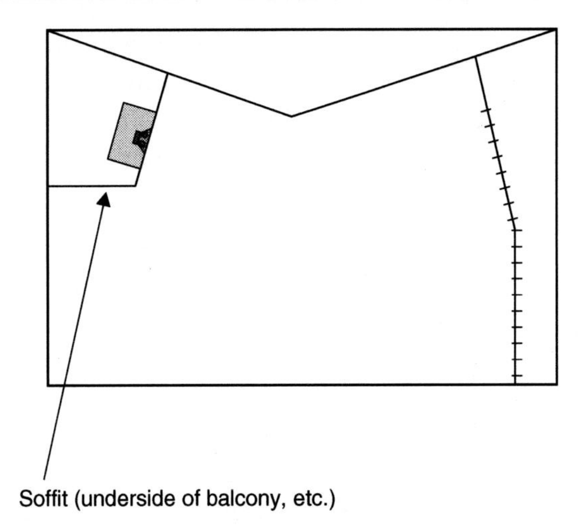

Figure 13.8:
Soffit-mounted loudspeakers. In the example shown, the loudspeaker is mounted above the soffit. The alcoves below are often used to house tape recorders. In general, the idea is not too good, both because the high mounting of the loudspeaker causes the high frequencies to arrive at the ear at a less than optimal angle, and because the alcove below can become resonant (the Italian word for ceiling is 'soffitto').

frequency acoustic efficiency of between 3 and 6 dB compared to the same loudspeakers free-standing. Therefore, for the same SPL in the room, only a quarter to a half of the electrical drive power is needed. The reduced demand from the amplifiers results in a big cut in the working temperatures of the voice coils, which tends to both reduce power compression and distortion, and improves the transient response of the system at high SPLs. Essentially, for serious monitoring purposes, there is no real alternative to flush mounting if the flattest response is required.

Flush mounting also allows the cabinets to be very solidly mounted in a massive front wall, of which they may form an integral part. This type of mounting, if correctly done in order to avoid structural transmission of the sound (essentially by using heavy, rigid boxes in a heavy, rigid wall – see also Chapter 20), can improve stereo imaging by reducing to a minimum any unwanted response artefacts due to the cabinets physically moving under high levels of low frequency drive. This can be of benefit to the overall system transparency and to the stereo imaging.

13.3 Directional Acoustics

All the leading philosophies of stereo control room design are directional inasmuch as the rear half of the room is acoustically different from the front half. It has been found, almost universally, that such a practice is necessary for rooms to exhibit the most detailed stereo images, smoothest pressure amplitude responses, and best low level perception (at least whilst maintaining a pleasant acoustic to work in), but this poses a dilemma for the design of surround-sound rooms. The two front halves of good stereo rooms cannot simply be built facing each other, as this would leave the loudspeakers facing inappropriate far boundaries; generally, the front walls are solid, which rear walls tend not to be. The 'mirror imaged front-half approach' led to many poor designs of quadrophonic rooms in the late 1970s. Good surround rooms *are* beginning to emerge,[8–11] but the design concepts are still quite contentious. Chapter 21 will deal with this subject in much greater depth.

13.4 Scaling Problems

Size differences affect room responses greatly. Wavelengths in air, at room temperature, are functions of frequency and the speed of sound in air, and the physical requirements of dealing with sound propagation remains fixed to the wavelengths which do not scale with room size. Wavelength-to-frequency ratios remain fixed because the speed of sound remains fixed. Therefore, it follows that a small room will exhibit a greater reflexion density than a larger one of similar nature, because the sound will encounter more boundaries per second in a smaller room. What is more, in smaller rooms, the first reflexions will return with more energy because they have travelled less distance. Therefore, when control rooms are designed, the physical dimensions of the room dictate some of the treatments needed. It thus becomes impossible simply to scale many of the room philosophies beyond certain limits. A LEDE room depends on the existence of reflexion free zones, and a suitable time lapse between the arrival of the direct signal and the first returning energy from any room boundaries. If the room is too small, the required time intervals cannot be achieved, and the philosophy will fail to function. The room shown in Figure 12.5 faces the same problems, whereby the reflexions would arrive sooner and higher in level if the room size were to be reduced; hence some of the prime design targets could not be met.

Non-Environment rooms, however, do not depend upon such carefully timed psychoacoustic phenomena, but they do require considerable space for their absorber systems. The absorber sizes are wavelength related, so in very small rooms, especially if the structural walls are heavy and rigid, and hence highly reflective, there may be insufficient space in the room to allow a useful working area if the necessary size of absorbent trapping is installed. In other words, they could become all absorber – no room! At best, a telephone kiosk size area may result from a 3 m cube.

Very small listening rooms of any design will produce problems due to effects such as the ones mentioned above, but they can also complicate issues by imposing additional loading on the fronts of the loudspeakers. The designs of most loudspeaker cabinets assume a relatively 'free air' loading on the outer face of the diaphragms. In small rooms, especially in cases where the low frequency absorption is less than optimal, reflexions can return from the room boundaries with considerable energy, as they have travelled much shorter distances than in larger rooms. Their superimposition on the loudspeakers can affect the performance of low frequency drivers and, more especially, any tuning ports. Pressure zone loading may also have implications here.

13.5 The Pressure Zone

This subject was briefly touched upon in Chapters 5, 6 and 11 (Sections 5.8.3, 6.10 and 11.2.3). Below the frequency of the lowest mode that any room can support, there lies a region known as the pressure zone. Transition into the pressure zone is gradual, and depends to some extent on the Q of the lowest mode. Within the pressure zone, travelling waves cannot exist, and the whole room is pumped up and down in pressure as the diaphragms of any loudspeakers move forwards and backwards respectively.

Myths and misconceptions abound with regard to loudspeaker response in the pressure zone. Some textbooks say that below the pressure zone frequency, also known as the *room cut-off frequency*, '... the environment loads any sources (such that often) the effect of this loading is to reduce the ability of the source to radiate sound into the rooms, and so results in reduced sound levels at these frequencies'.[12] Other texts state 'In rooms that are both physically and acoustically small, the pressure zone may be useful to nearly 100 Hz'.[13] Elsewhere, it can also be found in print that control rooms, for example, should have their longest dimension at least equal to half a wavelength of the lowest frequency to be reproduced in the room. For a '20 Hz room' this would be around 8.6 m if the lowest useful axial mode was needed. The implication often seems to be that in the pressure zone there will be no significant loudspeaker response. Indeed, the concept is reinforced by the use of the term 'room cut-off frequency' to mark its upper boundary. The low frequency cut-off can be calculated simply from:

$$f_{\text{cut-off}} = \frac{\text{speed of sound in metres per second}}{2 \times \text{longest dimension of room (in metres)}} \qquad (13.1)$$

The origin of the concept of requiring the loudspeakers to face the longest dimension of the room, although it is not strictly true, is clearly shown in Figure 4.17. In this figure, the modal support of the combined loudspeaker/room response is considerable, showing almost 10 dB of 'gain' over the anechoic response. It can also be seen that the response below the lowest mode falls back to the level of the rolling-off anechoic response. Given that at 30 Hz the perception of a reduction in SPL of only 4 dB can be that of a halving of loudness (especially below around 85 dB SPL), then the typical 6 to 10 dB reduction, as shown in Figure 4.17 between the modally supported response and the pressure zone (dashed) response, would result in the pressure zone response being perhaps only about one-sixth as loud as the modally supported response. These frequencies would, in fact, tend to be inaudible in a modally supported room at sub-80 dB SPL listening levels.

Obviously though, in rooms in which the modes are substantially damped, the difference between the modally supported and pressure zone regions will be greatly reduced. What is more, half space (2π) loading, such as is approximated by flush mounting the loudspeakers in a heavily damped (absorbent) room, will often serve to reinforce the output in the dashed line roll-off region of Figure 4.17, thus reducing the difference even more.

As for the concept of the pressure zone loading the loudspeakers, there are many things to consider. Perhaps if the room was *very* small, hermetically sealed, and possessed infinitely rigid walls, the statement would be true. Taken to an extreme, if the room were to be the same size as the cabinet loading the rear of the loudspeaker, then the room loading effect could be readily considered, because it would be *equal* to the cabinet loading. The reality is that the loading effect of the room is inversely proportional to the room/cabinet volume ratio. This means that a room with a cubic volume only ten times that of the loudspeaker cabinet would have only one tenth of the loading effect of the cabinet, which is hardly significant.

The mechanisms at work, here, are very interdependent. Below the resonance of a sealed box loudspeaker, the velocity of the cone falls with decreasing frequency in such a way that exactly offsets the tendency for the cone excursion to increase with falling frequency for any given input power. The effect is that the displacement stays constant with frequency. In the pressure zone, the pressure is *displacement dependent*, so the response of a sealed box whose resonance was equal to or higher than the frequency of the room cut-off (pressure zone frequency) would exhibit a flat response in the pressure zone. If the loudspeaker resonance is significantly *below* the pressure zone frequency, then the response will tend to rise with falling frequency until the resonance is reached, when the response will flatten out. Ported loudspeakers will tend to exhibit a 12 dB/octave (second order) roll-off below resonance (as opposed to their normal 24 dB/octave [fourth order] roll-off) when the port resonance is below the pressure zone frequency, although loading effects on the ports can complicate matters.

However, all of this assumes that the room is sealed and highly rigid. Non-rigid (diaphragmatic) walled rooms, such as are found in many timber-framed houses and most purpose-designed sound control rooms, present a totally different set of circumstances. Most timber-framed structures are, to varying degrees, transparent to low frequencies. The pressure zone onset is then defined by whatever high rigidity containment shell exists without the inner structure. A flimsy garden shed may well equate to a 2π space at very low frequencies, being *rigidly* bounded only by the floor on which it stands. The classic pressure zone equation:

$$f_{pz} = \frac{c}{2L_R} \tag{13.2}$$

(which is merely another form of Equation 13.1) is therefore unlikely to yield an accurate representation of the upper limit of something sudden happening at the entry to the pressure zone; or at the lower limit of the room's normal response – the room cut-off frequency – whichever way one chooses to view it.

Clearly, in a very small room where the pressure zone *may* exist up to 100 Hz (a 1.5 m cubed control cubicle, for example), the first modes would need to be very heavily suppressed if they were not to introduce severe colouration on their onset. The situation in rooms in general, however, is highly variable, and a very thorough understanding of the structural details would be needed before any accurate pressure zone response prediction could be made.

Nevertheless, one or two statements can be made definitively. Dipole radiators, such as many electrostatic loudspeakers, *cannot* radiate sound below the room cut-off frequency. Only at very small distances from the front or rear of the diaphragm would any sound be perceptible. A dipole source (see Chapter 11) would merely paddle back and forth, without creating any net change in the overall pressure within the room. The short circuit round the sides of the diaphragm would cancel out (except in the very close field) any local pressure changes due to the opposite polarity that would exist on the front- and rear-radiating surfaces – see Figure 11.3(a).

The very fact that dipole loudspeakers are an exception in itself implies that the room cut-off, as it is known, does *not* mean that conventional sealed box or ported loudspeakers cannot radiate useful output below that frequency. In fact, one characteristic of the response in the pressure zone is that the perceived response not only exists, but it is also extremely uniform, both in level and with respect to position, because no modes exist to create spectral or spacial variations. Nonetheless, a relatively flat extension of the low frequency response into the pressure zone can only be achieved in very damped (absorbent) rooms.

For these reasons, no loudspeaker system can be absolutely optimised for low frequency performance without knowledge of the room in which it will be used. Good monitor loudspeakers must perform in real control rooms, and fulfill the job for which they were designed. Good domestic loudspeakers must give pleasing results in less controlled circumstances. And as was also mentioned in Section 12.2.1, some loudspeakers have different bass alignments depending on the nature of the typical house constructions in their principal markets. There are so many things that a room can do to modify a loudspeaker response that no simple set of idealised test specifications can truly represent a loudspeaker in normal use.

13.6 One System

It should by now be clear that the design of the control rooms and their corresponding monitor systems cannot be done in isolation, because the performance of either one is dependent upon the performance of the other. This is why many studio designers require some degree of control over the loudspeakers to be used in rooms of their design if they are to put their name to the overall result. Directivity, for example, is a loudspeaker characteristic which may dictate how a particular loudspeaker suits a given control room design. Too wide a directivity pattern in a small control room may induce excessive early reflexions. Too narrow a directivity pattern in a large room may make the stereo listening area unworkably small. The vertically aligned 'D'Appolito' layouts of the loudspeaker shown in Figures 13.9 and 13.10 for use in the Non-Environment type rooms have strange vertical directivity. Nevertheless, in the horizontal plane the drivers remain reasonably time aligned wherever one walks or sits in the room, as long as one keeps one's head at a relatively normal height. A highly absorbent ceiling is required for such loudspeakers because with the three drivers operating at the crossover frequency their reflexions from a hard ceiling (especially a low one) would be very much out of step, and could produce transient smearing. The floor reflexions are normally scrambled by equipment.

Laterally spaced drivers in the highly absorbent rooms can lead to a 'phasiness' whilst walking around the room, because the path lengths to the individual drivers change as one moves horizontally. If the upper frequency limit of the parallel drivers is low enough, and thus the wavelengths are long enough, this problem is less consequential, although some such horizontally mounted pairs of loudspeakers operate up to 800 Hz, which is not a good idea in a highly controlled room. In a more lively room, however, the reflected energy will tend to mask this problem. The concept is shown diagrammatically in Figures 13.11 and 13.12.

Figure 13.9:
Capri Digital, Capri, Italy. The control room, designed by Tom Hidley, with the vertically aligned Kinoshita monitor loudspeakers (1990).

Figure 13.10:
The control room at Area Master Studios, (now Planta Sónica 2) Vigo, Spain, designed by the author, showing the vertical alignment of the drivers in the Reflexion Arts loudspeakers (1997).

13.7 Aspects of Small Control Room Designs

Gilbert Briggs, in his book *Loudspeakers*, published in the 1950s, wrote that loudspeakers are somewhat like boxers … 'In general, a good big one will always beat a good little one.' Well, the same can also be said for control rooms; when the size comes down, several things happen which conspire against a neutral sounding room:

1. Reflexions come back from wall surfaces *earlier* than in a larger room, because they have travelled less distance. This also means that they reflect more often, and hence small rooms produce a higher reflexion density than larger rooms.
2. Reflexions come back from wall surfaces *higher in level* than from similar wall surfaces in larger rooms, again because they have travelled less distance. This gives them a greater ability to interfere with the direct signal.
3. There is less space in the room to locate absorbent or diffusive materials or structures, so room effects (which are worse to begin with) will tend to be less able to be controlled than in larger rooms.
4. The resonant room modes (often referred to as standing waves) begin to separate at higher frequencies in smaller rooms, so the reduced ability to absorb or damp the room effects are exacerbated by the greater intrusion of the irregular response range into the low frequency range of the audible spectrum, as shown in Figure 4.15.

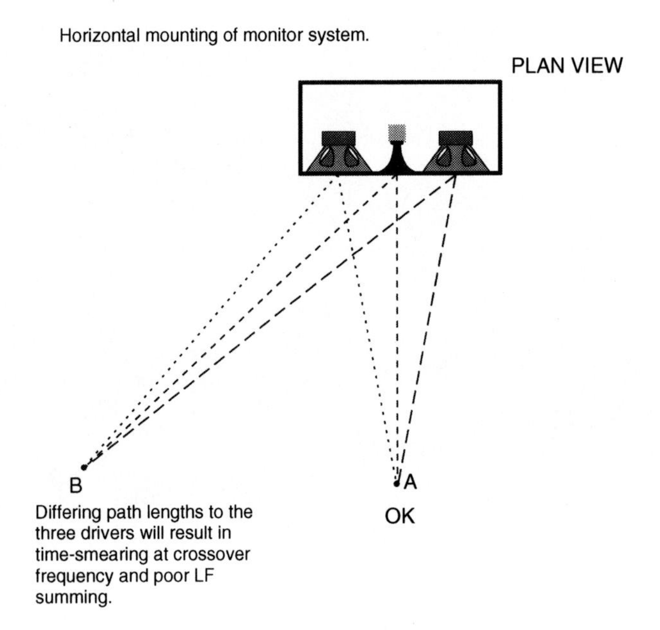

Horizontal mounting of monitor system.
PLAN VIEW
B
A
OK
Differing path lengths to the three drivers will result in time-smearing at crossover frequency and poor LF summing.

Figure 13.11:
Driver position considerations. At position A, all is well. The distance to each of the three loudspeakers is more or less equal. At position B, however, the distance to each loudspeaker is different. The path length to the left-hand bass driver is less than that to the horn, whereas the path length to the right-hand bass driver is longer. A sound emanating simultaneously from any two or more of the drivers will not arrive simultaneously at position B, so a flat frequency response would not be possible due to the phase shifts involved, unless, that is, the crossover frequency was so low that the wavelength at that frequency was significantly greater than the width of the box.

5. It becomes more difficult to work at a suitable distance from wide range, extended low frequency monitor loudspeakers, sufficient to avoid geometrical effects where the multiple loudspeaker drive units fail to gel into one integrated source … Highs can be heard coming from the tweeters, lows from the woofers, and small movements of the head cause changes in the perceived responses. (See Section 12.7.)

What should be clear from the above five points is that small rooms, in many cases even more so than large rooms, will tend to stamp their own individual ‘fingerprints’ on the sound of any given monitor system used in them. There is therefore a strong argument in favour of large control rooms, but economics and practical considerations have deemed that the majority of recordings are now undertaken in studios with relatively small control rooms … 24 to 40 m^2 being very typical. So, it would be useful to look at the above points in a little more detail.

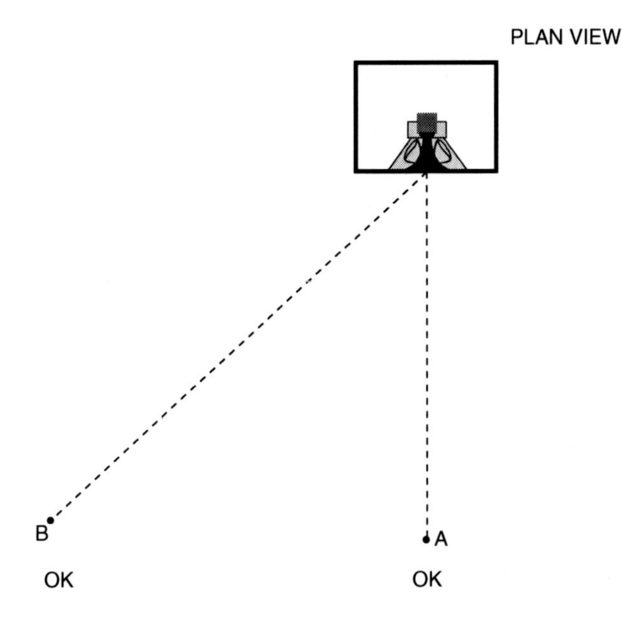

Figure 13.12:
Vertical mounting of drivers. With the drivers mounted vertically, the arrival times from each driver to a listener at position 'A' are essentially identical. If the listener then walks to position 'B', the arrival time from the different drivers remains equal. A flat response is therefore possible at both points A and B, and indeed at all points in between.

Reflexions returning from walls (more quickly and with more energy in small rooms) will tend to colour the sound of the monitors and smear the stereo imaging. Specular reflexions are the worst offenders, so they must be dealt with by means of either diffusion or absorption. In small rooms absorption is the only solution, for whilst diffusion may in some circumstances maintain the stereo imaging over a larger 'sweet' area, it nonetheless returns considerable energy to the listening position, and this *will* colour the sound. Diffusers are *statistically* diffuse, which means that when area-averaged they are equally diffuse at all frequencies within their design range. However, this does not mean that at any given point there will never exist hotspots at certain frequencies. In small rooms, where the listeners are inevitably close to the diffuse surfaces, such hotspots can be problematical. Angus[14] showed that the situation can, at times, be such that reflected energy from diffusers can be even more concentrated than from a flat reflective surface.

Adequate absorption, even if it does not remove *all* the reflected energy, will at least reduce the reflected energy and heavily damp the modes, making them broader in their frequency spread and lower in level, both of which tend towards producing less colouration. When the room is so small that the modal separation and entry into the pressure zone begins well into the lower octaves of the audible range, then heavy damping of modal activity is the only way to achieve a reasonably uniform frequency response.

The third point mentioned in the opening paragraph of this section was that in smaller rooms there was less space to install acoustic control contrivances. In the case of many absorber or diffuser systems, their lower-frequency cut-off points are a function of their depth. In other words, if a system needs, for example, 1 m of depth to be effective down to 70 Hz, then it requires that depth independently of the size of the room. Absorber and diffuser sizes cannot be scaled with the size of the room. These depths are fixed by the wavelengths of the frequencies, if 1m is needed to be effective to 70 Hz in a large room, then 1 m is still needed in a small room. Many studio owners complain when a designer asks to use 1 m depth of absorber on a rear wall when the room is only 4 m from front to back in the first place. Nevertheless, *physics dictates this, not the designer*, so a studio owner resisting the loss of floor space will pay the price of poorer response flatness in the room. Unfortunately, the reality is that to many of them, what they *see* is more important than what they *hear*, which is hardly an attitude that can be called professional, but it is rather widespread.

Point four is related to the scaling problem for absorbers and diffusers. Figure 4.15 clearly shows the identical modal pattern for two identically shaped rooms, but one having dimensions three times larger than the other. The modal patterns are a function of the relative room *dimensions*, i.e. the shape, and very little can be done to change this state of affairs. The perceived room response usually tends to follow the envelope of the modes and the spaces between them, so where the modes are closely spaced, a generally smooth response will be perceived. When the separation begins in the audible low frequency range, as is always the case in small rooms, the individual frequencies that coincide with the modes will be heard to ring on in the room, giving it a 'boomy' character on some notes, but not on others.

It is relatively futile to try to change these modal frequencies by angling the walls. Certainly, a small angle to avoid parallelism helps to reduce objectionable flutter echoes or their associated colouration, but as can be seen from Figures 5.7 to 5.10, even an angling of one wall by 1.5 m, between surfaces 10 m apart, has almost no effect on the pressure distribution at 70 Hz, and so will not change the modal frequencies significantly. The only viable solution is to damp the modes, by using the maximum practicable quantity of low frequency absorption on the side walls, the rear wall and the ceiling.

The last of the problems mentioned was the geometric consideration, relating to the spacial separation of the various drive units in a typical monitor loudspeaker. No studio for music recording can seriously call itself professional unless it has adequate full-range monitoring. Relying on small monitors that roll off rapidly below 60 Hz is not a professional approach. However, due to the physics of electroacoustics, loudspeakers which go down to 25 Hz, or thereabouts, and which can produce an adequate sound pressure level, tend to be relatively large. In a small room, large loudspeakers at short distances tend not to have flat responses, and the response tends to change with small movements of the head, which can be misleading.

In small rooms, it seems essential to flush mount the loudspeakers in the front wall of the room. This puts them at points of maximum distance from the listening position. It also avoids the low frequency response irregularities caused by the rear radiating waves reflecting from the front wall (from free-standing loudspeakers). These, as we have already seen, recombine with the direct signal with a phase relationship that causes peaks and dips in the combined response. In fact, flush mounting is the only realistic option for high-quality monitoring other than in free-field (anechoic) conditions. Otherwise, a maximally flat response cannot be achieved. Somewhat perversely, small control rooms tend to be the ones in which one is *least* likely to find flush mounted monitors, which is a pity, because well-designed small rooms *can* sound very even.

When looking for a site for a suitable control room, or when designing a building to house one, there are certain things which can be helpful towards later acoustic control. A ceiling height of at least 4 m is almost essential, because floors tend to be flat and hard. They are thus highly reflective to low frequencies, and must be opposed by a ceiling that is capable of dealing with the floor/ceiling modal activity. A floated isolation floor, which is usually required, may need around 20 or 30 cm, and the ceiling of the usual type of acoustic shell could require another 40 cm of height. If a finished ceiling height of 2.4 m was the goal, with a further 15 cm being needed for recessed lighting, then only 85 cm would remain from a 4 m original headroom for mounting acoustic devices to break up the floor/ceiling modes. Remember, when a room is already small, the ceiling height can be used to great effect to control the overall acoustics without losing too much floor space.

Rear walls, of necessity, do tend to need to take up a considerable area of floor space, because effective absorption (or diffusion) requires significant depth in order to deal with the lower frequencies. Somewhat like the floor/ceiling problem, the rear walls face front walls, which almost all tend to be reflective at low frequencies, so they must be adequately treated if a flat response is the objective.

13.7.1 Conflicting Requirements

One of the aspects of small control room construction that creates many problems is the need for sound isolation. If the need for sufficient isolation means the creation of a containment 'bunker' that is hard and heavy, then the bunker will be achieving its isolation by reflexion. External noises are kept *out* by reflexion, and internal noise is kept *in* by reflexion. The more reflective the containment shell, then the more the absorption that must be provided within the room to keep things under control, and space for this can be very hard to find in small rooms. In the rare cases where little isolation is required, then internal acoustic control tends to become much easier to achieve, because all of the sound that escapes from the room is effectively absorbed by the walls. Transmission through them or absorption within them leads to the same results – less energy reflected back into the room, hence better acoustic control.

These are some of the problems that seriously militate against the achievement of excellent monitoring conditions in small rooms. This is not to say that the objective cannot be achieved, but it can rarely be produced inexpensively, either in terms of money or of lost space. Again, somewhat perversely, money and space are often the two things that many people who want small control rooms do *not* have at their disposal, which leads to an enormous number of bad small control rooms.

13.7.2 Active Absorbers

Whether the future will see solutions to the small room problems is yet to be seen, but there is the possibility that we could soon see active absorber systems which will take up much less space than conventional passive absorption systems. Active absorbers use driven loudspeakers as energy sinks. The acoustic energy from the sound in the room is converted into heat in the voice coils of the absorber loudspeakers. At high frequencies, the technique is fraught with problems, but at low frequencies, practicable systems could be on the horizon. This sort of technology is rapidly advancing.

13.8 A Short Overview

Rooms form the final link in the sound chain from the loudspeaker to the ear. They are also the most variable links in the chain, and are the most difficult to standardise. In rooms with poor acoustics, no loudspeakers can be expected to perform well. Given rooms with reasonably controlled acoustics, the general tendencies are for rooms with some acoustic 'life' to sound more musical, and in some cases to support stereo images over a somewhat wider listening area. Very dead rooms allow perception of timbre and detail that is more accurate, but they are sometimes considered unsatisfactory for the enjoyment of acoustic music. They do, however, allow very precise stereo imaging and great clarity in the designated listening positions, and thus have their places in situations where quality control and attention to detail are important.

As highly damped rooms receive no 'support' from reflexions, they tend to require more power output capability from the amplifiers and loudspeakers as compared to more lively rooms of comparable size. On the other hand, because they contribute very little to the total perceived sound field, they are generally more consistent in their performance.

Considering the wide range of sound control room concepts and the great weight of experience that has been applied to their designs, the continuing existence of such variability of implementation suggest that there is no simple solution to the problem of room standardisation which is consistent with the provision of *all* the desired acoustics. Of course, the desired acoustic can be very personal – and most of the generally accepted control room philosophies have their partisan followers. Somewhat similarly, there is not one type of tennis racquet which is used to win all championships. Different weights, different string tensions and different designs suit different styles of play. No single room environment has been shown to be 'correct' for all purposes.

13.9 Summary

Control rooms should have a clarity of monitoring that is always one step ahead of the consumer systems.

Studio monitors are designed according to a different hierarchy of priorities compared to domestic hi-fi loudspeakers.

Most high-quality stereo control room designs rely on directional acoustics. The front and rear halves of the rooms are acoustically different.

Different designers have different opinions as to whether the front half or the rear half of control rooms should be reflective, and whether to use absorption or diffusion as the principal means of acoustic control.

For the best overall response, monitor loudspeakers should be flush mounted in the front wall. The increased reflexion density in small rooms makes adequate control all the more necessary. Flush mounting helps the low frequency control.

Because absorber and diffuser sizes are wavelength related, control can become difficult in small rooms which lack the available space for treatment.

Loudspeaker responses in the pressure zones are subject to many influencing factors.

For optimum performance, a loudspeaker and a room should be designed as one system.

At close distances, multi-way loudspeakers may not gel into a cohesive source.

At close distances, diffusers may produce reflective hot-spots.

References

1 Rodgers, C. A. (Puddie), 'Pinna Transformation and Sound Reproduction', *Journal of the Audio Engineering Society*, Vol. 29, No. 4, pp. 226—34 (April 1981)

2 Davis, Don and Davis, Chips, 'The LEDE Concept for the Control of Acoustic and Psychoacoustic Parameters in Recording Control Rooms', *Journal of the Audio Engineering Society*, Vol. 28, No. 9, pp. 585–95 (September 1980)

3 Haas, H., 'The Influence of a Single Echo on the Audibility of Speech', *Acoustica*, Vol. 1, p. 49 (1951). Reprinted in the *Journal of the Audio Engineering Society*, Vol. 20 (1972). This work was also investigated by Wallach, H. et al., *American Journal of Psychology*, Vol. 62, p. 315 (1949)
4 D'Antonio, P. and Konnert, J. H., 'The RFZ/RPG Approach to Control Room Monitoring', presented at the 76th Audio Engineering Society Convention, Preprint No. 2157, New York, (1984)
5 Newell, P. R., Holland, K. R. and Hidley, T., 'Control Room Reverberation is Unwanted Noise', *Proceedings of the Institute of Acoustics*, Vol. 16, Part 4, pp. 365–73 (1994)
6 Hirata, Y., Matsudaira, T. K. and Nakajima, H., 'Optimum Reverberation Times of Monitor Rooms and Listening Rooms', presented to the 68th Convention of the Audio Engineering Society, Hamburg, Preprint No. 1730 (March 1981)
7 Toyashima, S. and Suzuki, H., 'Control Room Acoustic Design', presented to the 80th Convention of the Audio Engineering Society, Montreux (Switzerland), Preprint No. 2325 (1986)
8 Walker, Robert, 'A Controlled Reflection Listening Room for Multi-Channel Surround', *Proceedings of the Institute of Acoustics*, Vol. 20, Part 5, pp. 25–36 (1998)
9 Holman, Tomlinson, *5.1 Surround Sound: Up and Running*, Focal Press, Oxford, UK (1999)
10 Various authors, 'Changing Rooms', *Studio Sound*, Vol. 43, No. 2, pp. 54–60 (February 2001)
11 Various authors, *The Proceedings of the AES 16th International Conference, Spatial Sound Reproduction* (1999)
12 Howard, David M. and Angus, James, *Acoustics and Psychoacoustics*, 2nd Edn, p. 301, Focal Press, Oxford, UK (2001)
13 Davis, Don and Davis, Carolyne, *Sound System Engineering*, 2nd Edn, p. 209, Focal Press, Oxford, UK and Boston, USA (1997)
14 Angus, James A. S., 'The Effects of Specular Versus Diffuse Reflections on the Frequency Response at the Listener', presented at the 106th Audio Engineering Society Convention, Munich (1999). Preprint No. 4938

Bibliography

Adams, Glyn, 'The Room Environment' (revised by John Borwick), Chapter 7 in *Loudspeaker and Headphone Handbook*, 3rd Edn (John Borwick ed.) Focal Press, Oxford, UK (2001)

Davis, Don and Davis, Carolyn, *Sound System Engineering*, 2nd Edn, Focal Press, Oxford, UK and Boston, USA (1997)

Howard, David M. and Angus, Jamie, *Acoustics and Psychoacoustics*, 4th Edn, Focal Press, Oxford, UK (2009)

Kutruff, H., *Room Acoustics*, 4th Edn, E. & F. N. Spon Publishers, London, UK (2001)

Newell, Philip, *Studio Monitoring Design*, Focal Press, Oxford, UK (1995)

Toole, Floyd E., 'Listening Tests – Turning Opinion into Fact', *Journal of the Audio Engineering Society*, Vol. 30, No. 6, pp. 431–45 (June 1982)

Toole, Floyd E., 'Subjective Measurements of Loudspeaker Sound Quality and Listening Performance', *Journal of the Audio Engineering Society*, Vol. 33, No. 1–2, pp. 2–31 (1985)

Toole, Floyd E., 'Loudspeaker Measurements and Their Relationship to Listener Preference: Part 1', *Journal of the Audio Engineering Society*, Vol. 34, No. 4, pp. 227–35 (1986)

Toole, Floyd E., 'Loudspeaker Measurements and Their Relationship to Listener Preference: Part 2', *Journal of the Audio Engineering Society*, Vol. 34, No. 5, pp. 323–48 (1986)

CHAPTER 14

The Behaviour of Multiple Loudspeakers in Rooms

Loudspeakers in different environments. Steady-state and transient response differences. Pan-pot laws. Fold-down compromises. Discrete and phantom image differences. Multiple surround formats. Sub-woofer options.

Chapter 11 discussed many aspects of the performance of loudspeakers in rooms. Principally, it was dealing with single loudspeakers, but a very important aspect of the performance of sound control rooms for music recording is the spacial performance in stereo or surround. This aspect of the performance of loudspeakers is difficult to specify, and in the case of a single loudspeaker, which is usually what loudspeaker specifications relate to, it is currently (2011) impossible to specify. Nevertheless, without knowing the basic behaviour of multiple loudspeakers it will be difficult to appreciate some of the more philosophical discussions that follow in later chapters. Let us now look at the behaviour of multiple loudspeakers in some detail. Firstly, however, we must begin by once again considering mono sources.

14.1 Mono Sources

Figure 14.1 shows a familiar set of response plots for a typical loudspeaker in an anechoic chamber. The individual plots show the response on-axis, and at various angles and directions off-axis. When moving horizontally across the front of such a mono source, it should be evident from the plots that the high frequencies will progressively reduce as one travels further off-axis. If one were to move behind the loudspeaker, then one would not expect to hear much top at all, and a glance back at Figure 11.5 will clearly show why.

If the high and low frequencies are equal in level on-axis, then it should be obvious that in the room as a whole, there will be more low frequency energy, because at no point do the high frequency levels exceed the low frequency levels. Conversely, the lows are present in *many* places at much higher levels than the highs. In fact, the areas of the high and low frequency patterns in Figure 11.5 give a reasonable idea of the relative proportions of the radiated power in each frequency band. Nevertheless, in an anechoic chamber, in an arc of $\pm 15°$ across the frontal axis, a highly uniform response can be achieved.

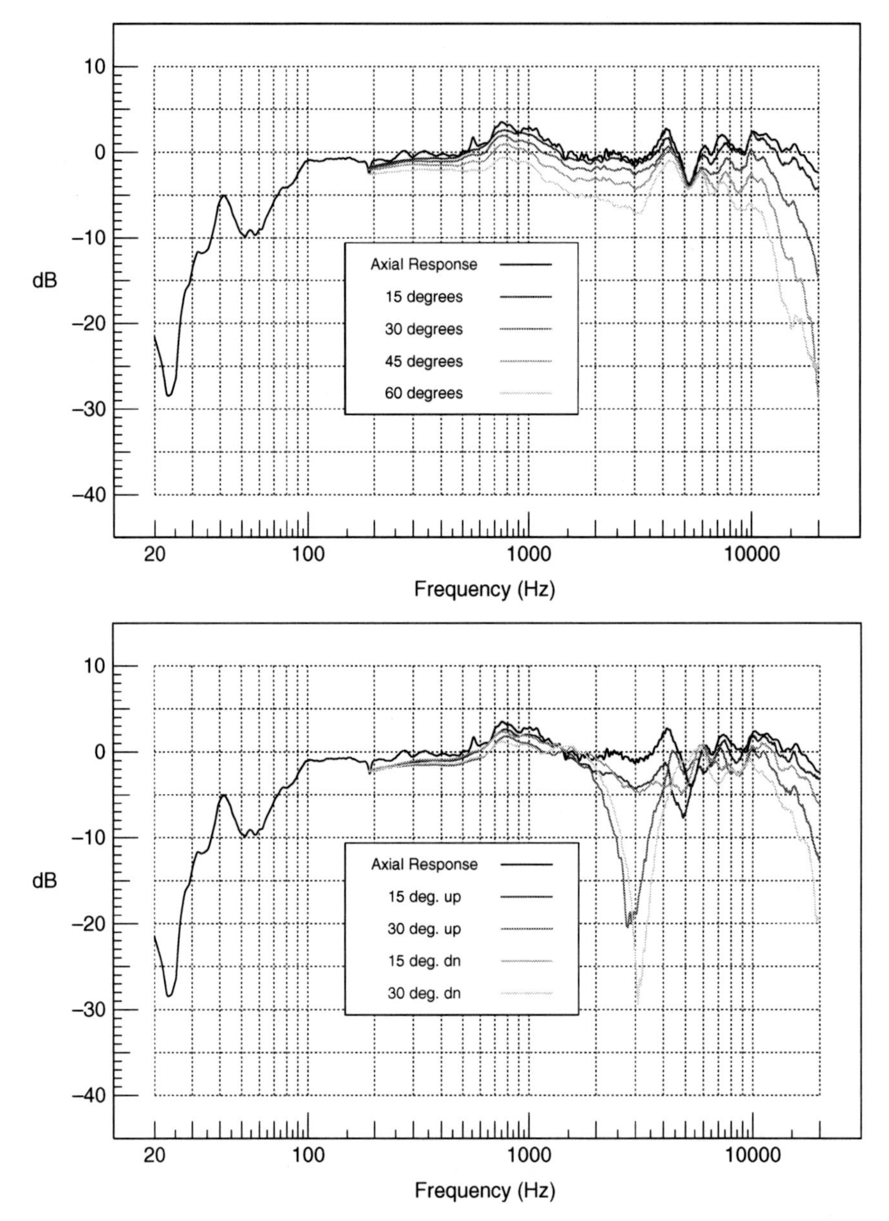

Figure 14.1:
Horizontal and vertical directivity plots of a Tannoy 'Reveal' loudspeaker.

If we now transfer the same loudspeaker from Figure 14.1 into a reverberation chamber, the diffuse field that the chamber gives rise to will serve to integrate all the responses from all the directions. As Figure 11.5 shows, the low frequencies are radiating in all directions, so they should also be radiating more total power into the room. Figure 14.2 bears this out, and shows the significantly higher level of low frequencies that would be perceived in the far

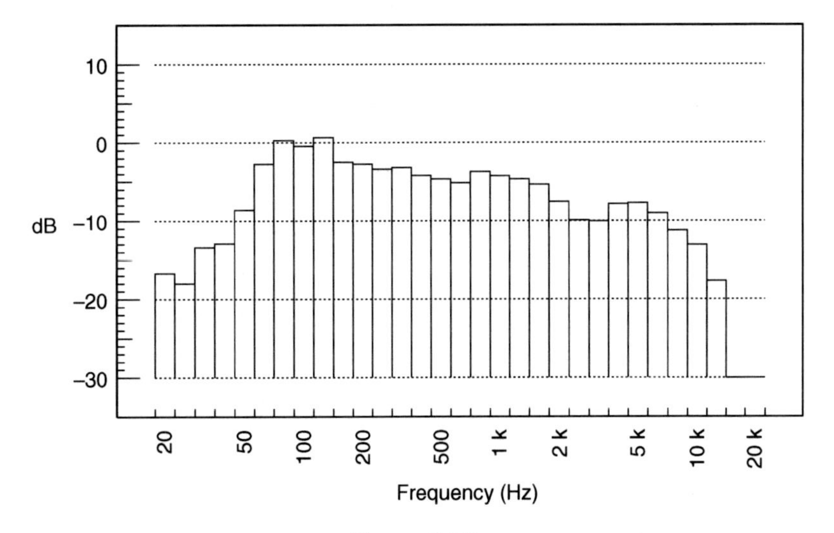

Figure 14.2:
Tannoy Reveal, total power response.

field at any point in a highly reverberant room. In fact, even on-axis at quite short distances, the response would still largely be as shown in Figure 14.2 because the critical distance in such a room, where the direct field is equal in energy to the reverberant field, is very short indeed. (See also Section 11.2.) Figures 14.1 and 14.2 were from actual measurements made by Dr Keith Holland in the anechoic and reverberation chambers, respectively, at the Institute of Sound and Vibration Research (ISVR) in Southampton, UK. All music listening rooms fall somewhere between these extremes, therefore the loudspeaker responses in those rooms will also fall between the same extremes.

14.2 Stereo Sources

Let us now consider two ideal point monopole sources, representing a stereo pair of perfectly flat, omni-directional loudspeakers. When reproducing a central mono image, the combined directivity would be as shown in Figure 14.3(a). For a single, mono source, the radiation pattern would simply be spherical and equal at all frequencies (as shown in Figure 14.3)(b), and thus the frequency response at any point in the room would be flat. Unfortunately, as shown in Figure 14.4, the off-axis frequency response of a phantom central image from a *stereo* source is anything but flat.[1] Considering the directivity plot shown in Figure 14.3 it could hardly be expected to be flat. Figure 14.5 goes one stage worse – a central, in-room image created by four sources in a surround system.[2] It can be seen that the situation can rapidly get out of hand. Only on-axis in an anechoic chamber can a phantom central stereo image mimic a discrete central image from a centrally positioned loudspeaker, and even then only as measured by a microphone, not by ear. In

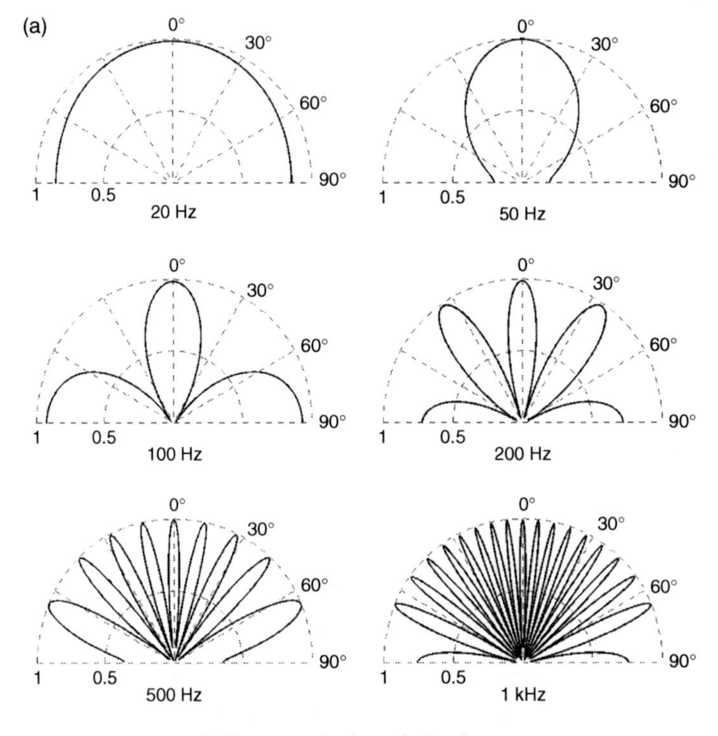

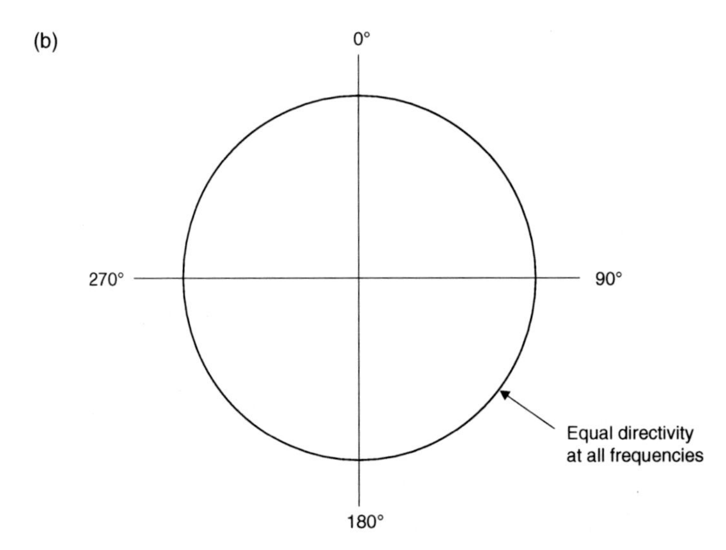

Figure 14.3:
(a, b) Discrete versus phantom source patterns: (a) Polar response of a pair of 'perfect' loudspeakers when producing a central phantom image and spaced 3 m apart.
(b) Polar response of a perfect loudspeaker in an anechoic chamber. The above plot represents the polar response of a perfect, discrete, mono source.

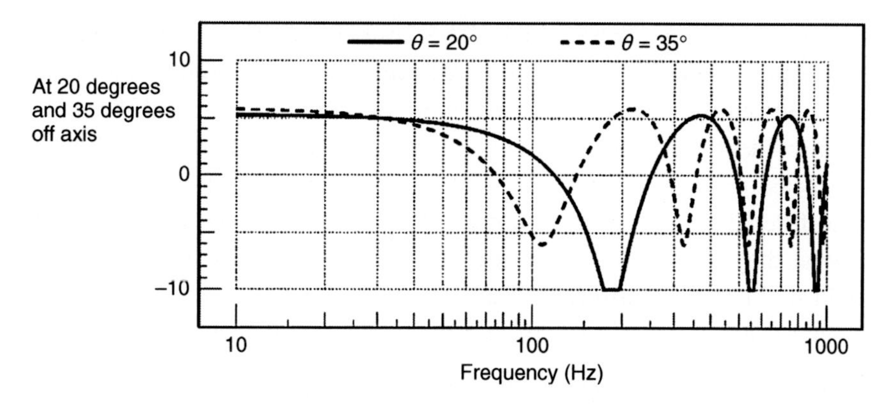

Figure 14.4:
Effect of comb filtering. Off-axis frequency response of a pair of 'perfect' loudspeakers when producing a central phantom image in an anechoic chamber.

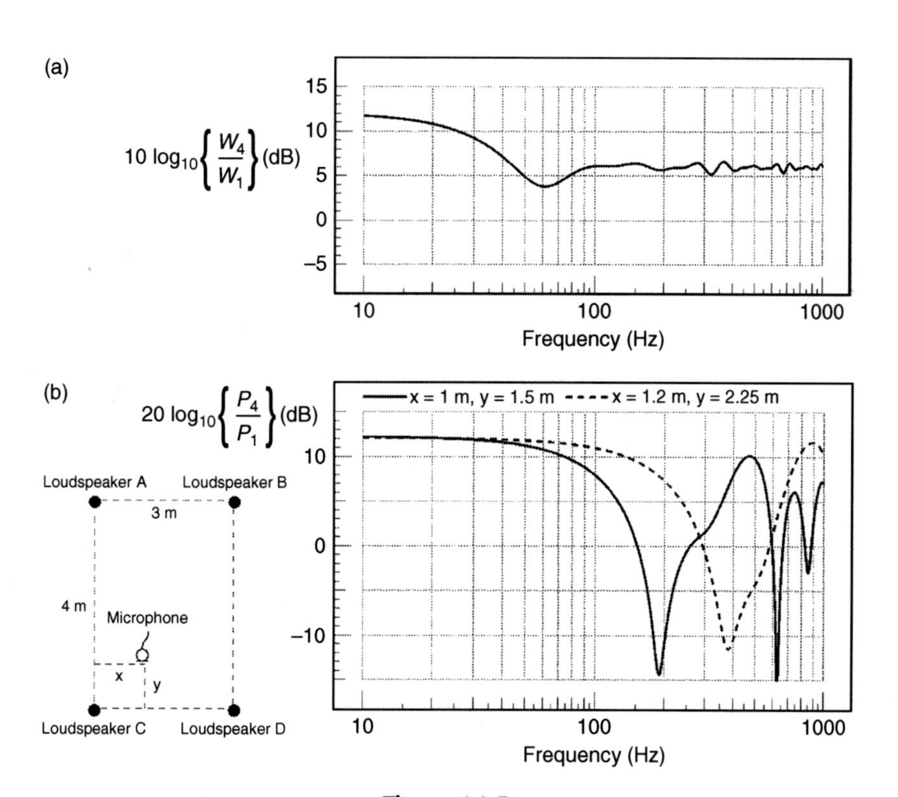

Figure 14.5:
Frequency response of four 'perfect' loudspeakers radiating the same signal: (a) Total power response in a highly reverberant room; (b) Combined frequency response of four omni-directional loudspeakers at two different positions in an anechoic chamber.

a reflective room, this can never be achieved, because all the off-axis radiation will be of the comb-filtered nature shown in Figure 14.4. All reflexions from points other than on the central plane between the loudspeakers will therefore also exhibit comb-filtered responses. It must be self-evident that if a single source radiates a flat response in all directions (we are still speaking about perfect omni-directional loudspeakers here) and a stereo pair radiates comb-filtered responses at all points off the central plane, then the reflexions will return what they receive (given perfect reflectors, of course). This means that the reflective field will exhibit a different frequency balance from a central mono source, such as a single central loudspeaker, than from a phantom central image from a stereo pair.

In rooms with significant modal activity, it must also be remembered that the modal pattern is very much dependent on the point(s) from which the room is driven. A central mono source has one drive point, whereas a phantom central image has *two* points of origin, separated by the distance between the loudspeakers, and *neither* point corresponds with the position of the phantom image. The two means of generating a central image can drive the room very differently, and thus cannot be expected to sound the same at *any* point in the room. Only on, or close to, the central plane in anechoic spaces will the discrete and stereo sources produce identical results, but as has already been mentioned, even these two cases may not be perceived to *sound* identical, the reason for which will be discussed in the following paragraphs.

14.3 Steady-State Performance

On 'steady-state' signals, the situation can be very different between anechoic and reverberant rooms. In anechoic conditions, the interference pattern from the left and right loudspeakers would sum by 6 dB on the central plane, and for a distance either side of it which would be dependent upon wavelength. The 3 dB of 'extra power' superimposed upon the 3 dB radiated power increase is gained at the expense of lower SPLs elsewhere in the room. Away from the central plane, the interference patterns would produce comb filtering, as shown in Figure 14.4, the nature of which would be position dependent. Off-axis there would also be additional low frequency power, which was the result of the additional radiation due to the mutual coupling. (See Glossary for Mutual coupling, if necessary.) That this is not perceived on-axis as solely an LF boost, but as an overall boost, can be considered to be a result of an overall directivity change when the two spaced drivers are operating together. As can be seen from Figure 14.3, on the central axis of the pair there is always a summation at *all* frequencies.

In highly reverberant conditions, a totally diffuse sound field would build up. As we are still considering perfectly flat omni-directional loudspeakers here, a similar frequency distribution would be radiated in all directions, and so the response at any point in the room would be the same. It would be a simple power summation of the two sources,

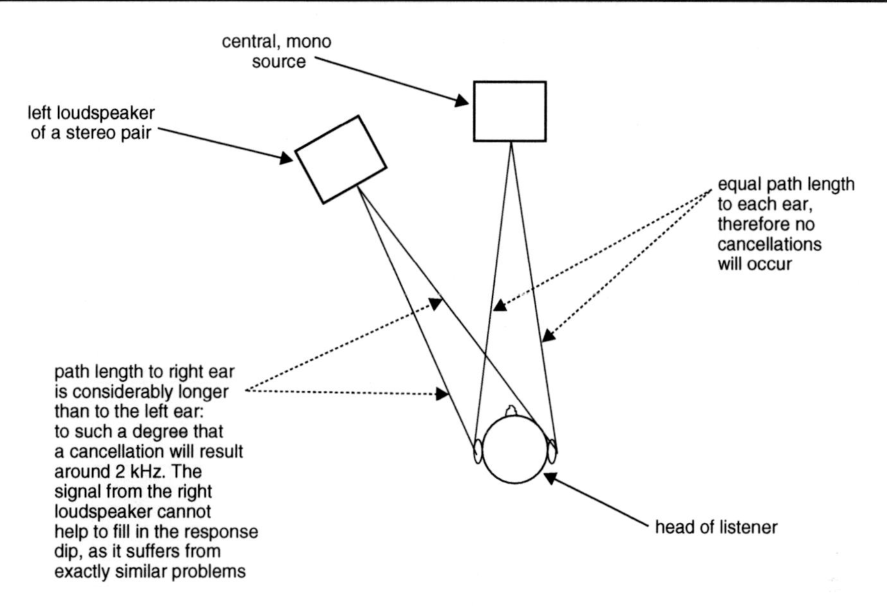
central, mono
source
left loudspeaker
of a stereo pair
equal path length
to each ear,
therefore no
cancellations
will occur
path length to right ear
is considerably longer
than to the left ear:
to such a degree that
a cancellation will result
around 2 kHz. The
signal from the right
loudspeaker cannot
help to fill in the response
dip, as it suffers from
exactly similar problems
head of listener

Figure 14.6:
Path-length anomalies for phantom central image.

producing a 3 dB increase in level plus whatever low frequency boost resulted from mutual coupling. Musically, though, things would sound very confused, with one note blurring into the next. In practice, many listening rooms of a reflective nature tend towards the 3 dB central summation, and some repercussions of this fact will be discussed in Section 14.5, but we should take a minute to look more closely at the area of 6 dB summation in the more anechoic conditions, because the width of the region is very frequency dependent. The region of perfect summation would only be around half a wavelength in width. At 20 kHz, this would be around 1 cm, but at low frequencies it would be many metres. At around 2 kHz, for a two-eared human being sat on the centre line, there would be an effective cancellation due to the spacing of the ears. This is due to the fact that the path length distances are not the same from each loudspeaker to each ear, as shown in Figure 14.6, which is another mechanism by which a centrally panned image, from a pair of loudspeakers, differs in the way that it arrives at the ears compared with the arrival of the sound from a discrete, central loudspeaker.

14.4 Transient Considerations

On the central plane of a stereo pair of loudspeakers, the transient *pressures* will also sum, producing a single pulse of sound 6 dB higher than that emitted by each loudspeaker individually. At all other places in the room, as the different distances to the two loudspeakers create arrival time differences, double pulses will result. This effect is clearly

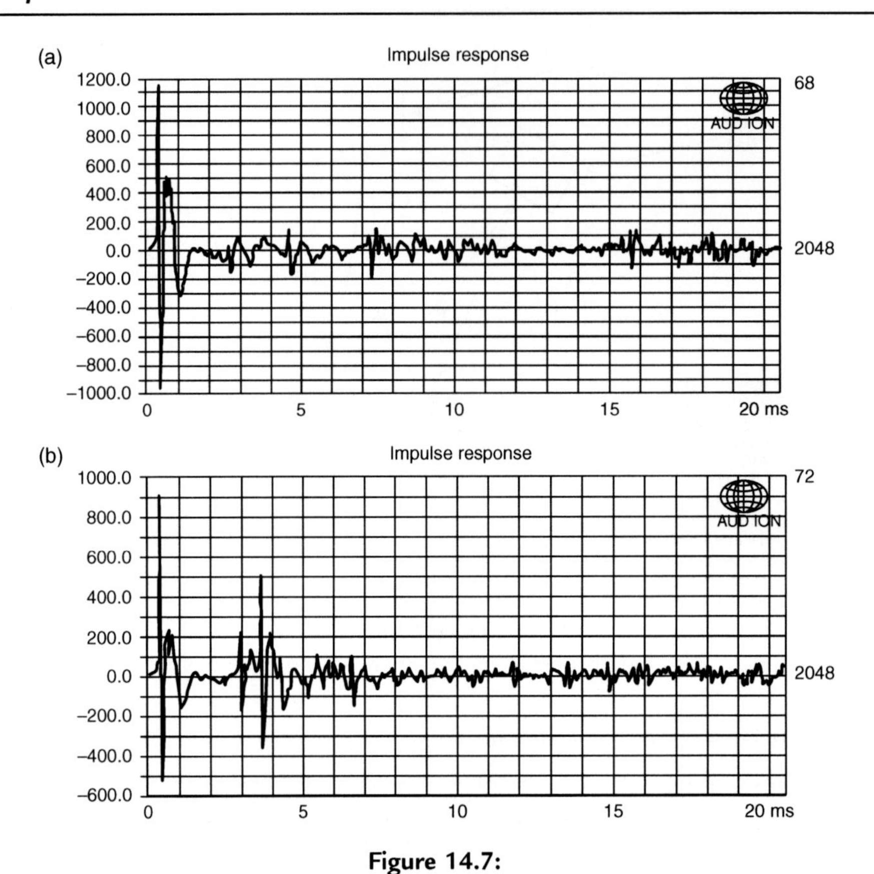

Figure 14.7:
(a) Impulse response received from a centrally panned image from a pair of loudspeakers. Measurement taken on the common axis of both loudspeakers (centre line). The response from a central, mono loudspeaker would be essentially similar; (b) Impulse response, as (a) but as received from a position 1 m behind (a) and 1 m to the left of the common axis (centre line). There are *two* clear impulses with the one arriving from the right-hand loudspeaker, about 3 ms later in time, and about 5 dB lower in level.

shown in Figure 14.7. Although it may seem to an observer on the central plane that four times the power (+6 dB) of a single loudspeaker is being radiated at all frequencies, to an observer off the central plane this would only be the case at low frequencies. At higher frequencies, the effect of the constructive and destructive superposition of the signals results in an average increase of 3 dB. The zones of coupling are shown diagrammatically in Figure 14.8. However, this still apparently leaves us with a 'magic', extra 3 dB of power on the centre line, which for transients cannot be described in terms of mutual coupling due to an increase in radiation impedance, because radiation impedance is a *frequency domain* concept, and transients exist in the *time domain*. We need to look at this behaviour in a different way.

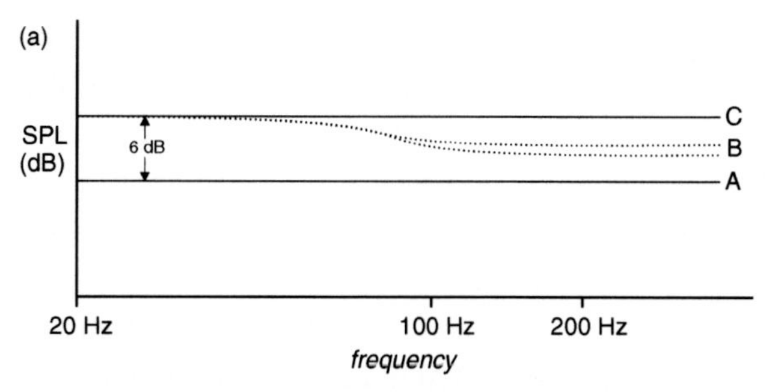

A: Response of single loudspeaker, anywhere in the room.

B: Response of a stereo pair of loudspeakers, each receiving the same input as "A", anywhere in the room, except on the central plane: precise response may be position dependent.
C: As in "B", but measured on the central plane only.

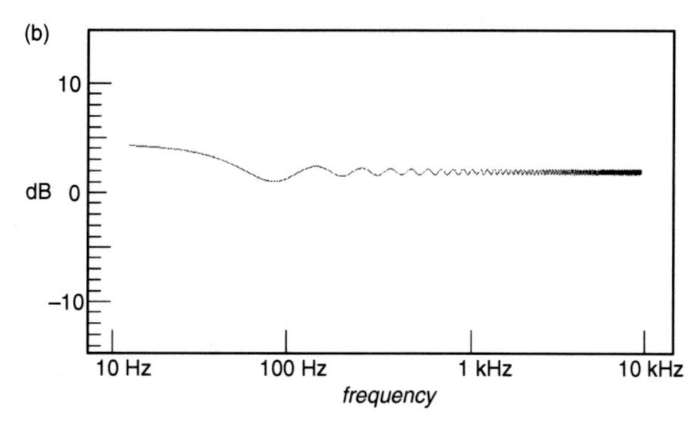

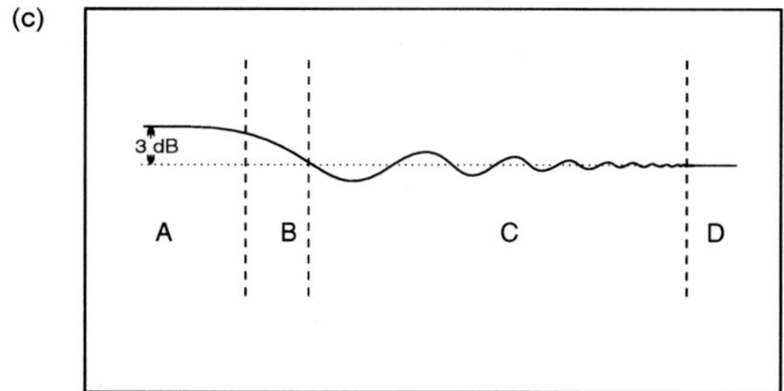

Zone A: Region where the separation distance between the loudspeakers is less than one-eighth of a wavelength, and where wholly constructive mutual coupling is effective.

Zone B: Region where the separation distance between the loudspeakers is less than a quarter of a wavelength, and where the mutual coupling is becoming less effective as the frequency rises.

Zone C: Region where the separation distance is *greater* than a quarter of a wavelength, and where the mutual coupling alternates, as the frequency rises, between being constructive or destructive.

Zone D: Region where the mutual coupling has ceased.

Figure 14.8:
Mutual coupling effects – omni-directional sources: (a) Pressure amplitude response in an anechoic room; (b) Frequency response of a pair of loudspeakers at any position in a reverberant room – combined power output; (c) Zones of loading/coupling – general response as in (b).

In the case of a perfect delta function (a uni-directional impulse of infinitesimal duration), the points of superposition would only lie on a two-dimensional, central plane, of infinitesimal thickness. As this would occupy no perceivable space, then no spacial averaging of the power response would be relevant. With a transient musical signal, which has finite length, there would be positive-going and negative-going portions of its waveform. At the places along the central plane where the transients crossed, they would not only meet at a point, but would 'smear' as they interfered with each other over a central area either side of the central plane. Around this central plane, the pressures would superpose, producing a pressure increase of 6 dB over a finite region each side of the centre line. (Doubling the *pressure* increases it by 6 dB.) As they crossed further, they would produce regions of cancellation, which would show overall power losses equal to the power gain in the central region of summation. The total power would thus remain constant when area-averaged. The above effect is shown in Figure 14.9, in which the average height of all the transients occurring at any one time in the room would be the same as that of a single transient emitted by one loudspeaker, though they would be doubled in number as there are two sources. Only where they interfered with one another would there be disturbances in their height, but there would be no extra total power in the room, just the simple sum of the power radiated by the two individual loudspeakers.

On transient signals therefore, because of their existence as separate bursts of energy, the performance of anechoic and reflective conditions differs only in that the reflective rooms will add an ever-increasing number of reflected energy bursts to the environment, though of ever decreasing energy, which will add to the overall perceived loudness. However, bearing in mind that a reflective/reverberant room must, by definition, be anechoic until the first

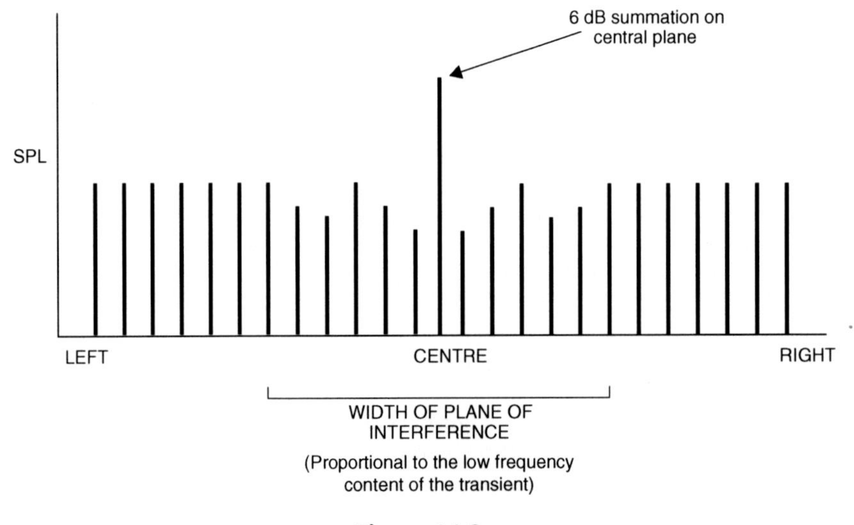

Figure 14.9:
Pulse superposition.

reflexion arrives, then dependent on the room size and the length of the transient burst, the subjective perception of the transient could change over time from the anechoic to the reverberant state. Continuous types of sounds (e.g. bass guitar notes) would be perceived more consistently in accordance with the type of room in which they were being reproduced. Perhaps it is little wonder then, that stereo perception can be so variable from room to room.

14.5 The Pan-Pot Dilemma

The perception of a 3 dB overall central summation (as tends to exist in most reverberant spaces) was at the root of the old pan-pot dilemma: should the electrical central position of a pan-pot produce signals which are 3 dB or 6 dB down relative to the fully left or fully right positions? Mono *electrical* compatibility of the stereo balance requires constant voltage, therefore each side should be 6 dB down (half voltage) in the centre in order to sum back to the original voltage when added *electrically.* (The pan-pots, or more fully panoramic potentiometers are potential [voltage] dividers, so, it is the voltages which must sum.) In the case of an *acoustic* stereo central image in a reasonably reverberant room, it is the *power* from the two loudspeakers which must sum to unity in the centre, which therefore requires a condition whereby the output from each loudspeaker would be only 3 dB down (half power) when producing a central image. Nevertheless, Figure 14.8(b) shows how below the mutual coupling frequency, the low frequency response would still approximate to the −6 dB requirement, because the power summation would tend to be augmented by a further 3 dB due to the mutual coupling effect between the two loudspeakers.

In stereo radio drama, where voices are often panned across the sound stage, −3 dB pan-pots would produce a uniform level as a voice was panned from left, through centre, to right. However, if the programme were then to be broadcast in mono, the voice would be perceived to rise by 3 dB as it passed through the centre position in the stereo mix, due to the *electrical* mixing of the signals. If the same task were to be repeated for a bass guitar, then in the mono broadcast it would still be subject to the same, uniform, 3 dB rises as it passed through the centre of the stereo mix. Somewhat inconveniently, though, when heard in stereo, it would be perceived to increase only in its low frequency content as it passed through the centre position on its way from left to right, due to the mutual coupling between the two woofers. The degree of boost, and its frequency of onset, would be dependent on the distance between the left and right loudspeakers, introducing yet another variable. It is therefore different for every different set of monitoring conditions, and cannot be compensated for in any standard mixing console.

As there is usually little dynamic panning of low frequency instruments in stereo music recording, a −3 dB centre position used to be considered optimum. In the case of radio drama, where mono compatibility for the majority of the broadcast listeners is a great priority, a −6 dB centre position was customarily required. Many mixing console manufacturers will produce consoles with different pan-pot laws for different applications, though they often opt for a −4.5 dB compromise, which produces only a 1.5 dB worst case error in either instance. This compromise fits nicely with most real-life rooms in which music will be heard, because they tend to exist in a region somewhere between reverberant and anechoic. In fact, in truly anechoic conditions, the acoustical sum on the central plane is identical to the electrical sum, at least in the region where the axial response holds true; but this is a central-plane-only condition. Elsewhere in the room, the summation approximates to the panning effect at all points in a more reverberant space, though with less confusion in the sound. The repercussions of these anechoic/reverberant and steady-state/transient phenomena create many problems for designers of subjectively reliable means of producing electrical fold-down systems for stereo/multi-channel compatibility, and for compatibility between large and small cinema mixing theatres. (See Section 21.12.1.)

14.6 Limitations, Exceptions and Multi-Channel Considerations

The theoretical concept of the perfect omni-directional loudspeaker therefore breaks down badly when the question arises of where to put them. Within any room other than one that is either perfectly anechoic or perfectly reverberant, boundary reflexions from the rooms in which they were situated would produce an irregular frequency response. Perfectly reverberant conditions would be useless for listening to music, as the perception of much detail would be swamped by the reverberation. Anechoic conditions, in which fine detail is most readily perceived, allow no reflected energy, so the only sound heard by a listener is that which passes directly from the loudspeaker to the listener. In such conditions there is no perceivable difference at the optimal listening position between a perfectly omni-directional loudspeaker and one that radiates a uniform frequency balance on an axis pointing directly towards the listener. A uniform response for ±20 or 30° off-axis allows for some movement about the central listening position and this is quite easy to achieve in practice. Therefore, in anechoic conditions, omni-directional loudspeakers would do nothing except waste power by radiating sound in unnecessary directions.

Just as the perceived response of a central image generated by two monopole loudspeakers will be different from that generated by a single, centre loudspeaker, as was shown in Figures 14.4 and 14.6, an image that is generated by *four* loudspeakers is likely to differ from an image generated by two. There could be a further mutual coupling rise of up to 3 dB as the number of sources was again doubled, as shown in Figure 14.5. With dipole loudspeakers though, this build up may be different, because they are pressure-gradient sources, coupling to pressure *nodes*, and not velocity nodes (pressure anti-nodes). Any bass lift due to mutual coupling will be greater in the case of typical multi-channel dynamic loudspeakers, as compared to electrostatics. This situation can pose even *more* problems for people trying to design electronic systems to 'fold-down' multi-channel mixes into stereo or mono, while still attempting to maintain the original musical balance.

As well as all the fold-down compromises caused by the effects of the rooms and combined loudspeaker directivity, it can now also be seen that the desired fold-down can also depend on the type of loudspeakers on which the music was mixed, as well as those on which it would be reproduced. To make matters worse, the compatibility of mixes is also affected by the fact that the optimum fold-down must take into account the frequency dependence of the directionality of human hearing. The subjectively desirable level of the sounds formerly in a rear loudspeaker may be considered excessively loud, or bright, when reproduced from a frontal direction after fold-down. This is because they may have been made brighter in the surround mix because the ears are much less sensitive to higher frequencies arriving from behind the head. Thus, when assessing claims about the degrees of compatibility of mixes from surround to stereo or mono, one must ask … 'Compatibility with what?' The electrical fold-down equation may differ very greatly from the purely acoustic, *or* the electroacoustic, *or* the perceived psychoacoustic fold-down requirements. The acoustic nature of the listening environment will also add its own variables – the room coupling. Once again, relatively anechoic (i.e. highly damped) monitoring conditions would seem to offer the fewest complications for the electrical fold-down requirements. What is more, as the spaciousness would be in the surround, there would be little need for a room to add any more if faithful reproduction was required.

So, even if it was possible to design perfect loudspeakers, the question would remain as to exactly in what sort of a room they would *behave* perfectly in stereo pairs? Only on the central plane in anechoic conditions, it would appear. In all other cases, the room will impart its influence upon the perceived response, and, even in anechoic conditions, there are aspects of the weaknesses inherent in stereophonic reproduction which mean that we cannot precisely reconstruct the sound from a single source by means of a phantom image generated by two sources. The situation is that we have imperfect loudspeakers *and* imperfect rooms, trying to reproduce a fragile concept.

This produces a strong argument in favour of three or five frontal loudspeakers, not only in terms of image stability when moving off-axis, but also from the point of view that fewer phantom sources and more discrete sources means that reflexion, absorption and diffusion will all be more uniform in frequency content. True, in real situations, the reflexion, absorption and diffusion will *not* all be uniform in their frequency response, but nonetheless, less confusing interaction from multiple sources for single image positions means, at the very least, a more predictable set of starting conditions. Unfortunately, this still presumes omni-directional, point source loudspeakers with perfectly uniform frequency responses, which do not exist.

The job of acousticians and electroacousticians to get the best out of any given set of circumstances is no easy task. Only by a very careful balancing of all the parameters can optimum end results be realistically hoped for, but parameters can only be balanced if they are understood, and many of the points being made in this chapter are *not* widely appreciated. The reason why no one principle of listening room design is universally accepted is because of the truly vast number of variables involved, of which the problems that have been discussed here form only a very small part. All of this does little, however, to change the perceptions in the minds of the public at large about room acoustics being a black art. In fact, in many instances, the object of an acoustician's work is to keep the variables within defined, acceptable limits, rather than to seek an as yet unachievable perfection. It is disappointing that in Internet forums, in hi-fi magazines, and even in some professional recording studios, there is a tendency to believe that there are simple solutions to even the deeper problems of multi-channel reproduction.

14.7 Surround in Practice

The drawings in Figures 14.10 to 14.23 show 14 surround formats, all of which are either already in use or have been proposed by people or organisations of sufficient respect to render them free of any 'crank' or 'minority interest' status. Even more elaborate systems are now being widely discussed, and the new Dolby 7.1 system splits the surrounds into Left, Right, Left-rear and Right-rear.

It must be understood that surround is not just a *little* more complicated than stereo, nor even twice as complicated – it is *vastly* more complicated. There is a saying 'Give a person enough rope and they will probably end up by hanging themselves'. Well, surround supplies an awful lot of rope, and the potential for people to get hung up in its complexities is enormous. Stereo supports one phantom sound stage between the two loudspeakers. Five-channel surround (let alone seven or more), by allowing panning between any individual pair or groups of loudspeakers, provides 26! Except in the centre of the loudspeaker arrangement in free-field acoustics, such as in an anechoic chamber (and even ignoring human hearing directional differences), surround systems will tend to result in different responses of timbre for any instrument panned across any of the possible different sound stages.

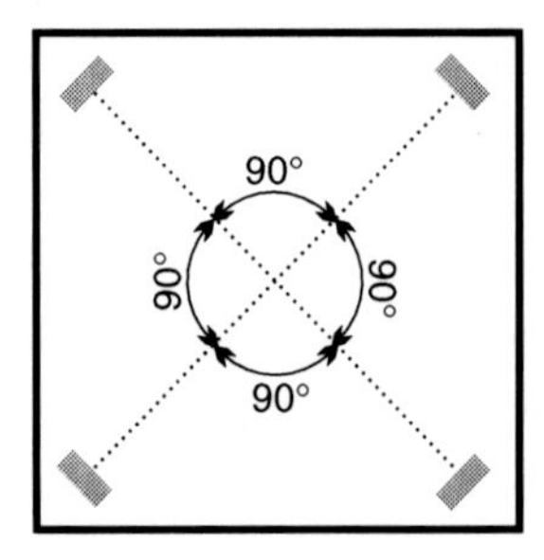

Figure 14.10:
4ch 'Quadrophonic', also used in basic Ambisonics.

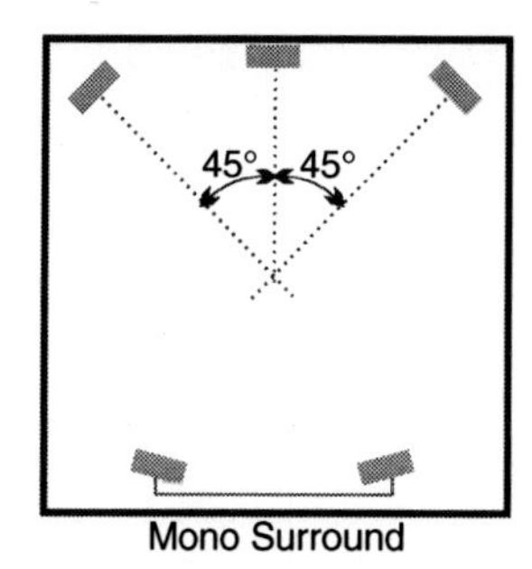

Figure 14.11:
Dolby Stereo 3/1 (Pro-Logic).

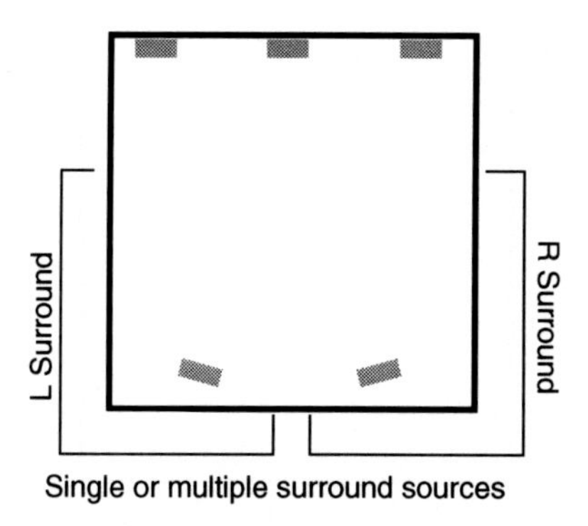

Figure 14.12:
5-channel 3/2.

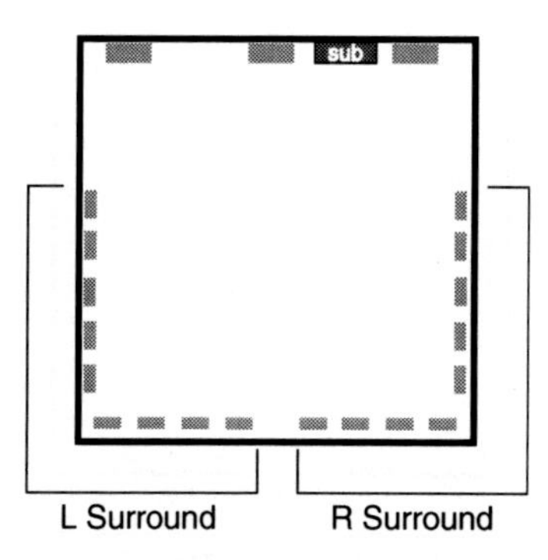

Figure 14.13:
Dolby Digital surround.

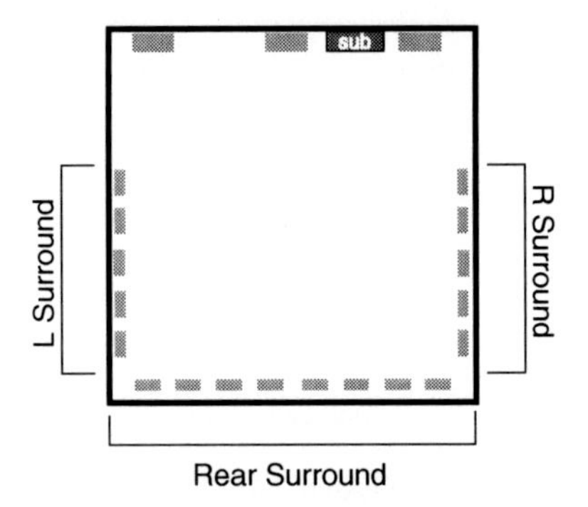

Figure 14.14:
Dolby EX 6.1.

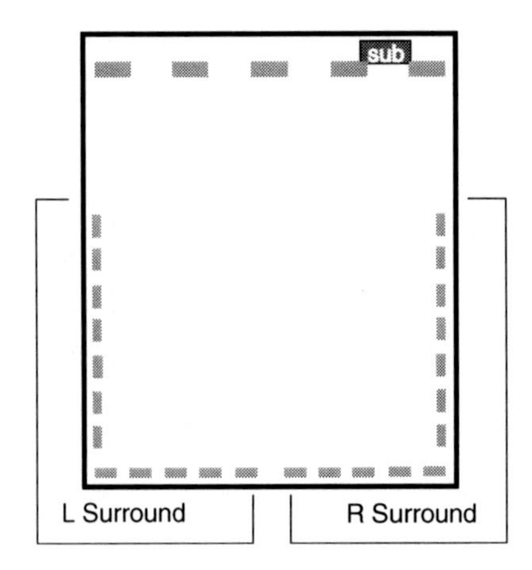

Figure 14.15:
Sony 7.1 (SDDS) (Sony Dynamic Digital Sound).

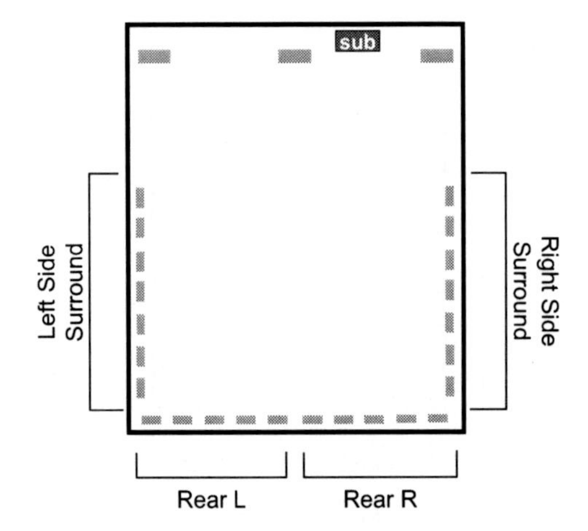

Figure 14.16:
Dolby 7.1.

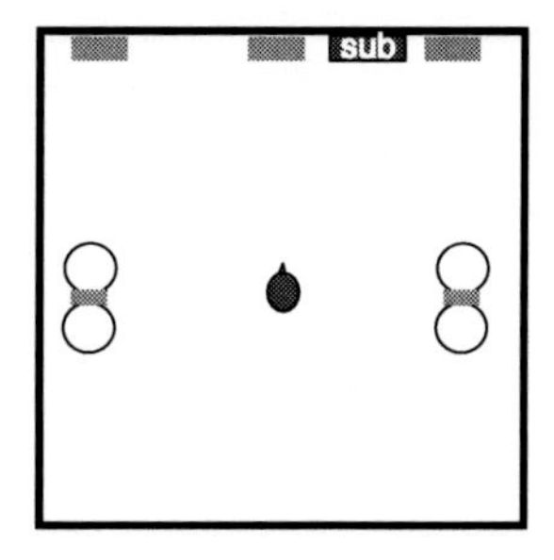

Figure 14.17:
5-channel surround using dipole side/rear loudspeakers.

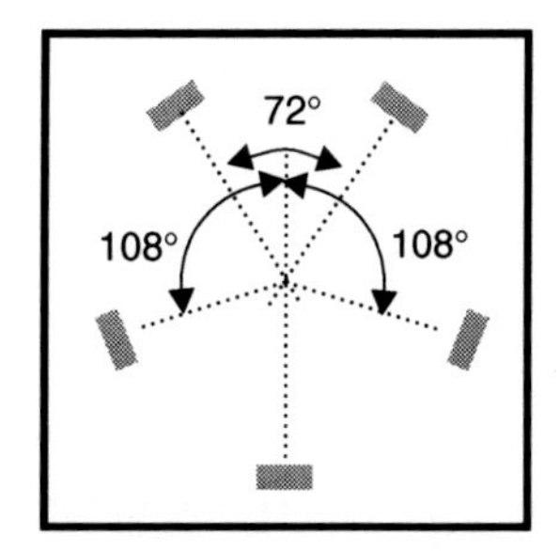

Figure 14.18:
Holman Pentagon. 'Pentagon' (Boston Audio Society *Speaker*, Vol. 22, No. 2, p. 31, May 1999) [by Alvin Foster].

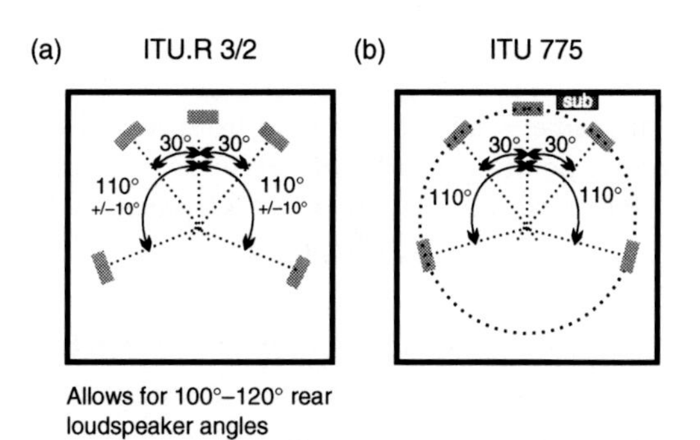

Figure 14.19:
(a) ITU.R 3/2; (b) ITU 775.

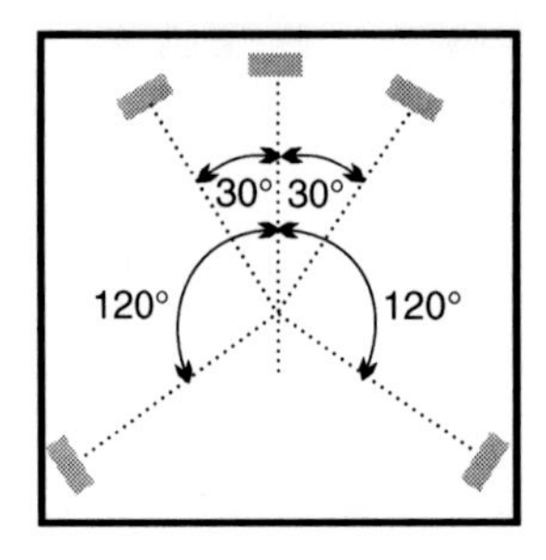

Figure 14.20:
HDTV study group — Japan, Preprint 4253, Copenhagen, 1996 AES. The surround loudspeakers are more distant than in the ITU specifications shown in Figure 14.19.

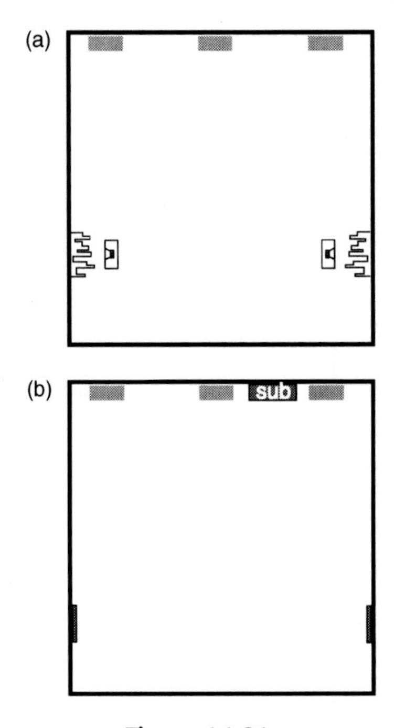

Figure 14.21:
Two variations on a similar theme — diffuse surround from two discrete sources: (a) Surround speakers mounted in room, pointing at diffusers on walls — as used by Bell and Dobson; (b) As at TOBIS, with diffuse surround sources — NXT panels — DMLs (see References 3 and 4, Chapter 21).

In fact, at very low frequencies, mutual coupling effects between the sound sources may result in different low frequency responses even in anechoic conditions, depending on the number of loudspeakers supporting any one phantom image. When the systems are transferred into a room, and boundary reflexions begin to affect the responses, the concept of any sound mixing personnel being able to predict the result outside of the actual mixing room becomes

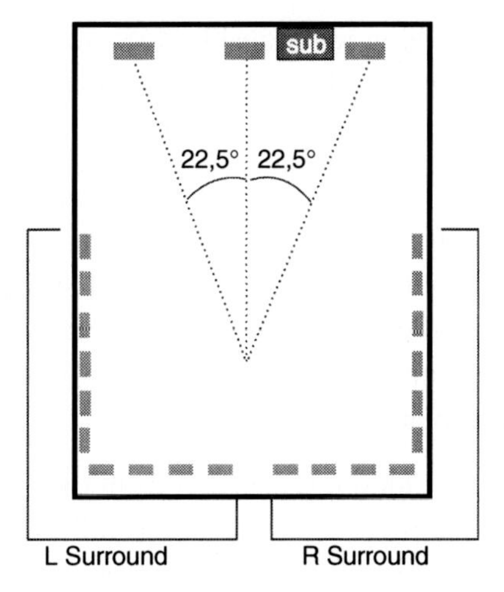

Figure 14.22:
THX requirement for film dubbing theatres — 45° subtended angle.

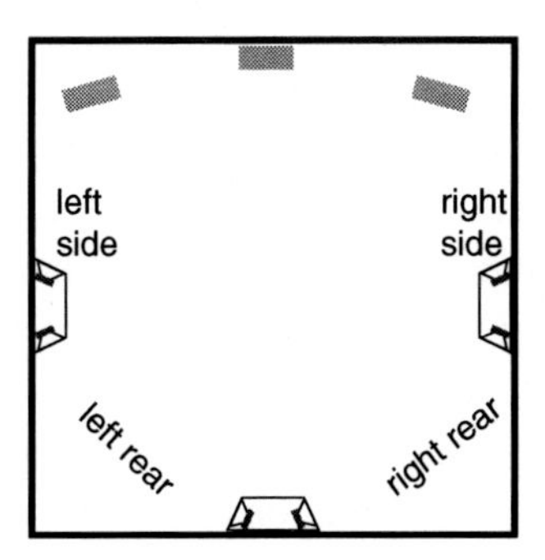

Figure 14.23:
Bipolar surround sources. Proposed by Fosgate,[3] using processors to decode 7 channels processed from 5.

a lottery. Despite all the magazine articles by producers and engineers who are excited about the creative possibilities of surround, the subject is a minefield, ready to trap anybody who does not know exactly where the mines are laid. Many people think that they *do* know about surround, but as James Moir reputedly pointed out many decades ago: 'If one thing is certain about acoustics, it is that if anything seems obvious it is probably wrong'.

The 14 layouts shown in Figures 14.10 to 14.23 indicate how difficult it has been to agree on any simple standard set of compromises. Many of the systems are specifically engineered to solve given problems, or sets of problems, but in doing so they each tend to create other problems. The fact that the minds of so many specialists have been put to work on the problems, only to result in chaos, suggests that not only is there no easy solution to the problems of surround, but that the simple, attractive, flexible, high-quality systems that the marketing people have been searching for probably does not and will not exist. It is *that* complicated. It therefore becomes very difficult to specify an environment for the optimum mixing of music in surround because the goals are still not clearly defined.

Even the drawings in Figures 14.10 to 14.23 are simplified inasmuch as many offer the options of whether to use full-range loudspeakers, or to use satellites and a sub-woofer, or even two, three or four sub-woofers. With these options, our 14 systems would now become 40 or more, each with its' own specific tonal differences. Their respective strengths and weaknesses would be dependent upon room conditions, programme material and how the mixing personnel chose to distribute the sounds.

Much orchestral and organ music definitely benefits from the low frequency 'wash' that can envelope the listeners when listening to a pair of stereo full-range loudspeakers. When such music is transferred to a loudspeaker system with satellite loudspeakers and a mono sub-woofer, the effect is largely lost. However, in rooms in which the low frequency response is less than very well controlled, the single source provided by a lone sub-woofer can often lead to a less confused low frequency response than would be experienced when using two full-range sources.

The question of 'to use, or not to use' sub-woofers is complex. Undoubtedly, they can offer the flexibility of being able to position them in a room such that the low frequency coupling can be optimised, but such positioning, whilst benefiting the flatness of the low frequency pressure response, may not be optimal in terms of time alignment. This is of particular importance in the context of surround sound when a front positioned sub-woofer may be both physically and temporally misaligned with the rear loudspeaker. The commonly held belief that there is no directional information in frequencies below 300 Hz is erroneous. The mis-location of sub-woofers can demonstrate this easily. When sub-woofers are placed in non-optimally damped rooms, it is advisable to mount them off-centre in order that they will drive the rooms asymmetrically. The central positioning, by virtue of the equal distance to each side wall, tends to cause symmetrical driving of the standing-wave field, which can concentrate reflected energy into narrow frequency bands. By contrast, the asymmetrical drive of the standing-wave field by a non-centrally positioned sub-woofer tends to create more peaks and dips in the response, but of lower magnitude, yielding a generally flatter overall response.

With such off-centre positioning of the sub-woofer, it is usually prudent to offset it to the right, because for orchestral music, the cellos and basses are normally positioned towards the right. For pop, rock or electronic music, where the bass guitar and bass drum are normally centrally panned, then whether the sub-woofer is offset to the left or to the right is of little consequence. However, when listening to orchestral music with the sub-woofer offset to the left, the non-collocation of the low frequencies with their respective instruments can often be noticeable.

Similarly, in surround systems, when a sub-woofer is positioned at the front of a room, with a full-range keyboard instrument panned to the rear, walking backwards and forwards in the room can often reveal very clearly the non-collocation of the different frequency bands. The coincidence of the arrival time of the full frequency range can have a great bearing not only on the punch and impact of a sound, but also on the room-to-room compatibility of the mixes. Mixes done in rooms with badly positioned sub-woofers can make it very difficult to judge the optimum tonal balance of the low frequencies, and it can be difficult in such rooms to make a mix that travels well to other rooms. The optimum balance of bass drum and bass guitar can be quite difficult to achieve in rooms with temporally offset sub-woofers. Furthermore, many sub-woofers use band pass enclosures that have notoriously poor time responses, and this tends to add to the time-smearing confusion.

The situation with multiple full-range loudspeakers is complex. In a five-channel system, the low frequency response of a signal panned to any individual source is reasonably predictable. However, when a low frequency signal is routed to two or more loudspeakers, the frequency response will be dependent upon how many loudspeakers it is routed to and the distance between the loudspeakers. The number of loudspeakers will determine the degree of low frequency boost due to the mutual coupling of the sources, and the distance between the loudspeakers will determine the frequency below which the mutual coupling boost will begin to take effect. Clearly, if total freedom of choice exists in whether to use single or multiple low frequency sources, then the degree of predictability of the response compatibility of the mixes with other systems will be poor.

It is more or less standard in serious stereo control rooms to refer mixes to a pair of full-range loudspeakers 2.5 to 4 m apart. At the time of writing, there is still no consensus about how to manage the low frequencies for surround music mixing. In the cinema world, the situation is well defined, with the more or less general use of a discrete low frequency effects channel. There is usually also little multi-source panning, except when the full power of the entire system is needed for effect, such as during an explosion. Most music is panned between no more than two frontal loudspeakers, which are usually capable of extending down to 40 Hz or so. In the TV and video world, a single sub-woofer tends to be the norm, though it must be said that when the sound is intended to accompany pictures in a domestic situation, it is rarely, if ever, given the sort of attention that it would receive in music recording.

Ironically, it is in the world of music recording, which is supposedly the flagship of the sound recording industries, that no consensus exists about how to deal with the low frequencies in multi-channel recordings. The lack of standards organisations, the great power of the vested interests of equipment manufacturers marketing their products, and the widespread ignorance of the difference between many of the low frequency options have all had their effect in delaying any general decision on how to handle the low frequencies of surround mixes. Nevertheless, perhaps the principal reason why chaos has ruled for so long is the fact that there *is* no simple option that will suit all types of music. Mono had one source. Stereo, from the outset, used the pair of low frequency sources. However, surround has a multitude of low frequency options, such as:

1. Five discrete sources all available as desired
2. Three full range channels at the front with reduced low frequency capacity in the rear channel
3. Left front and right front carrying the main panned music mix, with the centre channel principally for vocals or dialogue
4. Low frequencies handled by one sub-woofer
5. Low frequencies handled by stereo sub-woofers
6. Low frequencies handled by multiple, processed sub-woofers.

The problem is that all of the above options will sound different in any one room, and the way that each one will translate to a different system in a different room will also be different. From the point of view of producing control room and monitor systems with any semblance at all to a reference standard, first we need a standard. Without such a standard, the concept of surround sound cannot really be considered high fidelity.

14.8 A General View

The room effects become significantly more problematical as the number of simultaneously radiating sound sources increases. In fact, the problems can multiply so rapidly that with four, five or six sources, as are commonly advocated for surround systems, the complexity can become so great that there may be no solutions allowing delivery of the same degree of fidelity as can be achieved in better stereo rooms. If surround mixing is not carried out with full respect to the acoustic limitations, then trading quality for quantity may become a fact of life. The directional reflexion philosophies used in many stereo control room concepts will not function when 360° of sound sources are used. Many of the surround-sound problems can be solved by highly damped rooms in which all the ambience is in the surround, but this could lead to very unpleasant room acoustics in which to be when the music is switched off.

The subject of surround control rooms will be discussed at much greater length in Chapter 21, but this brief look at the situation will have at least prepared the way for the battles to come.

14.9 Summary

A central phantom image, produced by two spacially separated loudspeakers, creates a very different reflexion pattern in a room when compared to a central mono source. The two sources of the central image produce comb-filtered reflexions.

A central phantom image is also subject to a perceived dip around 2 kHz, due to the different distances from each loudspeaker to the two ears. A central mono source does not exhibit the same effect, so the timbre will sound different to a phantom image.

Pan-pot laws ideally need to be tailored to both the acoustics of the rooms in which the music will be used and to the programme material.

The theoretical concept of the perfectly flat omni-directional loudspeaker would appear to be of no advantage in practical circumstances.

Automatic electrical fold-down of surround mixes to fewer channels is fraught with compromises.

Five-channel surround is vastly more complicated than stereo, and can exhibit at least 26 different phantom sound stages between different loudspeaker pairs or groups. Stereo has *one* phantom sound stage.

At low frequencies, the mutual coupling between sources will be dependent upon the number of sources in use and the distance between them. The former will dictate the amount of boost, and the latter the frequency below which the boost will occur.

There are more than a dozen authoritatively proposed surround configurations. When one considered sub-woofer options, the number is close to 40.

Single sub-woofers should be slightly offset to the right of the centre-line.

Sub-woofers can produce considerable time smearing of the low frequency response, due to both the physical displacements and the common use of band-pass cabinets with inherently poor time responses.

The problem of surround compatibility from mixing room to listening room is considerable.

References

1 Holland, Keith R. and Newell, Philip R., 'Loudspeakers, Mutual Coupling, and Phantom Images in Rooms', presented at the 103rd Audio Engineering Society Convention, New York, Preprint, No. 4581 (1997)
2 Holland, Keith R. and Newell, Philip R., 'Mutual Coupling in Multi-Channel Loudspeaker Systems', *Proceedings of the Institute of Acoustics*, Vol. 19, Part 6, pp. 155–62 (1997)
3 Fosgate, James., designer of the Dolby Pro Logic II System

Bibliography

Borwick, John, *Loudspeaker and Headphones Handbook*, 3rd Edn, Focal Press, Oxford, UK and Boston, USA (2001). [In particular, Chapter 1.]

Newell, Philip, and Holland, Keith, *Loudspeakers – for music recording and reproduction*, Focal Press, Oxford, UK (2007)

Toole, Floyd E., *Sound Reproduction: The Acoustics and Psychoacoustics of Loudspeakers and Rooms*, Focal Press, Oxford, UK (2008)

CHAPTER 15

Studio Monitoring: The Principal Objectives

The functions of a monitor system. Basic points of reference. Differences in perception. Different balances of compromises. Desirable assets. Loudspeakers as information channels. Close-field monitoring. The NS10M.

15.1 The Forces at Work

From the previous chapter it should now be apparent that we are not dealing with perfection when we talk about music reproduction via loudspeakers. The late Richard Heyser (inventor of the Time Delay Spectrometry measurement system, and one of the audio giants of the twentieth century) said that: 'In order to fully enjoy the intended illusion of a recording, it is necessary to willingly suspend one's belief in reality. All recording and reproduction via two loudspeakers is illusory'.

The classical concept of studio monitoring has been 'The closest approach to the original sound', as the Acoustical Manufacturing Company stated in their advertising literature of the 1940s. David Moulton[1] pointed out that the loudspeaker itself, without formal recognition, has become the predominant musical instrument of our time. Music is being created *on* loudspeakers for playback *by* loudspeakers. Often there *is* no reality. Neither is there an original sound to attempt to make the 'closest approach to'. Given the disparity between many studio monitoring conditions, this must render the whole situation somewhat arbitrary.

It is also hard to get away from the fact that both the sound recording and sound reproduction industries are businesses to a greater degree than ever before. The truly professional 'irrespective of cost' or 'absolute attention to detail' approach to either one is now extremely rare. Reference was made in Chapter 1 to the building of The Townhouse studio in London in 1978, which in its day had the best of everything. Few studios could survive today if they were built at such real (i.e. inflation adjusted) costs.

Today, also more than ever before, the cost of advertising by the manufacturers who supply the professional industry, together with the cost of attending a large number of trade shows around the world, means that the sale price of much equipment must support all of these extra burdens. In fact, it is the case that the cost of components represents less than 15% of the sale price of much of today's professional recording equipment, far less than was the case in 1980. The upshot of this is that just because something seems to be expensive does not necessarily mean that it contains expensive components. The pursuit of true excellence is now in the hands of a *very* small proportion of the people involved in the music industries. The majority are in the pursuit of profits, and little else. Nevertheless, the principle aim of this chapter is to take the purist viewpoint as far as possible, and to try to make clear divisions between whether the compromises or shortfalls are based on the limitation of the science and technology, or on the reality of a market led world.

The main functions of studio monitor loudspeaker performance are directed towards enabling the users to detect things that may become problematical on the best domestic systems; to give the users the widest possible scope for artistic interpretation of what is being recorded and to be capable of producing the desired atmosphere in the creative environment. Somewhat surprisingly to many people, many well-used studio monitor systems do not rate particularly highly on the fidelity ratings used by many hi-fi reviewers and magazines. To lay persons this may seem disturbing, but they must appreciate that the function of such systems is not primarily to create wonderful sounds in the studios, but to enable the studio to be used to produce recordings that sound wonderful when played at home by the public. Although it will always be very worthwhile to aim for maximising the fidelity of a monitoring system, other requirements, such as large size to enable reliable production of high sound pressure levels at low frequencies, ruggedness in daily use, ease of servicing, consistency of performance, and many other factors, will also be high on the list of design priorities. A super-fidelity system that fails twice a week is of little use in studios.

That studio monitor loudspeaker systems are a class unto themselves is borne out by the fact that so few loudspeakers straddle the professional/domestic divide. Monitoring loudspeakers are rarely found in other environments, not because of price, as many 'high end' hi-fi systems cost more than many comparable monitor systems, but because they are highly specialised, highly evolved tools. They are only a means to an end. Indeed, one reason why many good monitor systems do not automatically become expensive domestic hi-fi is that they can often be unpleasant to listen to on poor quality programmes; they are too ruthlessly accurate, but it is their job to be so.

Conversely, the small monitors used in many home studio set ups, and which may be pleasant enough for domestic listening, often do not have the accuracy of resolution to make obvious during the recording process any build up of aggressive sounds, due, for example, to excessive signal processing artefacts. They have been designed to sound good, which is not much use when problem hunting. The result may often be that a recording which sounds acceptable on such loudspeakers may be less pleasant to listen to when heard on somebody's domestic high-end hi-fi. Recordings should not be released by a professional industry if they contain such badly monitored programme. People who work on poor or 'easy to listen to' monitors are merely burying their heads in the sand, taking a lazy way out, and leaving the record buying public to suffer the results of their failings; all because of a lack of professionalism being shown in the recording studios.

15.2 Where is the Reference?

A significant problem exists, because no one combination of loudspeakers, amplifiers, crossovers and rooms can be optimised for the most lifelike sensory illusions for all of the different styles of music or recording techniques. The lack of correlation between measured and perceived responses only serves to complicate matters further. There can also be complex convolutions of the different performance characteristics that manifest themselves in very programme dependent ways. It still cannot be said in general terms whether a ± 1 dB error in the pressure amplitude response is any more or less important than 1° in phase accuracy, or than an extra 0.1% of non-linear distortion. Even within the harmonic distortion performance, no absolute, programme independent ratio exists in terms of the relative importance of the individual harmonics. In fact, harmonic distortion measurements may be of little use at all, because the overriding source of unpleasant non-linear distortion would appear to be the intermodulation products, for which no common standards of measurement currently exists.

The aforementioned 'closest approach to the original sound' concept of a studio monitoring system has an inherently logical feel about it. It would seem to be self-evident that we should be able to set up an acoustic guitar, for example, in a good recording environment, listen to the recording in an acoustically neutral control room via a selection of loudspeakers, and choose the most realistic loudspeaker as a good monitor. The reality, however, is very different.

The different frequencies produced by the acoustic guitar radiate from many parts of the instrument. The highs tend to come from various parts of the strings. The lows come principally from the large surface of the body and the resonance hole. The 'mids' emanate from the complex vibrational behaviour of the body in a way that is quite spacially diffuse. What this means is that the sound-field around the acoustic guitar is very complex, and what we perceive when standing in front of it is further dependent upon the reflexions in the room and the fact that we have two ears. Once we try to collect this sound via a microphone, the fact that the microphone has a directivity pattern quite unlike our two ears means that the electrical signal that it generates will not be representative of what *we* hear from a similar point in front of the guitar. Perhaps an obvious solution would be to use a pair of microphones in a dummy head, but such a collection system only sounds natural when the pinnae (outer ears) on the dummy head are identical to those of the listener, and when listening on headphones. This is because if we play back the recording via loudspeakers, the inter-aural cross correlation that is inherent in the two signals coming from the dummy head would be processed a second time when the sounds from the loudspeakers entered our own ears. It would thus pass through *two* sets of outer ears, which can be avoided only by using headphones.

Another problem with the sounds from the loudspeakers is that the spacial distribution of the sound source is entirely dependent upon frequency. The lows come out of the woofers and the highs come out of the tweeters. In fact, in many cases, all the high frequencies in a stereophonic loudspeaker reproduction are generated from two points, only a couple of centimetres across (the two tweeters), which is quite unlike the distributed high frequency generation by the instrument.

Furthermore, if the control room in which the loudspeaker is situated has the same characteristics as the recording room, the room characteristics will be heard in double dose; once through the microphone and once from the excitations by the loudspeaker. Clearly, this is not accurate monitoring. However, if we listen in an anechoic room, the lack of reflexions will clearly leave the loudspeakers as being the only sources of the sound, which again is quite unlike the natural ambient 'surround' sound heard when stood in front of the acoustic guitar in the recording room.

If we were to take the acoustic guitar into an anechoic chamber, place a high-quality pair of measuring microphones in front of it, and monitor the sound via a pair of loudspeakers in another anechoic chamber, we could, perhaps, finally achieve a degree of realism via the loudspeakers. Under these circumstances, the loudspeakers with the widest frequency responses (pressure amplitude responses) the most accurate phase responses and the lowest non-linear distortion content would probably show themselves to be the most realistic in their reproduction of the original sound. Nevertheless, if the loudspeaker directivity was taken into account, then in a more lively control room they could give rise to reflexions with strange frequency balances, which would render the overall perception to be much less natural.

Even in the anechoic circumstances, however, two people may not agree about which loudspeakers were most accurate. The fact that the sound-fields from the loudspeakers are different to the sound-fields from the actual instruments means that the shapes of the pinnae of the listeners (as explained in Chapter 2) will ensure that the reception of the different sound-fields will create different responses in the ear canals of the different listeners. In other words, what reaches the eardrum can depend on the frequency distribution of the source in *spacial terms*, as well as the frequency distribution in terms of the pressure amplitude in the vicinity of the outer ear. In a test carried out in a studio in London in the 1980s,[2] two respected recording engineers failed to agree by 3 dB in the '6 kHz upwards' range about the similarity of a monitor system to a live acoustic instrument in the studio. The question was not about what sounded right, but simply what high frequency level sounded most similar to the live source.

During research at the Institute of Sound and Vibration Research in Southampton University in 1989,[3] a wide range of listeners took part in listening tests which were directed towards identifying characteristic sounds from mid-range horn loudspeakers. The question was asked for each of nine different sounds, 'To which of the four reference loudspeakers does the test sample loudspeaker sound most similar?' The tests were blind, and the sounds were a selection containing recordings of both transient and resonant nature. On some sounds agreement was almost unanimous, but on one particular sound and one particular sample, out of the first 11 listeners, two people thought that the sample sounded more like reference A, one like B, five likc C, two like D and one person thought that it did not sound like any of them. These were mainly people with some experience in listening critically to music, yet they could not agree on which loudspeaker sounded most similar to which. Had the question been reversed, the test loudspeaker could have been the reference sound. In that case, the question could have been asked, 'Which of the four loudspeakers is most similar to the reference sound?' The results would have been equally diverse.

Concurrently with the above tests, research was being done for Canon, who were assessing the value of a concept for a wide image stereo system.[3] The test set up was designed to detect spacial preferences by feeding artificial reflexions via combinations of 14 loudspeakers selected via a six position switch. Many participants complained about switch malfunctions when no change was heard. When the results were published, it was revealed that the switch 'faults' were non-existent. The 'faults' always occurred in the same places for the same listeners when the tests were repeated, but the 'faulty' switch positions were different for each person. The implication was that people not only had some very well defined spacial image *preferences*, but that the *sensations* of the images differed greatly from person to person, also.

In the first of the two ISVR tests described above, the anechoic conditions and the relative uniformity of the directions of the different sounds meant that the differences perceived were amplitude and phase differences in the *frequency* domain. In the second case, all the loudspeakers used for the simulated reflexions were nominally identical, so the differences perceived were in the *spacial* and *temporal* domains. Another outcome of the first of the tests was that some of the loudspeakers deemed to sound similar in the overall analysis showed harmonic distortion differences as great as 20 to 1. In other cases, loudspeakers with very similar distortion characteristics were considered to sound very different.

Loudspeaker systems as they exist today are far from perfect. Figure 15.1 shows a series of pressure amplitude response (frequency response) plots for a group of well-respected loudspeakers that are commonly used in recording studios. From these plots alone, and taking into account the fact that the loudspeakers all have different directivity characteristics, it should be apparent that their reflexions should sound different, and it is not hard to appreciate that they *do* sound different. It should also be easy to appreciate that because the response differences are so widely distributed, traditional equalisation systems could not adjust them all to a flat response.

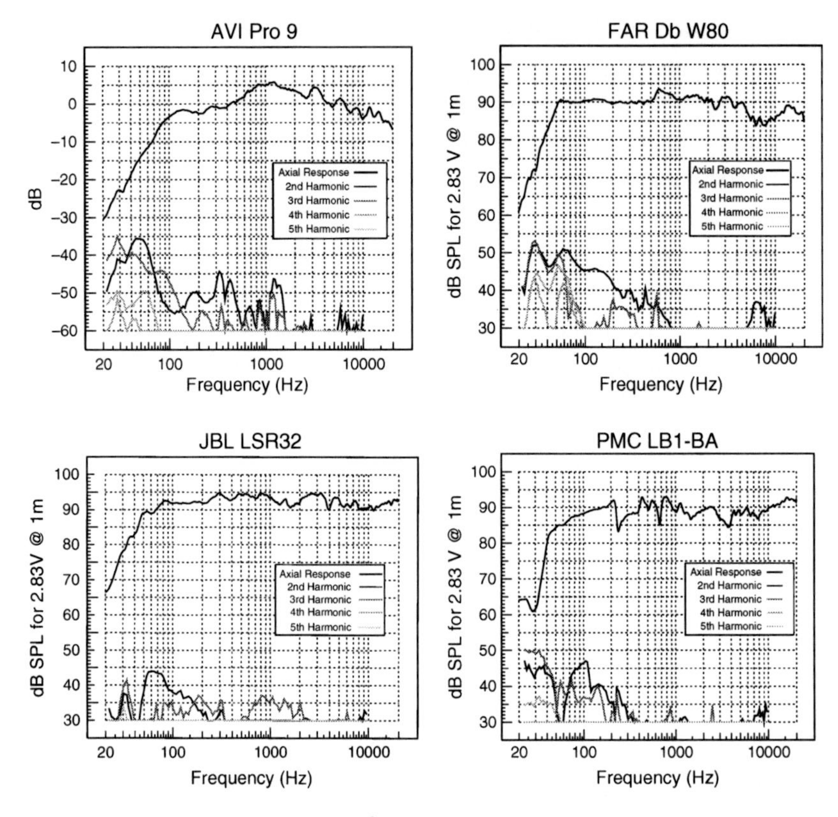

Figure 15.1:
Pressure amplitude responses of four loudspeakers, all ostensibly designed to perform more or less the same task.

In the light of what has been presented in this chapter so far, it should be apparent that we do not have loudspeakers that exhibit perfect responses in *any* of their characteristics. Neither do we have a sufficient degree of perceptual consistency from one person to another to form an 'expert listener' panel to select a 'best' loudspeaker. Even if we did, we would no doubt get different results when the loudspeakers were auditioned in different rooms and on various types of music. The choice of studio monitoring systems therefore tends to depend to a considerable degree on the ability of the users to get the desired results when using them.

In an interview in EQ magazine in June 1993, George Massenburg, one of the world's pre-eminent recording engineers and producers, said 'I believe that there are no ultimate reference monitor systems, and no "golden ears" to tell you that there are. The standards may depend on the circumstances. For an individual, a monitor either works or it doesn't … Much may be lost when one relies on an outsider's judgement and recommendations'. Also, it is worth repeating (from Section 2.4.1) that about a hundred years earlier, Baron Rayleigh, one of the true giants of acoustical research, stated 'The sensation of sound is a thing *sui generis*, not comparable to our other sensations. Directly or indirectly, all questions connected, with this subject must come for decision to the ear, as the organ of hearing, and from it there can be no appeal.' Rayleigh summed up the situation so beautifully before loudspeakers had even been invented!

As Toole and Olive wrote,[4] 'A loudspeaker isn't good until it sounds good. The traditional problem has been: how do we know what is good? Whose opinion do we trust?' Although they then go on to describe listening tests which when well devised and controlled can yield very consistent results, they also acknowledge that when considering anything but the finest of loudspeakers in ideal listening circumstances (which may be very different from the ideal working circumstances in a control room), assessments may be based more on the balance of positive and negative attributes as they are revealed by different kinds of music, rather than on absolute responses. There are many loudspeaker designers and researchers who would strongly argue that 'a loudspeaker is a loudspeaker is a loudspeaker', and that a good loudspeaker should be universally usable, but the evidence presented in the previous few paragraphs is hard to argue against.

15.3 Different Needs

There is a fundamental difference between classical and rock music recording. It is generally true to say that reproduction of classical and acoustic music in the control room seeks to emulate the original acoustic performance. Conversely, the *live* performance of rock, and especially electronic music, seeks to emulate the original studio *recording*. In classical music, classically recorded, the original performance exists as a real entity in time and space. Much popular music, however, never exists as a single integrated performance in any one time and place until the final mix is first heard in the control room. It is created *on* loudspeakers for performance *by* loudspeakers, and so no *natural* sound balance ever exists. The concept of the 'real' reference is therefore lost.

Even in the recording of acoustic instruments for rock music, a recorded bass drum sound, for example, rarely sounds like the actual drum. The recording of bass drums has now become highly stylised, perhaps to comply with what a recorded bass drum is now expected to sound like. Even with a listener's head inside the bass drum, one could not hope to hear the 'acoustic' version of the recording. Notwithstanding the risk of hearing damage, the gross distortions and compressions that the ear would superimpose on the sound at such high SPLs would make any comparison of the real versus recorded sound invalid. And even *that* is not taking into consideration the change in the acoustic properties of the bass drum when a human head was placed inside it as opposed to a microphone. Much modern music therefore simply has no acoustic counterpart, even though it may be made up entirely of acoustic instruments.

Consequently, in the field of modern music recording, we are rarely, if *ever*, dealing with absolutes in terms of loudspeaker performance. We tend to be dealing largely with illusionary perspectives. Given that all loudspeakers are far from theoretical perfection, our choice of loudspeakers may be dictated by the most important aspects of their individual characteristics relative to any particular music or instrumentation. This is somewhat reminiscent of the discussion in Chapter 6, relating to the suitability of halls for orchestral recordings also sometimes being favourable to specific styles of music or instrumentation. As we are dealing with music, whose purpose is to bring pleasure to its listeners, anything that enables or inspires the people involved in the recording to reach new emotional heights must be valid. In fact, if we can for a moment split the recording process from the mixing process, there are some studios which are liked for their inspirational capacity during the recordings, but which during mixing would lead to unrepresentative results. Nonetheless, if they can lift a musical performance, then they have a valid function, but such idiosyncratic studios must be considered to be more of a performance stage (control room included) than a general reference for what is 'right' about a balanced mix.

15.4 What is Right?

Traditionally, the pressure amplitude response has been the most important aspect of loudspeaker performance. It has been relatively easy to measure, and it tends to relate well to the perception of a natural sound. Unfortunately, as can be seen from Figure 15.1, even maintaining a response accuracy of ±3 dB in an anechoic chamber is not easy, and, as any recording engineer knows, being able to adjust a graphic equaliser by ±3 dB over the full frequency range can result in a very wide range of different timbres. From this it can be correctly deduced that all the loudspeakers of Figure 15.1 sound different, even on-axis in an anechoic chamber. It is now also quite a widespread view amongst mastering engineers that a flat frequency response is not their prime concern when choosing a monitor system, although the *smoothness* of the response is highly valued.[5] Low distortion and transparency are also highly valued attributes, together with a good transient response. Frequency response, it would seem, as long as it does not deviate excessively from a gently changing curve, is something to which the ear adapts quite easily. In fact, as the frequency balance is so variable from one home to another, the provision of tone controls to make some desired compensations is almost universal on domestic reproduction equipment. Nevertheless, a reasonably flat frequency response is a necessary part of good studio monitor systems if we are to avoid deviating too far from a common reference point.

The transient response of a loudspeaker is something which had not really received due attention until the 1980s. This will be discussed in more detail in Chapters 19 and 20, but essentially, the time response of a system (which defines its transient response) is a function of the convolution of the pressure amplitude response and the phase response. The three are related by the Fourier Transform (see Glossary).

Over 150 years ago, Ohm, and later Helmholtz, carried out experiments that appeared to show that the ear was 'phase deaf'. Their experiments had been carried out on sine waves, which are *very* steady in nature, and not representative of normal musical programmes. Nonetheless, quite unbelievably, their work is still quoted in some circles as implying the relative unimportance of phase responses in terms of audibility.

Phase responses have an *enormous* bearing upon the waveforms of transient signals, changing considerably the way in which the shock of the transient excites the ear. Again, an argument had been put forward implying that if the waveform distortion changed the envelope only within the integration time of the ear, then the effect would not be perceived, but once again, our perception systems seem to be far more subtle than often previously believed. The concept of integration time states that if the ear samples sounds in periods of time of length x, then it would not matter whether ten units of a sound occurred simultaneously, or whether each was separated by a period of one-tenth x, or even any combination in between. The ear would still perceive ten sound units within that time window of length x, so the perception of any combination would be the same in all cases. In Figure 11.29, the two waveforms shown differ only in the phase relationship of the component frequencies. If the whole figure represented waveforms of less than the integration time of the ear, then classic theory predicts that they should sound the same, but they do not. The difference is more easily perceived in highly damped (non-reflective) listening conditions, but it does tend to disappear in conditions that are more reverberant. In Professor Manfred Schroeder's own words: 'In listening to the two waveforms an astonishingly large difference is perceived ... Too bad Ohm and Helmholtz were not able to listen to (AM and QFM) signals. They might have been very hesitant to formulate their phase law'.[6]

Non-linear distortions should also be minimised in an ideal monitor system, because the natural timbre of an instrument can be either enhanced or degraded by non-linear distortions. One reason why valve amplifiers should not be used to drive monitor systems is that the typical second harmonic distortion that they often produce can be quite musical in nature, but that could mislead the recording personnel into thinking that they had a musical sounding *recording* when they in fact did not. They would merely be listening to it through a musical sounding amplifier. (A more in-depth discussion of the audibility of non-linear distortion will be found in Chapter 19.) So, although a flat frequency response and a good phase response (and hence a good transient response), low distortion and an off-axis response which does not cause unduly coloured reflexions are all desirable attributes of a monitor loudspeaker, they nonetheless *do not* define the whole sonic character of a loudspeaker.

An interesting proposal was put forward by Watkinson and Salter in 1999,[7] which addressed a hitherto undefined aspect of loudspeaker performance. They proposed treating the loudspeaker as an information channel of finite capacity that can actually be measured as an equivalent bit-rate. It would seem obvious that when listening to audio data compression codecs, for the purpose of assessing their audibility, a judgement of the degree of audibility could only be made when listening through loudspeakers of the highest resolution. When listening via a conventional, domestic telephone, for example, it would be impossible to hear whether the signal being transmitted was 24-bit linear, 12-bit linear, or coded by AC3, MP3, or any other reasonable data compression codecs.

Conversely, if an audio signal was passed through a loudspeaker via a system for finely reducing a bit-rate, then the loudspeakers that rendered the smallest bit-rate reductions to be audible would be those with the highest resolution. A possible scale of resolution could be drawn up for different loudspeakers. The tests would need to be carried out in stereo, because some spacial aspects of codec signal degradation are not evident in mono. This concept would seem to be of value, because one reason why so many bad recordings are for sale in the shops is that they have become aggressive sounding due to the build up of signal processing artefacts which were not clearly audible on the loudspeakers in use during the mixing stage. In fact, one reason for the boom in the work for mastering houses has been caused by the increase in the proportion of studios that currently use totally inadequate monitoring systems. It also goes some way to explain why the mastering engineers are valuing transparency, low distortion and a good transient response as being more important than a dead flat amplitude response.

Essentially, a good monitor system is one on which great recordings sound great, average recordings sound average and bad recordings sound bad. This is not necessarily in agreement with the requirements for domestic hi-fi systems, whose prime function is to please the listeners. Really, nobody in their right minds would want to hear an unintentionally bad sound for the purposes of seeking enjoyment. In domestic circumstances, if a bad recording sounds less bad than it really is, whilst a good recording still sounds good, then for the purpose of enjoyment such a loudspeaker system would be well received. A studio monitor loudspeaker system, however, does not exist for the purposes of giving enjoyment. It exists to tell the recording personnel if they are achieving their desired objectives on the recording medium itself, and not only when heard via the monitors.

The aforementioned Watkinson and Salter presentation also addressed the fact that most rectangular box loudspeakers produced diffraction artefacts (as discussed in Chapters 4 and 11) which had a tendency to mask some of the loss of spacial sensation that could result from the use of signal processing devices with less than adequate electronic signal paths. This again adds to the argument that the principal monitoring system in a control room should be flush mounted. In general, and *only* in general, Table 15.1 outlines some of the performance compromise differences that have developed over the years between the needs of the rock/pop and classic/acoustic recording worlds. Ideally, of course, a loudspeaker should excel in all the performance characteristics, but size and cost considerations can give rise to prioritisation.

Table 15.1: Areas of design parameter priorities on two-loudspeaker stereo perception for rock/pop and classical/acoustic recording. Generalisations argument that the principal monitoring system in a control room should be flush mounted

Rock	Classical
High output SPL capability, e.g. 120 dB @ 3 m.	Lower maximum output SPL, e.g. 105 dB @ 3 m.
More tolerant of mid-frequency time/phase distortion if no acoustic source exists.	Minimum mid-frequency time/phase distortion for good localisation and natural timbre.
Extended low frequency performance at high SPLs, holding the response relatively flat as far down as possible.	Can tolerate generally lower low frequency SPLs and tailing-off responses, as long as smooth roll-offs exist in terms of both amplitude *and* phase.
Require relatively dead rooms with minimal lateral reflexions in order to support strong phantom, amplitude-panned images. Axial response more important than off-axis response.	Require rooms with lateral reflexions, hence need to have smooth, even, directivity in order to achieve due sense of spaciousness.
Whilst harmonic distortion should remain as low as can be achieved, this should not be done at the expense of transient headroom. The stylised, highly transient sounds of electronic music will often be more audibly tolerant of harmonic distortion than of amplitude limiting.	Require low harmonic distortion because the more steady sounds and subtle ambient information of classical music, along with the lower general levels of transients, make harmonic distortion perception much more of a problem than high level transient headroom.
A second set of smaller loudspeakers are considered *de rigueur* for overall musical balance assessment, and also as a reference to how the recordings may be expected to sound in cars, at home, and on the radio.	No such similar need has arisen, and the reference to a smaller set of (usually) poorer quality loudspeakers has not evolved.

15.5 Close-Field Monitoring

It became apparent in the 1960s that the sound as mixed in recording studio control rooms did not always travel well to the then new transistor radios. As it was via such radios that most people decided which singles to buy, the success or failure of a single was seen, rightly or wrongly, to be dependent upon how it sounded on the radio. Whether an extra 2 dB on the level of the lead guitar would really make the difference between the single reaching No. 1 or only No. 18 is a moot point, but the competition between the record producers, *and* their inherent insecurity, drove the studios to provide a typical radio loudspeaker, either in or on the mixing console. By the early mid-1970s, the Auratone Sound Cube had become a *de facto* reference, and it was reasonably generally agreed that it was easier to judge vocal and reverberation levels more easily on the small loudspeakers. Judge them, that is, in terms of how they would typically sound in domestic circumstances. In those days, most control rooms were still put together on an *ad hoc* basis, and only rarely were they designed from scratch. Somewhat perversely, we have returned to a situation whereby the vast majority of control rooms are once again not properly designed, so it is little wonder that close-field monitoring is still in widespread use.

Undoubtedly, the close-field monitors, with their restricted frequency range, do give a good representation of typical domestic reproduction, which takes place largely on loudspeakers of a generally similar size or smaller. Listening close to or within the critical distance also helps to remove a degree of room-to-room variability. Unfortunately, though, this does not tell the whole story about the recording, and boosts applied to a recording at 70 Hz whilst monitoring via loudspeakers that roll-off at 50 Hz may result in gross effects at 35 Hz. These can be deemed totally undesirable when the music is heard on a full range, truly high fidelity domestic system. Once again, in many cases, the mastering engineers are now expected to save the day.

It would seem to be incumbent upon a professional industry to deliver a well-balanced product to the marketplace. This will yield results correspondingly appropriate to the systems on which they will be played. It does *not* seem professional to deliver products that are exposed as lacking by the purchasers of audiophile systems. In other words, the quality heard from a recording should be proportionate to the quality of the system on which it is played. All too often, however, due to inadequate monitoring systems in the studios, the audiophiles in their homes are the first to realise just how awful some recordings are.

No studio can reasonably call itself professional unless it can provide the ability to monitor to a high degree of resolution over the great majority of the audio bandwidth. The close-field monitoring is therefore a useful adjunct to, but not a substitute for, a high-quality full range system, although the precise choice of whether the close-field system should be of high or moderate quality depends to some extent on the type of music being recorded and its intended market. Few people monitor classical recording through a pair of NS10Ms, yet some respected rock/pop producers and engineers claim to be able to mix solely by reference to them.

15.6 Why the NS10M?

During the twentieth century, probably no other loudspeaker saw such widespread use in the field of contemporary music recording as the Yamaha NS10M and its derivative, the NS10M Studio. The original NS10M was conceived as a domestic hi-fi ‘bookshelf loudspeaker’ (see Glossary), but it was a commercial failure as such. In the early 1980s, however, it was seized upon by the music recording world. As a result of this, and after many early problems due to insufficient power handling and excessive high frequency response for studio use, Yamaha launched the more specialised NS10M *Studio*, which continued in production until late 2001. The loudspeaker was small enough to carry around from studio to studio, which is largely what producers and engineers did with them in the early days, yet it packed enough punch when connected to an adequate amplifier to reasonably replicate the sound character of a typical studio monitor of the day. Its use in the close-field helped to standardise things even more, by removing the room-to-room variability to a large degree. At the same time, the frequency range resembled quite closely that of typical mid-priced hi-fi systems, which is little wonder, because that was the purpose for which it was originally designed.

The general consensus seemed to be that the NS10M had a hard sounding mid-range, a distortion performance that was not particularly good (though this is not borne out by measurements), a low frequency response that was adequate (whilst not being by any means 'extended'), and an output capability that suited the close-field monitoring requirements of most studios. The NS10M can certainly deliver a low frequency punch that in its early days was not common from a loudspeaker of such a small size. It is a loudspeaker unable to resolve fine detail, and the transparency of its sound does nothing to impress people. Yet, despite all of these criticisms, these loudspeakers were used, in their day, to get a musical balance on pop and rock music by probably more people than any other type of loudspeaker of their era. For many people, they worked well for the assessment of the relative instrument balance on electric music, though they would tend to listen to other systems to judge the individual timbres of the instruments.

It has been extremely difficult to find users of NS10Ms who could explain what characteristics they possessed which rendered them so widely accepted. Some very highly respected engineers and producers have used them to create some outstanding mixes over the years, without ever having been able to fully explain why. However, in Sections 19.10 and 19.11 we shall attempt a denouement.

15.7 General Needs

Notwithstanding the idiosyncrasies of close-field monitoring, in general the main monitor systems do show a tendency towards greater acceptance in line with improving objective performance measurement. Of course, the value-for-money aspect distorts the direct comparison of acceptance in terms of numbers sold, but the trend is reasonably clear. Direct comparisons are also made difficult by the tendency for some studio designers to favour certain brands and models of loudspeakers, and thus the market itself is not necessarily deciding things. This is totally understandable, because if a designer is trying to guarantee a performance specification of a control room as a whole, then the imposition of an arbitrary monitor loudspeaker by the client is not likely to be acceptable. Unfortunately, the current situation can still be summed up by another quote from Toole and Olive[4]: 'We make accurate technical measurements that we have difficulty in correlating with listener evaluations, and then compound the problem by making subjective evaluations that are unreliable. It is very difficult to make progress under such circumstances'. Vested interests in the marketing of the manufactured products also do little to help the search for true fidelity. Ultimately, we must use our experience and trust our own ears, but again this will be discussed much further in the later sections of Chapter 19.

The rise of Internet downloads and the use of mobile telephones as a means of listening to music via 'ear buds' and 'microscopic' loudspeakers has introduced a new factor into the monitoring equation. There are a growing number of people in the recording industry who see this type of reproduction as the be-all and end-all of their creative work. They seem only to be concerned with how their recordings will sound on these systems, but the perception of the sound from these types of systems is highly variable. When used in the street, with the background noise of traffic, there is often only a very small dynamic range available between losing the detail of the music in the traffic noise and risking hearing damage. The music is almost universally data-compressed, again reducing the overall fidelity. Making masters which are compatible with both mobile telephones and high-quality loudspeaker reproduction may, in many (most?) cases not be possible, and 'dumbing-down' a mix so that it can excel on the former is a concept which many engineers choose not to do. Nevertheless, financial pressures and a desire to be accepted can be powerful forces to resist, and studio designers are now being asked by some clients to install monitor systems which facilitate the making of mixes which will sound good on ear-buds. Optimising final mixes just to sound good on headphones of micro-loudspeakers is an unwise path to follow unless the only goal is to produce recording for a short-term popularity and a limited market. Who knows what the new fashion will be in two years time?

This sort of practice is taking the worries that Toole and Olive spoke about (quoted in the previous paragraph but one) to a new extreme of uncertainty. Following the short-term market fads would soon lead to chaos and we could forget the concept of professional recording. If such a philosophy was to succeed, and music was mixed to the lowest common denominator, then the rest of this book would become largely irrelevant because the maximisation of quality would no longer matter.

15.8 Summary

Loudspeakers are a long way from perfection. Their imperfections group together in different ways, making some loudspeakers suitable for some specific purposes and other loudspeakers more suitable for other purposes.

The sound fields radiated by loudspeakers in no way represent the sound fields of the acoustic instruments whose sound they might be reproducing.

Subjective preference differences are also an obstacle to finding a standard reference loudspeaker.

Time domain responses are now accepted as having much more importance than was previously thought to be the case.

A monitor system can also be thought of as an information channel of finite capacity, and can be measured as such.

Rock and classical music recordists have tended to choose different criteria as being most important for their specific monitoring needs, but that is not to say that convergence is not possible when more is understood.

Relying only on small, close-field monitors is a risky practice.

Optimising recording systems to match low quality, mass market listening fashions is not recommended. Reference standards become lost.

References

1 Moulton, David, 'The Creation of Musical Sounds for Playback Through Loudspeakers', presented at the Audio Engineering Society 8th International Conference, 'The Sound of Audio', Washington DC (1990)
2 Newell, Philip, *Studio Monitoring Design*, Focal Press, Oxford, UK, Chapter 22 (1995)
3 Newell, Philip, *Studio Monitoring Design*, Focal Press, Oxford, UK, Chapter 5, pp. 65–68 (1995)
4 Toole, F. E. and Olive, S. E., 'Subjective Evaluation' in: Borwick, J. (ed.) *Loudspeaker and Headphone Handbook*, 3rd Edn, Focal Press, Oxford, UK and Boston, USA, Chapter 13 (2001)
5 Newell, Philip, *Project Studios*, Focal Press, Oxford, UK, Chapter 8 (2001)
6 Schroeder, Manfred R., 'Models of Hearing', *The Proceedings of the Institute of Electrical and Electronic Engineers (IEEE)*, Vol. 63, No. 9, pp. 1332–1350 (September 1975)
7 Watkinson, J. and Salter, R., 'Modelling and Measuring the Loudspeaker as an Information Channel', presented to the 'Reproduced Sound 15 Conference' of the Institute of Acoustics, Stratford-on-Avon, UK (November 1999)

Bibliography

Borwick, John, *Loudspeaker and Headphone Handbook*, 3rd Edn, Focal Press, Oxford, UK (2001)
Colloms, Martin, *High Performance Loudspeakers*, 5th Edn, John Wiley & Sons, Chichester, UK (1997)
Eargle, John M., *Loudspeaker Handbook*, Chapman & Hall, New York, USA and London, UK (1997)
Newell, Philip, *Studio Monitoring Design*, Focal Press, Oxford, UK (1995)

CHAPTER 16

The Non-Environment Control Room

Lack of consistency in current recordings. The concept of the Non-Environment. Discussion of criticisms. Spaciousness – on the recording or in the reproduction. The origins of stereophony. Perception variables. Benefits of the Non-Environment approach. Comparison with stereo microphone techniques. Traditional standards. The origin of the Non-Environment concept.

In this chapter, we will discuss an example of a control room philosophy which is generally representative of the types of stereo control rooms which use hard, reflective front walls and maximally absorbent rear walls. Chapter 17 will put forward an alternative concept, that of the Live-End, Dead-End (LEDE) control room, which uses an absorbent front half and a reflective/diffusive rear half to the rooms. That chapter is written by a specialist in LEDE control room design, in order to keep a balanced viewpoint from the perspective of the book. By placing these two chapters together, readers will be able to judge for themselves the pros and cons. What is more, as *both* systems are in use in highly sophisticated professional studios, it should become apparent why the subject of control room acoustics can be so controversial, with partisan supporters of the differing philosophies all ostensibly seeking more or less the same objectives via some very different means.

What follows in this chapter is an updated version of a paper entitled 'A proposal for a more perceptually uniform control room for stereophonic music recording studios' which was presented in 1997 to the 50th anniversary conference of the Audio Engineering Society, in New York, generally outlining the philosophy behind the Non-Environment rooms.[14]

This chapter also discusses the evidence that many of the roots of monitoring inconsistency lie not only in some erroneous and outdated beliefs relating to reverberation time, but also in attempts to extract more from stereophony than it can simultaneously supply.

16.1 Introduction

Judging by the wildly different tonal balances on many recordings for sale to the public, the state of room-to-room compatibility of the listening conditions in recording studio control rooms still appears to fall far short of what a professional industry should by now be achieving.

Digital recording has brought its share of problems, but one problem that we must credit it with removing is the variability inherent within the recording media. We may still have differences in the sonic performance of A to D and D to A converters, but the consistency of the digits which leave the recording studios and arrive in people's homes is now largely guaranteed. The vagaries of analogue transfer to (and recovery from) vinyl disc or magnetic tape have largely been consigned to the past, so the excuses for much of the previous variability can no longer be used in defence of the current situation. Some variability or inconsistency in the spectral balance of recordings is no doubt within a range of artistic interpretations by recording staff, because it can be dependent upon what they feel is appropriate for any given track, but careful listening to much of the available recorded material will soon reveal that most of the variability is *not* intentional. The source of the problem certainly lies, to a large degree, in the variability of monitoring conditions in the control rooms of the recording studios in which the music is mixed.

There is a general over-reliance on inexpensive, close-field monitoring loudspeakers, and one of the reasons for the existence of this state of affairs is the lack of faith in the widely disparate range of combinations of large studio monitors and different control room acoustics. As the differing points of view as to what is 'right' have continued to exist amongst many experienced designers of both the loudspeakers and the rooms, it is little wonder that the recording staff have continued to show uncertainty about what *they* feel comfortable with. They have, in many cases, opted for highly personalised solutions that work for them, individually, on the most usual types of music that they record.

Studio designers are all aware of the problems inherent in the use of small, close-field monitors, such as the inability to produce the lower frequencies at appropriate levels, or the inconsistency in room positioning, and hence the uncertainty about how the room modes will, or will not, be driven. There are also the attendant problems of reflexions from the mixing console and other related equipment. What is more, it is true that except for a very few types of commonly used small monitors, their ability to resolve fine detail is sadly lacking. As a result of this, noises, bad edits, the operation of gates, the clashing of phase distorted effects artefacts (which often lead to undue harshness in the sound) and a host of other problems, all too frequently add themselves to the irregularities of spectral balance which occur due to entire octaves of the musical frequency range being left unmonitored.

16.2 Sources of Uncertainty

We cannot, and nor should we, be too dictatorial about all that occurs in an artistic industry. There needs to be a range of concepts of what is right in order to accommodate the individualities of different producers, musicians and listeners alike. All should be free to make their decisions for themselves, but those decisions can only be valid as long as they are aware of the decisions that they are making, and that they are not being led into them by ill-conceived control room monitoring conditions. In many cases, the designs have been based on a long held belief that a control room should possess a 'reverberation time' (RT) which approximates to an average of domestic listening conditions. RT is, of course, not an accurate concept for use in most small rooms, and certainly is not applicable to highly absorbent rooms, but nonetheless, RT_{60}, T_{60} or whatever other decay description has been used as a design goal has frequently had some domestic point of reference.

In turn, many other design principles have been based on aspects of stereo perception in the rooms in which the commercially available recordings will eventually be played. This shows admirable concern for the record buying public who keep the music industry alive, but it is also apparent that much of the good intent has been misguided. In trying to take into account the extraordinary number of factors that are involved in the domestic listening process, and especially in the light of the fact that some requirements for the optimal reproduction of different types of music and recording techniques are mutually exclusive, the efforts have frequently not achieved their goals. By trying to take into account so many of the variables in the reproduction end of the chain, the production/record end has itself suffered a lack of certainty that has unfortunately served to introduce even more uncertainty into the reproduction end of the chain.

At the 8th International Conference of the AES, in 1990, Floyd Toole presented a paper 'Loudspeakers and Rooms for Stereophonic Reproduction'.[1] The abstract began thus:

> *Stereophonic reproduction attempts to reconstruct, in the minds of listeners, replicas of the timbral and spacial effects of acoustical events that have occurred at earlier times and other places. It matters not whether the 'live' event consisted of musicians in a natural acoustical environment, or a multi-track creation monitored in a control room. In all cases, musicians and production personnel presumably heard a stereophonic reproduction that met their artistic and technical expectations. Assuming that the necessary information has been preserved in the recording, a replication can be successful only to the extent that the loudspeakers are capable of reproducing the appropriate sounds, and that the listening rooms are capable of conveying those sounds to the ears of listeners. Variations in loudspeakers and rooms create many difficulties in achieving this goal. Although it has been traditional to consider the loudspeakers and room as separate entities, this approach is no longer justified. The loudspeakers, room and listener comprise a system within which the sounds and spacial illusions of stereo are decoded, and they must be considered together.*

The above first part of Toole's abstract is a lucid, concise, and powerful summing-up of a complex situation in real life. The last sentence of the above quotation is very significant: 'The loudspeakers, room and listener comprise a system within which the sounds and spacial illusions of stereo are decoded ...' Decoded! If they are being decoded, then in the production process they must have been, in some way, considered to have been encoded. No encode/decode process can be expected to work optimally unless the decoder can track the encode process, and in order to do this, the encode process must be known, but in reality, the monitoring (encode) conditions during mixing are rarely known to the listener at home.

Recordings are not sold to the public with instructions about on which type of loudspeaker, and in which size, shape, or other property of room they should be auditioned. There is also absolutely nothing expected from the foreseeable future that will be likely to reduce the range of domestic listening equipment and conditions. Furthermore, music is *not* the most important thing in the lives of the majority of people, so it will continue to be normal for people to buy houses not for the acoustics of their rooms, but for a multitude of other priorities, and then find appropriate rooms in which to listen to the music of their choice. Different loudspeakers will suit different rooms, different types of music, different recording techniques and media, different budgets, and different personal tastes, relating to precisely what people like to hear in order to achieve the most enjoyment that they can from the music of their choice during its reproduction. There is, therefore, adequate justification in having available a good choice of reproduction equipment.

Taking many of these variables into account, there will be the audiophiles who choose their system and listening conditions with great care, and who may well enjoy optimising them for their favourite types of music, recording techniques and storage media. The most appropriate loudspeakers and listening conditions for rock music recorded on an analogue system, will, in all probability, be different from those most suited to the enjoyment of middle/side, stereo microphone technique recordings of digitally recorded classical music. However, even within the latter, quite highly defined set of recording conditions, there will be a wide range of available and appropriate reproduction systems and environments, and all will sound different. There is an argument for the case that different recording styles should be mixed in control rooms optimised for their own specific characteristics. However, the variation within the sub-groups is so great that to standardise on one arbitrarily chosen type of loudspeaker, at a given distance and at a given level, will only transfer many of the vagaries of the reproduction environment into the production end of the chain. This is because the decision made as to which equipment and conditions to use for the production process may be largely based on the results on similar equipment in the reproduction end of the chain, and this could lead to very volatile standards as fashions change.

What is proposed here is an abandonment of the attempts in the control rooms to try to accommodate so many of the variables in the reproduction systems, and to concentrate on a streamlined version of the more fundamental aspects of the production and quality control needs. This will allow recordings to be monitored in more detail and with more consistency. It will also, with the knowledge and skill of the recording staff, make it easier to predict what the results will sound like in different listening environments. This approach would appear to be far more realistic than attempting to mimic one set of variables with an average of another set of variables.

16.3 Removing a Variable

In 1994, a paper was presented to the UK, Institute of Acoustics[2] (IOA) entitled 'Control Room Reverberation is Unwanted Noise'. The paper put forward the concept of the Non-Environment rooms, which sought to provide monitoring conditions as close as could be achieved to free-field conditions. These rooms can reduce the decay time of reflexions and modal energy to such low levels that the perception of many recording defects becomes much easier. The paper also contained a discussion of the majority of the other widely used control room acoustic control philosophies. It noted the fact that, due to the sensitivity of human hearing systems, most attempts at producing optimised decay conditions for music monitoring had yielded control rooms that sounded subjectively very different, and tended to lead to different musical conclusions when mixing the same piece of music in different rooms.

Figures 16.1 and 16.2 show the general concept of the Non-Environment approach.[3,4] It can be seen that the side walls, the rear wall and the ceiling are made as acoustically dead as possible to as low a frequency as possible. The front wall is hard, heavy and reflective, and the floor is also hard. These two surfaces, together with the hard surfaces of any equipment

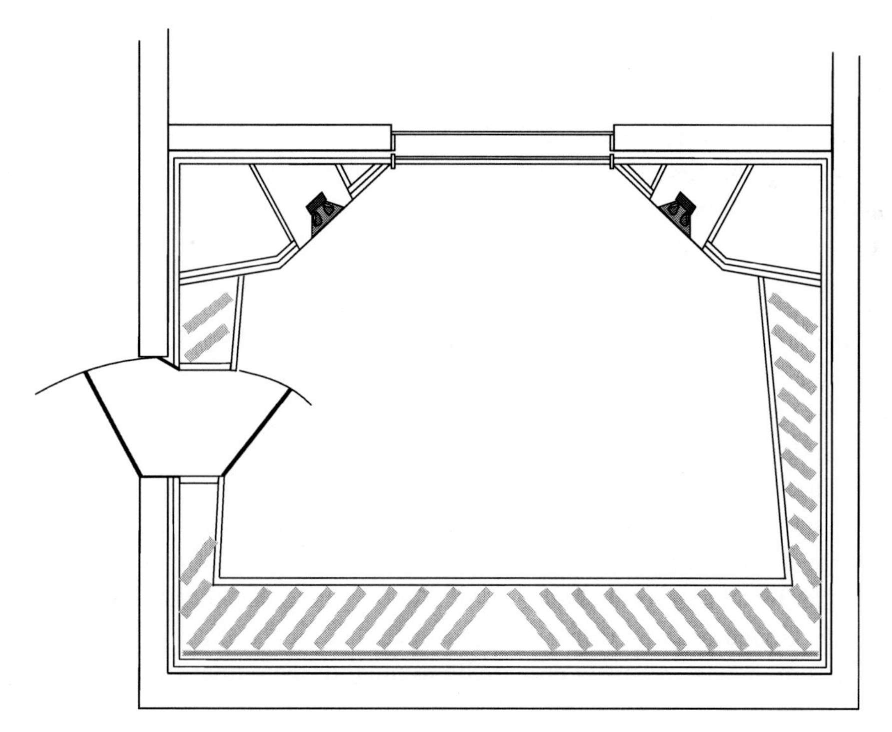

Figure 16.1:
Plan of Non-Environment control room. Shaded areas are wide-band absorbers.

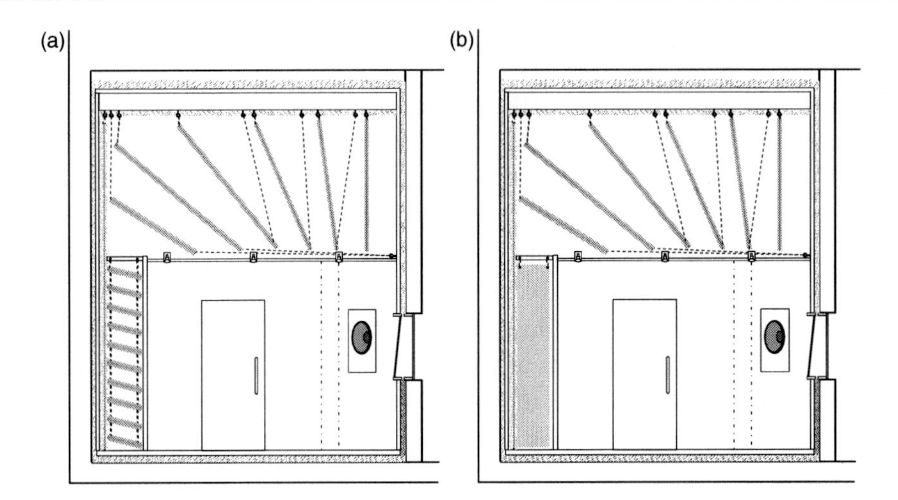

Figure 16.2:
Side elevation of Non-Environment control room: (a) Horizontal rear absorbers; (b) Vertical rear absorbers.

that may be facing the listener, provide a degree of acoustic life for sounds produced *within* the room, to alleviate any sense of being in an anechoic chamber. The loudspeakers are mounted flush in the solid front wall, and so are not actually *in* the room, but form a part of one of its perimeter surfaces. The front wall provides a large baffle against which the loudspeakers can push, thus aiding the efficiency and uniformity (flatness) of the low frequency radiation. The flush mounting also removes any response irregularities caused by cabinet edge diffractions, or by path length differences between the direct waves and the low frequency reflected waves from the wall behind free-standing loudspeakers.

Except for the floor and any equipment placed within the room, the monitors face something approximating a hemi-anechoic chamber. The acoustic conditions provided by the room are thus dependent upon whether a sound is produced within the room, or from one of its boundaries. In the two cases the overall decay characteristics of the room would be very different. From the monitoring direction, the reflexion problems from recording equipment can be dealt with by angling the equipment such that reflexions pass away from the listener and into an absorbent surface. If this cannot be done directly, then the offending surfaces can be protected, either by an absorbent shield, or by a streamlining device that will deflect the incident waves around or away from the object such that they will not reach the listening position. The aim in these rooms is principally to monitor the output from the loudspeakers, and nothing more.

By means of these techniques, rooms can be built which can achieve very high degrees of room-to-room compatibility by virtue of their relative absence of monitoring acoustics.

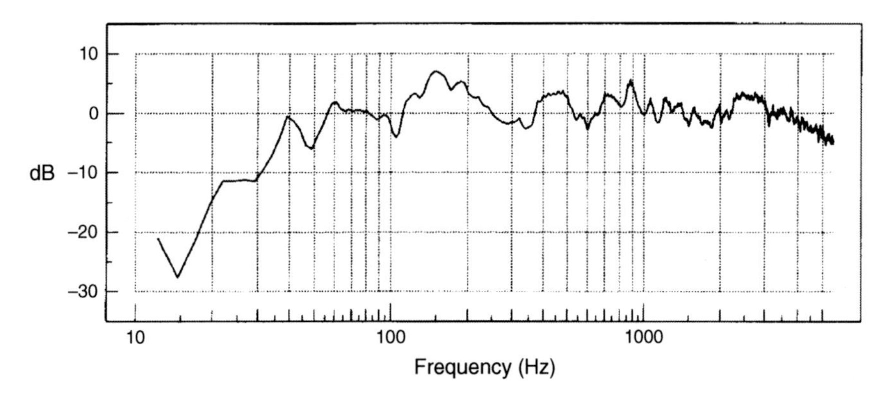

Figure 16.3:
Frequency response function of monitor loudspeaker at 2 m in a small Non-Environment room.

The studio designer Tom Hidley has been pursuing techniques of controlling the room modes down to frequencies as low as 10 Hz for his Hidley Infrasound rooms,[5] but some of the processes involved in achieving this very low frequency absorption lend themselves to the control of the more 'audible' low frequencies in much smaller rooms. This would seem to be important, because so many of the control rooms currently in use around the world are in the 25–35 m^2 region, and it has typically been this range of room size which has suffered so badly from inter-room incompatibility. Whilst the small Non-Environment rooms of different shapes and sizes have different ambient characteristics for general speech and noises produced *within* the rooms, (due to the different nature of the reflective materials and the different reflexion times in different sizes of rooms), they all have remarkably common monitoring characteristics. These characteristics are essentially those of loudspeakers, modified by whatever small ambient aberrations remain.

Figure 16.3 shows the frequency response function of one monitor loudspeaker at a distance of 2 m in a relatively small Non-Environment room. Figure 16.4 shows a measurement of a similar loudspeaker at 3 m in a large Non-Environment room. The two plots are remarkably similar considering the different room sizes. Figures 16.5 and 16.6 show the step responses of the monitors in the small and large rooms respectively.

In brief, the rooms are made highly absorbent at the mid and high frequencies by the use of conventional fibrous absorbent materials. Low-mid absorption is achieved as the waves pass through arrays of fibrous-lined ducts, formed by solid waveguide panels. The lowest frequencies are addressed by means of air-damped constrained-layer panel absorbers and 'deadsheet' membrane absorbers, which effectively line the room with a heavy, acoustically-dead, semi-limp bag. The overall control is provided by the whole system of absorption, but for the purposes of this discussion, the concept can be likened to an anechoic chamber, with one wall replaced by a hard wall in which the loudspeakers are mounted (effectively

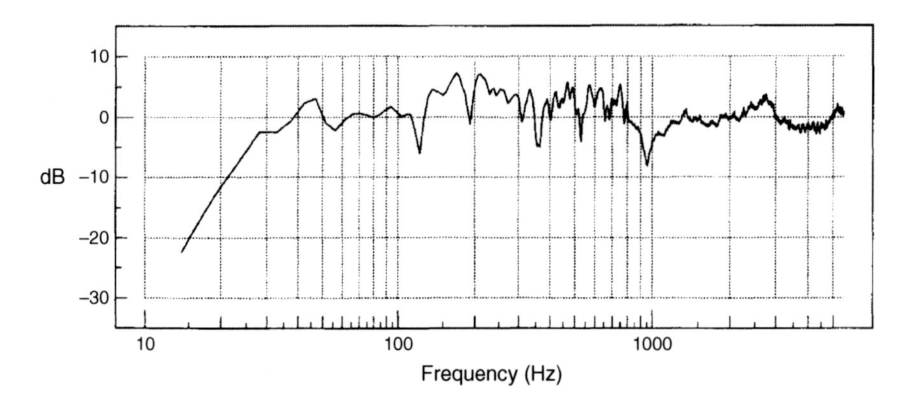

Figure 16.4:
Frequency response of similar monitor loudspeaker to that shown in Figure 16.3 at 3 m in a large Non-Environment room.

a hemi-anechoic chamber if not for the hard floor). The floor, in rare cases, may have openings at the front and rear of the room for the utilisation of under-floor absorption. This can go some way to reducing the floor induced response disturbances.

Irrespective of size or shape, such a termination will be highly uniform compared to that provided by more conventional rooms. The low frequencies may vary *somewhat* with room size and proportions, but with suitably effective absorption they are likely to do so to a considerably smaller degree than with most other current control room designs. What is more, the response perturbances caused by the rooms are likely to be at lower relative levels to the direct sound than is the case with most other rooms. This was one of the main benefits being proposed in the IOA paper.[2] Due to the reduction of room artefacts, the lower level

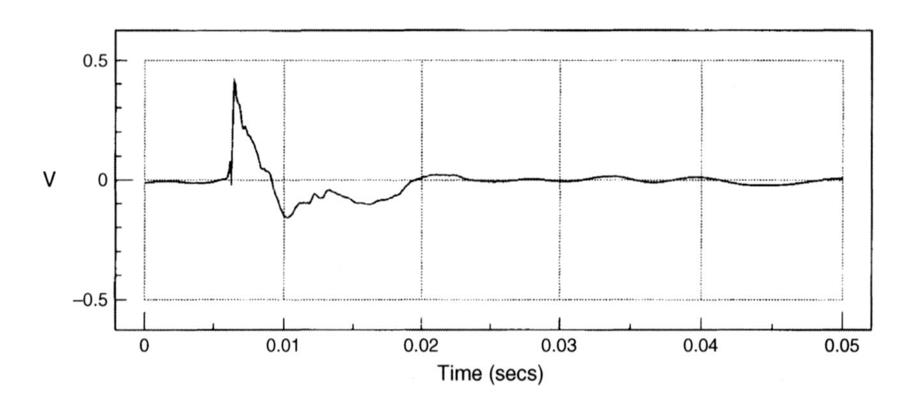

Figure 16.5:
Step response of monitor loudspeaker at 2 m in a small Non-Environment room.

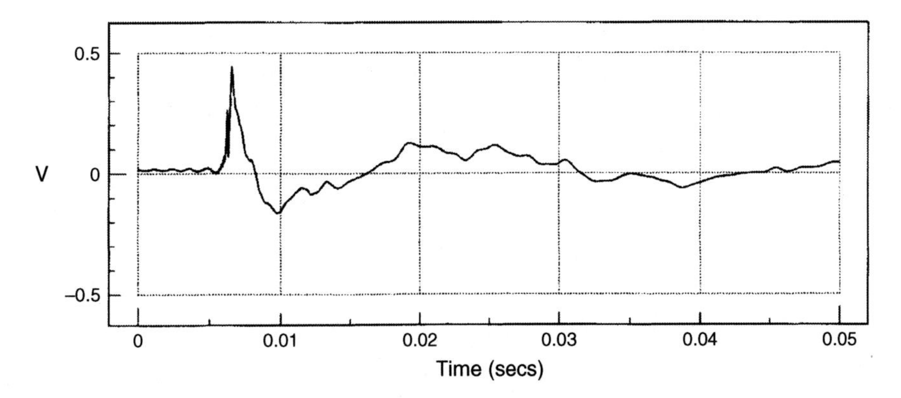

Figure 16.6:
Step response of monitor loudspeaker at 3 m in a large Non-Environment room.

details in the recorded music can be more readily perceived, and any unwanted aspects of the recordings, such as the audible operation of gates, can be dealt with before they become embarrassingly evident to the more discerning members of the recorded-music-buying public.

16.4 Limitations — Real and Imaginary

Over the years, a number of criticisms have circulated about the room concept being discussed here. Some of these comments have had substantial grounds to support them, such as the abandonment of the concept of having a 'domestic' decay time, but others have been based on misconceived theories. Examples of the latter type are comments such that the lack of modal support will produce rooms that are subjectively lacking in bass, and that an over-dead monitoring acoustic will lead to the excessive use of reverberation when mixing. The lack of modal support would only produce bass-light mixes if the decay time at the middle and high frequencies remained typical of more conventional control rooms. This was the case with some of the control rooms of the 1970s and early 1980s, where the excessive use of 'bass traps' was incorporated into rooms which still possessed significant decay times at higher frequencies. The Non-Environment rooms, however, are all-trapped, not just bass-trapped. A person who is used to working in a more lively room may initially be unaccustomed to the low decay time, but it is usually rapidly adjusted to, and the clarity and impact that the low frequencies possess is a revelation. If a recording is considered to be sounding too dry, then it is because a dry sound is what is on the recording medium. If it is considered desirable to remedy this, then either the mixes can be given more reverberation according to taste, or the recording acoustics or microphone location can perhaps be changed. Despite the criticism sometimes heard that working in low decay time environments leads to excessively reverberant mixes, experience has clearly shown that it is not so.

The fact is that when artificial reverberation is added, even in a totally dead room, it is unlikely to become excessive when played in a more reverberant control room because the decay responses of *any* reasonable control room will be short by comparison to the reverberation effects that are usually applied to recordings or mixes. One thing that is often noticed, though, is just how clear the reverberation tails in the recordings can be heard in very low decay time rooms. It is very useful to be able to monitor these carefully, because synthetic reverberation can produce some undesirable decay tail artefacts, which all too frequently go unnoticed in many control rooms. In low decay time rooms, the sound of the rooms in which the microphones were placed, or the different effects processor that had been used on a recording, become clearly recognisable to a degree that is normally only detected on headphones. What is more, every different conventional control room will produce different perceived ratios of short-lived versus resonant, quasi-steady-state sounds, certainly beyond the critical distance. The transient/percussive/consonant sounds tend to fall off at 6 dB per doubling of distance from the near-field of the source, because their life is too short to excite much modal resonance, but the quasi-steady-state signals may be supported by the different modal decay characteristics of the different rooms. This is a fact of life in all but the most acoustically dead monitoring conditions. It therefore suggests that relatively dead monitoring acoustics will allow more consistency in the perception of the transient versus steady-state balance of the sounds.

16.5 Spacial Anomalies

Three of the more substantial criticisms of the low decay time monitoring conditions are that they lack a sense of spaciousness; they are not representative of 'normal' listening conditions; and that in the smaller rooms of this type they fail to support an adequately wide area of stereo imaging. The last of the three points will be dealt with in some detail in Sections 16.7 and 16.8, but let us first consider the question of spaciousness. An *accurate* rendering of spaciousness can only be achieved by multiple, lateral reflexions, arriving from the directions and with the inherent delays that are appropriate to the performance space, whether that space was real, or imaginary. A less accurate sense of spaciousness, which is perhaps a more realistic goal, can still only be achieved by reflexions coming from a direction other than that of the stereo loudspeakers. Any sense of spaciousness (in the enveloping sense) is not therefore inherent in a conventional stereo recording. It will be dependent in its nature upon the reproduction acoustics. It can never be truly representatively 'monitored' at the time of mixing. Introducing an arbitrary set of reflexions into the mixing environment tends only to confuse matters. Surround sound helps us to tackle this problem somewhat more reasonably.

Spaciousness and the perception of detail tend to be mutually exclusive. This is true whether it is in the performance space, the microphone technique or the reproduction chain. Orchestral conductors hear more detail from their rostra than the audiences hear from the seats in the auditoria. The conductors need to hear the detail to be able to do their job, but most audiences like to hear the all-enveloping sound from the auditoria because it pleases them. Distant stereo microphone arrangements, such as spaced omni's, produce a greater richness of sound than close, multi-microphone techniques, but the latter can produce more fine detail, and perhaps more dynamic impact. The choice of which technique to use will be a creative decision by the people responsible for the recording. For critical monitoring, however, where the same compromises exist, it would seem that experienced personnel could far more realistically achieve their aims in rooms in which they could hear the fine detail, and *then* interpret how things would sound in a more reflective space. This would seem preferable to monitoring in rooms in which they could hear a spacious sound, but could only guess at what problems may lurk in the low level detail, masked by the 'spacious' reflexions of the room. In any case, it would not be too difficult a task to introduce suitable reflectors into a relatively acoustically dead room for a final and more spacious auditioning of the end result; once, that is, any problems in the finer details had been monitored and resolved.

The criticism about not being representative of domestic listening conditions would appear to be irrelevant. To date, all too many rooms that do attempt such domestic commonality often fail to produce the intended compatibility in the end result. Averages in themselves need not be representative. The average of the integers from 1 to 10 does not represent, even within $\pm 20\%$, more than 2 of the 10 integers (5 and 6). The majority of the integers would not be closely represented by the average. The worldwide range of domestic listening conditions is far too wide for any 'average' control room to represent. Motor cars and headphones, which now form a large part of the international listening environment, are also not represented by any average room. In fact, none of the normal arguments for control room specifications have much relevance for cars, headphones, or a wide range of domestic loudspeaker listening. What this seems to suggest is that we ought to know more about what is actually on the storage medium. This needs to be known (and recorded in a more predictable manner) in order for the disparate reproduction systems to be able to make more reliable attempts to decode the intentions of the recording personnel. The effect of possible reproduction environments must be deduced from the audiological and psychoacoustic cues in the recording, and how they will relate to the various listening conditions. In other words, the recordings should allow the maximum to be reliably extracted from them, without bias to any particular set of reproduction conditions, unless, that is, the recordings are being made for some highly specific purpose, such as television commercials or big screen cinema.

16.6 Solutions

By taking the control room acoustics out of the recording chain, the emphasis of the burden of monitoring uniformity shifts on to the loudspeakers. As loudspeaker performance has been converging faster than room performance, this simplifies the task of producing more compatible control room monitoring. Furthermore, much of the effort in loudspeaker design research has been involved with the amelioration of the problems caused by a typical loudspeaker/room interface. Loudspeakers designed for monitoring in Non-Environment rooms can concentrate on the optimisation of axial response performance, with less emphasis needing to be placed on the directivity problems well off-axis. This removes many compromises from the design process.

It is often the constraints of producing loudspeakers with a smooth, wide-angle directivity over a wide frequency range which restricts the choice of drivers in a monitor system. The fewer restraints that there are on driver choice, the easier it is to choose drivers for their sonic neutrality, low non-linear distortions, achievable SPL and many other parameters that the usual need for off-axis directivity control frequently does much to compromise. Simpler monitoring systems of excellent ability to reveal fine detail, and work at high SPLs, can be more reasonably priced. This can therefore spread the availability of more neutral monitoring conditions to a greater proportion of the industry. An affordable means of achieving a more uniform performance from the middle order of recording studios would be likely to have more effect on the recording industry's overall output than would be achieved by seeking to refine, even further, the upper echelon of elite studios, though that work should continue for its own valid reasons. One of the great benefits of Non-Environment rooms is that the techniques are not expensive, and apply with only minimal changes to control rooms of all sizes.

In the rooms being described here, phase responses become very important, because the absence of reflexions in the overall sound allows the detection of phase characteristics which even a single lateral reflexion can render inaudible. Many of these phase products, which are at the root of the harshness of many modern recordings, often go unnoticed, and hence also go uncorrected when using low resolution monitors in conventional rooms. With the absence of room characteristics in the monitor chain, the use of high-resolution monitor systems makes it much easier than is currently usual to achieve not only the desired timbral balance of individual instruments, but also the desired balance *between* the instruments. It also makes evident any non-linear distortions and the effect of any poor acoustics in the original recording spaces. The degree of openness and spaciousness *contained within the recording*, such as characteristics of 'transparency' and 'depth', can also be more easily assessed. (For further discussion of the audibility of phase, see Chapter 13.)

16.7 Stereo Imaging Constraints

Let us now turn to the other major point that has been raised in relation to these control rooms; their stereo imaging. Figure 16.7 shows a typical stereo perception area from a pair of loudspeakers situated in an anechoic chamber. The area is a function of geometry, so its actual size is determined by the distances between and from the loudspeakers, at least up to

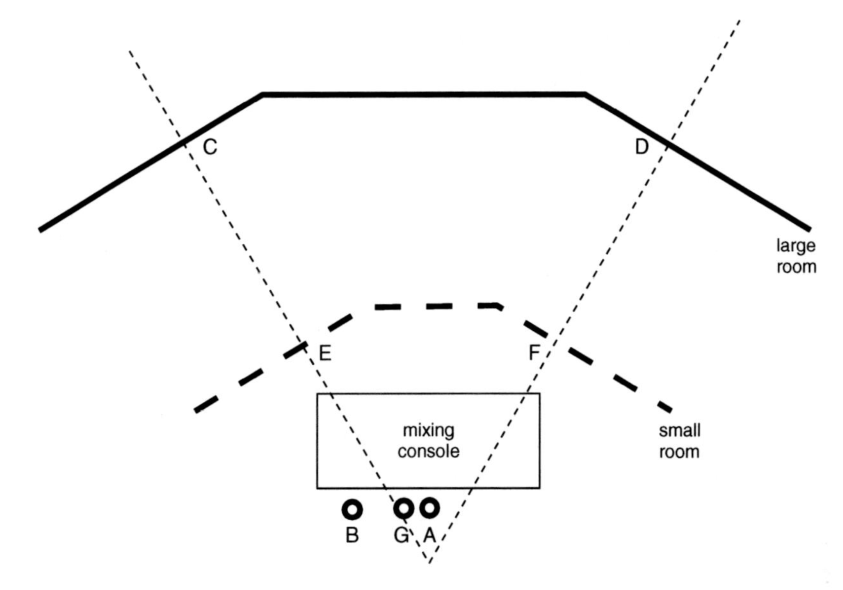

Figure 16.7:
Typical stereo perception areas for large and small rooms.
Moving from A to G in either situation would not significantly change the relative path length between the large-room monitors, C and D, or the small-room monitors, E and F. However, in the small room, moving to position B would make the relative distance to the monitors E and F 2:3. On the other hand, moving to position B would only change the relative distances to C and D by about 17%, quite different from the 50% in the case of the small room. Large rooms, therefore, tend to produce a wider and deeper listening area for critical use.

a point where the inter-channel delays become so great as to make stereo perception impossible. In large rooms with, say, 4 m between the loudspeakers and 5 m to the mixing position, the area available for stereo perception is sufficiently large to cover all the persons likely to be working behind the central 3 m or so of the mixing console. As the above dimensions decrease, so does the area of good stereo localisation. In very small rooms, and at close listening distances, the area of true stereo perception is perhaps only large enough for one person to appreciate comfortably. However, this should perhaps not be seen as a limitation of the room, but a clearer than normal demonstration of how two-loudspeaker stereo *should* behave. If clarity must be traded for spaciousness, then perhaps what we really need is ambient surround channels, rather than compromised two-loudspeaker stereo.

Moving from A to G in either situation would not significantly change the relative path length between the large room monitors, C and D, or the small room monitors, E and F. However, in the small room, moving to position B would make the relative distance to the monitors E and F 2:3. On the other hand, moving to position B would only change the relative distances to C and D by about 17%, quite different from the 50% in the case of the small room. Large rooms, therefore, tend to produce a wider and deeper listening area for critical use.

16.8 The Concept of Stereo as Currently Used

If we look back at the early history of stereo, there were two significant attempts at the reproduction of a 'solid' sound; 'stereos' being the Greek word for solid: a wall of sound, in other words. The first experiments relevant to the development of current stereophonic sound recording and reproduction took place in the 1930s, by Snow, Fletcher and Steinberg at Bell Laboratories in the USA, and by Alan Blumlein at what was to become EMI in the UK. The Bell scientists worked towards the reproduction of the originally recorded wavefront on a macro scale in the listening area, by using multiple spaced microphones and multiple loudspeakers. Blumlein, realising that a two-channel distribution system was all that would be commercially practicable in the then foreseeable future, considered the Bell proposals to be too much to ask of a domestically realisable system. He therefore opted for the implementation of a system relying on a set of psychoacoustic criteria that could reproduce, in the area of *a stereo seat* (Blumlein's own words), a realistic frontal sound stage using only a two-channel record/reproduce process.

The work at Bell Laboratories envisaged the likelihood of the use of at least three loudspeakers for reproduction, which they quite rightly considered superior 'by eliminating the recession of the centre-stage position, and in reducing the differences in localisation for various observing positions'. In the 1970s, 1980s and 1990s Michael Gerzon[6] put forward much work on new proposals for the three-speaker reproduction of stereo, some of which were totally compatible with two-channel recording systems. Although these more advanced proposals of Gerzon's would cover a considerable listening area, the early proposals of Bell Laboratories were still quite narrow, subtending an angle of only 35° at the listening position, so this was aimed at reproduction in larger spaces, such as in cinemas, where the listeners could be at some distance from the loudspeakers.

At EMI, Blumlein's aim was only to produce acoustical signals in a limited space around the head of one listener in a 'stereo seat'. This was intended to form an accurate virtual image of the source, by means of reproduction via two loudspeakers subtending an angle of 60° at the listening position. Blumlein's system constitutes the basis of what is now the well-established procedure known as Intensity Stereo, which inferred that simple level differences between the spaced loudspeakers would create both the necessary level and phase differences at the ears of the listener to produce a stereo image. This only occurs if each ear hears both loudspeakers, which is one reason why the stereo perception via headphones of a loudspeaker-derived mix can be so different, as no such inter-aural cross-talk exists with headphones. Shufflers can go some way to resolving this headphone problem, but they can also introduce problems of their own, such as position dependent frequency responses.

It was indeed possible for Blumlein's system to produce stable images between the loudspeakers by choosing suitable level differences between the left and right loudspeakers.

(The concept of 'left and right' is important here, because the effect is an aspect of human perception: the image supporting ability is not an inherent property of a pair of loudspeakers. The failure to fully appreciate this was one of the reasons for the failure of the many quadrophonic systems of the 1970s, where the assumption was often made, wrongly, that panning between a front/back pair of loudspeakers, on one side of the head only, would produce an analogous effect, which it does not.) The Intensity Stereo system is the one that the pan-pots of most mixing consoles employ, and which must surely be used in over 99% of all current recording processes. It is the implementation of Bauer's Stereophonic Law of Sines.

There is nothing limiting in the way that Non-Environment rooms present the stereo images, as the images perform exactly as one would expect them to perform, according to the way that the Intensity Stereo system was envisaged and implemented. (Incidentally, the Intensity Stereo referred to here has nothing to do with the psychoacoustic theories claiming intensity differences to be the key factor in localisation; here it merely relates to the level differences at the loudspeakers.)

Much work has been done in control room design to try to expand the area in which stable stereo imaging can be achieved, and the provision of certain lateral reflexions *can* serve to reinforce stereo localisation. Davis referred to 'Haas Kickers'[7] which are strong reflexions appearing after a suitably reflexion free period, and which help to maintain imaging. However, in many such ways, the means of supporting a wider stereo listening area are not the development of the concept of Intensity Stereo, but are psychoacoustic 'tricks' to help to extract more than the system inherently is capable of supporting. If a property is not inherent in the recording, then perhaps the enhancement techniques are best left for the final listening rooms, and not the control rooms. The problem in this is that the techniques tend to come at the price of compromises that must be made in other areas of monitoring. This latter point can be disturbing, as in the term 'control room monitoring' the words 'monitoring' and 'control' imply some sort of reference to a standard, which can hardly be the case if varying techniques are used to support the insupportable. What is more, if the control and monitoring are not defined at the recording stage, then to what standards do the domestic equipment manufacturers design their sound *reproduction* products? In the 'Studio Monitoring System' and 'Control Room' surely we must aim at some sort of tighter reference if the present unacceptably large range of end-product sound qualities are to be brought to a more repeatable equilibrium. Having said that, the current situation does provide a great deal of work for the 'Mastering' industry.

16.9 Conflicts and Definitions

There are a number of factors in studio monitoring which directly contradict domestic hi-fi requirements. Studio monitors are usually required to show up flaws and problems in the sound. They have an analytical requirement that is not normally necessary when listening to music solely for pleasure. Control rooms are for quality control as well as for the assessment of compatibility with the outside world. They are also, of course, used as creative environments, and that is a further aspect that makes its own demands. However, in almost all cases, the quality control function is degraded when attempts are made to imitate arbitrary domestic conditions, or to support artificially the stereo image stability over a wider area than was ever envisaged when the concept was formulated. It would thus seem that a very logical way to control the 'encode' side of the recording process is in rooms which simplify, to the greatest extent, the accurate monitoring of the signal which is being captured by the recording medium. Once there is a more reliable definition of the encode side of the system, it gives the manufacturers of, and the listeners to, domestic music reproduction equipment a better reference from which to make their own choices and decisions about how to get *their* desired 'best' *out* of the recording. The wider the tolerances are at the 'encode' side of the system, then the less consistent will be the ability of the reproduction systems to decode faithfully what the artistes and producers intended the listeners to hear. Arbitrarily designed control rooms do not aid the search for better standards of reproduction, because they are dependent upon far too many variables.

Toole highlighted the above point very forcibly in Section 2.4 of reference[1]: Reflexions and Absorption of Sound – Effects in Time and Space. This is not a simple subject, because:

1. The sounds radiated from loudspeakers in different directions are not the same;
2. The frequency dependent absorption properties of reflecting surfaces are not the same;
3. Listeners respond differently to sounds of different frequency;
4. Listeners respond differently to sounds of different temporal structure, e.g. impulsive or sustained;
5. Listeners respond differently to sound arriving at different times relative to the direct sound;
6. Listeners respond differently to sounds arriving from different directions;
7. Listeners respond differently to sounds in the presence of reverberation;
8. Listeners have many different perceptual responses; and
9. All of the preceding interact with each other and, to some extent, with the recording that is being auditioned.

That these interrelationships exist in domestic situations is incontrovertible, but surely all efforts should be made to remove as many of them as possible from the control rooms. The Non-Environment approach goes a long way towards achieving the lowest realistic number of room related variables.

Most domestic listeners want to hear music in a way that is pleasing. This is a valid requirement as they are seeking enjoyment, and they are at liberty to manipulate the above variables to suit their own requirements. However, what is pleasing should not be confused with what is on the recording medium. Stereo spaciousness can be very pleasing, but its presence in a domestic environment, *or* if created in a control room of any given design, is by no means necessarily an inherent property of what is on the recording. The use of early reflexions and reverberation can increase the stereo listening area, enhance the stereo listening pleasure, and extend it beyond the normal 'stereo seat' position,[8] but such techniques often compromise the detection of fine detail in low level signals, which in a monitoring situation risks allowing problems to pass by unnoticed.

In most truly professional studios, control rooms have already tended towards being less reflective than domestic listening rooms, undoubtedly because of a number of the above-mentioned reasons. Many professional recording personnel tend to prefer a more direct sound, even when listening for pleasure, as reported by Flindell et al.[9] In the paper 'Subjective Evaluation of Preferred Loudspeaker Directivity' they noticed that when their listening test results were separated into groups of naïve and professional listeners, the preferences of the two groups were very different. A few of the professional listeners even preferred frequency contoured reflected energy, which mimicked the conditions frequently encountered with the more directional loudspeakers in many control rooms. Many of the naïve listeners strongly favoured the spaciousness and extra high frequencies in the reflected sound, which were more typical of omni-directional (or multi-directional) loudspeakers in conventional rooms. No doubt there is a considerable degree of conditioning influencing the results for the professional listeners. Spending much time working in the conditions in which they do perhaps makes them more accustomed to hearing direct sounds. On the other hand, it is equally possible that as they are accustomed to listening for detail, such habits travel home with them.

The record and reproduce (studio and home) ends of the recording process have always been making their different demands, and it does *not* logically follow that the listening environment should be the same in both situations. Again quoting from Toole's paper[1]:

> *Strong reflected or diffused sounds from behind can seriously impair the clarity of the virtual sound images between the loudspeakers. Even at what appear to be safe distances the same can be true if reflecting or diffusing surfaces are large. A simple test is to reproduce monophonic pink noise at equal levels through both loudspeakers. For a listener on the axis of symmetry, the result should be a compact auditory image midway between the loudspeakers. Moving the head slightly to the left and right should reveal a symmetrical brightening, as the acoustical cross-talk interference is changed, and the stereo axis should 'lock in' with great precision. Start close to the loudspeaker and then move further away. It would seem a fundamental (minimum?) requirement that one should be able to find a stereo axis, and hear a clear centre image, in any position where critical judgements are made.*

If the new generation of cross-talk cancelling binaural and three-dimensional simulation systems is to be truly successful, a 'clean' acoustical path to the ears may be an absolute necessity. If a listening room garbles the crosstalk itself, it will most certainly garble the cancellation.

In a paper in the *Journal of the Audio Engineering Society*[10] in 1986, Jack Wrightson wrote:

> *The problem in the context of studio monitoring is that, regardless of the conditions, the room/monitor–loudspeaker combination places its imprint on all that transpires. For this reason a control room should be neutral, it should add as few sonic colourations as possible to the sound generated by the monitor loudspeakers. In this context, poorly designed loudspeakers should exhibit their flaws; well-designed loudspeakers should demonstrate their assets. The aural purpose of a control room is to provide the best possible free-air representation of the signals carried by the studio's audio system.*

Surely, the above conditions are most ideally met by the Non-Environment rooms of the type being discussed here. In this type of room, the conditions for neutrality and room-to-room compatibility would seem to be considerably greater than for any other concept of control room currently on offer. The number of variables in Toole's list in the previous paragraphs is significantly reduced.

1. Off-axis anomalies play little part in the proceedings.
2. Loudspeaker design is simplified.
3. As most reputable monitors have reasonably flat on-axis responses, the perceived difference when mixing with different monitors should be less than is all too often currently the case.
4. Reduced room decay time prevents the masking of low-level detail, an important factor in the 'quality control' process.
5. Reduced room decay minimises confusing timbre colouration caused by the room.
6. Reduced room reflexions enable precise stereo imaging, albeit over an area which is a function of room size and monitoring geometry.
7. Reduced room reflexions allow the detection of unwanted phase anomalies which can result from the over use, or inappropriate use, of effects processors.
8. Minimising room effects allow the various persons in the room to perceive the same musical balance between the instruments.
9. Reduced room effect allows the clearer perception of the ambience of the recording spaces or the use of artificial reverberation effects, and hence it is more easy to judge their appropriateness, or otherwise, to the recording.
10. Reduced room effect gives a greater possibility of working in other rooms of similar nature on a single recording project, even if the rooms are physically quite different, with a minimum of acclimatisation to the new location.

If the greatest price that must be paid for these advantages is a more restricted stable stereo imaging area in the smaller rooms, then it would seem to be a *small* price. When a mix is being built up, the desired timbre of an instrument can sometimes need to be changed in order to avoid masking by other instruments as they are introduced. Similarly, the optimum balance between the individual instruments can change. Just about the only thing that is usually *static* during the build up of a mix is the position of the instruments in the stereo panorama. Even in the smallest rooms where the stereo imaging will be true over perhaps only the space of one seat, then that seat will almost certainly always be available for occasional reference. Nothing about the imaging will suddenly change due to the dynamics of the mixing process.

In very small rooms, however, one should also consider the fact that the monitoring loudspeakers are often forced into positions where they cannot possibly subtend an angle of 60°, or less, at the monitoring position. This in itself will degrade the stereo imaging stability, irrespective of the type of room in which they are being used. Nevertheless, to subtend an angle of less than 60° in a small room would be likely to put any mixing personnel, other than the person on the centre line, outside of the loudspeaker pair, i.e. to the left of the left loudspeaker, for example. This situation would be less desirable, overall, than the less stable imaging produced by the greater subtended angle created by the wider spacing of the loudspeakers. Any comparisons of the stereo imaging in rooms of different design concepts should always take into account any differences in subtended loudspeaker angles, or the comparisons would be irrelevant.

Obviously, for the studios involved in the production of radio dramas or the like, where much more movement of the sound images is likely, the order of monitoring priorities *may* be somewhat different. In those cases, the greater use of dynamic panning, plus the possibility of having more people involved in the mixing process, may lead to a requirement for a large listening area over which the stereo sound stage was more stable. Perhaps this *would* take priority over the need for more absolute knowledge of the timbre of the sounds, however, the title of the paper on which this chapter is based did state '... for Stereophonic *Music* Recording Studios'.

If one is recording music for retail sale, and one *must* make monitoring compromises, then surely it is better that if there is one thing likely to be less easy to constantly monitor it should be the one thing which is *least* likely to vary. As not much tends to change in the static panning of a mix, then the stereo panorama is a good candidate for less constant monitoring. It should also be remembered that in larger rooms, the small 'sweet-spot' problem does not exist, and in the low decay time rooms the true imaging is better than in many other rooms, with all their attendant individual characteristics. Non-Environment rooms show stereo as it is recorded. If stereo is not enough, over two loudspeakers, then it is the *format* that should be criticised, not the rooms that show its failings. Surround sound systems are addressing this limitation to good effect.

Also noted in Toole's paper[1] were studies by Kuhl and Plantz[11] and Kishinaga et al.[12] Kuhl and Plantz, using only professional sound engineers as listeners, found that for dance and popular music, *plus* voice and radio drama, the preferences were for monitoring that was essentially the direct sound from the loudspeakers. At home, the majority of these same listeners, if listening to symphonic music, preferred a more reflective environment. Kishinaga et al. concluded from *their* investigations 'that in designing a listening room, optimum arrangement of absorbing and reflecting materials differs depending on the purpose of listening'. Recording/quality control and listening for enjoyment are very different purposes. Toole went on to say 'some recordings are clearly better matched to certain styles of reproduction than others. The situation (standardised listening conditions) would appear to be far from resolved'.

Indeed so, at the '*decode*' or reproduction end at least, where tastes and preferences lead to different conditions for maximum enjoyment of the music. However, if these same variables are allowed to affect the *encode* process in the studio control room, then it only leads to chaos in trying to decode-to-taste any set of non-standard encodings. Again quoting from Toole: 'In studio monitoring the general rule is to provide listeners with a sound-field that is predominantly direct. In these conditions, the principal impression of direction, image size and space are those that can be provided by the stereo signal itself'.

Surely, this is all that we can aim for in the studios. If we concentrate on what is on the recording medium, then the provision of a more consistently monitored product will allow the music-buying public to optimise their own listening conditions to suit their own pockets and preferences. Trying to guess what these conditions may be does nothing but harm to the encode process, and leads to absurd magnifications of the problems at the decode end. This being the case, the monitoring of the stereo in the Non-Environment rooms, without any enhancement or embellishments for greater enjoyment, would seem to be ideally suited to the production of recordings to a more consistent standard of reference, which should in turn make life easier for mastering facilities and the manufacturers of domestic equipment. Whatever that equipment may seek to achieve, its design and production would be made much easier without the often unintentional variability of the recorded material, affected as it is by the vagaries of current control room monitoring.

16.10 A Parallel Issue

In 1986, Stanley Lipshitz published a paper[13] on the subject of the spaciousness and airiness of different techniques of recording using spaced microphone techniques. The following quotations are taken from that paper, and many parallels can be drawn between the lack of detail and false spaciousness of spaced microphone techniques, and the loss of detail perception associated with the false spaciousness which results from anything less than 'direct' monitoring.

On perceived spaciousness:

> *I believe that spaced-microphone techniques are fundamentally flawed, although highly regarded in some quarters, and that coincident-microphone recordings are the correct way to go. The 'air' and 'depth' so valued in spaced-microphone recordings are shown to be largely the artefacts of phasiness due to the microphone spacing, and not acoustic ambience at all.*

> *I shall try to make a strong case for the use of single-point (i.e. coincident) stereophonic microphone techniques in preference to widely spaced microphone configurations.*
> *I am aware that I am treading on dangerous ground here, in that an aesthetic judgement is called for when attempting to rate stereophonic recordings as 'good' or 'bad'.*
> *Often it is the case that the more 'ethereal' the sound images appear, then the better the system is appreciated. Such systems can be regarded, however, only as attempts at pseudo-stereophony.*
> *I consider such blurring to be a defect, although I will admit that some people like soft-focus (photographic) lenses.*

On stereo reproduction:

> *The problem of freeing the listener from the 'stereo seat' by enlarging the region within which the image remains reasonably free from distortion is, in my view, a reproduction related question rather than one bearing directly upon the recording technique.*
> *If more than two transmission channels are available, one can do much better.*
> *For such reproduction systems (for example Ambisonics) an acoustically dead listening room would be preferable. It is my belief that as more sophisticated reproduction systems become available, the correct trend will be toward more anechoic listening environments.*

On the psychoacoustics of stereo:

> *Of primary concern is the fact that the ear on the side of the earlier loudspeaker need not receive the louder signal, and indeed at low frequencies does not! So the inter-aural level differences produced at low frequencies do not always reinforce the image produced by impulsive sounds. Sometimes, the low frequency image pulls in the opposite direction from the image of the transient, broadening and smearing the overall image.*
> *So we must consider stereo hearing as distinct from natural hearing, and actually quite unnatural – it is in fact an artificial creation.*

And, on the impact of modern recording technology:

> *The last few years have seen a dramatic improvement in our ability to accurately record, distribute and reproduce musical signals, and the benefits of this digital technology are now available to consumers in their homes.*
> *What is on the master tapes is now laid bare without the masking effects of the earlier technology, and what the consumer can now hear is frequently unpleasant.*
> *I feel that the source material (not referring to electronic music here) is now the weakest link in the chain from the artist to the listener, and that improvement here requires an enlightened reassessment of what goes on in the process of capturing the original sound and reproducing it through two loudspeakers.*

All of the above quotations from Stanley Lipshitz would seem to point to the need for detailed and direct monitoring as the only means of 'hearing into' what is *really* on the recording medium, and that spaciousness should, as discussed elsewhere in this chapter, be an aspect of the final reproduction environment. For detailed monitoring, it would appear that spaciousness and the resolution of fine detail are largely mutually exclusive. It should be recognised, however, that the authors (Newell and Holland) may possess a sensory bias towards the more detailed types of monitoring, as they admit to having a general dislike for soft focus photography: but also, it would seem, does Stanley Lipshitz.

So ended the AES conference presentation, but perhaps before closing this chapter it would be informative to look at how the concept initially developed. Firstly, however, this must be set in the context of the circumstances of the time.

16.11 Prior Art and Established Ideas

The traditional approach to control room design had been to achieve a smooth frequency response in a room with a decay time that was deemed representative of domestic, end-user, listening rooms. Apart from the standard decay time range relating to these outside-world conditions, there were two other proclaimed reasons for the choice. The first was the desired compatibility of mixes when work was transferred between different control rooms, and the second reason was the perceived general comfort of the personnel working in a 'typical' ambience.

Nevertheless, different specifications were drawn up which sometimes were excessively broad. The European Broadcasting Union (EBU) specifications, for example, are so broad as to be all but worthless; at least for serious music recording purposes. The fact is that the political problems incurred in trying to settle on a specification that 28 countries could agree to is something from which only 'men in grey suits' could be expected to gain any satisfaction. Why such specifications would be all but worthless if applied to commercial music control rooms is that many rooms can be built, entirely within the specifications, which are all very subjectively dissimilar. This does nothing to address the control room-to-control room compatibility problem, and if control rooms can be so dissimilar to each other, how can they *all* relate to some common concept of the outside world?

Furthermore, even if rooms *are* built to an acceptably tight standard, then this still offers little overall chance of achieving any significant level of compatibility as long as the choice of position for the monitor loudspeakers remains entirely arbitrary. Inevitably, if one moves a loudspeaker to different positions in any room other than an anechoic chamber, the cumulative response of the direct and reflected sounds will be different for each position. This renders flush mounting of the monitor loudspeakers almost mandatory if compatibility from room-to-room is desired. At least if one mounts the loudspeakers flush in a wall there is little risk that people will unwittingly move them from their preferred location, and hence perhaps also move the performance of the room/loudspeaker combination out of its desired range.

The compatibility issue, both between rooms *and* to the outside world is well addressed by the Non-Environment concept of control room design. The principle behind the idea is that any variables that do exist between rooms are acting upon a very small amount of room ambience. There is therefore less subjective variability in the room response because there is so little room response for the variables to act on, at least from the direction from which the monitors 'speak'.

This concept moved many such rooms outside the more accepted, legacy-standard parameters for room design, but experience has shown that working in this way has *not* produced mixes that subsequently sound 'wrong' when played in rooms that are more conventional. (At least not beyond any subjective or objective 'wrongness' inherent in the loudspeakers and/or the rooms in which they are being auditioned.) Surely, this *should* be the case. When a mix is done in a room which is 'somewhere within' the accepted specifications, it can sound either similar, better, or worse in a domestic room, depending upon how the control room's own inherent built-in misbalances coincide with the idiosyncrasies of the rooms in which the music is later heard. Ideally, a control room should add *nothing* to the recorded sound, or its monitoring conditions can only be considered arbitrary.

Much of this of course is enshrined in the philosophy of 'near-field' monitors, where loudspeakers of various types are used at close ranges. By this means, the characteristic sounds of a clutch of loudspeakers can become well known to a great number of recordists. If these are used at close range, particularly inside the critical distance (beyond which the direct sound sinks below the reflected sound), then even if such loudspeakers are not particularly *well liked*, at least they would be *well known*. Under such circumstances, room-to-room transfers of work can be judged with minimal variation. Unfortunately, such monitors frequently lack good transient accuracy, rarely do they have the transparency available from the better, bigger systems, and almost universally they fail to be able to monitor the full frequency range of the recordings. They usually also lack the necessary dynamic range to be able to assess properly low-level details in the presence of peak level sounds. Nonetheless, there is little in the concept of the close-field monitoring approach which in any way conflicts with the philosophy of the 'Non-Environmentalists'. What the Non-Environment approach offers is the possibility of enjoying the positive aspects of the close-field approach without its restrictions and limitations. Effectively, it extends the size of the close-field to that of the entire room. In many cases, the critical distance lies outside the boundaries of the rooms. This is especially so in the case of small rooms, and this is a very important point.

When people look for convenient buildings in which to site studios, they are often forced into compromises. Access to the toilets may be a problem, for example, requiring a corridor where the control room would ideally be sited. Therefore, control rooms do not always take priority in any absolute sense, and they must often be sited less than optimally. To try to get a standardisation of size and shape is totally impossible. As we have discussed previously, different rooms of conventional design will have differently perceived acoustic characteristics if the sizes, shapes and materials of construction are not identical. Hence, if standardisation is impossible, and the variability in room geometry is inevitable, the control rooms can never be considered to be performing to any tight reference standard *unless* the monitoring acoustics can be reduced to almost zero, because it is only zero which can be multiplied or divided by a range of variable factors without changing the result.

The Non-Environment approach advocated here adequately addresses the above problems, and it does so in rooms which remain comfortable for people to work in for hour after hour, week after week. In practice, the much-vaunted greater working comfort of the more traditional control rooms never became an issue, and furthermore, reliable monitoring added its own comfort factor. Until this option (and its close relatives) became available, acousticians had not been getting anywhere near as far as the industry needed in terms of achieving these compatibility levels. The whole close-field monitoring concept was borne out of the recording staff taking the problem into their own hands, and dealing with it in the only way that they knew how; sitting close to the loudspeakers. Given the fact that they chose to suffer the limitations and privations of such monitoring techniques, rather than work with endless room-to-room 'correction factors', was a strong statement of their dissatisfaction with the general state of monitoring affairs.

16.12 The Zero Option – the Origins of the Philosophy

In fact, it was the compatibility problems that prompted Tom Hidley's early retirement at the end of the 1970s. Until then, it was widely held that the one-third octave pressure amplitude response in the room was the strongest governing factor in the assessment of a room's 'rightness', despite the fact that it took almost no notice of the phase response, and hence the transient response. The whole 'room voicing' concept by means of graphic equalisers is now thoroughly discredited *and passé*, but Tom Hidley, amongst many others, learnt that one the hard way.

It was a chance set of circumstances in the early 1980s that drew him towards the Non-Environment concept. A previous client of his called him just after his 'retirement' and asked if he could make some changes to an older room. One current catchphrase at that time was that rooms were 'over-trapped'. The bass was not being perceived in the correct balance with the mid and high frequencies. Nobody seemed to be able to clearly define exactly what was wrong, because the spectrum analysers were reading flat, but it was a feeling that had gained some widespread acceptance. After making some modifications to the room, one of its subsequent users was Stevie Wonder, who, of course, is blind. It was noticed that when he was speaking about the sound, he kept pointing to the loudspeakers, but the places to which he was pointing were not consistent with the actual loudspeaker locations. Realising that something was amiss here, Tom Hidley thought that there must still be an excess of reflexions. He decided to *increase* the room's absorption, even though this was totally contrary to the suggestion that they were already 'over-trapped'. Nevertheless, the move was greeted with general approval, so he went back into his self-imposed retirement with quite a lot to think about.

A couple of years later, he was talked into visiting a company in Japan who wanted two studios of similar style to some of his earlier rooms in Tokyo. Hidley was a reluctant participant in all of this, because the compatibility impasse had still not been resolved. He eventually agreed that he would design *one* control room as the client wanted, but that the other would be designed to his own, latest, but still untested thinking. However, he imposed the additional condition: whichever of the two finished control rooms was least liked by the musicians and engineers would have to be demolished, and rebuilt according to the design of the preferred room. The 'winning' room was that of his first Non-Environment design, but it was radical and different, and was not immediately familiar to all its users. Nevertheless, it was also in Tokyo that designers such as Sam Toyoshima and Shozo Kinoshita were operating, and Toyoshima was also a strong advocate of very low decay times, hard front walls, flush mounted monitors and maximally absorbent rear walls (see also Section 13.1.5). Japan was therefore fertile ground for this sort of concept, which was almost diametrically opposed to the LEDE concept that was sweeping through the USA and was beginning to gain some acceptance in Europe. This was also the time of Tom Hidley's first use of the vertically aligned Kinoshita monitors, as shown in Figure 13.9.

Hidley soon came out of his retirement to produce more rooms, encouraged by what he had heard in Japan, but many of these rooms were monstrous by the standards of the day, as the low frequency aspects of the absorption techniques were very space consuming (see Figure 13.7). In the meantime, the author had been trying to achieve consistency, in his own way, by pointing rather directional monitors at an absorbent back wall absorber, but the problem remained of what to do about the variation in the omni-directional low frequencies between rooms of different shapes and sizes. A means for more effective low frequency control was badly needed.

Coincidentally, Tom Hidley was also looking for more effective low frequency absorption, but this time for the infrasonic region. He was designing his '10 Hz' control rooms for BOP TV in Boputatswana. Some joint research was therefore proposed that culminated in the one-tenth scale modelling work by Luis Soares at the ISVR. During the course of this work, Keith Holland was also coming to the end of his doctoral research on mid-range horns, and there came a time when both he and Luis Soares needed more measurements from real studios.

A weekend trip to London was arranged, in order to visit five different control rooms, each having a different designer and design philosophy. There was only one of Hidley's Non-Environment control rooms in London at that time (1989), and neither the author, Holland nor Soares had seen or heard one before. After the five control rooms had been visited, measured and listened to, including a then current Newell design, the consensus was unanimous that the Non-Environment room at Nomis was closest to what all three people present agreed that a control room *should* be. Luis Soares went back to his one-tenth scale modelling of the absorption systems, but it was not going well. (The scaling of complex interactive systems is notoriously difficult.) What was needed was a full-scale model, but there was no room at the ISVR (Institute of Sound and Vibration Research) where such a model could be built.

A flexible client for a studio design was subsequently found in Liverpool, UK, and in the new year of 1991 a full-scale model was begun, in which Holland and Soares could continue their experimental development work prior to the studio opening for business. (Purely coincidentally, the studio was called The Lab, as it was built in an old chemical analysis laboratory.) The control room was built to the same general principles as Hidley's Non-Environment room in London, though the shape was different, the size was different, the building shell was different, the materials of construction were different and the monitor system was different, although conceptually compatible to a large degree with the Kinoshitas in Nomis.

When everything was finished, as shown in Figure 16.8, Holland and Soares went to Liverpool with a car full of test equipment from the ISVR. Once all had been confirmed to be performing as intended, it was time to play some well-known CDs. Everybody in the room was stunned by the detail which could be heard in the music, but what was more, those in the

Figure 16.8:
A complete room. 'The Lab', at the Liverpool Music House (UK), in 1991. In this case, the floor is wooden parquet blocks.

room who knew Hidley's Non-Environment rooms were hearing something that they recognised. The degree of compatibility between Nomis and The Lab was remarkable, and yet there was hardly anything the same in the construction materials or the equipment. What was clear was that it was the *concept* that was giving rise to the compatibility, not the details: hard front wall with built-in monitors, hard floor and all other surfaces maximally absorbent, which could be trimmed to taste to add any desirable extra life.

The control rooms were still rather large, though; The Lab being about 48 m^2 and Nomis much larger (see Figure 18.1). The pressing problem was how to make the concept work in smaller rooms, which could not afford the loss of space for the absorbers. The breakthrough came with the use of membrane absorbers in front of panel absorbers, which provided reasonably useful wideband absorption in a depth of only about 30 cm, although 80 to 100 cm was necessary if more ideal results were required. It was like lining the surfaces of the rooms with a limp bag. Although there were some differences between the large and small rooms, the concept proved to be sufficiently robust to retain the compatibility in the music mixing. However, it was not until about ten rooms had been built could it be reasonably assumed that good compatibility could be achieved from a wide range of shapes and sizes. This was very much an acid test, as the first ten control rooms referred to ranged from 80 m^2 down to 12 m^2 (240 m^3 to 30 m^3), although the very small room was not surrounded by a heavy, rigid containment shell, and therefore some of the low frequencies escaped to the corridor and lounge areas. The finished room is shown in Figure 16.9.

The first obstacle faced was that the rooms did not sound quite like the previously accepted 'professional' rooms, but eventually people began to realise that the mixes were translating well to the world at large. The usual problem of taking something home to listen to it, then going back to the studio and remixing it, was becoming a thing of the past. Three years later, people were moving between the control rooms with absolute confidence and only minor compatibility problems. Since then, the number of rooms using this concept has multiplied, both for music mixing and in cinema control rooms for the mixing of soundtracks, as described in Chapter 21. At the other extreme of the size range, Figures 16.10 and 16.11 show a control room constructed in a space of only 17 m^2 in a reinforced concrete 'bunker', which was a very different situation to the starting conditions for the room shown in Figure 16.9. However, there was an available ceiling height of around 4 m, and so the primary low-frequency absorption was due to the sandwich structure of the 'box' (the basic construction as shown in Figure 5.5) and the 1m, or more, of absorption in the ceiling.

It was the success of the compatibility of these rooms, both between themselves *and* with the outside world, and not only on conventional home systems, but also on headphones and in cars, which prompted the writing of the AES paper on which the first part of this chapter was based. It was the 'The Zero Option': if something is going to be acted upon by a variable multiplier, then the smaller that that something is, the smaller will be the variation. If a room is

Figure 16.9:
A very small Non-Environment control room: Noites Longas, Redondos, Portugal. The room was built in a shell of only 3 m×4 m, with a sloping roof, but the acoustic situation was helped by the fact that the containment walls were not very rigid. The low frequency isolation to the lounge, behind the control room, was not considered to be very important. The principal performing room is shown in Figure 5.23. Both rooms were floated, and situated on the top floor of a house.

going to impose its character on the sound of the monitors, then the smaller the room sound is to begin with, the less it can affect the monitoring, even if it is variable from room to room. It is the concept of close-field monitoring, but with big monitor performance. Some construction and performance details can be found in Appendix 1, which follows Chapter 26.

Of course, the rooms had their critics. 'Boxes full of Rockwool … and the end of acoustic design' was the comment of one designer in the international recording press. 'Like every other neat theory … I guarantee that it is an oversimplification', was the reaction of another. Such scepticism is perhaps healthy, as it is only out of debate that ideas are usually developed. Nevertheless, all this is of no interest to the musicians and recording staff who neither know nor care about room acoustics. They just want rooms which make their daily work easier, more consistent, more rewarding, and which can help them to relax into their work with the confidence that hearing is believing. In the opinion of the author, the Non-Environment approach can most consistently deliver these things.

(a)

(b)

Figure 16.10:
(a) and (b) Two views of a control room under construction in a space of only 17 m^2, but with around 4 m height, surrounded by a concrete isolation bunker.

Figure 16.11:
The finished control room. Neo Musicbox, Aranda de Duero, Spain, 2008. (See also front cover.)

16.13 Summary

If recording personnel are choosing to work in the close-field of small monitors, despite suffering the limitations of small loudspeaker boxes, then this begs the question as to why not extend the close-field of the large monitors by reducing the room decay time from the monitoring direction.

This can be done, *and* without unduly deadening the room ambience from the direction of the people within it.

Highly damped rooms also exhibit less spacial variation of the sound, which is normally caused by the room modes.

Domestic listening conditions are very variable. It would seem unwise to allow the concept of average listening conditions to influence unduly the control room acoustics – the result of such influence is usually a lack of certainty in the mixing environment.

If conditions cannot be standardised at the mixing stage, then how can hi-fi equipment be made to match the reproduction response to that of the production environment?

The Non-Environment rooms have maximally absorbent rear walls, side walls and ceilings.

The monitor loudspeakers are flush mounted in a hard front wall, which is reflective to the sounds of people in the room, but from which the musical sounds from the loudspeakers cannot reflect, because they are propagating *away* from it.

The floors are generally hard.

Such rooms, by the absence of their monitoring ambience, are very consistent in the monitoring character from room-to-room.

The rooms should not be confused with the 'bass trapped' rooms of the 1970s and 1980s, which could lead to bass-heavy mixes. The Non-Environment rooms are 'all-trapped'.

Spacial sensations created in a control room, which are not an inherent part of the recording, cannot be considered part of a true monitoring process. Any spaciousness enhancement should be left to the final reproduction room acoustics.

Spaciousness and the perception of detail may be mutually exclusive. For critical monitoring it is probably better to aim for the more accurate perception of detail.

In Non-Environment rooms, loudspeaker design is simplified because less importance needs to be attached to off-axis responses.

Characteristics such as 'transparency' and 'depth' can also be more easily assessed without the confusion of room ambience.

Stereo, as we currently know it, was conceived as a sensation for a 'stereo seat'. Control rooms that compromise their clarity of monitoring in order to try to broaden the stereo perception area are trying to get something from two-loudspeaker stereo that it was never intended to provide.

Surround sound systems are the solution to more spacious stereo.

The low decay-time control rooms tend to achieve the lowest number of room-related variables of any of the current control room philosophies.

There is a lot of learned opinion that leans towards lower room decay times for more accurate monitoring conditions.

References

1 Toole, Floyd E., 'Loudspeakers and Rooms for Stereophonic Sound Reproduction', Proceedings of the AES Eighth International Conference, Washington DC (1990)

2 Newell, P. R., Holland, K. R. and Hidley, T., 'Control Room Reverberation is Unwanted Noise', *Proceedings of the Institute of Acoustics*, Vol. 16, Part 4, pp. 365–73 ('Reproduced Sound 10' Conference, Windermere, UK, 1994.)

3 Newell, Philip, 'The Non-Environment Control Room', *Studio Sound*, Vol. 33. No. 11, pp. 22–9 (November 1991)
4 Newell, Philip, *Studio Monitoring Design*, Focal Press, Oxford, UK (1995)
5 Stark, Eric, 'The Hidley Infrasound Era', *Studio Sound*, Vol. 37, No. 12, pp. 52–6 (December 1995)
6 Gerzon, Michael, 'Three Channels, The Future of Stereo?', *Studio Sound*, Vol. 32, No. 6, pp. 112–21 (June 1990)
7 Davis, Don and Davis, Carolyne, *Sound System Engineering*, 2nd Edn, Howard Sams, Indianapolis, IN, USA (1987)
8 Moulton, D., Ferralli, M., Hembrock, S. and Pezzo, M., 'The Localization of Phantom Images in an Omni-Directional Stereophonic Loudspeaker System', AES 81st Convention, Preprint No. 2371 (1986)
9 Flindell, I. H., McKenzie, A. R., Negishi, H., Jewitt, M. and Ward, P., 'Subjective Evaluations of Preferred Loudspeaker Directivity', AES 90th Convention, Preprint No. 3076, p. 6, Paris (1991)
10 Wrightson, Jack, 'Psychoacoustic Consideration in the Design of Studio Control Rooms', *Journal of the Audio Engineering Society*, Vol. 34, No. 10, pp. 789–95 (1986)
11 Kuhl, W. and Plantz, R., 'The Significance of the Diffuse Sound Radiated from Loudspeakers for the Subjective Hearing Event', *Acoustica*, Vol. 40, pp. 182–90 (July 1978)
12 Kishinaga, S., Shimizu, Y., Ando, S. and Yomaguchi, K., 'On the Acoustic Design of Listening Rooms', presented at the 64th Convention of the Audio Engineering Society, Preprint No. 1524 (November 1979)
13 Lipshitz, Stanley P., 'Stereo Microphone Techniques … Are the Purists Wrong?', *Journal of the Audio Engineering Society*, Vol. 34, No. 9, pp. 716–35 (September 1986)
14 Newell, Philip R. and Holland, Keith R., A Proposal for a More Perceptually Uniform Control Room for Stereophonic Music Recording Studios, presented at the 103rd Convention of the Audio Engineering Society, New York, Preprint No. 4580 (1997). [Reference 14 can be found at the beginning of the chapter.]

CHAPTER 17

The Live-End, Dead-End Approach*

The aims of the concept. The reflexion-free zone. Means of achieving the 'live' element. The need for a uniform decay rate. The practical needs of modern control rooms.

An earlier studio control room design philosophy than that of the Non-Environment room is the so-called Live-End, Dead-End (LEDE). As with the Non-Environment approach, this divides the control room into areas of contrasting acoustic function in an attempt to provide a listening environment that makes the room's own acoustics subservient to those of the monitored programme.

The quote attributed to Jack Wrightson[1] in the account of the Non-Environment room (Section 16.9) provides an equally valid reference point for the LEDE philosophy:

> *The problem in the context of studio monitoring is that, regardless of the conditions, the room/monitor–loudspeaker combination places its imprint on all that transpires. For this reason, a control room should be neutral; it should add as few sonic colourations as possible to the sound generated by the monitor loudspeakers. In this context, poorly designed loudspeakers should exhibit their flaws; well-designed loudspeakers should demonstrate their assets. The aural purpose of a control room is to provide the best possible free-air representation of the signals carried by the studio's audio system.*

The aspiration for an LEDE room is thus to create a neutral monitoring environment. Where this approach differs from the Non-Environment approach is in the definition of neutral. The aim for the room is to create an environment in which the 'context' information is coming from the monitored programme material, not the room in which you are listening, while retaining certain aspects of the room's natural acoustic. The things that make an environment non-neutral are, principally, the early reflexions (which give a feeling of warmth or timbre) and the build up of unconnected sound that combines with direct sound to create the impression that there is a tonal imbalance.

*This chapter was written by Tim Goodyer, and was based on a series of conversations with the studio designer David Bell, an experienced exponent of Live-End, Dead-End control room techniques. Many hundreds of control rooms have been built to the specifications of this concept over a period of around 30 years, so this book would have been seriously lacking if an authoritative description of LEDE principles had not been included.

17.1 First Impressions

Any room lends some form of context to the listening experience. This remains true even if the principal objective is to avoid giving it any colouration – this is true by definition since it is a different experience to listening outdoors in a field. If you do nothing other than provide the listener's eyes with a room to look at, this translates into an expectation of how the room should sound that is either fulfilled, or not, by the listening experience.

The way in which the ear is able to distinguish between, for example, a wooden drawing room and a stone basement is through the nature of the room's early reflexions – their frequency balance and how closely spaced they are in time – and the way the reflexions are changed by the surfaces they encounter. A good impression of the size of a room is obtained as you walk into it, through the first sounds that you hear. From experience you have a model in your mind of how large the room is and how it 'sounds'.

17.2 A Window of Objectivity

If a reflexion-free zone is produced so there are no reflexions in the desired 6–30 ms period (in a practical sized room, this is around 20 ms) it is possible to eliminate the usual colouration of a room acoustic and listen instead to that coming from the loudspeakers, whether it has been created in a real acoustic environment or with a digital reverberation unit.

Consequently, the design objective is to create a reflexion-free zone whereby the sound travels directly to the listener's ears, and then there is as long a gap as feasible (within the constraints imposed by normal room sizes) before any reflected sound arrives at the listening position. When those reflexions do arrive, they should be as diffuse as possible – that is, there should be as many as possible and they should be unrelated to each other. Depending on the size of the room and the available budget, this is achieved by mounting the loudspeakers in the monitor wall (again, in common with the Non-Environment room). The geometry and broadband absorption at the monitoring end of the room are used to eliminate early reflexions at the listening position; and by returning energy to the room in a non-destructive manner. The use of absorption in the front halves of the rooms gives rise to the 'dead' element of the LEDE description.

We are trying to model an idealised energy-time response within the room (see Figure 17.1). The intensity of the direct sound and the early reflexions can be calculated, and their relative levels managed through acoustic treatment of the listening room. Through this, the overriding reverberation characteristic will be that of the monitored system, as the colouration offered by early reflexions will only be those created in a studio's performing room or those coming from sound processors in the recording system. Achieving this involves consideration of where the sound energy enters the room space; how it travels to the monitoring person's ears and how the room reacts with the energy introduced into it to give the listening context.

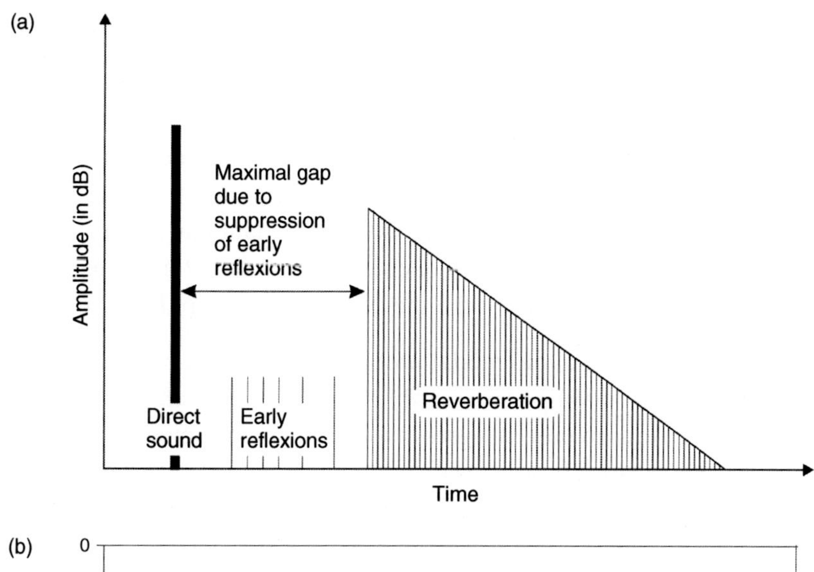

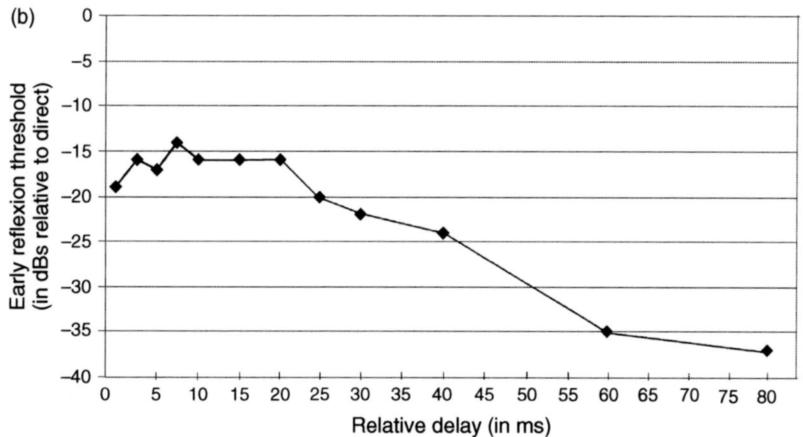

Figure 17.1:
Idealised energy/time response for a 'Live-End, Dead-End' control room: (a) Maximising the gap between the direct sound and the onset of 'reverberation' (or clustered late reflexions) by suppressing the early reflexions; (b) The minimum amount of reflexion suppression required for a reflexion-free zone.

The first order improvement in the operation of a loudspeaker is to mount it in a wall such that it works into half space; there is no fringe or edge effect produced by the loudspeaker, and there are no early reflexions from the monitor wall or anything in close proximity.

Early reflexions are destructive (or colouring) to the sound in that they are coherent with the direct sound, and that due to differing path lengths they may arrive out of phase with the direct sound. Because they are coherent and delayed, they interfere with a range of frequencies at multiples of their fundamental frequency, giving comb filtering effects. This causes aberrations in the perceived frequency response of the loudspeaker. By suppressing early reflexions within the room, the nature of the sound heard at the listening position is the nature of the sound coming from the loudspeaker, not any characteristic of the room.

It is necessary to retain subsequent reflexions, however, because a control room is finite in size (and generally small in acoustic terms) and complete absorption of all sound energy across broad frequency bands is not possible. It is, therefore, important that the returned energy from any reflexions that do occur is even with frequency. Thus mid and high frequency energy must be returned into the room to match the inevitable bass energy build up.

Careful design of the room will avoid modal problems as far as is possible, but the isolation requirements of real studio buildings necessarily imply bass level energy retention within the room (using massive walls to prevent egress of sound into unwanted areas – the performing room(s) even if nowhere else). Thus diffusion is employed in the rear half of the room to return energy in as even and non-coherent a form as possible.

If energy is returned to the room unmanaged, it is likely to cause similar effects as with early reflexions. Managing this energy is achieved by most designers by using proprietary diffusion products. These work by setting up interference between the arriving and reflected sound by the use of differing lengths of 'pockets' or 'wells' (see Figure 4.32) giving different delay times. This is similar to the way a film of oil on water produces different coloured reflexions, as internal reflexions in the oil film interfere with the light. A diffuser produces multiple, unrelated (termed non-specular) reflexions of sound that are returned to the room in different directions to the arriving sound, but that retain their energy and frequency content. The energy is therefore retained in the room at the original frequency, but without the ability to recombine destructively.

It can be shown that the intensity of direct sound decreases as the square of the distance from the source. The intensity of early reflexions can be similarly calculated, taking into account the additional path lengths and the characteristics of the reflecting surfaces, and can be shown to be inversely proportional to the square of the distance (and hence also time) from the source.[2] Absorption effects are highly frequency dependent but may be calculated by the use of the absorption coefficient for a given surface (see Tables 4.1 and 4.2, for example).

The Non-Environment approach to design would have it that the room contributes little or nothing to the monitoring mix. The aim is that all effects of the room are removed by very efficient, very broadband absorption throughout. Adherents to the LEDE philosophy have felt that it is very difficult to achieve highly uniform absorption of a wide range of frequencies. The bass end in particular is very difficult. If it were possible to allow all the bass energy to leak away, it would be more practical. However, a control room is frequently sited next to a performance space, and therefore it is necessary to contain the bass in order to make a normal studio operable; thus, a residual amount of bass remains in the room and builds to create perceived tonal anomalies.

Sound can be diffused by a number of means, and a number of different factors can determine effective diffusion. Given the space, diffusion can be maintained down to the order of 100 Hz. An additional benefit here is that there is a reasonable amount of bass absorption offered by the nature of the diffusers used at higher frequencies, in that the trapped air acts as elastic absorption.

The secondary effect of the diffuse high-end and mid sound energy that remains in the room is that it creates an even decay, producing a more neutral room. Hence the 'live' element in the LEDE design.

Taken as a whole, the frequency response of the reverberant field is shaped by broadband absorption, some bass absorption (of a broadband nature to maintain a smooth frequency response), and the maintenance in the room of some diffuse high and mid frequencies. The reverberation time of the space can therefore be made even with frequency.

Room modes can, and should, be minimised, though they cannot be completely eradicated. Modal effects can also be made more tolerable by arranging the principal listening areas to be free of modes. Any asymmetry in the room can make it difficult to predict modal behaviour, although there is a school of thought that suggests asymmetry in the front end of a room is a good thing because an irregular modal response is less perceptible than a regular one.

The net result is an early-reflexion-free period in which the timbre and acoustic qualities of the recording environment can be monitored, and there is a context against which to evaluate this in that the monitoring space is even in reverberation time. This has additional benefits of broadening the area over which the stereo field can be maintained. It is optimised at the main listening position for numerous reasons – principally that the listener is stereo-symmetrical with the loudspeakers, which is ideal, and stereo-symmetrical with the room, so that any artefacts are even on both sides. Any phase information is also symmetrical. Figure 17.2 shows a practical realisation of the LEDE concept.

17.3 Working and Listening Environments

The modern control room, however, is no longer a place where one technician in a brown coat makes hits out of the efforts of distant musicians working in a different environment. In many of the world's music studios, 95% of the creativity now takes place in the control room, and is rarely the work of a single person – necessitating the creation of a suitably sized listening area with a common listening characteristic. Everyone involved needs a similar, common experience of what is being recorded.

Figure 17.2:
Rear end of a 'Live End-Dead End' room showing the diffusers. BBC, Maida Vale.

Modern control rooms are built to impose huge amounts of acoustic control, and their acoustics have to be robust to the way in which they are used. This is important, as significant amounts of additional equipment are often brought into the acoustic space, and this can lead to significant livening of the rear of the control room. In the LEDE philosophy, a number of semi-randomly distributed 'workstations' of equipment distributed around the rear third of the room introduce further diffuse sound to a diffuse field, rather than detracting from a designed-dead acoustic. It is less destructive to perturb further a diffuse energy field than to introduce reflexions into an area that is intended to be dead – and it has been found in certain studios that there is a limit to the amount of additional equipment that can be brought in before the monitoring environment is compromised. (This will be discussed in much more detail in Chapter 18.)

The practicality is that it is necessary to accommodate this because a lot of recordings are made this way. The operational considerations have to include the control room offering a working environment as well as purely a monitoring environment. Typically, this may involve creating drum patterns, creating MIDI sequences, manipulating samples of audio and layering keyboard tracks in modern music.

There is a further, less well-defined, criterion: the idea of a 'comfortable' working environment. This is not analogous to working in a domestic room, but rather a room in which to assess realistically the elements going into the creative process with respect to the variety of listening environments that exist outside of the studio. These range from living rooms and headphones, to cars and nightclubs, to optimised listening rooms and single-speaker radios. In common with other design philosophies, the objective is to allow you to evaluate how your work is likely to sound when it leaves the studio environment and enters the ears of the 'real world'.

It is folly to try to create a 'representative' monitoring room because there is such a broad spectrum of listening environments and situations to address. Instead, what is required is to produce a non-colouring monitoring environment to assist in assessing the relationship of the work to these other environments.

It is important to note the difference between the listening and monitoring processes here. It is quite possible to listen to a sound, whether it is a familiar voice or a symphony, in a coloured environment and recognise the elements that characterise it. This is possible as the brain de-convolves the sound, in order to distinguish between the direct sound and the artefacts added by the environment. As such, it is possible to 'listen' in a wide variety of acoustic environments without being aware of either the effects of those environments or the systems through which one is listening. Through this mechanism, sounds become readily recognisable in a wide variety of different listening conditions. 'Monitoring', however, is not consistently possible in many such environments, as these artefacts may conceal aspects of the programme or modify them in ways that make objectivity impossible.

This again is why there should be no early reflexions – even of a controlled and diffuse nature – in a monitoring room. In contrast, this is quite acceptable in a listening room if it makes the listening experience more enjoyable.

17.4 Summary

The object of the design approach is to provide a neutral monitoring environment.

The principal difference to the Non-Environment approach is in the definition of how neutral 'neutral' should be.

The first design objective is to create a reflexion-free zone (RFZ) around the monitors and the listener(s).

The rear half of the room is designed to be diffusively reflective in such a way that the overall room response has an even decay rate at all frequencies of interest.

This is usually achieved by the installation of proprietary diffusers based on specific mathematical sequences. (See also Section 4.7.)

Room modes should be minimised as far as practicably possible.

The technique is also said to be beneficial in providing a better musical ambience for musicians to work in. This is considered to be important now that so much of the *playing* of the instruments takes place in control rooms.

The installation of considerable amounts of equipment in the rear section of a room is considered less noticeable in its effect on the diffusely reverberant acoustics of a Live-End, Dead-End control room than on the acoustics of a room with a maximally absorbent rear wall.

Modern exponents of the Live-End, Dead-End control rooms aim for a comfortable working ambience, and *not* a domestically representative decay time.

References

1 Wrightson, Jack, 'Psychoacoustic Considerations in the Design of Studio Control Rooms', *Journal of the Audio Engineering Society*, Vol. 34, No. 10, pp. 789–95 (1986)
2 Angus, James A. S., 'The Effects of Specular Versus Diffuse Reflections on the Frequency Response at the Listener', *Journal of the Audio Engineering Society*, Vol. 49, pp. 125–33 (2001)

Bibliography

D'Antonio, Peter and Konnert, John H., 'The RFZ/RPG Approach to Control Room Monitoring', presented at the 76th Convention of the Audio Engineering Society, New York, USA, Preprint No. 2157 (October 1984)
Davis, Don and Davis, Carolyn, *Sound System Engineering*, 2nd Edn, Chapter 9, Focal Press, Boston, USA, Oxford, UK (1997)
Davis, Don and Davis, Chips, 'The LEDE Concept for the Control of Acoustic and Psychoacoustic Parameters in Recording Control Rooms', *Journal of the Audio Engineering Society*, Vol. 28, No. 9, pp. 585–95 (September 1980)
Davis, Chips and Meeks, Glenn E., 'History and Development of the LEDE Control Room Concept', presented at the 72nd Convention of the Audio Engineering Society, Anaheim, USA, Preprint No. 1954 (October 1982)
Wrightson, Jack and Berger, Russ, 'Influence of Rear-Wall Reflection Patterns in Live-End, Dead-End Recording Studio Control Rooms', *Journal of the Audio Engineering Society*, Vol. 34, No. 10, pp. 796–803 (October 1986) (See also Section 13.1.3.)

CHAPTER 18

Response Disturbances Due to Mixing Consoles and Studio Furniture

Mixing consoles as acoustic obstructions. Equipment rack sizes and placement. Computers and their monitors. Sofas. Siting of monitor loudspeakers.

One basic limitation relating to the carefully calculated modal spread of some of the preferred ratios of room dimensions, discussed in Chapter 12, is that they all presume hard-walled, rectangular, empty rooms. The response of a loudspeaker in a carefully designed control room tends also to be calculated in terms of the *empty* room response. Once a large amount (or in some cases even a small amount) of equipment is placed in a room, both the modal response and the transfer function from the loudspeakers to the listening position will be altered, and sometimes drastically so.

No audiophiles would be likely to place their dining room tables and sideboards between themselves and their hi-fi loudspeakers, yet recording engineers, from necessity, do much worse things. They place large, resonant, reflective, absorptive, diffusive, diffractive objects (mixing consoles) between themselves and the monitor loudspeakers, and often, without due care and attention, position effects racks and computer monitors in places where disturbance of the monitor responses is effectively guaranteed to occur.

18.1 The Sound of Mixing Consoles

Historically, remarkably little thought was given to the way in which the physical presence of a mixing console could disturb the acoustics of a room, but as the better control room responses improved, sometimes great differences were noticed in the sound of the monitoring when an older console was replaced by a newer one. Size *is* important. In general, the smaller the mixing console, relative to the size of the room, the less will be the response disturbance. Somewhat perversely, out of ignorance and/or disregard for the monitoring quality, the business people who frequently dictate what happens in studios often also choose to put the largest mixing console that they can afford into a room, in the belief that it will impress the clients. This has a greater tendency to happen in smaller, less professionally operated studios, which are often the ones that are already marginal in terms of the quality of their monitoring.

When digital consoles were first being discussed, many of their designers predicted the 1 m^2 (or less) control surface, with the rest of the electronics being housed in a machine room. To many studio designers this seemed to be a definite step in the right direction, but the use of such systems has not become widespread. The aforementioned 'macho' sales value of the large console is one reason, despite being a triumph of marketing over sound quality. However, the other principal argument against the use of such small surfaces has come from the recording engineers, who need rapid access to all controls during large and pressurised recording sessions. There is often simply no time to leaf through menus in order to gain access to the required control. In small, single operator studios, which usually do not deal with large groups of expensive musicians on tight schedules, the slower access time of small control surfaces is often no great disadvantage, but these are frequently precisely the studios where ignorance exceeds experience and they opt for unnecessarily large consoles in order to appear more 'up-market'.

The worst offenders in terms of disturbing the monitor response are the mixing consoles with closed backs which extend from the meter bridge to the floor. The rear surfaces of such consoles are normally vertical, and they can form resonant spaces between the consoles and the front walls. They also obstruct the free expansion of the low frequency sound waves under the console, which causes undesirable amounts of diffraction, resulting in the subsequent disturbances in the frequency response, transient response and the stereo imaging. What is more, the large backs are rarely adequately damped, so they can rattle and resonate, thus both adding to and subtracting from the desired flat response. Some mixing consoles, it must be said, bring disgrace to their manufacturers by the sheer degree of disregard that they have paid to the acoustic intrusion of their products.

Metal panels underneath many mixing consoles can also be heard to ring in response to drum beats. Investigators at Vigo University, Spain, found that the decay time of an expensive control room which they measured, with otherwise excellent acoustics, was dominated by the resonance of the tubular steel frame of its ostensibly professional mixing console. It is interesting to note that when AMS/Neve launched their 88 R mixing console in late 2000, they made a point of advertising its minimal use of enclosing panels and the shallow inclined rear surface, both designed to minimise the acoustic disturbance to the control rooms. This is something for which they should be applauded, despite it having taken them over 35 years to correct their earlier oversights.

Studio designers have reported that significant differences in the response of control rooms, of very similar size and with identical monitor systems, have resulted from the installation of different mixing consoles. The tests have been carried out with the noise or music sources connected directly to the monitor systems, to avoid the possibility of the differences being due to the monitor circuitry of the consoles. Considerable changes of monitoring quality have been apparent when one mixing console has been changed for another of very different shape.

In fact, in some cases where a room has had a poor reputation for its monitoring, even though all electroacoustic consideration suggested that the room should be good, a change of mixing console from one that caused severe obstruction to the sound waves to one of smaller size has improved the monitoring by a significantly noticeable degree. This has even been the case when the replacement console was of lesser quality in terms of its electronic signal path. It has been rather unfortunate that some of the best mixing consoles, electronically speaking, have been the worst offenders in acoustic terms.

In general, a large mixing console needs to be installed in a large room, or its presence may dominate the acoustics in a small room to such a degree that the problems encountered cannot be remedied. There is a tendency for studio owners to want to fill their control rooms with all sorts of impressive looking equipment, but the temptation should be resisted or the monitoring quality will suffer. The installation of excessive quantities of equipment should be avoided.

It is also usually wise to remove all metal panels that are not necessary, and to add self-adhesive damping materials to the panels that must remain. Typical damping materials are the ones used in the automobile industry. However, before permanently removing any panels, the mixing console manufacturers should be consulted because it can, in some circumstances, change the air flow such that hot spots can develop in places where the convective air flow has been disturbed.

Another beneficial addition is an absorbent screen placed behind the mixing console. This could be in the form of a frame with heavily pleated curtains, or fabric-covered and filled with 20 cm of open-cell foam, cotton felt or mineral wool. Such a device is shown in Figure 18.1. Once again, care must be taken when positioning the screen to ensure that none of the ventilation slots or holes on the rear of the desk is obstructed or overheating may result. Such screens can also be useful as stands for the small monitor loudspeakers, because mounting the small monitors a little distance away from the console is usually preferable to mounting them directly on the console itself. Low frequency colouration is usually greater when a large plane surface, such as the control surface of a mixing console, is positioned directly below the loudspeakers. Figure 18.2 shows the effect of desktop mounting on the transient response, and neither colouration nor transient smearing is beneficial in terms of monitoring quality or stereo imaging.

18.2 Equipment Racks

There are many things done in control rooms these days which are for convenience much more than they are for accurate monitoring. In many studios, the monitors are mounted higher than ideal in order to allow a greater width of window, and an effects rack is mounted behind the mixing position and set at an angle for best visibility of the controls. Figure 18.3 shows the way in which the monitor height and effects rack angle combine to cause strong reflexions to return to the back of the engineer's head, typically 10 or 15 ms after the first pass of the

Figure 18.1:
Acoustic screen to suppress console reflexions (indicated by the arrow). A freestanding screen can be mounted behind the mixing console, spaced away by 15–20 cm to allow free ventilation if necessary. The screen should be around 20 cm thick, and should consist of an outer, rectangular frame filled with 60–80 kg/m^3 open cell foam slabs or wedges. Such a screen effectively eliminates reflexions between the console and the front wall or window. The screen also provides a better mounting location for close-field monitor loudspeakers than the meter bridge of the mixing console. A layer of foam on the meter bridge, immediately in front of the loudspeakers, helps even more (Nomis, London, UK, 1989).

direct sound, and only very slightly reduced in level. Strong comb filtering can result, and stereo image smearing can be considerable. It is often the case in such control rooms that the large monitors are not considered to sound very good. Figure 18.4 shows a control room where all equipment furniture is kept low, and is angled such that reflexions pass away from the mixing console. Perhaps such a system is not always quite as convenient as the rear-rack/keyboard-table approach, but if good monitoring conditions are desired then certain small compromises must often be made in terms of convenience.

18.3 Computer and Video Monitoring

Another case where convenience triumphs over good monitoring is when a mountain of computer and video monitors is placed such that they not only cause undesirable reflexions, but also obstruct the direct path from the loudspeakers to the engineers' ears. When studio owners and operators are questioned about this practice, the answer is often that it must be all right because of all the photographs seen in recording magazines showing famous studios elsewhere employing similar arrangements. The truth is that when photographs are being taken for magazines, the studio personnel often want to get the maximum publicity value from showing off as much of their equipment as possible. What is more, a photograph may be taken

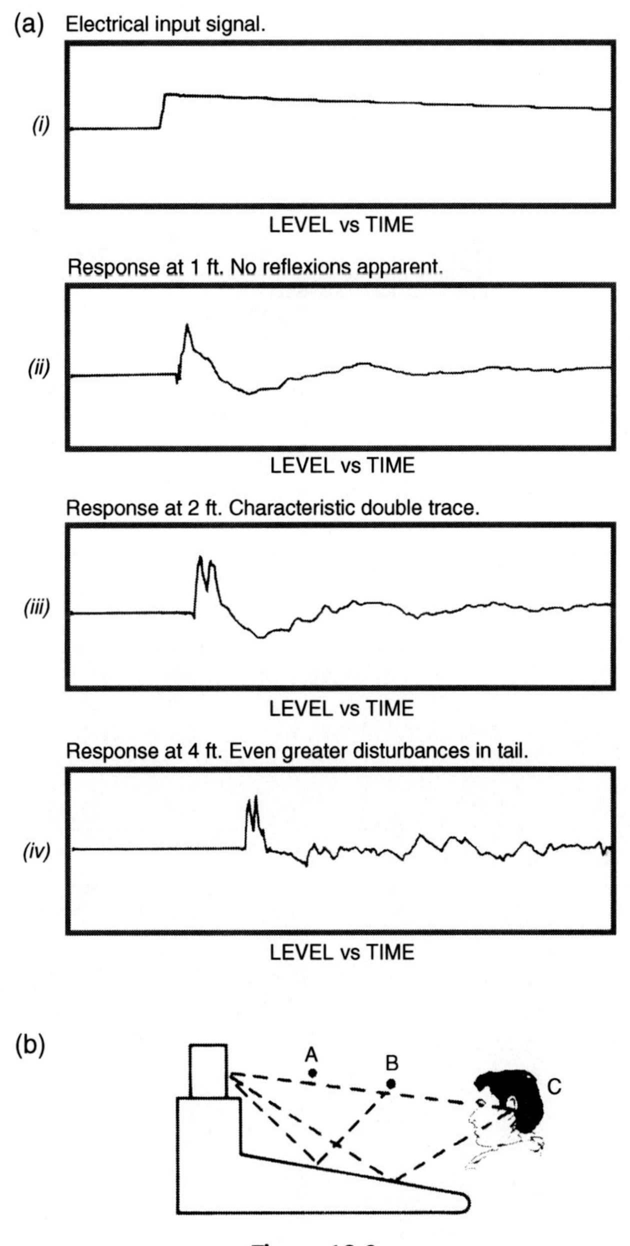

Figure 18.2:
Close-field monitor response: (a) The effect of the reflexions from the top surface of a mixing console on the transient response of a small loudspeaker placed on the meter bridge; (b) Loudspeaker directivity is too narrow to produce reflexions at position 'A'. However, reflexions are apparent at positions 'B' and 'C', with differing ratios of direct to reflected path lengths, hence they produce the different composite transient waveforms as in (a) (iii) and (iv). Positions 'A', 'B' and 'C' relate to the 1, 2 and 4 foot plots in (a) (30, 60 and 120 cm respectively).

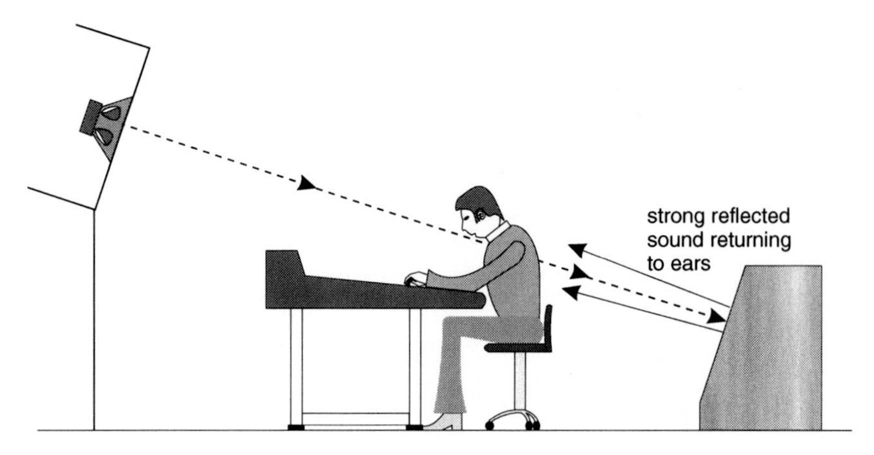

Figure 18.3:
In many cases, an effects rack placed behind the engineer, for easy access, also performs the function of an acoustic mirror, returning strong reflexions to the prime listening positions. This is especially problematical when the main monitor loudspeakers are mounted high.

Figure 18.4:
Hitokuchi-Zaka studios, Tokyo, Japan. In this control room, designed by Shozo Kinoshita, all furniture is kept below ear height, and is positioned to avoid confusing reflexions. Although the monitor loudspeakers are mounted relatively high, the size of the room ensures that the monitoring angle is kept low.

during vocal overdubbing, for example, but the situation with equipment clutter may be much better organised during serious mixing. The phase of the work during which the photographs were taken is rarely noted in the captions. The truth, also, is that many bad practices go on even in expensive studios. Ignorance is rife these days.

Nevertheless, there *is* definitely a trend towards allowing equipment to dominate many proceedings, and this reality must be recognised. However, the principal aim of this book is to take the purist point of view. Many studio owners and operators have simply never fully understood the repercussions of many of their actions, and do not even stop to consider what a pile of close-field loudspeakers and computer monitor screens may be doing to the main audio monitoring. Only by carefully pointing out the effects, and demonstrating them if possible, can they be put in a position to make their own balanced decisions about the priorities. Certainly, whilst overdubbing a guitar or making a quick recording for a TV jingle, convenience of operation may take precedence. However, when setting up to mix what is likely to be a million-selling CD, the record-buying public really deserve a little more consideration, because they are perhaps going to be listening to the results of the session for 30 years or more.

When possible, it is ideal from an acoustic point of view to mount video monitors in the front wall, but this should be done in such a way that no resonant cavities are formed, and no diffractive edges are too close to the loudspeakers. The new flat screens seem to be better than large cathode-ray video screens, because of their reduced physical presence and the ability to lie them down or move them into less intrusive positions when audio monitoring of a high resolution is required. Wherever possible, video monitors are best kept to the sides of the mixing console if they cannot be mounted in, or very close to, the front wall.

Hopefully, as computer-based recording equipment develops it will become more acoustically friendly, but one of the problems has been that many of the people who have been developing the equipment have not been 'audio people' in the traditional sense. Large amounts of money have been involved and the mathematical input has been enormous. Such developments are not made just for the benefit of a relatively few high-quality studios. They imply mass sales in order to get a return on the financial investments, so they lead to products that are largely aimed at a mass market. Acoustical niceties therefore tend to be well down the list of priorities. It has been a general fact of life since the early 1980s that dedicated specialist manufacturers for the professional recording industry have been becoming a dying breed. It therefore becomes an encumbrance upon the studio operators to find acoustically acceptable solutions for the appropriate mounting of their monitor screens.

18.4 Sofas

Sofas can be very useful in control rooms, not only to sit on, but also because they are generally good low frequency absorbers. Their physical depth makes them effective even when placed against walls. Generally speaking, fabric covered sofas are preferable because the leather and plastic covered varieties can be reflective at high frequencies, although their irregular shape usually makes the reflexions rather diffuse. From a non-acoustic point of view, the comfort that a sofa provides for producers and musicians is often greatly appreciated, and they are generally well received.

In larger rooms, where there is sufficient space between the mixing console and the front wall, it is sometimes possible to place a sofa in such a position that it can take the place of a purpose made acoustic screen to reduce the rear-panel reflexions from the mixing consoles. Sofas in this position can also have the beneficial effect of reducing floor reflexion response dips at the mixing position.

18.5 Effects Units and Ventilation

It is always best in a sound control room to avoid having anything higher than about 1m. The meter bridge of the mixing console may have to exceed this for operational reasons, but racks should be kept to a height of around 90 cm in order that things placed on top of them should remain below 1 m. Computer monitors tend to break this rule, but their positioning for minimal acoustic disturbance has been dealt with in Section 18.3. When things begin to exceed 1m in height, disturbances to the stereo imaging are likely to result. Metal racks should be avoided at all cost, unless they are of an open-frame, skeletal nature. Wood is the preferable material, of sufficient thickness not to resonate under normal conditions of use.

Ventilation can be provided by grilles, semi-open backs and spaces between the equipment, but fans should not be used as they inevitably contribute to the ambient noise in a most undesirable way. Obviously this type of ventilation is not suitable for densely packed equipment, but a high packing density is often a necessary result of inadequate space in the control room itself. Again, it is often the fact that people tend to build control rooms in spaces that are too small, which in turn forces equipment into spaces that are too tight, and undesirable measures then need to be taken to stop things from overheating.

18.6 Close-Field Monitors

There is an obvious conflict of priorities when it comes to the siting of small loudspeakers in the vicinity of a mixing console. Ideally, the main *and* secondary monitors both need to subtend the same angle in the area of the head of the recording engineer. If the subtended angles were the same, the small loudspeakers would need to be positioned directly in line between the main loudspeakers and the engineer. It would seem to be reasonable common sense that placing such obstructions in the direct line of sight to the principal monitors could do nothing other than disturb their response. However, the question may arise as to which ones *are* the principal monitors during different phases of the recording, overdubbing and mixing processes.

It is common to see the meter bridge area of a mixing console so loaded with different small loudspeakers and computer monitors that the large monitor loudspeakers are all but *invisible* from the mixing position. This is clearly an absurd situation, but it has been known to exist in some large and well-known studios. Quite how these situations arise in supposedly professional studios is hard to understand, except perhaps as an attempt to please all the clients by having all their favourite small loudspeakers on display, but it often drives a vicious circle. The main monitors sound less than good because of the first obstructions, so they are relied upon less. More small references are therefore introduced, which make the main monitors sound even worse, so the small monitors are relied on still more.

Another aspect of the fight for position is that in many cases the small loudspeakers are laid flat so that they obstruct less the view of the large monitors. As Figure 18.5 shows, many manufacturers are now clearly stating the intended orientation of the loudspeakers. When the drivers are aligned vertically, lateral movement of the engineer across the mixing console will cause no change in the relative distances from the high and low frequency drive units. If the loudspeakers are laid on their sides, and the drivers no longer share a common vertical axis, phasing effects may be heard during lateral movement and the off-axis colouration will be greater than when aligned vertically. This can be very easily demonstrated by listening to pink noise. Figure 13.11 shows the general principle, which can give rise to a further lack of compatibility between large and small monitors. Even if the two are highly compatible when listening exactly on the centre line between left and right, when the small loudspeakers (other than dual concentrics) are laid on their sides it is almost impossible to expect to be able to maintain

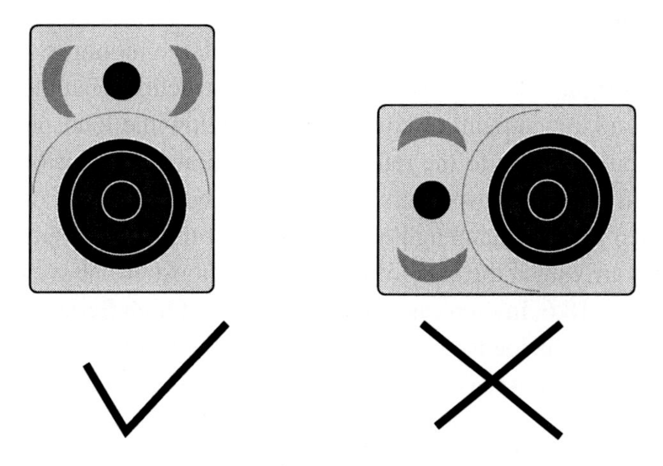

Figure 18.5:
Correct use of small loudspeakers. More and more loudspeaker manufacturers, such as Tannoy, Quested and Genelec are now publishing drawings, such as those above, in their manuals for small loudspeakers, instructing their users to keep the drivers in a common vertical plane. It matters nought that there are thousands of studios using such loudspeakers on their sides, the practice is WRONG (see also Figures 13.11 and 13.12).

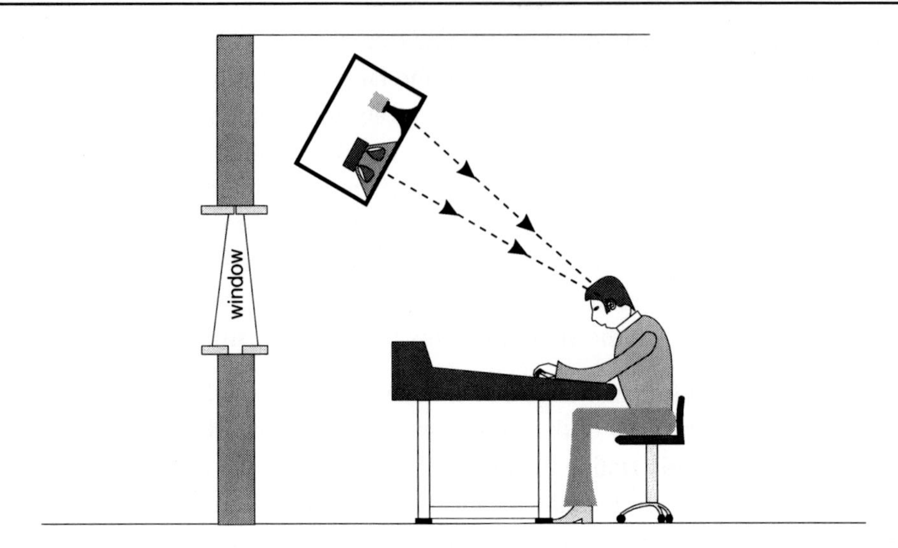

Figure 18.6:
When monitor loudspeakers are mounted at a steep angle, the high frequencies, in particular, arrive at the ears from an angle totally inappropriate for the perception of an accurate frequency balance. High frequencies will tend to be under-perceived when looking at the equalisation controls on the mixing console. What is more, unless the ceiling is highly absorbent, the low frequencies may suffer undue augmentation due to the proximity of multiple room boundaries.

that compatibility even when one moves only 5° or so off-axis. The practice of laying the small loudspeakers on their sides is usually a rare attempt to reduce the obstruction of the main monitoring, yet it is unwittingly done at the cost of the close-field monitoring qualities.

One attempted 'solution' to this problem is to mount the large monitors sufficiently high that there is a clear line of sight to them above the desktop obstructions, but there are drawbacks to this practice. Firstly, a large amount of equipment just below the loudspeaker axis will still inevitably cause colouration due to the reflexion, diffraction and diffusion from surfaces previously 'hidden' from the loudspeakers, so little may be gained. Secondly, when the monitors are mounted high, and an engineer is looking at the control surface of the mixing console, the sound is arriving at the ears from a position almost 90° above the normal listening axis, as shown in Figure 18.6. In such cases, the perception of high frequencies will be different from when directly facing the loudspeakers. This leads to the tiresome experience of constantly needing to look up at the loudspeakers after every adjustment, in order to audibly check the effect. If this is not done it can lead to incorrect conclusions about the equalisation.

18.7 General Commentary

All the above effects are rather subtle, but cumulative. Nevertheless, even if some of them are only marginally noticeable in isolation, the difference between a control room where they have all been taken into account and one in which no such consideration has taken place can be very noticeable indeed. The difference is not even always apparent for what it is. It is rather that well thought out control rooms, in which many small details *have* been taken into account, tend simply to induce a feeling of confidence in the people using them. This no doubt results from a general awareness of the better quality of the monitoring conditions and the lack of confusing cues.

Although easy access to equipment is a very important part of the functionality of a control room, the tendency to box-in the listening area by portable racks is absolutely asking for inaccurate monitoring. This practice will affect the responses of the small loudspeakers also. Placing so many reflective surfaces in the vicinity of the ears is simply inviting trouble.

Most studio personnel have not had the opportunity to hear a well-designed control room with only its monitor system installed. This is unfortunate, because if more people could hear what the rooms were capable of sounding like *before* the rest of the equipment was installed, it may inspire them to take a little more care about the acoustic consequences of the equipment positioning. It may also inspire them to ask the equipment manufacturers a few more questions about their design philosophies, which would also do the whole industry a great favour.

18.8 Summary

Few audiophiles would put their dressing table and a few chests of drawers between themselves and their hi-fi loudspeakers, yet that is no less reasonable than that which occurs in most recording studios.

Mixing consoles should be proportionate in size to the control rooms in which they are situated. A mixing console that occupies a great proportion of the space in a control room will inevitably negatively affect the sound quality of the monitor system.

Large mixing consoles should not be chosen for their ability to impress clients, because a price will be paid in terms of monitoring accuracy. Acoustically speaking, the rule for mixing consoles is 'the smaller, the better'.

If the panels on the back and underside of mixing consoles can be removed, or treated with self-adhesive damping materials, it is often wise to do so.

The monitoring character of control rooms has been known to change significantly with the change of a mixing console.

Absorbent screens behind the mixing consoles can be advantageous to the monitoring.

Equipment racks should be kept low.

Slanted equipment racks behind the recording engineer's head should be avoided.

Computer monitors should be kept to the side of the mixing console during serious mixing sessions. Ideally, they should be positioned well out of the way of the line of sight to the loudspeakers.

Convenience should not be allowed to dominate decisions at the expense of monitoring quality.

Sofas are generally beneficial, either at the back of a room or between the mixing console and the front wall.

Close-field monitors are better placed on stands behind the mixing console, rather than on the meter bridge. (But see also Chapter 19.)

Small loudspeakers should be mounted with the common axis of the drive units in a vertical line. Despite the label on the front of the NS10M Studio suggesting a horizontal (landscape) position, this is not a good idea. Images are less stable and more 'phasey' when drivers are offset horizontally.

Mounting the main monitors high up is not a good idea. The acoustic centres of the monitors should be sited low enough to reduce desktop reflexions and to keep the timbre constant for people moving around the room. (See also Subsection 20.7.1.)

Bibliography

Newell, Philip, 'The Acoustics of Mixing Consoles', *Studio Sound*, Vol. 33, No. 8, pp. 59–63 (August 1991)
Newell, Philip, *Studio Monitoring Design*, Chapter 14, Focal Press, Oxford, UK (1995)

CHAPTER 19

Objective Measurement and Subjective Evaluations

Description of objective tests. Discussion of relevance. Assessment of results. Subjective views of close-field monitoring. Analysis of the NS10M. Inter-modulation distortion.

The history of audio has had a frustrating tendency to produce 'undefinables' in terms of quality differences. The hi-fi press has invented a whole vocabulary of adjectives to describe sound qualities, but without objective verification there is no way to know if the terms mean the same thing to different people, or not. Spaciousness is one obvious example. In the stereo reproduction of music, spaciousness is a highly valued attribute, yet no unit of measurement exists for its quantification. Transparency is another example: we think that we know when we hear it, but can we prove it?

19.1 Objective Testing

In 1998, Dr Keith Holland began a series of loudspeaker tests, carried out at the Institute of Sound and Vibration Research (ISVR), for the magazine *Studio Sound.* His format can be followed here in order to present a series of objective measurements in a logical form. In a summary article, after about 20 tests had been performed on small loudspeakers, *all* intended for close-field or mid-field studio monitoring purposes, it was pointed out that no two responses of any aspect of the performance of any two of the loudspeakers were the same. People are often heard to ask why so many loudspeakers sound so different when they all measure the same. Well they may seem to measure the same when reading the manufacturers' sanitised literature, but as Dr Holland pointed out, they do not *sound* the same simply because they do not measure the same.

Figures 19.1–19.4, 19.6, 19.9 and 19.10 show the anechoic responses of three different loudspeakers. In each case (except 19.3(c)) the (a) plot is that of a JBL LSR32, the (b) plot is that of a Westlake Audio BBSM-5 and the (c) plot is a Tannoy System 600 A. The pedigree of the manufacturers is beyond reproach, yet they have chosen three radically different physical layouts. The JBL is a conventional vertical 3-way design, the Tannoy is a 2-way concentric design, and the Westlake is a 2-way design using horizontally spaced low frequency drivers with the tweeter mounted above and between them.

19.1.1 Pressure Amplitude Responses

Figure 19.1 shows the on-axis pressure amplitude responses and the distortion components up to the fifth harmonic. The axial frequency response was measured with the loudspeaker mounted on top of a pole in the large anechoic chamber at the ISVR. The microphone was a Bruel and Kjaer type 4133 mounted on a pole 1.7 m away from the loudspeaker. The measurement amplifier was a B & K 2609. The distance of 1.7 m was chosen in order to be distant enough to avoid geometrical and near-field effects, but close enough to achieve a good signal-to-noise ratio. Where relevant, the microphone was placed between the geometric centres of the low and high frequency drivers, although the true 'acoustic centre' of many loudspeakers is not well defined. Pink noise was used as the test signal to ensure a good signal-to-noise ratio throughout the audio bandwidth. The input signal was measured at the loudspeaker input terminals, thus eliminating power amplifier and loudspeaker cable effects, and the frequency response was derived from the output of the microphone pre-amplifier and the reference input signal.

Strictly speaking, the complete frequency response should show the pressure amplitude response *and* the phase response, but the phase responses have been omitted because of the notorious difficulty of interpreting them on the logarithmic scale that best suits the interpretation of the pressure amplitude plots. The phase plots have been replaced by 'acoustic source' plots, which will be described in Subsection 19.1.4. The measurements were analysed with a 16,384-frequency line resolution, from 0 Hz to 20 kHz, after averaging 160 records.

19.1.2 Harmonic Distortion

The harmonic distortion was measured by feeding the loudspeaker with a swept sine wave input. The sweep lasted for one minute, during which time it swept logarithmically from 10 Hz to 10 kHz. The signal level was adjusted to give the equivalent level of 90 dB SPL at 1 m distance. A test signal analysis was carried out to ensure that any harmonics it contained were below the noise floor. Only harmonics above −60 dB (0.1%) are displayed.

For reference, Table 19.1 shows a comparison of distortion figures in decibels and percentages.

19.1.3 Directivity

The directivity plots show the horizontal and vertical off-axis responses. The measurement techniques were similar to those used for the axial pressure amplitude responses. The microphone position remained the same, but the loudspeakers were turned on the pole in increments of 15°.

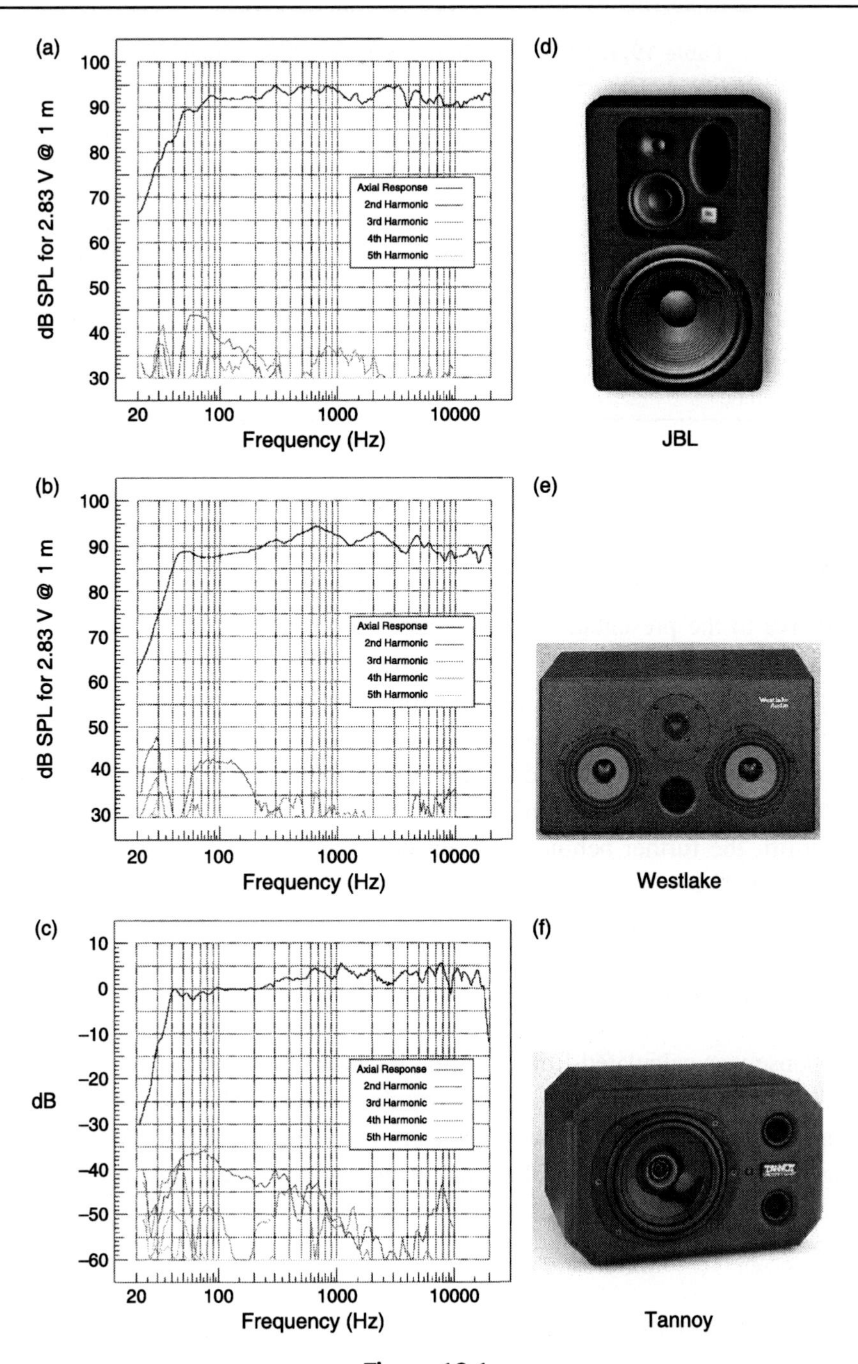

Figure 19.1:
On-axis frequency response and harmonic distortion of three loudspeakers with physically very different driver layouts: (a) JBL LSR-32; (b) Westlake BBSM-5 and (c) Tannoy System 600 A; (d) The JBL LSR-32, with 300 mm low frequency driver; (e) The Westlake BBSM-5, with 2×125 mm low frequency drivers and (f) The Tannoy System 600 A, with 150 mm low frequency driver and coaxial tweeter.

Table 19.1: Equivalent harmonic distortion quantities

100%	−0 dB
30%	−10 dB
10%	−20 dB
3%	−30 dB
1%	−40 dB
0.3%	−50 dB
0.1%	−60 dB
0.03%	−70 dB
0.01%	−80 dB
0.003%	−90 dB
0.001%	−100 dB

19.1.4 Acoustic Source

As an alternative to the presentation of the phase responses, acoustic source plots are shown which were derived from the phase responses. The group delay was calculated as a phase slope (the differential of the phase response relative to frequency) and the results were derived from the multiplication of the group delay by the speed of sound. The acoustic source was plotted as an equivalent distance in metres that each frequency apparently emanated from in relation to the front baffle. It will be seen that the steeper the low frequency cut-off, the further behind the front baffle will be the apparent acoustic source of the low frequencies.

19.1.5 Step-Function Response

The step response was calculated from the time integral of the inverse Fourier transform of the on-axis, high-resolution, complete frequency response (i.e. the pressure amplitude *and* phase responses). The step-function response shows the equivalent effect of the application of a DC voltage directly to the loudspeaker, and gives a good representation of the transient performance.

19.1.6 The Power Cepstrum

The power cepstrum is the Fourier transform of the logarithm of the pressure amplitude response. The resultant chart has units of time versus non-dimensional decibels. Full details of the methodology can be found in the reference at the end of this chapter.[1] Cepstrum analysis evolved from the world of seismology, as a means of finding echo patterns in conditions of very poor signal-to-noise ratios. Some of the more or less equivalent terms are listed below, all of which are anagrams:

spectrum – cepstrum
frequency – quefrency
magnitude – gamnitude
phase – saphe
filter – lifter
harmonic – rahmonic

In this strange, pseudo-dimensional domain, 'high-pass filters' become 'long-pass lifters'.

Of particular value is the fact that reflexions and echoes show up very clearly as spikes in the power cepstra, even though these may not be apparent from the disturbances that they cause in the frequency response plots. Edge diffractions and diaphragm termination effects can be separated from the confusion of the general response. The displacement of a spike along the quefrency axis represents the delay in milliseconds separating the direct sound from the reflexion. The height of the spikes shows their relative strength. From the cepstrum display, one can readily compute the possible positions of the sources of the reflexions.

19.2 The On-Axis Pressure Amplitude Response

The on-axis pressure amplitude response is generally considered the most important of all loudspeaker specifications. For studio monitoring purposes it is widely accepted that this response should be as even and smooth as possible, although gradual roll-offs above 8 or 10 kHz are sometimes specified. Ideally, in many cases, relative uniformity of response down to 20 Hz would be desirable, but this is only achieved in rare circumstances, partly because of the physical size necessary to generate 20 Hz at realistic monitoring levels. It is therefore not considered a fault if smaller loudspeakers fail to reach down to 20 Hz.

Although the frequency of roll-off and the rate of roll-off can have a great effect on the character of the low frequency sound, no consensus exists as to exactly where and how the roll-off should occur (i.e. the frequency and the slope). What is more, it is difficult to specify the necessary anechoic response of the loudspeakers because the rooms in which they are used, and the position within those rooms, not to mention whether the loudspeakers are mounted on meter bridges or on stands behind the mixing consoles, *all* seriously affect the in-room frequency response.

Looking at Figure 19.1 it can be seen that all three loudspeakers maintain a frequency response within ±4 dB (from an arbitrary medium frequency) between 60 Hz and about 19 kHz. The

JBL maintains a response ±2.5 dB from 70 Hz to 20 kHz, which is quite remarkable for *any* loudspeaker. The low frequency roll-off is approximately third-order (18 dB per octave). The Tannoy maintains its frequency response to within ±4 dB from 40 Hz to 19 kHz, which is again a commendable performance, but the roll-off rates are very rapid at each end of the spectrum. At low frequencies, the roll-off is sixth-order (36 dB per octave [each 'order' representing 6 dB per octave]). The Westlake also maintains ±4 dB, this time between 40 Hz and 20 kHz (and probably beyond) and exhibits a fourth-order (24 dB per octave) roll-off at low frequencies.

On closer inspection, the responses show definite differences of design approach. The JBL clearly tries to adhere to a flat response, and the low frequency roll-off is the most gradual of the three. The relatively large (300 mm) woofer and the 50 litre cabinet allow this with reasonable sensitivity (93 dB at 1m for 2.83 V). With these dimensions, the low frequencies do not need to be 'forced' by special tuning alignment or electrical boosts. The Tannoy (Figure 19.1(c) and (f)) displays a very gradually falling low frequency response, but there is a general trend towards the response rising with frequency. A smoothed plot would show a distinct inclination or 'tilt' in the response. As this is a free-field anechoic response, the implication is that if the loudspeaker were mounted on a meter bridge, the constraint of radiating angle would tend to level the tilt by reinforcing the low frequencies. The flatter low frequency response of the JBL suggests that it was intended for use on stands, with free space around it.

The Westlake response shows a pronounced peak around 650 Hz with the response tending to fall either side of this frequency. The tilt up and tilt down at either side must give a different timbral response to the JBL and the Tannoy. The response of the JBL at 1 kHz is the same as at 100 Hz and only one decibel more than at 10 kHz. The Westlake, however, has a response that at 1 kHz is about 5 dB up on the responses at 100 Hz and 10 kHz. It is somewhat like putting a broad 5 dB equalisation lift between 300 Hz and 1 kHz, which anybody involved in recording will know to be very audible indeed. The Westlake also shows evidence of a bump at 50 Hz, which is probably the tuning port acting strongly to augment the low frequency response from the pair of 125 mm low frequency drivers in their approximately 30 litre enclosure. The side-by-side mounting of the woofers suggests that the cabinet was intended to be used 'landscape', with the longest axis horizontal. This also suggests that it was designed for meter bridge mounting, which would tend to augment the bass response by virtue of the flat reflective surface below the loudspeaker, and where the sort of boost shown in Figure 11.26 could be expected. Mounted on stands, though, it is difficult to see how the 650 Hz peak would not dominate the response. Nevertheless, having said this, there are many other loudspeakers with generally similar responses.

The high frequency response shows a gradual roll-off of the type that is commonly found in large monitor systems. Time and experience has suggested that a roll-off of the high frequencies can lead to a more compatible match with the outside world. The two probable reasons for this are the psychoacoustic differences when working at SPLs louder than the domestic norm, and that at these levels less ear fatigue (and threshold shift) takes place if the high frequencies are reduced.

The Tannoy uses a 150 mm woofer in a 13 litre enclosure, and it is driven by built-in amplifiers and equalisation circuitry. The low frequency response is typical of such a system; the bass extension continuing with electrical assistance before a protection filter circuit cuts the sub-40 Hz frequencies abruptly, to prevent driver damage at high SPL.

From the three frequency response plots shown in Figure 19.1, it can clearly be seen that different philosophies and design approaches have been used in each case. It should also be apparent that the three will all sound markedly different if used under similar circumstances. The manufacturers will all claim flat responses, although clearly, the mounting condition will have considerable effect on the low frequency response, and the intended mounting conditions should be taken into account before commenting on the publicised responses.

Strictly speaking, as explained earlier, we should be referring to the pressure amplitude responses, here, as we are not dealing with the phase response, the two of which constitute the true 'frequency response'. However, for the sake of recording studio convention, if not academic convention, the term frequency response has been used because of its greater familiarity.

19.3 Harmonic Distortion

Non-linear distortions, such as harmonic distortion, intermodulation distortion and rattles, are ones in which frequencies are produced which were not present in the input signal. On pure tones, a non-linear device initially produces harmonics of the fundamental, but sum and difference products can also be produced by intermodulation when reproducing complex signals such as speech or music. Rattles generate totally unrelated frequencies. In Gilbert Briggs[2] book of the 1950s, *Sound Reproduction*, he opened his chapter on intermodulation distortion with a quote from Milton: '... dire was the noise of conflict', which quite well sums up its subjective quality.

The audibility of non-linear distortion and its subjective effects are not well documented. The mechanisms of distortion can give rise to subjective effects that are not attributable to the distortion *per se*. An amplifier giving 1% of distortion would (except for a few idiosyncratic, valve, 'hi-fi' devices – which tend to produce even-order harmonics) not really be considered to be hi-fi, yet 1% is a very acceptable figure for a low frequency loudspeaker. James Moir, of the BBC, computed a graph of harmonic distortion detectable on single tones,[3] which showed that at 400 Hz 1% of second or third harmonic distortion is undetectable. At 60 Hz, 7.5% of third harmonic distortion was undetectable, and that at 80 Hz, even 40% of second harmonic distortion was undetectable. Magnetic tape recorders, even of the professional kind, have traditionally been aligned to produce not more than 3% total harmonic distortion at maximum operating levels. One of JBL's most famous high-frequency drivers, the 2405, was designed for use above 6 kHz and has a rated distortion of 7%. According to Colloms,[4] MacKenzie suggested that for loudspeakers, a maximum of 0.25% harmonic and intermodulation content for the range 200 Hz to 7 kHz was desirable if high fidelity was the goal.

Of the loudspeaker distortions shown in Figure 19.1, the JBL exhibits remarkably low distortion for a loudspeaker of its size and output capability. At a worst case 60 Hz, the second harmonic distortion reaches only 0.56%, but has fallen below 0.1% (−60 dB) by 200 Hz. The third harmonic distortion peaks at 1% at 32 Hz but remains below 0.2% (−55 dB) at high frequencies. The Westlake BBSM-5 is also an excellent performer. It is much smaller than the JBL, which usually implies higher distortion at low frequencies, but the second harmonic distortion is only around 0.5% at 100 Hz. True, this rises to 4.5% at 30 Hz, but this could be considered out of the range of operation of such a loudspeaker, and it is still well below Moir's detectability threshold. The total distortion between 200 Hz and 6kHz at 90 dB SPL is generally less than −55 dB, or 0.2%, well within MacKenzie's detectability limits. The Tannoy remains below 1% of total harmonic distortion except for its lowest octave. Distortion in the 2–5 kHz range is commendably low at around 0.1%, or −60 dB.

Once again, as with the pressure amplitude responses, the harmonic distortion performance of all three loudspeakers is well within the accepted standards for professional use. Nevertheless, in some cases, the distortion figures are different by up to 20 dB, and although they are all below the threshold levels for detection on single tones, the effect on highly complex music signals is still hard to quantify. On transient signals, which make up a great proportion of music, it can be difficult to detect moderate distortion. However, on string sections, especially when reverberation is present, much lower levels of distortion can be noticeable. The main point of this section, though, has been to show that the distortion products of the three loudspeakers are quite distinctly different from each other.

19.3.1 Intermodulation Distortion

Practically anybody who knows anything about sound amplification and transducer systems will have heard about harmonic distortion. It is one of the fundamental response parameters that have been customarily measured since the early days of audio. It might therefore come as a shock to some people that many of these measurements may all have been in vain and that by making them we have largely been wasting our time. (That is, from the point of view of the end-user as a means of ranking loudspeaker sound quality, as opposed to its use as a research and development figure.) It has been mentioned elsewhere in this book that little direct correlation has been found between absolute harmonic distortion figures and perceived audio quality, at least not below a certain, surprisingly high threshold. In the opening paragraph of Section 19.3, Briggs' reference to Milton's 'dire was the noise of conflict' may be the key to why harmonic distortion measurements have been so frustratingly inconclusive in their subjective correlations.

'The noise of conflict'! How apt this statement may turn out to be. Non-linear distortions refer to any distortions that contain frequencies that were not present in the input signal. If a sine wave is distorted, then in order to take on its new form, it *must* contain other frequencies. That is *how* it can develop its distorted waveform. When an amplifier clips, the resultant output, if viewed on a spectrum analyser, will show products at 2, 3, 4 times, etc. the fundamental frequency of the sine wave which it is amplifying. Similarly, the progressive non-linearity from loudspeakers produces distortion, which gradually increases with level. However, music signals are not sine waves. They contain many frequencies simultaneously, and what is more, unlike sine waves they vary with time.

When two frequencies are being amplified by a device that is not perfectly linear, it will produce not only the harmonics of the two input frequencies, but also the sum and difference frequencies. For example, for 1000 Hz and 1100 Hz there would be outputs at 1000 plus 1100 Hz (2100 Hz) and 1100 minus 1000 Hz (100 Hz), and these are only the first-order products. If we take two frequencies which are musically related by a perfect fifth, such as 1000 Hz and 1500 Hz, the first few *harmonic* distortion products would be 2000 Hz and 3000 Hz (second harmonics) and 3000 Hz and 4500 Hz (third harmonics), etc. Relative to 1000 Hz, the harmonics at 2000 Hz, 3000 Hz and 4500 Hz are the octave, the fifth above the octave and the second above the double octave. Relative to 1500 Hz, the harmonics at 2000 Hz, 3000 Hz and 4500 Hz would be the fourth, the octave and the fifth above the octave. The harmonics of both frequencies are therefore musically related to both fundamental tones. This is not surprising, because all musical instruments produce harmonics naturally – playing tunes with sine waves only, would be rather boring.

On the other hand, if we look at the *intermodulation products*, they would produce not *multiples*, which are precisely what harmonics are, but sum and difference tones, such as $f_1 + f_2, f_1 - f_2, 2f_1 - f_2, 2f_2 + f_1$, etc. These combinations can produce inter-modulation frequencies that are *not* in any musical way related to the fundamentals. What is more, when a complex musical signal is being produced or reproduced, the complex spectral spreading of the intermodulation products begins to look *not* like an enrichment of the harmonic structure of the music – which is what at least the lower order harmonic distortions add – but something which more resembles the addition of *noise*... 'The noise of conflict'.

When we are measuring harmonic distortion, we are measuring, or at least we are trying to measure, the degree of non-linearity in a system. In the case of a loudspeaker, the effects are due to such things as the non-linearity in the restoring forces applied by the suspension systems, or the non-linearity in the behaviour of the magnetic fields under different drive conditions. We are measuring the effects of things that are not perfectly symmetrical in their behaviour. These are the things that give rise to non-linear acoustic outputs. If they were not there, there could be no harmonic distortion, so, if there *is* harmonic distortion, it implies that there *are* non-linear mechanisms in the system, and these must, in turn, give rise to intermodulation distortion products.

At the end of this chapter, we will look at intermodulation distortion more specifically, but the main point to be understood, here, is that the harmonic distortion figures are better only to be viewed as a guide to what else may be happening in an *un*musical way, and that we are *not* measuring. Direct comparison of harmonic distortion figures between different devices can therefore be very misleading in terms of their sonic performances. Valve amplifiers which may (or may not!) be deemed to sound 'better' than transistor amplifiers of much lower harmonic distortion are an obvious example.

19.4 Directivity – Off-Axis Frequency Responses

In an anechoic chamber, when listening on-axis (directly in front of the loudspeaker), the off-axis frequency response is the one aspect of loudspeaker performance that is irrelevant. However, in environments where reflexions exist, the off-axis response is more likely to influence the room sound than the axial response. The total power response, as described in Chapter 11, is the factor that is most likely to drive the reverberant field. Figures 19.2 and 19.3 show the horizontal and vertical off-axis responses of the same three loudspeakers, as shown in Figure 19.1.

The horizontal off-axis response of the JBL LSR32 is extremely well behaved. Indeed, the LSR in the name refers to 'Linear Spatial Reference', because the loudspeaker is designed especially to yield a smooth reverberant response in typical rooms. Even up to 60° off-axis, the response differs from the on-axis response only by a gradual high frequency roll-off. The vertical directivity is similarly well controlled (Figure 19.3(a)), except for the sharp dip around 2 kHz in the '30° up' direction. To maintain this sort of response with non-collocated drivers implies superb engineering and forethought. Almost nobody is going to listen from 30° above the loudspeaker, and the reflexions from that direction are likely to be innocuous in well-designed listening rooms. In fact, in many control rooms the ceiling is relatively absorbent, so the energy travelling 30° up is not likely to contribute *anything* significant to the room sound. The crossover has no doubt been carefully designed to push that cancellation node into the least problematical direction.

The Westlake exhibits an entirely different approach. The 30° up dip in the frequency response of the JBL is due to the crossover design and the physical path length difference problems outlined in Figure 13.11. The Westlake uses a horizontal driver alignment, so if we look at the *horizontal* directivity of the BBSM-5 (Figure 19.2(b)) we notice the

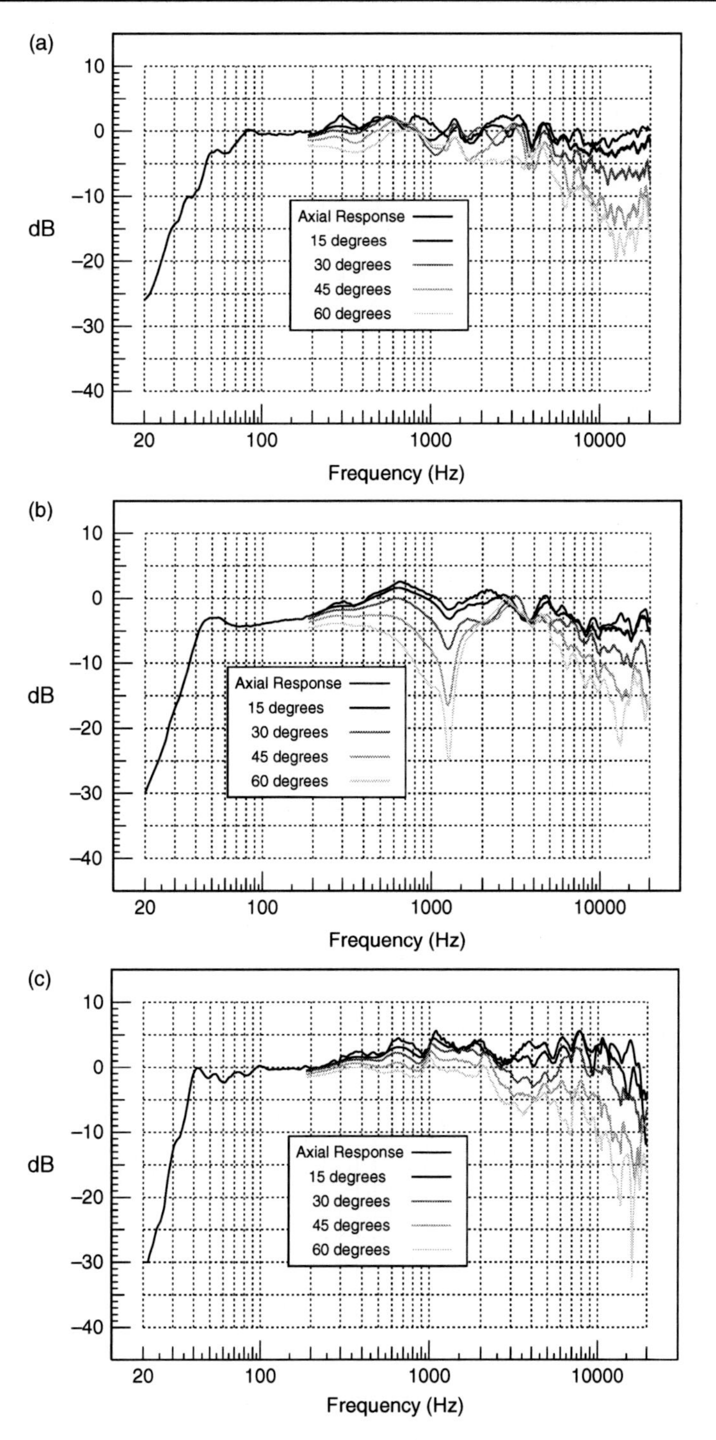

Figure 19.2:
Horizontal directivity: (a) LSR-32; (b) BBSM-5 and (c) System 600 A.

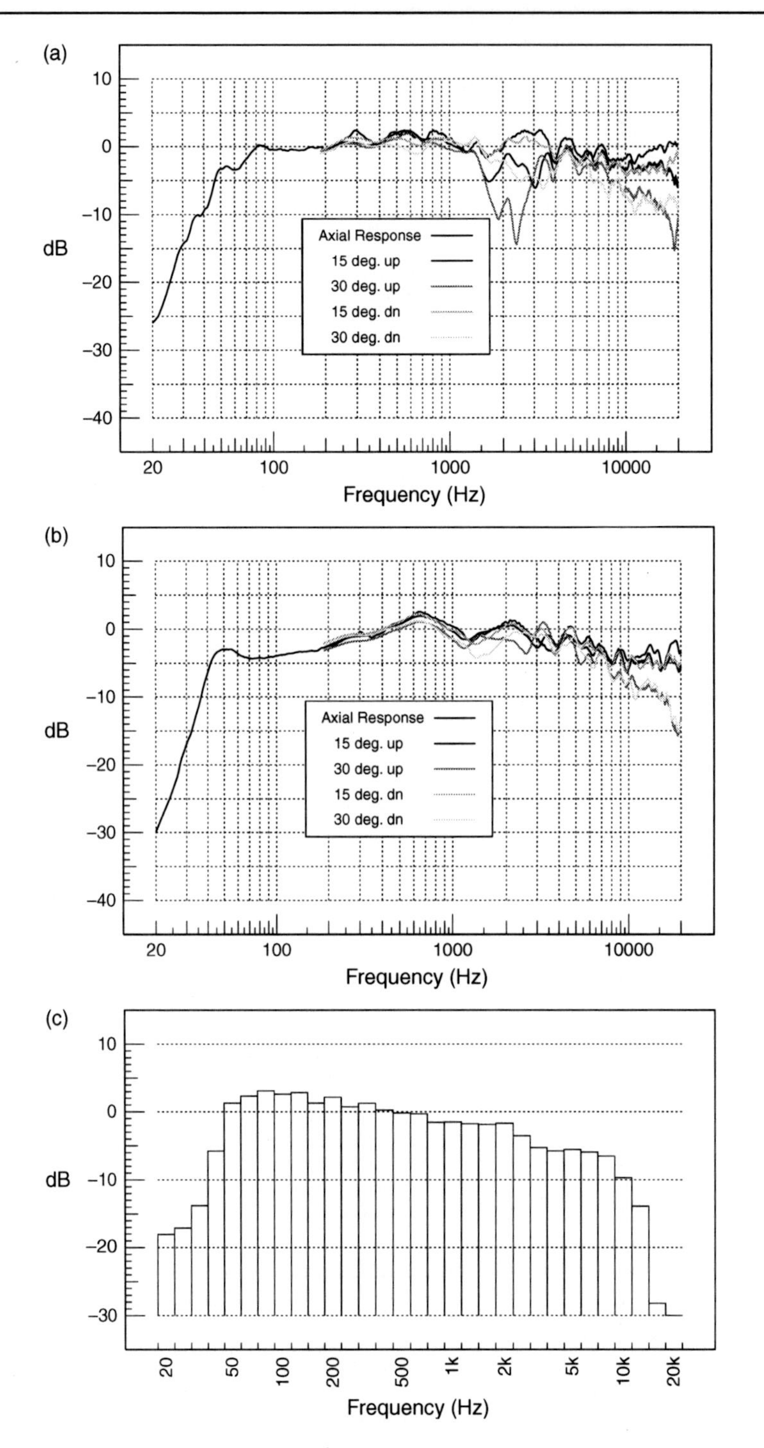

Figure 19.3:
Vertical directivity of JBL and Westlakes plus total power response of Tannoy 600 A: (a) LSR-32; (b) BBSM-5 and (c) Total power of System 600 A.

cancellation dips that are evident in the *vertical* directivity plots of the JBL. The notches are quite severe, but they may be more pronounced than normal due to the on-axis dip at the same frequency, 1200 Hz. The vertical directivity of the Westlake is excellent due to the low mounting of the tweeter, which puts it no more than about 10 cm above the line passing through the centre of the bass drivers. At the crossover frequencies therefore, when all three drives are operating together, they are more or less horizontally in line. This type of arrangement is good from the point of view of not obstructing the main monitors when the small ones are mounted on stands behind a mixing console, or on the meter bridge, but it does cause the response to change with horizontal movements of the listener, as described in the previous chapter. Sharp response dips of this nature do not tend to be very audible, which is fortunate because they are an unavoidable result of spaced driver geometry, but the Westlake dip is quite broad.

In fact, the JBL allows the front panel, which holds the tweeter and mid-range driver, to be rotated through 90° for horizontal (landscape) mounting of the cabinet if desired. However, the sheer size of the cabinet, almost 40 cm in width, would still make it rather too large for meter bridge mounting if the main monitors were not to be obstructed. The vertical notch at 30° up would still remain because it is related to the mid-range and tweeter interference. Where the loudspeakers are used as the principal monitors, of course both options are available.

The Tannoy avoids all of these problems by means of its dual concentric construction. The tweeter fires through the centre of the woofer, which it uses as its horn flare. The concentric design means that the horizontal and vertical directivities are equal, because there is no physical offset of the drivers which can create cancellation at the crossover frequency. The Tannoy off-axis response shows a gradual fall in upper-mid frequencies that is perhaps slightly more than is the case with either the JBL or the Westlake. However, the Tannoy exhibits a slightly rising *on*-axis high frequency response, whereas the other two have somewhat falling responses. There is an old hi-fi trick, described in Chapter 11, which is to boost slightly the on-axis high frequency response to try to flatten the total power response at the listening position, compensating for the 'darker' reverberant room response caused by the *off*-axis roll-offs. The subjective effect of this is highly room dependent. As the horizontal and vertical directivity plots of the dual concentric design are identical, and are both as shown in Figure 19.2(c), Figure 19.3(c), shows the total power response for the Tannoy System 600 A, which is very smooth indeed. The on-axis rising response no doubt contributes to this.

The price to be paid for the use of dual concentric drivers is that the tweeter horn is being modulated at high levels of low frequency output, and the driver design is more complex. The former may lead to a rather idiosyncratic, level-dependent sound, and the latter may force other design compromises that would perhaps not be desirable had the coaxial requirements not existed. Nevertheless, Tannoy are one of the true long-term players in the studio monitor industry, with over 60 years of fully professional acceptance, and they perhaps know more about dual concentric driver design than anybody else. Indeed, Tannoy registered the description Dual Concentric in the 1940s. If the System 600 A is mounted horizontally or vertically, it performs in an identical manner.

19.5 Acoustic Source

The acoustic source represents the distance behind the baffle from which the different frequencies appear to emanate. Obviously a big change in the acoustic source at low frequencies will cause them to arrive noticeably later than the higher frequencies, which causes the time-smearing of transients. As sound travels at around 340 m/s, each metre of displacement will cause an arrival delay for the affected frequencies of about 3.4 ms. The acoustic source shift is due to group delay in the electro-mechanical filters, which results from the electrical crossover roll-offs *and* the associated roll-offs of the drivers and their associated ports and cabinets. For any given frequency, the steeper the roll-off, the greater the group delay. (See Glossary.)

Figure 19.4 shows the acoustic source plots of the three loudspeakers. As mentioned in Section 19.2, the JBL has a third-order low frequency roll-off, the Westlake a fourth-order roll-off and the Tannoy a sixth-order roll-off. Not surprisingly, it can be seen that at 40 Hz, the JBL has an acoustic source almost 2 m behind the baffle, the Westlake 2.4 m and the Tannoy over 3 m, the latter representing over 10 ms of signal arrival delay for the low frequencies relative to the high frequencies.

The audibility of these time shifts is not well understood, but the consensus is that the sound is more natural if the group delays (the phase shifts) are kept to a minimum. The late Michael Gerzon, one of the most respected audio investigators of the twentieth century (co-inventor of the Soundfield microphone, Ambisonics, the Meridian Lossless Packing data compression system, and the first proponent of dither noise shaping), was of the opinion that for natural sound reproduction, the minimisation of phase shifts down to 15 Hz was more important than the maintenance of a flat amplitude response down to much higher frequencies. Figure 19.5 illustrates the idea diagrammatically, with the low order roll-off of the (a) curve tending to sound more natural, despite rolling-off from a higher starting point than the (b) curve. In general, the subjective audibility of the low frequency roll-off order in small loudspeakers still needs much further investigation.

19.6 Step-Function Responses

Figure 19.6 shows the step-function responses of the three loudspeakers. The perfect step-function response would be as shown in Figure 19.7. For the tail not to slope

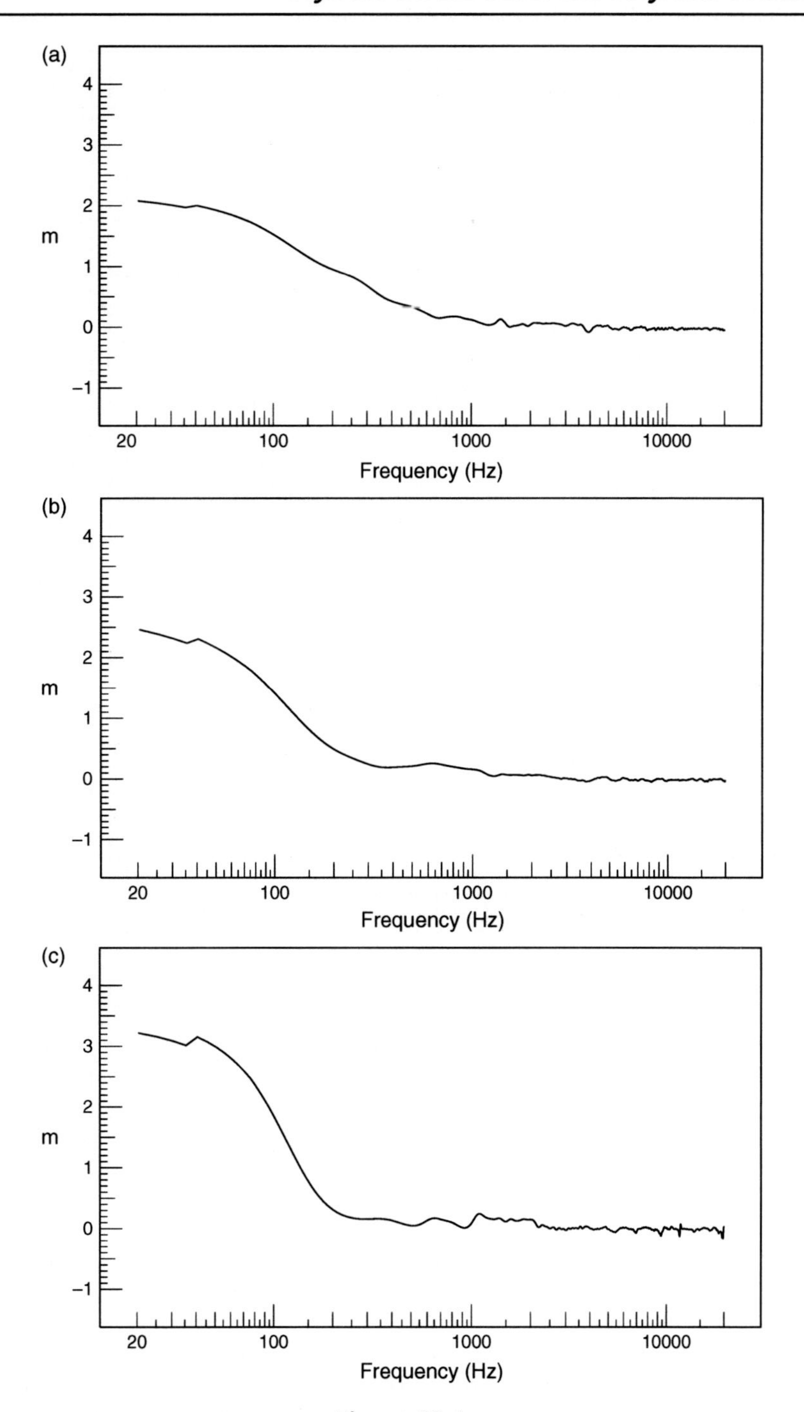

Figure 19.4:
Acoustic source plots: (a) LSR-32; (b) BBSM-5 and (c) System 600 A.

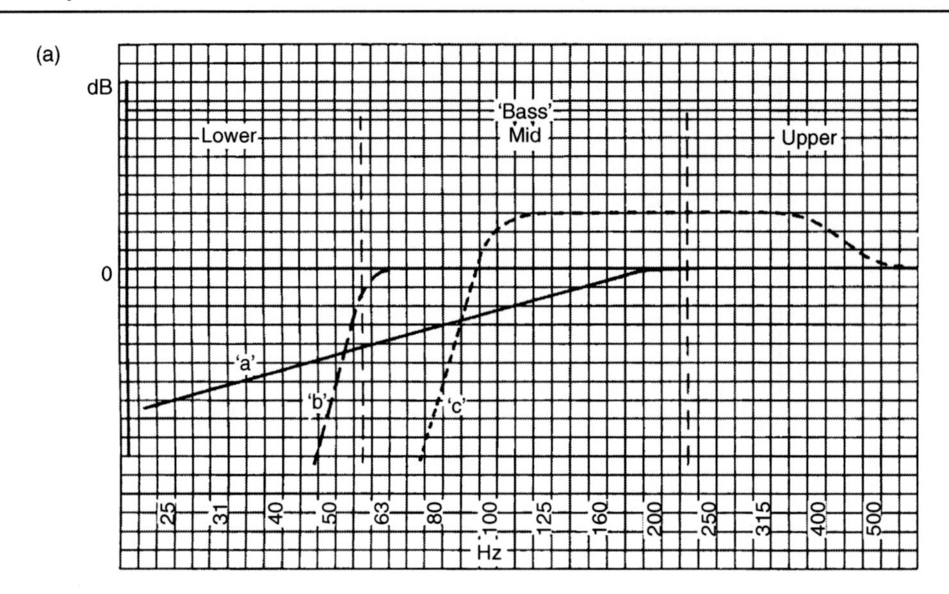

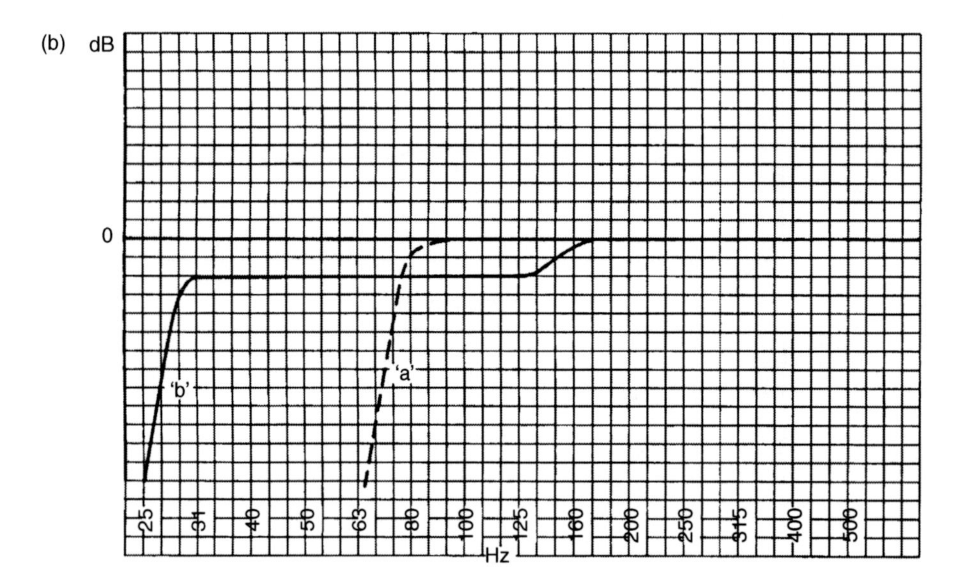

Figure 19.5:
(a) The plots in this figure show three very different bass responses. Despite also imparting very different tonality or timbre to the music, they could nonetheless all be perceived as having subjectively the same 'quantity' of bass, as related to mid and top, if their responses above 500 Hz were all the same, although it would be highly dependent upon the nature of the musical signal. In general, the rock world tends to favour something akin to the 'b' line, the classical world the 'a' line, and many cheap domestic systems the 'c' line; (b) Plot 'a' is typical of many good quality small, free-standing loudspeakers, such as are frequently employed as close-field monitors, on top of or immediately behind the mixing console. Plot 'b' would be typical of a large, high quality, built-in monitor system.

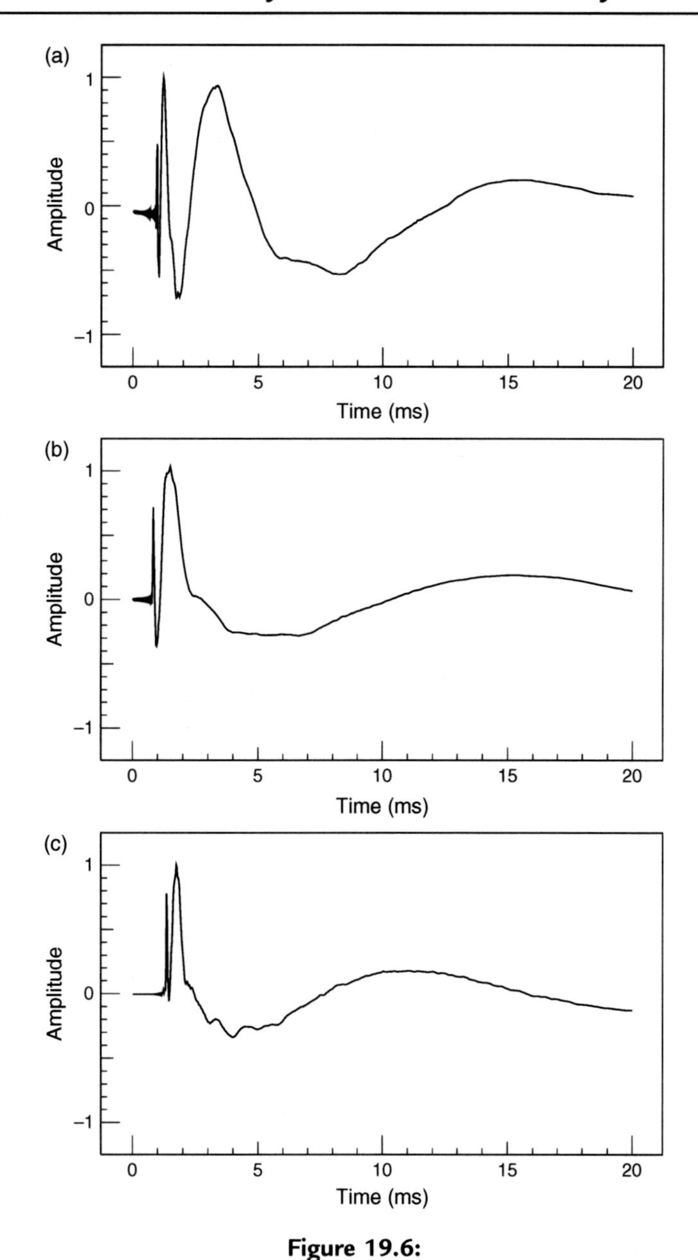

Figure 19.6:
Step-function (Heaviside function) responses: (a) LSR-32; (b) BBSM-5 and (c) System 600 A.

downwards, the response would need to extend flat to DC, so the steepness of the decay slope is a function of the low frequency response of the loudspeaker system. It could only be horizontal if the loudspeaker box was sealed and placed in an airtight room. The spike at the beginning of the JBL response in Figure 19.6(a) is the tweeter, which can be seen to respond a fraction in advance of the mid range drive unit, which is represented

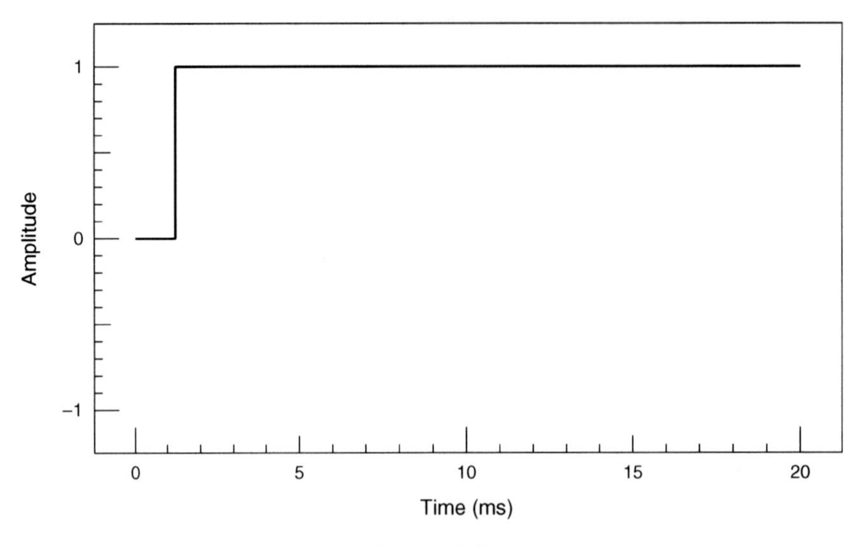

Figure 19.7:
A step function.

by the second, higher peak. The third, more rounded peak, at around 3.5 ms, is the woofer. This non-synchronicity of the arrival times is due to the physical alignment of the drivers and the group delay due to the crossover frequency and slope. The manufacturers considered the subjective effect to be too small to warrant complicating the design. All arrive within about 2.5 ms of each other.

The Westlake shows in Figure 19.6(b) a more compact response. The tweeter can still be seen to speak in advance of the woofers, though only by about 500 microseconds. The response can be considered good. The time separation of the drivers at the crossover point also manifests itself as a kink in the acoustic source plot, Figure 19.4(b), at around 800 Hz. The Tannoy shows a better step-function response, this time with only about a 200 microsecond delay, which is probably inaudible. However, for comparison, Figure 19.8 shows the exemplary step-function leading edge of two Quad Electrostatic loudspeakers, which shows what *can* be achieved.

19.7 Power Cepstra

The power cepstrum responses highlight problems due principally to surface irregularities and diffraction. The power cepstrum plot of the JBL, in Figure 19.9, shows almost no evidence of reflexions, which suggests that the stereo imaging should be good due to the lack of physically-separated secondary diffraction sources. The cepstrum plot of the Westlake shows evidence of echoes at 0.2 and 0.5 ms, and these may be responsible for the slightly irregular frequency response in the mid-range, due to interference between the direct and

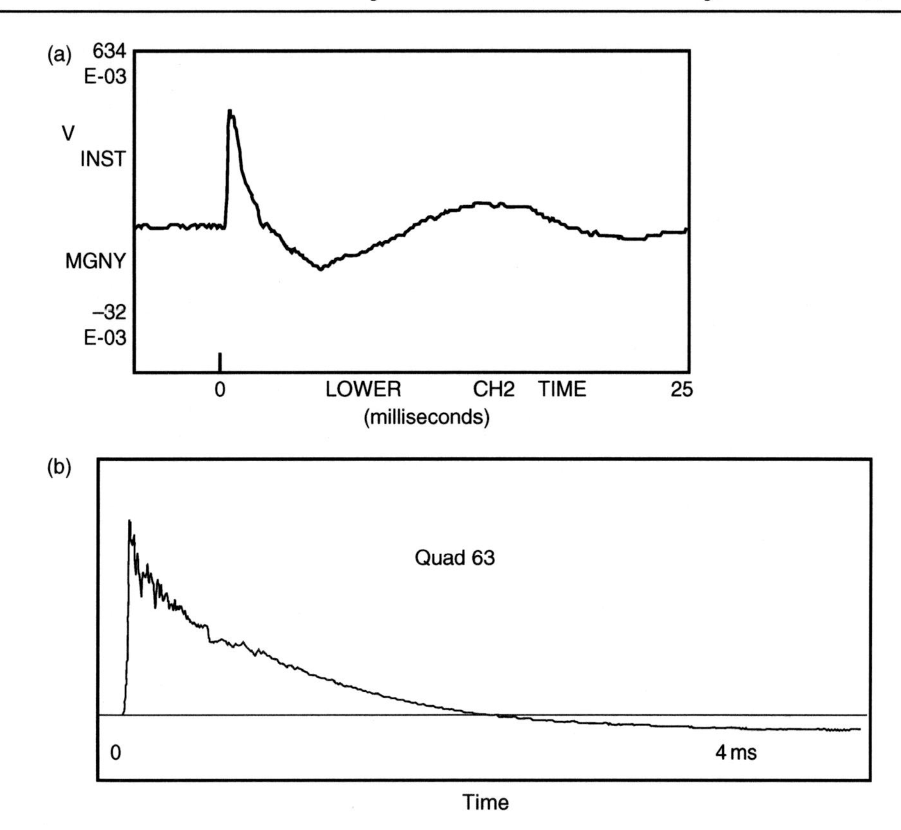

Figure 19.8:
(a) Step function response of a Quad Electrostatic Loudspeaker (ESL); (b) Step function response of a Quad ESL63, but on a much shorter timescale than (a), showing an almost perfect leading edge.

secondary sources (see Figure 11.20). The Tannoy shows evidence of reflexions at 140 and 280 microseconds (0.14 and 0.28 ms) which may be due to the discontinuities in the horn flare where the metal phasing plug meets the coil gap of the low frequency cone. Again, these could be the source of some of the on-axis mid-range frequency response irregularities.

19.8 Waterfalls

Figure 19.10 shows the waterfall plots for the JBL, the Westlake and the Tannoy. Waterfall plots display time, frequency and pressure amplitude in a three-dimensional form. The most noticeable difference between the plots is the greater low frequency overhang of the Tannoy. Although the decay of the Westlake ultimately falls below the –40 dB level more rapidly than the JBL, there is much more initial energy in the decay below the 100 Hz region. The steeper roll-off of the low frequency response of the Tannoy results in more ringing energy in the filters. The more resonant a filter, the sharper will be its Q and the longer it will

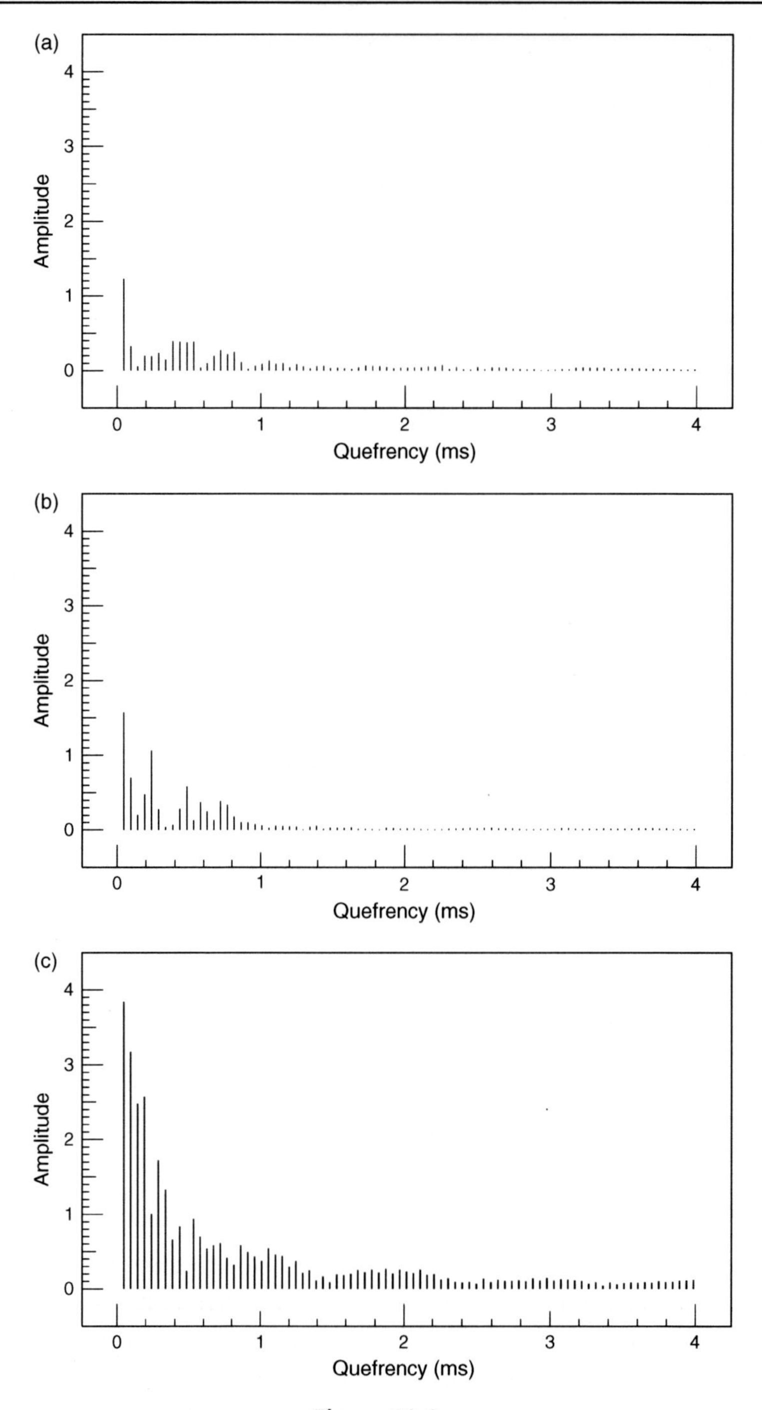

Figure 19.9:
Power cepstrum: (a) LSR-32; (b) BBSM-5 and (c) System 600 A.

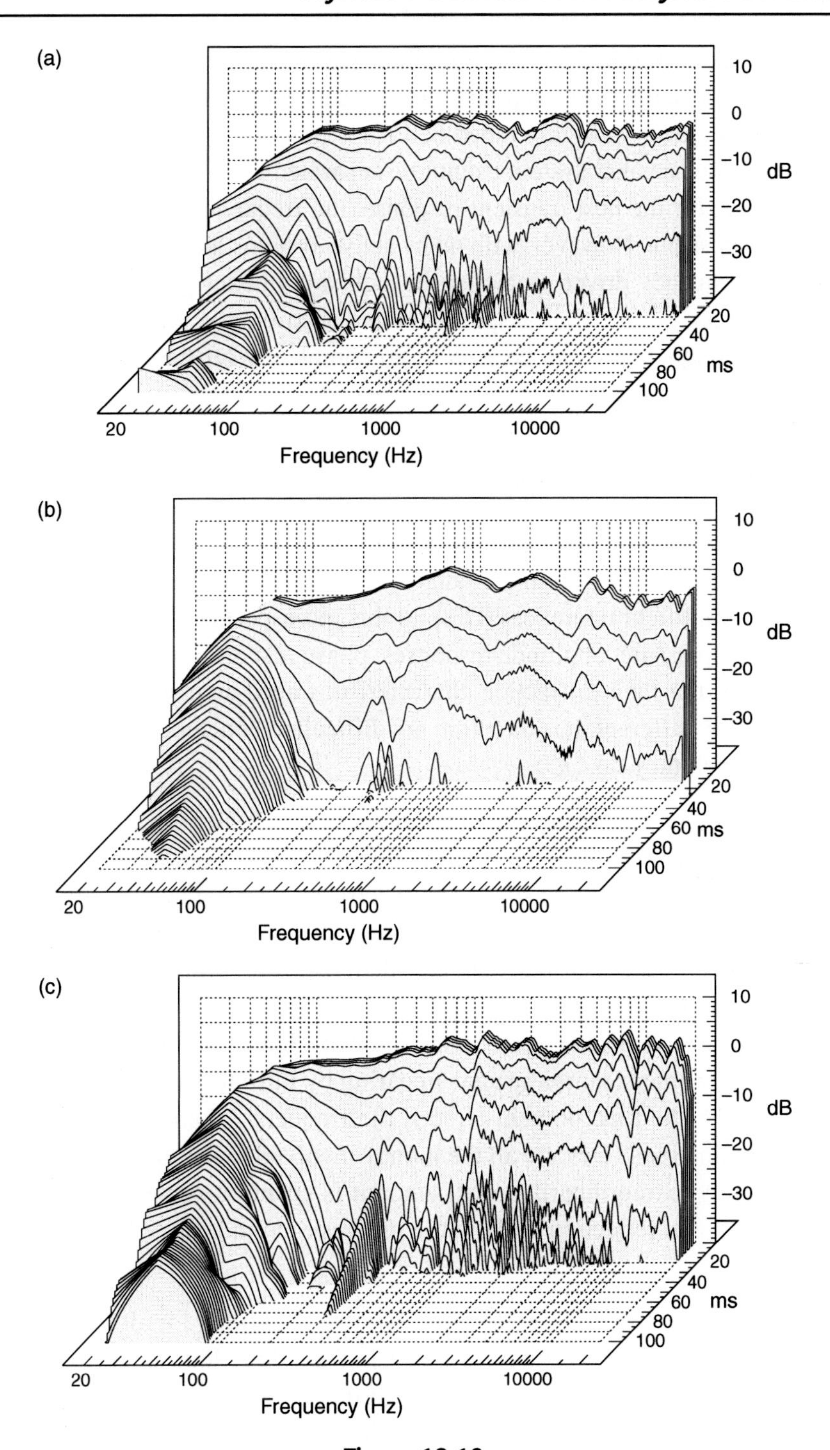

Figure 19.10:
Waterfall plots: (a) LSR-32; (b) BBSM-5 and (c) System 600 A.

ring; and it makes no difference whether the filter is electrical, mechanical or acoustical. When the note is more defined, or the filter slope is steeper, the result will be a longer 'ring-on' time, or resonant hangover. A highly tuned bell rings longer than an old metal barrel. The time response of a loudspeaker system cannot be separated from the roll-offs in its frequency response. Therefore, for the best transient response, the frequency response must be as wide and as flat as possible. However, with many active systems like the Tannoy System 600 A, the need for low frequency driver protection in such small boxes gives rise to the need for steep protection filters, but the time response suffers accordingly. The time response of the Tannoy, at low frequencies, can be seen to go completely off the scale.

19.9 General Discussion of Results

The three loudspeakers under discussion in this chapter were chosen because they are all good performers. They are all well in the range of loudspeakers typically used in professional studios as close or mid-field references. Despite this, no aspects of their performances match. Neither their pressure amplitude responses, phase responses, time responses, harmonic distortion responses, nor their diffraction characteristics need an expert to separate them. They are all obviously different. It is therefore not difficult to conclude that the loudspeakers all *sound* different; which they do.

A parallel exists with microphones, which are more or less loudspeakers in reverse. Nobody would pay €3000 for an old Neuman microphone if a €200 Shure could be equalised to sound the same. The characteristic sounds of electro-mechanico-acoustic transducers are highly complex combinations of the convolution of their time and frequency domain responses, along with the spacial responses, which also affect their overall responses in non-anechoic conditions. One great problem for investigators looking into the audibility of various aspects of loudspeaker performance is that of separating combined effects. Where a diffraction problem causes spacial smearing by introducing secondary sources, then what is being heard? The time domain effects, the frequency domain effects, the spacial effects, or the combinations? Despite all the money involved in the music, hi-fi and recording worlds, an absolutely extraordinarily small amount is spent on subjective/objective correlation. This is a sad reflection on how the professional industries have been usurped by the business people.

The current situation is that there is no loudspeaker that is optimal for all rooms. This fact can be clearly seen from the on and off-axis responses of the three loudspeakers studied in this chapter. No loudspeaker is optimal for all music. Of the three types of low frequency responses shown in Figure 19.5(a), one would not normally choose to play classical music through 'b' or disco music through 'a', if both options were available, because they would simply sound much more appropriate in the reverse order.

Obviously, no loudspeaker can sound optimal in a bad room, though the Tannoy would probably sound better than the Westlake because of its very smooth total power response, as shown in Figure 19.3(c). All the designs are affected very differently by their circumstances.

High definition monitor systems with low distortion, good transient accuracy, flat and extended frequency responses, high SPL capabilities and decent sensitivity are expensive to make and require expensive components. They also tend to be large. Commercial realities therefore force compromises on loudspeaker designs, and the wide range of loudspeakers that are available reflects the need for choice if best-fit compromises are to be economically matched to circumstances. Unfortunately, the people who choose the loudspeakers are rarely fully familiar with the reasons for the compromises. They hear a loudspeaker in one set of circumstances, presume that the sound quality is the responsibility of the loudspeaker, alone, and then wonder why they sound different in another environment. This is why for the highest quality of reproduction, the loudspeaker system and the room should be designed as one entity.

All this makes it even stranger that a loudspeaker of seemingly dubious quality, such as the Yamaha NS10M, has found favour as a workhorse in such a wide range of environments, but with close inspection, reasons *can* be found. In fact, it may help to put some of the subjects discussed so far in this chapter into a useful context if we try to correlate the subjective and objective characteristics of a loudspeaker as well known as the NS10M. After the cessation of production of the loudspeaker in 2001, a series of interviews were made on behalf of *Studio Sound* magazine, and were published in August of the same year.[5] What follows in the next section includes some excerpts from those interviews, which will hopefully throw some light on to why the NS10Ms were so widely used. Section 19.11 will then try to match their performance characteristics to the comments of their users.

19.10 The Enigmatic NS10

As previously mentioned in Section 15.5, in the 1960s it was customary for 'pop' record producers and engineers to check what a mix would sound like on a 'transistor radio loudspeaker' before approving it. It was generally believed that what people heard on the radio had a big influence on whether they bought a record, or not. Radio *was* a big influence on sales, and the consoles such as the Neves of the late 1960s had small, built-in loudspeakers, which in many cases could be used as a radio reference. As awful as these things were, and to a large degree they were not really representative of anything, they did sometimes help to get a more appropriate balance between guitars, vocals and reverb levels. They were very inconsistent though, and recording engineers were always complaining about them. People then moved on to making a Compact Cassette and playing the results back on a domestic machine, either inside or outside the studio. Somewhere in the early 1970s, there appeared a new 'reference' loudspeaker, the Auratone 'Sound Cube'. These relatively tiny loudspeakers could produce prodigious output levels for their size when driven by Crown D150 amplifiers, or similar. Their small size also caused very little obstruction of the main monitors if they were mounted on the meter bridge of a mixing console. The fact that they were single driver loudspeakers ensured a smooth phase response through the vocal region, and this no doubt contributed to their usefulness, because crossovers in general simply will not sum their outputs to give an accurate reproduction of the input.

The Auratones reigned supreme for several years, although not everybody used them or liked them of course, but in the early 1980s, there came a challenger, the NS10M. Designed as a home hi-fi loudspeaker, it was generally badly received by the critics, but this little loudspeaker had a rock-and-roll punch which caused many people in the recording industry to take note. It was also reasonably robust, but the thing that puzzled so many rival loudspeaker manufacturers for the subsequent 20 years was just *why* it was so popular. What characteristics did it have that led so many people to have confidence in the fact that the mixes that they did on them would translate well to the street? A number of well-respected people from the music industry were therefore asked to give their opinions about the NS10s, in the hope of getting a little more insight into what made these loudspeakers so special.

The first person to be approached was Bob Clearmountain, one of the great exponents of the NS10s. He said that in the early 1980s his favourite mixing loudspeaker was the KLH17. The problem with them was not their sound, but their fragility; they just would not withstand the punishment of studio use. He was then introduced to the NS10 in its early domestic form, and was impressed by the way that the mixes travelled. The only problem was that the mixes lacked top outside of the studio, so toilet paper placed over the tweeters was used as an acoustic filter, to reduce the high frequencies perceived during mixing. Apparently, this worked well, because his mixes then did not surprise him when played elsewhere. When the NS10M 'Studio' version came out, the toilet paper was no longer necessary, and the NS10Ms were then his mainstay for many years.

What had puzzled the interviewer for years was how INXS's *Devil Inside* could sound so well balanced on wideband monitor systems, in good rooms, when Bob had claimed to have only monitored the mix on NS10s. One problem with the use of close-field monitors, only, is that people often have a tendency to make changes to the bass and bass drum levels and equalisation, according only to the limited frequency range that they can hear. It is a common complaint from mastering engineers that if these have been wrongly judged on small monitors, there is often nothing that the mastering houses can do to repair the situation such that things will still sound well balanced on full-range systems; be they in a home or a studio environment. Bob solved the mystery by paying great tribute to the producer Chris Thomas and the recording engineer David Nicholas. It seems that the multi-track recordings, which *had* been checked on large monitors and had been made in very professional studios, were of such quality and free from problems that Bob's job was mainly to balance the relative instrument and effect levels. The quality control had already been done for him. He recognised that he was in the fortuitous position of normally only working with high-quality artists and recordings. Principally he is a mixing engineer, rather than a recording engineer. He trusted the NS10s for so many years because they were the tools that helped him do his job. 'They helped to pay my rent', he said, 'because I could have confidence in the work that I did on them. I'm not a very technical guy, I'm a music mixer, and I use the tools that I can work with best. *Why* they work is not too important to me'.

The next person to be spoken to was Alan Douglas (the Grammy winning engineer of the *Riding with the King* duet album of B. B. King and Eric Clapton), on the day before he was starting recording with Annie Lennox. He said that he thought that NS10s sounded hideous, but they had a rock-and-roll sound, which if you got used to it could lead to good results from a mix. Alan said that the Quad amplifier/NS10M combinations, often used in the UK, was something that he found very difficult to deal with, but the NS10s on a large Crown amplifier was something that he had achieved good results with. Like Bob Clearmountain, Alan's preference at the time of the interview was for KRKs, but he also had used the Auratones many years previously. 'If you can get a mix to sound good on an Auratone, then you know you're really in business', he said. He preferred the KRKs because they could still, despite their size, get a good control room 'buzz' for the musicians. At home he liked to listen on Tannoys, but found it difficult to work on them in a studio. On the other hand, the loudspeakers that he *could* work with in the studios he did not like for home use. Such are loudspeakers!

Nick Cook had spent much of the previous ten years as the head of Fairlight's European operations, and as Sales Director of Amek. From the mid-1970s though, he was a busy recording engineer, working in some of Britain's best facilities.

> *I always used to check the core sound on the large monitors, then do the equalisation and balancing on the NS10s. They are bright and harsh, and the old ones without the 'loo' roll led to too little top on the mixes. The change from the old domestic NS10M to the NS10M Studio version caused a short period of confusion, because the reference changed, but the sound had essentially the same character and people soon adjusted.*

Nick believed that the recording and mixing processes require very different loudspeakers.

> *During the recording you need to hear the instruments in detail, whilst at the same time being able to inspire the musicians to great performances. The subtleties of fine balance are not important at that stage. By the mixing stage you should already have well recorded multi-tracks, so the mixing process becomes more a question of balance.*

Nick also believed that one asset of the NS10 over many other of its contemporary loudspeakers, when it was first used in studios, was its ability to accept a 'solo'd' bass drum without expiring. 'They also sounded good when sitting on the top of SSL consoles, and once their reputation had been established in such a position, and they had been used on "top" recordings, the industry in general saw them as a reference'. Nick also commented on the solidity of the NS10 cabinet construction, which ensured that the cabinet, itself, was not audible. There was thus, no boxy sound, which helped it to have more of the character of built-in monitors. Furthermore, the solid cabinet did not tend to excite mixing console resonances, which could be a problem with some of the NS10s more flimsy rivals. 'What I've also noticed', he said, 'is that the general sound of recordings has changed over the years, and some of the reason for that could have been that the most popular mixing loudspeakers begin to set their own standards for what is perceived to be right'. What you become accustomed to becomes a *de facto* reference in the absence of any clearly defined industry reference.

Now over to the man from the 'Beeb'. Although Chris Jenkins had worked for SSL for the previous 20 years, he was BBC trained and spent some years with Virgin. He has been generally well respected as a musician, recording engineer, maintenance engineer and console designer. Surely, *he* must have some insight into the popularity of the NS10M. 'I used to be able to perceive a lot of dynamic detail on the Auratones', was his first comment. His technique was to get a clean recording by using the main monitors, then switch to the NS10s or Auratones to get a balance between the instruments. 'Once you learn to trust the balances that you get on a certain loudspeaker that you can rely on, then from the engineering point of view there is really no need to know why – you just get on with it. I tended to prefer the Auratones to the NS10s', he said, 'because I got more of a sense of the dynamics; perhaps due to the fact that it was a single loudspeaker with no crossover'. Somewhat surprisingly, not only could a person as musical *and* technical as Chris not put his finger on the 'why', but he did not even seem to care. Only the results counted.

The next person to be approached was Mick Glossop, the producer-engineer of artists such as Frank Zappa, Van Morrison and Sinead O'Connor. Mick was continuing to use both NS10s and Auratones. 'I don't really discuss what or why with other people, because it is only the results that I get that count', he said. I then read him the quote from George Massenburg, mentioned elsewhere in this book. To recapitulate, Massenburg had said 'I believe that there are no ultimate reference monitors, and no "golden ears" to tell you that there are. The standards may depend on the circumstances. For an individual, a monitor either works or it doesn't … Much may be lost when one relies on an outsider's judgements and recommendations'. 'Absolutely!' was Mick's reply. In fact, the George Massenburg quote had been tested on several other people during these interviews, and they all agreed that he had (as is his tendency) hit the nail right on the head.

Mick went on to say:

> *If things sound too good on the loudspeakers, then it makes you lazy and you don't work on the music. You then take things home and realise that you could have, and should have, done*

a lot better. I use large monitors, from time to time, to check the bottom end, but often I do whole recording sessions on NS10s. I really don't like them at home, though; they're for work only. I have often noticed that during the mastering, I need to add something around 2.8 kHz, because I tend to undermix those frequencies when using NS10s. I still use Auratones to set the bass and vocal levels in a track. I can also hear distortions better on the Auratones, and they have a lot of detail when quiet. I can get a good balance on Auratones which works well when switched to NS10s, but I won't use the Auratones from scratch. I dread having to change monitors, because I have become so used to the ones which I use. If I had to change, I would probably go to KRKs, but I also really like the Questeds, I can work well on those. I use large monitors when available as a low frequency check, but if I don't know the room that they're in, it can be misleading. Essentially, though, I want to put all my energy into the mix, and I don't want to have to waste time thinking about the loudspeakers.

Finally, here are the comments from the London-based songwriter/engineer/producer/studio owner, Michael Klein.

What I like about working on the NS10s is that they make the mid-range very clear and prominent. This is normally where many instruments are fighting for the same space. The NS10s allow me to concentrate on getting the mid-range finely balanced, and once that is done, the basis of a mix is usually well established. I wouldn't record on them, though, and I certainly wouldn't want them at home, but for mixing, they are a great help.

So, with all those words of wisdom from such esteemed recording engineers, where does that leave us? Well, it seems that none of them are the slightest bit interested in loudspeaker design – they just want to use the ones which work for them. They all also appear to agree with George Massenburg's statement. They almost universally saw the NS10M as a mixing tool, not a recording monitor, and it is perhaps in this area where so many small studios miss the point. They see top name engineers using NS10Ms for mixing but fail to realise, firstly, that they often use large monitors during recording and, secondly, that they are immensely experienced people who can rely on that experience to interpret what they are hearing. They are perhaps not taking what they hear to be gospel, but are interpreting what they hear in the light of their experience.

The people interviewed all used the monitor systems that worked for them, and they all tended to agree that they wanted to use loudspeakers that made them work hard at a mix. Interestingly, there was a general rejection of a very popular brand of powered loudspeakers because they did not make bad mixes sound bad enough. This was summed up by Alan Douglas's comment that 'If you can make a mix sound good on Auratones then you *know* you're in business'. Not one person spoke about frequency response, or hardly any other technical aspect of performance. They all spoke in subjective terms, even though some of them have deep technical knowledge. However, this lack of accurate, descriptive feedback to the manufacturers has not helped the further development of monitor systems, and that communication gap between users and manufacturers still exists.

As the author of this book, I suppose that I ought to add my own comments, from my points of view as a former recording engineer and producer, and now, principally as a studio designer. My career as a mainstream recording engineer and producer was largely coming to an end when the NS10 was first introduced, but, in the years since, I have always been around studios and recording personnel. In fact, understanding the needs of other engineers is perhaps a greater part of my job now, than ever it was before. My own opinion of *any* monitor is not that it should sound nice, but that it should scream in your face when things are wrong. This philosophy is borne out by the number of top producers and engineers who, when relaxing at home, absolutely do not want to listen to the loudspeakers that they work with. They *want* to hear the problems *smoothed over* when listening for pleasure, at home. (Though, as always, there are some exceptions.)

What has been discussed in this section are the subjective opinions about why the people interviewed have elected to use the NS10Ms. What has not been discussed, however, are the aspects of the NS10M's performance that may be responsible for its widespread use. The following section will now attempt to highlight those aspects of performance, in objective terms, which correlate well with the subjective comments. The work is based on a paper presented to the Reproduced Sound 17 Conference of the UK Institute of Acoustics in November 2001.[6]

19.11 The NS10M – a More Objective View

The original NS10M was conceived as a domestic hi-fi loudspeaker for bookshelf mounting. As such, it was not a great commercial success, and neither was it very well received by the international hi-fi press. However, it was readily adopted by many recording personnel as a close-field studio monitor for rock/pop mixing. It effectively took over the mantle that had largely been carried by the Auratone 5C 'Sound Cube'. Despite the output of the Auratone being prodigious for its size and era, its limitations had led many users to seek other loudspeakers with higher output levels and wider frequency ranges. Nevertheless, many of them still sought loudspeakers that exhibited the more valued characteristics of the Auratones. The NS10s were widely considered to fill that need.

The original NS10M fell short of the requirements on two counts; firstly, it was still somewhat lacking in output capability and secondly, it was considered to have an excess of high frequencies. The former problem gave rise to the need for frequent driver replacements, whilst the latter was commonly solved by the fixing of a piece of toilet paper over the tweeters. The old tale about the discussions over which brand was most appropriate, and whether one sheet or two was required, was not a joke; such discussions actually did take place. Yamaha subsequently dealt with both problems with the introduction, in the mid-1980s, of the 'NS10M Studio', hereinafter simply referred to as the NS10M.

19.11.1 Specifications and Measurements

The NS10M is a two-way loudspeaker consisting of a 180 mm paper coned low frequency driver and a 35 mm soft domed tweeter, all in a 10.4 litre sealed box. The crossover is second-order passive with asymmetrical turnover frequencies and in-phase connected drivers. The frequency range is quoted as 60 Hz to 20 kHz, the sensitivity is 90 dB for 1 W at 1 m, and the maximum (peak) input power is rated at 120 W. The crossover frequency is 2 kHz and the nominal impedance is 8 Ω.

Measurements carried out by Dr Keith Holland at the ISVR revealed a frequency response with a deviation of ±5 dB over the range from 85 Hz to 20 kHz under anechoic conditions. This would hardly seem to be impressive in itself, but closer examination shows that this deviation is due to an inverted 'V' characteristic response shape rather than the irregular wiggles exhibited by some loudspeakers.

Figure 19.11 shows a selection of nine of the 36 waterfall plots published in Reference 6, and reproduced in Appendix 2 at the end of this book. The nine plots represent the last eight of the alphabetical order of the 36 plots, plus the Auratone. The two outstanding features of the Auratone and NS10M are the inverted V response shape and the very rapid response decay over the entire frequency range. Both of these characteristics are largely due to the sealed box nature of the designs. We will return to this point in the concluding part of this section, but suffice it to say that of the 36 waterfall plots depicted in Appendix 2, the only other loudspeakers exhibiting a similarly rapid response decay were the ATC SCM20A, the AVI Pro 9 and the M&K MPS-150.

Figure 19.12 shows the step-function responses of the same nine loudspeakers. All of these are very good compared with many typical monitor loudspeakers of 25 years ago. The Auratone exhibits the more exemplary rise because of its single driver nature. The separate peak of the tweeters responding early can be seen in most of the other plots. The Yamaha shows a better than average step-function response, which is a good indicator of its transient performance. The rise time is very short. The response tail is also well damped, which corresponds with the rapid decay shown in the waterfall plot.

An electrical input signal having a step response is shown in Figure 19.7, but for anybody not familiar with it, a battery connected to the loudspeaker terminals via a switch can also be used as a crude source of a step-function. Rise time, simultaneous response of all drive units in a system, and ringing in the decay tail are things that step-functions show up well.

Figure 19.13 shows the harmonic distortion performances of the same nine loudspeakers as in Figures 19.11 and 19.12. Again, neither the NS10M nor the Auratone are bad performers. This is made even more emphatic when one considers that the other seven loudspeakers (and the other 34 if one looks at the full presentation in Appendix 2) are all of reputable make and are all designed for professional use.

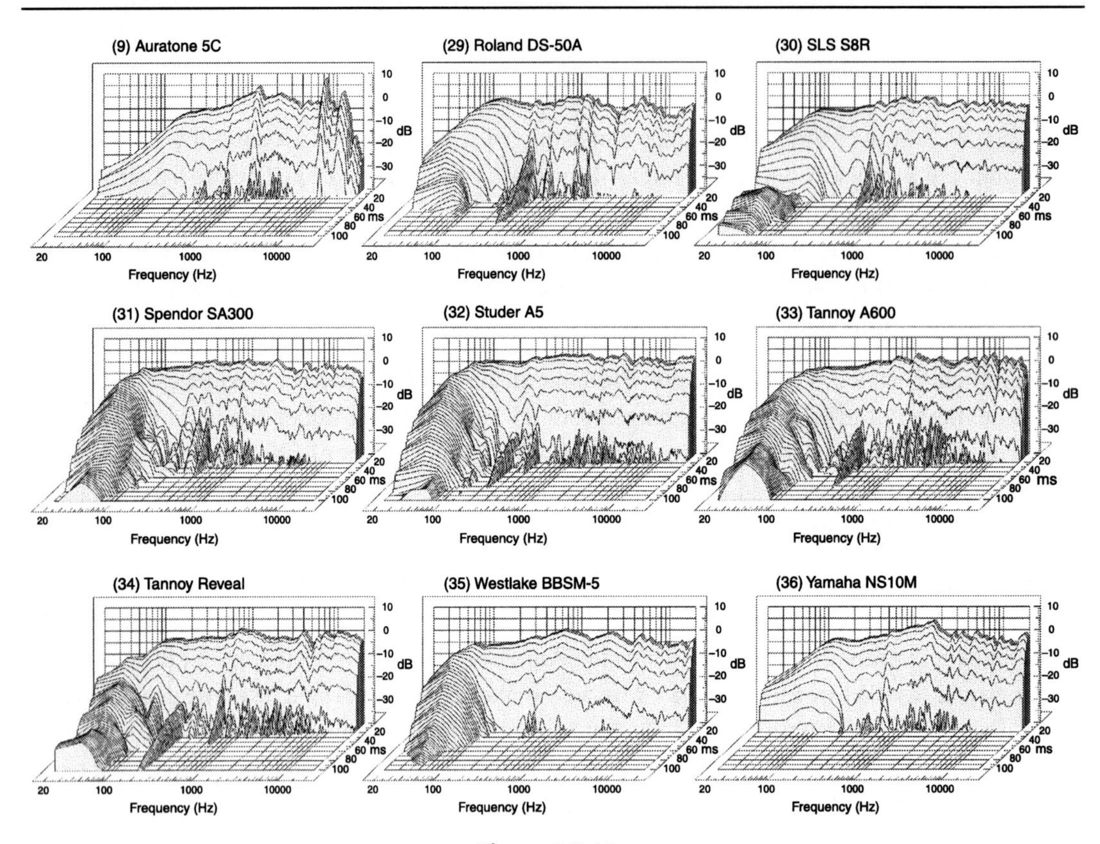

Figure 19.11:
More waterfall plots. Note the rapid decay of the 5C and NS10M.

19.11.2 Discussion of Results vis-à-vis Subjective Perception

It is widely considered that a reference monitor loudspeaker should exhibit a relatively flat frequency response. However, it should be remembered that the loudspeaker and its mounting in a room are part of one system. It is the frequency response of that *system* which really needs to be flat in order for the recording personnel to perceive an accurately frequency balanced representation of the music being recorded. The free-field response of the loudspeaker is not what is heard in a control room. The following four figures may help to clarify this point.

Figure 19.14 shows the response curves of an idealised loudspeaker of approximately similar size to the NS10M, both in free-field conditions and flush mounted. Figure 19.15 shows the response of an NS10M suspended in the open air, about 4 m from the nearest reflective surface. The response wiggles are due to the reflexions from the nearby surfaces, but the overall shape can be seen to be very similar to the free-field response shape in Figure 19.14. Figure 19.16 shows the response of an NS10M mounted on top of the meter bridge of

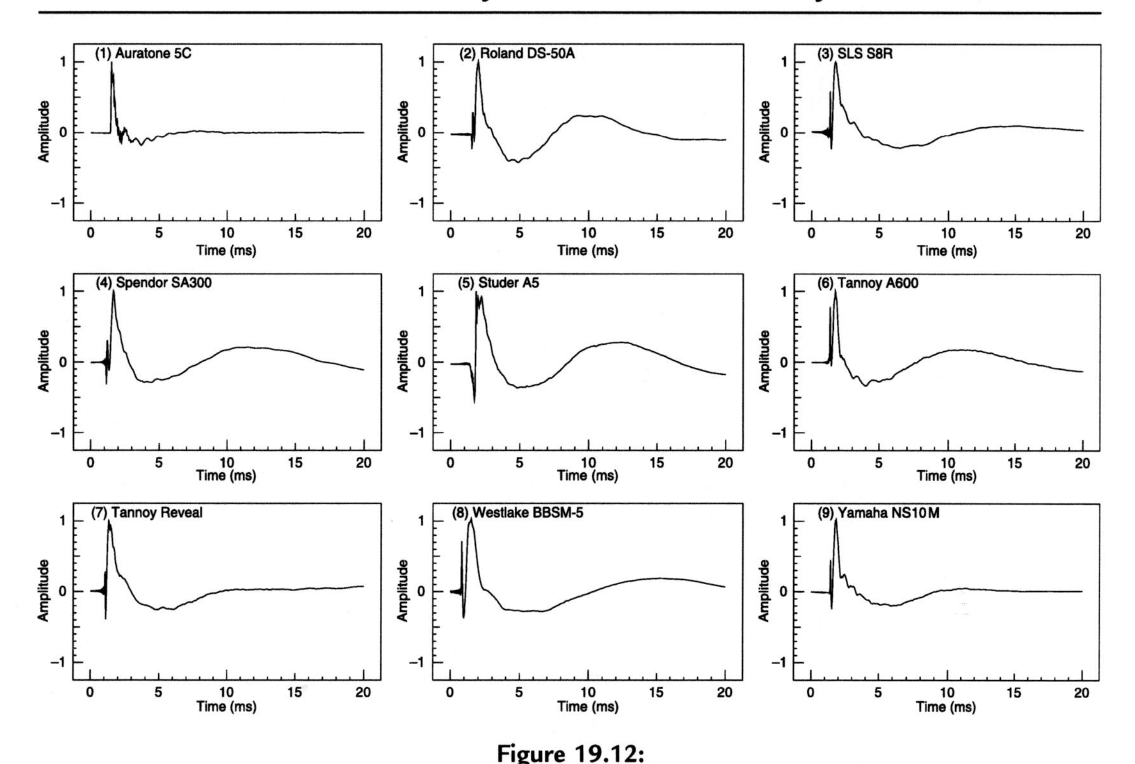

Figure 19.12:
Step responses. Note the rapid damping of the tails of the 5C and NS10M.

a mixing console, both suspended in mid-air. Additional comb filtering of the response is evident due to the proximity of the top surface of the mixing console, but the overall trend is that of the flattening of the bass response. Figure 19.17(a) shows the response of an NS10M on top of the meter bridge of a mixing console in a room typical of many recording studio control rooms. Despite the extra irregularities due to boundary reflexions, the overall trend of the low frequency response shape is in the direction of the flush mounted response shown by the dashed line in Figure 19.14.

In Section 19.10, Nick Cook was quoted as saying, 'They also sound good when sitting on top of SSL consoles'. Although he was referring specifically to SSL consoles, he actually acknowledged that they worked well on other consoles, also. It is worth noting, though, that the structure and shape of the console on which they are mounted *will* affect the response, and perhaps it is likely that the more solidly built consoles, like the SSLs, will colour the sound less than would be the case with lighter, more resonant consoles. The implication is also that the NS10Ms first established their reputation on the better consoles, on which they also tended to sound best, and that once their use had been established in the 'first division' they were then seized upon by the lower echelons. Nevertheless, Figures 19.14–19.17 do seem to reinforce the concept of the NS10M,

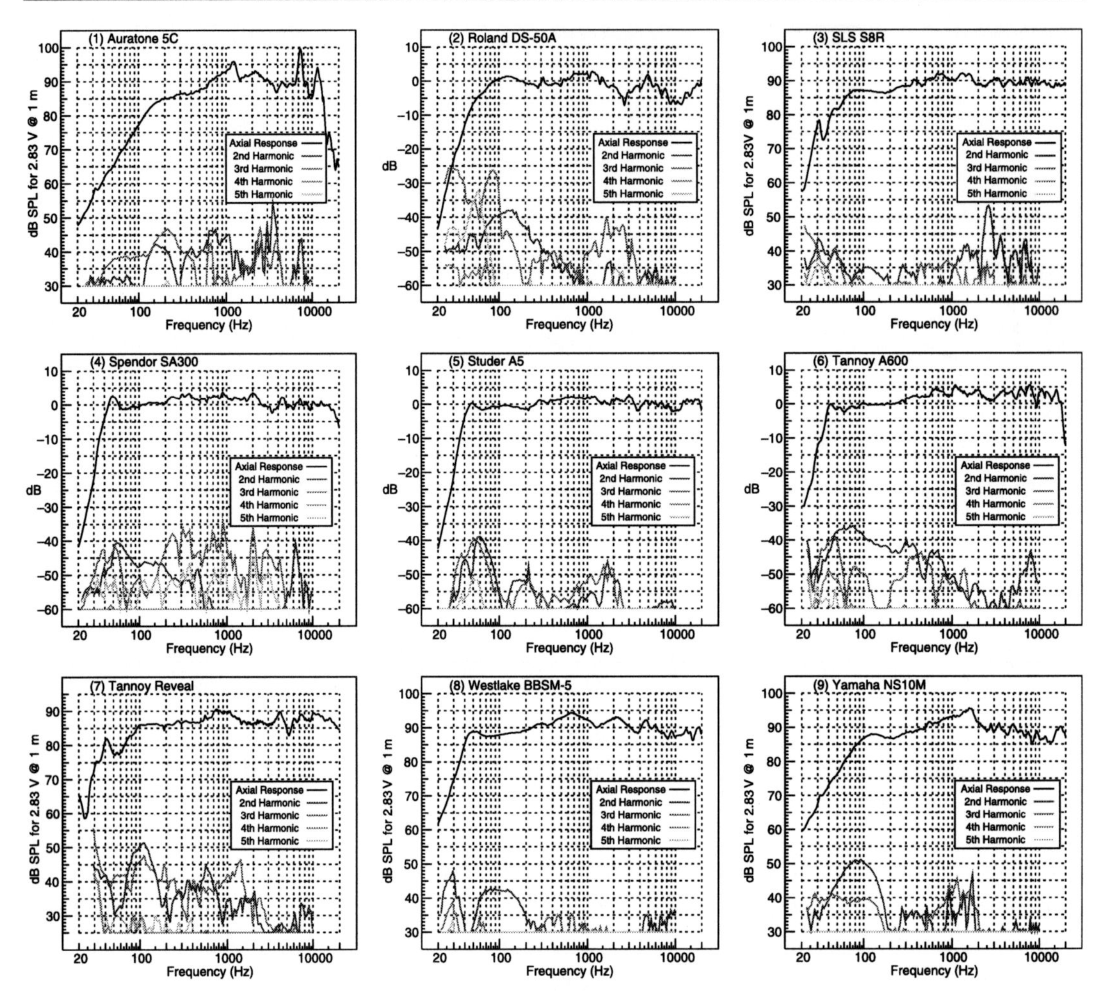

Figure 19.13:
On-axis frequency response and distortion.

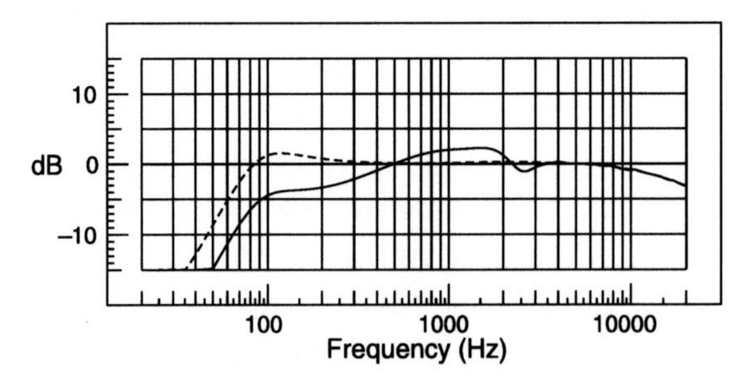

Figure 19.14:
Response of idealised loudspeaker: flush mounted (---) free-field (—).

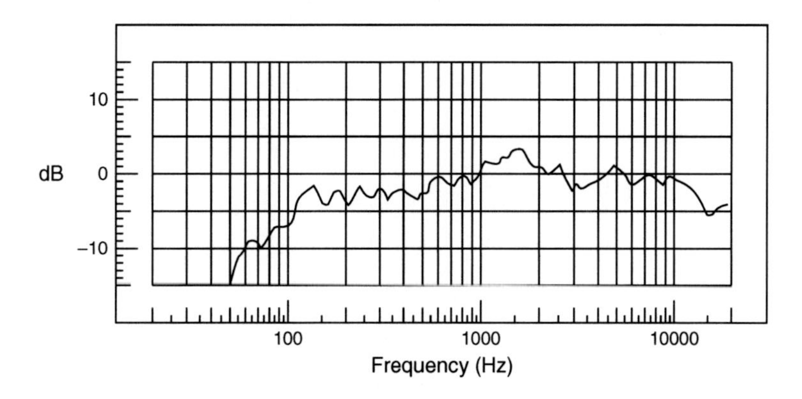

Figure 19.15:
Response of NS10M, flown from a crane, 4 m above the ground and 4 m from the nearest wall. Measurement taken outdoors.

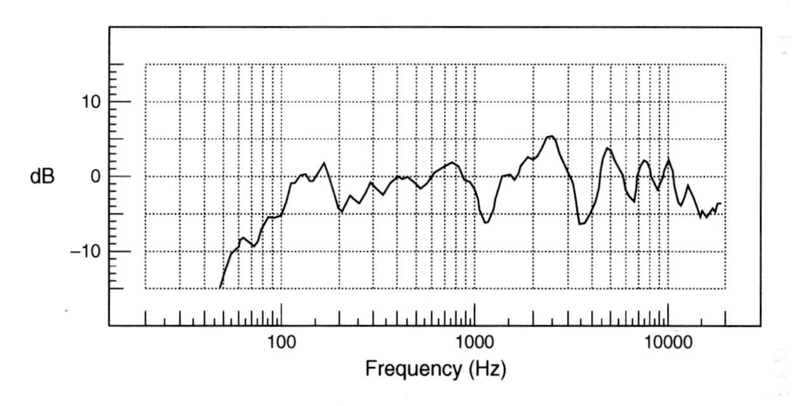

Figure 19.16:
NS10M, mounted on the meter bridge of a mixing console, both flown from a crane, outdoors, 4 m above the ground and 4 m from the nearest wall.

plus a mixing console *and* a typical control room, yielding an overall frequency response of a nature that many recording personnel appear to need to hear. That the mixing console plays its part in the response is perhaps reinforced by the number of people who work daily with NS10Ms but who do not choose to use them at home – where large mixing consoles are usually conspicuous by their absence.

The mid-range response peak, which is clearly observable around 1.7 kHz in Figures 19.11 and 19.15, appears to be responsible for the 'harsh' description that is often referred to when discussing the NS10M. This could objectively be considered a negative asset. However, during the interviews quoted in the previous section, Michael Klein said 'What I really like about the NS10Ms is that they make the mid-range very clear and prominent.

(a)

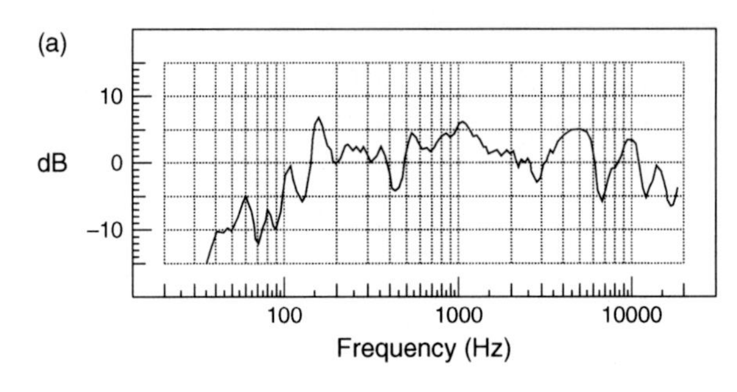

(b)

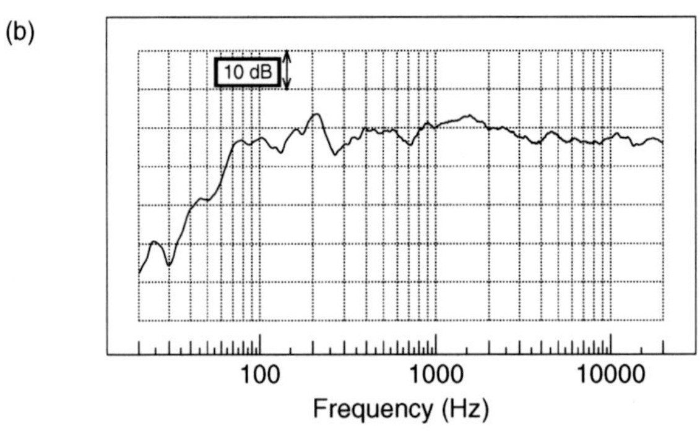

Figure 19.17:
(a) Pressure amplitude response of an NS 10M, on top of the meter bridge of a mixing console, in a room of about 30 m^2, containing material typical of a small studio control room;
(b) Actual response of an NS10M in a small studio, placed directly against a wall, with *no* mixing console beneath it. In effect, the bookshelf mounting for which it was originally designed.

This is normally where many instruments are fighting for the same space. The NS10Ms allow me to concentrate on getting the mid-range finely balanced, and once that is done the basis of a mix is usually well established'. Many people would no doubt echo Michael's comments. His words again highlight how the NS10M has been seized as a tool to help to get a job done.

Certainly in terms of frequency response, the NS10M appears to have a free-field response which, when in the typical surroundings of a recording studio, gives many people what they need in order to get a job done. The relatively low distortion no doubt also helps.

But what of the time response? Let us now turn again to the waterfall plots of Figure 19.11. Again, during the interviews recounted in Section 19.10, several people referred to the 'rock and roll punch' or the 'rock and roll sound' as Alan Douglas was quoted as saying. This clearly relates to the rapid decay of the NS10M. Two other things can also be said to result from the time response. The first is that the rapid decay is reminiscent of many good, large monitor systems in well-controlled rooms. Such systems often have cabinet resonances tuned way down below 30 Hz, and they are usually without any protective filtering in the audio frequency band. The tuning ports and protection filters that are typically *inside* the lower bass region on smaller loudspeakers give rise to the low frequency ringing which is typical of most of the plots shown in Figure 19.11. The NS10M has neither tuning ports nor protection filters. The tightness of the bass, however, *can* be influenced by the amplifiers driving the NS10Ms, and powerful amplifiers with extended low frequency responses should be used if the full potential punch is to be realised. High instantaneous current delivery is also important.

The second point is that the rapid decay of the low frequencies from the NS10M is less likely to cause confusion by distorting the time responses of the bass drums and bass guitars. One repeated complaint from many mastering engineers is that people who mix on a variety of small monitors often get the bass-guitar to bass-drum ratio wrong. As these exist in the same frequency range, an inappropriate balance between the two can often not be resolved by equalisation (or any other process) at the mastering stage. It is probable that fast decays are less likely to lead to such erroneous relative balances. After many investigations there is now growing evidence to support this argument, because the obvious low frequency time-response differences shown in Figure 19.11 will inevitably convolve themselves with the musical instrument sounds. Determining which one is contributing what, during the mixing, may be all but impossible in the case of the loudspeakers with the longer decay tails.

19.11.3 Conclusions

From the above investigations, it would appear that the following statements can be made:

1. The free-field frequency response of the NS10M gives rise to a response in typical use that has been recognised by many recording personnel as being what they need for pop/rock mixing. The principal characteristics are the raised mid-range, the gentle top end roll-off (which is typical of many large monitor loudspeakers), relatively low distortion and a very short low frequency decay time. The latter is aided by the 12 dB/octave low frequency roll-off of the sealed box design.
2. The time response exhibits a better than average step-function response, which implies good reproduction of transients.
3. The output SPL is adequate for close-field studio monitoring with good reliability.
4. They appear to mimic, in many ways, the characteristics of the good, large monitor systems (within their limited range) typically used in pop/rock recording, and hence they are recognisable to many recording personnel in terms of overall suitability for their needs.

Of course, the information presented here will only be deemed truly worthwhile if it can be used in the design of future loudspeakers for studio use. General acceptance of any such loudspeakers is, however, not merely a technological challenge. Widespread acceptance requires widespread exposure, which implies mass manufacture with good worldwide distribution networks and an affordable price. These are non-technical realities, which nonetheless affect the choices in today's recording industry.

A strong implication from the data presented here is that loudspeakers that exhibit a flat free-field response will *not* have a flat response characteristic when placed on top of a mixing console in a control room. Many of the manufacturers of active loudspeaker systems provide significant ability (via d.i.p. switches and the like) to contour the low frequency response to the mounting conditions, yet it is remarkable in how many studios the switches are set 'flat' in a misguided belief that this provides the flattest response, even when mounting conditions dictate otherwise. However, one alarming result of looking at all 36 waterfall plots in Appendix 2 is the enormous variability in the low frequency *time* responses.

Another great controversy is whether it is wise to place the loudspeakers on the meter bridges or whether they should be on stands just behind the console. The latter system does tend to give a more open stereo imaging and less comb filtering, because of the reduction of desk top reflexions. On the other hand, one then loses the bass reinforcement provided by the desk top, as shown in Figure 19.16. There are obvious compromises being made here. Clearly, though, it would seem that for optimum mounting *behind* the console, a design with a little more bass than the NS10M would be desirable. This subject is discussed further in Section 20.7.2.

It would therefore seem probable that the NS10Ms have so frequently been placed on the meter bridges because that is where they have been found to exhibit their flattest overall response, even if some other aspects of their performance are compromised. The NS10M, almost certainly found a waiting gap in the studio monitoring market which it was reasonably well suited to fill. It had many of the characteristics needed for the then relatively unconsidered (in 1982) task of serious close field monitoring in rock/pop music studios. Nevertheless, whether by design or accident, it has made its presence felt in the music-recording world perhaps like no other loudspeaker to this day.

19.12 The Noise of Conflict

As discussed in Section 19.3.1, intermodulation distortion is probably the real enemy – the hidden enemy – that we fight when we try to reduce *harmonic* distortion. Dr Alexander Voishvillo, the Russian electroacoustician, has been a driving force behind the search for a means of quantifying and comparing intermodulation distortion responses. His work is recommended reading for anybody who seeks to know more about the subject.[7–10]

Total harmonic distortion gives the output response of a system to a single frequency input when the input frequency is filtered out. Spectrum analysers can separate the different harmonics, and sweep frequency analysers can display individual harmonics as a swept function of the input signal. Such plots are shown in Figure 19.1. Nevertheless, at any given time, the stimulus is a single frequency, and the measured products are multiples of that frequency. They are the very harmonics that would be produced by any acoustic instrument if a musician were to attempt to play a single frequency. The harmonics give rise to the tone colour (timbre) of an instrument. Harmonics are therefore musical sounds; in fact, they are the essence of rich musical sounds.

Why then do we measure harmonic distortion? The answer seems to be, 'Because we can'. There is no doubt that it is a measure of the non-linearity in a system, but work over many years has shown it to correlate only poorly to perceived loudspeaker quality. Whilst it is true that the same mechanisms that give rise to harmonic distortion are the ones that are responsible for the intermodulation distortion, the latter is difficult to measure, because when every frequency modulates every other frequency, the variables seem to be endless.

There is also no magic ratio between harmonic and intermodulation distortion, because the intermodulation distortion depends on:

1. The absolute level of the signal
2. The bandwidth of the signal
3. The complexity of the signal
4. The peak-to-mean ratio of the signal
5. The waveform of the signal
6. The interaction between the above and a number of other factors, also.

Some instances may occur, in simple cases, where it could be said that, for example, the intermodulation products were three or four times the level of the harmonic products, which in many cases would be typical, but on complex musical signals, any seemingly fixed relationship tends to break down badly. Intermodulation distortion in a non-linear system is therefore frequency dependent, level dependent, waveform dependent … in fact, it is very difficult to devise any simple test signal that could yield a realistic description of how the intermodulation performance of two systems could be compared. In the words of Dr Voishvillo:[10]

> *Since the dynamic reaction of a complex non-linear system such as a loudspeaker cannot be extrapolated from its reaction to simple testing signals, such as a sweeping tone, the thresholds expressed in terms of loudspeaker reaction to those signals (total harmonic distortion [THD], [individual] harmonics, and two-tone intermodulation distortion) may not be valid.*

That intermodulation distortion is the number one enemy of loudspeaker designers is probably true, whether too many of them realise it or not. However, with such a characteristic nature that is constantly changing with the music, any useful presentation of its quality and quantity, numerical or graphical, must have a correlation to the psychoacoustic perception of the problem. Until now, no such credible presentation system has been widely accepted. So far, all attempts have been flawed, so intermodulation problems often tend to be ignored.

Although the venerable Gilbert Briggs had so clearly identified it as a great problem by the early 1950s, the intervening years have not yet come up with any adequate quantification method.

The thing that must be borne in mind is that the intermodulation distortions (IMD) of complex signals tend towards being a modulation related noise. It is a little like trying to listen to a good hi-fi system whilst somebody is just outside the window using a chain-saw, whose noise production is dependent on the level and spectral density of the music, though not directly proportional to either. With IMD, transparency is lost, low-level detail is buried, and the sense of effortless reproduction cannot be achieved. Brass bands and choirs are great victims of intermodulation distortion. Their reproduction via loudspeakers can be very disappointing if one is accustomed to hearing them live. Whilst all the sources emanate from different points in space, the intermodulation products are low, but when they are all mixed together and passed through a pair of loudspeakers, the intermodulation becomes obvious. This need not be a fault of the loudspeakers alone, because microphones are also often responsible for intermodulation distortion, and the electronic systems and digital converters can further add their products.

It must be remembered that the air, itself, is also non-linear at very high SPLs, and where such levels exist, such as in the throats of horns (be they trumpets or loudspeaker horns), intermodulation can be very evident. This is more evident in PA systems rather than at the levels normally experienced in recording studios, but nevertheless, in very large control rooms it can be a problem. Figure 19.18 shows the propagation distortion from a source at 1m distance. Figure 19.19 shows the propagation distortion, for the same SPL at the measurement point, from a source at a distance of 5m. There is a strong implication here that for the same SPL at the listening position, the lower source SPL generated by a close-field monitor may produce significantly less IMD than that from loudspeakers at 5m distance, unless, that is, the loudspeakers at the greater distance generate significantly lower levels of non-linearities. In many cases, the exact opposite is the case, where the large monitors actually produce *more* non-linear distortion, level for level, than the ones used at close-range. Furthermore, where air propagation distortion is concerned, the greater distance travelled at high SPLs also gives rise to more distortion, so the generation of high SPLs from considerable distances seems to be a viable proposition only in studios with extremely linear monitor systems.

Figures 19.18 and 19.19 were derived from a system of multitone testing, where the device under test is injected with a whole series of tones, simultaneously. The basic concept of multi-tone testing dates back to 1913, but for the measurements shown in Figures 19.18 and 19.19 a special series of 20 frequencies were used. These frequencies were chosen such as to maximise the separation of the intermodulation product frequencies, to help to see more accurately the spread of the problem. Inappropriately chosen frequencies could produce intermodulation products at the coincident frequencies, in which case some masking of the problems would take place. The multi-tone testing has shown that the total IMD can often

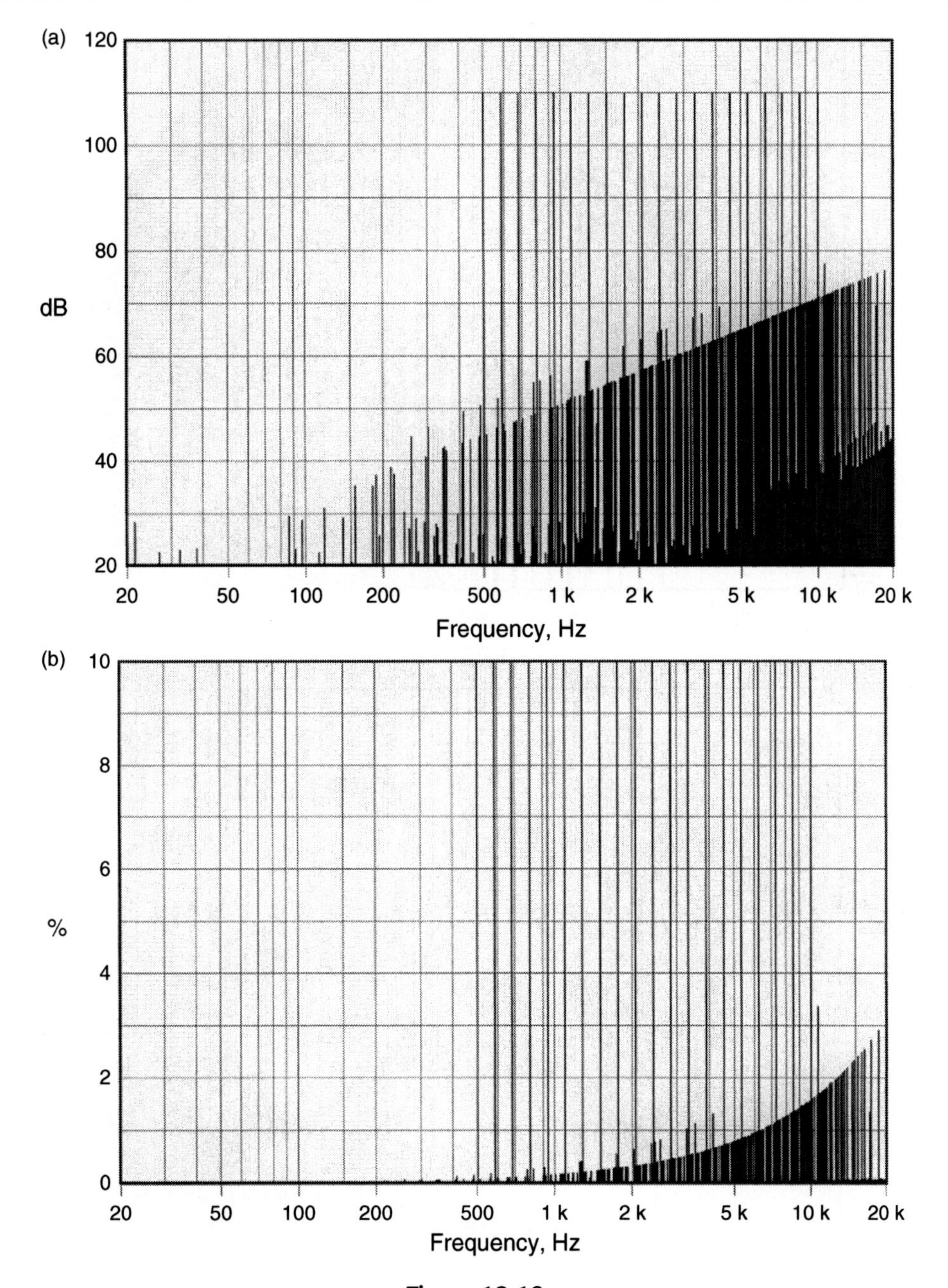

Figure 19.18:
Multi-tone testing of intermodulation distortion. Propagation distortion in spherical wave, SPL 110 dB at 1 m from the source. Radius of the source 0.5 m. First graph (a) is dB SPL, second graph (b) is per cent IMD to fundamentals (courtesy of Dr Alexander Voishvillo).

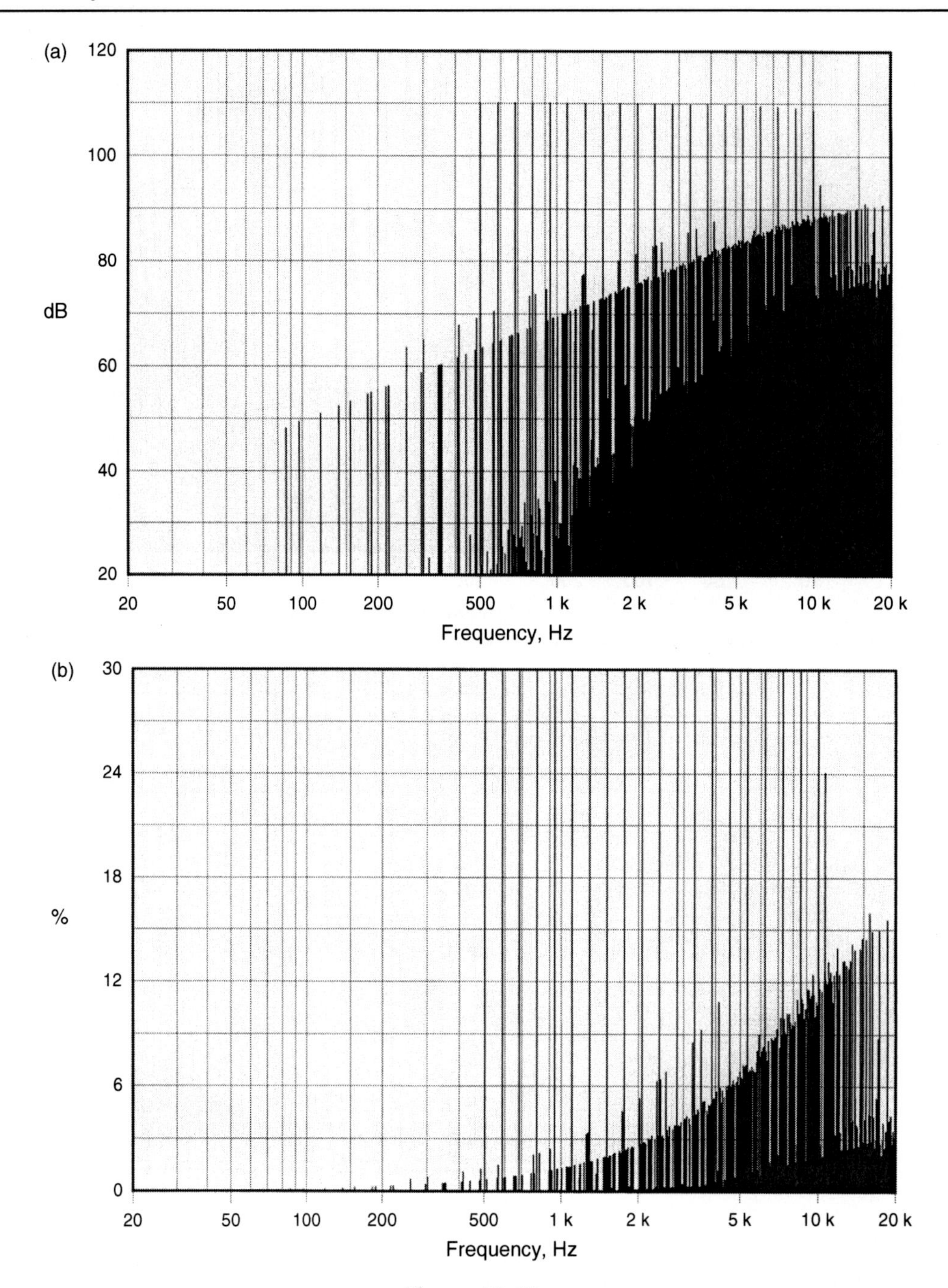

Figure 19.19:
Intermodulation distortion at higher level. Propagation distortion in spherical wave, SPL 110 dB at 5 m from the source. Radius of the source is 0.5 m. First graph (a) is dB SPL, second graph (b) is per cent IMD to fundamentals (courtesy of Dr Alexander Voishvillo).

Figure 19.20:
The multi-channel monitoring system in the private studio of Eugene Czerwinski, the founder of Cerwin Vega Inc. The use of one-track-to-one-monitor is said to reduce drastically the levels of intermodulation distortion, and to increase greatly the level of perceived openness and neutralness in the reproduced sound (photograph courtesy of Eugene Czerwinski).

be around four times greater than the measured THD. The implication is that in many cases, what people frequently believe to be harmonic distortion of music is actually, predominantly, intermodulation distortion. This helps to explain why there have been such poor correlations between measured THD and perceived distortion. The perceived distortion is more likely to be the IMD, which is neither being measured, nor adequately quantified.

What is more, various studies have shown that IMD can be irritating at only 10% of the equivalent THD levels because of its inharmonic nature. Therefore, producing four times more of something, which is ten times more irritating, *could* suggest that the IMD problem is 40 times greater than the THD problem. It has even been shown, by using a fifth-order static polynominal model, that non-linearity of the type $2f_i + f_j$ can be 34 dB higher than the fifth harmonic distortion in the same system.

Voishvillo's explanation for the extreme clarity that has been reported from the monitor system shown in Figure 19.20 is that not only are the independent recordings of the instruments not sharing any of the loudspeakers, but that they are also not sharing the same 'air conduits' between the loudspeakers and the ears. (The 16 loudspeakers are fed from separate tracks of a multi-track recorder.) At low SPLs, this idea of air conduits would violate the concept of the linear superposition of acoustic waves, but the non-linear interference concept has already been proved by experiment at high SPLs.

Normally, the harmonic distortion is measured by means of a fixed level signal, either static in frequency or swept. However, the non-linear distortions that relate to what we hear are the complex interactions within a complex, time-varying signal, and THD correlates poorly with this. Whilst THD is presented as a *percentage* of the signal, it could be that we are sensitive to *absolute* levels of IMD. We are accustomed to the harmonic density of instruments increasing as the musicians play louder, but somebody vacuuming in the room can be disturbing for the hi-fi listener whether the music is loud or quiet.

Anthony New published some interesting data in a pair of articles entitled 'THD Is Meaningless'[11,12]. Albeit taken from radio frequency amplifiers, he showed plots of the intermodulation products of two, three and four simultaneously applied tones, with a level about 65 dB above the noise floor. On the two tone plot, four intermodulation products could clearly be seen above the noise floor. With three tones, nine intermodulation products were clearly visible, whilst with four tones over 30 intermodulation products were evident. The IMD products increase rapidly as the number of stimulating frequencies increases, because the IMD products begin to intermodulate between themselves. One can easily imagine the result for a complex musical signal – a random-type modulation noise, perhaps only 40dB below the level of the music. His arguments are quite convincing.

To say that THD measurements are meaningless is perhaps going a bit too far, but to say that they are meaningless, in themselves, is probably fair comment. It is well known that many valve amplifiers produce much greater levels of second harmonic distortion than the better quality transistor amplifiers, without sounding distorted. A quick look at the tendency towards second order non-linearity generation may explain why this is so. In fact, the second order non-linearity produces no odd-order, unpleasant intermodulation of the $f_1 \pm 2f_2$ nature. The products of the second order non-linearities tend to be fewer in number and lower in level than the higher order products. As just discussed, the higher order products can increase rapidly in both their density and cumulative level. They also tend towards a random noise type of frequency distribution, which the music cannot mask. (Note the black mass in Figure 19.19(a).)

When valve amplifiers produce second harmonic distortion it can be pleasant in moderation, but perhaps what is even more pleasant about such amplifiers is the *lack* of higher odd-order intermodulation products. This is perhaps also true of Class A amplifiers in general, which are free from crossover (zero-crossing) distortion artefacts at any level. In fact, it would appear that harmonic distortion is not a separate entity to intermodulation distortion, but is a special case of intermodulation distortion. If multiple tones are being used for intermodulation testing, and they are individually sweepable and all brought together in frequency, then when they blend together they will produce only harmonic distortion. Perhaps in this way it is easier to visualise that the same non-linear mechanisms give rise to both forms of distortion.

In electro-mechanical systems, such as loudspeakers, the non-linear restoring forces in the suspension systems are further sources of intermodulation and harmonic distortions. However, the deflexions of these suspensions under musical drive can be complex in the extreme, and they bear little relationship to the electrical factors that give rise to non-linearities in amplifiers, for example. This may explain why 1% of second harmonic distortion may be excellent sounding from one device whilst being intolerable from another. It is therefore not the level of second harmonic distortion, in itself, which we are comparing, but the existence of other non-linear artefacts resulting from the different non-linear mechanisms or systems. Such mechanisms can also give rise to phase distortions of a dynamic nature, which can affect the stereo imaging. Another particularly nasty source of loudspeaker intermodulation distortion is magnet BL (force factor) non-linearity[13], which can give rise to many more intermodulation distortion problems. Once again, though, these highly audibly unpleasant products may not be anywhere near adequately exposed by simple THD measurements.

Very much more work still needs to be done on the reduction, and the meaningful measurement, of IMD. One reason why there has been a lack of motivation by many commercial manufacturers of audio equipment has been that, because of the lack of accepted specifications, they have been reluctant to pursue expensive improvements that cannot be proved in the sales literature. A meaningful objective/subjective correlation measure would be a powerful stimulant to a global attack on the IMD menace. Nevertheless, it is probably true to say that the perceived clarity in any audio system is usually a direct result of the absence of intermodulation distortion.

19.13 Summary

The on-axis anechoic frequency response of a loudspeaker is not what will be heard in a control room.

Loudspeakers have many different physical layouts and will exhibit different directivity characteristics because of this.

Different on-axis responses may suit different mounting conditions.

Harmonic distortion figures can be misleading, but they are an indicator for the existence of intermodulation distortions, which are probably the real culprits for any lack of musicality.

Steep low frequency roll-offs give rise to group delays that smear the time response of a loudspeaker.

Just as microphones of different designs cannot be equalised to sound identical, nor can loudspeakers.

Many recording engineers and producers use different types of loudspeakers for work and for pleasure.

Loudspeakers with anechoic pressure amplitude responses similar to the general shape of that of the NS10M tend to flatten when placed on the meter bridge of a solid mixing console.

Above all, it would appear that the users of Auratones and NS10s have been using them because of their very rapid low frequency decay characteristics, which tend less to confuse bass guitar and bass drum relative levels, for example.

Of the non-linear distortions, harmonic distortion is the one that is most measurable, but intermodulation distortion is that which is most offensive.

Intermodulation distortions tend towards the character of noise, whereas harmonic distortions, alone, free of the presence of the IMD that usually accompanies them, may even enrich some music. However, neither signal degradation nor enhancement is desirable in monitor systems.

Air is also non-linear, so the propagation of high SPL for long distances can be expected to add to the other non-linear distortion.

Multi-tone signals, although dating back to 1913, are perhaps the most likely candidates for future IMD measurements.

IMD results from the interactions of complex time-variant musical signals on complex non-linear systems. It is very hard to quantify meaningfully.

References

1 Holland, Keith R., 'Use of Cepstral Analysis in the Interpretation of Loudspeaker Frequency Response Measurements', *Proceedings of the Institute of Acoustics*, Vol. 15, Part 7, pp. 65–72 (1993)
2 Briggs, Gilbert, *Sound Reproduction*, 3rd Edn, Wharfedale Wireless Works, Bradford, UK (1953)
3 Moir, J., 'Doppler Distortion in Loudspeakers', *Wireless World*, pp. 27–28 (April 1974)
4 Colloms, Martin, *High Performance Loudspeakers*, 5th Edn, John Wiley & Sons, Chichester, UK, p. 376 (1997)
5 Newell, Philip, 'The NS10M', *Studio Sound*, Vol. 43, No. 8, pp 54–56 (August 2001)
6 Newell, Philip R., Holland, Keith R. and Newell, Julius P., 'The Yamaha NS10M: Twenty Years a Reference Monitor. Why?', *Proceedings of the Institute of Acoustics, Reproduced Sound 17*, Vol. 23, Part 8, pp. 29–40 (2001)
7 Voishvillo, Alexander, 'Nonlinear Distortion in Professional Sound Systems — From Voice Coil to the Listener', presented at the 'Reproduced Sound 17' Conference of the Institute of Acoustics, Stratford-upon-Avon, UK (November 2001)
8 Czerwinski, Eugene; Voishvillo, Alexander; Alexandrov, Sergei and Terekhov, Alexander, 'Multitone Testing of Sound System Components – Some Results and Conclusions, Part 1: History and Theory', *Journal of the Audio Engineering Society*, Vol. 49, No. 11, pp. 1011–48 (November 2001)
9 Czerwinski, Eugene; Voishvillo, Alexander; Alexandrov, Sergei and Terekhov, Alexander, 'Multitone Testing of Sound System Components – Some Results and Conclusions, Part 2: Modeling and Application', *Journal of the Audio Engineering Society*, Vol. 49, No. 12, pp. 1181–92 (December 2001)
10 Voishvillo, Alexander, 'Assessment of Loudspeaker Large Signal Performance – Comparison of Different Testing Methods and Signals', presented at the 111th Convention of the Audio Engineering Society, Contribution to the Task Group SC-04-03-C, New York, USA (December 2001)
11 New, Anthony, 'THD Is Meaningless, Part 1', *audioXpress*, pp. 36–40, USA (January 2001). www.audioxpress.com
12 New, Anthony, 'THD Is Meaningless, Part 2', *audioXpress*, pp. 54–61, USA (February 2001)
13 Dodd, Mark, 'The Transient Magnetic Behaviour of Loudspeaker Motors', presented at the 111th Convention of the Audio Engineering Society, Preprint No. 5410, Munich, Germany (September 2001)

References 11 and 12 were first published in *Electronics World*.

CHAPTER 20

Studio Monitoring Systems

The concept of the system. Common weak links. Amplifiers. Cables. Crossovers. Filter slopes. Loudspeaker cabinets and drive units. Magnet and diaphragm properties and performance. Horn loading.

Ideally, the main control room monitoring system in any studio should be of high resolution, full audio bandwidth, low distortion and should be able to provide sufficient sound pressure level at the mixing position for all forms of music recording likely to be undertaken in the studio. As discussed in the previous chapter, there are people, some of them with great skill, flair and experience, who choose to work principally on their favourite mid or small sized monitor loudspeakers, but that does not negate the need for a high-quality system. If they choose *not* to use the large monitors, then it is their prerogative to do so. Nevertheless, it should be incumbent on all professional studios to have available a full range monitor system for persons who wish either to work on them as standard practice, or to use them to check for things which are outside of the range of performance of the small monitors.

Although studio monitor system design is a huge subject, well beyond the scope of this book (see Bibliography), this chapter discusses many of the aspects of the component parts of a monitor system which need to be taken into consideration, both in terms of the functionality of the entire control room, and the way that they may influence overall results. For example, the characteristics of a cabinet which influence the way that it needs to be mounted, or the effects of loudspeaker cables which may dictate the siting of the amplifiers. It needs to be well understood that loudspeaker systems are not self-contained, independent entities which can simply be wheeled into a control room and be expected to be ready for use as soon as the acoustic construction has been completed. In fact, there are several aspects of studio monitor design that *cannot* be considered independently of the control room design as a whole.

20.1 The Constituents of the System

The monitor chain really begins in the monitor circuitry of the mixing console, because everything heard in the control room will pass through the console's monitor level control. The circuits which feed that control, the input selector circuits, are therefore the true front end of the monitor system. From the mixing console we then need audio cables to the inputs of the crossover if the loudspeaker system is active or to the amplifiers if the system is passive.

In the former case we would then pass to the amplifiers and on to the loudspeaker drive units via suitable loudspeaker cables. In the latter case, the amplifiers would connect to the high-level crossover via suitable loudspeaker cable, and thence to the loudspeaker drive units. The drive units (drivers) will be mounted in some form of enclosure, which will modify their free-air response, and from the cabinet mounted drivers the air in the room will couple the loudspeaker vibrations to the ears, via any obstructions that they may find in their way. This entire pathway describes a studio monitor system.

In a monitor system there are therefore many different component parts, and not only the parts themselves determine the overall system response. The interfacing must also be considered, because interactions can take place between inputs and outputs which must be carefully and suitably matched if response disturbances are not to result in signal degradation. The interfacing cables can also be sources of interference and general signal degradation if not chosen appropriately, although the better professional equipment tends to be much more tolerant of cabling than a lot of the domestic hi-fi equipment due to the generally more robust circuits in the input and output stages. Nevertheless, care still needs to be exercised when selecting appropriate cabling.

Let us therefore consider the needs and pitfalls as a signal passes from the output busses of a mixing console on its way to the ears of the recording personnel. In this way, we can discuss the options available and any problems as they arise. The description of the monitor system in the previous couple of paragraphs seems simple enough, but the realities are somewhat more complicated. Inevitably, this will be a chapter of some considerable length, because the options and dilemmas are multifarious.

20.2 Console Monitor Circuitry

On completion of a control room it is customary for studio designers to play a range of recordings which are well known to them in order to confirm, aurally, that all is performing according to expectation. As mentioned in Chapter 18, the introduction of a mixing console into a room can cause considerable acoustic disturbance, which in the past has often been confused with a change in sound due to the mixing console electronics. Nevertheless, the monitor circuitry of many consoles leaves much to be desired, and so there are *two* ways that a mixing console can degrade the monitoring performance of a control room: acoustically and electrically.

The classic test for mixing console electronic transparency is to connect a signal source directly to the power amplifiers in the case of a passive loudspeaker system or to the input of the electronic crossover if the system is multi-amplified. In either case, care should be taken to ensure that the signal source (CD or DVD player, for example) has outputs that are suitable for driving the crossover or amplifier inputs correctly. The next stage is to listen carefully to a selection of recordings, and then listen again with the signal source connected via the 'tape returns' of the console monitor circuitry. If any difference is heard in the signal clarity, stereo width, transparency, stereo imaging, frequency response or any other aspect of the sound quality, then one should begin to worry about the console monitor circuitry. All too often, a difference *is* heard, and in some cases it is not a subtle difference. Many very expensive mixing consoles are not immune to this problem. It should be noted that such tests can only be truly meaningful if the signal source is of very high quality. If, for example, a cheap CD player is used, it may itself limit the reproduction quality to such a degree that the lesser degradation due to the console would not be evident, so any D to A converters used in these tests should also be of the highest quality.

There seems to be a tendency amongst many mixing console manufacturers to cut costs in the monitor circuits, in the misguided belief that as they are not in the direct signal path of the recording they are somehow less important than the mixing busses in terms of minimising noise and maximising quality. Of course the mixing bus outputs are of great importance, because the final programme *will* pass through them, but unless the monitor circuitry is of equal quality, *or better*, then one has no way of knowing the precise quality of what is being recorded.

Some mixing consoles of reputable manufacture have been shown to be seriously lacking in terms of monitor circuit quality. Sometimes as little as an extra €6 or €7 per stereo monitor channel could significantly increase the quality of the sound that passes through them. Unfortunately, the competition between the manufacturers is so fierce that the cost of improvement to six stereo monitor inputs (mix out, tape returns, etc.), which could add €100 to the price of a console, is a cost which they would rather avoid. Believe it or not, consoles costing €100,000 or more have been subject to such miserly cost cutting. This seems totally disgraceful, but it is a reality of the de-professionalisation of the music recording industry. Even studios that *are* trying hard to be professional are often subject to performance degradation due to hidden cost cutting of a nature such as described above.

In fact, one reason why manufacturers have got away with such shoddy practices is because many of the monitor systems used in current studios are of too poor quality to reveal the problems. It has been demonstrated clearly in studios with excellent monitoring that differences that can be plain for all to hear can be rendered totally inaudible on loudspeakers such as NS10s. This must *not*, however, be taken as a criticism of the NS10s, *per se*, because they were never intended to be used as the principle quality references. Their value has been in achieving appropriate musical balances, and on certain styles of music they have shown themselves to be useful tools for such work. However, it does emphasise once again the need for high resolution monitoring in a professional studio, irrespective of whether they are going to be generally used for mixing, or not.

Clearly, it is what is on the main mixing busses which goes to the recording medium, but all the decisions about the correctness or suitability of the sound will be made via a monitor system which begins with the pick up of the monitor signal *from* those mixing busses. The dials and gauges on a racing car do not directly affect the performance of the car, but they do influence its performance via the information which they feed to the driver. Monitoring is somewhat similar, the ultimate performance of a mix or a recording is affected by the information that the monitors feed to the musicians and recording personnel during the recording or mixing process.

Sometimes, it must be said, the lack of attention to detail in the monitor circuits of quite expensive mixing consoles can be disgraceful. Flimsy printed circuit tracks that are badly screened and of very unequal length, in what may be ostensibly balanced circuits, are often typically to be found. The 'balanced' circuitry is also often found to be designed around single op-amps, which cannot be used to derive truly balanced inputs. The consoles also often fail to come with adequate explanations about the circuit topologies or the recommended way of unbalancing them if it is necessary to do so; for example, whether the negative (cold) connections of the inputs or outputs should be grounded or left floating. This is relevant because not all amplifiers in studio use have balanced inputs. Inappropriate connection can be infuriatingly difficult to trace, because often no normal test equipment will show any obvious fault condition or response anomalies. However, correct termination can transform the monitoring clarity. Monitoring systems do need to be considered as single systems, and not a sum of individual parts, but what goes on inside a mixing console monitor section is rarely within the control of a studio designer. In fact, investigation and improvement is often actively discouraged by studio owners, who fear that the guarantees (warranties) may be rendered void if their brand new mixing console is modified. Manufacturers are often reluctant to make modifications at a studio's request because it first necessitates admission that all is not as good as it could be with the stock item.

It really cannot be over-emphasised that one can no longer presume that high-quality mixing consoles have been built with the highest fidelity in mind. More often, they tend to be built for the maximisation of the characteristics for which they are individually best known, and to meet the points on the price-versus-performance curves which their marketing departments consider most profitable. Even consoles costing over €500,000 are not immune to political interference in their design. Therefore, it is down to the studio owners and users themselves to make sure that the monitor circuitry is not limiting the performance of the entire monitoring chain, and to take whatever action they deem to be appropriate if it *is* found to be doing so. As previously mentioned, a high-resolution monitoring system and high-quality sound sources (very high-quality D to A converters, also) will be needed for any tests, or the failings may go unnoticed, as they so often do in daily use.

20.3 Audio Cables and Connectors

In general, the higher quality mixing consoles have well-designed balanced outputs; and most professional monitor amplifiers and electronic crossovers have well-designed balanced inputs. They are usually very tolerant of the type of cable that is used to interconnect them. Nevertheless it is good engineering practice to keep cable runs to a minimum between the console outputs and the crossover or amplifier inputs, and to avoid running the cables in parallel with other cables such as data cables or power cables which may be likely to induce interference into the monitor signals.

It tends to be the case that many less expensive consoles, with less advanced output circuitry, are more sensitive to the type of cable used to feed the main monitor system. Perversely, the more money that is spent on the mixing console usually means that less money needs to be spent on the cable. Although good quality cable should always be used, esoteric cables have often been shown to exhibit little sonic benefit when used with high-quality consoles. It should be remembered that professional audio systems interface at 10 to 20 dB higher levels than domestic equipment, precisely to reduce the effects of interference. Together with the lower impedances of the inputs and outputs, this helps to make interfacing much less critical than can be the case with domestic hi-fi.

There are those who argue that the reason why no difference is usually noticeable between good quality and esoteric monitor interface cables is that one is merely adding a few metres to a monitor circuit, when the mixing console itself may contain dozens of metres of 'normal' cable from the monitor modules to the jackfield (patchbay) and to the output connectors. Hundreds and even thousands of metres of cable can be used in some large consoles. Some studios have been known to re-wire entire consoles with oxygen-free signal cable, and improvements in sound quality have been claimed, but few opportunities exist to compare directly the modified and unmodified consoles under the same, controlled conditions. In any case, the sheer quantity of cable within the console is likely to swamp the benefits of esoteric cables connecting the console to the main monitors. In the case of cables feeding the close-field monitors, they are most likely to be too short for subtle cable differences to show much effect due to the low impedance balanced circuitry typical of most good monitor systems. However, there is some evidence to suggest that the cable quality becomes more critical as it handles more signals that are complex, so the monitor interface cables could be more critical than those connecting a single channel module, but this is difficult to prove.

As with the cables, good quality connectors should also be used, but there is little to suggest that any significant benefits result from the use of special connectors. Good quality, well-made soldered joints are an obvious requirement.

20.4 Monitor Amplifiers

There can be no doubt that power amplifiers can sound noticeably different to one another. However, what cannot be said is that there is any absolute order of quality. Clearly, a poor amplifier will surely always be worse than a good one, but amongst good ones, things can become very dependent upon circumstances. A well-documented test took place at the Tannoy factory in Scotland in the early 1990s.[1] Intending to recommend a standard amplifier with a new range of four loudspeakers, they had difficulties in making a choice. Eventually they opted to ask some respected persons to carry out blind testing. Four amplifiers were chosen for the comparisons. What had been expected was that one amplifier would be chosen as being superior to the others by the 'golden ears', but the reality was very different. In repeatable tests, one of the amplifiers was chosen to be the best on two of the loudspeakers, one on another and a third amplifier on the remaining loudspeaker. Only one of the amplifiers was not chosen to be the best on any of the four loudspeakers.

Loudspeakers present an enormous range of different load conditions to the amplifier outputs, and amplifier performance can be influenced by the loading to a considerable degree. Loudspeakers with high level, passive crossovers vary enormously in the terminations that they present to the amplifiers. Some loudspeakers (principally for hi-fi use) have impedance corrected crossovers and can present an almost perfect 8 Ω load to the amplifier. At the other extreme, monitor loudspeakers such as the Kinoshitas have nominally 4 Ω impedances, but can drop to as low as 0.8 Ω at certain frequencies and under certain drive conditions. Given their enormous power handling of around 1000 W, under certain drive conditions the Kinoshitas can call for as much as 100A from the power amplifiers. This is way in excess of what many power amplifiers can deliver, even if they *are* rated at 1000 W into 4 Ω, which according to the specifications alone should be sufficient to drive the Kinoshitas (shown in Figure 20.1). The simple I^2R formula cannot be applied to difficult loads, because reactive currents can be enormously underestimated. In the above case, it would suggest:

$$\begin{aligned} W &= I^2R \qquad (20.1) \\ 1000 &= I^2 \times 4 \\ \frac{1000}{4} &= I^2 \\ 250 &= I^2 \\ 1 &= \sqrt{250} \\ I &= 15.8\text{A} \end{aligned}$$

less than 16% of the true peak current demand.

(a)

(b)

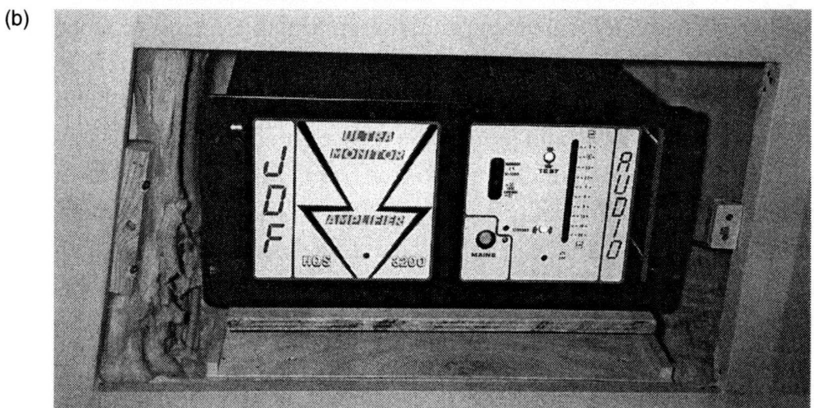

Figure 20.1:
Kinoshita '28 Hz' monitor at Masterfonics, Nashville, USA, and a typical amplifier. (a) The vertical 'D'Appolito' mounting configuration gives rise to much less response change with position when moving around in a control room. The concept allows the simplicity of a two-way system to be a practical solution for large monitors, as well as small ones; (b) A JDF amplifier, of enormous output current capacity, of a type often used with the Kinoshita monitors.

Loudspeaker sensitivities also vary widely. Two studio monitors commonly in use in the 1990s were the large, 3×15 inch+horn UREI 815, and the small ATC SCM10. The sensitivity of the UREI is 103 dB at 1m distance for 1 W (more accurately 2.83 V) input. The sensitivity of the SCM10 is 81 dB at 1 m for 1 W input. The effect of the sensitivity difference is that in order to sound as loud as a 1 W input into the UREI, the SCM10 would need 158 W, which is +22 dB relative to 1 W. If both loudspeakers were to be driven by a 100 W amplifier with the intention of listening at a sound pressure level of 90 dB SPL at 2 m (equivalent to 96 dB at 1 m in free-field conditions), which is quite a reasonable monitoring level, then very different demands would be made from the amplifier. In the case of the UREI, the typical power demand from the amplifier would be less than a quarter of a watt. (1 W would produce 103 dB at 1 m, hence 6 dB less at 2 m.) For the same SPL, the ATC would need to take 32 W from the amplifier.

Amplifier distortion figures are usually quoted in relation to maximum output power prior to the onset of gross distortion. Their performance at very low output power is rarely specified, but it can have a great bearing on their sound quality under different circumstances. In the case of the ATC loudspeaker in the previous paragraph, signals 20 dB below the normal level would, during quiet passages for example, need just over a quarter of a watt. The UREI, producing a similar SPL would require only just over 2 *milliwatts.* The performance of different 100 W amplifiers at the 2 milliwatt level can be very different indeed. At such levels, crossover distortion in particular can become a very significant proportion of the output, and can produce rather nasty audible effects. When one considers some mid-range horn loudspeakers with sensitivities in the order of 110 dB at 1 m for 1 W, then the power requirements can be surprisingly low. When playing quieter passages of music, 20 dB below a 90 dB normal listening level at 2 m from the source, which is a perfectly reasonable situation in a studio, the power needed would be only 400 *microwatts.* This represents only 0.4 milliwatts. When serious listening is taking place whilst drawing only *microwatts* from the amplifier, Class A amplification, which is free from crossover distortion, would seem to be a wise choice, although it must be said that there are other amplifier classes which are also free from crossover (i.e. zero-crossing) distortion these days (see Bibliography – Self [2009]). It is doubtless that many high efficiency mid-range horn loudspeakers have been judged to sound harsh and distorted merely because they are reproducing distortion which their drive amplifiers exhibit at very low levels.

One very great advantage of the multi-way amplification of monitor systems is that amplifiers can be matched much more accurately and appropriately to the individual loudspeakers. The lack of crossover components between the amplifier and the loudspeaker voice coil also gives the amplifier much more authority over the movement of the diaphragm. Nevertheless, a loudspeaker drive unit is still a reactive device (see Glossary), even without passive crossover components, so simple resistive loading of the amplifier is still not achievable. However, a drive unit alone provides a much more comfortable termination to the amplifier than when fed via a high-level passive crossover. Although the aforementioned constant impedance crossover networks also provide an easier load for the amplifier to deal with, they have often been rejected on subjective listening. What is more, at very high power levels, they can become very difficult and expensive to realise. They can also dissipate considerable power in their compensation resistors, and the drawbacks can soon outweigh the advantages.

For the highest quality of monitoring systems it is better to consider the amplifiers carefully in terms of the power requirements, the distortion content at relevant power levels and their ability to drive the required loads. It is usually of little use selecting a 'favourite' amplifier from some other set of circumstances and then specifying it in untested circumstances. The results are likely to be less favourable than from the use of a carefully chosen amplifier for the specific purpose.

Another matter for consideration when selecting amplifiers for studio use is mechanical noise. It is now generally considered wise to mount the amplifiers as physically close as possible to the loudspeakers, in order to reduce loudspeaker cable length to a minimum. The mounting of noisy, fan-cooled amplifiers in remote racks is no longer accepted as good engineering practice. The reasons why will be dealt with in the following section. The use of quiet amplifiers, without cooling fan noise or mechanical transformer buzzes, is desirable, unless they are mounted in ventilated enclosures behind flush mounted loudspeakers. Noisy, fan-cooled amplifiers inside a control room are an absurdity, as they add to background noise levels and hence reduce the dynamic range of the monitoring chain of which they form a part. Figure 20.2 shows convection cooled amplifiers mounted below the loudspeakers in a position where loudspeaker cables can be kept to 2 m, or less, and where the clip and protection lights are clearly visible. The openings above and below the amplifiers allow an ample supply of air to circulate through the heat sinks. The cavity behind is highly acoustically damped, to prevent any resonance problems colouring the monitoring.

The close coupling of the amplifiers to the loudspeakers reduces cable impedance effects and allows better damping of loudspeaker resonances. One often sees damping factor figures of 500 or more in the specifications war amongst amplifier manufacturers, but this figure alone is rather meaningless. The damping factor is the ratio of the load impedance to the amplifier output impedance. Amplifiers can actually be made with zero output impedance, which would theoretically give an infinite damping factor, but in practice they may damp the loudspeaker no more than any other amplifier. This is because the loudspeaker cable has resistance, inductance and capacitance, which lumped together form an impedance to the signal flow. This forms part of the amplifier output impedance from the point of view of the loudspeaker. Even 0.1 Ω of loop impedance in the cable feeding an 8 Ω loudspeaker, connected to an amplifier with infinite damping, would result in an effective damping factor of no more than $8 \div 0.1$ (or, in other words, 80). There is little to suggest that damping factors greater than 40 or 50 are directly audibly beneficial in themselves. For rapid transient response, a high instantaneous output current and a fast slew rate (the ability to make rapid voltage changes – normally measured in volts per microsecond) tend to have more influence on the sound quality than very high damping factors. When judging

(a)

(b)

(c)

Figure 20.2:
Sonobox, Madrid, Spain. The amplifiers can be seen mounted below the loudspeakers, to ensure the shortest practical cable lengths: (a) The finished control room (2001); (b) loudspeaker and amplifier mounting during construction; (c) The cavity behind the amplifiers must be very well damped to avoid any resonant effects.

amplifiers it is essential to avoid the type of false syllogism which goes, 'Cats have tails, my Rottweiler has a tail; therefore my Rottweiler is a cat'. Unfortunately, such conclusions are drawn on a daily basis. An amplifier's performance must be considered in its place as a part of a monitor system. It must be understood that no one amplifier is the best in all circumstances.

Valve (tube) amplifiers, which many music lovers value for home listening, are not usually advisable in recording studio monitor systems. The 'nice sound' that they can impart to the music is something that *they add* to a recording, and therefore the recording itself does not necessarily sound so nice without the help of the amplifier. The principal objective of monitoring is to assess what is on the recording medium, which needs transparent and neutral amplification. When recordings are good, the sound should be good; when recordings are bad, the sound should be bad. At home, when listening for pleasure, one does not really want to hear how bad a recording may be, so valve amplifiers may well be valid, but in the studio it is essential to *know* when one has a bad sound.

The question of the amplifiers used in 'active' monitors, i.e. loudspeakers with built-in amplifiers and electronic crossovers, is something of a two-edged sword. It can cut both ways. Certainly, in an integrated system the manufacturers are in control of everything. There is no loudspeaker cable problem to consider, and the whole electronics package can be designed to provide the correct acoustic response in front of the loudspeaker. In fact, it is the case with many small loudspeakers that their extended response and high output SPLs are *only* achievable by using dedicated electronics packages with in-built response correction and overload protection. The distortion produced during low frequency overloads in a passively crossed over system tends to send much of the distortion products into the tweeters, resulting in harsh sounds and the need for the frequent replacement of blown diaphragms or coils. By means of the use of separate amplifiers, distortion products are constrained within the bands that produce them, and therefore cannot affect components in other frequency bands.

The negative aspect of the built-in amplifiers is that in today's marketing driven industry they have probably been built to a 'just good enough' quality. The user cannot upgrade the amplification system if he or she decides to spend more on the electronics than the manufacturer has allocated to them. It is an unfortunate but absolute fact that almost all such monitors are built with a competitive price higher in the order of design priorities than the absolute sound quality.

20.5 Loudspeaker Cables

The number one 'golden rule' about loudspeaker cables is, 'the shorter the better'. An enormous amount of pseudo-science clouds this subject, but a few simple facts can be clearly stated. All cables exhibit a mixture of resistance, inductance and capacitance. Resistance is a bad thing, and should be minimised. It wastes power, reduces the ability of the amplifier to control the low frequency cone movement, and forms a potential divider with both the amplifier output impedance *and* the loudspeaker input impedance. In the former case, it affects the control of back-e.m.f.'s (see Glossary) and in the latter it can cause modification of timbre if the crossover and/or drive units have non-uniform impedance/frequency characteristics.

Inductance is also a bad thing. Its effect is frequency dependent, and series inductance will act as a low pass filter, gradually attenuating high frequencies. Inductance can also be a means of contamination by radio frequency interference, which can enter the amplifier via the loudspeaker cables and cause grittiness in the sound. Low cable inductance is essential for high-quality monitoring systems. Capacitance, however, is rather innocuous. It can, if rather large, cause instability in some amplifiers, but such instability is usually seen to be more a shortfall of the amplifiers' ability to tolerate capacitive loads rather than being a problem due to the cable capacitance itself. Hence, if cable construction can trade inductance for capacitance, in general the lower inductance option would be preferable.

The negative effects of loudspeaker cable are exacerbated by length, and by the width of the frequency range that they are handling. Long cables from a remote amplifier room to full-range (passively crossed over) loudspeakers are a worst case. In some cases where 5–10 m of cable is run to a loudspeaker with an internal passive crossover, cables of special construction costing typically €2500 each are known to be in use. Moving the crossover away from the loudspeaker and close to the amplifier can help matters. By running separate cables from the crossover to the low frequency and high frequency drive units, the length of cable that handles the full frequency bandwidth can be restricted to the short section from the amplifier to the crossover. It also takes the crossover away from any magnetically hostile loudspeakers. Little will be improved in terms of low frequency damping, but the higher frequencies can be rendered more transparent.

The low frequency currents can often exceed the high frequency currents by thousands of times. Under such circumstances, magnetostriction effects, the attraction and repulsion between the individual cables in a pair, and the dielectric charge migration may, in high power systems, superimpose artefacts of the enormously greater low frequency currents on the high frequency signals in cables carrying full frequency band signals. By separation of the cables to the low and high frequency drivers, such superimposition can be greatly reduced. Another, more usual variation on this bi-wiring theme is where the crossover remains in the loudspeaker cabinet but the common inputs to the high and low pass sections are separated. Individual cable pairs are then run from the high frequency and low frequency filter inputs, back to the amplifier. However, the advantage of removing the crossover from the proximity of the loudspeaker drive units is that the possibility of magnetic induction effects, which can interact between the motor systems and the crossover inductors, is effectively eliminated.

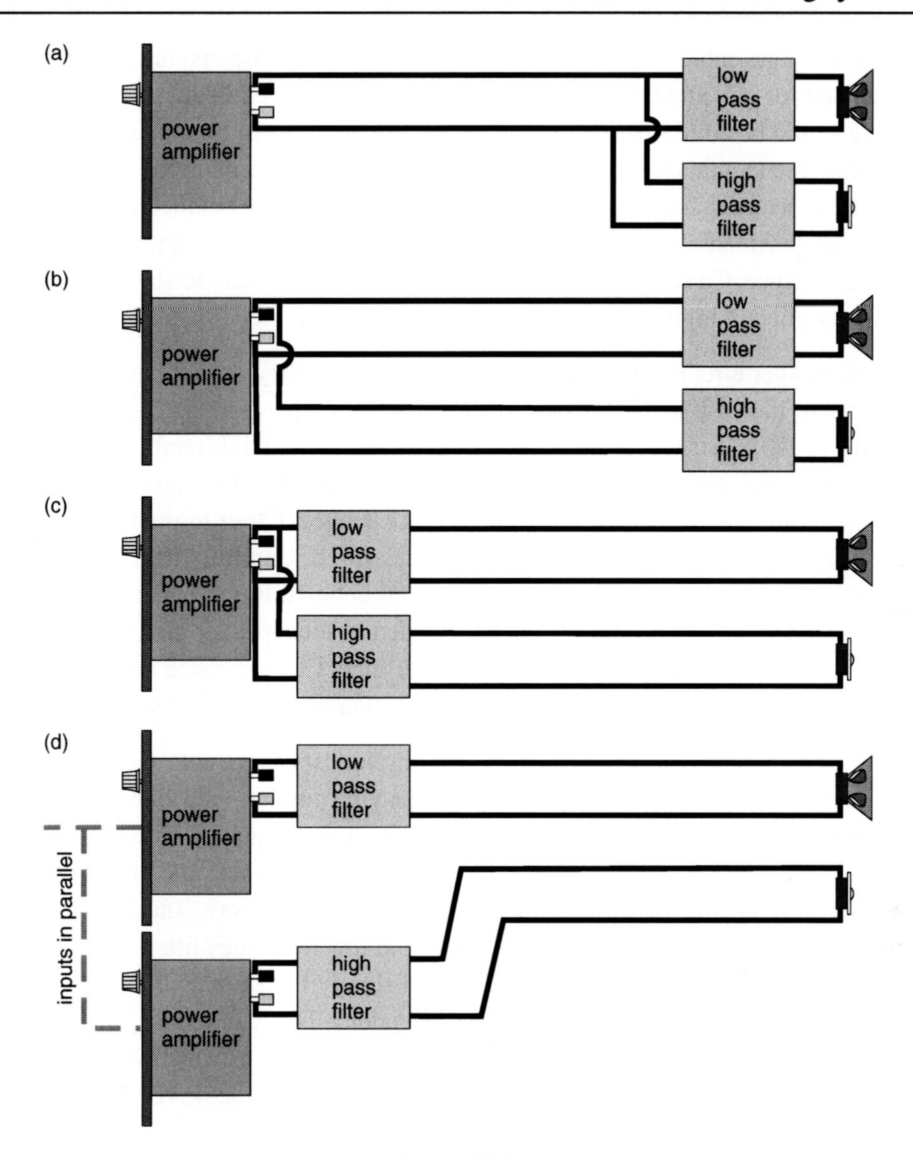

Figure 20.3:
Some wiring options for passive crossovers: (a) full-range, direct; (b) conventional bi-wiring, minimising the length of cable carrying full-range current; (c) bi-wiring with crossover filters remote from the loudspeakers; (d) amplifier *inputs* in parallel, poly-amplification, still using high-level, passive crossovers.

Cable conductor material is a source of continual debate. There does seem to be evidence that silver has subtle sonic benefits over copper, and that oxygen-free copper, especially of the long-crystal type, has benefits over standard copper. However, when loudspeaker cable lengths are kept to 2 m or below, and when the cables only carry band-filtered signals to individual drive units, the evidence for sonic differences between the conductor materials is scarce, especially if adequate cable cross-sections are used. For 8 Ω loudspeakers, 2.5 mm^2 cable is recommended for mid and high frequency drivers and 4 mm^2 cable for bass drivers. At 4 Ω, 6 mm^2 cable would be advisable for low frequency drivers. It must be remembered that whilst doubling the length *and* cross-section may more or less maintain the same resistance value, doubling the cross-section can do nothing to lower the series inductance, which will increase proportionately to the length. Only very close coupling of the conductor pairs can reduce inductance. Coaxial cables can be useful here, and ordinary RG59 video cable can be very usefully applied to high frequency drivers as long as current demands are not too high.

Fuses in loudspeaker circuits are absolutely to be avoided. They do not have constant resistance; it will increase with temperature. The fuses can therefore act as limiters, increasing the series resistance as the drive current increases, and thus tending to oppose it. The fuses also have a time constant, which means that their own thermal capacity will ensure that the temperature changes always lag the current changes. Close to the blow current, non-linear distortion of the order of 4% has been measured. Carefully designed electronic protection circuits in dedicated amplifiers are a much better solution when loudspeaker protection is needed, but relays are also to be avoided in loudspeaker circuits because of a tendency for their contact resistance to change over time. Contaminated contacts can also act like diodes, introducing non-linear distortion into the loudspeaker drive signal.

Figure 20.3 shows four means of driving and cabling loudspeakers with passive crossovers. The first (a) is the most usual, but is also the worst from the point of view of cable performance. The second (b) is simple bi-wiring, which removes the low frequency current stresses from the cables carrying the high frequencies. The third (c) removes the crossover from the electromagnetically hostile environment of high power loudspeakers. The fourth (d) uses separate amplifiers to drive independently the high and low frequency filters and loudspeakers. Although in case (d) the two amplifiers receive and supply the same *voltage* drive, the *current* drawn from each of them is frequency dependent. This can reduce intermodulation distortion. It is known as poly-amplification, which is only one step away from bi-amplification and multi-amplification, where the crossovers are placed before the drive amplifiers.

20.6 Crossovers

Ideally it would be useful to have a single loudspeaker drive unit which possessed a response from 20 Hz to 20 kHz, flat to within a decibel, and capable of giving sufficient acoustic output for studio monitoring purposes.

Unfortunately, the laws of physics dictate that this cannot be so, because the requirements for low frequency radiation and high frequency radiation are very different. A high power, low frequency loudspeaker needs to be large if it is to be efficient, whilst a high frequency loudspeaker (of high *or* low power) needs to be small. We are therefore stuck with a situation whereby different frequency bands need to be handled by different drive units. The electrical dividing networks that are used to feed the appropriate frequency bands to each drive unit are commonly known as crossovers, although 'frequency dividing networks' is a more representatively accurate term. The term crossover relates to the way in which the electrical filter slopes superimpose at the dividing frequency, rather than describing the function that they perform (i.e. the filter slopes crossover, or overlap).

It was probably the mid-1970s before the full implications of the inherent problems in crossovers and their application to loudspeaker systems began to be realistically addressed. Until then, standard theoretical filters were largely used, without consideration for the practical repercussions of filter use. No filter can cut on or off abruptly. A filter rolls off, from a flat response to a slope determined by the time constants of the reactive elements of the network – the inductors and capacitors. The slopes tend towards regular numbers of decibels per octave, normally in multiples of 6 dB, which is a direct result of the behaviour of not only inductors and capacitors, but also of many mechanical systems, such as the loudspeaker drive units themselves.

The signal, therefore, does not suddenly switch from one drive unit to another as the crossover frequency is passed. There is an overlap where the slopes cross over, hence the name. If the slopes are set so that the high and low frequency sections are 3 dB down where they cross over, then they will tend to sum to full power at that frequency. There are also other types of crossover in common use which are 6 dB down at the crossover frequency. The voltages sum to unity and they can provide a flat axial response, but the power response would be 3 dB down at the crossover frequency. The first option would initially seem to be the logical solution, but group delays inherent in filter circuits cause time-shifts that prevent minimum-phase summation, rather like the room reflexion problems discussed in Chapter 11. In neither case can these time-shifted dips be equalised in the frequency domain, so monitor equalisers cannot help the situation.

The time shifts, which are related to the phase shifts, also tend to steer the combined wavefront from a loudspeaker system in directions other than at a normal to the front baffle plane. This is because a loudspeaker drive unit receiving a delayed signal, with respect to another drive unit in the same system, will behave as though it were farther away from the listener. It is like having two drive units on a baffle that is inclined with respect to the listener, so the wavefront radiates at a normal to the imaginary, inclined baffle. The axis of the flat frequency response will therefore not be directly in front of the loudspeaker, but inclined upwards or downwards. Many things about loudspeakers are not as obvious as most people tend to believe. Crossover responses can affect the directivity pattern of the combined loudspeaker system. This is why many hi-fi loudspeakers are seen to have staggered front baffles. The low frequency loudspeakers are the furthest forward, to compensate for the delayed signal that they receive. As the delay tends to increase with the filter slope and reducing frequency, a steep slope low frequency filter will cause considerable delay.

Many loudspeaker drive units have irregular responses outside of their normal frequency range of use. It is therefore advantageous to use reasonably steep filter slopes to avoid such ragged responses from superimposing themselves on the total signal. A 6 dB/octave crossover would mean that, with a 3 dB down point at the crossover frequency, the output would only be about 9 dB down one octave away from the crossover frequency. If a drive unit had a 5 dB response peak one octave away from the crossover point, then its response would only be 4 dB (9 dB − 5 dB) below the wanted signal level, and so the irregularities in the response could clearly be audible. Alternatively, a 72 dB/octave slope would solve the problem completely, but would introduce other problems, such as greater group delays and possible ringing problems from the high Q filters needed to achieve such a slope. Crossover slopes are therefore chosen as compromises, and there are indeed many options.

20.6.1 Passive Crossovers

The function of a crossover is to split the frequency bands and send only the appropriate frequencies to each drive unit − lows to the woofers, mids to the squawkers, highs to the tweeters, etc. Traditionally this was done with passive networks of inductors and capacitors placed between the amplifier output terminals and the loudspeaker drive units; usually inside the loudspeaker cabinet. Typical, historical networks are shown in Figure 20.4. The values of components were often chosen presuming that the loudspeakers were electrically like resistors, but in fact that is not the case, because they exhibit resistance, inductance and capacitance. Their input impedance (resistance plus reactance) therefore tends to vary with frequency, and two typical impedance curves are shown in Figure 20.5. The traditional crossovers were thus usually mismatched to the load impedance, and response flattening often took place with the 'adjust on test' addition of components.

Very few large studio monitor systems still use passive, high level crossovers. They are usually either relatively poor performers or extremely expensive. The overwhelming modern consensus is that passive high-level crossovers (as opposed to passive low-level crossovers − see later) are not appropriate for top quality studio monitoring. The Japanese TAD and Kinoshita monitors are still in current use at a high professional level, but the latter, with suitable amplifiers to drive the very complex impedance presented by the crossover components and drive units, cost around €50,000 *each*: €100,000 for a stereo monitor system. The electrolytic capacitors that the crossovers use have a life of about 10 years, and can be subject to changing values with age. For the people who like these monitors − and there are many − if they deliver the desired sound in small PA system quantities they will pay the price if they can afford it, but this type of approach is not really a modern day engineering solution.

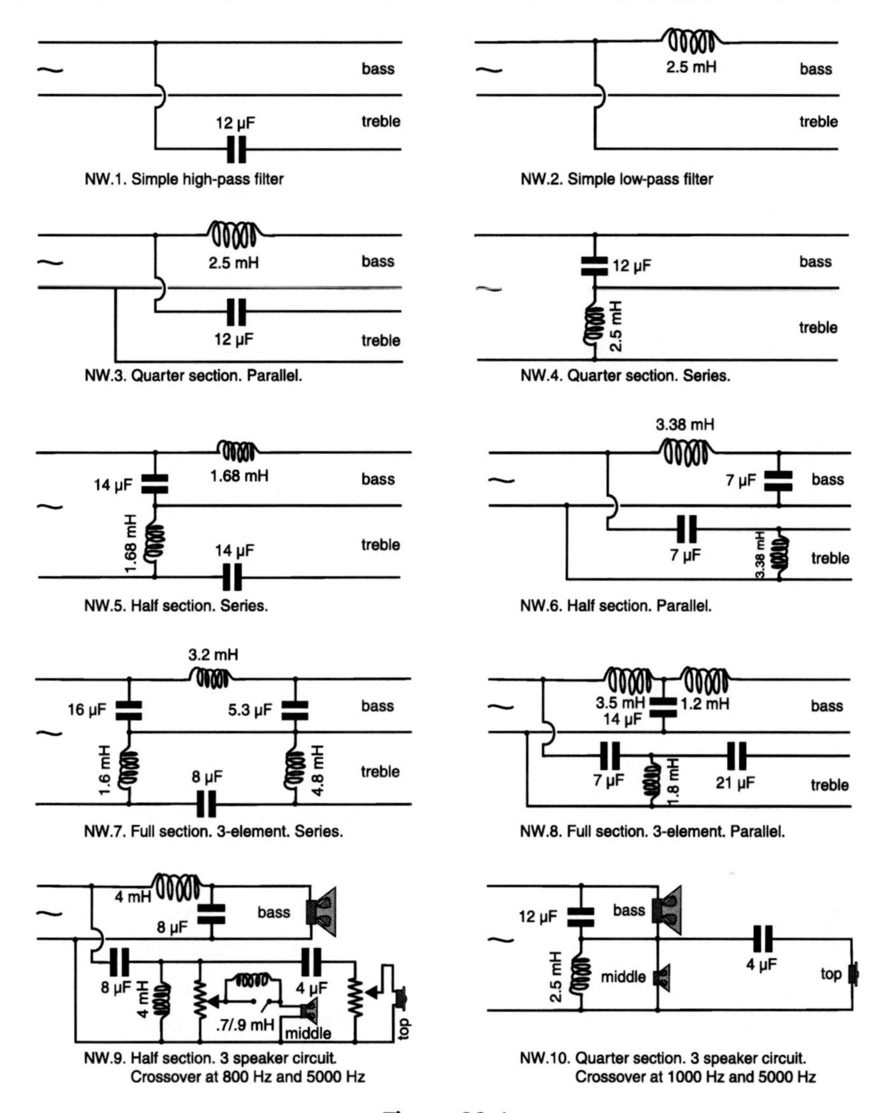

Figure 20.4:
Traditional crossover networks. Typical constant resistance crossover circuits with values for 15 Ω speakers and 1000 Hz crossover, except NW9 that is for 800 and 5000 Hz, and includes two volume controls and a switch for 2-way/3-way operation.

The problems are endless in the design of large monitor systems with passive crossovers. Amplifiers tend to need to be enormous, because, if they clip, the distortion products of the lower frequencies (which are usually the first to overload) are higher frequencies, and they pass straight into the high frequency drivers, often with very spiky waveforms. This can, and often does, produce premature failure of the high frequency drivers. In addition, the drive signals are passing one way through the filters whilst the back-e.m.f.'s (electro-motive-forces),

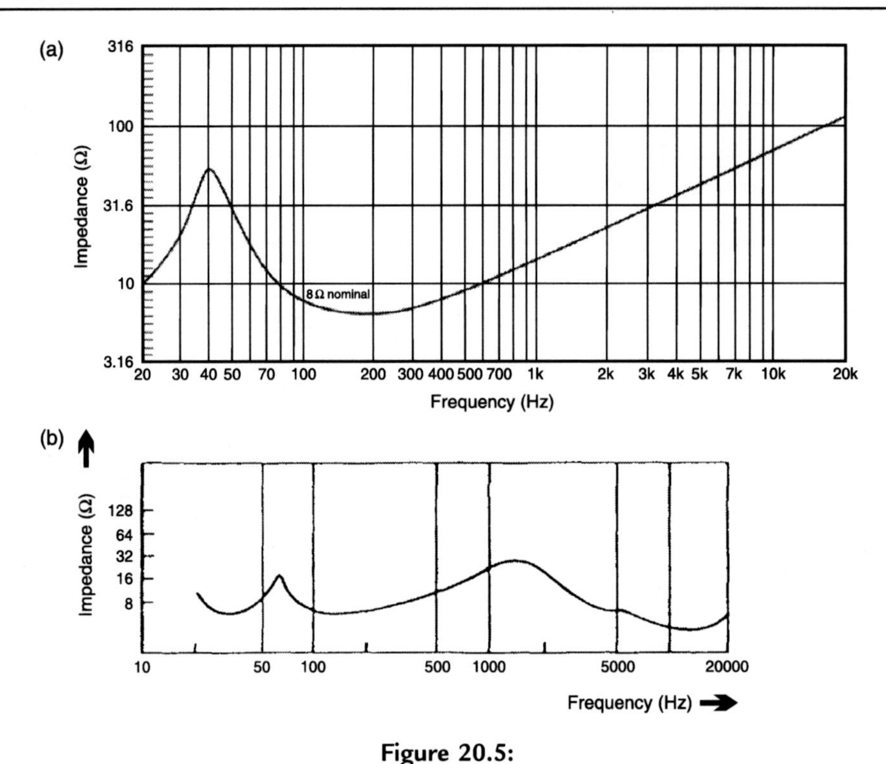

Figure 20.5:
Loudspeaker impedance curves: (a) Low frequency driver. Typical impedance curves for 8 Ω LF driver. Impedance shown on logarithmic scale (after Eargle); (b) A complete, full range loudspeaker system. KEF Model 104, impedance versus frequency.

created by the drive unit and filter resonances, are passing the other way through the filters. The complex impedances caused by this superposition can be almost beyond calculation; they are very signal dependent and they can put great strain on amplifiers. This is why the choice of amplifier can be so system-specific when driving passively crossed-over loudspeakers.

As mentioned in the previous section, there are passive crossover designs that can fully compensate for the impedance variations with frequency. In one design, described by Colloms,[2] it is possible to maintain a 4 Ω input impedance, ±0.5 Ω, between 20 Hz and 12 kHz. The three-way design uses a total of 11 inductors, 15 capacitors and four resistors, all to try to make the loudspeaker appear to the amplifier like a pure resistor. The possibility of being able to put so much between the amplifier and the loudspeaker drive units without negatively affecting the sonic performance would seem to stretch the imagination. To try to do this on a high power system, such as would be needed for studio monitoring, would be approaching the ridiculous. With capacitor values up to 600 microfarads, component drift with age would tend to lead to a very unpredictable response within a short period. This is perhaps why the Kinoshita solution was to let the input impedance vary, and look for a mega-amplifier to drive the 100 A currents that the loudspeakers occasionally demand.

The concept of the passive crossover can also be used at low levels, ahead of the power amplifiers, but they are rarely used. This is because once one has added the almost obligatory output driver stage (needed to cope with driving metres of cable, a range of amplifier input impedances, and balanced/unbalanced input/output configurations), it has already become an 'active' crossover, so one may as well incorporate the filter into the feedback circuitry. This then constitutes a typical, modern, active crossover, which offers many advantages.[11]

20.6.2 Active Crossovers

Active crossovers can, by virtue of their feedback loops, remain very stable whilst also offering almost infinite possibilities of filter slopes. State-variable filters, even if they do drift, can cause the high and low pass filters to drift in unison, hence lumps or dips in the total response are avoided. Active filters can also easily incorporate delay compensation for driver mounting offsets, such as when a horn driver is set behind the woofers. Response tailoring is independent of the loudspeaker impedance complexities. The list of advantages in favour of active crossovers and multi-amplification is impressive:

1. Loudspeaker drive units of different sensitivities may be used in one system, without the need for 'lossy' resistive networks or transformers. This can be advantageous because drive units of *sonic* compatibility may be electrically incompatible in passive systems.
2. Distortions due to overload in any one band are captive within that band, and cannot affect any of the other drivers.
3. Occasional low frequency overloads do not pass into the high frequency drivers, and instead of being objectionable, if slight, may be inaudible.
4. Amplifier power and distortion characteristics can be optimally matched to the drive unit sensitivities and frequency ranges.
5. Driver protection, if required, can be precisely tailored to the needs of each driver.
6. Complex frequency-response curves can easily be realised in the electronics, to deliver flat (or as required) acoustic responses in front of the loudspeakers. Driver irregularities can, except if too sharp, be easily regularised.
7. Complex load impedances are simplified, making amplifier performance (and the whole system performance) more predictable.
8. System intermodulation distortion can be dramatically reduced.
9. Cable problems can be significantly reduced.
10. If *mild* low frequency clipping or limiting can be tolerated, much higher SPLs can be generated from the same drive units (*vis-à-vis* their use in passive systems) without subjective quality impairment.

11. Modelling of thermal time constants can be incorporated into the drive amplifiers, helping to compensate for thermal compression by the drive units, although they cannot eliminate its effects.
12. Low source impedances at the amplifier outputs can damp out-of-band resonances in drive units, which otherwise would be uncontrolled due to the passive crossover effectively buffering them away from the amplifier.
13. Drive units are essentially voltage-controlled, which means that when coupled directly to a power amplifier (most of which act like voltage sources) they can be more optimally driven than when impedances are placed between the source and load, such as by passive crossovercomponents.
14. Direct connection of the amplifier and loudspeaker is a useful distortion reducing system. It can eliminate the strange currents that can often flow in complex passive crossovers.
15. Higher order filter slopes can easily be achieved without loss of system efficiency.
16. Low frequency cabinet/driver alignments can be made possible which, by passive means, would be more or less out of the question.
17. Drive unit production tolerances can easily be trimmed out.
18. Driver ageing drift can easily be trimmed out.
19. Subjectively, clarity and dynamic range are generally considered to be better on active systems compared to their passive equivalents (i.e. same boxes, same drive units).
20. Out-of-band filters can easily be accommodated, if required.
21. Amplifier design may be able to be simplified, sometimes to sonic benefit.
22. In passive loudspeakers used at high levels, voice-coil heating will change the impedance of the drive units, which in turn will affect the crossover termination. Crossover frequencies, as well as levels, may dynamically shift. Actively crossed over loudspeakers are immune from this effect.
23. Problems of inductor siting, to minimise interaction with drive unit voice coils at high current levels, are nullified.
24. Active systems have the potential for the relatively simple application of motional feedback, which may come more into vogue as time passes.

Conversely, the list of benefits for the use of passive, high-level crossovers for large studio monitors would typically consist of:

1. Reduced cost? Not necessarily, because several limited bandwidth amplifiers may be cheaper to produce than one large one capable of driving complex loads. (The loudspeaker may be cheaper but the system more expensive).
2. Passive crossovers are less prone to being misadjusted by studio engineers. True, but in a *professional* studio one ought to be able to expect a degree of responsibility in the use of professional systems. However, passive systems have a tendency to mis-adjust themselves with age.

3. Simplicity? Not really, because very high quality passive, high-level crossovers can be hellishly complicated to implement, not to mention the amplifiers which are needed to drive them.
4. Self-containment of the loudspeaker system: but this is probably only interesting for hi-fi enthusiasts who wish to use them with their favourite amplifiers.
5. Ruggedness? No; because the electrolytic capacitors (necessary for the large values) are notorious for ageing, and gradually changing their values.

Clearly, the advantages of active, low-level crossovers completely eclipse those of passive, high-level crossovers, yet it was only around the late 1980s that dedicated active crossovers began to be seriously used on large-scale monitor systems. Prior art used stock electronic crossovers and perhaps these caused some delay in the acceptance of totally active designs because they normally were made with fixed slopes on all the filter bands. It took some time before people generally began to accept the need to buy a specific, bespoke crossover with a monitor system, which could be relatively useless in any other application. There was still a mix-and-match mentality towards component parts, each of which was expected to function as a 'standalone' device. Once attitudes like this have become established it can be very difficult to introduce new concepts. In fact, it took a long time before self-powered, actively crossed over *small* monitors could establish their place in studio use. Domestic resistance to their acceptance has been even more pronounced. Established practices die hard, and they can be remarkably difficult to change, even in the face of clearly superior technology.

20.6.3 Crossover Characteristics

Crossovers are more than simple filters. Due to the group delays that exist in filter circuits, the outputs of the various filter sections are time-shifted by virtue of the phase shifts which are inextricably linked to the roll-offs. The result is that when the outputs are recombined, either electrically or acoustically, there are often non-minimum phase response irregularities in the frequency responses. Essentially, these have to be lived with, because they cannot be corrected by analogue means. Digital crossovers *can* be made to provide summing outputs, but the sonic benefits are not necessarily worth the effort and digital crossovers, unless they employ high sampling rates (96 kHz or more), will restrict the resolution in such a way that could make it difficult to accurately monitor some analogue or high sampling rate digital recordings. What is more, a three-way stereo crossover used with an analogue mixing console would require two A to D converters and six D to A converters. If these were to be of the highest quality (and in a high quality monitoring system they could be expected to be nothing less) the cost of the unit could be exorbitant. Using anything less than the finest converters would make a mockery of trying to monitor *recordings* made through the best converters. This subject is discussed further in Section 20.6.5.

Crossover output summation is a complicated affair. First-order (6dB/octave) crossovers can sum perfectly to reproduce a square wave applied to their inputs. This is because the 3 dB hump, which the in-phase outputs would be expected to produce at the crossover point, is exactly cancelled by the 3 dB dip which results from the ±45° phase shifts of the high and low pass section outputs. The step response of a 6 dB/octave crossover is shown in Figure 20.6. When the outputs of a two-way crossover using traditional second order filters (12 dB/octave) are connected in-phase, they produce a smooth phase response but a dip in the on-axis amplitude response. When connected out of phase (reverse polarity), the axial amplitude response is flat, but the phase response is kinked. As the phase *and* amplitude response must be flat in order to reproduce a transient accurately (see Section 11.5.3 and Glossary – Transient response) second-order crossovers cannot normally do so, no matter which way round they are connected. The effect on a step-function is shown in Figure 20.7. However, driver offsets at higher frequencies of crossover *can* sometimes reintroduce a better transient response, but this may require filters with asymmetrical turnover frequencies, such as used in the NS10M.

Third-order (18 dB/octave) and fourth-order (24 dB/octave) crossovers also cannot accurately reproduce transients, such as step-functions. However, due to the steeper slopes, the amplitude and phase response irregularities are confined to much narrower frequency bands, and hence are much less audible. The Butterworth (constant power) third-order crossovers produce slopes that are 3 dB down at the crossover point. The voltage outputs are in phase-quadrature (90° out) at all frequencies, by virtue of being shifted ±135° in each direction. In fact, they are 270° out of phase when connected 'in-phase'.

Another class of filter that is very popular with studio monitor designers is the fourth-order (24 dB/octave) Linkwitz-Riley crossover. These consist of two, second-order Butterworth sections in cascade. The in-phase connection gives a flat voltage summation, and can give a flat amplitude response on axis, but the nature of the filter's total frequency response (phase and amplitude) is that it must give rise to the 3 dB dip in the total power response at the crossover frequency. This is not a practical problem in most control rooms, because in highly damped conditions the total power response is not of as much importance as the on-axis frequency response.

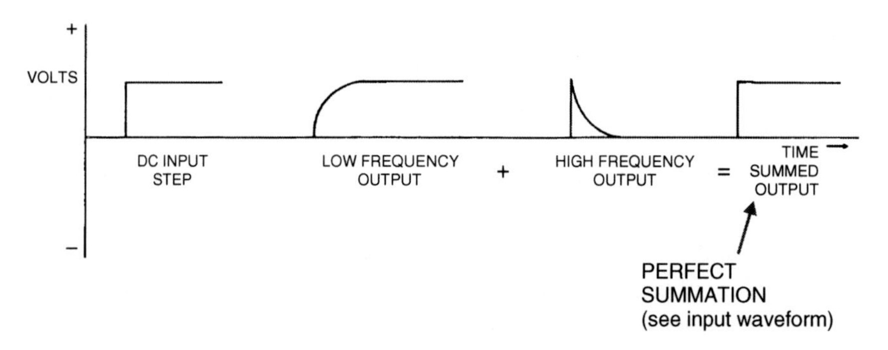

Figure 20.6:
First order (6 dB/octave) crossover step function summing.

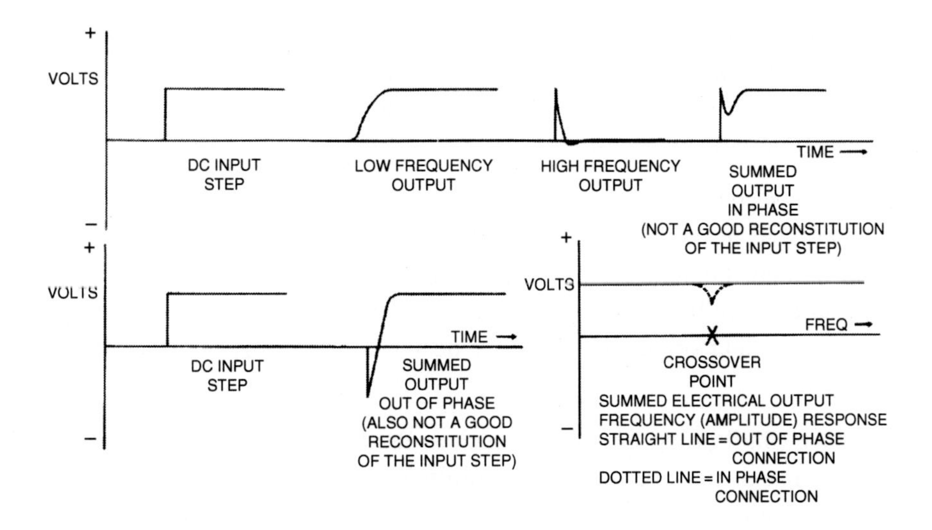

Figure 20.7:
Typical electrical step-function responses for 12 dB/octave crossover. Note that in-phase or out-of-phase, the summed output is not a true replication of the input waveform.

The range of available filter slopes and shapes is enormous, and detailed analysis of crossover design and performance is way outside the scope of this book. Readers needing a more in-depth study should refer to the Bibliography at the end of the chapter (especially Self [2011]). However, a few important characteristics *can* be considered here. Things will be discussed in terms of active, low-level crossovers and multi-amplification, because in the twenty-first century it appears to be very hard to justify the use of high-level passive crossovers in serious, new, studio monitor designs. Although there are some known and trusted monitors still in use with passive crossovers, it is unlikely that more will arrive. Despite the fact that the multi-amplified monitors took some time to fully establish themselves (monitoring has always been a rather conservative subject, and perhaps quite rightly so), once they took hold the benefits became so obviously apparent that older techniques are now rarely considered in new designs.

20.6.4 Slopes and Shapes

At least in analogue designs, the 6 dB/octave (first-order) Butterworth crossovers are the only ones that can truly reconstruct a transient waveform at their combined outputs. The main problem with their use is that their attenuation rate is so slow that out-of-band interference between the drivers can be a big acoustic problem. What is more, the slow rate of filter attenuation is not much use in protecting high frequency drivers from the damaging low frequency inputs. The only common use for 6 dB/octave filters is in 'half-crossovers', such as when a naturally tapering high frequency response of a mid-range driver is matched by an electrical filter which progressively brings a tweeter into circuit. There is another use for 6 dB/octave slopes in some complex, mixed slope, linear-phase/constant voltage crossovers, but the need for one driver to exhibit a flat response up to three octaves, or so, out of range makes such designs rather impracticable.

The 12 dB/octave (second-order) filter provides a good polar response, and has been used in many famous passive designs such as the Yamaha 'NS10M Studio'. However, in many cases of their implementation, the problems involved with getting the transient responses adequately accurate have been a handicap. In passive designs, this type of filter was popular because it was cheap and simple, but in active designs, they hardly reduce the price or complexity at all.

The 18 dB/octave (third-order) crossovers, such as the Butterworth, are widely used, despite the asymmetric off-axis performance which results when using non-coincident drivers. Perversely, it has nearly always been the case with such designs, that they have *not* been used with coincident drivers.

The fourth-order Linkwitz-Riley (24 dB/octave) crossovers have come to be the generally preferred form, although their design is rarely straightforward in high-quality monitors. The electrical filters are usually designed to have a 24 dB/octave 'target response' when superimposed on the electroacoustic response of the drive units. The sequence of events for the design of crossovers for high-quality monitors therefore tends to be:

1. Select the drivers for their sonic performance.
2. Measure their responses around the crossover region.
3. Design a filter that will give the desired 'target function' response in the room when feeding the appropriate driver.

In fact, the days of the use of standard electrical filters in crossovers are long gone. It is now almost *universally* recognised that the response shapes of the drivers themselves should be taken into account as part of the crossover filter design.

20.6.5 Digital Crossovers

Digital crossovers have seen increasing use as time has advanced. They are particularly attractive because of the relatively easy implementation of almost any amplitude response, phase response, delays, driver compensation and even room compensation. However, the big limitation to digital crossover use in studio monitoring systems is the cost of the converters. In any monitoring chain, the quality of response should be equal to or greater than the response in any other part of the system. If this is not so, then one cannot monitor the quality of the recording chain. It is difficult to measure to 140°C with a thermometer which only goes from 0 to 100°C. Likewise, if a studio purchases ultra-high-quality converters for any part of its recording or mixing chain, then the use of anything less in the monitor chain is an absurdity. One would not be able to hear how good the recording was, due to the limitations of the monitor converters. Surely, this ought to be self-evident, but it often appears to get overlooked.

As mentioned previously, but worth repeating here, a stereo, three-way digital crossover has two inputs and six outputs. For use with analogue mixing consoles, which at the time of writing *still* occupy the majority of 'high-end' studios, this would need two ultra-high-quality A to D converters, and *six* ultra-high-quality D to A converters. With a digital input, the six D to A converters would still be required to drive the power amplifiers. Even if digital crossovers are used with a digital mixing desk and amplifiers with digital inputs, then one must still ask questions about *their* (the amplifiers') converter qualities, which are almost certainly likely to be less than the best.

Likewise, whatever sampling rates or bit-rates are in use, the electronics of the monitor circuitry should always be equal to or greater than anything anywhere else in the chain. For example, one could hardly expect to hear any benefit of 192 kHz sampling through a crossover based on 48 kHz sampling. These things all need to be carefully considered before choosing digital crossovers as an option for high-quality monitoring systems.

20.7 Loudspeaker Cabinets

The cabinets in which loudspeaker drive units are mounted are an important part of the loudspeaker as a whole. When an electrical drive signal enters the voice coil of a loudspeaker, the coil is attracted or repelled, depending on polarity, by the magnet system. However, the force which is produced by that motor system has no way of 'knowing' whether it is being resisted by the stiffness of the loudspeaker suspension system, the load imposed by the air in front of the diaphragm, or the spring which is formed by the air inside the loudspeaker cabinet. All it sees is the total impedance to its motion. That is why (as in many loudspeaker design books) the cabinet can be modelled by electrical symbols of capacitance, inductance and resistance.

A small loudspeaker cabinet will behave like a stiffer spring than a large one. Therefore, for two drive units of equal free-air resonant frequency, the stiffer spring of the smaller cabinet will force the resonant frequency upwards to a greater degree than would be the case for the identical driver in the larger box, hence restricting the low frequency response, which tends to roll-off below resonance. This can only be brought back down again by using a heavier cone, which will then need more power to move it. For an equal acoustic output over a given frequency range, smaller boxes need more power to drive them. Smaller boxes also usually have a greater difficulty in dissipating the heat from the voice coils, so the greater heat produced and the greater difficulty in getting rid of it can mean that smaller loudspeaker boxes are more prone to thermal compression effects when reproducing high levels of low frequencies. The thermal compression results from the rise in resistance of the copper or aluminium voice coil with increasing drive current. This increase in resistance prevents the coil from drawing as much current from the amplifiers, for any given output voltage, than a cold coil. The effect is exactly like that of a compressor. This concept is discussed further in Section 20.8.2. Whilst it is true that circuitry could be used to model the compression and partially compensate for it, this would require more voltage from the amplifiers, so even larger amplifiers may be necessary, probably needing more power still, so efficiency reduces further.

There is, effectively, an 'eternal triangle' of size, efficiency and frequency range. Whenever size is reduced, then for a flat response, either the efficiency of the loudspeaker system or its low frequency extension must also reduce. In most cases, both reduce with size. Therefore, it is not that one cannot generate very low frequencies from small boxes; it is just that one cannot do it efficiently. Hence the sensitivity difference between the large UREI 815 and the small ATC SCM10, described in Section 20.4. For high levels of extended low frequencies, from reasonable amplifier power, loudspeakers must be large. This is somewhat like the fact that a double bass needs to be much larger than a violin. The need for large low frequency sources will be discussed further in the following section on loudspeaker drive units.

Large cabinets have large areas of cabinet walls, and with high internal pressures being generated by the drive units it can be difficult to prevent the walls from vibrating. The problem with panel vibrations is that they act as secondary radiation sources, and the secondary radiation can seriously colour the sound by adding to or subtracting from the direct radiation from the driver(s). Panel vibration therefore needs to be minimised, which calls for high rigidity and damping. Usually, that double requirement means that there will need to be considerable mass – the cabinets will be heavy. For commercial reasons, monitor loudspeakers intended for large-scale sales often tend to be lighter than acoustically optimum. Transport costs and handling difficulties would make very heavy cabinets more expensive and less manageable, and hence less likely to sell in large quantities. Conversely, some specialised studio monitors which are solely intended for flush mounting, and which were not intended to be mass-market products, never need to be compromised by marketing considerations and so can be enormously heavy, to good advantage.

20.7.1 Cabinet Mounting

The difference in cabinet weights and rigidity leads to two common approaches to mounting the loudspeakers in a monitor wall. The lighter cabinets are usually more subject to panel vibration problems, which can pass into a control room structure if rigidly mounted. This can lead to vibration passing through the structural materials at speeds much faster than through the air. These vibrations can re-radiate from parts of the structure and form secondary sources, sometimes arriving at the ear before the direct sound.[3] The usual answer to this problem is either to resiliently mount the loudspeakers, on rubber for example, or to try to substantially damp the cabinet walls by encasing them in concrete. The much heavier, specialist loudspeakers, because of their highly damped massive nature, need less special attention, as their external cabinet vibration levels are usually insignificant.

Many hi-fi enthusiasts mount their loudspeakers on spikes, which bite into a wooden floor. In some cases, this is to prevent recoil effects if the loudspeakers are mounted on carpet. When the bass drivers punch in and out, there *can* be cabinet reactions which, if mounted on carpet, could cause the whole cabinet to shift forwards and backwards sufficiently to cause phase modulation of the tweeter response, hence blurring the high frequencies. Spikes penetrate the carpet to get a solid fixing on the floor below. Spikes are also sometimes used directly on wooden floors to avoid rattles, which may occur due to the imperfect contact between the flat underside of the loudspeaker (or its stand) and the floor boarding. Spikes are therefore often a solution for light to medium weight cabinets. With very heavy and well-damped cabinets, the problems do not usually arise.

Whilst on the subject of loudspeaker mounting, a little should be said about the height at which they should be positioned in a wall. The general tendency amongst most careful designers is to site the mid-range units approximately 145–148 cm above the floor. This is a mid position between the ear height of a small person sitting down and a tall person standing up. It minimises the variability in frequency balance as people move around the room, or move out of their chair to a standing position. It also ensures that the sound enters the ears from a natural direction during most operations. The practice of mounting monitor loudspeakers high is to be discouraged, because in a normal, slightly nose down working position for the people at the mixing console, the sound from high-mounted monitors tends to arrive at the ears from a direction perpendicular to the top of a person's head, as shown in Figure 18.6. Both high frequency perception and stereo imaging can be badly affected, and the reflexions created by the working surface of the mixing console can muddy the sound.

It should be recognised that differences in control room sizes and design philosophies *can* modify the requirement to some degree, but there is now great consensus amongst designers that the 145–148 cm (4 foot 9 inches to 4 foot 10 inches) mounting height, with the monitor cabinet baffles vertical (i.e. no inclination) is the best compromise. The concept is shown diagrammatically in Figure 20.8.

The horizontal placement of the cabinets can be influenced by window positions and sizes, and the need to maintain the chosen subtended angle at the listening position. However, one thing which is almost standard practice is to focus the monitors at a point about 60 to 80 cm *behind* the principal listening position. This barely compromises the stereo at the prime listening position, but it significantly enlarges the area in which people can sit side-by-side at the mixing console and still more or less hear the same stereo panorama. Figure 16.7 shows the general idea.

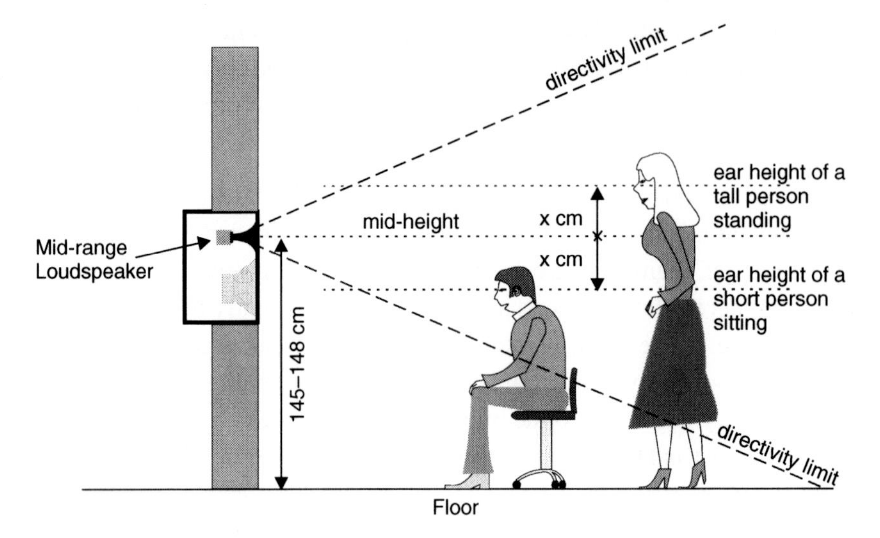

Figure 20.8:
By mounting the loudspeakers vertically, and *not* inclined, the height of the mid-range loudspeakers can be set between the ear heights of a small person sitting and a tall person standing. Only by mounting the loudspeakers vertically can this condition be maintained throughout the room. Such mounting is also less likely to cause reflexions from the upper surface of the mixing consoles.

20.7.2 Cabinet Concepts

Loudspeaker cabinets are huge subjects in themselves, and they are very much a part of the loudspeaker electroacoustic system. They are not merely convenient boxes in which to mount the drive units. A good understanding of the concepts is very useful in aiding the better application of the most appropriate choices. The main point to bear in mind is the way that the loudspeaker mounting conditions have a large effect on the overall system performance, and that the mounting conditions are, essentially, a part of the loudspeakers themselves. The Bibliography lists sources where this subject is taken much further.

As few, if any, recording studios now work on large monitors alone, the mounting conditions for the small loudspeakers need some consideration, or their performance will not be optimal. Figure 20.9 shows the adverse effects commonly found when mounting small loudspeakers on console meter bridges. Mounting the loudspeakers on stands, behind the console, can go a long way to reducing such reflexion problems; however, the low frequency response of the two mounting systems can be very different. Depending upon the nature of the structure of the console, the physical contact between the flat surface and the loudspeaker can either reinforce the low frequencies if the console is rigid, or reduce the low frequencies if the console is significantly resonant. As so many different situations exist, it is hard to specify specific rights and wrongs. Nevertheless, a very good way of

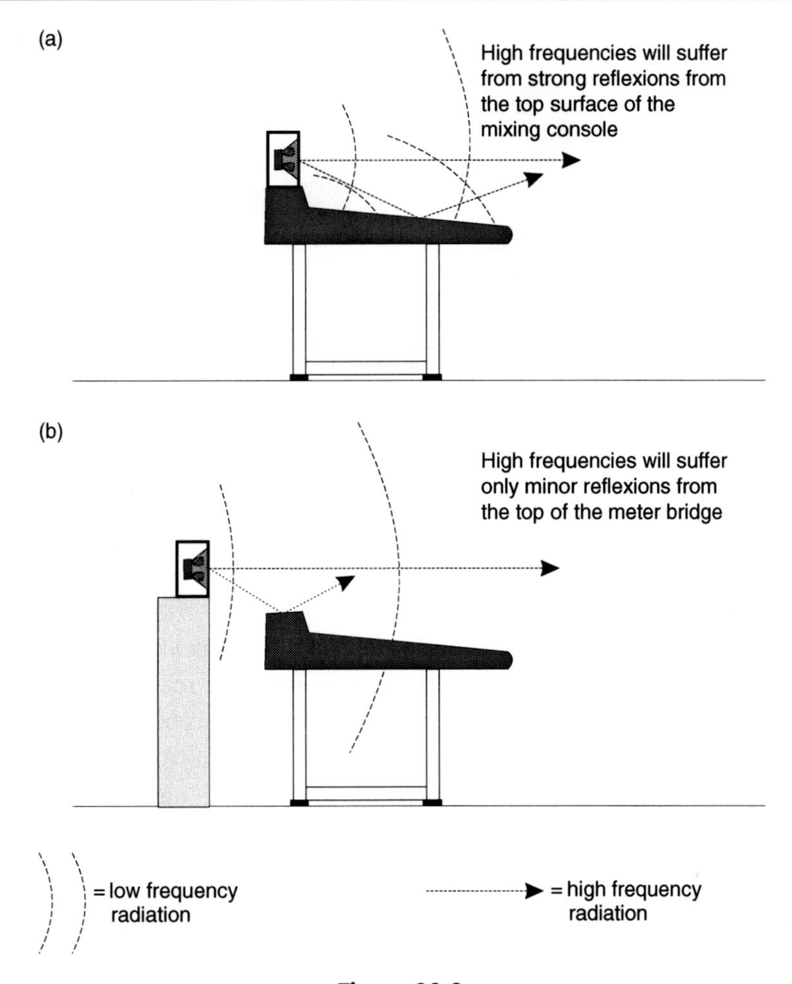
(a)
High frequencies will suffer from strong reflexions from the top surface of the mixing console
(b)
High frequencies will suffer only minor reflexions from the top of the meter bridge
= low frequency radiation
= high frequency radiation

Figure 20.9:
Small loudspeaker mounting: (a) Small loudspeakers mounted on top of the meter bridge will suffer low frequency reinforcement because the low frequency waves are not free to expand. They are constrained by the close presence of the mixing console (this may or may not be a bad thing); (b) Low frequencies will be more free to expand, so they will tend to be weaker than for loudspeakers mounted on the meter bridge. Situation (b) is generally preferable unless the loudspeakers are somewhat lacking in bass.

determining the best mounting position for small loudspeakers is to have two people hold one loudspeaker each, against their chests, and walk from a distance of about 1m behind the console towards the listening position, as shown in Figure 20.10. If the loudspeakers are finally placed simultaneously on the meter bridge, any sudden change in the timbre of the sound could be attributable to structural or cavity resonances in the console. In many cases, as the people bring the loudspeakers closer to the console, great changes can be noticed in the clarity of the sound and the stereo imaging. Colouration often increases the closer the cabinets are to the meter bridge, due to reflexion and diffraction effects.

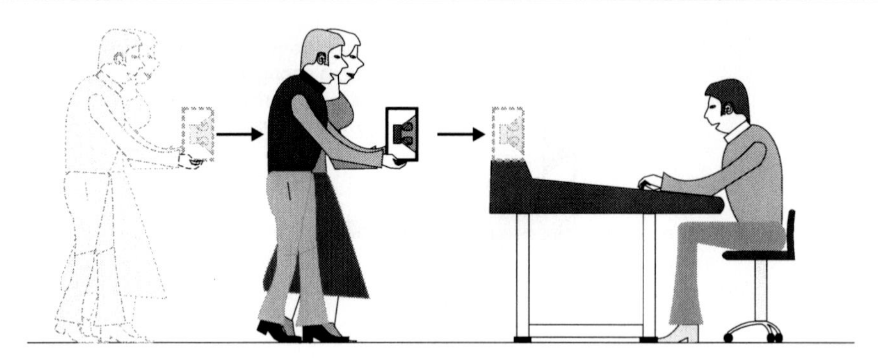

Figure 20.10:
Two people can hold one loudspeaker each, walk towards the mixing console from a position a metre, or so, behind the console, and finally put the loudspeakers on the meter bridge. Care should be taken to ensure that the two people (holding the left and right loudspeakers) move in unison. Any changes in frequency balance, colouration, or stereo imaging should be clearly heard by the listener sat in the engineer's position.

By using this method, an optimum position for the small loudspeakers can usually be easily chosen. However, distances of more than 30 cm or so from the rear of the console may necessitate the covering of the back of the console with an absorbent material, to prevent potentially disturbing reflexions from the console returning to a front window, for example, and reflecting back to the listening position. As the mounting of an absorbent screen behind the console is also generally good practice from the point of view of the main monitoring, it should not be seen as a disadvantage to the more distant mounting of the small loudspeakers. It must be borne in mind, though, that moving the small loudspeakers from a distance of 1 m to a distance of 2 m from the listening position could require 3–6 dB more drive – up to four times the amplifier power – due to the doubling of the distance from the listening position.

20.7.3 Mounting Practices and Bass Roll-Offs

In small control rooms, the proximity of the loudspeakers to the front wall can cause a bass boost, as discussed in Chapter 11. In the case of reflex enclosures, the rise can sometimes be compensated for by the blocking of the tuning port(s). The consequent reduction of the low frequency output from the port can often offset the rise due to the proximity of a wall. This is a perfectly legitimate exercise, because it is the response perceived at the listening position which is all-important. However, it will tend to change the slope of the ultimate low frequency roll-off, and it may affect the ventilation and heat loss from the voice coils.

A loudspeaker on an open baffle board will suffer a low frequency roll-off from a frequency set by the cancellation around the sides of the baffle, and hence by the baffle size. For practical sized baffles, this will begin from quite a high frequency, but the roll-off will only gradually fall to 18 dB per octave. In the case of a sealed box, the frequency where the roll-off begins can be reduced compared to the same drive unit on an open baffle, and the slope will tend towards 12 dB per octave. With reflex tuning of a similar sized box, the roll-off can be delayed until still lower frequencies, but once the roll-off begins, it will attain a slope of 24 dB per octave.

Different types of loudspeaker cabinets therefore have different characteristic low frequency roll-off slopes, and hence, not surprisingly, different characteristic sounds. This is even less surprising when one considers that the slope of the roll-off also affects the time response, with the steeper slopes giving rise to more signal delay in the region of the cut-off frequency. (See Figure 19.11.) The fastest transient responses are exhibited by the loudspeakers with the shallower roll-offs. Work carried out in 2001 at the Institute of Sound and Vibration Research, in the UK, has shown that above 30 Hz, the subjective colouration due to steep roll-offs can be considerable. However, the roll-off rate itself is not the whole story. Bank and Wright[4] have noted that despite the bass reflex and transmission line cabinets both exhibiting a 24 dB per octave roll-off, the subjective perception of their low frequency responses can be quite different. It is thus necessary to appreciate how some seemingly simple changes in loudspeakers can have some disproportionately complex results.

Intuitively, blocking the port on a loudspeaker would suggest that the drive units would be better protected from low frequency overloads by virtue of the pressure changes inside the cabinet during high excursions. In fact, at resonance, a loudspeaker in a vented (reflex) enclosure moves much less than a similar driver does in a sealed box of equal size. Down to the resonant frequency, a vented enclosure actually protects a driver better than a sealed box, although below resonance all control is lost. This is why small, ported loudspeakers intended for high-level use often have steep filters to protect them from excessive drive below their useful operating frequency band. Despite their widespread use in modern actively driven (built-in amplifier type) loudspeakers, such filters do colour the sound of the low frequencies, and can cause severe deterioration of the transient response. Figure 19.4 shows the effect of cabinet tuning and low frequency alignment on the acoustic source positions.

20.8 Loudspeaker Drive Units

The drive units (or motor systems) of loudspeakers are at the core of any studio monitoring system. The needs of low frequency and high frequency sound radiation are very different, and therefore for high-quality loudspeaker systems more than one drive unit is required to cover the entire audible frequency range. Low frequency drivers need to be large, because they need to move a large volume of air. Most dynamic loudspeakers are volume/velocity sources, where the acoustic output is the product of the two. So, at low frequencies, where cone velocities are small, the volume of air which needs to be moved is proportionately larger than at high frequencies, where the vibrational velocities are much higher. The acoustic requirements of a conventional high frequency loudspeaker require it to be small, in order that the high frequencies will be well distributed. If the radiating surface were to be large, then the parts of the sound arriving from different parts of the surface could have considerable phase shift at high frequencies, where wavelengths are short. At 20 kHz, the wavelength is around 1.7 cm, so if the difference in path length from the ear to two different parts of the radiating surface were only 8 *millimetres*, the radiation from the two places would arrive at the ear 180° out of phase, and so would cancel.

20.8.1 Low Frequency Driver Considerations

In general, larger low frequency drivers tend to be preferable to smaller ones for any given radiated output level. If the volume-velocity product is to be the same, an 8 inch (20 cm) diameter loudspeaker would need to travel much further than an 18 inch (45 cm) loudspeaker for each cycle of the music. An 18 inch loudspeaker cone has a surface area of 254 in^2, whereas an 8 inch loudspeaker cone has a surface area of only 50 in^2. In order to displace the same amount of air, the smaller cone would have to *move five times* the distance of the larger cone, and would need to travel at five times the speed in order to displace the same amount of air in the same amount of time. The difference has several implications.

When higher frequencies are simultaneously being radiated from a cone that is undergoing low frequency excursions, frequency modulation can take place due to the Doppler effects (see Glossary). Obviously, a large cone moving slowly over short distances will produce far less Doppler shift (frequency modulation) than a small cone moving rapidly over greater distances. Furthermore, a small cone punching its way violently through the air will create far more turbulence at its edges than a large cone moving slowly. The implication of this is that spurious noise generation is far less likely in the case of the larger cone. Another point to consider is that loudspeaker suspension systems and magnet systems can be notoriously non-linear, and the degree of non-linearity tends to be proportional to the distance travelled. It is thus generally easier to keep non-linear distortions lower in low displacement large cones than in high displacement small cones for similar acoustic outputs.

There *are* limits to woofer size, however, because the cones can be very difficult to keep rigid once their diameter increases beyond 45 cm (18 inches). The tendency is therefore to use more woofers of the same size when output needs to be augmented, rather than a single driver of greater size. Some woofers designed to operate to relatively high frequencies – say 500 to 1000 Hz – are specifically designed to decouple concentrically the outer parts of the cone from the centre section as the frequency rises. Such decoupling prevents the phase cancellation effects described earlier, when the source size becomes large with respect to a wavelength of the frequency being generated. Ideally, by 500 Hz, the radiating proportion of a large cone would need to be only the central portion of about 25 cm in diameter. If such internal cone decoupling cannot be achieved, then the only solution is to cross-over into a different drive unit of smaller diameter. If this is not done, the large cone will begin to emit the higher frequencies in a narrow beam, which would give rise to undesirable off-axis frequency responses and a narrower usable listening area.

20.8.2 Efficiency and Sensitivity

These two loudspeaker parameters often get confused. Efficiency is the degree to which a loudspeaker can convert its total electrical input energy into acoustic output energy. The sensitivity describes the acoustic output of a loudspeaker, on-axis, at a given distance (usually 1m) for a given electrical input (usually 1 W, or 2.83 V [the voltage equivalent of 1 W into 8 Ω]) in its operational frequency range.

Sensitivity and efficiency are important in that their absence creates problems. Inefficient, insensitive drivers need a lot of electrical power input in order to achieve an adequate sound output. A woofer with a 500 W power rating and 0.5% efficiency will, at full power, dissipate 99.5% of its input power – or 497.5 W – as heat within its voice coil and magnet system. Hence, the blackened magnets and chassis heat sinks to be found in many high power loudspeakers: the black colour helps to radiate the unwanted heat into the air. A loudspeaker with a sensitivity of 85 dB for 1 W at 1 m, dissipating 400 W, would produce the same acoustic output as a loudspeaker with a sensitivity of 95 dB for 1 W at 1m dissipating 40 W. The 10 dB difference represents a ten times power difference and losing 40 W to the atmosphere is much easier than losing 400 W. Consequently, for the same given output power, the 95 dB loudspeaker would run much cooler than the 85 dB loudspeaker. The principal significance of this is that a typical copper voice coil resistance increases by about 0.4% per °C. As the resistance rises, the ability to draw current from a typical amplifier reduces, so more drive voltage is needed to maintain the output from a hot coil when compared to a cold coil. As most amplifiers do not know the temperature of the coil that they are driving, the usual result is that as the voice coil temperature rises, the acoustic output falls. This is known as thermal compression. A temperature rise of 250°C will double the coil resistance, and many inefficient loudspeakers do work at temperatures of 250–300°C. Three or four dBs of compression under such circumstances would be typical. In conventional voltage driven loudspeakers, there is no total solution to thermal compression, so if a loudspeaker will be subjected to moderately high-level sustained outputs, the compression can only be avoided by using drivers of greater sensitivity, or fan cooling the chassis.

Many of the woofers used in commercial monitor systems are of relatively low sensitivity, because of a commonly held view that amplifier power is cheap, whereas powerful magnet systems are expensive. Almost nothing is now free of cost restraints, and the price at the point of sale is the god. Nevertheless, over the years, the cost of the extra electricity consumption needed to drive the low sensitivity loudspeakers will also add up to a considerable amount, but hidden costs such as these are usually ignored, despite being very real. In fact, in the long-term, and after the receipt of many electricity bills, it *may* be realised that amplifier power is *not* cheap. What is more, in many countries, obtaining high current electricity supplies is difficult and expensive, and in hot climates, extra power for the amplifiers means extra power for the air-conditioning to take out the heat that the amplifiers are pumping into the control room. In many circumstances, these problems are not trivial.

20.8.3 Magnet Systems and Cone Materials

When Zambia became an independent country in the 1970s, a civil war broke out. Zambia produced around 80% of the world's cobalt, an important constituent of most high-quality loudspeaker magnets at that time. Many of the mines became flooded during the war, and consequently the price of cobalt on the world market rose enormously. So much so, in fact by about 2000%, that the majority of the world's loudspeaker manufacturers began to look for magnet materials other than the Alnico (*al*uminium/*ni*ckel/*co*balt [with copper and iron]) which was becoming so expensive. Other magnet materials, such as Ticonal and Alcomax, also contained cobalt (the former also with titanium), and so suffered large price rises.

The industry generally turned to a range of barium or strontium ferrites, which were ceramic, non-conducting materials. They were also cheap. However, the physical shapes and sizes needed for the new ferrite magnets often meant totally redesigning the motor systems. A serious problem with ferrite magnets, which was not at first appreciated, was due to their non-conducting nature. This was the inability of the magnetic material to provide an electrical short circuit to eddy currents derived from the interaction of the magnetic fields of the coil and the magnet itself. The flux consequently moves in a series of jumps, rather than smoothly. The result can give rise to a form of programme-modulated noise, somewhat similar to modulation noise on magnetic tape recordings. If the magnetic material is conductive, then any magnetic field changes need to generate large eddy currents in the electrical short circuit provided by the metallic magnet, and are thus smoothed out.

For many years, audiophiles have extolled the virtues of the older magnets, and only recently has the explanation become more commonly understood. Ferrite magnets can therefore exhibit a 'bit-rate' that limits their resolution. Hence it is not only the weight saving which has spurred on many manufacturers to use the newer rare-earth magnet materials, such as neodymium/iron/boron and samarium/cobalt (yes, back to cobalt again, but in smaller quantities). In fact some of the newer magnet materials *demand* high sensitivities if they are for high output use, because they tend to demagnetise at high temperatures. The use of ferrites led to so much bad design partly due to their tolerance of high temperatures without demagnetising. Notwithstanding all of this, there *are* some good loudspeakers which employ ferrite magnets, although they also tend to employ special mechanisms to overcome the problems.

Cone materials can also significantly affect resolution, and pulped paper cones can still give a good account of themselves from this point of view. Many modern materials have been chosen for their physical properties in relation to their ability to perform as ideal pistons, but some low-level hysteresis losses within the materials can produce a rather similar type of programme-modulated noise that can result from the use of poorly designed ferrite magnets. Despite the measured performance of some ferrite-magnet/composite-cone low frequency loudspeakers being highly impressive, the subjective performance of some of the metal-magnet/pulp-cone loudspeakers is often considered to be more 'open' and 'transparent', probably due to the lower levels of programme induced artefacts. However, the degree of subjective differences can be heavily dependent on the frequency range being handled by the woofer. When the upper limit is restricted to 200 Hz or so, the subjective differences are greatly reduced, both for the magnetic material and cone material choices.

20.8.4 High Frequency Loudspeakers

The physics of high frequency sound reproduction demand that the surface area of the generating diaphragm should be small. There is now an enormous range of high frequency loudspeaker types, but in recording studio use, the general tendency is for the use of cones, domes, compression-driver/horn combinations, ribbon loudspeakers and Heil air-motion transformers. Somewhat rarely, Manger drivers are also to be found (see Section 20.8.5.4). One of the greatest controversies is usually in the choice of hard or soft materials for the domes. Hard domes may be made from titanium, aluminium, beryllium, copper and plastics of various natures. Phenolic-doped fabrics have also been employed. Soft domes are usually made from fabrics, suitably doped with rubbery materials which impart to them the desired stiffness and damping.

The precise choice of high frequency drive unit in any given system is highly subjective, with some manufacturers adopting specific types with almost religious fervour, whilst eschewing the favourite choices of other manufacturers. Marketing also has a great deal to do with this, where one manufacturer wants to impose on its products a recognisable brand image, which it can then sell for all it is worth. Unfortunately, this often does little for the benefit of the professional user.

Domes, cones and ribbons are usually relatively low sensitivity devices, and because they need to be small for sound radiation purposes, their ability to lose heat is often restricted. Sensitivities in the 90 dB for 1 W at 1m are typical, and ferro-fluid is often used, these days, to aid voice coil cooling. However, the stability of long-term performance of ferro-fluids is still not accurately determined, and may be environmentally dependent. The result is that at high levels, the effectiveness of cone, dome and ribbon drivers is limited, and at levels of

105 dB or so, many of them are reaching the limits of their linear operation. For higher levels of use, the only solution is either to restrict the high frequency driver to very high frequencies and use more crossover points, or to use horn loaded tweeters, yielding sensitivities as high as 108 dB for 1 W at 1 m. The disadvantage to the latter case is normally only cost, because little controversy exists about the use of horns above 7 or 8 kHz.

It also should also be noted that, contrary to much widely accepted mythology, dome tweeters do not exhibit better sound dispersion at high frequencies. The dome shape gives many people the idea of a section of a spherical radiator, but a dome can move in one plane only, and often tends to act more like a ring diaphragm, exhibiting a rather strange directivity. This subject will be dealt with in more detail in Section 20.8.5.2. Once again, as is so common in electroacoustics, things are often very different to what 'common sense' would tend to suggest.

One of the greatest influences on the sound from high frequency drivers is the design concept of the entire loudspeaker system. As discussed in Chapter 19, there is no universally accepted standard high frequency response for monitoring systems, and the subjective sound of two different tweeters may be dictated as much by the way that they are being used and the frequency response contour of the entire system as by the inherent properties of the high frequency drive units themselves. It has even been reported how adverse criticisms have been made of high frequency drivers operating above 6 kHz, when the source of the subjective harshness was found to be due to distortions in the low-mid-frequency drivers, operating only *below* 1.2 kHz.[5] One has to be very careful before pronouncing judgement on almost any part of a monitor system, because the roots of many 'obvious' problems are very often not obvious at all.

20.8.5 Mid-Range Loudspeakers

This is a controversial subject; of that, there is no doubt. In many ways, this is not surprising, because it covers the range of maximum hearing sensitivity and the most significant range of musical instrument timbre. In general, for use in medium to large size studio monitors, the choice lies between cones, soft domes and horns, although hard domes do occasionally find use, especially in medium-sized systems. The mid-range of audio frequencies can probably be considered to span the range from 300 Hz to 5 kHz, which covers about four octaves. The whole range is rarely covered by a single drive unit, because the physics of sound reproduction usually dictates that a crossover occurs in the region.

A 500 Hz wavelength is around 67 cm, which is larger than the diameter of most low frequency loudspeakers. This means that low frequency loudspeakers are generally small compared to all of the wavelengths that they are radiating. By 4 kHz, however, the wavelength is only about 8 cm, which is small by comparison to the diameter of the typical cone or dome loudspeakers that would be needed for high level radiation (110 dB SPL+) of signals in the

500 Hz region. Because of the off-axis phase cancellation for 8cm wavelengths being generated by 20 cm diameter cones, the directivity would tend to become too narrow for studio use. There are therefore problems in covering the whole mid frequency range with a suitable drive unit for high level, high-quality monitoring, although the horn-loaded Kinoshitas go from about 500 Hz to 20 kHz with a single, though phenomenally expensive, compression-driver/horn combination. (See Figure 20.1(a).)

20.8.5.1 Cone drivers

Cones and domes are both used in domestic high fidelity loudspeakers covering the entire mid-range, but such small units can rarely be capable of dissipating the waste heat in high-level systems. Cones have always been a popular choice for mid-range drivers, and they have the rather fortunate tendency to break up in such a way that the higher frequencies are often decoupled from the outer edges of the cone, and thus tend to radiate from the smaller diameter central section. They can thus maintain their wide directivity as the frequency rises. Figure 20.11 shows two small mid-range cone drivers covering the mid frequency range in a high power monitor system, but here they have been fitted with 'noses' and mounted in a shallow waveguide. The combination of the nose and waveguide lead to what is really a form of horn loading, which helps to augment the sensitivity of the system and also helps to control the directivity, both of which are desirable assets. Such a system of mounting provides

Figure 20.11:
The Genelec 1035B monitor loudspeaker. Note the 'noses' in the centre of the small mid frequency cone loudspeakers, and the sculpted waveguide into which they are set.

the option of rotating part of the baffle through 90°, thus allowing the use of the cabinet either horizontally or vertically. Generally, the restriction of the use of cone drivers in high-level studio monitor mid-range use is because of the limit to the amount of heat that can be dissipated by the relatively small coils that tend to be used. Theoretically, however, there would seem to be no reason why small cones cannot be driven by larger than normal diameter voice coils, as has been pointed out by Watkinson.[6]

Colloms[7] has noted that some cone materials can lead to unpleasant distortions at high levels due to gross compression and rarefaction stresses at the neck of the cone. This effect is most noticeable with certain types of plastic cone materials. He has also noted that the choice of cone material requires serious subjective testing, as well as objective analysis, because standard mid-range cones generally operate in a controlled break-up mode. This inevitably adds a degree of colouration to the sound which is characteristic of the particular material employed. 'A sonic signature of the chosen materials technology', as he puts it.

20.8.5.2 Dome drivers

Hard dome drivers, such as the beryllium dome used by Yamaha in the NS1000 loudspeaker, can produce outstandingly low distortion levels, but once they do break up, at high levels, they can do so rather aggressively. In high level monitoring, the *soft* dome has been fashionable now since around 1980.

The serious soft dome mid-range concept dates back to the work of Bill Woodman at ATC in England in the late 1970s. His 75 mm diameter unit was revolutionary in its time, and it could cover the range of about 400 Hz to 5 kHz with high output level, low power compression, low distortion and reasonably high efficiency. The dome was made of a doped soft fabric with a coil attached to its perimeter. A dual suspension system, somewhat unusual in previous domes, ensured a good pistonic performance. This dome was also used in early Quested monitoring systems. Stanley Kelly later designed a 100 mm unit for use in the Discrete Research 'Boxer' systems. ATC and PMC also had considerable success with their monitor systems using soft dome mid-range units in the 1990s and beyond.

The soft domes spawned an entirely erroneous, widespread belief that they had significantly better sound dispersion than cones. As previously mentioned in Section 20.8.4, their hemispherical shape suggested to many people that the sound radiated as in Figure 20.12(a), but in fact, as they move in one plane only, their action is pistonic, as shown in Figure 20.12(b). However, once they enter break-up, the *centre* of the dome uncouples from the high frequencies, contrary to the behaviour of a cone loudspeaker, because the dome is driven from its edges. The rather undesirable result of this is that the dome becomes a ring radiator, which lends to response beaming at high frequencies, because an annular radiator behaves like separate displaced radiators with a space between them, the centre of the ring being the null point.

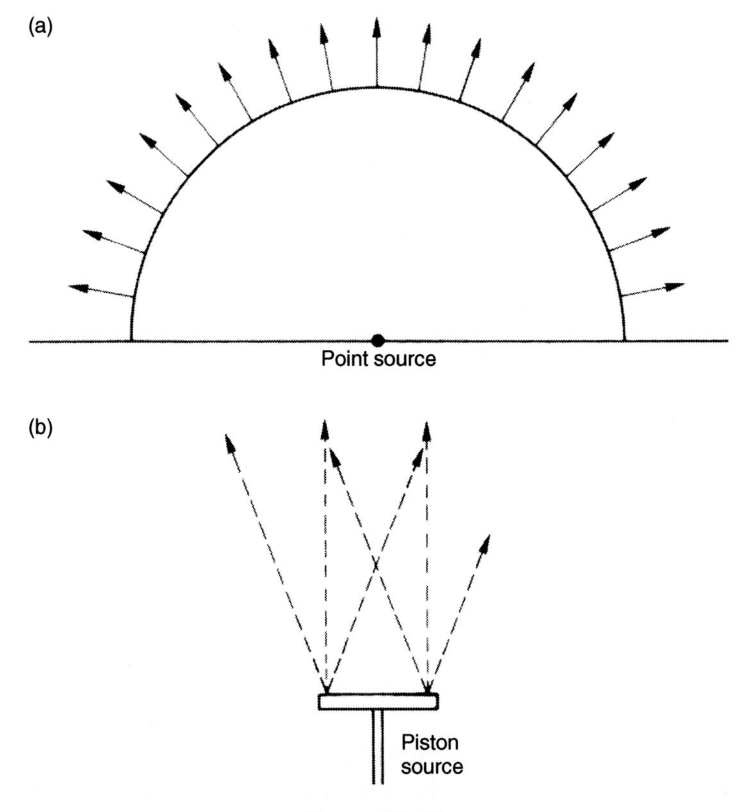

Figure 20.12:
(a) Spherical expanding wave with all particles moving away from each other; (b) single plane, pistonic movement, characterises radiation from dome loudspeakers. Typical directivity is that of equally sized cone loudspeakers.
The radiation shown in (a) would be typical of that from a pulsating sphere, and to some degree from many horns. A dome, however, although appearing like a hemisphere, does not pulsate. It moves backwards and forwards, in one plane only, and hence acts more like the piston shown in (b). At higher frequencies, if the centre of the cone decouples from the outer diameter, the driven portion acts like a ring diaphragm and radiates only from its perimeter. Radiating pistons produce particle motions that can interfere with each other, producing interference patterns and subsequent lobing of the polar pattern plots.

The more expensive dome radiators can produce excellent results, but they need careful attention to many aspects of their design. The lightweight domes are usually quite transparent to sound, so unless the cavity behind the dome is well damped, the rear radiation can colour the sound. Adequate damping requires expensive magnet systems, and these are sometimes hollowed and/or tapered to reduce resonance effects. Cheap 'look-alike' copies should be avoided, because the production of high-quality mid-range dome loudspeakers is difficult and expensive. The success of the better domes has largely been due to the ability to use large voice coils and large magnet systems, which have yielded high sensitivity and good heat dissipation. Good design also requires critical damping of the dome, itself, with viscous plasticisers.

The radiation shown in (a) would be typical of that from a pulsating sphere, and to some degree from many horns. A dome, however, although appearing like a hemisphere, does not pulsate. It moves backwards and forwards, in one plane only, and hence acts more like the piston shown in (b). At higher frequencies, if the centre of the cone decouples from the outer diameter, the driven portion acts like a ring diaphragm and radiates only from its perimeter. Radiating pistons produce particle motions that can interfere with each other, producing interference patterns and subsequent lobing of the polar pattern plots.

The enormous Community M4 loudspeaker has been used as a dome in some custom designed monitor systems. It was really designed as a compression driver, but it has seen use without its phase plug and mounted on a vestigial horn. Meyer, in some of their mid-field cabinets, has also used dome drivers at the throat of horns for the upper mid-range/high frequency part of the spectrum. In fact, most of the high power dome units employ a very short horn around the dome, which, although not appearing like stereotype horns, are horns nonetheless.

20.8.5.3 Mid-range horn loudspeakers

Perhaps nothing in the discussion of studio monitoring loudspeakers excites so much argument and passion as the question of the use of horn loudspeakers. For many people involved in recording, they are 'The Devil Incarnate', yet for many other people they are their optimal choice. It has to be said that many badly designed horn systems have been marketed over the 70 years, or so, of their use. Compared with the domestic hi-fi and public address markets, high-level studio monitoring is a very small market for loudspeaker manufacturers. It is therefore of little surprise that it has been forced to borrow much of its technology from the research done for the exploitation of the larger markets. Many large loudspeaker systems for studio use have directly adopted components from public address systems, and mid-range horns and compression drivers have been no exception. Despite the prime necessities of the two uses often being very different, many large-scale monitor systems have historically used unmodified components designed for sound reinforcement in concert halls and stadia. If the use of such systems has led to the widespread opinion that horns are undesirable as mid-range units in studio monitoring systems, it is not surprising. Then again, nor is it necessarily true.

Carefully and *specifically* designed horns for studio monitoring systems can be excellent performers, at low levels as well as high levels. Shown in Figure 21.3 is a monitoring system with a purpose designed mid-range horn/driver combination that does not possess a horn-like sound. However, there are limits to horn design which must not be transgressed if a natural sound is to be the result, and these can place restrictions on the shape and frequency range of the highest quality horns. In general, the rules are as follows:[8–10]

1. The distance from the diaphragm to the mouth of the horn should not exceed 35 cm, or thereabouts. (Magnet system design can determine the distance from the diaphragm to the throat, typically 0—10 cm.)
2. Horn mouths should flare as smoothly as possible into the front baffle, avoiding sudden discontinuities.
3. The flare rate determines the cut-off frequency, so to flare from a 25 bmm (1 inch) throat to a smooth encounter with the front baffle in no more than 35 cm of length requires a highflare rate, such that cut-off frequencies below 1 kHz are difficult, if not impossible, to achieve if the highest sonic resolution is the goal.
4. To minimise internal disturbances that can give rise to both on and off-axis response anomalies, all corners, angles and obstructions in the horn should be avoided.
5. Discontinuities in the throat region should be avoided, which makes the design of dual concentric arrangements somewhat difficult if they need to have gaps for the low frequency voice coils.
6. The compression driver throat should not diverge from the flare angle at the throat of the horn. The transition from driver flare to horn flare should be smooth. Therefore, replacing one driver with another of different internal throat flare will not give fair comparisons, because they cannot *both* be optimally matched.

The last of the points mentioned above means that not all 1 inch-exit drivers will match all 1 inch (25 mm) horn throats. The smoothness of flare rate across the junction is of great importance if honky sounds are to be avoided. In the past, some very expensive monitor systems have made disgraceful cross-sectional changes at the driver/horn junction. It was traditionally believed that changes in cross-section that were small compared with the wavelengths involved were inconsequential, but cepstrum analysis, as described in Chapter 19, has revealed otherwise.

Mid-range horns and their associated drivers can give an effortless reproduction at high SPLs which many other systems can have difficulty in matching. A typical mid-range horn would have a sensitivity of around 110 dB for 1 W at 1 m. This means that a mere 10 W would produce 120 dB at 1 m, and the typically huge magnet systems associated with such efficiency have no trouble in dissipating the few watts of waste heat. Thermal compression therefore barely comes into the question. Conversely, when listening at a distance of 2 m to music at a peak of 85 dB SPL, much of the musical signal would, in quiet passages, be in the order of 70 dB SPL. If 1W would produce 110 dB SPL at 1 m, then it would produce around 104 dB SPL at 2 m. Thirty decibels represents a power ratio of 1000:1, so 30 dB down on 104 dB would be 74 dB, and this SPL could be achieved with an input to the horn driver of one *milliwatt.* Seventy dB SPL would therefore be produced at 2 m with less than 400 *microwatts* input. As mentioned in Section 20.4, it has to be asked, how many mid-range horn systems have been damned by the relatively high levels of crossover distortion present in many amplifiers when producing only microwatts?' One cannot blame horns for faithfully reproducing the distortions that they are fed with.

It has been shown[8–10] that horns above about 40 cm in length can give rise to 'hornlike' sounds because reflexions from imperfect mouth terminations can become separated in time from the direct signal sufficiently to make the reflexions audible. Abrupt changes in angle anywhere around the mouth of a horn, as it meets the baffle, will almost certainly also give rise to reflexions. Many poor sounding horns have sharp angles at their mouths, so the poor sound is only to be expected. Furthermore, 'Constant Directivity' horn technology *relies* on the diffraction from abrupt angular changes in the flare, so it must also be concluded that as diffraction within horns produces reflexions, the Constant Directivity horns will be unlikely to yield the most audiophile results. However, they were not designed to do so, so their misapplication should not be judged as a criticism of the horns themselves, but of the persons misapplying them. Don Keele devised the constant directivity horns for speech reproduction, but some others have hijacked his ideas without his approval. The choice of compression driver is also highly critical, and not only must a good driver be chosen, but it must also be a good driver which precisely matches the horn.

In 99% of cases, or even more, the above rules have not been respected in mid-range horn application to studio monitor systems. That such misapplications have been made by some of the world's most famous loudspeaker manufacturers has done little to dissuade many (most?) people from thinking that the whole concept of the use of mid-range horns in studio monitor systems is taboo. The *real* facts show that the technology of horn loudspeakers still has much to offer, even at the very highest quality levels.

In fact, another great benefit of horns is that they radiate sound in a way very similar to that which would come from the ideal, point-source, pulsating sphere. They *do* radiate sound as shown in Figure 20.12(a). Despite the mouth size, the throat is the effective source. The sound waves leave the horn mouth in the form of a section of a spheroidal wavefront, free of the off-axis response dips exhibited by the cones and domes when the wavelengths become small compared to the radiating area. As frequencies rise, the polar pattern tends to gently narrow, rather than splitting into lobes.

Mid-range horns often receive a bad press, but this is so often due to misunderstanding and misapplication. When correctly applied they can achieve truly outstanding results. People suggesting otherwise are merely displaying their ignorance.

20.8.5.4 Ribbons, Heil air-motion transformers and Manger drivers

The basic concept of a ribbon loudspeaker is shown in Figure 20.13. The diaphragm is usually pleated in one direction to give it rigidity, and when an AC current flows through the metallic ribbon, it moves backwards and forwards as a whole, with no bending or flexing except at its extremes where it is coupled to its mountings. Modern technical advances have led to 'printed circuit' sheets, which behave similar to ribbons, but the traditional ribbons are usually made from aluminium, which is a good compromise between resistance and mass.

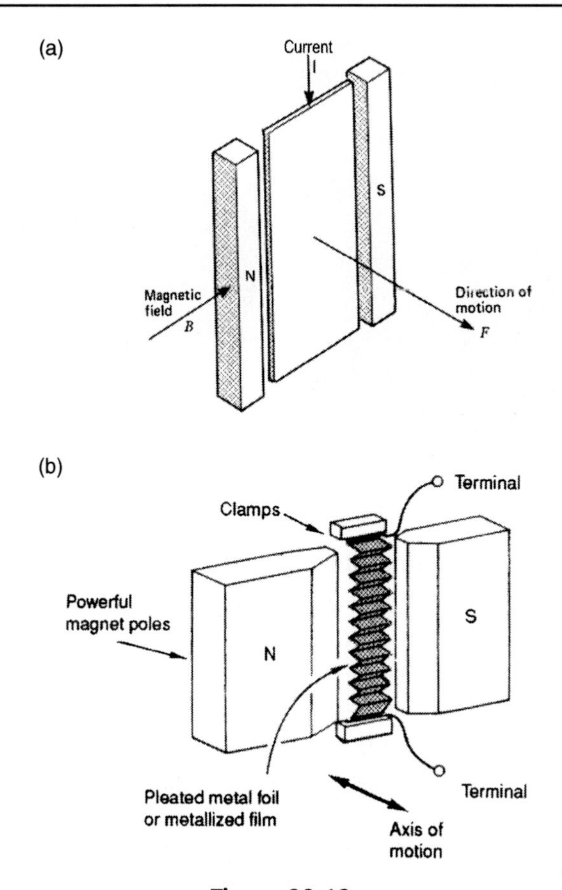

Figure 20.13:
A ribbon driver. (a) The basic concept of a ribbon driver. The magnetic field resulting from the current flowing through the diaphragm in the direction of the arrow reacts with the static magnetic field from the permanent magnets at either side, giving rise to a force, and hence a movement of the diaphragm, in the direction shown. (b) A more practical realisation of the concept.

Ribbons have always had a good reputation for sound quality, ever since the modern ribbon emerged in the 1950s, although as a concept they pre-date the moving coil loudspeakers. Their main drawbacks have traditionally been fragility and difficulty in manufacture, as with ribbon microphones, but some modern designs are proving to be much more robust. The American company SLS has made great use of ribbon loudspeakers above 2 kHz.

Often confused with ribbon loudspeakers are the Heil air-motion transformers. Their outward appearance is somewhat similar to a ribbon, but instead of moving backwards and forwards they are driven by a longitudinal, lengthening/shortening movement, somewhat like playing an accordion. The air is therefore pumped in and out of the folds as the current flows through a conducting track which gives rise to different magnetic field directions on adjacent folds. The diaphragms are typically made of p.t.f.e. or polythene.

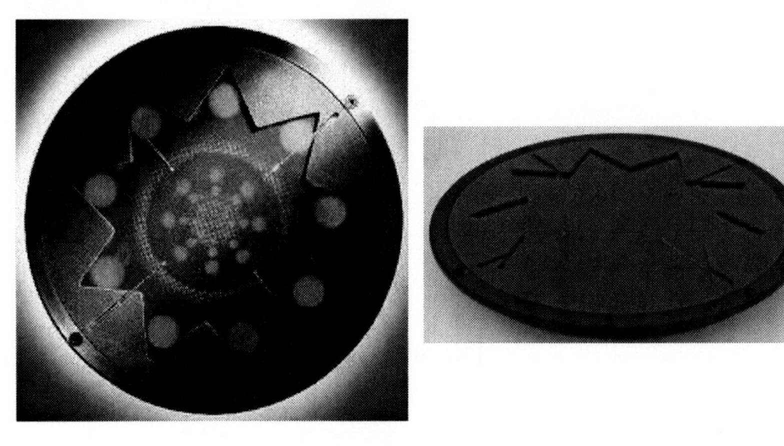

Figure 20.14:
The Manger driver.

They are called air-motion-transformers because there is a ratio of about 4 to 1 between the velocity of the air particle motion in and out of the folds and the movement of the pleated membrane itself. As with the ribbons, the magnet system must be large so that the whole membrane can fit into the magnetic field. In general, their high-frequency distortion is greater than that of many other mid-range devices, but the distortion products tend to be above 10 kHz, and their audibility is debatable. The German company ADAM has made great use of this technology, but despite their marketing names, they are *not* ribbon transducers.

Like the Heil devices, the Manger Sound Transducer is also a German invention. It is a bending-wave transducer which uses a flat diaphragm with a serrated termination which mechanically implements the necessary delays to create a phased array. Quite remarkably, the usable frequency range extends from about 80 Hz to 35 kHz, and the driver has been successfully used in a large monitor system from 800 Hz upwards, with a remarkable purity of sound and considerable robustness (see rear cover of book).

The lack of colouration in the sound from these drivers, together with the exceptionally fast transient response, makes them ideal candidates for monitor loudspeakers. Whilst they *are* moving coil devices, the voice coils have a diameter of 19 cm, and so the diaphragms are driven from the perimeter. The Manger drivers exhibit several of the desired characteristics of electrostatic loudspeakers, but with a peak output capability of around 114 dB at 1m. Figure 20.14 shows two views of the Manger driver, which is about 20 cm in diameter.

20.9 Review

This long chapter has been a relatively in-depth discussion about many of the aspects of studio monitor systems which affect the decisions about their use and installation in high-quality recording studios. They are highly complex systems with much interaction between their component parts, to such a degree, in fact, that artistry will always come into their design, especially because the end results must ultimately face subjective assessment. A very great deal of what is said about studio monitor systems is based on myths and hearsay. The triumphs of ignorance and marketing over electroacoustics and psychoacoustics are multitude. With the popularisation of the recording industry, and the extravagant marketing claims of commercial manufacturers, it is hardly surprising that many facts get buried in hype, or even get flatly denied. However, the truth still exists, and the people who want to hear it are perhaps the ones who will be better prepared to make high-quality recordings in high-quality studios.

20.10 Summary

The monitor systems of a recording studio consist of the entire electrical, electro-mechanico-acoustic, and acoustical signal path from the monitor circuits of the mixing console to the ear of the listener.

A professional studio should have available for its clients a full-range, high-definition monitoring chain.

Some mixing-console monitor return circuits are sonically questionable. Comparing a signal source via the monitor returns, and then directly into the crossover or amplifier inputs, often reveals the shortcomings of the mixing console.

Test for sonic purity must be made with very high-quality signal sources and converters, because if the source is not of the highest resolution, any subsequent loss of resolution may not be noticed.

The monitor system is the window through which to view all that is recorded.

Given the wide range of loudspeaker specifications and performance, power amplifiers should be carefully matched to the circumstances of use. On high sensitivity loudspeakers, the amplifier performance below one milliwatt of output power may be highly relevant.

Multi-amplification is less demanding on almost all of the component parts of a monitor system, and is an excellent way of minimising intermodulation distortion.

It is unwise to use valve amplifiers for monitoring, because any pleasantness which they add to the sound is not a property of the recording. The recordings should be monitored for what they are.

The cables from the amplifiers to the loudspeakers should ideally be kept below 2 m in length.

The minimisation of resistance and inductance in the loudspeaker cables should be a prime goal.

Fuses should not be used in loudspeaker circuits, because they can add distortion and act as compressors.

The avoidance of the full frequency band currents being passed through any single loudspeaker cable is generally beneficial.

Crossovers tend to introduce non-minimum-phase group delays into a monitor chain. The design of good crossovers is not trivial, and they tend to be specific to each type of loudspeaker system. 'Off the shelf crossovers are usually not suitable for the highest quality monitor systems.

Active crossovers are usually preferable to passive crossovers for a multitude of reasons.

Digital crossovers may seem to be an inviting technical solution to the problem, but when the cost of very high sampling rate, high bit-rate, and highly sonically neutral converters is considered, they can be prohibitively expensive. A cheaper digital crossover whose converters limit the monitor chain neutrality is not a sensible option.

For a given low frequency extension, loudspeaker cabinet size and efficiency are proportional – more size; more efficiency.

Differences in loudspeaker cabinet construction can require different conditions of mounting.

A widely accepted rule is to mount the loudspeaker cabinets with the baffles vertical, and the acoustic centres about 147 cm above the floor.

The different types of cabinet – open baffle, sealed box, bass reflex, etc. exhibit different rates of low frequency roll-offs, and different transient responses.

For low distortion at high levels, large radiating surfaces that move slowly are preferable for low frequency loudspeakers. Doppler and intermodulation distortions tend to be reduced.

Low sensitivity loudspeakers tend to suffer more from thermal compression.

Magnet and diaphragm/cone materials can exhibit significantly different sound qualities.

Mid-range horn/compression-driver combinations, if applied within strict limits, can give very high-quality sound reproduction.

The bad name associated with many horn loudspeakers has been a result of misapplication and non-compliance with the rules.

References

1 Newell, Philip, *Studio Monitoring Design*, Chapter 6, Focal Press, Oxford, UK, pp. 74–5 (1995)
2 Colloms, Martin, *High Performance Loudspeakers*, 5th Edn, John Wiley & Sons, Chichester, UK, pp. 262–8 (1997)
3 Rodgers, Carolyn Alexander, Ph.D. Thesis, 'Multidimensional Localization', Northwestern University, Evanston, IL, USA (1981)

4 Bank, Graham and Wright, Julian, *Loudspeaker and Headphone Handbook*, 3rd Edn, Borwick, John (ed.), Focal Press, Oxford, UK, p. 307 (2001)
5 Newell, Philip, *Studio Monitoring Design*, Focal Press, Oxford, UK, p. 227 (1995)
6 Watkinson, John, 'Transducer drive mechanisms' in Borwick, John (ed.), *Loudspeaker and Headphone Handbook*, 3rd Edn, Focal Press, Oxford, UK (2001)
7 Colloms, Martin, *High Performance Loudspeakers*, 5th Edn, John Wiley & Sons, Chichester, UK, p. 213 (1997)
8 Holland, Keith R., Fahy, Frank J. and Newell, Philip R., 'The Sound of Midrange Horns for Studio Monitors', *Journal of the Audio Engineering Society*, Vol. 44, No. 1–2, pp. 23–36 (Jan/Feb 1996)
9 Newell, Philip R. and Holland, Keith R., 'Do All Mid-Range Horn Loudspeakers Have a Recognisable Characteristic Sound?', *Proceedings of the Institute of Acoustics*, Vol. 12, Part 8, pp. 249–58 (1990)
10 Holland, Keith R. and Newell, Philip R., 'Axisymmetric Horns for Studio Monitor Systems', *Proceedings of the Institute of Acoustics*, Vol. 12, Part 8, pp. 121–28 (1990)
11 Self, Douglas. Active Crossover Design, Focal Press, Oxford, UK (2011)

Bibliography

Borwick, John, *Loudspeaker and Headphone Handbook*, 3rd Edn, Focal Press, Oxford, UK (2001)
Colloms, Martin, *High Performance Loudspeakers*, 5th Edn, John Wiley & Sons, Chichester, UK (1997)
Duncan, Ben, *High Performance Audio Power Amplifiers*, Newnes, Oxford, UK (1996)
Eargle, John, *Loudspeaker Handbook*, Chapman & Hall, New York, USA (1997)
Newell, Philip, *Studio Monitoring Design*, Focal Press, Oxford, UK (1995)
Newell, Philip and Holland, Keith, 'Loudspeakers – for music recording and reproduction', Focal Press, Oxford, UK (2007)
Self, Douglas, *Audio Power Amplifier Design Handbook*, Focal Press, Oxford, UK (2009)

CHAPTER 21

Surround Sound and Control Rooms

Cinema surround – its origins. TV surround – the differences. Music-only surround – its needs. Stereo and surround – the compromises. Perception and surround. Fold-down complications. Rear channels – the different concepts. Low frequency options. Close-field options. Study of an actual design. Dubbing theatres. Room compatibility issues. The X-curve. Room equalisation: the concept and the problems.

At least as far as the music-only market is concerned, the concept of surround sound seems to have slowly descended into a mire. In 2011 it has become very difficult to find high-fidelity surround recordings in any commercially available format. What exists has largely become only limited-interest, special-order-only items. Whatever is currently available as soundtracks to video and film releases is almost invariably data-compressed in a lossy manner (meaning that all the original quality of the recording *cannot* be reproduced), and hence does not comply with the concept of high fidelity. What is more, professionally, the market for music-only surround control rooms has largely disappeared. Whatever surround rooms that are currently being built in recording studios are mainly in the form of post-production rooms, with arbitrarily placed loudspeakers, and often with no means of even reproducing SACDs or DVD-As (96 kHz, audio-only DVDs). Sub-woofer positioning and calibration is haphazard, and response uniformity is almost non-existent. Remote gain controls are even found for sub-woofers in professional facilities, so that low-frequency levels can be 'adjusted to taste' between one recording and another, or to 'reference' recordings which, themselves, were probably mixed in equally arbitrarily calibrated rooms. Attitudes have degenerated to a point where there is almost no reference for making surround mixes, and domestic listeners are left to their own devices to get whatever they can out of a recording.

Nevertheless, this chapter will look at the different surround concepts available, both for music-with-picture, general cinema, *and* music-only, and an attempt will be made try to understand what *could* be achieved if only the music industry were interested. It should become obvious that perhaps one reason for the currently sad state of affairs is that many of the concepts were over-simplistic in the way that they were 'sold' to both the industries and the public. The reality of trying to achieve *true* high fidelity surround sound is something which is very difficult and expensive to put into practice. Unfortunately, this did not coincide well with a post-2000 world which was heading in the direction of asking for 'more, and cheaper' rather than for 'better quality', even if the latter cost only *a little* more.

Historically, it was the cinema industry which led the way in the serious use of surround sound. Walt Disney's animation film 'Fantasia' featured multi-channel surround audio 30 years before the record companies made their first, albeit short-lived, forays into quadrophonics in the early 1970s.

21.1 Surround in the Cinemas

Dolby did the industry a great service many years ago by clearly specifying their intentions for their cinema systems and issuing adequate guidelines to mixing personnel. They did not try to establish a system that would do all things for all people. Instead, they chose systems that would, within a restricted range of use and in reasonably controlled acoustics, give a good representation to the cinema-goers of the film directors' creative wishes. They defined the reasonable limits of surround reproduction at an early stage, and history has shown them to have been remarkably insightful given the chaos of quadrophonics from which they developed their early systems.

A few important facts regarding mixing for Dolby cinema surround are given below.

1. The most important signals will usually be coming from the front, because the action is on the screen, and the directors do not want people to be distracted from it by important sounds coming from rear locations. This would tend to excite the human reflex response of turning to face the sound, which may be warning of danger in real life, but it is not much use in a cinema, except in IMAX, perhaps.
2. The mixes will be played back in theatres which comply with predetermined acoustic and electroacoustic requirements (although the degree of compliance is somewhat variable)
3. Due to the sheer cost of film production, mixing will usually be carried out by skilled, experienced, and knowledgeable personnel, who will almost invariably limit their creative exuberance to getting the very best out of what they have to work with. They will not get too carried away creating wonderful effects in the mixing room which may be detrimental to the experience in the theatres. The success of a film tends to ride on how many people see the first screenings in the cinemas and recommend the experience to their friends. Creating effects which are not universally obvious for all the audience, or at worst are confusing, does not help this situation.
4. The objective is usually to create a one-off, big performance, and it generally means that high-quality programme material and equipment are available to achieve the high standards that are normally expected.
5. The whole concept of the various Dolby surround formats (and others which follow generally similar guidelines) is to deliver a balanced programme to a group of people, with no particular 'sweet spots'. This they have proved themselves very capable of doing, over many years and around the world.

21.2 TV Surround

TV and video, at first consideration, seem to need the same general requirements as cinema because once again the final audience will be sat in front of a screen with moving pictures, and will be surrounded by (usually) five loudspeaker channels. However, there are other things to consider with regard to domestic 'viewing'.

1. Reproduction will usually *not* take place with SPLs of over 90 dB.
2. Background noises in the listening environment will usually be higher than in cinemas.
3. Because of the two above points, dynamic ranges will be much more restricted.
4. Many broadcasting companies have their own set of standards, and some times are forced by international agreements to follow standards that are not always beneficial for the sound quality.
5. Reproduction will be presumed to be largely on poor-to-medium quality domestic loudspeaker systems, and getting a reasonably good sound for all is fundamental if the revenue of advertisers is what keeps the TV channels in business. The high impact of cinema audio is neither a requirement nor a practicable concept for TV mixing.

The above, and many other points, generally mean that rooms designed for TV or domestic video production will be much more compromised in terms of overall sound quality. It is no doubt that it is the existence of so many restrictions which often leads to a lamentable attitude about sound quality from many people in the world's TV industries.

One therefore cannot easily mix for surround TV in a room designed for big-screen, high dynamic-range productions. Neither would it seem wise to try to mix audiophile quality music recordings in rooms designed for TV/video sound mixing, where many of the necessary subtleties for music mixing would almost certainly not have been considered during their design. This does highlight the fact that surround room design becomes very specific to the goals of the reproduction circumstances, and gives some insight into why the variations shown in Figures 14.10 to 14.23 have come into existence.

21.3 Music-Only Surround

Now we come to the *big* question; rooms for mixing music-only, high fidelity surround. Unlike in the cinema, TV and video worlds, there are no set standards. In rooms for high-quality *stereo* music mixing there are several schools of thought. There are the 'Live-End, Dead End' types of rooms (see Chapter 17), and earlier designs by people such as Jensen, which were intended to give a clean first-pass to the sound from the loudspeakers, with a diffuse room decay from the rear, and to varying degrees from the sides. The 'Non-Environment' concept (see Chapter 16) attempts to make the room as anechoic as possible to the loudspeakers. Only the floor and *front* wall are reflective in order to give life to speech and actions in the room, thus avoiding the creation of an uncomfortably dead ambience in which to work. Sam Toyoshima and Eastlake Audio both opt for something that is rather similar to the Non-Environment concept, but with a careful distribution of reflective surfaces (principally for the higher frequencies) scattered about the room to add a touch of extra ambience. The rear walls are highly absorbent in both cases. It has to be said that well-designed and well-constructed types of *all* of the above rooms, and others besides, in the hands of skilled recording staff can produce stereo recordings of the very highest quality. None of them are perfect though.

As designers have their own individual hierarchies of priorities, each type of room has its own special strengths, such as the resolution of ultra-fine detail, the ambient 'feel' of the room, the low frequency behaviour of short duration sounds, or other aspects of performance. Nevertheless, all of the better rooms of each type, when the recording staff know the rooms, can produce results which travel well to domestic situations. However, what they all have in common is that they are bi-directional: the front halves differ from the rear halves. In no case can one simply transfer the monitor loudspeakers to the rear wall and still have a room that sounds as good. Yet, this is exactly where rear loudspeakers are needed in a surround room that is intended to have fully symmetrical five-channel monitoring. Therefore, if all of the very best stereo mixing rooms need bi-directional acoustics, this would seem to lead to only two possible conclusions.

1. In surround rooms with bi-directional acoustics (like the good stereo rooms), the rear loudspeakers responses will not be as good as the front loudspeaker responses.
2. In surround rooms that do have fully symmetrical acoustics and monitoring, the frontal stereo cannot be as good as in the best bi-directional stereo rooms.

It would seem that the first choice would be the better compromise for high-quality music mixing, because even in a surround mix the frontal sound stage is the most important in 95%, or more, of recordings. To compromise our current high-quality stereo for a more enveloping sound is to trade quality for quantity. The second option described above is reminiscent of an old comedy on British television many years ago, about the textile industry, called 'Never Mind the Quality, Feel the Width'. Some surround room designs seem to be saying, 'Never mind the quality, feel the space'. Whether or not that is a backward step is perhaps a subjective issue.

In the classical recording world, the orchestral *layout* is a fundamental part of the music, which is designed to deliver the composers' emotions *to the audience.* Nobody *within* the orchestra hears the true, intended balance of instruments, so having the orchestra wrapped around the listener does not seem to be a worthwhile goal. This suggests that for most classical recordings the surround channels will be delivering purely ambience, and only rarely for special effects will they be reproducing the direct signals from musical instruments. In most classical works where off-stage sounds *are* used, they are usually intended to be ethereal in quality, so they will be of an ambient nature, which is well suited to surround in the bi-directional rooms. These rooms may not be appropriate for a close-mic'd wraparound orchestral recording though, because of the lack of 'symmetrical' monitoring.

Based on all the above, it would appear that the design of surround mixing rooms for music-only recordings should not differ fundamentally from the design of rooms for Dolby cinema mixing. The object should be to try to achieve the best that *can* be achieved from the rear channels without compromising the response of the frontal stereo, be it two, three, or five channels. The only significant difference in design may be in the choice of the front wall materials in the absence of a screen. In surround rooms, the front walls still need to be solid to act as baffle extensions for the loudspeakers, but they should also be irregular to break up any specular (discrete) reflexions of the sound striking them from the rear loudspeakers.

The most high fidelity (i.e. highly faithful to the original) surround sound reproduction can only really take place in anechoic surroundings, but even the craziest of surround audiophiles are unlikely to have anechoic chambers for domestic listening. Nevertheless, it would seem to be a professional attitude for the music industry to try to generally monitor in conditions that are better than could be expected in domestic reproduction because, eccentric as many audiophiles are widely deemed to be, it is hard to deny their right to expect that the recording standards should live up to their equipment standards. There is no justification for the widespread television attitude in a professional music recording industry, but there now seems to be ominous pressure (such as from MP3) to lower many of the general quality aims. This is a risky situation though, because if compromise *is* forced on the end result, enthusiasm will be lost at the production end of the chain, and that would be ruinous for creativity. Standards erosion must be resisted!

To maintain the quality levels of *stereo* hi-fi in commercial surround recordings requires vigilance and discipline. Producers and engineers are likely to be disappointed if they expect to be able to do whatever they want with the distribution of instruments whilst hoping to produce results of a quality equal to the current quality of the best stereo. (And to be even more disappointed if they expect equally balanced reproduction in most people's homes.)

There is a huge carrot dangling in front of studio designers to produce 'the ultimate quality' in symmetrical surround mixing rooms. Time will surely find the rooms lacking, though. This is because achieving fully symmetrical five-channel monitoring with no compromises to the level of quality of the frontal two-channel stereo; *and* to have the rooms also usable for multi-format mixing (cinema/music/TV, etc.), basically cannot be done. Control room design for surround mixing is therefore a very format-based concept. Until we get more rationale than the situation shown in Figures 14.10 to 14.23, the difference in the concepts of surround control rooms will lead to great confusion, with people being tempted to do the wrong things in the wrong rooms; or even the wrong things in any room.

21.4 An Interim Conclusion

From many points of view it would seem that the best answer for the future of surround is for everybody to follow the basic five-channel cinema format. It has many things going for it.

1. It is proven.
2. It is widespread.
3. It was created by professionals after a great deal of knowledgeable and rational thought.
4. It *can* do (or is capable of doing with little adaptation) almost all that surround can reasonably be expected to do.
5. Much of what it will not do is not possible anyway, which becomes clearer when the finer points of surround are better understood.
6. It requires no separate systems for music and cinema.
7. It is more room-tolerant than many other systems.
8. It does allow some flexibility in the choice and location of the rear/side loudspeakers.

It is worth remembering there are many things that one can *dream* of doing on other systems, but there is not much that one can actually do *well* which cannot be supported by the current cinema formats.

21.5 The Psychoacoustics of Surround Sound

The relatively stable phantom images of conventional stereo only function well when they subtend an angle of about 60° centred on our noses. This fact was grossly under-appreciated during the quadrophonic era of the 1970s. An enormous number of people (the author included) attempted to pan musical signals to all points around the room on quadrophonic pan-pot 'joysticks'. It was a great source of mystery to many people why the front—left/rear—left panning only stabilised when at fully front or fully rear positions, with all points in between yielding images which flip-flopped from one extreme to another at the slightest movement of the head. In fact, it is only in the horizontal frontal listening arc where we generally have good localisation. Over this arc, it can be *very* good, with many people being able to resolve differences in position to an accuracy of 1° for some sounds. When a loudspeaker is placed *behind* the head, the accuracy of localisation is reduced, and phantom images between two or more loudspeakers cannot be stable. The perceived frequency response will also be different from that perceived from similar loudspeakers in front of the listener.

In surround listening, whether to live instruments or recordings via loudspeakers, it is a fact that perceived tonality will change according to source position. However, in many people's heads, there seems to be an idea that if *symmetrical* monitoring conditions can be achieved, then this will lead to more ideal surround sound monitoring. A big problem arises when surround mixes monitored under such circumstances are heard in stereo or mono. What was perceived as correct when heard from behind may be inappropriate in both level and frequency balance when folded down to fewer channels and heard from a frontal direction.[1] In such cases, it is hard to see why distributed rear loudspeakers, such as those used in the cinema world, would be at any great disadvantage to discrete loudspeakers, because neither relate perfectly to the perception of a folded-down mix.

It has been recognised that the ambient or discrete-instrument type of surround mixes are better served by correspondingly distributed or discrete loudspeaker sources at the rear, but the existence of these conflicting concepts is largely due to the fact that it has also been widely recognised that five channels are not enough to provide the best of both worlds. Ten channels are seen to be a minimum for truly flexible surround sound systems, but it is deemed too unwieldy to be commercially acceptable.

When used for ambient mixes, the two-discrete-loudspeaker option for the rear channels often does not work as well as some of the other rear channel loudspeaker options (see also next section). Therefore, the whole concept of discrete rear loudspeakers does not appear to have much in its favour (except for the fact that market pressures are pushing for it), but as we shall see later, it can sometimes win by default.

Dolby cleverly recognised these conceptual weaknesses in the 1970s when they introduced 'Dolby Stereo', which was really a four-channel surround format (see Figure 14.11). In those days of analogue-only technology, the quadrophonic vinyl-disc systems used rather poor phase encoding techniques to try to put four channels on a stereo disc. Dolby saw the nonsense of it all, but then effectively used a similar phase encoding technology to produce not the left—front/right—front/left—rear/right—rear of quadrophonics, but a left—front/centre—front/right—front/single-channel-rear format. They put one of the loudspeaker channels at the centre—front location, which stabilised things immensely, and split the mono rear signal between several loudspeakers, widely distributed, which could give quite a spacious ambience. For what it was, the system worked well. Once again, therefore, the cinema people moved one jump ahead of the music industry on the subject of surround. The advent of this system could have saved quadrophonics, but the music business was already largely fed up with it, and a generation was to pass before they again attempted to re-invent the rather useless 'square wheel' under a new name — surround. With digital technology, the cinema world soon moved on to three front channels and distributed *stereo* surround, which represented yet another march forward.

21.6 Rear Channel Concepts

It has been shown that multi-channel surround is more realistic than two channel surround (surround here meaning the channels other than the front channels), although two surround channels can be made more effective as ambience channels by using diffuse source techniques rather than two, simple, discrete loudspeakers. Tomlinson Holman proposed another concept for domestic use; single dipole sources to each side and slightly to the rear of the listener (see Figure 14.17).[2] These sources presented their nulls to the listener, who would receive only reflected energy from the room loudspeakers, and hence a more diffuse sense of spaciousness. However, this concept will only work in rooms with reasonably reflective surfaces. In a dead room, the surround channels would almost disappear.

David Bell successfully used single discrete loudspeakers as the rear channels in small post-production rooms by hanging them from the ceiling, facing away from the listening position, but pointing directly at proprietary, wall-mounted diffuser panels to reflect energy back into the listening area.[3] The author successfully used distributed mode loudspeakers, which are naturally diffuse sources, in a film laboratory screening room that met Dolby specifications.[4]

These things are all ultimately subjective, but a large body of opinion believes that discrete instruments occasionally played through the above diffuse surround systems suffer less loss of realism than two-channel ambience played through a pair of discrete rear loudspeakers. The consumer market may ultimately dictate the *de facto* 'standards' to be used, but the diffuse sources referred to in the previous paragraph could well be more tolerant of listening room acoustics than the fully discrete, five identical loudspeaker option which seeks to achieve ostensibly identical monitoring from five locations, whether the sources are perceived to sound identical, or not.

Unfortunately, it seems that the consumer end is becoming ever more chaotic. The majority of people just assemble their systems and hear whatever they hear. In some ways, the concept of domestic 'fidelity' to an original recording quality appears to be vanishing. In fact, some of this is probably due to the lack of reliable guidelines for the production end of the industry. Just as the public got tired of the format wars of quadrophonic vinyl discs, it seems to be getting equally tired from the lack of clear leadership from the (now dwindling) surround-music-production industry.

21.7 Perceived Responses

Clearly, the five different surround-channel systems so far discussed will all tend to produce different responses, both measured and perceived, at the listening position(s). So, let us now consider the responses of the various approaches in different acoustical conditions.

21.7.1 The Simple Discrete Source

Taking the simple, single loudspeaker first, the response at the listening position will be dependent upon the mounting conditions and the nature of the surfaces that face the loudspeaker.

Let us assume that we have put a full, five-channel system of identical loudspeakers in an anechoic chamber, mounted at the recommended points on the perimeter of a circle. An omni-directional measuring microphone at the listening position would pick up the same frequency response from each source. On the other hand, a listener at the same position as the microphone would perceive less high frequencies from the rear loudspeakers than from the front loudspeakers, due to the pinnae (outer ears) being more responsive to high frequencies from a frontal direction. As there would be no reflexion of sound, the rear loudspeakers could be perceived as single discrete sources. Nevertheless, these conditions would be the ones under which the overall flattest response could be expected from all the loudspeakers.

In a stereophonic type of control room with a dead front end and a live rear end, the situation would be somewhat different. At high frequencies, the frontal loudspeakers would project directly into the ears of the listener, *and* excite the reflected field from the rear, but the rear loudspeakers would do neither. The perception of the rear loudspeakers would tend to be dull and lifeless. The situation at low frequencies would depend on the mounting conditions. If all the loudspeakers were flush mounted, the front loudspeakers would, hopefully, face a rear wall with enough low frequency diffusion to ameliorate the effect of standing wave resonances, but the rear loudspeakers would face a front wall which would typically be quite solid and reflective at low frequencies. Response disturbances could be expected due to the strongly reflected waves interfering with the direct waves.

If the loudspeakers were free-standing, then many of the same results would obtain, except that a less uniform bass response could be expected. The omni-directional low frequency radiation would travel behind the front loudspeakers and reflect off the front wall, again causing response irregularities due to the interference of the direct and reflected waves. The rear radiation from the rear loudspeakers would hopefully have a less disturbing effect due to the presence of the rear wall diffusers, but the forward moving radiation would still suffer interference from the front wall reflexions.

Somewhat differently, in a room with a wideband reflecting front wall and absorbent rear wall, such as described in Chapter 16, the rear loudspeakers would subjectively sound brighter than in either the anechoic or the dead front-end conditions, due to the high frequencies reflecting from the front wall and back to the listener from a forward direction. A measurement microphone at the listening position would show a less flat mid-range response than in a room with an absorbent front end, due to the interference of the direct and reflective waves. At low frequencies, the mounting conditions would give rise to different responses from front and rear loudspeakers, due to the rear absorber wall providing a less effective baffle extension to the loudspeakers than that enjoyed by the front loudspeakers (see Chapter 11). This could be legitimately equalised to some degree, but not precisely. For free-standing loudspeakers, the rear channels may be perceived to be flatter at low frequencies than in a room with a reflective/diffusive rear wall, due to the relatively non-reflective, adjacent rear surface.

Tomlinson Holman, and others, have proposed that rooms for surround should be of generally lower decay time than conventional listening rooms, both to avoid colouration of the recorded surround and to help to ameliorate the variable low frequency response problem. Less ambient 'help' from the room is needed for surround sound because the ambience is usually already recorded in the surround channels. Holman and others have also suggested that the room should be made more acoustically uniform by the careful and appropriate distribution of reflective/diffuse materials on the surfaces of the walls and ceilings. These are eminently sensible suggestions for a more uniform perceived response from all of the loudspeakers, but they cannot avoid the criticism that the frontal loudspeaker response, which is usually of prime importance in the vast majority of music recordings, cannot be as good as in the finest, acoustically bi-directional stereo rooms. However, the proponents of the ideas *were* talking about surround listening rooms, and not about stereo compatible control rooms. The compromising of the frontal responses to the benefit of rear responses may well be valid in the context of some surround-only rooms, but it must be accepted that a trade-off exists. The degree to which this trade-off is beneficial or otherwise may be heavily dependent upon programme material and personal tastes. We are also being driven towards mono sub-woofers.

21.7.2 The Multiple Distributed Source

This is the Dolby cinema approach. In anechoic conditions, the main difference between this and the discrete source concept is that it would still be perceived as a distributed source, because that is exactly what it is, (although the precedence effect could have a tendency to pull the image towards the nearest loudspeaker of a group). Depending upon the precise distribution of the loudspeakers, it could be the case that the distributed sources would sound brighter than the single discrete rear sources, because some of the loudspeakers could be expected to be more directly pointing towards the ear canals, as can be seen from Figure 14.13. The response of a measuring microphone at the listening position in anechoic conditions would be less flat than from the single discreet source because of the constructive and destructive interference due to even minor path length differences from the different sources. Subjectively, however, this may not be a problem at all, but a response down to 20 Hz could not be expected from the use of multiple, smaller loudspeakers.

In the case of either extreme of stereo control room design, whether the front walls be reflective at high frequencies or not, the response from multiple distributed sound sources would tend to be less different from that which could be expected from a discrete rear source in the different rooms. Surface mounting of the multiple distributed sources is normal, because flush mounting tends to be rather structurally complicated, and multiple free-standing sources tend to be too much of an obstruction to everyday work and general activities.

Subjectively, the distributed system seems to work very well, especially for ambient and special effect surround. When the rear channels are fed via a signal delay, as in the Dolby system, the effects can be very lifelike, because for any sounds in all the channels the precedence effect will tend to ensure that front-originating sounds cannot be pulled back into the surrounds. This can be effective even when a listener is closer to one of the surround loudspeakers than to the most distant frontal loudspeaker, as can be the case in many cinema and home theatre installations, especially where an audience of more than one person can often be the norm. For home use, though, the multiple distributed surround loudspeaker concept is cumbersome.

21.7.3 Dipole Surround Loudspeakers

This option is shown diagrammatically in Figure 14.17. Clearly, in anechoic conditions this choice would be somewhat of a nonsense, because with the nulls facing the listener (see Figure 11.3(a)) and no reflexions returning, not much would be heard. In a 'stereo' style of control room, the side of the dipole facing the hardest surface would give rise to most of the high frequencies. Obviously, therefore, the most suitable type of room for using such a system would be one with relatively evenly spread reflective surfaces, such as tends to be found in many domestic rooms. This is not too surprising, because Tom Holman initially proposed this technique for domestic use, where it does have a lot of potential, but it is hard to see how a flat frequency response could be expected at the listening position from such an arrangement using the purely reflected energy from arbitrary boundary conditions.

21.7.4 Diffuse Sources

For the purpose of this discussion, the true diffuse sources such as the DML (Distributed Mode Loudspeaker) and the approach of a discrete loudspeaker pointing at a wall-mounted diffuser can be lumped together (see Figure 14.21). They could both be expected to deliver a relatively flat response to the listening position, almost despite the nature of the room acoustics. The DML behaves rather differently from other loudspeakers in that the initial SPL drop with doubling of distance tends to be more like 3 dB rather than the conventional 6 dB. (Due to the source area.)

Diffuse sources have wide radiation patterns over an extended range of frequencies, and tend to suffer less from the effects of room resonance and standing wave interference. The DMLs do tend to suffer from a rather curtailed low frequency response, but they can be so advantageous as ambient sources that the extra effort involved in adding a common sub-woofer to them would seem to be well worthwhile.[5] Essentially, however, they consist of a radiating surface made from a material which exhibits a very dense modal activity spread throughout its surface. They are energised by moving coils, but not in the sense of a conventional magnet and chassis system. Despite being a mass of resonances, the early part of the impulse response is remarkably rapid. Since the late 1990s these loudspeakers have been causing some reassessment of conventional thinking, and in many areas they have been well received, especially in the creation of ambience effects.

Obviously, the discrete loudspeaker pointing at a wall-mounted diffuser will behave more like a conventional source at low frequencies, where the radiation pattern tends to be that of an omni-directional compact source. Its response in this region may differ from that of the DML, but the low frequency responses of surround systems in general are something of a minefield, so perhaps that is what we should now look at in more detail.

21.8 Low Frequencies and Surround

Figure 21.1 shows a typical layout for a Dolby Digital theatre. This is the archetypal 5.1 system, where the 'point-one' (0.1), or low frequency effects channel, is fed to a dedicated sub-woofer system. In cinema mixing for Dolby Digital, DTS (Digital Theatre Systems) and SDDS (Sony Dynamic Digital Sound), what goes to this channel is determined by the mixing engineer. However, in the Dolby Stereo analogue system (which despite its name is a matrix *surround* format), the sub-woofer is fed from a low frequency management system in the processors, somewhat like in domestic 'home theatre' systems.

It will be noted from Figure 21.1 that the sub-woofer is set off-centre. This is done to try to avoid driving the room symmetrically, where the tendency would be to drive fewer modes more strongly due to the equal distance from a centre loudspeaker to the two side walls. The off-centre arrangement tends to produce more response peaks and dips but of lesser magnitude than would be the case for a symmetrical drive. The degree to which the off-centre location of a sub-woofer can be detected by ear is usually a function of the upper frequency limit. Below 80 Hz it is generally very difficult to detect the source position, but as the cut-off frequency rises, the low frequency source position can become more noticeable. Below 50 Hz, localisation is impossible.

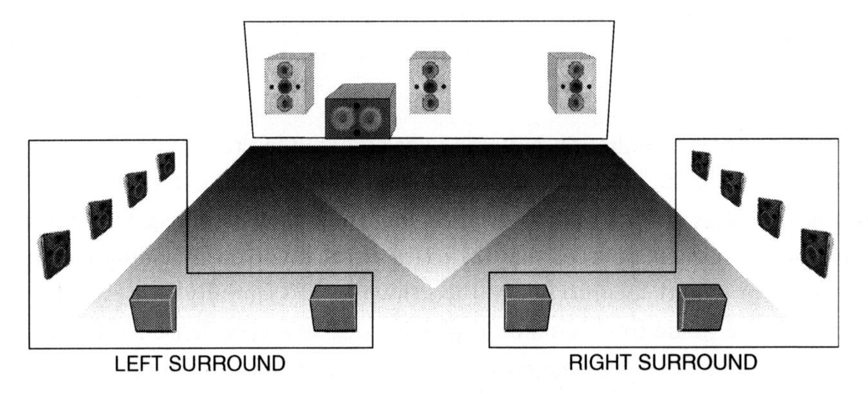

Figure 21.1:
Three-dimensional conceptualisation of a Dolby Digital theatre.

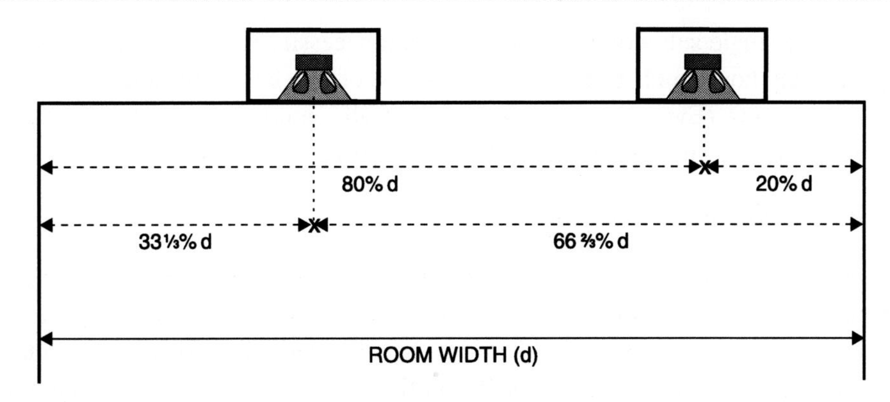

Figure 21.2:
Dolby recommendation for siting two sub-woofers; one of them one-fifth of the room width from one side wall, the other one-third of the room width from the other side wall. This not only avoids the localisation of a single sub-woofer towards one side of the room, but also avoids the symmetrical driving of room modes by the central placement of the sub-woofer(s).

To eliminate noticing the off-centre source location whilst maintaining an asymmetrical room drive, Dolby now recommend the use of two sub-woofers, fed from the same electrical signal. One should be placed one-third of the distance across the room from one side wall, and the other placed one-fifth of the distance across the room from the opposite side wall, as shown in Figure 21.2. The use of one or two large sub-woofers in such cinema installations takes into account the fact that cinemas, whilst being designed to at least *reasonable* acoustic criteria, are not so heavily controlled as music control rooms. Fewer low frequency sources are therefore more practical than three or five full-range sources. They also address the need for a large area, relative to the size of the room, to be covered by a respectably even sound-field, so that the paying customers all receive their money's worth. The existence of sweet seats and poor seats, acoustically speaking, would not be commercially viable. Large sub-woofers also offer the power required for explosions, and other sound effects that would tax full-range loudspeakers to their limits in larger theatres.

There is thus little in common between the dedicated 'low frequency *effects*' (LFE) channels of digital cinema and the low frequency *extension* (LFE) sub-woofers used in Dolby Stereo systems. In fact, just to confuse matters further, the high definition television standards have another LFE definition – Low Frequency Enhancement! The first and third of these LFEs are discrete channels; the second is not. However, the first is obligatory for reproduction because it may contain signals which are both important and *not* appearing in the other channels, whereas the second and the third do *not* contain 'essential' material (just some extended LF) and so their reproduction is optional. None of this confusion helps the general acceptance of surround sound. Somewhat disgracefully, there is no compatibility between the three LFE concepts. This is an insult to the general public!

Domestically, the single sub-woofer is a commercial necessity resulting from the great reluctance of many households to accept the presence of five large full-range loudspeakers in domestic living rooms. (Even if they *could* find sonically good places to put them that did not block a doorway or cut half the light from a window.)

21.8.1 Music-Only Low Frequencies

In the mixing of surround sound for music only there is no such luxury as specifying the means of playback and the environment in which it will be heard, let alone any means of enforcing compliance with any specifications. Sub-woofers, therefore, tend to add a complication to the mixing environment if traditional, full range monitors are not used. They add another crossover point, and mono sub-woofers lack the undoubted benefit which stereo bass can add to certain music, especially in terms of the spaciousness, which is, somewhat perversely, surround sound's *raison d'être.*

Nevertheless, the situation still exists that a mix done on one concept of system is likely to sound quite different when played back in a different control room, when the number of sub-woofers can vary from zero to four, and the full range loudspeakers (if used) can vary (normally) between two, three and five. In fact, the four-channel (no centre-front) option is another possibility, and does still see use, albeit in a slightly modified geometry from the old quadrophonic (square) layout. (See Figures 14.10 and 14.11.) At the time of writing (2011) some people are still mixing to a four-channel format which resembles the five-channel format of Figures 14.19 and 14.20, but without the centre-front loudspeaker.

It would seem that the optimum arrangement for a music-only control room for the highest quality monitoring would use five full-range loudspeakers and no sub-woofer. However, Chapter 14 discussed the behaviour of multiple loudspeakers in rooms, and from that discussion it can be understood that one must be careful not to route individual low frequency signals to any more channels than absolutely necessary if playback compatibility problems in other rooms and on other systems are to be minimised. Moreover, only in rooms with well-controlled acoustics can the full-range, five-channel option be heard to be superior to the single (or double) sub-woofer option. In rooms with less ideal low frequency control, the single sub-woofer option can reduce the variability of a sound as it is routed to different locations (which at low frequencies would tend to drive different modes to different degrees). Whether such poorly controlled rooms should be in use for serious surround mixing is a moot point, but the fact is that many of them *are* used for such purposes. It should also be added, here, that attempting to mix on a sub-woofer/satellite system using bass-management can be very risky. These systems can often be so far from the reality of what is on the recording medium that they could hardly be considered to be 'monitoring' anything other than their own idiosyncratic sound.

21.8.2 Processed Multiple Sub-Woofers

Some of the larger manufacturers of professional monitoring loudspeakers have accepted the fact that there is now a tendency for many people to mix all forms of music recordings in rooms with little acoustic treatment, especially at low frequencies where treatment can be both expensive and space consuming. Companies such as JBL and Genelec have put much effort into the design of low frequency loudspeaker systems which are intended to deal with room problems in such a way that, whilst not reaching the standards that can be achieved in well-controlled rooms, certainly improve the conditions of monitoring when compared to *conventional* loudspeakers in poorly treated rooms.

One such system, described by Toole,[6] was headlined on the cover of the AES Journal as 'Adapting to Acoustic Anarchy in Small Spaces'. In the paper, he described a system of multiple sub-woofers, spaced in various ways around the room and equalised both individually and globally. By this means, some unwanted modes can actually be cancelled, thus improving the time response of the room. The pressure amplitude flatness and distribution around the room can also be greatly improved, even in relatively poor rooms.

In most cases, perhaps only one resonant mode will be particularly troublesome below 100 Hz. One of Toole's techniques for dealing with such relatively simple problems involves locating a pressure node (a point of minimum sound pressure) of the offending mode, and placing the sub-woofer at that point, where it will only weakly couple to the mode. (This of course presumes the use of a monopole sub-woofer, which is a volume/velocity source. A dipole sub-woofer would require being located at an antinode [a point of maximum sound pressure], as dipoles are pressure *gradient* sources.) Obviously, some phase adjustment will have to be available because moving the sub-woofer nearer to or farther away from the listener will affect the overall time response of the system. The other technique involves using two sub-woofers, one placed on each side of the node, because on one side of it the pressure is falling whilst on the other side the pressure is rising. Therefore, two woofers, one on each side of the node and connected in parallel will destructively drive the mode, thus greatly reducing its effect.

The next step involves using four sub-woofers below 80 Hz, and Toole claims that research has shown that there seems to be no benefit in using *more* than four sub-woofers. These are fed via signal processors which, as mentioned previously, both individually and globally equalise the loudspeakers. Obviously, with systems such as those being described here, the low frequency signal must be mono. Nevertheless, Toole claims with considerable justification that if first class acoustic conditions are neither available nor achievable for practical reasons, then fast, flat, mono bass below 80 Hz may be greatly preferable to non-flat, resonant bass in stereo. He also makes a strong point of the fact that any type of response correction by sub-woofer can only be effective below about 100 Hz *and* when the troublesome modes are reasonably well separated. Above 300 Hz the loudspeaker dominates the response, but in the gap from 100 Hz to 300 Hz acoustic treatment tends still to be the only viable solution. Fortunately, in this range, control measures tend not to be so bulky as treatment below 100 Hz, but it definitely requires something more than merely sticking some foam on the walls.

The fact is that it still remains difficult to achieve 'true', compatible low frequencies for surround mixes, which is the reason why so many palliative measures are offered from so many sources. Concepts which may approach reality are so far from end-user playback environments that mixing often becomes something of a lottery. Different styles of mixing also suit different low frequency loudspeaker arrangements, so that it adds yet another set of variables. The optimum way of dealing with the low frequencies for high fidelity, five-channel music systems is a problem that still has not been solved at audiophile level.[7]

21.9 Close-Field Surround Monitoring

Not unlike the way that many people resorted to the use of small, close-field loudspeakers in an attempt to escape from the problems of stereo monitoring variability from studio to studio, the use of satellite loudspeakers, on stands in the close-field, and a common sub-woofer, has found widespread use for multi-channel mixing. In this case, though, one of the driving forces behind the choice was the lack of purpose-designed surround control rooms with adequate full-range monitoring. The reasons for this dearth of facilities have been:

1. Lack of clear guidelines/standards for the design of music-only surround rooms.
2. Because of 1 (above), there has been a corresponding lack of people willing to invest in the building of dedicated surround rooms, which may be short-lived in use if the 'wrong' layout is chosen.
3. As the ideal needs of surround rooms and stereo rooms are not entirely compatible if the highest performance is required, people have not been willing to compromise their stereo rooms whilst surround sound has remained only a challenger to the market supremacy of stereo.
4. Good dedicated surround rooms require the commitment to a considerable and long-term investment, and the recording market has shown no clear intention of being prepared to pay significantly more for surround mixing than for stereo facilities. For many studio owners, only earning the same rate as for stereo recording does not warrant taking the risk of investing on such a shaky basis.

Thus many commercial surround mixing rooms are, in fact, nothing more than stereo mixing rooms in which a satellite/sub-woofer system has been installed, perhaps with the addition of a few acoustic contrivances to help to (or at least to appear to) control a few acoustic irregularities. The fact that this can be passed off as professional mixing is due partly to the fact that the mixes will be expected to pass through a surround mastering facility in order to make them sound like the perceived, accepted, current norm. However, it can also be got away with because of the fact that the situation in the domestic playback circumstances is variable to the point of absurdity. If the mixes are not up to the highest standard, then who is going to know? If nobody knows, then nobody is likely to complain, but is this a professional attitude? The impression given is more that it is all a bit shoddy.

Manufacturers of domestic equipment and programme material have done little to help the situation. Ludicrous situations have arisen whereby the promotion of audiophile quality DVD 'A' discs have only been found to be viably marketable by making relatively cheap DVD video players read them digitally, whilst only passing the output through D to A converters of the lower resolution used for the DVD videos. In many cases it has been almost impossible to tell apart the compressed audio of the ordinary DVD audio channels and the high sampling rate/high bit-rate of the potentially vastly superior DVD A when passed through the cheap converters.

It may well be that this is not too different from using common master recordings/mixes for Compact Cassette and CD release. Those who choose to buy the appropriate playback equipment get the appropriate results … hopefully. On the other hand, Compact Cassettes never claimed to be superior to CDs, but the marketing of the surround formats certainly contained many implicit suggestions that a whole step forward was to be expected from DVDs *vis-à-vis* stereo CDs. This is certainly not the case for DVD A when played on DVD video systems. In fact, it would take a whole step forward in the world economy before people would be able to afford five loudspeakers, five amplifier channels and all the associated processors and converters which were of equal quality to the *two* of everything required for stereo. The development and specification of professional music surround facilities has been greatly hampered by the badly conceived marketing hype that has tried to force a new medium on a public who were not exactly crying out for it.

21.10 Practical Design Solutions

Figure 21.3 shows a control room which was designed by the author for high-quality, music-only surround use. By discussing some of the design options and choices it will be possible to highlight some of the concepts and compromises that have been touched on in the previous sections of this chapter.

The room was required to be principally a stereo room that could be used for high-quality surround recordings and mixing. It was considered to be very important for surround use to have three, full-range, flush mounted monitor loudspeakers, all at the same height. The reasons for setting the mid-range loudspeaker drivers of all three loudspeakers at a height of about 147 cm above the ground was discussed in Subsection 20.7.1 and

(a)

(b)

Figure 21.3:
A music-only surround room: Producciones Silvestres, Catalonia, Spain:
(a) The empty room; (b) With a view of the forest.

illustrated in Figure 20.8. The intention was to make no compromise which could reduce the quality of the two-channel stereo. The wall on the right-hand side of the room contains a window with a view of the local woodland, and at the rear of the left-hand wall is a glass door which leads to the studio rooms. However, the directivity of the loudspeakers was considered to be narrow enough not to expect problems from these surfaces.

As the stereo was of great importance, no effort was made to change the bi-directionality of the room acoustics, and so the rear wall was made maximally absorptive, in accordance with the chosen control room philosophy. The side walls were made relatively absorbent, except for the two windows and the glass door. The glass was all of the 12 mm laminated variety, and the windows were angled quite steeply upwards, to attempt to persuade any reflected energy to head in the direction of the absorbent ceiling. It must be remembered that even plane surfaces give rise to a certain degree of scattering at high frequencies, but from these positions it was considered improbable that too much energy would return to the listening position from any of the loudspeakers. Had the rear loudspeakers not been needed, the windows would also have been angled towards the back of the room, to tend to direct the reflexions from the front loudspeakers into the rear trap. This was really the only significant compromise that was made to the room design for the benefit of the surround performance. The front wall was made with an irregular surface of stone, to help to reduce any tendency for specular reflexions with the rear loudspeakers, although this in no way compromises the frontal stereo performance.

The centre loudspeaker, it should be noted, should be connected to powered-up (switched-on) amplifiers, even when not in use during stereo recording and mixing, to avoid loudspeaker resonances from affecting the response of the other loudspeakers. If the amplifier is not connected to the low frequency drivers, or is connected but switched off, the loudspeaker cones and the tuning ports would be free to resonate at their natural frequencies. With power to the amplifier, the very low output impedance acts as a brake on the movement of the loudspeaker cones, holding them rigidly. The port resonance is less of a problem because the receiving/radiating area is smaller, and the excitation tendency (at the sub-20 Hz tuning frequency) is less likely. In general, loudspeakers should never be left in control rooms if the amplifiers to which they are connected are not switched on, because they can affect the sound from nearby loudspeakers both by absorption and coloration, due to their resonant tendencies.

Anyhow, what we have described so far is really a 'three-channel stereo' room with full range monitoring from 20 Hz to 20 kHz, built to the principles discussed in Chapter 16. To convert this into a surround room we therefore need to add a system of suitably chosen and mounted rear loudspeakers, and, in this case, the general consensus was to set the rear loudspeakers at around 120° either side of centre–front.

One hundred and ten degrees, or less, would have obstructed doors and windows to an unacceptable degree, which again highlights the fact that ideal surround mixing rooms are best built as such, and should not be compromised by access to studio rooms or views of the forest. Again, the cinema people have a better approach – dedicated mixing rooms – but the tight budgets of the music industry tend to require rooms of more flexibility in use. Nonetheless, in this case, nobody really believed that positioning surround loudspeakers 10° aft of 'normal' would significantly alter the perception of the sound in such a room.

21.10.1 The Choice of Rear Loudspeakers

Here, the concepts discussed in Section 21.6 can be considered again in the context of a specific room. The overall room design was as shown in Figure 21.3. It should be obvious that symmetrical monitoring would not be possible. The front loudspeakers are set into a solid, very rigid front wall, whereas the rear loudspeakers, even if identical to the front cabinets, would perform differently because they would be set in absorbent surroundings (see Figures 16.1 and 16.2). They would not enjoy the low frequency loading provided by the front wall acting as a baffle extension. The effects of such loading differences were discussed in Chapter 11, and whilst it is true that the reduced low frequency loading could be equalised in the feeds to the rear loudspeakers, this could require up to four times the amplifier power to do so. This may or may not be a great problem, but the intransigent problem is that of the first reflexions from the opposing wall.

The rear absorbent trap is designed to minimise the effect of the reflected energy from the front loudspeakers from interfering with the direct signal. With the rear-mounted loudspeakers *facing* a solid front wall, nothing can effectively be done to prevent the response irregularities caused by the reflected wave, and no conventional equalisation could flatten the response. The mid and high frequencies would also face different terminations at the front and rear of the room. Therefore, even notwithstanding the differences in perception in terms of the frequency balance of signals arriving at the ears from the front or from the rear, the loudspeaker/room combination itself could not deliver identically balanced signals to a measuring microphone at the listening position.

To enable such a symmetrical monitoring condition to exist would, in the opinion of all the people concerned with the design of this room, have required unacceptable compromises to be made to the stereo performance. This also applied to the frontal sound-stage performance of a surround mix, which was also considered to take precedence over the rear channel performance. The option of identical loudspeakers all round was therefore abandoned.

The dipole option, as described in Subsection 21.7.3, in this case would result in virtually zero sound coming from the rear because of the highly absorbent rear wall. Little would arrive directly from the loudspeaker to the listening position because of the null in the plane of the baffle. Almost all of the audible output from the loudspeakers would therefore be by reflexion from the front wall, which would produce no *surround sound* at all, only confused stereo. Not surprisingly, this option was also rejected.

The multiple loudspeaker (cinema) choice was rejected because this was a music-only studio. Despite the fact that the cinema technique has much to offer to music-only mixing, the music industry in general has not woken up to this fact. The option was therefore rejected on the basis of lack of acceptance by the clients, but *not* from a system-engineering viewpoint.

The single diffuse arrangements, such as the DMLs or loudspeakers pointing towards wall-mounted diffusers, were considered carefully. The DML option was finally rejected due to the lack of low frequency response unless very large panels were used. The option of using smaller DMLs with a common sub-woofer was rejected on the grounds of unnecessary complexity. The much larger panels were rejected because of worries about large reflecting surfaces at the rear of the room creating problems with the frontal stereo, although their hemispherical directivity over a wide frequency range could have allowed the panels to be angled such as to minimise this effect. Unfortunately, this would also have taken up valuable space in a smallish control room of just under 50 m^2, but this option remained under discussion until the final choice was made. It was certainly a serious contender.

As the rear of the room was relatively dead, acoustically, and the designated mixing/listening area was so small (6 m^2 or thereabouts) not much benefit was seen from the option of pointing a loudspeaker at a diffuser panel on each sidewall. Ultimately, the decision was made to use a pair of single, conventional, discrete loudspeakers, but effectively only by the rejection of all the other options. (Although see note at end of this section.)

The owner then consulted his clients, and nobody seemed to be intending to use the rear channels for bass guitar or bass drums. Therefore, in order to minimise any interference with the pure stereo use of the room, it was convenient to use relatively small loudspeakers that could be mounted on stands at the rear of the room. The actual model of loudspeaker was chosen for its sonic compatibility with the front monitors. Their ability to produce around 108 dB SPL at 1 m, down to 70 Hz, was considered sufficient in a room where nobody listening seriously would be more than 4 m from them, hence around 100 dB at the listening position from each surround loudspeaker would be guaranteed. The peak response was about 6 dB higher.

It was also acknowledged that certain clients might wish to use their own choice of satellite and sub-woofer systems; and that the studio may in the future purchase its own such system. In this case, the rear loudspeakers could be moved on their stands to a closer position in order to serve as the rear loudspeakers for the satellite system. The thinking behind the two choices of monitor system was that, as in much stereo recording, the large, full range monitors could be used during the recording process, to track down any distortions or noises and to check the low frequency balances. The satellite systems could be used for the 'domestic reference' and for those who desired to mix on them. The large system could also be used to 'vibe' the musicians (or other personnel) when necessary.

Although no claims are being made that the system described above is definitive, the description and the discussion about the thinking process that led to its final design can perhaps be useful to help to outline the options and typical compromises that go into the design of surround sound control rooms. (In 2006, two years after the opening of the studio, the rear loudspeakers were changed to relatively large DMLs [60 cm × 40 cm], mounted in the rear corners of the room.)

21.11 Other Compromises, Other Results

Figure 21.4 shows a small screening room in a film laboratory using DML panels for the surround. In this instance the compromises produced different results, because the needs were different. The room had to meet the specifications for Dolby Digital, and hence the surround sound-field over the area of 26 seats could not be allowed to vary by more than 3 dB. The 3 dB drop per doubling of distance in the close-field of the DMLs, together with their wideband hemispherical directivity pattern, made them an excellent engineering choice. Their diffuse nature, and the relative inability of the audience to localise them audibly, made them a good psychoacoustic choice. The fact that they were readily received for the overall natural impression of the surround tracks also made them a good subjective choice.

In the case of this screening room, the typical choice of multiple surround loudspeakers, as normally used in cinemas, was rejected because it was considered that it would be difficult to prevent the audience from localising the surround sound on the nearest loudspeaker. This was due to the narrower directivity angles of typical cinema surround loudspeakers and their extremely close proximity to the seats in this room. It was also considered problematical to get the required evenness of coverage unless a very great number of loudspeakers were used, again because of the directivity of the typical surround loudspeakers.

Also, in this room, the seating was all *much* closer to the surround loudspeakers than to the front loudspeakers, and the precedence effect suggests that the source of any sound routed to the front *and* the surround loudspeakers (which in fact is a relatively rare thing in cinema mixes) would be localised in the surrounds. However, as mentioned earlier, in the Dolby processors there is always up to 100 ms of delay to the surround feeds. This ensures that in any theatre where the difference in the distance to the listener from the surround and front loudspeakers is less than about 30 m, the sound will always be localised in the front loudspeakers.

(a)

(b)

Figure 21.4:
Front and surround monitoring in the Dolby Digital screening room at the Tobis film laboratories, Lisbon, Portugal: (a) Front monitor distribution; (b) The DML panels, high on the side walls, used as diffuse surround sources.

Once again, the cinema people did their homework and came up with a carefully conceived standard. However, the delay, which works well in the cinema, could limit the options for music-only surround if lead instruments were to be put in the rear channels. With ambience in the surrounds, the effect of the delay can be advantageous, but obvious timing difficulties would be encountered if an ensemble were to be split across front and rear sources.

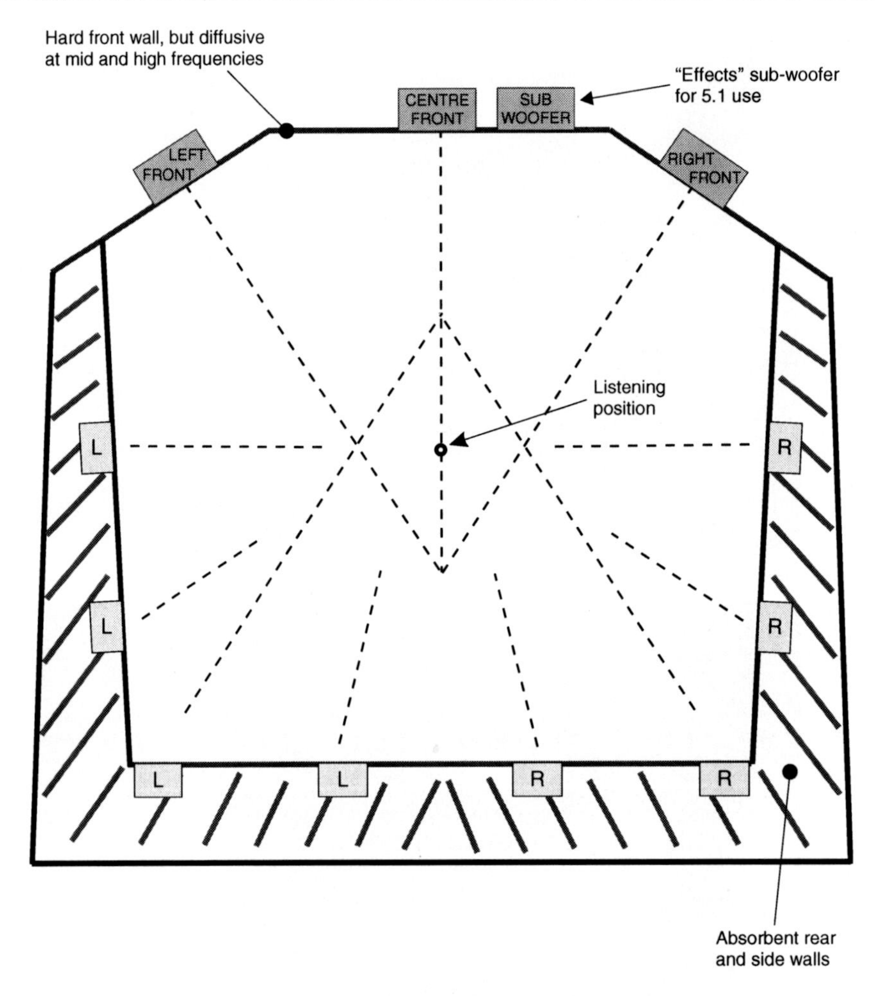

Figure 21.5:
Non-Environment concept using multiple rear sources. Versatile 5.1 channel, full-range surround monitoring system, with multi-format monitoring capability. Left and right surround channels split between two groups of four loudspeakers, or, alternatively, perhaps with diffuse radiators, such as distributed mode loudspeakers. All loudspeakers full-range, except the optional 'effects' sub-woofer.

Nevertheless, a room for ambient surround mixing could be extremely effective if built according to Figure 21.5. The only precaution necessary would be to ensure that any *reverberation* fed to the surround loudspeakers was fed via a short delay, otherwise it may arrive *before* the direct sound, but as this is relatively standard practice, it should cause no extra complications.

In the design of a surround-sound control room, therefore, many things need to be taken into account; it is not a simple exercise. However, the process is exacerbated by the lack of consensus or standards in the music industry about precisely what the goal is intended to be. The quality of the frontal stereo, be it two or three channels, is still of prime importance in most cases, so it is likely that most rooms used for surround mixing will be unlikely to compromise this merely to benefit the rear channels, from which our hearing is anyhow much less discriminating. The question of 'how to surround' then becomes a function of the likely type of programme and the basic philosophy of the room acoustics. It is a very complex subject, which is made no easier by the electroacoustic complications created by multiple sources, as described in Chapter 14.

The fact is that the psychoelectroacoustics of surround is a far more complicated subject than is the case with stereo, and not least because of the much increased number of possibilities for making poor choices based on lack of a thorough understanding of the problems. Furthermore, the choice of 'only' five channels is not sufficient to provide the optimum ability to compromise to the multiple and often conflicting requirements without significant loss of performance. An interesting development of the concept is shown in Figure 19.20, which can take the clarity of perception to an entirely new level. Very significant reductions in the level of intermodulation distortions is also reported, which is hardly surprising.

21.12 Dubbing Theatres

Currently, the most impressive surround environment is probably the cinema. Traditionally, rooms for mixing cinema soundtracks ('dubbing theatres' as they are usually referred to in British English) have borne little relationship to music mixing rooms because the perception of sound whilst watching a picture on a large, distant screen is different from both the perception of sound in a small room whilst watching a small screen, or listening to the sound without picture in any room. Some of the reasons for this are explained in more detail in the following sections, but there is a tendency these days for cinema theatres (i.e. the public performance theatres) to become smaller and acoustically drier. As a consequence of this, cinema dubbing theatres, whilst still being large, are now often very close in size to some of the smaller theatres, so this offers the prospect of excellent compatibility between the mixing and reproduction environments.

In these smaller, drier cinemas, provided that they have good quality loudspeaker systems, the perception of detail in the soundtracks is considerably greater than it has tended to be in larger, more reverberant theatres. In any professional industry it would seem to be a requirement that the professionals stay one step ahead of the consumers, and many older dubbing theatres can now be shown to be more coloured in their acoustics than many of the newer theatres for public screening. This is leading to a trend for more precise dubbing theatres, although the cinema industry is still very conservative, so changes tend to happen rather slowly.

Figure 21.6 shows the acoustic control measures inside a low decay-time dubbing theatre (Cinemar Films, in Milladoiro, Spain). The ceiling consists of three layers of materials supported on two independent systems of wooden beams. Progressively, from the inside, each subsequent layer is less absorbent and more reflective, so the inner layers absorb more and the outer layers isolate more. The inner layers, below the first set of beams, consist of felts and deadsheets, the intermediate layers are of plasterboard and deadsheets,

(a)

(b)

Figure 21.6:
The dubbing theatre at Cinemar (Milladoiro, Spain) under construction: (a) showing the front monitor wall; and (b) the distribution of surround loudspeakers.

and the outer layers are of plasterboard, heavier deadsheets and plywood. In the case shown, the inner ceiling is rather low in terms of conventional width/length/height ratios for Dolby theatres, but in acoustic terms, if the ceiling is very absorbent, then it is really not there, so its height is somewhat irrelevant. Dolby now generally have no problem with this concept.

Figure 21.7 shows the level versus distance of the above-mentioned room in one-third octave bands, measured at 1m intervals between a point 2 m from the screen until 40 cm from the acoustically transparent fabric covering the rear absorber. As can be seen, the modal activity is minimal, and the tendency towards a low frequency build up in the vicinity of the rear wall only begins to become evident below the 31.5 Hz band. Despite its size, this room is still only intended for use in the zone behind the mixing console, but, as mentioned earlier, perceptions of sound can change with the size of the picture and the distance to the screen (as described in Subsection 21.12.3), so for this reason the room needs to be big. The completed room at Cinemar is shown in Figure 21.8. The owners had chosen this concept specifically because of their own experiences in a similar room which they had frequently rented (which was already fully approved by Dolby) in which they could make fast and reliable decisions during the mixing process, and the subsequent excellent translation of the work from the studio to the cinemas. That room, which differs principally in its provision of variable acoustics in the front half of the room, is shown in Figure 21.9.

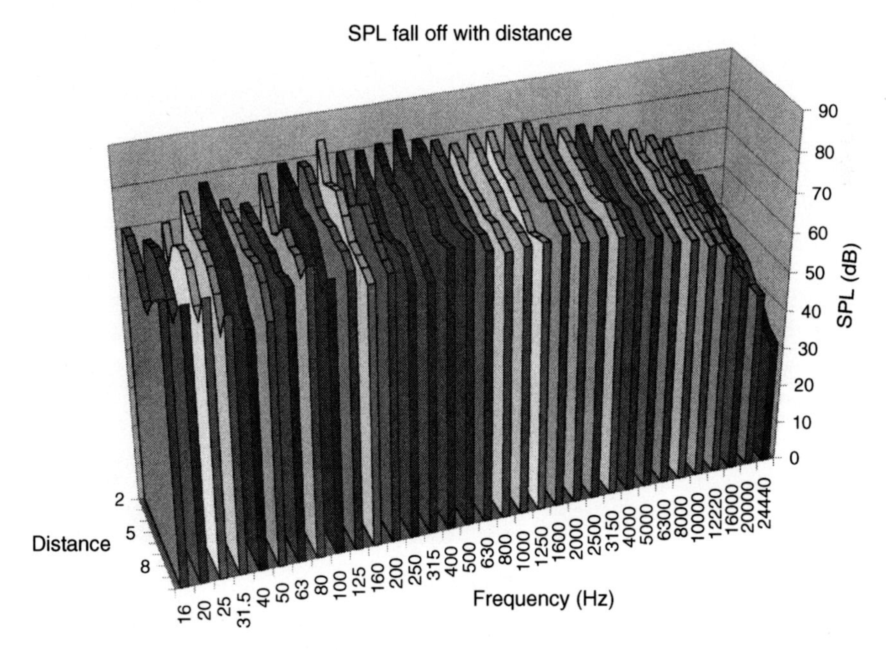

Figure 21.7:
The measurement at 10 m is only 40 cm from the face of the rear 'trap'. Except for a small rise around 100 Hz, all of the frequency bands continue to fall into the trap above 40 Hz, showing how the rear wall of the room acoustically disappears.

Figure 21.8:
The completed dubbing theatre at Cinemar Films, Milladoiro, Spain (2006).

Figure 21.9:
The dubbing theatre at Soundtrack (Sala 15), Barcelona, Spain, showing the rotating acoustic control devices in the front half of the room, as outlined in Figure 7.9(a)(ii) and (b). The front half can be made more acoustically bright by rolling up the carpet and rotating the triangular panels to expose a reflective or a diffusive surface. In this mode, up to about 25 string instruments can be recorded to picture.

Relatively large rooms tend to give rise to mixes which are more compatible with large-screen cinemas compared to those made in smaller rooms. However, without the limitations imposed by the need for a relatively even coverage over a large audience area, a dubbing theatre with few seats can take advantage of the fact that only a limited area of the room requires an even sound coverage. This allows better optimisation of the response over the critical listening area, yielding more detailed perception of the sounds being mixed, and therefore enabling the decision-making process to be both faster and more reliable. Furthermore, without the need for the main loudspeakers to exhibit very wide directivity for the even coverage of the audience, lower-powered and more precise mid and high frequency loudspeakers can be used.

There are some people who worry that a lower decay time in the mixing theatre, compared to a typical commercial cinema, would lead to incorrect decisions being made about the appropriate levels of reverberation to add to a soundtrack, but experience has shown that the problem does not exist in practice. Once again, the extra precision has been shown to make decision-making much easier, hence saving time and effort, easing the work-load on all concerned and allowing them to concentrate more on other important matters. These concepts were controversial in the world of music recording in the 1980s, but experience again showed that low decay times and fast transient responses were very necessary if things such as low-level coding artefacts and digital processing errors were to be noticed early enough to prevent the public release of flawed recordings. No effect was noticed regarding decisions about the application of artificial reverberation.

The surround loudspeakers shown in Figure 21.6(b) are 10 JBL 8340A, standard cinema surround loudspeakers. They are each driven by separate amplifiers, which allows not only their precise level settings, but also the simple switching by the Dolby processor between 5.1, 6.1 and 7.1 modes of operation. Obviously, where little reverberant integration of the loudspeaker outputs is occurring within the room itself, any imbalances in the sounds arriving from each loudspeaker would be more noticeable than normal, so individual amplifiers help considerably here.

21.12.1 Room-to-Room Compatibility

In general, once reverberant colouration is added by the room to the output from the loudspeakers, many changes take place in the perceived sound. Although some degree of reverberation is often considered to be desirable in public cinema theatres, in order to give a more uniform sound character over the entire audience area, in the mixing rooms the reverberation can mask low level details and other problems and noises on the recorded soundtrack. Room sizes, also – surprisingly to many people – can affect the perceived frequency responses. Therefore, if a film is mixed in a room of different size *and* reverberation/decay time to a room in which it is subsequently shown, the *frequency response* of the loudspeaker systems may need to be adjusted if the most similar *perception* of frequency balance is to be achieved. However, this is a totally separate and additional issue to the one of reverberant colouration changing the sound character.

Acknowledging this fact, in the 1970s, the cinema industry began investigations into room-to-room compatibility, which later led to the standardisation of the 'X-curve' as the frequency response characteristic to which the soundtracks would be mixed and shown. The X-curve, (described further in Subsection 21.12.2) is not a clearly defined line: it is a line which is banded by considerable tolerance. Consequently, when a dubbing theatre is being certificated by Dolby, the approximate curve is achieved by measurement, but the final response adjustment is made by ear, with the Dolby engineers using a range of well-known test recordings to make the definitive decisions about frequency balance. There are too many variables in this equation to work solely to prescribed measurements but unfortunately, in public cinemas, such care and attention in the set-up is rarely applied.

Usually, as the room sizes increase, the high *frequency* response of the loudspeaker system must be reduced if a flat *perception* of the sound is the goal. Also, as the reverberation or decay time of the room increases, the high frequency response once again tends to need to be reduced. Consequently, a large reverberant room will tend to need a significantly more attenuated high frequency response as compared to a small, dry room. But there is also a *level-related* question which affects cinema studios or theatres of different sizes: 'Why does 85 dBC at the listening position in a small room tend to sound louder than 85 dBC at the listening position in a large room?' After all, SPL is SPL; or is it? The question will be discussed further in Section 21.12.3.

As yet, there have been no definitive answers to this question, but there are some very different small/large room characteristics which may give rise to such perceptual differences. Imagine that we are in a small room, listening 4 m from the loudspeakers and 3 m from each side wall, which by definition if they are not anechoic will produce some reflexions. The early reflexion paths will probably differ from the direct signal paths by less than 5 m – perhaps much less – so they will be heard about 10 ms after the direct signal and with a frequency balance which will depend solely on the nature of the reflective surfaces. Air absorption at high frequencies tends to be around 1 dB at 10 kHz for every 5 m travelled, and so for short distances it can be ignored for first-order reflexions.

If we now go into a larger room where we are listening at 12 m distance from the loudspeakers and 8 m from each side wall, the situation becomes very different. The first reflexions may arrive from the surfaces (presumed here to be of a similar nature to those in the aforementioned small room) with a delay of around 25 ms, arriving via pathways of 8 or 10 m more than the direct path. So, even the very first of the first order reflexions may be around 2 dB down at 10 kHz compared to the direct-signal/reflected-signal balance in the smaller room. The reflexions in a larger room will also be separated more in time, both from the direct signal and from themselves, than the reflexions in a smaller room.

We therefore have a situation where the reflexion density is higher in a smaller room (i.e. the reflexion arrival times are more closely spaced) and the reflexion levels will be higher relative to the direct signal because they have travelled less distance, and so have expanded less (that is to say, their intensity will be higher than for comparable reflexions in a large room). What is more, the reflexions in a small room will be brighter sounding because the longer the reflexion path the more air absorption will take place at higher frequencies, so the reflexions in a larger room will tend to suffer a greater high frequency roll-off relative to the direct signal.

To briefly recapitulate the situation:

1. In a small room, the reflexion density will be greater than in a larger room; that is, the reflexions will *arrive closer in time* to the direct signal, and to themselves, tending more to reinforce the perceived loudness of the direct signal.
2. In a small room, the relative reflexions (i.e. first-order, second-order, third-order, etc.) will *arrive more closely in level*, with less difference between their relative levels than in a larger room.
3. In a small room, the reflexions will arrive with a greater high frequency content due to suffering less air absorption than the respective reflexions in a larger room of similar acoustic treatment.

A consequence of all of this is that the direct-to-reflected level difference in a small room will tend to be less than the respective proportions in a larger room. Therefore, if we set a level of 85 dBC at the listening position in a small room and in a large room, with both rooms having similar surfaces and absorption characteristics, there will be a higher proportion of reflected energy in the small room (compared to the direct energy) than would be encountered in a larger room, and the reflexions would also exhibit more high frequency content, giving rise to the differing sensations of loudness in the large and small rooms. The differences in reflexion density, reflexion level, and reflexion frequency response are powerful cues which human auditory systems can use to extract information about the distance to the source of the sound. Anthropologically speaking, a loud sound which is perceived to be close tends to suggest a greater potential danger than a sound which is equally loud but perceived to be more distant. The closer event requires a more immediate reaction, hence a greater perception of loudness can be a greater stimulus to act, and thus in some cases a benefit to survival. The above characteristics of a sound-field undoubtedly give rise to different sensations of loudness, but what has so far not been produced in any standardised form is a correction adjustment for level versus room size. A suggestion for a possible solution to this problem will be discussed in Section 21.12.3.

In anechoic chambers, the effect of room size on perceived sound level with distance should be zero, at least within the frequency ranges where the rooms are truly anechoic. However, it will be shown that when we listen to sounds which are associated with a moving picture, the visual cues can indeed introduce perceptual differences which can affect our opinions of precisely how loud a sound may be.

21.12.2 The X-Curve

Cinema theatres, for the public performance of feature films, tend to be rather different from domestic rooms which are used for listening to the high fidelity reproduction of music. Almost anybody who has ever been to a large cinema will know that the sound in the cinema is rarely reproducible at home. In large rooms, the perception of the sound is different from the perception in small rooms, so the cinema industry has, since its earliest days, tended to mix film soundtracks in rooms which were representative of, or in effect were, cinema theatres.

Optical soundtracks have been the ever-present means of recording sound to film, and even Dolby Digital film soundtracks still have optical, analogue tracks alongside them, for back-up in the event of a digital system failure. Aspects of film reproduction such as the projector slit height, electrical noise-filters, negative/positive print-loss filters, loudspeaker characteristics, losses through the perforated projection screen, and even air absorption losses in the theatres (almost 1 dB for every 5 m at 10 kHz) all meant that the maintenance of a given frequency response from a soundtrack mixed in one room and played back in a significantly different room would not be a likely outcome. Therefore, if the dubbing theatres (the mixing studios) were generally similar to the public screening environments, the mixing engineers could at least mix and equalise the soundtracks in a reasonably representative environment in order to achieve the most natural or most desirable audio quality. This historically meant using significant high frequency boost, to compensate for all the losses in the optical reproduction chain, but care had to be taken to avoid the boost leading to distortion. The high frequency roll-off of the reproduction did, however, serve a useful function as a form of noise (hiss) reduction in the days before Dolby noise reduction came on to the scene. Perhaps, somewhat surprisingly, the significant high frequency roll-off in the overall reproduction did not sound as dull in the typical, large, reverberant cinema theatres of those days as a visual inspection of the frequency response would suggest. In practice, the 'Academy Characteristic' as shown in Figure 21.10 was the general response curve for cinema reproduction systems. A roll-off began around 1.5 kHz, which fell to about −18 dB at 8 kHz, so the HF attenuation was quite severe.

It has been traditional to split the sound reproduction chain into the A-chain, from the playback head to the output of the projector; and the B-chain, from the output of the projector to the audience. What we are principally dealing with when we discuss modern cinema equalisation (as modern A-chains are essentially flat) is the B-chain, but historically the A + B chains could even be 20 dB down at 8 kHz, and this could lead to excessive distortion when equalised to sound natural in the cinemas, or it could lead to excessive noise if the recorded levels were reduced in order to avoid the distortion. There was a very fine line between excessive noise and excessive distortion.

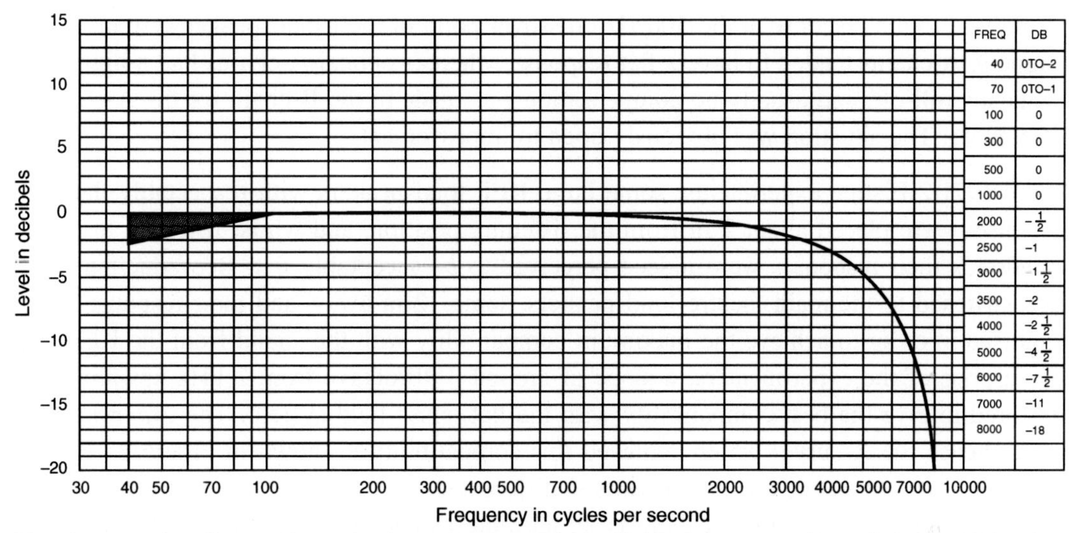

Figure 21.10:
1948: 'Academy Characteristic' for Altec Lansing systems.

In 1971, during the mixing of 'A Clockwork Orange', experiments were made in Elstree Film Studios, near London, to work with a wider range response. This film was the first to use the Dolby A-type noise-reduction on all its pre-mixes and mixes, and ways were being sought to exhibit this to its full advantage. The state of affairs then current is shown in Figure 21.11, which shows the responses of nine Hollywood dubbing theatres in 1974,

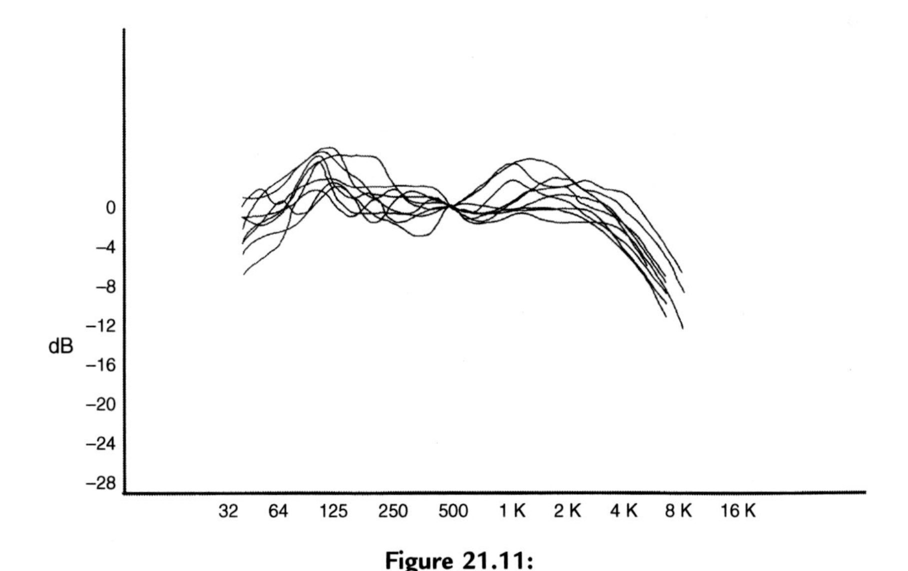

Figure 21.11:
1974: Nine Hollywood dubbing stages — B-chain only (normalised at 500 Hz).

normalised to the response at the 500 Hz crossover point of their loudspeaker systems. Also, magnetic A-chains tended to have a flat response, whereas optical A-chains had a high frequency roll-off, and usually no compensation was made in the B-chain for the A-chain differences. The situation was obviously open to improvement.

In the 1971/2 Elstree experiments, some large KEF loudspeakers with respectably flat responses were placed about 2 m from the listening position, and whilst listening to a flat soundtrack their responses were indeed perceived to *be* acceptably flat. Reproduction was then switched to the main theatre loudspeakers, some 12 m away, and their response was equalised in order to achieve the maximum compatibility with the sound from the KEFs. To the surprise of the experimenters, a slope of around 3 dB/octave from 2 kHz upwards appeared to give the best compatibility with the flat, close-field system. The *experimental* curve (hence the X-curve) which achieved this compatibility is shown in Figure 21.12, as drawn in 1972. Repeating the experiments in different sized rooms led to the family of curves shown in Figure 21.13, and still there is no absolute explanation of why this should be so, although Ioan Allen[8] has suggested that it could be due to some psychoacoustic phenomenon involving far away sound and picture, or perhaps the result of increasing reverberation which generally follows room size. No doubt the reflexion density also plays a part, as mentioned in Section 21.12.1.

The tendency was for the roll-off to be greater both in large rooms and in rooms with longer reverberation times. A small, low decay time room tends towards needing a flat response, whereas a large room with a generous reverberation time would need considerable high frequency roll-off in order to sound subjectively similar in equalisation to the small, dry room. Remember though that in the world of cinema *dialogue* is pre-eminent, and

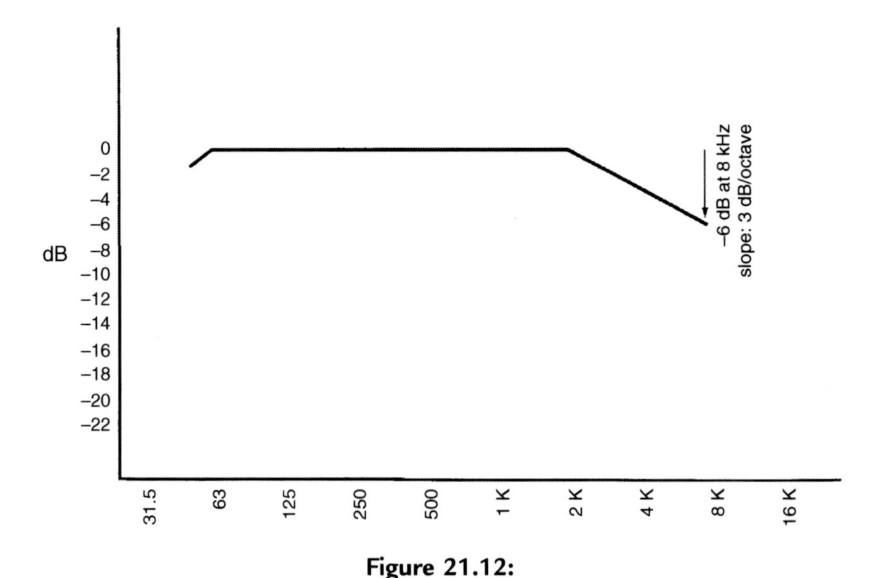

Figure 21.12:
1972: The first wide-range B-chain characteristic — later called the X-curve.

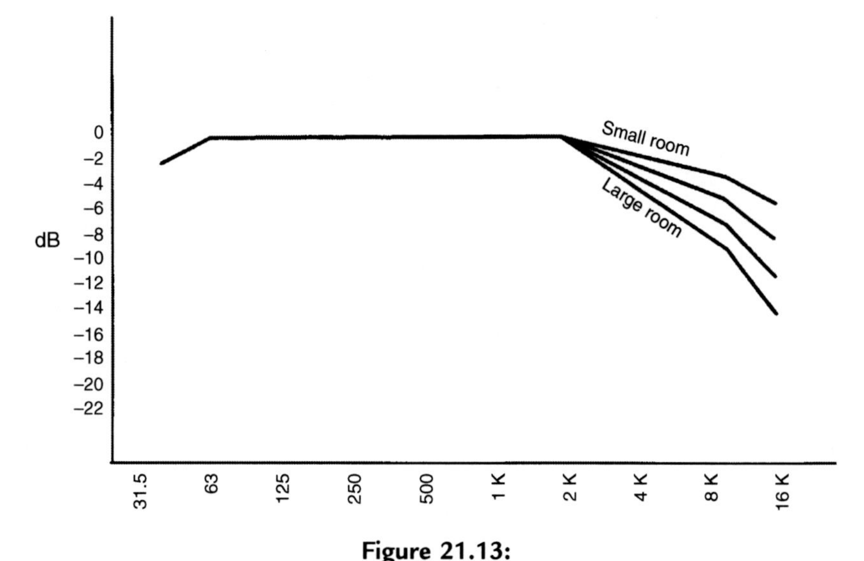

Figure 21.13:
SMPTE 202M — 1982 — Corrections for theatre size.

its intelligibility is all-important. The short sounds of dialogue do not receive reverberant enhancement in terms of loudness, so the natural sound via a relatively flat *direct* response from the loudspeakers may be more important than the flatness of the combined loudspeaker/room response, because it is the direct response which gives rise to the clear perception of dialogue. Hence, in drier rooms of *any* size, the overall curve tends to be flatter as it does not have to take into account the normally sloping reverberant response which tends to fall as the frequency rises. In large reverberant rooms, it was considered that the direct response may have to be made excessively bright if attempts were made to achieve a flat response to pink noise in a naturally bass-heavy room, and obviously the general tonal nature of the dialogue needs to reasonably match the rest of the soundtrack. The X-curve is therefore not a fixed curve, but it is a curve with upper and lower tolerance limits, which allow it to be tailored to different rooms in order to maintain the overall perceptual compatibility. Figure 21.14 shows the recommendations published in 1998, as extended to 16 kHz and with a second knee at 10 kHz, but the final adjustment still needs to be made by ear by experienced people. Figure 21.15 shows the typical equalisation for the surround loudspeakers in order to give good compatibility with the X-curve response of the behind-the-screen system. Note how with increasing room size (distance from listeners) and increasing reverb, it is the *turnover* frequency that is adjusted, *not* the slope. Subjective assessment had shown that the surround loudspeakers tend to need a brighter characteristic than the screen channels (i.e. less HF roll-off), which could be due to the public being closer to the surround loudspeakers, and hence more in their direct field.

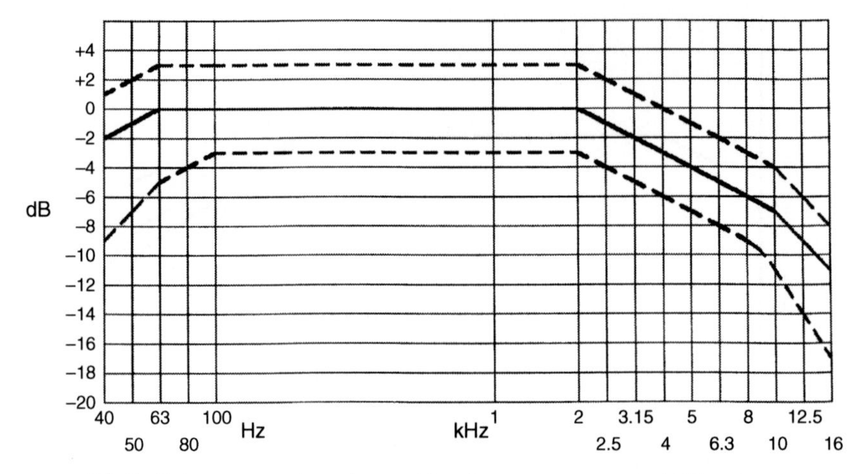

Figure 21.14:
SMPTE 202M – 1998 – X-Curve extended to 16 kHz with second break point at 10 kHz.

The set up of cinema monitoring systems is clearly not trivial, and as Allen pointed out 'if material were to be mixed in a small room, in a large theatre it would have to have information about the content as it varied between short-duration (speech) signals, and long duration (music and possibly effects) to have perfect playback translation'.[8] Large dubbing theatres are therefore still the only viable mixing environments for high-quality cinema soundtracks, because the perception of the sound between large rooms and small rooms varies so much.

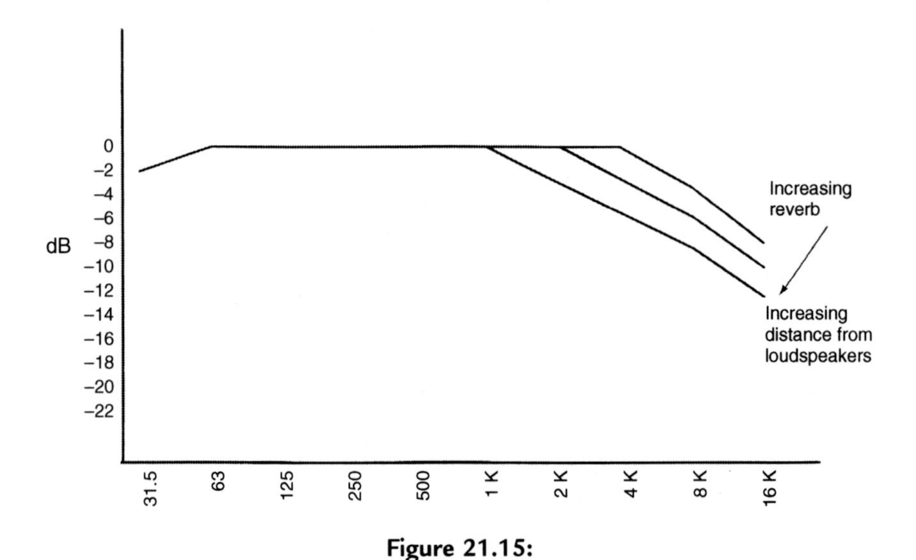

Figure 21.15:
Adjustments for room size and distance to surround loudspeakers – adjusting the turnover frequency as opposed to adjusting the slope.

Of course, human perception of the relative balance of highs and lows also depends upon level, as shown in Figure 2.1, so it is essential in film production that the mixing and the public performances should take place at the same levels if the equalisation compatibility is to be maintained. After the introduction of digital soundtracks, 18 dB of headroom became available above the normal 85 dBC reference level. Many film directors began to abuse this by using it throughout the film – (if the plot is weak, turn the level up!) – which led to many cinema owners turning down the volume from the standard setting. The level reduction in the cinemas was partly due to the avoidance of distortion from some marginal reproduction systems, but also because of complaints from the cinema-goers about excessive level. Unfortunately, one result of turning the level down in the cinemas is that the quieter parts of the dialogue can become unintelligible at the reduced levels. The general situation is still rather arbitrary, and discussion about the standards continues, especially in the light of the arrival of digital cinema.

21.12.3 Sound Level versus Screen Size

It was mentioned previously that in anechoic chambers there should be no difference in the perception of a sound from sources at different distances, at least not until those distances became so great as to introduce significant, high frequency, air absorption losses. Nevertheless, in 2008, at the University of Vigo, Spain, experiments were carried out in a hemi-anechoic chamber which did show that people expected to hear a lower sound level as a screen was moved further away from them, step by step. This test was part of a series of tests relating to screen size and distance versus the 'appropriate' sound level.[9] The conclusions were that:

1. At a fixed *distance*, as the screen size increases, the sound level necessary for a realistic combination of sound and picture also *increases*.
2. For a fixed *screen size*, as the distance from the viewer increases, the sound level necessary for a realistic combination of sound and picture *decreases*.
3. For a fixed *viewing angle*, where the screen size increases as the distance from the viewer increases, the sound level necessary for a realistic combination of sound and picture also *increases*. This concept is shown diagrammatically in Figure 21.16.
4. For a fixed *sound level*, as the screen size increases, *the amount of low frequencies* necessary for a realistic combination of sound and picture also *increases*.
5. Little evidence was found of any connection between the screen distance or dimensions and the appropriate level of *high* frequencies.

One of the authors of the paper (Christian Beusch) proposed the curve shown in Figure 21.17. This was developed from a curve that had been proposed in an earlier paper.[10]

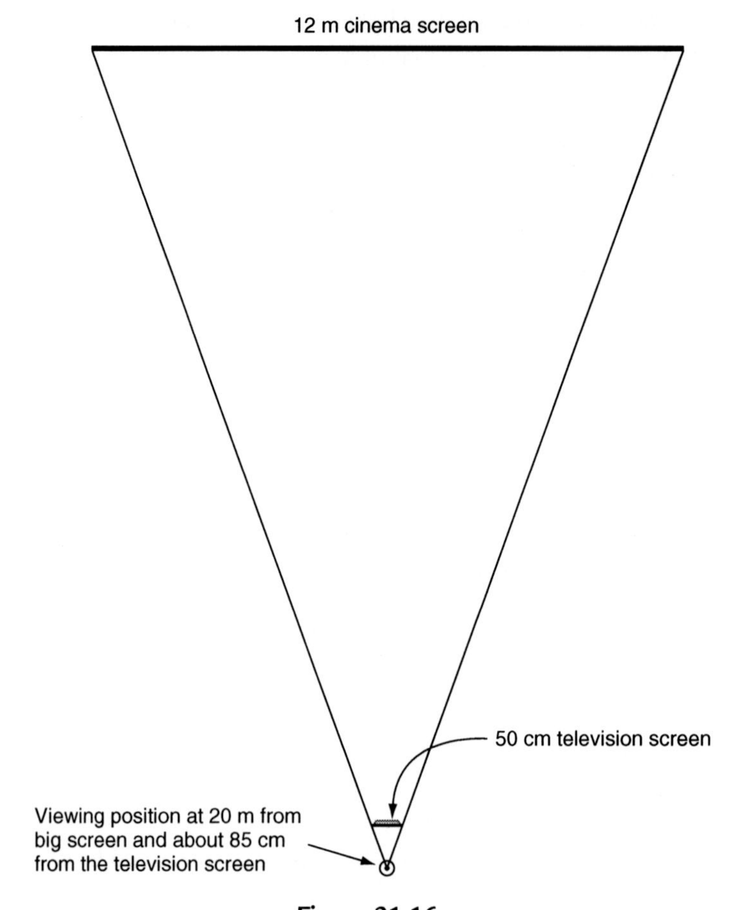

Figure 21.16:
Same relative picture size for the viewer; same viewing angle; different distances from the screen. It should be self-evident that the appropriate sound levels for a 'natural' perception would be different in each case.

A look back at Figure 21.16 will put into a 'common sense' perspective the findings of points 3 and 4, above. It is not difficult to imagine a battle scene in an action film, taking place on a 12 m wide screen at 20 m distance in a big room, and for explosions to reach a very exciting 115 dBC. However, if, as shown in the figure, the scene were to be repeated at a viewing distance of 85 cm on a 50 cm (20 in) television screen, nobody in their right minds would wish to watch it with 115 dBC of accompanying low-frequency rumble down at 20 Hz. In nature, small things do not make loud noises. What is more, as mentioned towards the end of Section 21.12.1, loud noises at close distances are usually a sign of danger, and so can be psychologically disturbing. The curve shown in Figure 21.17 has been experimentally tested and it does tend to suggest that there is no one calibration level that suits all sizes of screens at all distances. As the screen gets closer to the viewer, the 'natural'

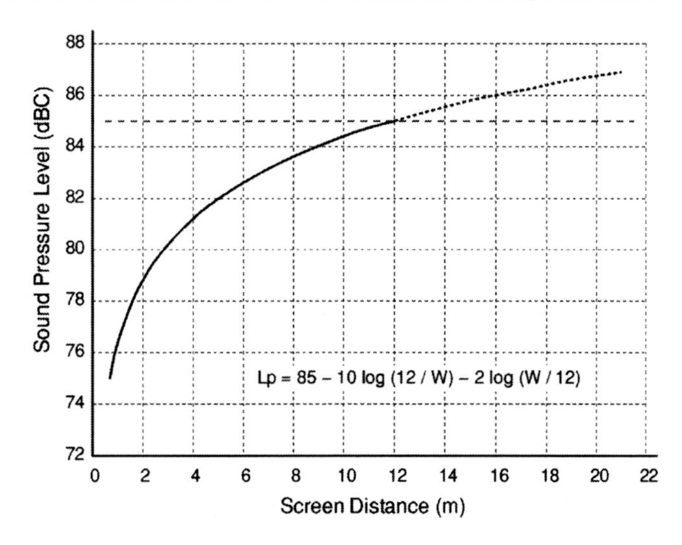

Figure 21.17:
Proposed curve relating screen distance to calibrated sound level for a 45° viewing angle.

sound level begins to fall off more rapidly. This is a clear indication of why many soundtracks sound 'too loud' in small cinema rooms. It also suggests several reasons why the cinema industry has always tended to mix in large rooms:

1. In smaller rooms the soundtrack may be mixed with insufficient level for large cinemas.
2. In smaller rooms the soundtracks may be mixed with insufficient low frequency content for large cinemas.
3. Because of the reflexion density question, mentioned in Subsection 21.12.1, the dialogue levels may receive greater reflective support in smaller rooms, and hence may be mixed with insufficient dialogue level for large cinemas.
4. There is also a question of dynamics. As the background noise in most cinema rooms will be around 30 dBA, if the soundtrack is mixed at a lower average level in a small room, the difference between the average level and the noise floor will be less than in a large room, and so the dynamic range of the mix will be reduced.

The accumulation of these factors relating to human perception suggests that it may always be necessary to mix *in* big rooms *for* big rooms; which is exactly what the cinema industry has always tried to do, although without ever clearly explaining the underlying reasons why. So, it seems that large dubbing theatres will still be required for as long as feature films are shown in large, public cinemas.

21.12.4 Room Acoustics and Equalisation

When the X-Curve was proposed in the early 1970s, not only the public cinemas but also the dubbing theatres where the soundtracks were mixed were in a rather poor state of inter-room compatibility. The introduction of the X-Curve as a response goal, and a standard means of measuring it within the rooms, certainly brought a measure of order to the chaos, but what was known in 1971 was a long way short of what is known in 2011. Measurement and analysis techniques have advanced greatly, as have the knowledge of psychoacoustics and the means of introducing corrective measures. In 1971 there was no such thing as a parametric equaliser. What *were* just becoming available however were third-octave real-time analysers and third-octave equalisers. There was also a school of thought developing which held that the ear 'heard' in critical bands of one-third-octave width. If that were the case, then if the energy in each one-third-octave band was equal, each band would be perceived as being equally loud. We now know that this is not the case, but in the early 1970s, to many people, one-third-octave analysis and equalisation appeared to be the solution to the room-to-room compatibility question. In fact, the use of third-octave equalisation became very widespread, not only in cinema rooms but also in recording studios and live concert sound, but it has since been largely abandoned by all except the cinema world. The problems with loudspeaker equalisation in general are discussed in Section 23.1.1, but in any case, one-third-octave equalisation, at least for the purpose of sound system alignment, is simply too coarse to function as intended.

The simple fact is that *rooms cannot be equalised.* In any room other than an anechoic chamber, the response at any position will be different, and in most cases *significantly* different. If we begin with a flat source, such as an excellent omni-directional loudspeaker radiating wideband pink noise, the response will be different at all places in the room. The acoustic waves expand three-dimensionally and subsequently interact with their own reflexions and the room resonances. If, therefore, the response is different at all places, it must follow that there is no equalisation setting which can correct for anything except for one point in space at a time. This can only be considered to be *local* equalisation, and not *room* equalisation.

It cannot be over-stressed that a measuring microphone and analyser combination has, in many cases, very little similarity to an ear/brain combination. There is now a very strong body of opinion which asserts that human beings tend to 'hear through' rooms. This is best explained by considering the case where the source of sound is 'natural', which for discussion purposes could be someone playing a baritone saxophone, which is a harmonically rich instrument covering a wide range of frequencies. If the instrument was being played at the front of the room, then to a listener walking around the room it would be evident that it was the same instrument playing. Taking measurements at many places around the room would show that the responses at each place would be different, and in some cases grossly different. The degree of difference would be totally out of proportion with the degree of perceived difference by the listener, who would merely hear the *same* instrument, but modified by a different room acoustic from place to place. *Never* would the listener be fooled into thinking that a different instrument was playing, not even if listening whilst blindfolded. In all cases, the first sound heard for any new note or change in tonality would be the first arriving sound (the direct sound), which by definition must arrive anechoically because it arrives straight from the instrument, whereas all reflexions and resonances suffer arrival delays. It is the first arrival of any sound that the brain 'locks on to', and *that* sound has a characteristic 'fingerprint' to which all the subsequent sound is referenced. This is why acoustic, natural sounds *sound* natural. A listener could say that an instrument was perhaps not sounding as good as it could in a given room, but it could never be said that the instrument sounded *unnatural*.

If we now change the situation to one in which the source of sound is a loudspeaker playing a flat, anechoic recording of the same baritone saxophone, and given that the loudspeaker has roughly similar directivity characteristics to the instrument, there is no reason to expect much difference in the overall assessment of the sound in the room. However, if we were to now 'correct' the response for the room, by playing pink noise through the loudspeaker and measuring the response from our 'prime' listening position, everything would begin to change. First of all, it is almost unthinkable that the analysing equipment would show a flat response from the unadjusted system given all the reflexions and resonances involved. The tendency would be (given current thinking and practices) to equalise the system for a flat response. The application of the equalisation would distort (change) the frequency response of the loudspeaker system. As the loudspeaker would be the source of *all* the sound, it follows that the direct sound first arriving at the listener would change; the frequency balance of the reflexions would therefore also obviously change; the relative balance of the room resonances would change; and the overall reverberation characteristic would change. What is more, the relative effect of all of these changes would be different at all places in the room, and none of them would be the same as when the baritone saxophone was being played live. How, one could well ask, could this 'corrected' sound be more natural than the instrument itself? Patently, it cannot *be* so.

Let us now consider taking the saxophone on a tour of ten cinema theatres, all with rather different acoustics. In each room, the overall sound will be different, but the fact that the saxophone was always the same one (or at least a very similar one) would be very evident. If we then made a second tour with the excellent loudspeaker reproducing the recording of the saxophone, we could to a large degree experience a similar result. *If*, however, we then made a third tour, before which the loudspeaker had been equalised to give a flat response at a designated measuring/listening position in each room, the result would be that the recording of the saxophone sounded significantly more different that when it was reproduced flat. The fact is that with ten different equalisations in the ten different rooms we would be listening to ten different sources. When the sources are different, it is virtually impossible for the in-room responses to be the same. What is more, when the *ten sources* are different, the ear/brain has *ten different references*, and so perceives *ten different sounds*, and not the same sound in ten different rooms, which is what it *should* be doing. Brains understand this, even if only sub-consciously, but spectrum analysers do not! The situation addressed in Section 11.5 is not equalisable. *Rooms* are not equalisable.

If room-to-room compatibility is the goal, *the sources must all be the same*. If rooms have *gross* acoustic problems, they must be addressed by *acoustic* solutions. If the sources *are* the same, and the room problems are *not* gross, then room-to-room compatibility is almost assured. Equalisation is unnecessary except for any desired changes to the overall response.

A study of twenty rooms around Europe in 2010, all Dolby certified, indicated the damage being done by the current one-third-octave equalisation regime.[11] Figure 21.18 shows the results of the measurements at 2 m from the screens, and at two-thirds distance from the screen to the rear wall, (which is the standard calibration position), for 11 dubbing theatres. Figure 21.19 shows similar measurements for nine commercial cinema theatres of various sizes. In effect, the measurements at the 2 m distance broadly represent the direct sound from the loudspeakers. It can be seen very clearly from the figures that an enormous amount of damage is being done to the integrity of the direct sounds in order to try to force the far reverberant field measurements to conform to a predetermined standard by means of 'room' equalisation. Given the differences in the 2 m measurements in Figures 21.18 and 21.19, there

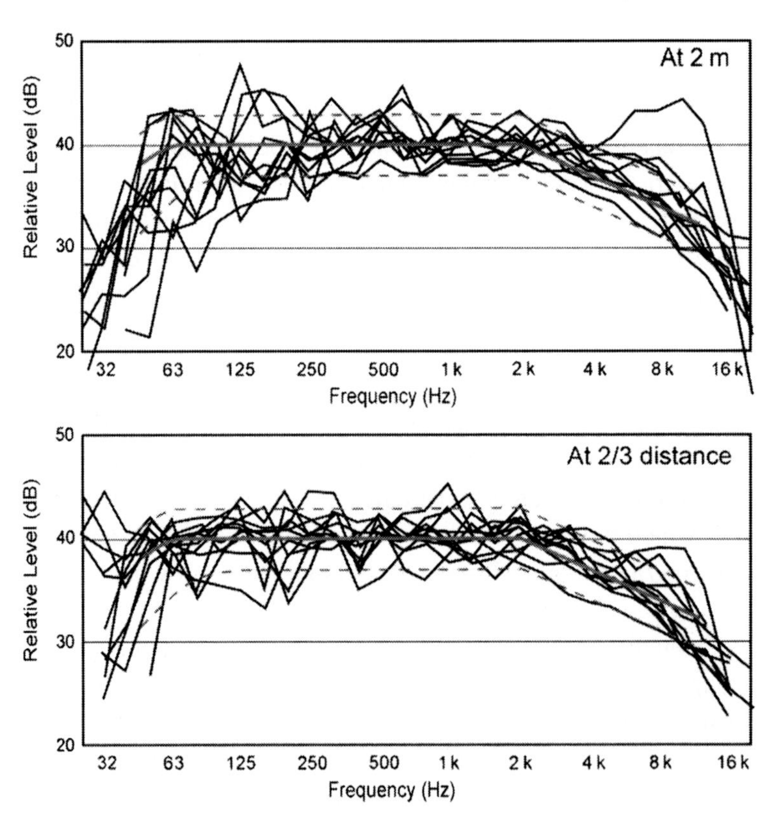

Figure 21.18:
Overlaid, third-octave responses of 11 Dolby Digital dubbing theatres.

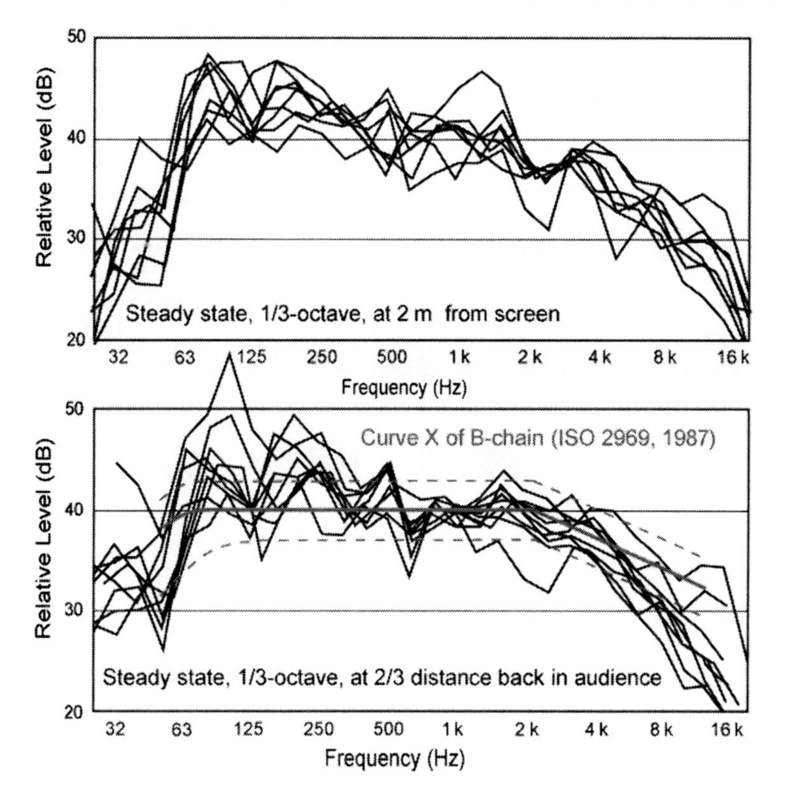

Figure 21.19:
Overlaid, third-octave responses of nine Dolby Digital public cinemas.

is *absolutely no hope* that such different direct-sound responses could sound similar. As has already been explained, if the direct sounds are *not* the same, then all hope of overall compatibility of the sounds from room-to-room is lost. The problem is that the pink noise measurements are simply lumping together the direct sound with all of the reflexions and resonances and general reverberation. The analyser cannot separate these components, *but the ear can*!

By contrast, Figure 21.20 shows the responses at 24 different positions in a music club.[12] In all cases the source was the same loudspeaker, which had a relatively flat response. Although there is no great difference in the degree of variability between the responses shown in Figures 21.18, 21.19 and 21.20, the tendency is for all the responses of the first two figures to sound different, but for those of Figure 21.20 to sound very recognisably as coming from the same source, which indeed they were. This relates back to the aforementioned discussion about the baritone saxophone. If the *source* is the same, the perception in any part of any reasonably controlled room is that of the same source being heard in different acoustic conditions. The brain knows what part of sound is the source, and what is the room.

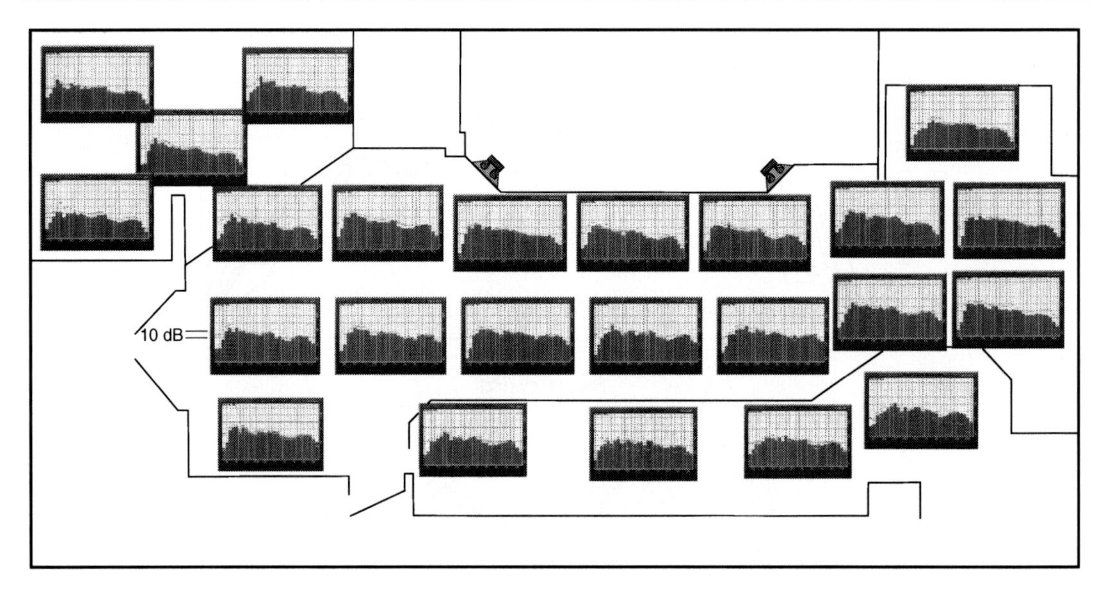

Figure 21.20:
Third-octave responses at 24 different places in a music club. In all cases, the source was the same loudspeaker mounted in the same position. The responses appear to vary considerably but the general impression of the sound was the same at all locations. The position of the plots on the drawing relate to their position in the club, the shape of which is outlined.

Conversely, even when the reverberant-field measurements are reasonably similar, if the frequency balances of the sources are different, the perception may be of different sources, which, in effect, is what they are. If the baritone saxophone excites disagreeable resonances in certain rooms, then for every place where there may be an unpleasant 'honk' on certain notes, there would be other places in the room where those same notes were weak. This situation applies whether we are referring to the real instrument or its reproduction via our appropriate loudspeaker. In the latter case, one could equalise the loudspeaker to reduce the resonance, but by doing so it would virtually kill the note at other places where it was already weak, and also upset the attack of the instrument at *all* locations. This is *not* a correction!

The situation is that for room-to-room compatibility, the direct sounds must be the same and rooms must be reasonably well damped, and not excessively reflective. If the rooms are *not* reasonably well controlled, equalisation cannot make them much more compatible, one with another. The only truly viable use of equalisation is to compensate for loading effects, such as where the left and right loudspeakers receive more low frequency loading from the side walls than do the centre loudspeakers, or to introduce desired corrections to the *direct* response from the loudspeakers. The acoustic losses due to the screen being in front of the loudspeakers can also be legitimately compensated for by equalisation, but it is customary to consider the screens as a part of the loudspeaker systems because they *can* have an effect on the overall directivity of a loudspeaker. In other words, the directivity of the loudspeaker, alone, cannot be presumed to be the same when placed behind a perforated projection screen. Obviously (or at least it should be obvious), the loudspeakers should exhibit a smooth and uniform directivity pattern which is adequate for the intended coverage area. If this fact is *not* respected, the probability is that the direct sound will *not* be flat in all the areas of the room and that the reflexions from the surfaces of the room will not have smooth frequency responses. It is this failure which gives rise to the generally uneven spectrum of the reverberant responses in many rooms, and which it turn calls for the 'absurd' equalisation which is customarily found to have been applied in so many cinemas after 'aligning' the rooms by means of pink noise measured in the far, reverberant fields of the rooms.

A more complete argument can be found in the papers referred to in the above text.[11,12]

21.12.5 Dialogue Levels and Room Equalisation

As previously mentioned, it has been traditional to mix cinema soundtracks in large mixing theatres in order that the many variables involved in transposing a mix from the studios to the public cinemas could be minimised. However, the more recent needs for compatibility with smaller theatres and subsequent DVD releases have further highlighted the problems of the compatibility of playback in rooms of different sizes. The X-curve has long been used for cinema loudspeaker playback, and has also (at least in principle) employed different high frequency roll-offs depending on room size and decay time, but another room-size related problem – that of the perceived level of the dialogue varying relative to the music and effects – has led many people to ask whether a further compensation needs to be defined and applied.

In general, as rooms become larger they tend to exhibit a decay time which rises at low frequencies. They also tend towards their first reflexions arriving later than those in small rooms, and the subsequent reflexions are more separated in time. It has been noted by many people that the relative level of the dialogue in a film soundtrack which has been mixed in a large room can seem excessively loud when reproduced in a small room. The situation is complicated to define, due to the number and the interaction of the variables involved. However, Allen, in his 2006 paper,[8] demonstrated how the reverberation in a room would develop in response to pink noise, and three figures from his paper are reproduced as Figures 21.21(a), (b) and (c). Given a reverberation characteristic as shown in Figure 21.21(a), it can be seen from Figures 21.21(b) and 21.21(c) how the first arriving signal, having a flat response, is subsequently subjected to a reverberant build up which is moderate at mid frequencies but greater at low frequencies and less at high frequencies. The signal is effectively amplified, and the spectrum is modified by the reverberation characteristics of the room.

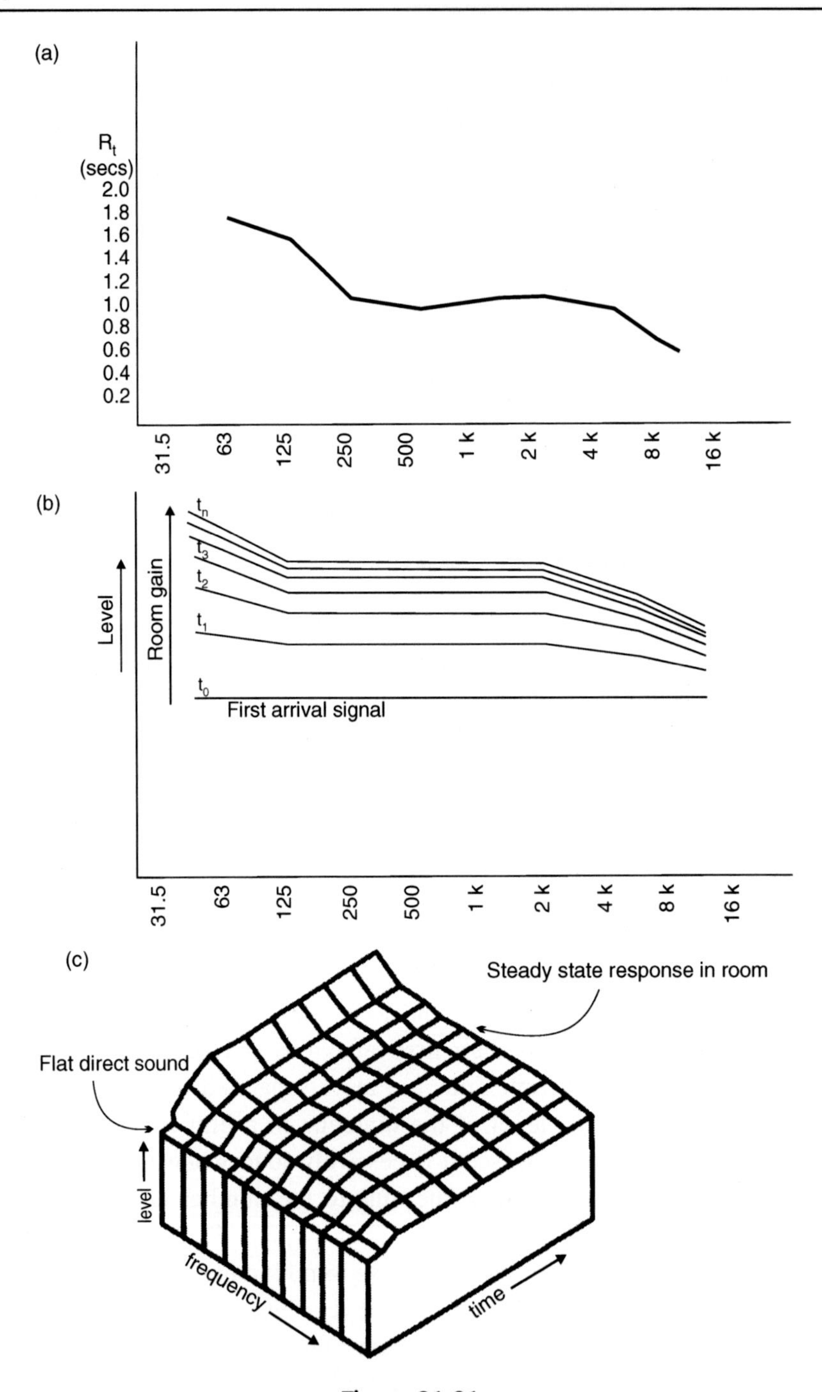

Figure 21.21:
(a) Typical medium to large size theatre reverberation characteristic.
(b) Pink noise build up over time in medium to large size theatres.
(c) Frequency response changes with duration of signal.

The short sounds of the spoken word are essentially in the range of 100 Hz–2 kHz, but their duration is too short to exhibit signs of reverberant build up. They are not steady-state signals like the pink noise, and so do not inject energy into the reverberant field for long enough to drive the reverberation to its full level. They are also above the frequency band that would suffer the greatest reverberation in a complex, wideband soundtrack. Consequently, in such cases, the low frequency build up could tend to mask the short sounds of dialogue and reduce the intelligibility as compared to dialogue at the same level but with less ambient or musical accompaniment. The natural tendency, therefore, is for the person making a mix for a soundtrack to elevate the dialogue levels when mixing in a room with a longer and low frequency-dominant reverberation time, as compared to mixing in a less reverberant room with less low frequency build up. This is due to the fact that despite the large and small rooms both being equalised flat, (but only, of course, to the steady-state pink noise signal) the time-smeared reverberant response exhibits a greater masking effect. As a result of this, when such a soundtrack is played back in a smaller, drier room, the dialogue levels may seem to be excessively high in terms of their relative balance with the rest of the soundtrack because the anticipated low frequency reinforcement does not occur. This is yet another complication to the large-room/small-room compatibility issue.

When a *room* is 'equalised' to be flat in response *to a pink noise signal*, with the equalisation of the loudspeaker system compensating for the reverberant build up at low frequencies, the time-history of the response would tend to be as shown in Figure 21.22, also taken from Allen's paper. The tendency would be for a direct signal to be reduced in bass and increased above around 2 kHz, leading to a generally thinner, harsher sound, which is exactly

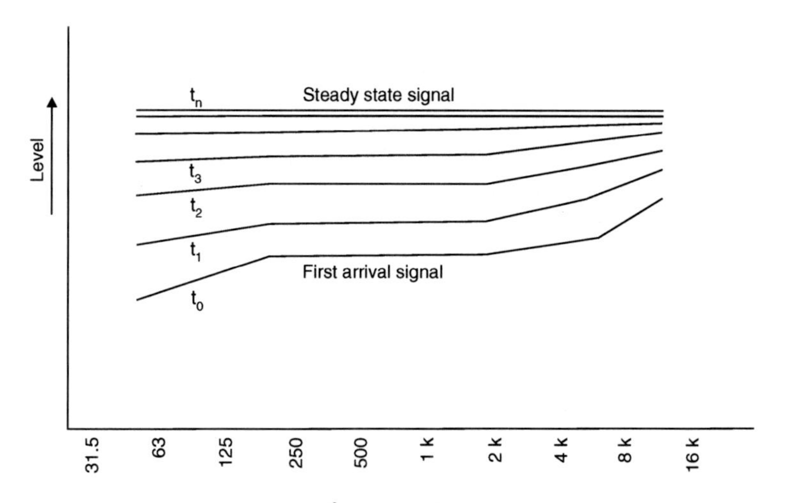

Figure 21.22:
What would happen to first arrival signal if pink noise was tuned for a flat steady-state response.

what has been reported by many people as being the nature of some of the sounds when auditioned in small rooms after being mixed in larger, more reverberant rooms. However, from the previous discussions it can be seen that the degree to which these effects may occur are highly dependent on the nature of the individual components of a soundtrack. The situation is by no means either clear or simple.

Nevertheless, in real life, if a human being were to speak in rooms of different sizes, the characteristics of the *direct* sound would *not* change. Therefore, to linearly distort (upset the frequency balance of) the direct sound in order to compensate for room effects, which although affecting more steady-state sounds do not make a significant change to the perception of the spoken word, can only be detrimental to the uniformity of the perception in different rooms. That is to say, once a large room has been equalised to the X-curve with pink noise, such that the direct response from the loudspeakers is similar to that shown for the first arrival in Figure 21.22, the re-recording (film soundtrack) mixers will probably equalise the dialogue to give it a more natural characteristic – in effect adding the inverse of the first-arrival response. When the soundtrack is played back in a smaller or drier room, in which the loudspeakers have been equalised with a flatter general characteristic, the dialogue may then tend to be perceived with a much 'heavier' sound. The concept is described more fully in Section 11.5. It follows from this that the older, larger, more reverberant dubbing theatres would not seem to be the ideal rooms in which to mix complete soundtracks for the generally drier acoustic conditions of modern cinemas.

A further indication from the above discussions is that even if the X-curve, or any other standardised equalisation, continues to be used, it would be better applied in the close field of the loudspeakers so that the direct sounds, in all cases, are more similar from room to room. It is in the similarity of the direct sounds where the compatibility lies; *not* in the similarity of the highly complex, far-reverberant-field responses.

Therefore, taking a mix from a large room, with a reverberation characteristic similar to that shown in Figure 21.21(a), and playing it back in a smaller room with a drier, flatter decay characteristic, we could observe the following:

1. The overall response of a large, more reverberant room with a significant rise at low frequencies (in the R_T) will consist of a direct signal which is bass light and, perhaps, slightly treble heavy. People mixing in such rooms will probably compensate for this effect by adding equalisation to the dialogue, and other short-lived sounds, which will restore the natural frequency balance of the direct signal. Subsequently, on playback in a smaller room which has been equalised with a flatter direct response, the dialogue will sound to have been boosted by the applied equalisation, which would not have been necessary in the flatter direct monitoring conditions of the smaller, drier room, and which therefore sounds excessive.

2. The dialogue will receive no significant reverberant support in either a large or a small room, so the effect of reverberation can be largely discounted when considering the dialogue only. However, the music and low frequency ambience would *not* receive as much reverberant support in a small, dry room as in a larger, more reverberant room, and so may sound weaker. Note that even if the levels of these signals had been boosted at low frequencies by the flatter direct signal from the loudspeakers in the smaller room, the faster overall decay may well still leave the dialogue more exposed.
3. The dialogue may receive more support from early reflexions in the smaller room due to the close proximity of reflective surfaces, because the reflexions arriving within 40 ms of the direct sound will reinforce its level. This would perhaps be noticed more on the short sounds of the dialogue than on the longer sounds of the music and ambience, because the latter two had received reverberant support during the mix whilst the dialogue had not.
4. Air absorption at high frequencies would be less in the smaller room, so any reflexions would tend to be brighter than in a larger room, once again giving a boost to signals which were being supported by the higher reflexion density.

From the differences in perception described above, it would appear to be self-evident that no simple application of a predetermined family of third-octave curves could lead to any uniform sound quality in the different rooms.

It should be noted that one of the main reasons why these points are so pertinent to cinemas, as opposed to most other forms of public performance via amplified sound systems, is that the cinema situation is somewhat unique in that there is no operator to adjust the sound level and equalisation for each performance. These characteristics are ‘frozen’ in the dubbing theatre, where the mixing personnel have only a *preset* monitor level control. Mixing is carried out at the (typically) Dolby reference level, which means that all mixing is carried out at the sound levels that would be expected in similarly calibrated cinemas. In principle, the projectionist in the public cinemas should leave the gain control at the calibrated setting (No 7) for all performances of all films, and he or she has no equalisation available if the dialogue lacks presence or is screaming at the audience. Conversely, in a discothéque, or other music venue, the DJ or sound operator will have level and equalisation controls available to adjust to taste any piece of music being played. It is the fixed nature of cinema projection which puts so much emphasis on the need for standardisation from the post-production to the public screening environments.

21.13 Summary

The lack of consensus about what format music-only surround should take has not been a help to the design of appropriate studios.

The most practical solutions for surround sound came from the cinema industry.

TV and video surround formats take into account their own sets of requirements.

Surround for music-only recordings would seem to serve little purpose if the frontal stereo channels are compromised.

The need for monitoring systems using five identical loudspeakers seems to be questionable, because the application and perception problems do not lead to symmetrical performance.

Five channels is a commercially imposed limit, which is only part way between two-channel stereo and what surround *should* be. Things get much better with ten channels, but it seems to be commercially impossible to implement.

Many surround (rear/side) formats have been proposed and tested. All tend to give different results, with their strengths and weaknesses suiting different music and circumstances. No outright winner has emerged.

In rooms where the acoustics have been designed to be as uniform as possible for the surround monitoring, the frontal stereo performance invariably seems to be compromised.

Conversely, in the better performing stereo rooms, the different concepts yield different responses and perception of the surround (rear) loudspeakers, and so provide no standard reference.

It is generally accepted that surround sound control rooms should have short decay times – perhaps the shorter the better until a point of ambient discomfort is reached.

The choice of separate low frequency sources or a single sub-woofer can be very dependent on the room acoustics. Perhaps the greatest degree of reality can be achieved by separate sources in highly controlled rooms.

'Processed multiple sub-woofers, in mono below 80 or 100 Hz, can offer considerable improvement in response uniformity when highly controlled room acoustics cannot be provided.'

'Three-channel stereo' rooms can be built with realistic rear loudspeaker responses which do not compromise the stereo performance. For 'stereo and ambience' surround mixes they can give excellent results.

Cinema dubbing theatres tend to need to be large because perception of the subjective sound character can vary greatly with room size.

As cinemas become smaller than they have been in earlier times, the decay times of both the cinemas and the dubbing theatres have been tending to go down.

The perception of high frequencies tends to increase as both reverberation time and room size increases, giving rise to the need for high frequency roll-offs in the reproduction systems of large, reverberant rooms.

The X-curve is now the recommended standard for the equalisation of cinemas and dubbing theatres, being appropriately modified to room size and decay time.

The surround loudspeakers follow a different equalisation curve to the behind-the-screen systems, changing their turnover frequencies rather than their rates of roll-off.

Sources must be as similar as possible to each other if room-to-room compatibility is the goal.

Rooms *cannot* be equalised.

Cinema mixing is intended to take place at the same sound pressure levels as it will be heard by the public in the theatres.

Different room sizes can give rise to the perception of different 'ideal' sound levels for any given picture.

References

1 Newell, Philip, 'A Load of Old Bells', *Acoustics Bulletin* (*Journal of the UK Institute of Acoustics*), Vol. 26, No. 3, pp. 33–7 (2001)
2 Holman, Tom, 'New Factors in Sound for Cinema and Television', *Journal of the Audio Engineering Society*, Vol. 39, No. 7–8, pp. 529–39 (July/August 1991)
3 Chase, Jason, 'Hi-Fi or Surround. Part Two', *Audio Media* (European Edition), Issue 92, pp. 122–6 (July 1998)
4 Newell, P. R., Holland, K. R. and Castro, S. V., 'An Experimental Screening Room for Dolby 5.1', *Proceedings of the Institute of Acoustics*, Reproduced Sound 15, Vol. 21, Part 8, pp. 157–66 (1999)
5 Bank, Graham, 'The Distributed Mode Loudspeaker (DML)' in Borwick, John (ed.) *The Loudspeaker and Headphone Handbook*, 3rd Edn, Chapter 4, Focal Press, Oxford, UK (2001)
6 Toole, Floyd, E., 'Loudspeakers and Rooms for Sound Reproduction – A Scientific Review', *Journal of the Audio Engineering Society*, Vol. 54, No. 6, pp. 451–476, June (2006)
7 Newell, Philip R., Holland, Keith R., 'Surround-Sound – The Chaos Continues', *Proceedings of the Institute of Acoustics*, Vol. 26, Part 8, pp. 135–147, Reproduced Sound 20 conference; Oxford, UK (2004)
8 Allen, Ioan, 'The X-Curve: Its Origins and History', *SMPTE Journal*, Vol. 115, Nos 7 & 8, pp. 264–275 (2006)
9 Newell, Philip R.; Holland, Keith R.; Neskov, Branko; Castro, Sergio; Desborough, Matthew; Torres-Guijarro, Soledad; Pena, Antonio; Valdigem, Eliana; Suarez-Staub, Diego; Newell, Julius P., Harris, Lara; and Beusch, Christian, The Effects of Visual Stimuli on the Perception of 'Natural' Loudness and Equalisation, *Proceedings of the Institute of Acoustics*, Vol. 30, Part 6, pp. 15–26, Reproduced Sound 24 conference, Brighton, UK (2008)
10 Newell, Philip R., Holland, Keith R., Neskov, Branko; Castro, Sergio; Desborough, Matt; Pena, Antonio; Torres, Marisol; Valdigem, Eliana; and Suarez, Diego, The Perception of Dialogue Loudness Levels Within Complex Soundtracks at Similar Overall Sound Pressure Levels in Rooms of Different Sizes and Decay Times, *Proceedings of the Institute of Acoustics*, Vol. 29, Part 7, pp. 125–139, Reproduced Sound 23 conference, Gateshead, UK (2007)

11 Newell, Philip; Holland, Keith; Torre-Guijarro, Soledad; Castro, Sergio; and Valdigem, Eliana, Cinema Sound: A New Look at Old Concepts, *Proceedings of the Institute of Acoustics*, Vol. 32, Part 5, Reproduced Sound 26 conference, Cardiff, UK (2010)
12 Newell, Philip; Holland, Keith; Newell, Julius; and Neskov, Branko, New Proposals for the Calibration of Sound in Cinema Rooms, *Presented to the 130th Convention of the Audio Engineering Society*, Preprint No 8383, London, UK (May 2011)

Bibliography

Holman, Tomlinson, *5.1 Surround Sound – Up and Running*, Focal Press, Boston, USA and Oxford, UK (2001)
Toole, Floyd E., *Sound Reproduction: The Acoustics and Psychoacoustics of Loudspeakers and Rooms*, Focal Press, Oxford, UK (2008)

CHAPTER 22

Human Factors

Needs of different personalities. Human expectations and fulfilment. Perceptual differences.

When designing recording studios, what must always be borne in mind is that the studios exist to capture artistic performances, and anything that the studios can positively contribute to those performances is desirable. The recording equipment is a product of the technology, which is in turn a result of the application of the relevant sciences. What makes studio design so fascinating is that it involves art, technology and science in such a way that they can harmonise to produce fabulous music. In the final analysis, however, it will be the subjective assessment of a recording studio, and of the recordings and mixes that it produces, which will determine whether any design is successful, or not.

22.1 The Ambiance of the Occasion

In general, if the objective aspects of a studio design are not up to standard, it is unlikely that the studio will satisfy its subjective requirements. As discussed in Chapter 1, there *are* cases where some very objectively poor studios can make successful recordings, but the successes are likely to be idiosyncratic and limited in scope. However, no studio, no matter how well designed, can be all things to all people.

For example, many musicians need an intimate feel in a studio. Perhaps they can only perform in a small room with little lighting, and do not wish to be stared at by the recording personnel. They prefer that their presence should be more or less ignored, and that only their music should be apparent. Conversely, there are artistes who need a big stage in order to perform at all. For them, the small studio lacks the majesty that their performances require. A big studio gives them a sense of occasion, and a group of onlookers in the control room only adds to their sense of importance. There is no right or wrong, or better or worse approach. They are equally appropriate to the personalities of the performers in question, and somewhat surprisingly, the stage persona of an artiste does not always allow one to guess what type of studio would suit them best.

There are musicians who arrive at a studio dressed very casually, and there are musicians who arrive at a studio in their full stage regalia. Some people often think that the latter case is rather pretentious, but it should be remembered that during much of the twentieth century the BBC required its *radio* newsreaders to wear evening dress, because it was considered to add to the sense of occasion and tended to make the presentation more formal. As the BBC was seeking to present an authoritative voice, the evening dress was considered to contribute to that goal. Similarly, if a musician needs to feel outgoing and important in order to perform, then whatever dress aids the achievement of the sensation is a valid recording accessory.

Recording engineers and producers are also artistes, so anything that makes their working lives easier will contribute, by allowing them to spend more time in a relaxed mood, and thus concentrating more on the music rather than on the equipment. By contrast, there are the 'techno' types whose level of creativity rises in proportion to the number of knobs, bells and whistles within easy reach, and who feel that *they* cannot perform on equipment 'below' their presumed status. No problem with that, either, if it goes towards improving the recorded results. Studio design, therefore, contains a considerable degree of psychology, like the design of a new football stadium, which can make a team feel at their best and play accordingly.

22.2 The Subjectivity of Monitoring

Ultimately, a recording engineer will choose to work on whatever loudspeakers and in whatever conditions are conducive to achieving the desired results. This may seem like a strange statement to make after 21 chapters on some of the finer aspects of studio and control room monitoring design, however, familiarity, culture, peer pressure, training, the success or failure of prior recordings and many other factors, all have a bearing on personal choice. A person who has spent many successful years working on one type of monitor system may be very reluctant to change their point of reference, even when confronted with a monitor system whose performance is demonstrably superior. This was clearly highlighted in Section 19.10.

There is also a widespread tendency for many very experienced recording personnel to use different loudspeakers for recording, mixing and listening at home. A typical pattern would be to use large, full frequency range, high-resolution monitors for recording, a not too good small loudspeaker for mixing, and something pleasant to listen to at home. At least that is a trend in the pop/rock music world. In the classical recording world there is usually much less use of the lower quality mixing loudspeakers, and the large loudspeakers do not need to be used to 'give a buzz' to the musicians, because the musicians are normally much less involved in the decision-making in the control room. The producers, engineers and conductors are also tending to look for classical sounds, so synthetic processing for creating special effects is generally eschewed. However, as we shall discuss later, classical recording is also not free from controversy.

One school of thought seems to believe that this situation is a result of the failure of the electroacoustics industry to produce a definitively accurate loudspeaker, but careful conversation with very experienced recording engineers and producers soon reveals that this is not the case. The situation is somewhat similar to a photographer having different lenses to choose from, calling variously for panoramic views, close-ups, or special effects lenses, depending upon perceived requirements. In rock recording, the big picture is needed during the recording phase. If a multi-track recording can be made such that each track has been well scrutinised for noises, distortions and tonal balance, then the mixing phase can be made much easier. During the mixing, however, it is often in the mid range where many instruments may be fighting for the same space. In such cases, a loudspeaker with a prominent mid-range can be useful to draw attention to, and to expose, the critical mid-range where vocals, lead instruments, and reverberation and effects may all require careful balancing.

Admittedly, it could be argued that a better or more controlled musical arrangement should have ensured that the conflicts never arose in the first instance, but that is not the reality of recording. Recording is a very human process, and it is surprising how many recording engineers and producers have almost no interest in how or why a given loudspeaker works. They only want to know if it works *for them* at any stage of a recording process. There is also a noticeable tendency for many people to have loudspeakers at home that are very different from those that they work on. At home, they want to hear the shine and the polish, not the nit-picking detail that they spend their working days trying to resolve. Conversely, there are hi-fi enthusiasts who wish to use recording monitors so that they *can* hear all the minor problems that the recording staff would rather that they did *not* hear, and did not intend that other people *should* hear.

So much of this boils down to what each individual *wishes* to hear from a recording, so we are very much in an area of subjective choice, and not only objective standardisation. People want to hear what they want to hear, for their own purposes. This can create a difficult situation, because the imposition of a 'perfect standard monitor', even if it were achievable, would not be enforceable. Ostensibly, a process of refinement is always in progress, and it is clear that the general upper level of monitoring was better in 2002 than it was in 1982. Nevertheless, looking at the recording industry as a whole, the same cannot be said. The 20 year reign of the NS10M would not have been possible as a *de facto* reference if the general trend was leaving it behind.

Improved loudspeaker/amplifier systems have appeared with corresponding increases in prices, which completely buck the trend of the reducing real cost of much recording equipment. The human response in the lower and middle echelons of the recording industry has usually been to scale the costs of the monitor systems to the costs of the whole recording set up. This situation has not led to a general improvement in monitoring quality, even when it is technically feasible to move several paces forward from what was previously in use. The fact is that the majority of loudspeaker manufacturers now begin with a price for a loudspeaker which they think that they can market, and then design a loudspeaker around that price.

Quite seriously, sound quality may be number four or five on the list of priorities, following price, appearance, lightness (for cheaper shipping), or features of competing models. Such things are then sold as monitor loudspeakers, which is a travesty, but a reality nonetheless. This situation cannot really be said to be encouraging.

22.3 Conditioning and Expectations

Toole and Olive have stated that:

> *It turns out that much, if not most, of the variability on commonplace opinions can be traced to differences in the listening experiences that led to the opinions. When these nuisance variables are controlled ... it is remarkable how consistent, and how similar, human listeners can be. We can be very good measuring instruments.*[1]

The problem faced by most studio designers is that they are often confronted by clients of very limited breadth of listening experience, and which has *not* been well controlled. Even some very experienced producers and engineers openly admit to dreading the necessity of changing from their usual monitor loudspeakers, even though there may be no objectively based reasons why their chosen monitors should be used as a reference.

The recording industry is a working entity which has to survive the onslaught of fierce (and often 'unfair') competition, financial constraints, shareholders' profit expectations and the demands of newly successful artistes, whose new found power may be totally disproportional to their knowledge and experience of the recording process. Perhaps above all, the industry has to face the barrage of sometimes quite ruthless advertising campaigns. The majority of people working in the industry are also trying to survive; they need to make a living. They are trying to do their own jobs in the middle of a fashionable and rather unstable business, where the ultimate paymasters, the consumers, are usually buying the products for purely nebulous reasons. Relatively few people in the industry are prepared to fight the avalanche of subjective opinion that drives so much of the decision-making, because they have families to support.

Therefore, whilst the comments of Toole and Olive are doubtless very factual, the chances of getting an industry to form any reasonable sort of consensus about such things as monitoring are perhaps too remote to be hoped for. Indeed, Toole and Olive themselves published a paper entitled 'Hearing is believing vs believing is hearing'. It related to the difference in results from tests on the same loudspeakers depending upon whether they were carried out blind or sighted. They also wrote, 'Judges of wine, and the like, are willing to put their reputations on the line in blind evaluations; it is time for the audio industry to do likewise'.[2] Again, whilst it is very hard to fault Toole and Olive's reasoning, it must be remembered that the audio and recording industries in general are not particularly logic or reason based. Peer pressure and marketing work to such an extent that many people convince themselves that they are hearing what they think that they are expected to hear from a given piece of equipment.

A recording studio designer therefore must take these aspects of human subjective variability very seriously, because they cannot effectively be changed when an industry is based on such fragile foundations. They must also be taken seriously because, even though they may be unreasonable in the context of what perhaps ought to be a more professional industry, the reputation of the designers may nevertheless stand or fall according to the effect of prior expectations. A potentially great studio may even develop a bad reputation due to the incompetence of its staff and owners.

The majority of reputable professional studio designers are experienced and knowledgeable people, and they usually have a considerable degree of influence over the wishes of their clients. However, when clients require designs that are too subjectively based, with too much disregard for the principles of good electroacoustic design, it would be foolhardy for the designer to go along too far with the wishes of the client. Unfortunately, and at times despicably, there are other people with vested interests who will readily remind all who are willing to listen to them about the nonsense committed by designers who *have* bent too far to clients' irrationalities, implying that bad design was due to the *designer's* ignorance. It can be remarkably difficult to shake off the bad reputation due to one imprudent design, even though a designer may have been in considerable disagreement with the client's ideas. Such errors of judgement can stick like glue, so it is usually wiser for a designer to refuse a commission, no matter how apparently lucrative, if it is considered to stray too far from his or her beliefs about how a studio *should* perform.

22.4 Lack of Reference Points in Human Judgements

Of course, it is well known that pop/rock recording personnel have a wide range of opinions about which monitor loudspeakers to use. This is not too surprising when one considers that so many sounds are so stylised that no obvious or natural reference exists, but the classical music world does not appear to have clear-cut judgements, either, despite the rather obvious and even ubiquitous existence of the 'real sound' for reference. The plethora of 'preferred' microphone techniques bears this out. Considering that they all would produce different sounding recordings of the same orchestra if used simultaneously, there must be some disagreement about which technique would be most 'right'. People can, and do, argue at length about this, and the views of Stanley Lipshitz on this subject can be found in Section 16.10. The obvious consequence of this is that if people cannot decide upon which type of microphone arrangement is right, then how can they be expected to decide upon which loudspeaker system is right?

Therefore, as different loudspeakers can tend to suit different microphone techniques (and *vice versa*), we begin to run into the 'chicken and egg' paradox, where one microphone technique can tend to make a person prefer a certain type of loudspeaker, or where one type of loudspeaker can lead a person to choose a certain microphone technique. This problem is not easily solvable, and a much more disciplined approach to subjective/objective evaluation would be necessary before the problem could even be reduced. Toole and Olive have stated:

> *We make accurate technical measurements that we have difficulty in correlating with listener evaluations, and then compound the problem by making subjective evaluations that are unreliable. Would any serious engineer make a voltage measurement with an unreliable voltmeter? Not likely! … If this industry is to be thoroughly professional we must put as much effort and intellectual investment into subjective measurements as we do into their technical counterparts.*[1]

Given the way that this industry is developing, however, such investment seems unlikely.

22.5 Studios and Control Rooms

That the design of the recording rooms is an art form is in little doubt. The rooms in which the musicians play are instruments, or at the very least they are *extensions* of the instruments being played, because often the originally intended tonal character of the instruments took into account a degree of ambient sound. If the rooms feel good to the musician and sound good when recordings are made via microphones, then the rooms *are* good. Of course, a degree of understanding of the scientific principles involved goes a long way to achieving a successful design, but the same also applies to instrument designers and manufacturers. Few people would expect that they could work at weekends in their garden sheds and produce an electric guitar like a Les Paul or a Fender Telecaster. These were the products of highly talented and experienced designers.

It can also be said that if a control room leads the recording personnel to achieve great mixes, then it too must *be* good, irrespective of its measured performance, but the situation here is somewhat different. Notwithstanding what has been discussed in the previous sections, there is still an implicit need for a control room to be a room of general neutrality, which allows the listeners to hear *the recording*. Unlike in the studio, where the ambience can affect the performance *and* the recording, in a control room the ambience is a property of listening in that room, and that room alone, and this can be further complicated by any non-uniformity or non-linearity of the monitor system response.

The fact that some exceptional and gifted people, such as those mentioned in Section 19.10, can make high-quality mixes on a pair of NS10Ms does not mean that a pair of these loudspeakers in any room will justify the title of 'a monitor system'. These people are working with a great deal of experience, and often have the luxury of beginning a mix from a very high-quality multi-track master, made by skilled engineers in good rooms and with excellent musicians, instruments and recording equipment. This is a long way from a pair of NS10Ms in a poorly treated room where budget priced equipment is used to record non-virtuoso musicians with good, but not excellent instruments. In these cases, excellent monitoring conditions could often make the difference between a mediocre recording and a very good one. The more true information that the monitoring can pass to the recordists, the easier will be their task. Unfortunately, this is often difficult to explain because poor rooms defeat many of the subtle characteristics of the better monitor systems. Laying on its side a loudspeaker which was designed to stand upright (see Figure 18.5) may not make any difference to the sound in a room that is already a chaotic mass of reflexions. Only when better basic acoustic conditions exist can many such important subtleties be detected. However, until a person has heard the benefit of the differences, their virtues can be difficult to explain.

Studio designers know this, but the industry is now largely in the hands of people for whom ignorance is bliss. However, quality has a tendency to rise to the surface and to be noticed. Although they are now only a small proportion of the total number of studios, the well-designed ones still form the backbone of the industry, and they will surely continue to do so. Nevertheless, it is also probable that human perceptions and misconceptions will also continue to rock the boat.

22.6 Summary

Studios exist to capture artistic performances. Anything in the studio environment that helps the artistic ambience is a positive contribution to the whole event.

Different characters and personalities respond differently to different environments.

Many recording personnel who work in various studios may use a familiar set of loudspeakers in the close-field in order to have a rather more fixed point of reference.

The choices of such loudspeakers can be very personal, and are often made instinctively rather than logically.

In the whole recording industry and its peripherals, above all, many people are trying to earn a living. It should therefore not be considered that all 'professional' decisions would be altruistically based for the benefit of better recording. Many decisions will in fact be based on how to make more money for the supplier of the service or product. This is the way of the world, but its repercussions should still be considered seriously in any decision-making processes.

No studio is, in itself, good. How it performs will be very dependent on the skill, experience and personalities of its staff.

It may be unwise for designers to undertake jobs which their experience tells them are too heavily compromised. The bad reputation which they may get from the results may be difficult to lose.

It is perhaps surprising that classical recordists, who invariably have a live sound source as a reference, choose to work with some very different loudspeakers which they each consider most 'right'.

Judges of wine, and the like, are 'willing to put their reputations on the line in blind evaluations; it is time for the audio industry to do likewise'.[2]

It can be difficult to demonstrate the advantages of monitoring via excellent loudspeakers if the room in which they are auditioned is limiting their performance. Bad rooms often tend to lead people to use poor monitor loudspeakers, because the benefits of the better ones are often difficult to perceive in bad rooms.

Well-designed studios still form the backbone of the recording industry.

References

1 Toole, Floyd E. and Olive, Sean E., 'Subjective Evaluation', in: Borwick, John (ed.) *The Loudspeaker and Headphone Handbook,* 3rd Edn, Chapter 13, Focal Press, Oxford, UK (2001)
2 Toole, Floyd E. and Olive, Sean E., 'Hearing is Believing vs Believing is Hearing: Blind vs Sighted Listening Tests and Other Interesting Things', 97th Audio Engineering Society Convention, Preprint No. 3894 (November 1994)

CHAPTER 23

A Mobile Control Room

The acoustic characteristics of a vehicle body. The problems presented and their means of control. Loudspeaker design for specialised conditions.

It would seem appropriate in at least one chapter to try to go through an entire design process. This way it will be possible to get an idea of the typical problems that arise, and the means available to solve them. Furthermore, if a mobile recording vehicle is chosen as the object of the exercise, it will allow us to look at another set of circumstances that have not been discussed in earlier chapters. So, let us now look at the concept, design and construction of a multi-track mobile recording vehicle, designed by the author at the end of the twentieth century.

23.1 The Problems to be Solved

During the spring and summer of 1998, Portugal was presenting Expo '98, a world fair of the oceans. The enormity of this exercise was to occupy much of the country's mobile recording and broadcast facilities, so the company Banzai Lda, which was engaged in many other recordings, decided to invest in a mobile sound recording vehicle in order to fulfil their own recording contracts.

The final decision to build a vehicle was only taken in early 1998, and the first recordings were to begin at the start of the following May. The three months available for the design and construction appeared to be too short to make an entirely new truck, so a search was made to find an existing vehicle that could be rebuilt to their specifications. They discovered that NOV, the Dutch broadcasting organisation, had a sound vehicle for sale, so one of Banzai's owners went to Holland to see it. The truck appeared to be suitable for their needs, and it had the necessary infrastructure, but internally it needed to be totally rebuilt.

This was quite an interesting challenge, because the request came exactly at a time when 25 years earlier, the selfsame designer had been building the first of the Manor Mobiles, to which the construction and dimensions of the NOV truck were not too dissimilar. The problems were therefore known intimately, and it was a great opportunity to see if it would now be possible to deal with the limitations that had been so seemingly insoluble in the past.

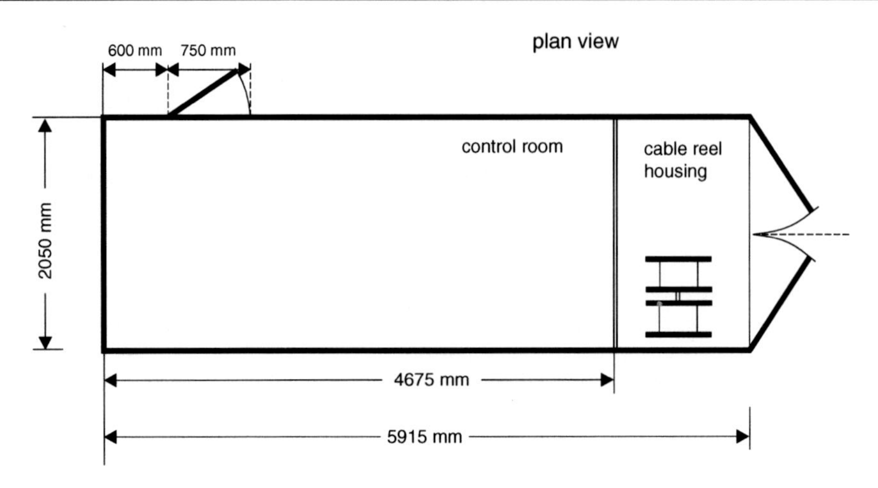

Figure 23.1:
Available space after removal of original recording equipment.

Monster vehicles were easier to deal with, but the mid-sized vehicles, with control rooms in the 5 to 6 m range, tended to be too small, the wrong shape, and too light from an acoustic point of view if a flat monitoring response was required.

The first Manor Mobile was built in a 20 ft × 8 ft × 8 ft (6 m × 2.4 m × 2.4 m) freight container, and the Banzai vehicle had a working space only slightly smaller, as shown in Figure 23.1. Both vehicles were made principally from glass-reinforced plywood (GRP) with heavily damped plywood floors on steel beams. The basic response of this type of shell, driven from an end wall, is shown in Figure 23.2. The Manor container was fitted with a rear bulkhead having a door and cable housing, so that the truck could be used with its heavy container doors open, as shown in Figure 23.3. The walls and were lined with 13 mm plasterboard to improve their damping and sound isolation, but the sheer size of the then current equipment precluded the use of much space for any resonator boxes – the traditional method of 'tuning' a room in those days.

The plasterboard/plywood walls did absorb *some* low frequencies, but they achieved much of their sound isolation by reflexion, so they tended to bounce a lot of energy around inside the room, creating the overall response shown in Figure 23.2. This was not ideal, by any stretch of the imagination, but the situation had been helped to some degree by the use of Tannoy Dual Concentric loudspeakers in Calrec/IMF transmission line cabinets. At least with these coaxial loudspeakers the 'point source' nature of the sound generation meant that most reflexions arrived at the listener with their phase and frequency responses intact. Loudspeakers with spacially separated drive units for different frequencies tend to produce reflexions where the different path lengths from the ear to the different drivers cause time and phase distortions to occur. Hence the 'don't lie your NS10s on their sides' rule.

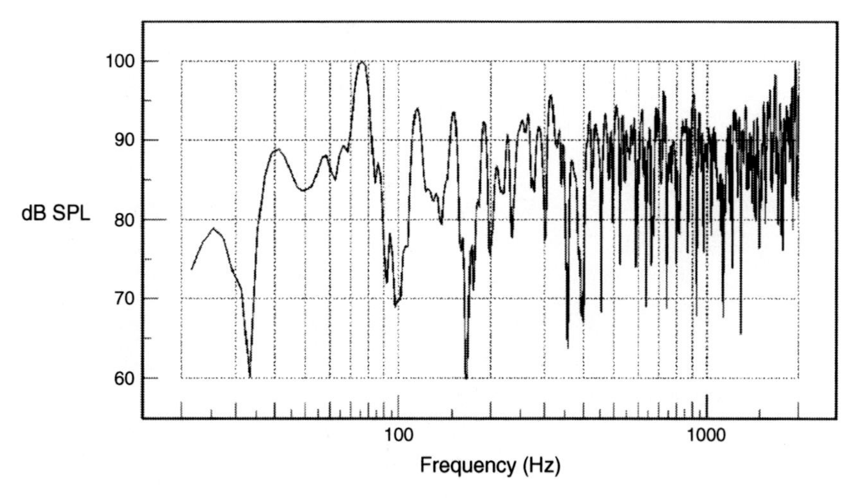

Figure 23.2:
Simulated frequency response of bare shell (45 cm × 200 cm × 200 cm).

The Manor Mobile was lumpy in its low frequency response, but no more so than the other vehicles around at the time. However, the crew knew it well, and were largely able to compensate. This was a huge improvement on the situation during the designer's early days with Pye Records, whose mobile unit *without* a control room in a vehicle had two Neve Series 80 consoles, each of which could be split in half for transport, or the two complete units could be joined together. It took four people to lift each section, and two people to lift each of the three sections of the 'portable' 3M M56 8-track tape recorders. All this, together with the large Lockwood/Tannoy monitors, would be set up in the best available space for each recording: dressing rooms, offices, hallways and the like. Obviously, this meant that the monitoring conditions varied wildly, so apart from listening for noises or distortions, and perhaps assessing the relative musical balance of a section of instruments, not too much could be done in the way of judging the timbre of an instrument, especially at the lower frequencies.

Nevertheless, this situation did have its compensations, such as recording at the 'Talk of the Town' in London, where the only available 'control room' was the chorus girls' changing room. During the various acts, they would come in, strip off, and change costumes. Concentration *definitely* suffered on such occasions.

Together with the monitoring variability, the great loss of time for rigging and de-rigging led Pye to build their first dedicated mobile vehicle in 1971, which went on the road just after the Rolling Stones' mobile. It was immediately appreciated that this gave a new consistency in monitoring conditions, which, even if they *were* wrong, at least they were always the same. The crews could learn how to come to terms with their 'wrongness' which was something which was very difficult when portable equipment was set up in a different room for each recording venue. In fact, when The Manor Mobile bought the Pye

(a)

(b)

Figure 23.3:
(a) The first Manor Mobile, in 1974, recording Queen at the Rainbow Theatre, London. The rear container doors are fully open, revealing the rear bulkhead and doorway. The huge fuel tank, beneath the trailer, was one of three, fitted during the 1973 oil crisis. The tanks gave the mobile a range from Oxford (its base in the UK) to the south of France, and back, without refuelling. Fuel was bought wherever available; (b) The inside of the vehicle, a week before going into operation in July 1973. On the right can be seen the world's first Ampex MM1 100 24-track tape recorder, bought by The Manor Mobile in the preceding January. The machine was a pre-production model, built in the Ampex laboratories, not the factory.

mobile in 1974, not only was it up-dated and expanded, but it was refitted into a new vehicle, very similar to the first Manor Mobile. This was specifically done to try to ensure the closest match of monitoring conditions between the two. The crews therefore did not have to remember which truck they were in, or which mental compensation to make.

23.1.1 Electronic Control Limitations

The trucks worked well for recording purposes, and built an excellent reputation, but they were difficult environments for mixing, which was hard work and uninspiring. In fact, the trucks had never been intended for mixing, but the lack of 24-track studios in those days often left no other recourse. When Virgin built The Manor Studios with Tom Hidley, in 1975, its equalised Westlake monitors became the new reference. The Manor's analysing equipment was taken into the trucks, and efforts were made to try to equalise the monitors, but it was simply not possible to achieve a flat response, the peaks and dips were too severe. Six decibels represents a four times power difference, so trying to correct a 6 dB response dip simply overloaded the amplifiers and loudspeakers (which were only rated at about 50 W, anyway) as it called for four times more power. Furthermore, no matter what equalisation was applied to the system, it always seemed to sound subjectively less natural, even if it did at times sound a little more uniform.

Of course, such efforts were futile. The majority of the response irregularities were caused by the reflexions in the room and the standing wave field. This was obvious from the fact that the low frequency character changed significantly as one walked around the room. Equalisation can *only* be used to deal with the minimum-phase problems, which are the response irregularities caused by driver roll-offs, the *loading* effects due to the proximity of room boundaries (*not* the reflexions themselves) and certain other response problems which are independent of listening positions. These, remember, are the effects that are superimposed on the acoustic signal at its time of generation. In these cases, correction of the amplitude response by equalisation will tend also to correct the phase and time responses.

However, *non*-minimum-phase problems are the ones such as those created by *multiple* reflexions, when the disturbance is caused by a mixing of the direct signal and time-shifted reflected energy. The arrival-time difference between the two signals produces comb filtering in the signal, as was shown in Figure 5.12. It is impossible to use conventional equalisers to compensate for such room effects, because any attempt to equalise the source signal to compensate for the combined signal will distort the frequency balance and the transient response of the direct signal. The integrity of the direct signal is super-important, because by definition, until the first reflexion arrives, *all* rooms are anechoic, so the direct sound is the first sound heard, and it must be accurate.

Any equalisation of the direct sound will upset the transient response, and hence the wavefront of the signal. Anybody with the vaguest knowledge of how to get sounds on synthesisers will know well how important the attack is to the character of a sound. Quite remarkably, in the early 1970s, although these principles were known in academic circles (and obviously to Robert Moog), they were largely unknown in the recording world, and it had only been a few years earlier that Altec had brought out their 'Acoustavoicing' system, precisely to commit equalisation atrocities to monitor systems.

The warning bells started ringing for the recording staff of the Virgin Studios group in late 1978. Their new Townhouse Studios One and Two, in London, had very similar control rooms. Studio Two had the first large SSL mixing console in the UK, and therefore had its own specialist staff because this mixing console was very different in concept and operation compared to the usual mixing consoles of the day. The monitor systems in the two control rooms were identical, but the equalisation settings on the monitor equalisers were done by different people, and those settings were very different, as was the sound. This was despite the response of both rooms reading flat on a 27-band spectrum analyser. Chris Jenkins, Andy Wild, Steve Cater and Malcolm Heeley, the maintenance staff, brought this to the attention of the author, who was technical director of the Virgin Studios group, at that time. It seemed that they each had their own method for adjusting the equalisers. It was decided to zero all equalisers and have each of five people equalise each room separately. After taking notes of the settings, it was found that there were ten different filter responses, five for each room as adjusted by five different people. Now, *one* response in *one* location in *one* room from *one* signal source has *one* inverse response, *not five*! It was beginning to appear that it would have been possible to find dozens of different equaliser settings which would all read flat on a spectrum analyser, yet which would all produce different sounding monitor systems.

The situation was reported to Tom Hidley, who had designed the control rooms and monitor systems. He said to try to find the setting with the minimum of large excursions from zero; the smoothest curve. This concept was then applied as the general principle for any monitor equalisation in the Virgin Studios group, but the general confidence in the value of monitor equalisers had been shattered. Since the early 1980s, neither Tom Hidley nor the author have used monitor equalisers, and this is the case for most designers these days.

These days, there is a proliferation of digitally adaptive 'room' control systems, but they also have their limitations. Firstly, the improvements at the 'hot seat', or in the chosen listening area, are gained at the expense of a poorer response in the rest of the room; see Figure 11.28. Secondly, many of them are based on 16 bit/48 K audio. When so much is now being done to introduce 24 bit/96 K audio, there would seem to be little purpose in trying to compare different high bit-rate/high-sampling-rate recording systems via monitors that only have 16bit/48 K resolution. It is on the monitoring system where one needs the most transparency of all, or judgements cannot be sensibly made.

23.1.2 Space Problems

Back to the situation with the Banzai mobile, their need was for a truck in which they could not only record, but also mix for television *and* CD releases. Hundreds of recordings in the old Manor Mobiles 1 and 2 had yielded much experience about the problems of the response in this type of basic shell, but, as has been discussed in the previous paragraphs, correcting this by monitor equalisation was out of the question. The Banzai truck was also seemingly too small to install sufficient conventional *acoustic* control systems, so how could a flat monitoring response be achieved? Fortunately, acoustic control measures had moved a long way forward from the use of the Helmholz resonator boxes of 30 years previously, but they are still area and depth dependent. In general, the surface area covered by the absorbers controls the amount of absorption for any given frequency, and the depth of the absorbers will determine the lowest frequencies that can be absorbed; although some trade-off is possible between the two.

Conventional acoustic control measures, as discussed in Section 13.7, tend to be wavelength dependent. They are not vehicle-size dependent. Tape One, in the late 1970s, had an excellent vehicle, but it was twice the size of the Manor Mobiles, and was almost half filled with acoustic control systems. In the case of the Banzai vehicle, such techniques (as they cannot be scaled) would have resulted in a truck that was all absorber – no control room! Clearly, this approach was out of the question. What was needed were solutions that could produce a respectably flat response in the control room, especially around the listening position, but also as uniform as possible elsewhere. Electronic solutions were not acceptable, and any acoustic solutions could not be space consuming to the extent of anything more than about 10% of the empty internal volume of the control room section of the vehicle. This was a challenge.

23.2 The Vehicle

After the purchase of the NOV truck by Banzai, it was taken from Holland to England, where Kustom Konstructions had been engaged to make the conversion. The truck was stripped out in London, then taken to Swanwick, near Southampton, where the carpentry work was carried out. The brief was to make a mobile control room that was suitable for recording *and* mixing, with the provision of 24-track analogue and 24-track digital recorders. Before the actual design of the vehicle began, the mixing console had been specified as a Euphonix CS3000, so a suitable space for this had to be allocated. There also needed to be space for limiters, compressors, effects processors, two television monitors, a jackfield, the analogue 'audio tower' for the mixing console and a communications system.

Once stripped of its former recording equipment, the vehicle was left in the state previously shown in Figure 23.1. A heavy bulkhead separated the 'control room' from the cable reel housing. This was a heavy sandwich construction of GRP (glass-fibre reinforced plywood) and expanded polyurethane foam; in all approximately 10 cm in thickness. The bulkhead was well-damped and provided excellent sound isolation between the two sections of the vehicle. It was decided to site the audio tower in the cable reel section, as the tower had seven fans and produced much noise and heat.

The Euphonix console was chosen primarily for the rapidity with which it could be reconfigured: ideal for recording television programmes or concerts with several acts performing in rapid succession. An acoustic benefit was that the console itself was only a digital control surface, and was very slim. Slim-line consoles generally cause much less acoustic disturbance to any control room, but in very small rooms, large consoles can cause acoustic chaos, so the compact nature of the Euphonix was very well suited to the application under discussion here.

The side walls of the vehicle were made from a sandwich of GRP, mineral wool and aluminium, the latter of which formed the outside surface. The ceiling was aluminium, with a 10 cm soft foam lining. The floor was double skinned, with cable ducts running within it, accessible by heavy hatches. Its top surface, which was entirely above the height of the wheels – thus providing a plane deck – was made of 30 mm plywood over a heavy framework of 10 cm T-section steel, with 15 mm plywood below. Underneath the floor were compartments for the air-conditioning equipment, a fuel-burning heater, power stabilisers, isolation transformers, and the batteries for the 24-volt reel-motors, standby lighting, and air-conditioning control.

23.3 Acoustic Discussion

The available space in the control room section was 2.05 m × 2.05 m × 4.67 m. There was a perforated wooden ceiling in the original vehicle, at a height of 2.05 m, above which, and below a lining of 10 cm foam rubber, ran air-conditioning ducts and the cables for the lighting systems. It was decided to adopt a 'Non-Environment' approach to the room design, which called for an absorbent ceiling. The open-cell foam was left in place, but the wooden ceiling was removed, and was replaced by 'stretch' fabric over an open wooden frame, into which new lights were mounted. The carpet was removed from the original floor, and was replaced by 1.5 cm of solid beech planking.

A monitor wall was constructed in front of the bulkhead, into which would be set the monitor loudspeakers, televisions, amplifiers, crossovers, console dynamics displays, communication systems and the main electrical breaker panel. The finished wall is shown in Figure 23.4. It was made from a framework of 5 cm × 5 cm timber, faced with 13 mm plasterboard, a 5 m/m^2 deadsheet, and 19 mm MDF (medium density fibreboard).
The television monitors were to be covered by a sheet of heavy, non-reflecting glass, so that there would be no cavities in the finished wall surface that could resonate or cause other acoustic disturbances. In these types of designs, the front wall acts as a baffle extension,

Figure 23.4:
The basic framework of the Banzai monitor wall, with the lateral membrane absorbers also visible.

and also as a means of providing life to the speech of the occupants within what could otherwise, in this case, be a rather uncomfortably dead acoustic in which to work for long periods. Three of the surfaces of the 'Non-Environment' were thus in place; the hard floor, the sturdy, reflective monitor wall and the absorbent ceiling. The question at this stage was how to achieve the maximum acoustic absorption on the remaining three surfaces, consistent with using the minimum amount of space for the treatments.

As far as the internal acoustics are concerned, transmission *through* the walls could be considered equivalent to absorption *within* them, as in either case the energy would not be reflected back into the room. The basic shell provided isolation from inside to outside of about 30 dB, though somewhat less at the lower frequencies, and the internal acoustic control measures would be likely to add another 10 dB. Experience suggested that this would be adequate for most purposes, as mobile recording vehicles are rarely used in situations where they are, themselves, the most significant source of noise. A monitoring level of 90 dB SPL inside the vehicle would thus yield an SPL of around 50 dBA at a distance of 3 m from the outside surfaces. Few recordings are likely to take place in areas where these sorts of leakage levels would cause offence, and in many cases the use of much higher internal levels (100 dB, or so) would also not be expected to cause any nuisance. After due consideration of the above figures, it was decided not to attempt any further isolation, because it would be space consuming. Furthermore, it could also be prejudicial to the internal acoustic control, by reflecting energy back into the listening area. Such had been the effect of the plasterboard lining in the first Manor Mobiles.

From an internal acoustics point of view, a lightweight, well-damped shell is desirable, as the whole surface area of three walls and a ceiling can be used for absorption. If some of the very low frequencies are allowed to re-radiate into the outside air, it effectively makes the room acoustically much bigger at low frequencies than it is physically. Remember, at around 30 Hz the threshold of hearing is about 65 dB SPL. Five to ten decibels up on this would be the minimum level that could provoke complaints in most countries, so only 20 dB of isolation at those frequencies would still allow monitoring at 95 dB SPL with impunity. With the monitors of the truck eventually being sited only about 120 cm from the listening position, 100 dB SPL at 1 m was considered excruciatingly loud. Usually, there is little point in providing more isolation than necessary, because it also adds weight, which is another important consideration in a mobile recording vehicle. Of course, the minimum amount of energy reflected back into the vehicle is also important for minimising response disturbances.

The mixing console of 56 channels was only about 5 cm narrower than the inside of the vehicle, thus little treatment was possible on the lower halves of the walls unless the desk was to be 'permanently' built in. The lower 1 m of each side wall was carpeted, though little of these surfaces would remain visible after the installation of equipment racks (wooden), and the digital and analogue tape recorders. The upper metre of each wall was kept free from any equipment or obstructions, and was treated as shown in Figure 23.5 with membrane absorbers. These consisted of 5 cm deep, wooden frames, running back 250 cm from the monitor wall. The 5 cm space was then filled with an absorbent cotton waste felt, known as A1, which is self-extinguishing and has a density of around 60 m/m^3. The frame was then covered in PKB2, which is a 3.5 m/m^2 deadsheet, heat bonded to a felt similar to the A1. The whole system was then covered by a wooden frame, in turn covered in a stretch Lycra material.

This type of damped membrane absorber is quite effective at low frequencies, and the 2 cm felt surface is an effective absorber at high frequencies. Given their position in the vehicle, *vis-à-vis* the monitor loudspeakers, they receive most of the sound energy at near grazing incidences, which tends to augment their absorption. (See Figure 5.20.) Their primary purpose was to prevent early, lateral reflexions from disturbing the monitoring clarity, as well as to aid the overall low frequency control of the vehicle. The extra absorption that they provided would also improve the sound isolation through the side walls.

The remaining problem was what to do with the rear wall, which is usually the most troublesome surface in any control room.

23.3.1 Rear Wall Absorber

As can be seen from Figure 23.1, between the side entrance door and the front wall of the vehicle (the rear wall in terms of the control room layout), there exists a distance of about 60 cm. Ideally, from an acoustic point of view, this wall should be 100% absorbent at *all*

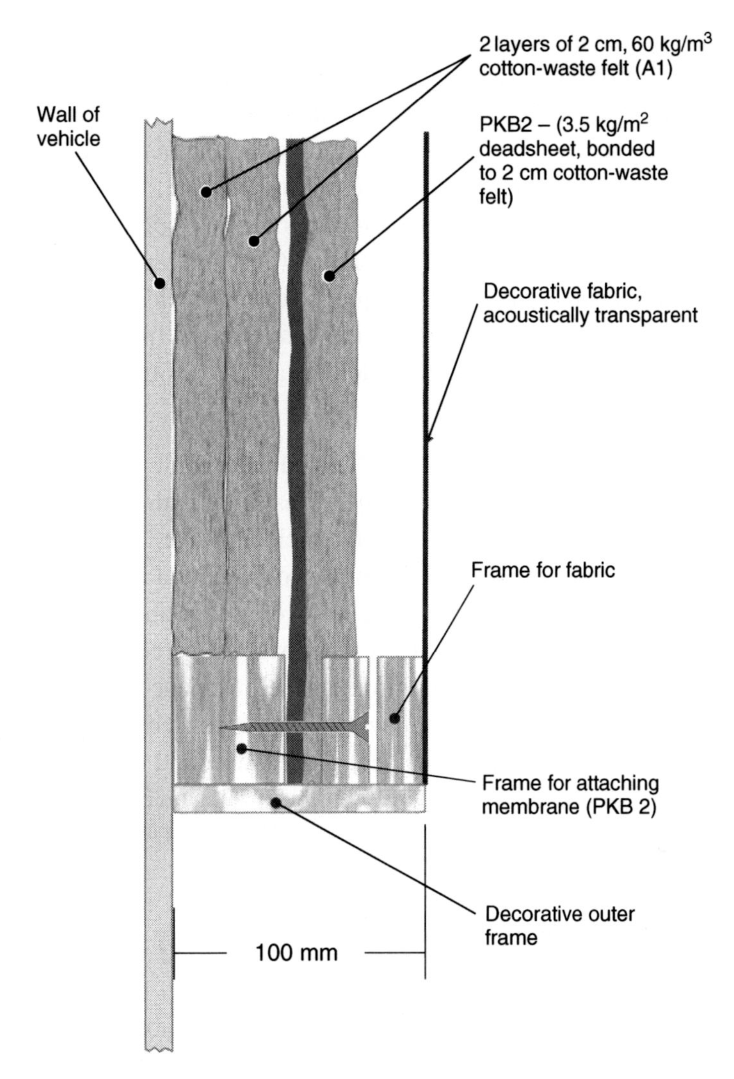

Figure 23.5:
Side wall absorber system.

frequencies, in order to realise the most uniform response. In most vehicles, this wall, if untreated, causes the greatest disturbances to the low frequency response at the monitoring position, by supporting (in the long dimension) isolated and strongly resonant axial modes in the standing wave field. As usual in these circumstances, there was a conflict between the space required by the acoustic design and that required to house all of the necessary equipment. The 24 V charging and control systems of the original vehicle were situated in this region, as was the control system for the air-conditioning. The job of rewiring and relocating these systems would have been a major task, so it was decided to keep them in the same general region. There was still, also, the matter of where to site the jackfield.

Bearing these points in mind, a multi-element absorber system was installed at the rear of the room, utilising the full 60 cm of available depth between the door and the end wall. The construction is shown in Figure 23.6. The end wall, the side walls and floor *within* the absorber, were made in the same way as the membrane absorbers on the main side walls of the vehicle. The front face of the space (towards the mixing console) was fitted with an open wooden frame, which provided support and mounting for the air-conditioning and 24 V control systems. These were mounted in the two top corners. (See Figures 23.7 and 23.8.) As the jackfield was essentially a mass of holes, it was judged that it would not be very reflective, so it was decided to arrange the mounting of this in the centre of the frame. This was convenient both in terms of ease of access for patching, and its position directly over the central cable duct in the floor.

The remainder of the space was then fitted with angled waveguide panels. The small volume, above the jackfield and between the control panels was fitted with ten small, suspended rectangles of PKB2, at an angle of about 30° to the incident wave. The larger spaces, at either side, were each fitted with five panels of 10 mm plywood, slotted into wooden guide rails and supported on 10 cm of foam rubber. The panels were set at about 30° from the vehicle front/back centre line, and were mirror imaged. They were covered on one side by a layer of LA5, a deadsheet material of 5 m/m^2, to damp the panel resonances, after which each side was covered in A1. The whole rear absorber system thus consisted of porous absorbers, lined duct absorbers, waveguides to break the incident plane wave, and membrane absorbers. The large jackfield and its attendant masses of cabling further helped in the overall absorption and diffusion.

23.3.2 Frequency Breakdown

The operation of this type of wideband absorption system can be explained as follows. Frequencies below 80 Hz, or thereabouts, are largely lost by absorption within and transmission through the walls of the vehicle. Releasing 70 dB SPL (linear) at 50 Hz may not be acceptable within a building, but to the outside world it will generally be lost in background noise, and represents something in the region of only NR35, which is unlikely to cause disturbance. The SPL will also tend to reduce rapidly with distance from the vehicle, because, unless parked in a narrow street, the waves are free to expand in the outside air. Frequencies in the region of 50 to 200 Hz are absorbed to a large degree by the membrane absorbers, and by the wall structure of the vehicle. The 200 to 500 Hz region is addressed by the absorption within the system of 'waveguide' panels. These consist of cotton-felt-covered, damped, plywood panels of 130 cm × 50 cm. The lined ducts, which they form, generally soak up a very large proportion of the frequencies above 500 Hz.

(a)

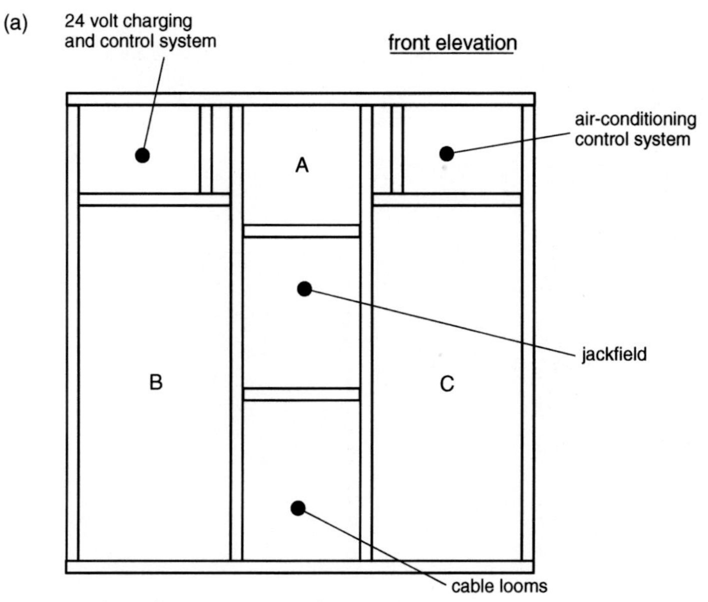
24 volt charging
and control system
front elevation
air-conditioning
control system
A
jackfield
B
C
cable looms
A – Space for small waveguide/absorber panels
B,C – Space for large waveguide/absorber panels

(b)

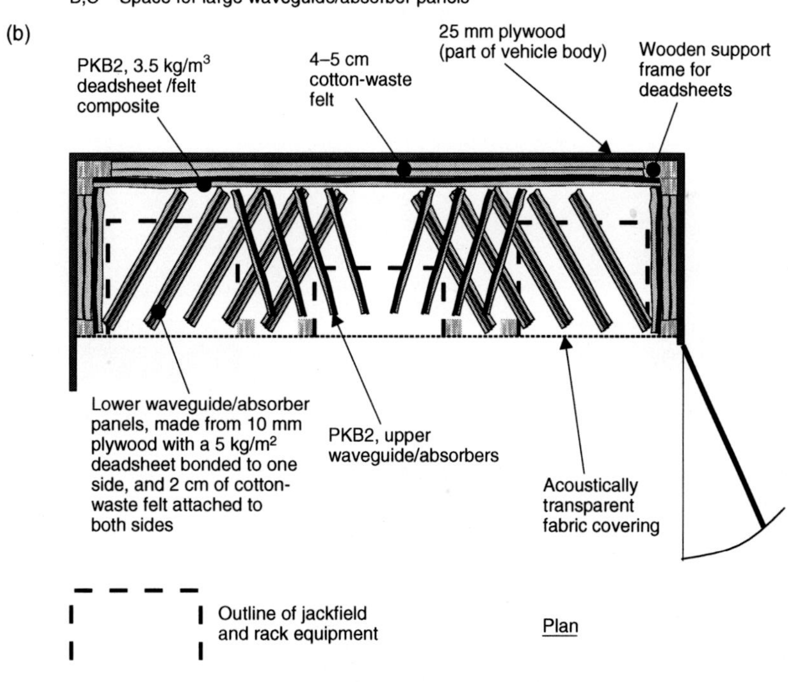
PKB2, 3.5 kg/m3
deadsheet /felt
composite
4–5 cm
cotton-waste
felt
25 mm plywood
(part of vehicle body)
Wooden support
frame for
deadsheets
Lower waveguide/absorber
panels, made from 10 mm
plywood with a 5 kg/m2
deadsheet bonded to one
side, and 2 cm of cotton-
waste felt attached to
both sides
PKB2, upper
waveguide/absorbers
Acoustically
transparent
fabric covering
Outline of jackfield
and rack equipment
Plan
Ceiling and floor of "trap" treated as the walls

Figure 23.6:
Rear wall absorber system: (a) Allocation of space; (b) Plan view.

Figure 23.7:
Exposed rear-wall trap (absorber) system, with the central jackfield mounting installed, and the cable access hole visible below.

Figure 23.8:
The completed rear-wall absorber system.

The angling of the ducts, to form waveguides, also tends to increase the effectiveness of the membrane absorbers by steering the incident wave and causing it to strike the membrane absorbers at angles that are more conducive to absorption, and by confounding the attempt of the weakened reflected waves to re-enter the listening area. Work by Walter,[1] carried out at the Institute of Sound and Vibration Research, Southampton, UK, has demonstrated the degree of this waveguide effect. Figures 23.9 and 23.10 show examples of the waveguide effect of such systems. This data is previously unpublished.

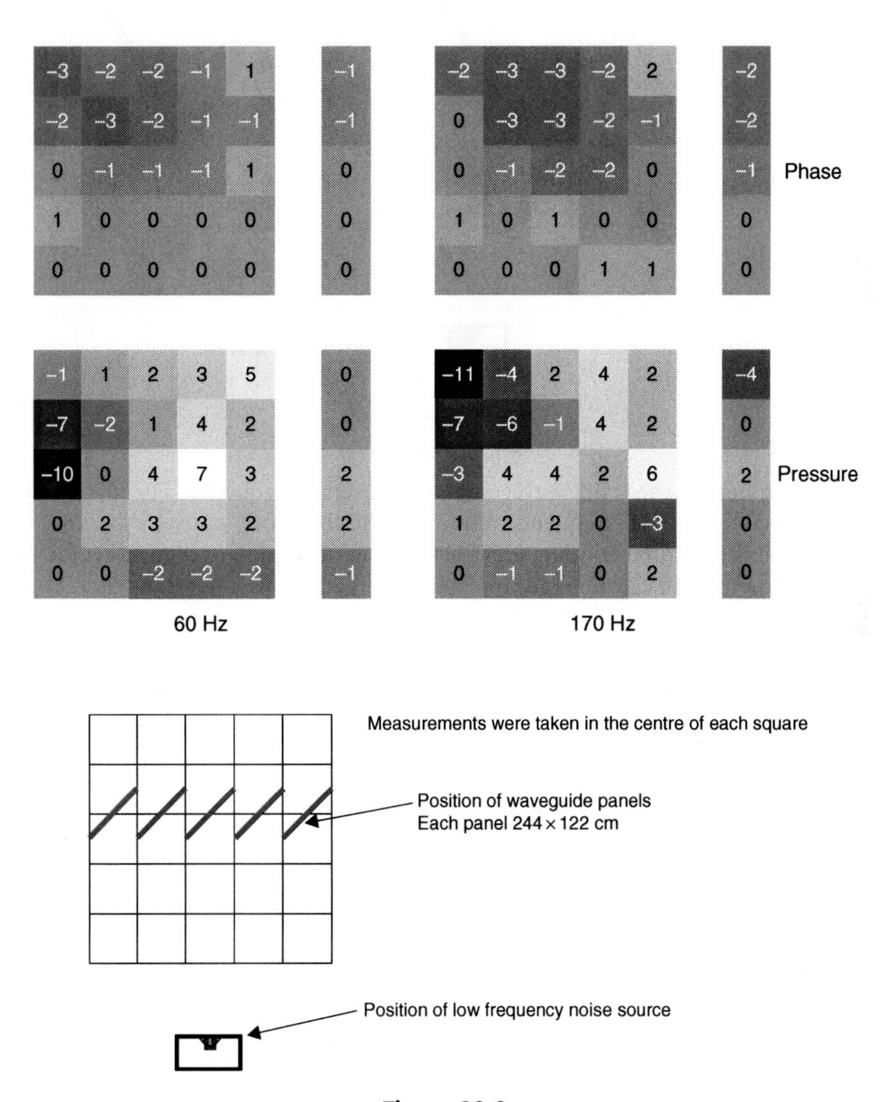

Figure 23.9:
Effect at single frequencies. Typical maps showing distribution of phase and amplitude at two different frequencies. Panels with one layer of felt, 45° angle and 30cm spacing. The numbers in the squares of the phase measurements are in radians, and on the pressure measurements in decibels.

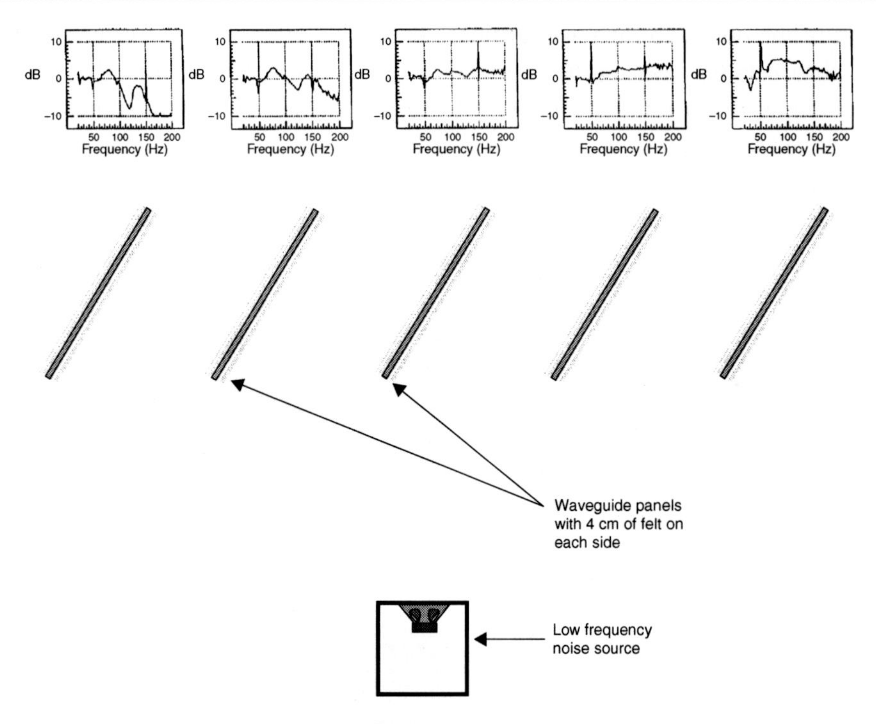

Figure 23.10:
Pressure distribution behind the 'trap' waveguides – SPL versus frequency. The low frequency noise source was corrected for a flat frequency response. The five plots, above, show the deviation from a flat response at the corresponding locations on the other side of the waveguide panels. The steering effect of the panels can be seen, removing energy from the left-hand side of the array, and adding energy to the right-hand side. In the tests, the low frequency source was further away from the panels than is indicated in the drawing.

By this stage, we finally had an acoustically controlled shell, but we still had the problem of how best to drive it. What was needed was a high-quality, small monitor system which could deliver around 110 dB SPL and which was suitable for very close range listening. The practical layout for the vehicle necessitated that the mixing console should be across the width of the truck, and should be as close as possible to the monitor wall. This resulted in each side wall being within 1 m of the ears of the recording engineer, with a floor about 1 m below, and a ceiling about 1 m above. The distance from the listening position to a line drawn through the centre of the two loudspeakers was also about 1 m. The brief for the design required a peak monitoring level capability of approximately 105 to 110 dB SPL and that the response should be sufficiently flat for serious mixing for CD releases.

23.3.3 Side Wall Reflexions

Placing the loudspeakers in the end of what is effectively a closed tube is not a particularly easy means of achieving a flat response. In fact, mobile recording vehicles through the ages have been plagued by the problems thus created. The first problem to arise is due to the proximity of the side walls. Given a 2 m × 2 m front wall, there is little space to get an adequate stereo separation, so inevitably the loudspeakers must be sited more or less adjacent to the side walls. At higher frequencies, where the wavelengths are less than about four times the distance from the centre of the loudspeaker to the wall, severe comb filtering can result from wall reflexions, as shown in Figure 5.12. In fact, and perhaps even more disturbingly, the side wall reflexions can produce severe smearing of the stereo image, not only because of the displaced phantom source of the reflexions (the mirrored room effect) but also because, with very close walls, the reflexions can return within the 0.7 ms time window before the Haas effect begins to apply. Such reflexions can be ruinous for the clarity of perception[2,3] as the Haas effect/Precedence effect (of the first arriving signal dominating) does not apply.

The solution for the above problem is to position an absorber system alongside the loudspeaker. The absorber must be effective down to frequencies where wavelengths will be sufficiently long that the reflected wave will return to the listener substantially in phase, and will not cause interference except to boost the output. Everything above this needs to be absorbed so that the reflexions also do not mar the stereo imaging. For a path length difference between the direct and reflected waves of around 30 cm, this would mean effective absorption down to 150 Hz or thereabouts. In small spaces, this may not be easy to achieve, but geometry can help the situation to some degree, as shown in Figure 23.11. Here, the loudspeakers are built

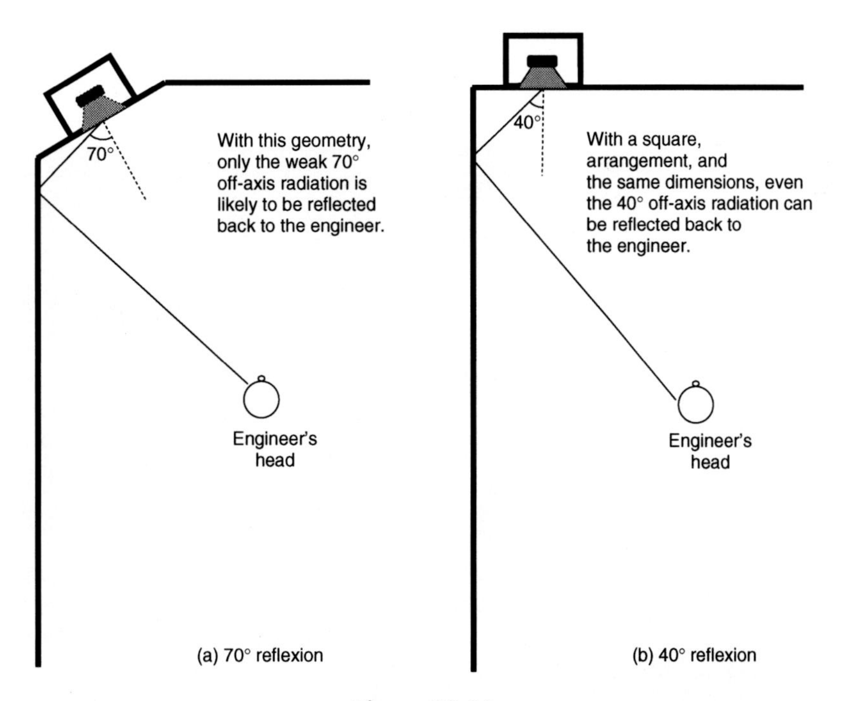
70°
With this geometry, only the weak 70° off-axis radiation is likely to be reflected back to the engineer.
Engineer's head
(a) 70° reflexion
40°
With a square, arrangement, and the same dimensions, even the 40° off-axis radiation can be reflected back to the engineer.
Engineer's head
(b) 40° reflexion

Figure 23.11:
Effect of angling the loudspeakers. With most loudspeakers, the 70° off-axis radiation is likely to contain little above 300 Hz, and is also likely to be lower in high frequency content than is the case for the 40° off-axis radiation.

into an angled section of the front wall, such that the side wall absorbers could only return energy back to the listener which emanated from 50° or 60° off-axis. The effectiveness of the absorber determines the strength of the reflected waves. However, in this instance, the engineer could be brought very close to the loudspeakers in order to take advantage, as far as possible, of the direct sound.

23.4 Close-Range Monitoring

In the design of a vehicle like this, it is inevitable that the listening position should be close to the monitors. Large monitor systems cannot be readily used in such circumstances, because at short distances all multi-element loudspeakers exhibit geometrical problems, where the drivers covering different frequency ranges are perceived as separate sources, except, of course, for coaxial designs. In fact, coaxial designs initially appeared to be the optimum engineering solution for this problem, but the owners of the mobile studio in question expressed a lack of satisfaction with the sonic qualities of the coaxial loudspeakers commercially available in the size and power ranges required.

The chosen solution was to use a pair of Quested Q405s, which had previously existed only as prototypes. These are shown in Figure 23.12, and in form are something of a hybrid between a d'Appolito array and a coaxial array. They were also small enough to avoid the above geometrical problems.

Figure 23.12:
The Quested Q405.

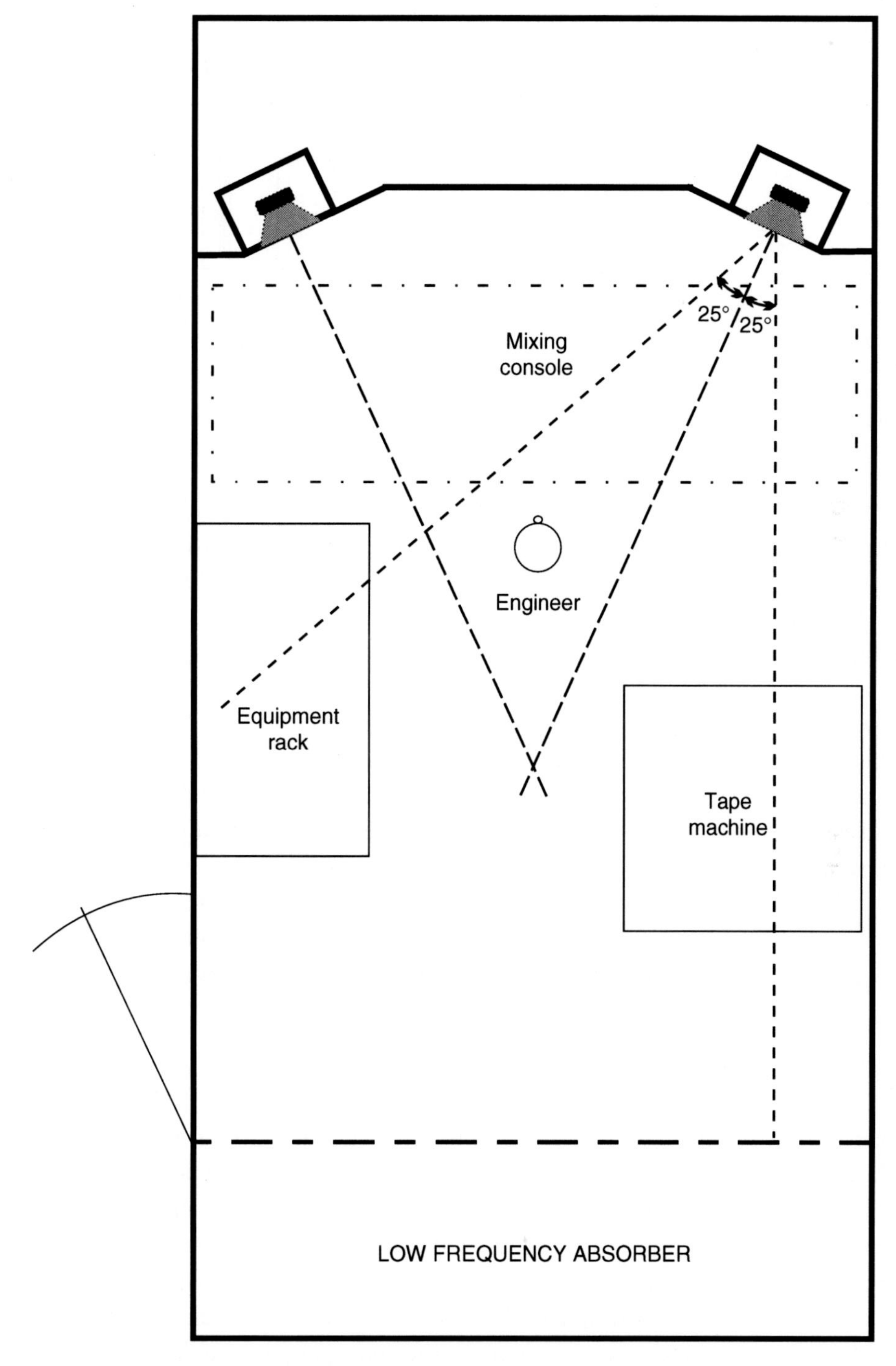

Figure 23.13:
Loudspeaker coverage area. More or less the whole working area of the vehicle is within the 0–25° response field of the loudspeakers.

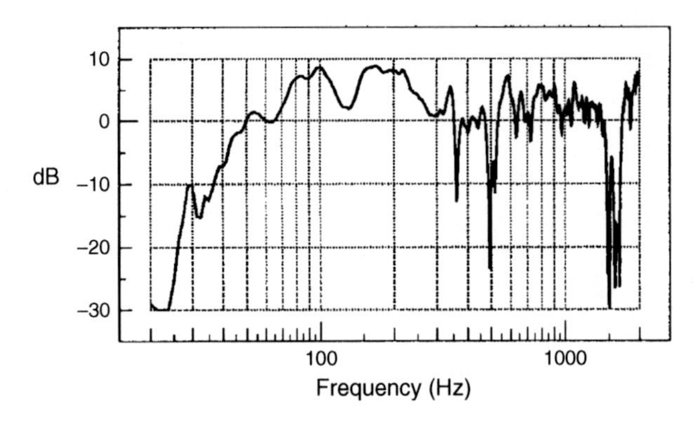

Figure 23.14:
Cancellation notch at 1.6 kHz from out-of-phase tweeters. Both loudspeakers driven, measured at 2.5 m from the back wall, on the centre line of the vehicle (10 drivers radiating).

At the crossover frequency of 1.5 kHz, all five drivers operate, and the slightly elongated quincunx layout ensures a reasonably symmetrical horizontal and vertical directivity. Of course, at angles of 40° or 50° off-axis, the one quarter-wavelength differences due to the different distances to the nearest and furthest pair would begin to give rise to response irregularities, especially in the vertical plane where the driver pairs are spaced further apart, but in this case no such problem could occur. In this type of vehicle it is simply not possible to remain within any practical working area and be more than 20° or 25° off the axis of either of the loudspeakers. This fact is illustrated by Figure 23.13. Figure 23.14 shows the cancellation at the crossover frequency with the tweeter polarity intentionally reversed, with both loudspeakers (left and right) radiating, and measured at a central position within the working area. The cancellation, and hence the interference pattern, remains remarkably constant, with clean, steep sides. If any multiple-source smearing were present, such clean cancellation could not be achieved or maintained. Any potential problems in using so many drivers were thus confined to large off-axis angles that are of no relevance in such a vehicle.

23.5 Directivity and Total Power

Most loudspeakers designed for free-standing in conventional control rooms are designed to have a flat axial amplitude response when so mounted, but due to the omni-directional directivity of the low frequencies, below around 300 Hz the total power response tends to increase as the frequency lowers, and the directivity patterns broaden. If the low frequencies spread in all directions, whilst the high frequencies beam out in a narrower band, then clearly more total energy must be radiated at low frequencies in order to maintain an even SPL on axis. (See Section 11.4.) However, due to room reflexions, the concept of a flat axial response from free-standing loudspeakers in a conventional room tends to be a contradiction in terms. In the case under discussion here, the loudspeakers were mounted in a rigid monitor wall, which due to the increase in radiation resistance provided by the constraint of the radiating angle, created a boost in the axial response below about 250 Hz (see Chapter 11). This variability of mounting conditions is why the manufacturers of many active monitor systems provide low frequency level and roll-off controls, exactly to equalise this sort of response boost.

Conventional rooms also allow a degree of lateral and vertical expansion of the sound waves, but in a truck of the dimensions being discussed here, and given the incomplete absorption (including transmission) of the low frequencies by the walls, floor and ceiling, the 'room' tends to act like a plane-wave tube. This has the effect of further increasing the low frequency radiation resistance, and boosting the output still further. The effect can clearly be seen from Figure 23.15. Nevertheless, this boost does provide a useful extension of headroom. Effectively, the increased resistive loading also increases the sensitivity of the loudspeakers, and because no non-minimum phase delays are involved in the response disturbance (such as is the case where reflexions combine with the direct signal after having travelled a significantly different path length) the problem lends itself to correction by relatively simple electrical equalisation. In these cases, the correction of the amplitude response will also tend to correct the phase response, consequently improving the time response. Corrective equalisation to lower the response elevation at low frequencies is further beneficial because it will increase the headroom by the same amount. This is an important factor to consider when trying to use the smallest practical source size to produce the required output capability of 110 dB SPL at 1 m. Remember, large monitor systems were out of the question because of the geometrical and near-field problems. Inefficient, small loudspeakers may also suffer from thermal compression problems, therefore *small and efficient* loudspeakers are ideal. Although this is normally a contradiction at low frequencies, the special circumstances of the mobile vehicle make it achievable.

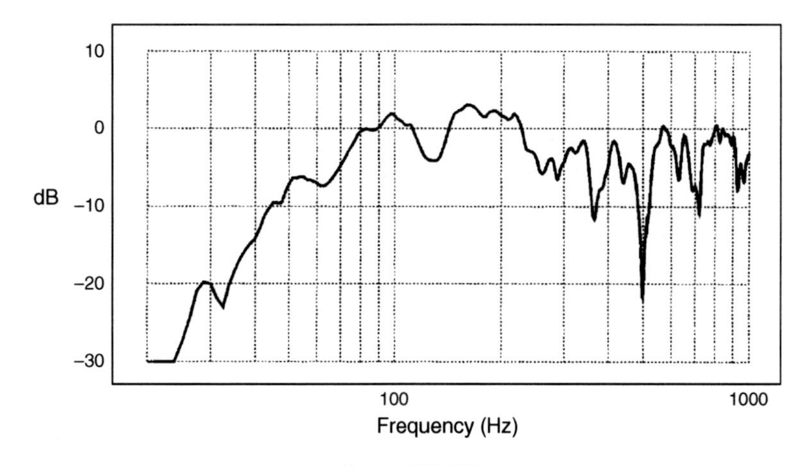

Figure 23.15:
Axial response of one loudspeaker without sub-woofer or equalisation. Note the general boost in the 80–200 Hz frequency range, due to the restricted radiating angle.

23.6 Attaching a Sub-Woofer

The above system, alone, was deemed to be lacking in bass for full-range monitoring purposes, so it was decided to employ a sub-woofer system to augment the lower frequencies, below 70 Hz. In general, the use of single sub-woofers on stereo systems for critical listening is not recommended, but again, in this case, the special circumstances not only demanded a single sub-woofer, but also allowed it to be used without the usual drawbacks. The sub-woofer was to be sited within a quarter of a wavelength of the main stereo pair at the crossover frequency. Indeed, if a *pair* of sub-woofers were to be used they would also, inevitably in this case, be located within a quarter-wavelength of each other, and hence would tend to act as an acoustically coupled pair for mono signals. The space saving benefit of using a single 15 in woofer far outweighed any acoustic disadvantages, and offered the advantage of reducing the low frequency variability of panned images. The sub-woofer would also tend to drive the room like a plane wave tube, without significant reflexion.

As the room was rather absorbent, even at low frequencies, any modal activity would be very weak. Furthermore, in a space so small, the only mode that could develop in the sub-woofer range would be the first, axial, front-to-back mode at around 57 Hz, below which a pressure zone would be in effect, giving a very smooth response. The use of the sub-woofer and its associated crossover would also prevent excessive low frequencies from entering the main stereo pair, thus easing their workload and increasing their reliability at high output levels. What is more, as the pressure zone exhibits an entirely minimum phase response, it can be electrically equalised, if required, without any detrimental effect. (See Section 13.5.)

23.6.1 The Appropriate Equalisation

In reality, it is impossible to predict the precise absorption characteristics of a mobile recording vehicle, or the exact loading effects created by the acoustic characteristics of the space. The problem is too complex, and too little is known of the inherent properties of the structure. However, because of the reasons previously described, it could be assumed that in the vehicle under discussion the response irregularities would be of a predominantly minimum phase nature, so an empirical method of evaluating the extent of the problems, and the required corrective measures, would be adequate.

After the installation, pink noise was put into the system, and a simple real-time spectrum analyser reading was taken. In highly damped rooms with low decay times and few reflexions, this is a viable technique. A high-quality, multi-band, stereo parametric equaliser was then used to correct the response before listening tests were carried out. The equaliser was then adjusted until the optimum subjective response was agreed upon by a group of recording engineers, listening to ‘known’ material. Subsequently, the equaliser was readjusted

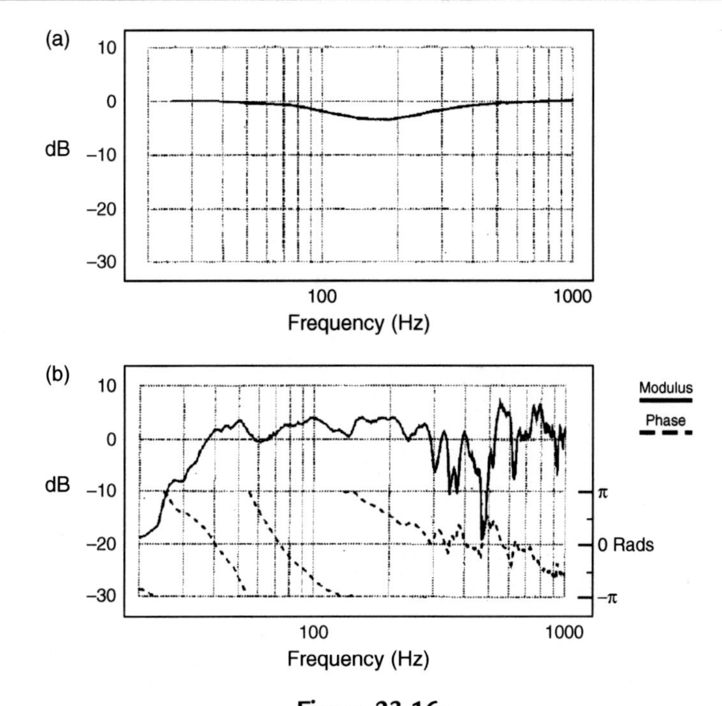

Figure 23.16:
(a) Response of equaliser; (b) Equalised frequency response with sub-woofer.

to the nearest simple, smooth response correction, before the system was auditioned for a final check of its sonic acceptability.

The next step of the operation was to record the pink noise from the output of the equaliser. The recording was then taken away for analysis, and the transfer function of the equaliser was plotted. The crossover used in this system had provision for the insertion of custom equalisation cards, and hence was a very practical choice for such circumstances. A card was subsequently inserted with the required characteristics to flatten the response. Its characteristics are shown in Figure 23.16(a).

23.7 Results

The front end of the completed 'room' is shown in Figure 23.17, with the sub-woofer just visible below the mixing console. The response of the final system in the recording vehicle is shown in Figure 23.18, which is a very commendable response indeed for a mobile control room. The ragged response in the 500 Hz region is due to reflexions from the top surface of the mixing desk, which are inevitable in such situations. The step-function response is shown in Figure 23.19. It can be seen that the tweeter comes in about 0.5 ms before the woofers, with

Figure 23.17:
The complete front half of the truck, with the sub-woofer just visible below the mixing console.

the sub-woofer showing about a 3 ms arrival delay. In fact, the sub-woofer was a Quested VS115 situated under the console about 0.5 m *ahead* of the main monitors. The VS115 uses a 12 dB/octave crossover at 38 Hz, which gradually augments and extends the falling response of the Q405s. Without the use of a sub-woofer, the 'power response' boost below 250 Hz, as shown in Figure 23.15, could be equalised down to flat, but the lower frequencies

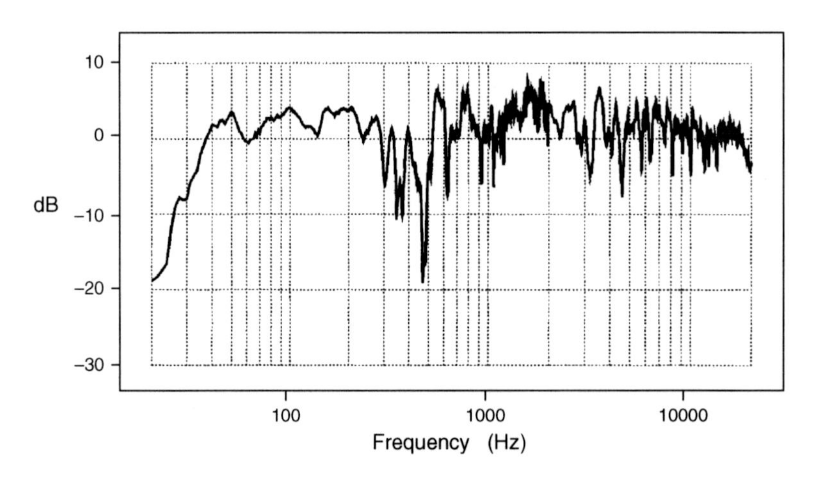

Figure 23.18:
Frequency response of finished monitor system (the aberrations around 500 Hz are from the inevitable console reflexions).

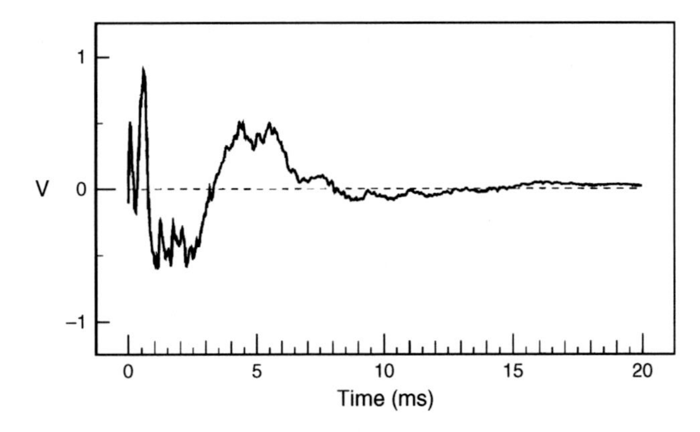

Figure 23.19:
Step-function response.

could not be boosted much, because the excursion limits of the 5 in (125 mm) drivers would have been exceeded below 60 Hz if using monitoring levels of more than 85 dB SPL. In this particular instance, the sub-woofer was perhaps the only option available for achieving an extended, flat response, and an output capability of over 100 dB SPL.

Of course, 12 dB/octave crossovers can never be correct in amplitude and phase at the same time. However, Quested had chosen to use them in the sub-woofers after conducting extendedlistening tests, and for this application, they are indeed very workable. The VS115 sub-woofer has a polarity reversal switch and a variable phase control. In this case, a reversed polarity and 90° phase shift produced the flattest amplitude response, but the phase inconsistencies can still lead to the errors in the time response. Such are the compromises with sub-woofers, which is why they tend not to be used in critical listening situations, but in the case being described here their use was a viable option in difficult conditions. The overall response in this vehicle is remarkably uniform considering the starting conditions, and, in fact, the 'room' performs better than many small, fixed rooms, although this is largely due to the less rigid structure of its walls.

23.8 Conclusions

The Banzai mobile was clearly a great improvement on the first Manor Mobile in terms of its monitoring acoustics, though it must be said that the recording paths have not significantly advanced. Although the digital control system and general reliability have seen great development, the basic signal circuits, from microphone to tape, have not. In terms of the acoustics and monitoring, however, it is interesting to compare what could be done in 1998 with what could be done in 1973.

1. Much more is now known about the psychoacoustics of monitoring, so more things can be taken into account during the design stages.

2. The importance of the accuracy of the direct signal is much better understood, so more care can be taken to deliver it intact.
3. Loudspeaker/room interactions are better appreciated, and the nonsense of using monitor equalisation to correct room reflexion problems has been largely laid to rest.
4. The specialised acoustic deadsheets, which were extensively used in the Banzai mobile, were not available in the early 1970s.
5. The rear wall trap systems were in their infancy in the early 1970s, and had never been applied to mobile vehicles.
6. In 1973, it would still be ten years or more to the birth of the 'Non-Environment' room concept.
7. Small loudspeakers with high output capability were generally not available in 1973.
8. There was no experience of the combination of acoustic control measures employed in the Banzai vehicle.
9. 'Slim-line', high-quality mixing consoles of the size of the Euphonix were not available.

There is no doubt that The Manor Mobile was the state of the art in 1973, but the electroacoustic performance of the Banzai mobile, 25 years later, was clearly superior. From a designers point of view it is all very well looking back and thinking 'if only…' but the nine points above emphasise the developments that the intervening 25 years had produced. All in all, it was a very interesting exercise to undertake.

23.9 Summary

Sound propagation in a mobile recording vehicle of conventional dimensions tends to be more like propagation in a tube rather than in a room.

Where no space is available for adequate low frequency absorbers, the necessary absorption can be achieved through transmission to the outside of the vehicle if reduced sound isolation can be tolerated.

Console shape, size and siting can be crucial factors in the overall acoustic performance of a vehicle.

It is a good idea to try to reduce as much as possible the reflexions from the end wall opposite the loudspeakers.

Because of the very close proximity of the side walls to the loudspeakers, effective absorption should be placed on those walls in order to prevent the very early reflexions from smearing the stereo imaging.

Because of the difficulties encountered in achieving adequate acoustic control, close-range monitoring is a wise option, but the appropriate small-size loudspeakers may lack an extended low frequency response.

Geometrical near-field problems can be ameliorated by the use of a concentric or semi-concentric driver layout.

Such a close-range monitoring situation fortunately lends itself well to low frequency augmentation by means of a sub-woofer.

References

1 Walter, Alistair, Private communication of internal ISVR document from which Figures 23.9 and 23.10 were extracted, with permission (1998)

2 Haas, H., 'The Influence of a Single Echo on the Intelligibility of Speech', *Journal of the Audio Engineering Society*, Vol. 20, p. 146 (1972)

3 Poldy, C. A., 'Headphones' in: Borwick, John (ed.), *Loudspeaker and Headphone Handbook*, 2nd Edn, p. 529, Focal Press, Oxford, UK (1994); 3rd Edn, p. 642 (2001)

CHAPTER 24

Foldback

The performing environment for the musicians. Constant voltage distribution systems. Impedance calculations. Stereo vs mono. Mixing by the musicians. Types of headphones. Advantages of different options.

A very important aspect of *any* recording studio is its monitoring environment, as this is the means by which all the studio proceedings are ultimately judged. However, it should not be forgotten that there is another very important monitoring environment; the foldback system. When musicians are using foldback, it is their world. The acoustic sounds of the studios, the real sounds of the instruments, or the sounds as heard by the recording engineers in the control rooms are all only peripheral concepts to a musician wearing headphones. What is of utmost importance to a performance at the time of recording is the ambience created for the musicians in the foldback systems. This subject was introduced in Section 9.3, which may be worth re-reading before beginning this chapter.

24.1 A Virtual World

In large studios, 'tracking loudspeakers' are sometimes used, as shown in Figure 24.1, but only certain types of room lend themselves to being optimally usable with this form of foldback. The system has always been popular because many musicians prefer not to use headphones if they can avoid it. Some of this is no doubt because our perception via headphones is different from our perception via our pinnae (outer ears), and musicians tend to like to perform in a familiar world. However, it could well also be that many musicians have shunned the use of headphones not only because they find them to be an unnatural 'world', but also because they have too often had to endure some pretty appalling foldback mixes.

To an alarming number of recording personnel, foldback is something via which a musician keeps in time and in tune to a backing track, and little else. To musicians, the foldback is their creative space. Badly balanced foldback will distort the musicians' perception, and nobody can be expected to perform with appropriate expression and feeling if they are not hearing a balance which is conducive to such a performance. A musician cannot be expected to build up a feel for a track if levels are going up and down, and instruments are cutting in and out as recording engineers adjust *their* balances, while hoping that the musicians can use

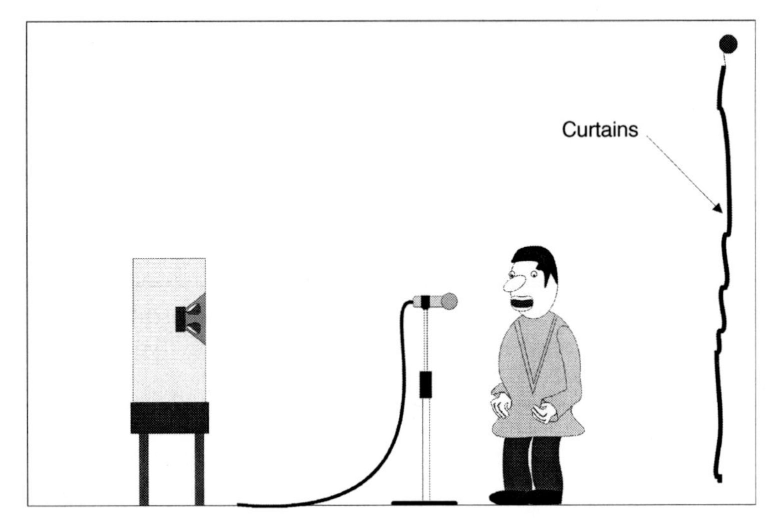

Figure 24.1:
Use of tracking loudspeakers. In this situation, a directional microphone is used to record the vocalist, with a loudspeaker providing the foldback. A typical arrangement would be for the loudspeaker to be placed directly behind the microphone and pointing towards an absorbent surface in order to control the amount of reflexions of the foldback signal returning to the front of the microphone.

such run-throughs for their own rehearsals. Musical performances can be fragile, and this fact should be taken into account before any atrocities are committed via the foldback.

24.2 Constant Voltage Distribution

The time-honoured tradition of providing foldback in many large studios has been to use a power amplifier as a constant voltage source, and to use medium impedance headphones to bridge the line. Such a system is depicted in Figure 24.2. The principle of such a system is to take a feed from the auxiliary output of the mixing desk and to feed this into the inputs of the amplifier. The amplifier gain control is usually set such that 0 VU from the desk would produce about 3 dB below clipping at the output terminals of the amplifier. For a typical amplifier producing 100 W into 8 ohms, the output voltage at full power would be given by the formula:

$$V^2 = WR \tag{24.1}$$

where:

W = power in watts
V = voltage in volts
R = resistance (or the impedance for a complex load) in ohms

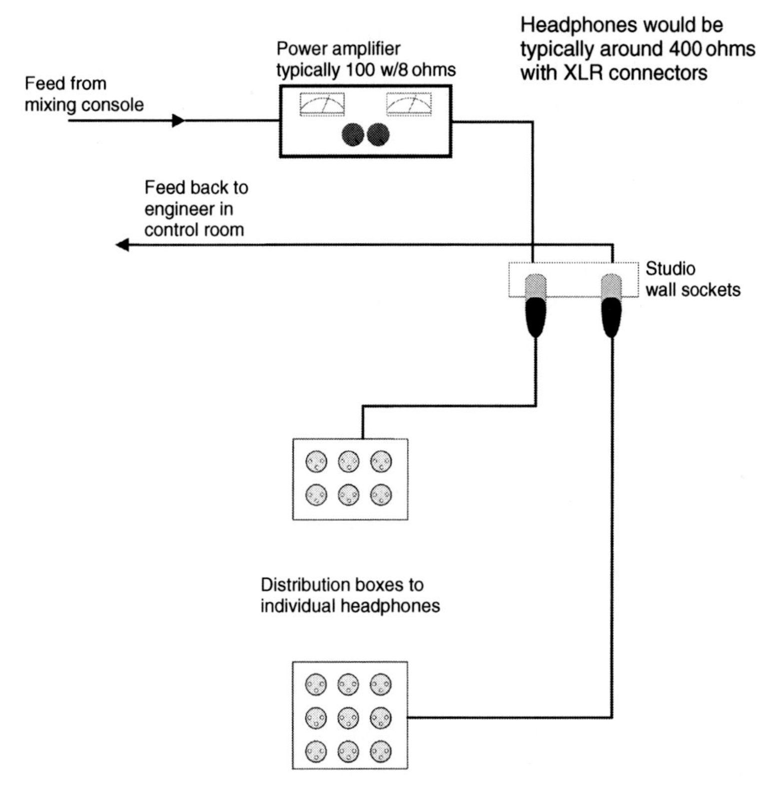

Figure 24.2:
Constant voltage foldback system. NB: if individual volume controls are used for each set of headphones, they should be potentiometers of about 600 ohms, linear track and should each be capable of dissipating at least 5 watts, or they will tend to burn up.

∴ For 100 watts into 8 ohms, the voltage would be:

$$\begin{aligned} V^2 &= 100 \times 8 \\ V^2 &= 800 \\ V &= \sqrt{800} \\ V &= 28.3 \text{ volts} \end{aligned}$$

To be 3 dB down on this would be half power, so:

$$\begin{aligned} V^2 &= WR \\ V^2 &= 50 \times 8 \\ V^2 &= 400 \\ V &= \sqrt{400} \\ V &= 20 \text{ volts} \end{aligned}$$

(Note: 3 dB down is half *power*, 6 dB down is half *voltage*)

The output from a typical power amplifier is referred to as a constant voltage source because when an input voltage causes an output voltage to appear at the loudspeaker terminals, differing load impedances do virtually nothing to cause the voltage to vary. At least, that is, until an impedance is reached which is so low that the amplifier power output becomes limited by its ability to supply current into the load, when either the voltage will begin to sag or protection devices cut in.

From the formula $W = V^2/R$ it can be seen that if the voltage remains fixed, then changing R will change W. In fact, reducing R will always increase W. That is, if the voltage remains fixed, if the resistance (or impedance) goes down the power will go up. The *power* output is thus dependent upon the load impedance. As more headphones are added in parallel to such a foldback circuit, the impedance of the load to the amplifier will drop, and it will be able to supply proportionately more power. (See Glossary for a discussion on resistance and impedance.) If headphones of suitable impedance are chosen, then it can be ensured that they cannot be overloaded if the inputs (from the mixing desk outputs) are inadvertently turned up excessively, or if a feedback should occur.

A typical studio headphone would be something like the Beyer DT100. Each capsule has a maximum power rating of 2 watts, and they are available in a variety of impedances from 8 ohms to over 1000 ohms. A common value for studio use is 400 ohms. It was calculated earlier that an amplifier rated at 100 watts into 8 ohms could give a maximum output voltage of 28.3 volts. If 400 ohms is substituted into our above formula, it can be seen that such an amplifier could deliver, at maximum:

$$\begin{aligned} W &= \frac{V^2}{R} \\ W &= \frac{28.3^2}{400} \\ W &= \frac{800}{400} \\ W &= 2 \end{aligned} \tag{24.2}$$

i.e. 2 watts (into a 400 ohm load).

Consequently, even if driven to maximum output voltage, the amplifier could not deliver enough power to destroy the headphones.

If six headphones were used, the combination of six 400 ohm loads in parallel would give a total impedance of:

$$\frac{400}{6} = 66.6 \text{ ohms}$$

So, once again, we can use the formula $W = \frac{V^2}{R}$

$$\therefore W = \frac{28.3^2}{66.6}$$

$$W = \frac{800}{66.6}$$

$$W = 12 \text{ watts}$$

Twelve watts into six headphones is still 2 watts per headphone.

Making the same calculation for 20 headphones, 20×400 ohm loads in parallel would give 400 ohms $\div$ 20 = *20 ohms*

$$W = \frac{28.3^2}{20}$$

$$W = \frac{800}{20}$$

$$W = 40 \text{ watts}$$

which, divided by the 20 headphones is, yet again, 2 watts per headphone.

It can thus be seen that if the headphone impedance is correctly chosen the amplifier will simply provide more current (and therefore more power) as the number of headphones is increased; at least until the maximum output rating of the amplifier is reached. However, below this limit, adding more headphones will not affect the loudness of any of the headphones already connected, nor will there be a risk of burning out headphones if other parallel devices are disconnected, because for any given input signal, the peak output voltage remains constant, and independent of load impedance.

Nevertheless, one precaution which is worth taking with such systems is to feed the foldback amplifiers via un-normalled jacks on the jackfield/patchbay so that they must be patched in when needed. All too frequently, people wire them via jacks that are normalled to the auxiliary outputs of the console. During mixdown, when the auxiliaries are being used to drive effects processors, headphones are often unintentionally left connected in the studio. When the effects processors are plugged into the half-normalled auxiliary outputs, the feeds to the amplifiers are not disconnected, so the high levels which are being sent to the effects units will also be feeding the foldback amplifiers. Although the amplifiers may not be able to overload the headphones with clean power, if the amplifiers do overload, then the distorted signal waveforms continuously running into the headphones may destroy the capsules after some time of such abuse. However, by making it necessary to physically patch the amplifier inputs into the auxiliary outputs when foldback is required, the headphones will automatically be disconnected when the auxiliaries are needed for other purposes, and damage to the headphones will be prevented.

With the type of foldback system under discussion here, all the feeds are sent from the auxiliary outputs of the mixing console. As all the headphones on one circuit share one amplifier, the foldback balance is thus entirely in the hands of the recording engineer. It is usual for the engineer to be provided with a headphone socket in the vicinity of the mixing desk, fed from the same amplifier output so that he or she will always be able to check, at any time, the exact balance and level that is being fed to the musician(s). In order to optimise this facility, it is the usually preferred method for the engineer to use identical headphones to the musicians.

Of course, not all of the musicians want to hear the same mix, so multiple foldback systems are often provided. Drummers may need to hear lots of bass but little drums in the headphones, whilst bassists may need to hear lots of drums and little bass. No matter how many systems are in operation, there would be a feed from each one back to the engineer, so that the balance can be carefully set on each channel in the sure knowledge that the musicians will be hearing the selfsame sound.

24.3 Stereo or Mono

Without doubt, a stereo foldback presentation is easier for musicians to work with compared to an 'above the head' mono jumble. In general, because of the spacial separation, individual instruments can be readily detected in a stereo mix even if their level in the mix is on the low side, or if they are in a frequency range which is shared by other instruments which are playing at the same time. Nevertheless, mono foldback is still common, especially when multiple mixes are needed, as there may not be enough auxiliary outputs or foldback channels to send stereo feeds to everybody. Mono mixes usually require much more delicate balancing if the musicians are to hear the detail that they need, and it is often wise not to send unnecessary instruments to the mix. Simplicity can be the key to clarity in mono mixes.

Some recording engineers, when overdubbing, send a stereo feed to the foldback which is derived from the monitor feed in the studio, plus a little extra boost on what is being recorded. This system can work well, but it must be done in such a way that any muting or soloing on the main monitors does not affect the foldback, or it can be very disturbing and frustrating for the musicians. It may also, at times, be an inappropriate balance, because perception via headphones can be very different to perception via loudspeakers. Once again, though, if the engineer has a headphone feed from the same power amplifier output that feeds the musician's headphones, then the appropriateness of the balance can readily be checked.

24.4 In-Studio Mixing

An alternative foldback philosophy to the one outlined in Section 24.2 sends line-level sub-group outputs to the rooms where the musicians are recording, and small mixers are provided for each musician, or group of musicians. Three variants of this concept are shown in Figures 24.3 and 24.4(a) and (b). By this means, sets of sub-groups such as drums, guitars, keyboards, vocals, brass and other such sections can be sent to the studio, where the musicians can set their own balance and level, at least to the degree that the subgroups can provide discrete feeds. Whilst this may sound like an ideal system, with much greater flexibility than with mixes sent from the control room, things may not be quite as beneficial as at first sight. The system tends to work very well in small studios where only one or two musicians are recording at any one time. However, with large groups of musicians there are potential risks of the band not playing together if they are all listening to different balances. To highlight this point, consider the case of a foldback system where the drums were fed to four foldback mixer channels; say bass drum, snare, tom-toms and overheads. Dependent upon whichever drum was prominent on any particular foldback mix, the individual musicians may be led to emphasise different beats. During larger recording sessions it is usually the producer or musical director who dictates which beats are to dominate; it is not a question of chance, determined only by an individual musician's choice of foldback balance. Setting the foldback sets the mood, and it is a particular skill of experienced engineers and producers to be able to use the foldback balance to drive the musicians to play in a desired way. Remember, if the musicians are all going their own ways with the foldback mix, then who is in control of the proceedings? Probably nobody! Quite literally, the musicians may not all be playing to the same beat. In some cases, this is not a problem, but in other cases it could be like an orchestra playing without a conductor and with no agreement as to what to emphasise or where.

It should also be noted that the musicians do not all necessarily know *how* to make an appropriate foldback mix, especially of a song with which they are not familiar. Foldback balances are really jobs for the engineers and producers to make, with frequent reference to the musicians to confirm their comfort with the balances. A problem with individual mixes is that without the ability for the recording staff to monitor the headphones in the control room, a foldback mix can be living a life of its own, which may or may not be relevant to the producers' wishes.

Many project studios have begun with a small, individual foldback mixing system on grounds of cost, and have augmented the number of mixers as the studios have grown. Often, the staff of such studios is totally unaware of the constant voltage system because they have never received any professional training. In truth, the different systems have their strengths and weaknesses, and when studios develop there is no reason why they should not offer both the options of local and remote mixing. Many experienced engineers have been appalled when they have gone into a studio and heard the mixes that some musicians were actually trying to work with. Conversely, many musicians have been appalled with what they have been sent from a control room. Getting the foldback balance right is a fundamental necessity for any good recording, so very careful attention should be given to foldback systems, because they form a very important part of the infrastructure of any studio.

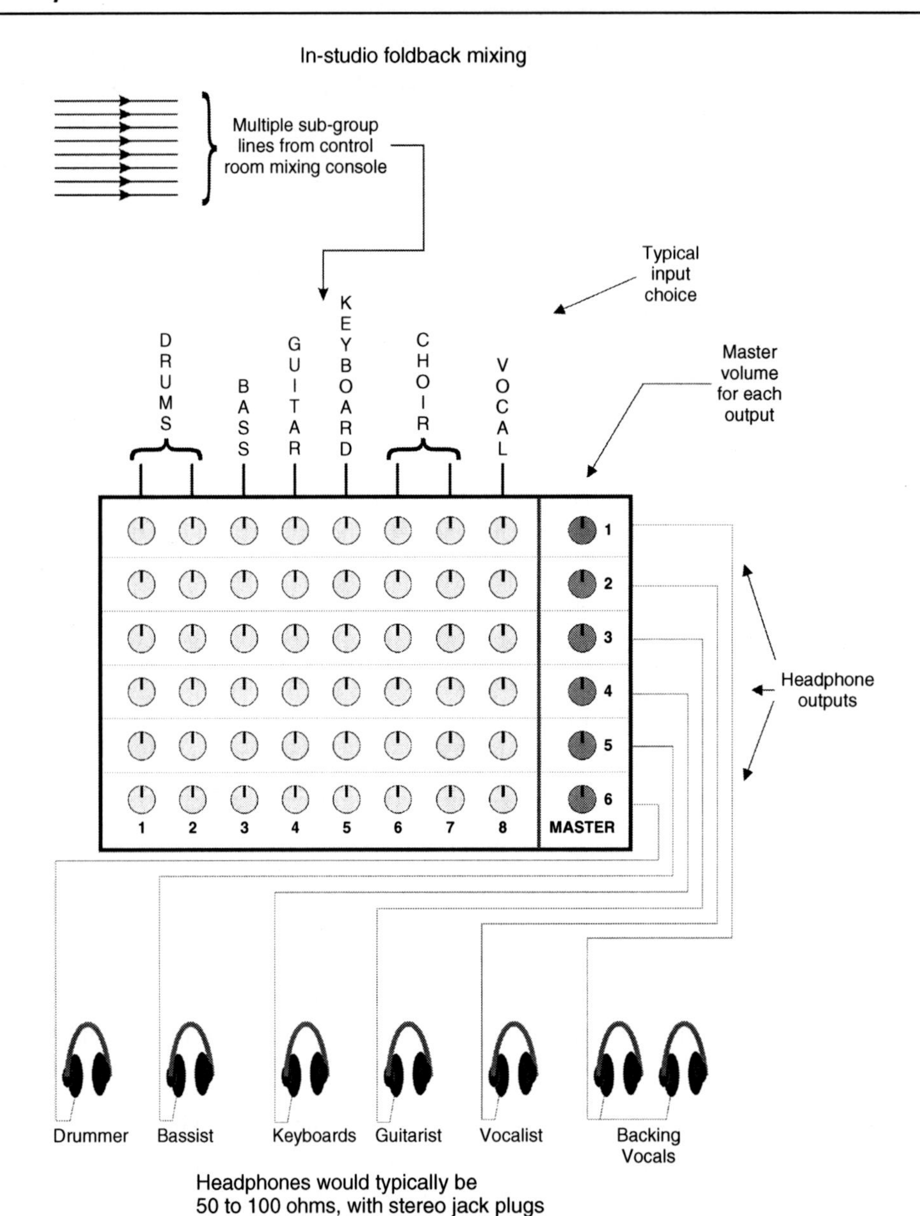

Figure 24.3:
In-studio foldback mixing. In this arrangement, single instruments or sub-groups of instruments are fed at line level into the studio. These go to the inputs of a mixer/amplifier matrix which allows each musician, or group of musicians, to set their own foldback mix and overall volume level. The system allows great flexibility in terms of each musician being able to choose an optimum mix, but it needs to be used by musicians who have some skill in mixing, and presumes that they know the score. Its drawbacks are that it splits the concentration of the musicians between mixing and playing, it denies the producer or engineer control over the foldback, and it isolates the engineer from knowing what the musicians are hearing.

(a)

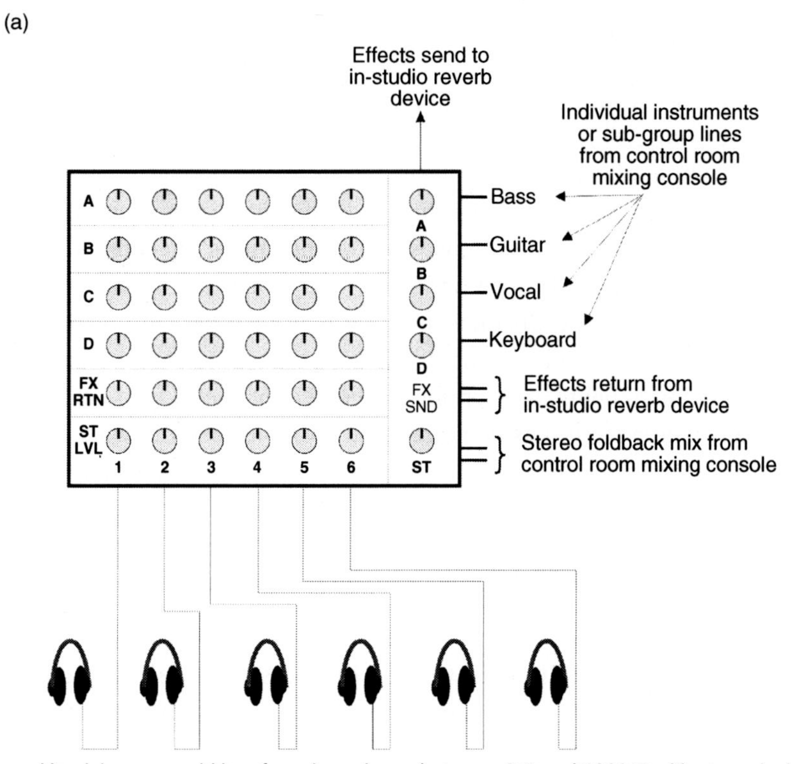

Headphones would be of any impedance between 8 Ω and 1000 Ω with stereo jack plugs

(b)

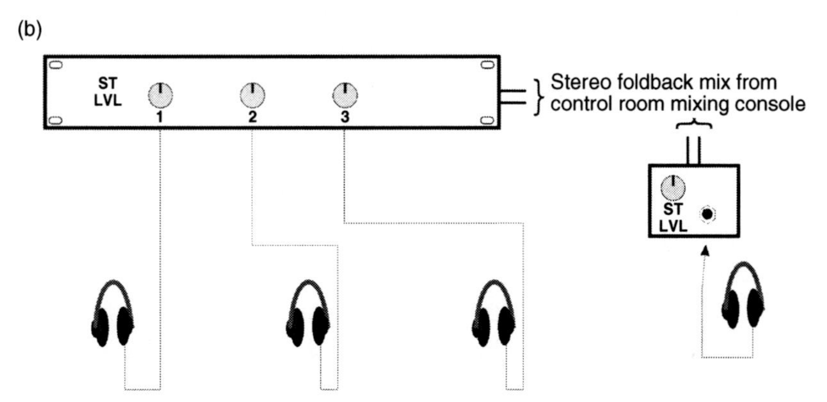

Headphones would be of any impedance between 8 Ω and 1000 Ω with stereo jack plugs

Figure 24.4:

In-studio foldback mixing (commercially available alternatives): (a) In this arrangement the stereo input level serves as the output level for each one of the headphones. Effects return level can also be used as a spare stereo input, for a stereo sub-group sent from the control room mixing console; (b) Single output and multi-output headphones amplifiers. Most of these devices come provided with a mono/stereo switch, and some have an alternative stereo input, which may be used both as effects return or as a second foldback mix input.

24.5 Types of Headphones

Where possible, groups of musicians on a common foldback feed should all be using identical headphones. If mixtures of different impedances are used, or mixtures of headphones with different sensitivities, the loudness in all the headsets will not be equal. This is why it is important for the engineer who is monitoring the headphone mix in the control room to be on the same circuit *and* to use identical headphones whilst making or checking a mix. However, the preferences of musicians towards open or closed headphones are another factor for consideration. Many prefer open headphones, because they feel less cut off from their environment, but there can be situations when their use is inappropriate. Drummers, or other musicians playing loud instruments, may prefer closed headphones. Open headphones provide virtually no isolation from outside noise, so a drummer playing with a kit in a live room, and producing a local 120 dB SPL, will need a potentially ear-damaging level of foldback if using open headphones. On the other hand, headphones with good ear seals may provide 30 dB of isolation, and so from a resulting ambient 90 dB, the drummer could be fed with a mix of instruments which could be heard above the drum level *without* risk to the ears.

In fact, one must always be careful about the use of excessive level in headphones. The level required in the ear canals for headphones to sound equally as loud as loudspeaker listening can be up to 6 dB higher. This is probably due to the loss of the contribution which the tactile sense and bone conduction provide as augmentation to the ear-canal-only sound, which means that the ear drum could receive four times the acoustic power from headphones than when listening to equally loud loudspeakers. This extra power can cause unexpected ear damage at levels which do not seem to be subjectively too loud. It should therefore be apparent that if a drummer wears open headphones, the levels necessary to overcome the acoustic 120 dB of the drums by an in-ear subjective 125 or 126 dB, would perhaps, in reality, be subjecting the ears to SPLs equivalent to around 130 dB from loudspeakers. This is way over any level which could be considered to be safe. For musicians, whose ears are their tools, the results could be disastrous. To put things into perspective, 130 dB represents over 1000 times the acoustic power to which European industrial workers would be allowed to be exposed, and whose ears are *not* so necessary for their work.

The above description shows clearly why closed headphones should be used to keep sound out, but vocalists often should wear closed headphones to keep the sound *in*. The problems can occur when overdubbing vocals. Percussion sounds in the backing track can produce the ubiquitous ‘tizzy tizzy tizz’ that one now hears everywhere from personal stereo systems. If this should leak into the vocal microphone, as it often does, then subsequent processing of the vocal, such as with delays, reverberation or other effects, will also process the overspill. Especially if the vocal is compressed during the mix, the overspill will be brought up in level and the problem will become worse. If a particular vocalist does not like wearing closed headphones it may still be better to work with them, but with one ear-piece partly off the ear, to allow some natural sound in, rather than to run the risk of an undesirable overspill entering the main vocal microphone. 'Click' tracks entering vocal microphones via headphones are also a bane.

24.6 Connectors

When the constant voltage type of foldback system is used it is common to provide distribution boxes for the headphones, say with six or eight outlets, at the end of a cable of around 4 or 5 m length which plugs into a wall panel. Outlets can be provided for each foldback channel, either as mono feeds, or stereo pairs, via a simple arrangement of wall sockets, as shown in Figure 24.5. If suitable cable is used, the boxes can be daisy-chained when more outputs from any given feed are needed. Heavy duty microphone cable will usually suffice for such purposes unless the number of headphones being driven from any single cable would be likely to draw more current than the cable could safely handle. If a tracking loudspeaker were to be used, then it would be better to plug it directly into the wall sockets, which should be wired to the power amplifier outputs with good loudspeaker cable.

The use of 'stereo jacks' on such circuits is not recommended. XLR type connectors tend to be a better choice. The problem with stereo jacks is that when they are inserted into the sockets they can sometimes short the jack connections together, and a mono jack mistakenly inserted would certainly short circuit the output of one amplifier, leading to possible damage. Jack plugs are better reserved for use with the individual mixer type of foldback systems. In any case, these systems usually benefit from the use of headphones of much lower impedance than the 400 ohms which is normal for the constant voltage distribution systems. The use of different connectors therefore prevents the erroneous connection of headphones of the wrong impedance into either of the systems. This is especially important in situations where both types of system are being used in one studio.

As a further precaution against inadvertent shorting of a power amplifier output, it may be feasible to wire a 15 ohm, 50 watt resistor in series with the feed on the headphone outlet wall panel. With only small numbers of headphones in use, this would cause only a minimal voltage drop. However, with 20 headphones in parallel, yielding 20 ohms in total, the 15 ohm resistor would rob almost half of the power, whilst at the same time restricting the amplifier to only half of its output into 20 ohms. It should also be remembered that under such circumstances the resistor, on music signal, could get very hot as it dissipates around 20 watts of heat. Its siting should therefore be carefully considered. If such a device were to be used for protection of the headphones, then a separate output, perhaps with a 'Speakon' connector, could be wired directly to the amplifier for use when a tracking loudspeaker is needed. The pros and cons of these systems need to be individually evaluated for each set of circumstances.

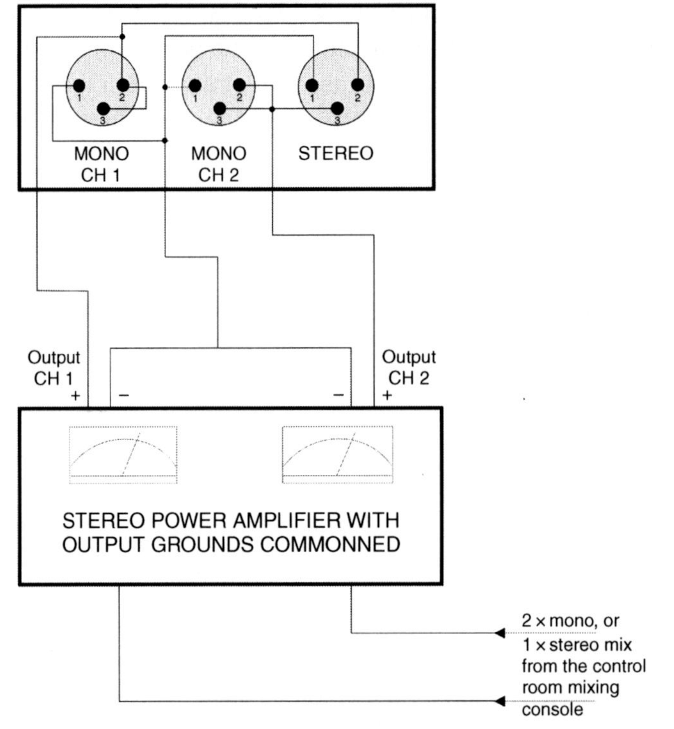

Figure 24.5:
Wall socket arrangement for a constant voltage foldback system. In the above arrangement, the headphones are wired with the −*ve* of each capsule connected to *pin* 1 of the XLR; the +*ve* of the left-hand capsule to *pin* 2, and the +*ve* of the right-hand capsule to *pin* 3. When plugged into the mono outlets (or into distribution boxes plugged into the mono outlets), the two capsules of each headset will be connected together, in parallel, from whichever mono channel of the amplifier they are connected. In the stereo outlet, *pins* 2 and 3 are connected to different amplifier channels, so the signal in the headphones will appear in stereo.

24.7 Overview

Whatever foldback system may be in use, the importance of providing the musicians with good foldback cannot be over-stressed. Musicians are not instrument-playing robots; they tend to be highly emotional human beings, which is probably why most of them *are* musicians. Creativity can be nebulous, with moods forming and dissolving on some very subtle bases. It should surely come as no surprise that a musician given a superb, spacious, inspiring foldback mix will probably perform more creatively than a similar musician receiving an ill-balanced mono mix via a pair of headphones of indifferent quality.

Even the tuning of vocals can sometimes be affected by foldback, with the relative loudness of a vocal in a mix being able to drive a vocalist sharp or flat. The feel of a song can also be varied, dependent upon which instruments are emphasised. A prominent snare drum, for instance, will tend to draw the musicians to follow that beat, whereas the emphasis of a hi-hat on an off-beat, with the snare subdued, could pull the track in a different direction. The direction of a recording can be greatly influenced by the choice of foldback instruments and their relative balance, but this can usually only be achieved by a producer or engineer dictating the foldback mix.

Nevertheless, many studio owners and operators are more under the influence of equipment manufacturers and dealers than under the influence of top professional studios and *their* working practices. The option to mix in the studio is readily supplied by commercially available boxes, which their manufacturers and dealers want to sell. These can be bought, plugged in with suitable leads, and 'off you go'. On the other hand, the constant-voltage/power-amplifier option is more labour intensive to install, and may not be as easy to assemble with ready-made items. However, for multiple headphone use, this option will certainly be cheaper for the studio, but it is not necessarily in the interests of the equipment dealers to suggest its use.

Of course, when recordings are being made by relatively inexperienced engineers, the reality is that they may not be too skillful in achieving sensitive foldback mixes, and when working alone, without a producer or an assistant, the delegation of the responsibility of foldback mixing to the musicians may be a great relief. What is more, in such circumstances, the musicians may well do a better job of it, so one cannot be too dictatorial here about which system is best. Nonetheless, at least if people are aware that the various possibilities exist then they have some options when selecting what to do when the time comes for expansion. Unfortunately, some mixing consoles do not have sufficient auxiliary outputs to make multiple stereo foldback mixes, so this must also be taken into account when selecting the most appropriate system for any given set of circumstances.

Foldback should never be seen as something of a secondary event in the process of recording; it is crucial. A parallel exists in live sound for concerts. In many countries, the struggle for the prestige of being the front of house (FOH) mixing engineer has left the job of stage monitor mixer being relegated, almost literally, to anybody who did not have another task to perform. In truth, the monitor mixing engineer is more likely to have an effect of the quality of the performance than the FOH engineer, for it is the monitor engineer who sets the conditions under which the band must perform. No band is going to play well with poor monitor mixes, and no front of house engineer is going to make a poor performance sound like a good one. In many instances, the calibre of the monitor engineer is more critical to the overall event than that of the FOH engineer. Stage monitors *are* foldback, and when considering foldback in terms of stage monitoring, perhaps the relevance of studio foldback mixes is easier to understand. In fact, when major touring artistes are travelling the world, their choice of monitor engineer is usually a matter of considerable importance on their lists of priorities. In many cases, the monitor engineer is almost considered to be like a member of the band. The relationships can be very close, because the artistes understand that the monitor mix can be like a lifeline that can either feed or starve a performance. Such is the power of foldback.

24.8 Summary

For the musicians, the foldback system is their working environment, and it can greatly affect their creativity.

Tracking loudspeakers are sometimes used instead of headphones.

Many recording engineers fail to realise the importance of a good foldback sound for the musicians.

Constant voltage distribution systems allow the use of a great number of headphones, all receiving identical signals and levels, controlled by the recording personnel.

Stereo foldback in headphones greatly eases the task of hearing different instruments in the mix. The spacial separation is a great advantage.

In many studios the musicians can make their own foldback mixes in the studio, but this runs the risk of them not all playing to the same backing track. What is more, the producer has no control over or knowledge about what each musician is listening to. Which system is best will depend on circumstances and the relative experience of the musicians and recordists.

Open and closed headphones each have their advantages and drawbacks. Excessive headphone levels can be ear-damaging, so care must always be taken to avoid the accidental sending of hazardous signal levels.

When different types of foldback systems are in use in one studio, using different impedances of headphones, it is advisable to use different connectors on each type of headphone to prevent damage by accidentally connecting the low impedance headphones into the high voltage circuits.

The use of jack plugs on high voltage circuits is to be avoided.

The importance of a stage monitor mix during a live concert is analogous to the importance of good foldback during a recording.

CHAPTER 25

Main Supplies and Earthing Systems

Safety and technical earths. Sources of interference problems. The need for low impedances. Mains filtering. The number of phases. Power conditioners. Balanced power.

25.1 The Ground Plane

Large professional studios often have 'technical earth' systems, where a dedicated earth is sunk into the ground, to which no equipment is connected other than the audio equipment. Sometimes this earth is isolated from other earths, and sometimes it is bonded to the earth which is used by the electricity supply company. The regulations about this vary greatly from country to country, and even from region to region, so in a book such as this, designed for an international readership, it is only possible to discuss things in general terms. However, it is always imperative to discuss earthing matters with a local, licensed electrician who is fully acquainted with the local regulations. There is an inherent problem in this, though, because many qualified electricians are totally unfamiliar with the concepts of technical earths, and even professional studios sometimes encounter problems with such negotiations.

Nonetheless, the object of the exercise is to provide a ground reference plane with a low impedance to earth, which acts as a common reference plane for all electrical signals. As the impedance of the surface of the planet is so exceptionally low, and its mass is so great, if one can get as low impedance a path as possible to earth, then whether the technical earth is, or is not, connected to an electricity company safety earth is usually not of too much purely technical relevance. The task is to get the audio system earth connected to any other earths in such a way that any common impedances are minimal. Figure 25.1 shows two different, right and wrong approaches which should be more or less self-explanatory.

Incidentally, the terms earthing and grounding, which frequently get interchanged, may also have different meanings in different countries. Often, they are one and the same thing, but, strictly speaking, the chassis of an aeroplane can be the ground for all the electronic systems, even though in flight the aeroplane is not connected to the earth. The ground, in general, is the common reference plane for the electrical/electronic circuits, whether or not it is physically connected to the earth. In fact, in some audio installations, the audio ground may be separated from earth by a suitable impedance.

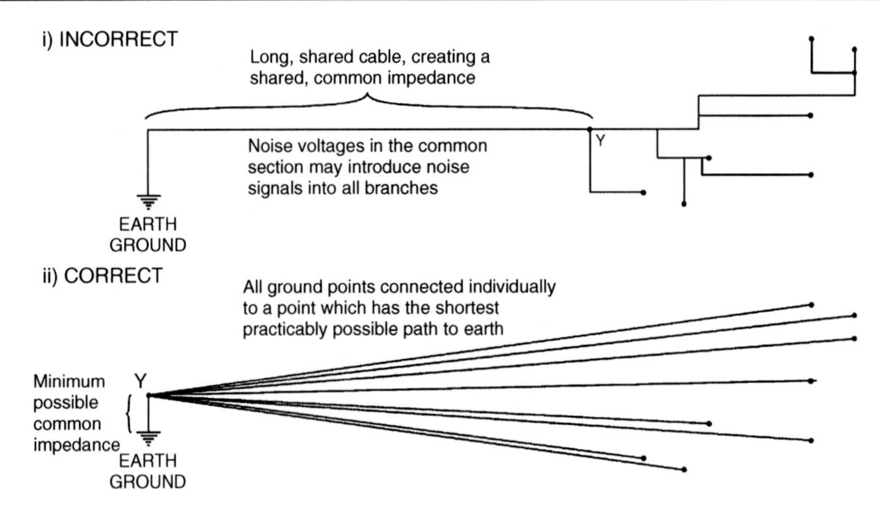

The section of the cable between point Y and earth, in each case forms the lower half of a potential divider between the individual earthing points and earth ground. The impedance of this section should be minimised, which means that the series components of resistance *and* inductance should be carefully considered. As we are dealing with AC, it is futile using a path of minimum resistance if the inductance is still allowed to remain high. In the case of power wiring, the close coupling of the individual cables for live and neutral can be used to help to cancel the inductance in each strand, but with single earth (ground) cables this option is not available, so short runs are the only real option.

Figure 25.1:
Grounding.

Normally, however, in order to overwhelm the effects of any differences in ground potentials due to leakage inductances and capacitances to earth, a low impedance connection is made from the ground plane to the physical earth of the planet. Furthermore, if all metal chassis are earth grounded, then they cannot become lethal in the event of a loose live wire touching them – a fuse or a breaker merely blows, and not a human being.

Geography, or rather local geology, can also play a great part in the effectiveness of earthing systems, and if a studio is sited in a region of poor ground conductivity then little can usually be done about it. However, when the audio system interfacing is carefully and correctly terminated, the necessity for an excellent earth is usually significantly reduced. On the other hand, when the earthing system is excellent, an audio system may stand a better chance of tolerating less than optimum audio interfacing without the manifestation of too many problems. Obviously, though, one should strive to get both aspects as good as one can.

The general level of electromagnetic interference has risen in recent years to a point where we now seem to be swamped with it. The amount of radio traffic is now enormous, which has polluted the air with electromagnetic signals of all sorts. Even in the earth, surprisingly large noise currents can flow near to railway lines where digital signalling and control equipment is used. In industrial areas, there can also now be enormous amounts of electromagnetic pollution in the air, in the ground and in the mains electricity supply. Staying clear of all this is not an easy task, but by paying due attention to each potential source, the effects can usually be reduced to insignificant levels. Unfortunately, the types of equipment often found in low budget studios tend to be more prone to external interference than top-of-the-line equipment. This problem is compounded by the fact that the level of knowledge about how to deal with the noises is less readily found in the smaller studios. This now often leads to situations where the studios which are least equipped to deal with these problems are the ones which most need to do so.

Good audio interfacing practices and good earthing practices can go long way to reducing interference problems, but they may be able to do little about mains-borne interference. The European Union and other authorities around the world have already introduced legislation which restricts both the amount of interference which equipment can generate, *and* its sensitivity *to* it, but this will take years to take real effect. In fact, due to the seemingly endless growth of the use of electrical and electronic equipment, much of the legislation may only serve to slow down the rate of growth of electromagnetic pollution, rather than to actually reduce it from present levels. The two cardinal rules are to ensure that studio equipment is supplied with the lowest practicably achievable source impedance for its electricity supply, and to keep all the interconnected equipment on the same phase. Figure 25.2 illustrates a right and a wrong way of installing power wiring if interference problems are to be avoided.

25.2 Low Impedance Supplies

Many pieces of equipment used in studios, such as most power amplifiers, are notorious for drawing current in surges. Figure 25.3 shows the typical way that an amplifier (other than the inefficient, constant output-stage current, Class A designs) will draw a large transient current from the mains in response to a loud bass drum signal being driven into the loudspeakers. The amplifier will initially draw current from the reservoir capacitors in its power supply, but once the voltage rails fall below the level where the peak voltage on the input rectifier exceeds it, large currents may begin to flow to replenish the reservoirs. The power drain from the electricity supply is therefore not continuous, but a continual series of high current pulses. As the supply impedance is in series with the amplifier, a potential divider will be formed, with the amplifier being the lower element. When the amplifier calls for a current surge, its mains input impedance will drop, and unless the power source is of a significantly lower impedance, the mains supply voltage measured at the wall socket will decrease as each pulse is drawn. Switch-mode power supplies can also be very demanding, as shown in Figure 25.4, creating the same voltage sags as shown in Figure 25.3.

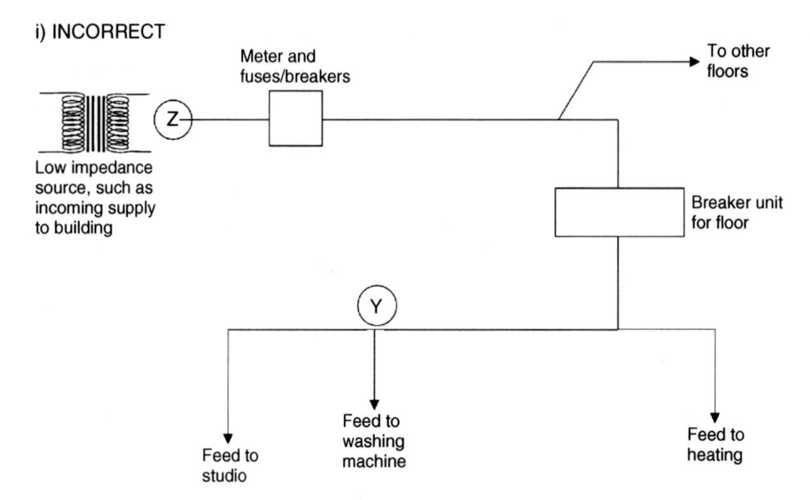

In the above arrangement, the long, shared section of the supply between points Y and Z serves to increase the impedance of the supply to the studio. Electrical loads on the other floors, together with current taken by the heating and other devices, will cause the voltage at Y to fluctuate as the current demand varies. Furthermore, interference caused by the washing machine motor will be superimposed on the supply line to the studio. The result would be a dirty and unstable supply to the studio when other electrical systems in the building were in use.

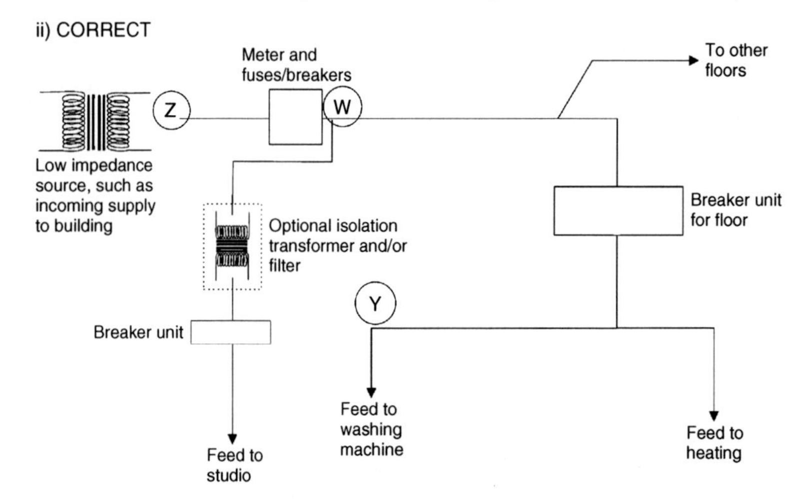

Here, the studio is fed directly from the meter at the incoming supply to the building, via over-sized cable. Whilst the voltage at point Y may still vary as other equipment in the building is turned on and off, the studio feed, from point W, will take advantage of the stability of the low impedance supply and will remain relatively stable. The section Z-W, which is of only minimum length and impedance, will form the lower half of a potential divider, Y-W-Z, with most of any interference from washing machine motors, or similar devices, being dissipated in the upper portion, Y-W. This solution may be seen as 'inconvenient' by the electrical installer, and unnecessarily expensive from a conventional power wiring point of view, but it may be the only way to ensure an adequate supply to sensitive recording equipment.

Figure 25.2:
Power distribution.

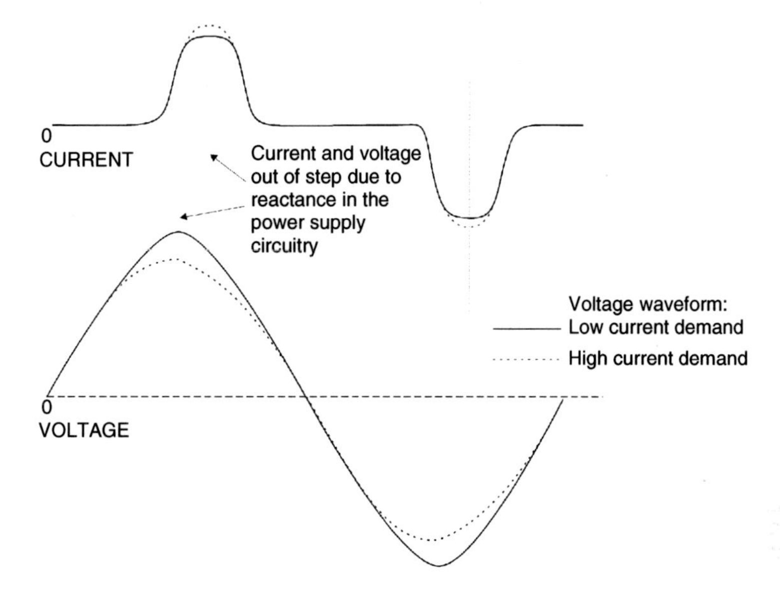

In the case of a stable, resistive load, such as an electric heater, the current demand would follow the sinusoidal waveform of the voltage, and both would be in phase. An audio power amplifier, however, presents a very different load to the supply. Due to the presence of inductance and capacitance in its power supply circuitry, it will present a partially reactive load to the supply, in which the current and voltage peaks will be displaced in phase. What is more, the current demand will *not* be sinusoidal, but will be more like the waveform shown above.

As the output stages of the amplifier supply power to the loudspeakers, they will deplete the charge in the reservoir capacitors. In a conventional power supply, the capacitors will be replenished via a transformer and bridge rectifier, but the bridge cannot begin to conduct until the voltage on the AC side exceeds the DC voltage on the capacitors. This means that as the AC voltage rises from zero, no current will be drawn until the voltage on the transformer secondary winding exceeds the residual voltage in the capacitors. The above figure shows no current initially being drawn, followed by a sudden rush of current once the DC voltage level is matched. Once the AC input voltage of the rectifier falls below the charge voltage on the capacitors the current then ceases, relatively abruptly. Note that as the current demand increases, the voltage falls, as shown by the dotted lines. The amount of voltage drop is dependent upon the supply impedance. A supply of zero impedance would suffer no voltage drop, which is why very low impedance supplies are often referred to as constant voltage sources.

The dotted lines in the current plot show the likely demand from a zero impedance supply, but the impedance of a poor supply would limit the current which can be drawn. This would cause the voltage to sag, and could cause the voltage on the studio power outlets to look something like the dotted lines on the voltage plot. This, in turn, produces harmonics on the supply voltage, which can extend into the hundreds of kilohertz region. Such harmonics can easily enter recording equipment, creating harsh sounds, and can play havoc with computer systems.

It is necessary to bear in mind that when current is drawn in this way, it is not as would be calculated by a simple V × I calculation. For example, a heater consuming 1 kW from a 230 volt supply would draw 4.35 amperes. However, if current was only being drawn 25% of the time, as could be the case with an amplifier, then 1 kW (or rather 1 kVA) would draw 17.4 amperes. From the point of view of an electrician, a 5 amp cable would suffice, because the time-averaged heating effect in both of the above cases is the same. The need for oversized cabling in studios is thus not due to cable heating or power consumption, but to avoid the voltage drop associated with the high current pulses, as these cause the voltage waveform distortion which gives rise to the intrusive harmonics.

Figure 25.3:

Current and Voltage interaction. Current demand cycle of a typical power amplifier under heavy drive conditions.

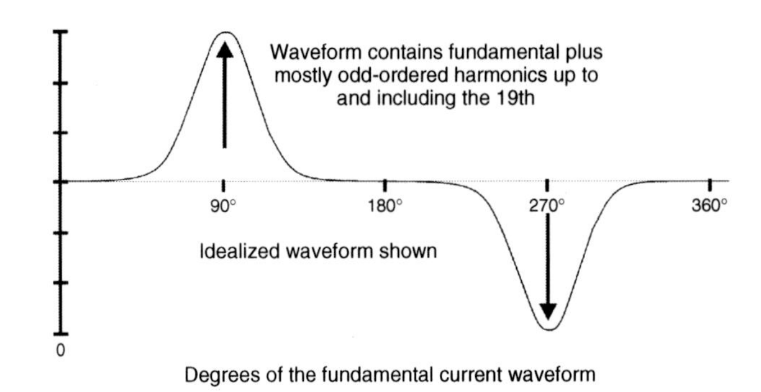

Figure 25.4:
Current and Voltage interaction. Typical high-peak-current waveform for switch-mode power supply input on mains power circuit (after W.H. Lewis).

It matters naught whether or not there is a power conditioner on the mains input to the studio, because these harmonics, which can extend up to quite high frequencies, are being generated locally, and so they will have the same effect as any externally generated harmonics entering via a non-conditioned mains electricity supply. Harmonics on the supply can lead to a whole raft of problems, which will be discussed as this chapter progresses. However, if the supply impedance is low enough it will tend towards acting as a constant voltage source, and within reasonable limits will continue to supply a clean waveform even when high current demands are made. This is exactly akin to an amplifier with a high damping factor acting as a constant voltage source to the varying load which is characteristic of most loudspeakers. Sonically, this means more punch in the bass response, but, unless the amplifier itself draws current from a low impedance supply, the punch may be lost if its own power supply is depleted by the transient signals, especially when working near to its power output limits.

The power supplies of many pieces of equipment are not very good at filtering out higher harmonics on the mains. In all transformers, there exist leakage capacitances between the windings, and these can act as easy routes for the high frequency harmonics to enter the signal circuits. In analogue audio equipment they can cause harshness in the sound, but, somewhat surprisingly to many people, in digital equipment they can cause glitches and even crashes. They can also introduce errors into control circuits, and jitter into digital audio signals. It is remarkable how many such problems can be solved by using a low impedance supply.

Wherever possible, a studio should draw its mains supply directly from the electricity company's input to the building, and not from any branch circuits. It should not share any cable run with any other part of the building, and the audio supply should not even share any cable runs with the studio's own air-conditioning units or refrigerators. It is also worth using the heaviest sensible gauge of wire, even if the fusing is much below this rating. This one can sometimes be difficult to get the electricians to understand, as, to many of them, power is power, and cables and fuses go hand in hand. Electricians may often frown strangely when asked to supply a fused, 50-amp feed with cable rated at 100 amps or more, but when the reasons are explained they are usually willing to comply. The situation is analogous to using loudspeaker cables of much higher current rating than is necessary for the continuous rated power handling of the loudspeakers. The rules, in fact, are identical: to supply the loudspeaker, or the studio, with the shortest, lowest impedance cables which can be practically realised.

Impedance is the mixture of resistance and reactance which is presented to an alternating current. It is the combination of the DC resistance, which can be measured by an ohm-meter, and the inductive and capacitive reactances. This can fool many people. A short, thin, closely intertwined pair of cables may possess an identical resistance to a longer, fatter, spaced-apart pair, but the former may possess a lower impedance than the latter. Inductance is an electromagnetic phenomenon which tends to cancel as the pair of wires forming the circuit are brought closer together, because the equal and opposite magnetic fields tend to cancel. The capacitance between the cables will increase as they are brought closer together, but this is a *parallel* phenomenon, whereas the inductive reactance is in *series* with the DC resistance. In practice, at power supply and audio frequencies, and especially with cables of typical studio lengths, the capacitive effects are negligible.

From the point of view of both power wiring *and* loudspeaker cabling, cables sharing the same outer jacket, and hence which are closely physically coupled, tend towards having the lowest impedances, and hence are best suited to studio use. In many countries, earth wires are also required to lie close to the live and neutral cables. It is hard to find specific data on this, but it is believed that the impedance of the earth wires tends also to be reduced by this method of dressing, and there is less chance of stray currents activating differential circuit breakers. Nevertheless, obtaining the lowest possible impedance path to the earth point, from each piece of equipment, is something that should be sought. The earthing system should also follow the same rules as the live and neutral feeds, in that the earth wire should share no common length of cable with any other building earth, as was shown in Figure 25.1(ii).

25.3 The Number of Phases

In some countries, anything other than a mono-phase connection would be prohibited by regulations which state that no pieces of single phase equipment connected to different phases can be installed within the distance of the spread of a person's arms. This is so that the two pieces cannot be touched simultaneously. On the other hand, in southern Europe, it is still not unusual to find 12-way plugboards installed inside studio equipment racks with four sockets connected to each of the three phases,[1] and these have been installed by licensed electricians. Even in France, it has not been unusual to find a multi-track tape recorder on one phase and its associated mixing console on another phase. Noise levels under such circumstances are rarely satisfactory. In some cases, to have been able to supply one phase to the studio with capacity enough to supply all the equipment would have meant rewiring the whole mains supply system within the building, which the owners were not prepared to do.

Electricity companies supply three phases because it is a much more efficient way both to generate and to distribute electricity in large quantities. Just as multi-cylinder piston engines run smoother than their similar sized, single cylinder counterparts, so three-phase generators run smoother and more efficiently than single-phase generators, in which the electromagnetic effects of each revolution are less balanced. When power can be balanced across three phases it requires less cable to distribute it, because any neutral cables carry composite currents which are not in-phase, and so do not produce the cable heating effect of the total current passing in three, single-phase, neutral 'returns'. More efficient generators and distribution mean lower costs to the electricity companies, and hence cheaper electricity for the users.

The United Kingdom is a relatively rich, industrial country, with plenty of electricity and a culture in which concepts of safety are deeply entrenched. It is quite normal for a single domestic dwelling (house or flat [apartment]) to have a single phase 100 amp supply. In Iberia, on the other hand, where electricity is in short supply, a whole house may be limited to a 15 amp, single-phase supply, and many commercial premises may only have a 20 amp, three-phase supply. Requests for more power can incur large installation costs and a heavy surcharge on each subsequent bill. The British requirement for a single-phase supply to all interconnected pieces of audio and/or video equipment is based on the premise that as potentials of around 400 V can exist between phases, then equipment insulation could be more likely to break down when interconnected equipment is fed from different phases, as compared to being fed from a single, nominally 230 V supply. What is more, the effects of any breakdown across different phases would be potentially more lethal than a breakdown on a single-phase supply. The Iberian philosophy seems to be 'That is all that we have got – take it, or leave it'. The French, no doubt, explain it with the ubiquitous 'Gallic shrug'. In fact, in the United States, there are many bi-phase supplies, derived from the two half-windings of a centre-tapped transformer, giving 120 V from each half; frequently with one half going to each of a pair of adjacent houses. Internationally, these things are far from standardised.

25.3.1 Why One Phase Only?

Anyhow, the science behind all of this is *not* a variable, and it is for good reasons that a single-phase supply is needed. If all the studio equipment were to be operated split across three phases, then all would be quiet as long as the three phases were supplying pure sine waves and were taking equal current. However, perfect sine waveforms on mains electricity supplies are somewhat rare, and studio equipment is not like the three balanced windings of a three-phase electric motor. It does not draw constant current because things are being turned on and off, such as tape machine motors, or have variable current demands such as most power amplifiers when music signals are passing through them. Current imbalances produce harmonics, and strange currents can be created in the common neutral wire. This subject is dealt with in great detail in two of the books listed in the bibliography at the end of the chapter. To quote from the Davis and Davis book (1997, p. 394) 'It can be seen ... that noise generated by the power system can be minimised if only a single-phase power system is used throughout the electronic equipment'. To quote from the Giddings book (1995, pp. 54–56), 'Another issue regarding AC power is the number of phases that are being used to drive the audio system. The ideal number is one, because of the capacitive coupling between most electronic equipment's case and the power supply within. This coupling creates a voltage fluctuation at the line frequency on the case, which is usually the ground reference for the electronics within ... Pieces of equipment on the same phase will have a similar oscillating ground-reference, so they tend to cancel. If a second piece of equipment operating on a different phase is interconnected with these, a small voltage difference exists in the ground references due to the phase differences, and can be picked up as common-mode ground noise by the input, and amplified ... The more often the signal passes between pieces of equipment powered by different phases, and the greater the gains of this equipment, the more the problem is compounded'.

Over many years now, the method of power distribution that has been used in many good quality studios is to power all of the audio equipment from one phase, all the single-phase air-conditioning, or heating, on another phase, and the lights, ventilation and general power (to the office, etc.) on the third phase. If one phase of the incoming supply is obviously cleaner or more stable than the others, then this one should be chosen for the audio. However, this should be periodically checked, as circumstances can change as a result of modifications in adjacent buildings.

25.4 Line Filters and Power Conditioners

As mentioned earlier in this chapter, it can be of little use conditioning the incoming power if there are pieces of equipment *within* the system which are introducing electromagnetic interference (EMI). Nevertheless, additional filtering of the supply ought to do no harm. The trouble is that so many EMI or RFI (Radio Frequency Interference) filters (or suppressors) fail to dispose of the problems, and merely result in dumping even more electrical hash on the earthing system. Figure 25.5 shows how this can commonly take place. In fact, the notion that uninterruptible power supplies will supply an absolutely clean supply is also often very misguided. Only units with internal output voltage feedback control will ensure a clean waveform, and such units often need to be of substantially higher capacity than initially assumed, or even they may cause more problems than they solve. The problem is that they rarely have sufficiently low output impedances to ensure freedom from the production of harmonics by mechanisms such as power amplifier transient demands. Some of these things can actually *cause* computer crashes unless they are connected to individual computers without any other equipment sharing the supply. (For readers unfamiliar with these mechanisms, a careful reassessment of Figure 25.3 could be useful.)

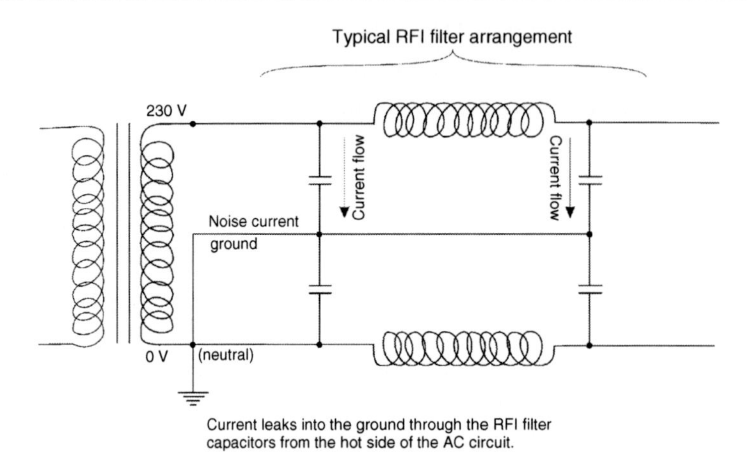

Figure 25.5:
RFI filters in unbalanced circuits. Radio frequency interference (RFI) filters can often remove much noise from the AC supply, only to dump most of it on the ground, which may be equally noise-sensitive: 'Out of the frying pan, into the fire!'.

25.5 Balanced Power

Figure 25.6 shows the balanced equivalent of Figure 25.5. The drawing was taken from an article by Martin Glassband,[2] the author of the 1996 amendment to the United States National Electrical Code (Sections 530–70 to 530–73).

It has long been known that supplying sensitive audio equipment from a balanced power source can resolve many noise problems. The essence is to use a centre-tapped transformer supplying two supplies of opposite polarity, each of half of the total voltage. For example, a 230 V supply is replaced by a centre-tapped supply, supplying two 115 V 'halves' balanced around the centre tap which is taken to earth. Figure 25.7 shows how difficult reactive current problems can also be nullified by balancing the power feeds. In some countries, to supply the standard wall sockets of a building in such a way may be prohibited by local regulations, but in many cases no problem exists if alternative types of sockets are used for the balanced power.

In the United States, the fact was recognised in the late 1980s that certain high technology systems were so sensitive that conventional electricity supplies were limiting their performance. In 1996, there was an amendment added to the US National Electrical Code which allowed audio-video and similar installations to be run from balanced power systems as long as certain specific conditions were met. In other countries, after due explanation, it may be possible to persuade the local electrical safety people of the necessity of power balancing. On the other hand, as such codes relate only to systems installed within the structures of buildings, it may be possible to use a balancing transformer as a free-standing unit, feeding the plugboards to which the audio equipment is connected. This is not something

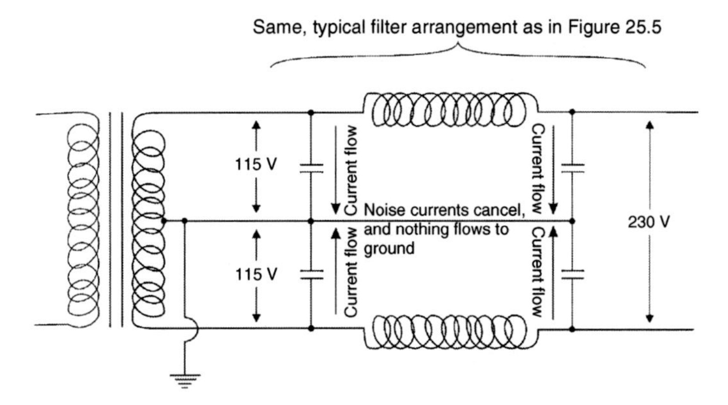

Figure 25.6:
RFI filters in balanced circuits. The grounded centre tap on the AC transformer provides grounding for the AC system and balances the leakage currents to ground, thereby eliminating capacitive leakage from RFI filters as a source of objectionable ground currents, and the consequential noise.

which should be undertaken without expert advice, however, because the use of isolation transformers might defeat the operation of differential circuit-breakers, which are essential safety features in many installations. Further protection will likely be necessary on the isolated side of the wiring.

There are now many installations in which balanced power has been used, and the benefits can be quite wide-ranging. Studios with historic problems of noise have reported over 15 dB less interference pick-up. Due to the ability of balanced power systems to null high frequency interference in a way that is almost unachievable in conventional systems, jitter has been reduced in digital circuits, allowing much cleaner audio signals. The measured reduction in jitter has sometimes been more than 30%. There is also an effect on capacitive interference, because the effects of an electric field are related to the square of the voltage. A nominally 230 V balanced system can only be 115 V away from ground, compared to the 230 V of an unbalanced system. (Or 60 V instead of 120 V in some countries.) Halving the voltage to ground therefore reduces the interference strength by 22, or four times. Balanced supplies can be very effective even in the most arduous of circumstances.

One reason why balanced power is so appropriate to smaller studios is that, unlike fully professional studios, they are very likely to have the sort of equipment mix which does not easily lend itself to optimal audio interfacing. They may be using mixed signal levels, many balanced to unbalanced connections, and much equipment that is only marginally engineered in the first place. Power balancing is also appropriate because lower budget studios are often sited in buildings where dedicated earths and low impedance mains supplies are difficult to arrange. They may also be in remote areas, at the end of long power lines, or on the 21st floor of a building where a good earth can be hard to achieve. Remember, though, never to attempt *any* electrical modification without the aid of a qualified electrician. Lives are at risk!

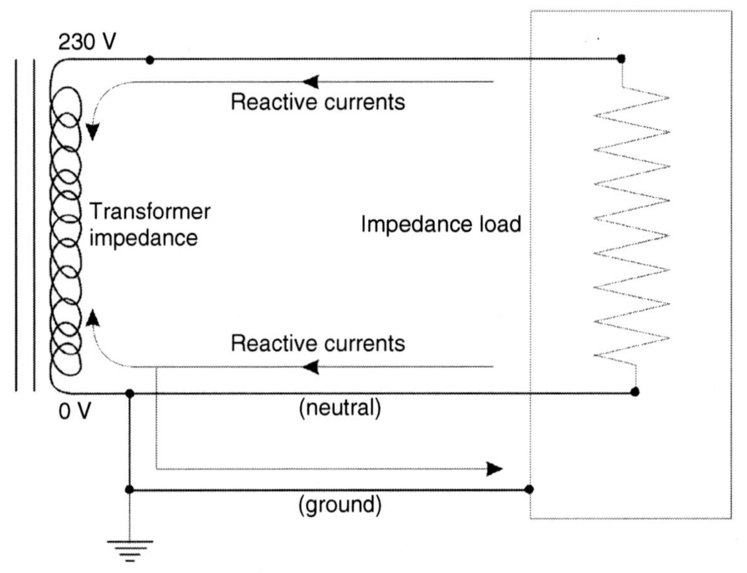

Reactive currents from non-linear loads are present on the ground in proportion to the AC transformer impedance

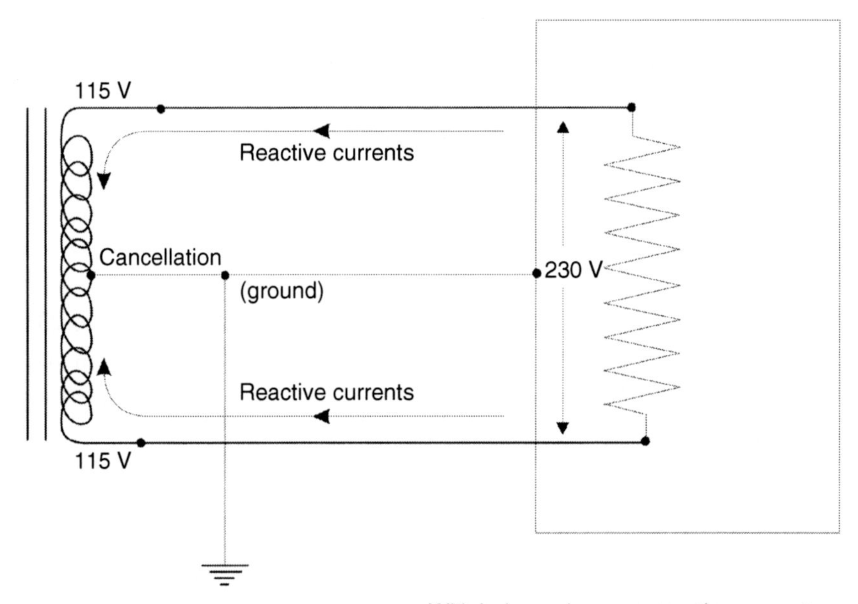

With balanced power, reactive currents null at the centre tap of the AC transformer, thereby eliminating reactive current in the ground as a source of interference in signal circuits.

Figure 25.7:
Comparison of behaviour of reactive load currents in unbalanced and balanced AC systems.

25.6 A General Overview

Whilst it is way outside the scope of this book to go into details of how to make special earthing or power supply arrangements, hopefully this chapter can at least stimulate awareness of the range of problems and solutions. Indeed, it would be totally irresponsible to encourage non-specialised personnel to attempt to make any changes on such safety related issues. Anybody who is particularly interested in the subject should refer to the books mentioned in the Bibliography at the end of the chapter, which explain very clearly just how deep and complex the subjects are.

The main point to be made in this chapter is that a good, clean mains supply and earth are not to be expected from a simple, domestic supply outlet. Many more computer crashes are caused by power problems that are ever realised, and software is often blamed when it is not at fault (although very frequently it is). Many noises are blamed on poor earths when they may actually be caused by poor mains, and *vice versa*. Many power conditioners do not provide the degree of cleanliness that their users believe they will deliver, and many uninterruptible supplies are not as free from mains-borne interference as they would lead their users to expect.

Once again, domestic and semi-professional equipment may be much more prone to disturbances by poor mains supplies and earthing arrangements than their fully professional counterparts, yet they may be the very pieces of equipment which are most likely to be used in less than optimum conditions. One thing which must be borne in mind, though, is that the fixing of poor mains supplies and earthing, to get the best out of semi-professional equipment, may be more expensive than buying professional equipment which may be more tolerant of poorer supply conditions.

A recurring problem these days is getting electricians to understand that the different considerations which may need to be taken into account when making recording studio installations are easily achievable inside the normal electrical regulations. Whether as a result of the shortage of highly qualified electricians to cope with the construction booms or the endless pressure on developers to keep costs down, it appears that electricians are gradually becoming 'installation technicians' rather than true electricians. The knowledge of their local regulations is often limited to that which they need to know for their routine work, and any request outside of that area of application is often greeted with a simple 'No; it cannot be done like that'. Two incidents which occurred in 2010, in highly developed parts of Europe, may serve to highlight the problem.

In one case, a reputable, small, electrical installation company was asked to do the principle electrical cabling in a private studio for a very successful record producer. Two people arrived to make the quotation and plans: one was the electrician who would actually carry out the wiring, and the other was his supervisor. They immediately went into the ritual of trying to make plans to put one air-conditioning unit on each phase; one power amplifier on each phase; the lighting for each of the three separate rooms on different phases, and so forth. It was explained to them politely that the phase distribution had already been specified by the studio designer, and that it should be respected, but the response was very negative. An attempt was made to explain that, even given the preferred distribution of the electrical installers, the use of current was so irregular for each device that there would be no advantage to be gained over the designer's distribution plans in terms of balancing the consumption over the three phases. It was put to the electricians that the neutral cable should be increased in cross-section, in order to help to stabilise the voltage on each phase when the current on the different phases was not well balanced. Quite unbelievably, the *supervisor* then said that this would have no effect because no current passed down the neutral.

Clearly, this revealed his lack of understanding of electrical engineering theory. Granted, in a three-phase motor, for example, there *is* no current flow down the neutral, and therefore, in many such cases, there *is* no neutral wire, but this is only possible because the three windings are accurately matched and they are all drawing equal current whenever the motor is running. However, as soon as perfect balance is lost due to different devices being on different phases, then current *does* flow in the neutral cables. In such cases, if there *was* no neutral, then the voltage would fluctuate, perhaps dangerously, as one phase began to draw either more or less current than the others. This problem was somewhat disastrously highlighted just after the opening of the studio shown in Figure 9.1. Housing development had taken place quite rapidly in an area with poor electrical infrastructure, and the local electricity supply company had delayed making the necessary upgrades. One cold, December night in 2000, when many households were probably using much electrical heating, a neutral cable burned out in the road. The voltages on one phase rose suddenly, and six houses lost their refrigerators, televisions and various other electrical appliances. The studio suffered serious damage to the mixing console and its power supplies. Fortunately, the electrical supply company soon paid for all the damage, but one has to wonder if it would have done so with so little hesitation had it not been for the fact that one of the people suffering damage was the then current mayor of the town.

The other incident which occurred in 2010 concerned a very large installation, of which a recording studio was only a small part. The whole installation was in the hands of an electrical contractor, who was responsible for arranging all the certification with the local town hall. At the planning stages the requirements for the studio had been given to the contractor, who initially accepted them. Nevertheless, when the installation was under way, the foreman of the installation tried to revert to 'normal' procedures. A long battle then followed, which even saw the architectural 'overlords' siding with the electricians. They flatly rejected the over-sized cables (and particularly the neutrals) which had been specified by the studio designers to go directly to the incoming breaker board without sharing any 'local area' breaker boards. The situation at times became very tense, and much of the 'special' studio installation was done on the understanding that it would probably have to be removed and replaced in order to pass the building inspections. In these circumstances, it is asking a lot of the studio owners to have confidence in the studio designers in the face of such 'specialist' opposition.

The first lucky break came when the main contractor began to fall behind schedule, and an electrical sub-contractor was brought in to deal with some of the work that was delayed. By chance, the boss of the small company of sub-contractors was of the 'old school', and had had a thorough education in electrical engineering. The owner of the studio explained some of the 'strange' requirements of the studio designers, and the engineer's reply was to the effect of 'Ah yes; I see. That makes sense'. This restored some of the client's confidence in the studio designers, but he then began to worry about the 'macro' installation in the hands of the main electrical contractors. Some months later, the intensive inspection of the electrical system was carried out by a team of consultants, before signing off the connection of the 0.65 megawatt transformer feeding the building. Some changes were asked for, but none of them related to the recording studio installation, which passed the inspection without a murmur.

In reserve, the studio designers had the number of the relevant EU regulation, No 331.6, which stated that if for any logical reason the phases of the supply could not be balanced, then the impedance of the supply should be reduced, which in practice means that the section of the supply cables should be increased. [It is a moot point as to whether or not all electricians fully understand the inverse relationship between 'supply impedance' and 'cable section'.] The information was all there, written in the norms, but there is a tendency these days for electricians only to learn the regulations which affect their normal, daily work. Things out of the ordinary receive a simple refusal. Dumbing-down is reaching epidemic proportions.

One of the roots of the problem is that electricians think in terms of *power distribution* – how much current is a cable supplying, and how much heating will it suffer. Studio designers, on the other hand, think in terms of the *supply impedance*. The latter is important to sink interference and to maintain stable voltages with varying current demands. The means of doing this have been discussed in Sections 9.5 and 25.2, but the reasons for them often need to be clearly explained to the electricians, who may worry that a large-section cable may also need a larger capacity circuit breaker. The electricity supply company may see this as an attempt to defraud them, believing that somebody was planning to draw more current than contracted at a later date. Nevertheless, a knowledgeable electrical engineer can usually explain the situation to everybody's satisfaction.

In the end, the improved results for the studio in terms of a clean and stable power supply usually do make it worthwhile to fight for the most appropriate electrical installation.

25.7 Summary

The precise nature of earthing systems can depend on local geology and regulations.

Even in the earth, interference currents can now be very high, especially in industrial areas or near electrical transport systems.

One of the best protections against 'earthing' problems is to ensure low impedance mains supplies.

Harmonics on mains power supplies can be very detrimental to system performance.

Studio power wiring should be taken from the primary inlet to the building. Sensitive and interconnected equipment should all be connected to the same phase.

Line filters and power conditioners can actually create problems if they are not installed with a full understanding of how they work.

Balancing the power can be highly beneficial in circumstances of high mains-borne interference levels.

No power wiring should ever be installed or modified without reference to a qualified electrician who is knowledgeable about local conditions and regulations.

Not all electricians seem to be aware of the regulations which permit the increase of cable section in cases which, for justifiable reasons, give rise to problems in the even distribution of the power consumption across multiple phases.

References

1 Newell, P. R., 'Namouche', *Studio Sound*, Vol. 36, No. 12, p. 48 (December 1994)
2 Glassband, Martin, 'The Origin of Balanced Power', *Sound and Video Contractor*, pp. 54–60 (September 1997)

Bibliography

'Sound and Video Contractor' [P.O. Box 12901 Overland Park, KS 66282-2901 USA] (September 1997). Issue dedicated to power quality

Giddings, P., *Audio System Design and Installation*, Focal Press, Boston USA and Oxford UK (1995). Previously published by Howard W. Sams. Indianapolis, IN, USA (1990)

Davis, D., Davis, C., *Sound System Engineering*, Second Edition, Focal Press, Boston, MA, USA (1997)

CHAPTER 26

Analogue Audio Interfacing

General interfacing problems. Old solutions. Modern developments. Polarity standards. The case for jackfields. Balanced–unbalanced interfacing. Grounding of signal screens.

When people plan their first studios of any significant complexity, they are often shocked once they find the true cost of the cables and connectors needed to interface all the equipment. They tend to count carefully the cost of the items of equipment when preparing their budgets, but when everything is ready for installation, and the reality of the interconnection costs confronts them, there is often not sufficient money left to do the equipment justice. It is always a great risk to cut corners on the audio cabling infrastructure because so many seemingly simple solutions can lead to traps. A simple lead, such as one connecting a phono (RCA) plug to stereo jack plug, may seem appropriate to connect the output of one device to the input of another, but there could be five or six ways of connecting those two 'simple' plugs. Although under many circumstances a signal will flow between the two pieces of equipment with *any* of the likely wiring arrangements, it is surprising to many people just how wrong some of those connections can be, and how they can degrade the sound without necessarily making any change in the measured frequency response.

Figure 26.1 shows a range of possible methods of connecting the two connectors mentioned above. In many cases, these get made in studios for specific purposes, but subsequently they get used for other purposes for which they 'look right', but are unsuitably connected. A typical example would be a cable which was perhaps being used to connect the unbalanced output from a CD player into the balanced tape return input of a mixing console, but which had originally been made to connect the transformer balanced output of compressor into an unbalanced input of a tape recorder. (The input and output function of each connector thus being reversed.) Whether or not it works in its new role is entirely dependent upon the type of balancing used in the input circuitry. The result of such a random interconnection may be noisy, it may be quiet, it may sound 'perfect' or it may sound degraded. The result of the incorrect wiring may be totally impossible to predict without knowledge of the precise type of circuitry to which it is connected, and which will not be described in most manuals for semi-professional equipment. The individual pieces of recording equipment in any studio really need to be brought to a common place of standardised termination if they are to be

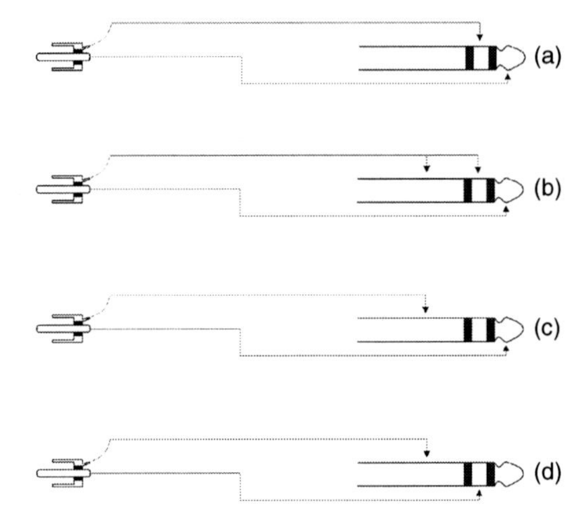

Figure 26.1:
Viable RCA (phono) to 3-pole jack configurations. Four possible wiring arrangements which are likely to be found in studios. (d) is an arrangement which may be found when using some mixing console insert jacks as 'direct' sends, or, in some cases, returns. All of the above have their specific uses.

flexibly and reliably connected together. The normal analogue way of doing this is via a 3-pole jackfield (or patchbay in US English).

We currently have a state of affairs whereby equipment manufacturers produce their equipment, often with little regard for how compatible it will be in its interconnection with other equipment. This is a more or less inevitable consequence of the evolution of much 'semi-professional' equipment out of the consumer market, where no interconnection standards were ever properly developed. In the realms of truly professional equipment there is less of a problem, because the professional equipment evolution has been more controlled.

26.1 The Origins of the Professional Interfaces

In the early days of recording, valve (tube) equipment *demanded* a strict code of interface. Valves could not easily drive long lines directly, so transformer matching was used. The lines were usually run at +4 dBm (0 VU) which was 1.223 volts into a 600 ohm load, a standard which had been adopted by telephone companies as the best compromise between noise and crosstalk in telephone systems. The jackfields themselves were also borrowed from telephone technology, where they were used in the switchboards of exchanges. Output transformers required accurate terminations if their frequency responses were not to be disturbed, so inputs were terminated with standard 600 ohm impedances. Partly to ease the loading problem of one output feeding multiple inputs (which was a more usual requirement in studios than in telephone systems), the studio industry began to terminate the outputs directly, and not rely upon a 600 ohm loading by the following input. This allowed 10 kohm 'bridging' input impedances to be used, and by the mid-1960s almost all professional studio equipment had nominally 600 ohm outputs and 10 kohm inputs; balanced, of course. Forty years later, at the top of the professional end, electronic balancing of inputs has become widespread, and the outputs have often tended to become unbalanced, although they are almost universally of very low output impedance; less than 50 ohms. However, there is usually no problem whatsoever in interconnecting any of the truly professional equipment produced over the last 50 or 60 years. Old style outputs will drive old or new style inputs, with at worst only the addition of a simple resistor, and old style inputs will accept new or old style outputs. About the only complication that arises is that it may still be necessary to confirm which pin is in-phase ([+] [hot] [live]) on an XLR type connector.

The Cannon XL connector ('L' for the 'latching' version of the original 'X' connector) was widely used in the 1950s as a professional audio connector, but it was found that there could be problems of poor contacts after many insertions. Cannon's fix was to use a resilient rubbery material in which to mount the contacts, so that they could be slightly offset and made to wipe on each insertion or withdrawal. The resilient (R) XL became the XLR connector, and it soon became, along with its equivalents by other manufacturers, very much a professional recording industry standard connector. In the United States, in the 1950s and 1960s, pin 3 usually seemed to be the accepted hot pin, with pin 2 being cold and pin 1 ground (earth). In Europe, pins 2 and 3 were frequently used in the reverse sense. In most cases, the male connector was used for outputs, but in the late 1960s, some recording equipment appeared with male input connectors and female output connectors, probably adhering to the concept that for safety (especially at higher voltages) outputs should not appear on exposed pins. Fortunately, this third 'standard' was rapidly abandoned, certainly for low level signals. Microphones have nearly all adhered to one standard pin configuration, with pin 2 producing a positive voltage in response to a positive pressure at the front of the diaphragm.

The AES (Audio Engineering Society) recommended standard is now for pin 2 hot, which would keep the microphone polarity-standard throughout the audio chain, but its adoption has not been universal. For some large manufacturers, with so many years of production behind them and so much equipment in the marketplace, to reconfigure their 'in-house' standard could lead to great confusion in the interchange of old and new equipment. JBL has suffered from this problem with their loudspeaker drive units, on which a positive voltage to the red terminal will produce a backwards (inwards) movement of the cone or diaphragm. The AES standard calls for a positive voltage on the red terminal to produce a positive pressure at the front of the cone; that is, an *outward* movement. JBL adhered to a very old but very logical standard. This stated that when the microphone diaphragm goes *in*, the loudspeaker cone should go *out*, because if the sound source was a voice, for example, it would be an *out-going* pressure from the mouth which would push the microphone diaphragm *inwards*, so the loudspeaker should follow the polarity of the source; not the microphone. However, few other companies followed this concept, but JBL produced such a large quantity of material that to change polarity after so many decades would be extremely difficult. There would be no way to inform all their users of the change, so old units would often be likely to be replaced by incompatible new ones, and total chaos could result, perhaps giving JBL products a bad name solely from the lack of information available to the people making the repairs. JBL compromised on this problem by making their newer *cabinets* adhere to the standard, so a positive voltage on the red terminal of the box will produce a positive pressure at the front of the loudspeaker, but the older style drive units remain as they were.

As previously stated, at least as all reputable microphones use pin 2 as positive (hot) there are usually no polarity problems in that department, except via incorrectly wired leads. In the case of a piece of equipment which is likely to be connected in different parts of the signal chain at different times, such as a tape recorder or an equaliser, then if it has balanced inputs and outputs, and if it is feeding and being fed from correctly connected balanced inputs and outputs, it really does not matter which pin is hot. As long as the inputs and outputs are in-phase with each other, the relative polarity of the signal will be maintained. The problems arise when a piece of equipment has XLR connectors on its main inputs and outputs, but jacks or some other similar connectors on intermediate inputs and outputs, such as insert points or compressor side-chains. In such cases, the polarity of the inputs or outputs on the jacks will be dependent upon whether pin 2 or pin 3 was 'hot' on the XLRs. Operators of mixing consoles need to know the phase relationship between external inputs and outputs and the jackfield, and this usually requires permanent and correctly polarised wiring of the equipment, which usually means that *all* inputs and outputs are brought to a standardised jackfield.

(Strictly speaking, the term 'polarity' should have been used throughout this discussion, and not phase, because a polarity reversal is a 180° phase change at all frequencies. However, the polarity reversal switches on almost all mixing consoles are labelled 'phase', and this terminological misuse has somewhat established itself, at least in many parts of the industry. Nevertheless, two signals with a relative phase shift of around 15° could be referred to as being substantially in-phase, but not in polarity. Polarity relates to phase relationships of 0° and 180° being 'in' and 'out' respectively.)

26.2 Jackfields (Patchbays)

It would seem to be a fundamental requirement of any studio which considered itself to be professional that the equipment should be wired to a central jackfield, and that this jackfield should be of a three conductor, tip, ring and sleeve type. In top line studios, this is *de rigueur*, but the home studio influence has cast some bad influences on many lower budget studios, and also on some others who should know better. The home studios and many project studios have often drawn their experiences, influences and equipment from various sources,

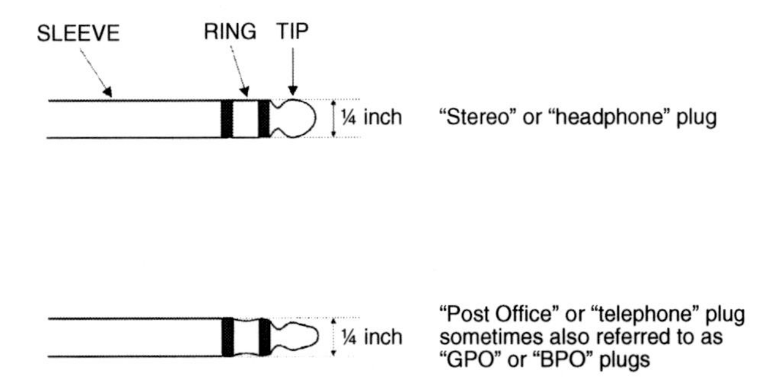

Figure 26.2:
Two types of ¼ in (6.35 mm) 3-pole jack plug. (Note: the 'bantam' or 'TT' jack plug is a 3/16 inch [4.76 mm] version of the GPO/BPO/telephone plug).

including home hi-fi, consumer recording equipment, live (stage) equipment and the professional recording world. This can lead to a very difficult mixture of balanced and unbalanced inputs and outputs, from −20 dBV to +4 dBV nominal signal levels (low level outputs tend to be less costly to produce) and both two contact and three contact connections.

There are three basic types of 3-pole jacks, as shown in Figure 26.2. First there is the 'stereo' jack, where the plug is like a mono ¼ in (6.35 mm) jack with an additional ring connector. This type is used on many headphones. As long as the plugs and the connectors on the jacks are kept clean, they can perform well, and cost considerably less than the more professional types.

(Incidentally, in all cases, the 'jack' is the female connector, and the 'jack plug' is male.) For use in jackfields, quarter inch BPO jacks are preferable; these being again of a ¼ in (6.35 mm) sleeve diameter, but the ring and tip of the plug are narrower. They were formerly known as GPO jacks in the United Kingdom, and telephone jacks elsewhere, but the name changed when the General Post Office in the United Kingdom became the British Post Office. These are more expensive than the stereo jacks, but come with a variety of very hard, low resistance and self-cleaning contacts, with the normalling (switching) contacts usually being made from hard metals that will not oxidise, such as platinum or palladium and occasionally other related metals. These are fully professional jackfields, good for 20 years of daily use. Switchcraft, the US connector manufacturer, developed a miniature version of these jacks, variously called 'bantam' jacks or 'TT' jacks. TT stands for 'tiny telephone' and bantam is, according to the Concise Oxford Dictionary, 'small but spirited', for example bantamweight boxers. Due to their compact size and good performance, the bantam jacks are gradually taking over from the ¼ in (6.35 mm) types, though the smaller plugs can be tricky to wire unless special cable is used. Some of the varieties need special tools for assembly, and may perhaps not be able to be disassembled or reused. They are certainly nowhere near as robust as their ¼ in (6.35 mm) counterparts, which may ultimately be a better choice for smaller studio use as they are easier to wire, reusable, and, in most cases, less expensive, but the massive jackfields alongside many large studio consoles require the compactness of the bantams.

In general, and somewhat perversely, domestic/semi-professional equipment is often far more problematical to connect together than the very expensive professional equipment. In most cases, a relative novice at recording could successfully connect together the equipment of a fully professional studio, but it usually takes a real professional to wire together a domestic studio, once, that is, it progresses beyond a hard-wired package, and this is where a properly wired jackfield becomes invaluable. Be warned, though; except for the most basic of patching, mono jack, 2-pole jackfields are taboo. They create nightmares in terms of grounding (earthing) which usually can never be correctly resolved for all combinations of cross-patching.

26.2.1 Balanced to Unbalanced Problems

Three-pole jackfields should be wired with the 'hot' (+ , in-phase) conductor of a balanced pair, or the signal wire of an unbalanced input or output, connected to the tip. The 'cold' (− , out of phase) conductor of a balanced pair, *or the screen of an unbalanced cable*, should be connected to the ring. The sleeve (ground) connector should only be used for the screens of balanced cables, and should *never* be connected to any conductor that forms part of an audio circuit, such as the screen of an unbalanced cable. There is a defective logic which is often applied which seems to imply that the difference between a balanced and unbalanced system is that there is no 'neutral', 'cold' connector on an unbalanced system. In fact, they both have a 'hot' connection, and they both have screens, but in an unbalanced system, the screen acts both as a 'cold' *and* a 'screen'. One of the main drawbacks of this is that any interference which the screen shields from the 'hot' wire will cause a current to flow in the shared cold/screen conductor, so it will superimpose itself to some degree on to the audio circuitry. Correct and incorrect connection procedures are illustrated in Figure 26.3.

In a three-pole jack system therefore, wired as discussed, a balanced to balanced connection will pass through the jackfield as though passing down one continuous cable. An unbalanced to unbalanced connection will do likewise, though without any contact being made with the sleeve (ground) of the jack. A transformer-balanced output signal, appearing between the tip and the ring, will automatically connect correctly to the 'hot' and 'cold' connections of an unbalanced input without any change in the grounding arrangements. In the case of an unbalanced output feeding a balanced input, the situation is entirely dependent upon the type of balanced input being used. Here, there are three basic possibilities, as shown in Figure 26.4. The first is the old, reliable transformer balanced input. This is true balancing, with many megohms of resistance to ground. The drawback to using transformers are that they are very expensive (if they are to do justice to high quality sound equipment) and they usually cannot pass very low frequencies, although 5 Hz or less is possible with some better quality devices. (In many low budget studio though, the restricted low frequency

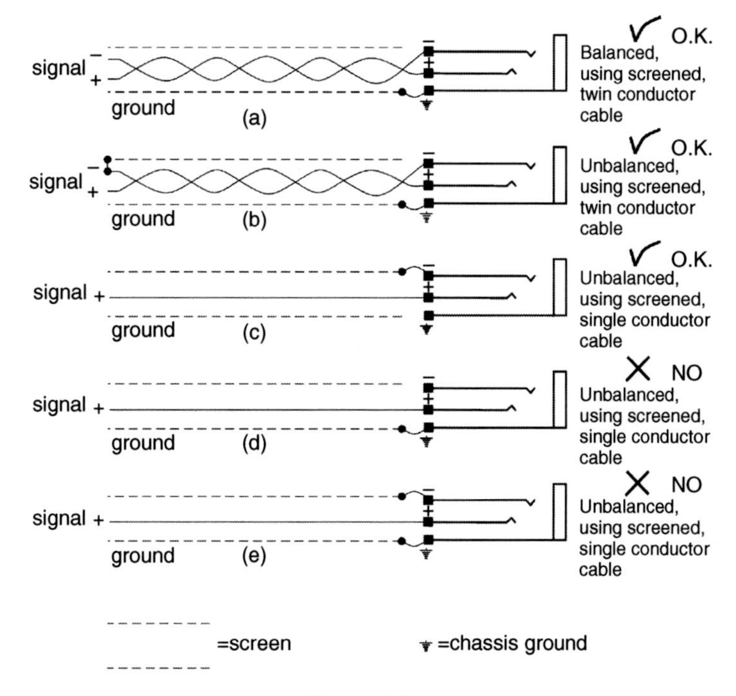

Figure 26.3:
Two- and three-wire connections to a jackfield.

response of the more moderately priced transformers is usually of lesser consequence, as the type of equipment chains normally found in such studios will also tend to have limited low frequency responses.) Secondly, and thirdly, there are two basic types of electronically balanced inputs, one of which *can* accept the 'cold' terminal being shorted to ground, and one which cannot. In the case of the type of balanced input which cannot accept its 'cold' being shorted to its *chassis* ground, care should be taken, and the manufacturer's advice should be followed when connecting such an input to a jackfield. This point is a good example of how, in such situations, a jackfield can be used to standardise all inputs and outputs, so that with only a few exceptions anything can be connected to anything else, in any order. Grounding arrangements can also be optimised, so that hum and other noise pick-up is minimised, though contradicting requirements *can* exist. Nevertheless, if a few high quality transformers are accessible via the jackfield they *can* be brought into good use when otherwise incompatible interconnections need to be made.

26.3 Jacks – Two or Three-Pole?

An absolutely ludicrous state of affairs exists in terms of the use of the mixture of two-pole and three-pole jacks. It is frequently the case that much equipment using ¼ in (6.35 mm) jacks gives no information whatsoever as to whether the inputs and outputs are unbalanced or

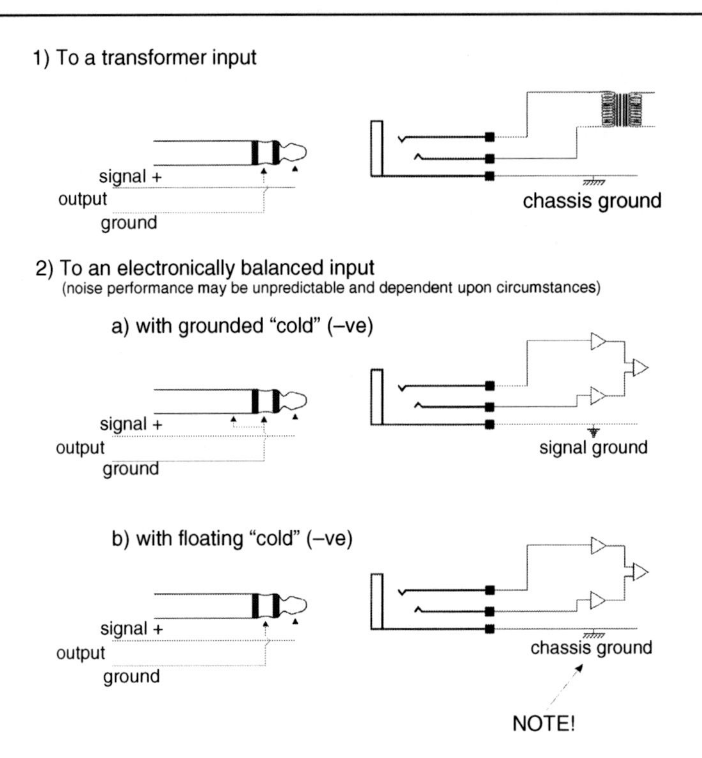

Figure 26.4:
Unbalanced outputs to balanced inputs.

balanced, and whether or not they use 2-pole or 3-pole jacks. Even many instruction manuals give scant information about the terminations. Looking down into the jacks with the aid of a pen-light is one way to find out if it is definitely unbalanced, which it *must* be if there are only two contacts on the jack. On the other hand, some types of equipment use 3-pole jacks as standard, so even though a circuit appears to be balanced it may still be wired unbalanced. Even removing the top and bottom panels from the equipment may not readily reveal the answer, as double-sided boards are often employed and the jacks may cover the tracks. The difficulty of discerning the precise nature of the jacks inevitably leads many people to a 'try it and see' approach.

In fact, there would appear to be no reason whatsoever, except for the excuse of saving an outrageously small amount of money, why 2-pole ('mono') jacks should *ever* be used on equipment intended for studio use, or even on musical instruments having line level connections and which are likely to be used in studios. However, this probably all goes back to the electric guitar and amplifier, with the historical standard use of mono jack plugs and high impedance leads. Early electronic keyboard instruments tended to follow the same standard, but the evolution of devices for use in conjunction with them conflicts badly with the professional studio wiring approaches. If only *all* new equipment used 3-pole jacks, wired in accordance with what was suggested earlier for jackfields, then it would almost all still be mono jack plug compatible, as the mono, 2-pole plug would always short the ring and sleeve together, hence, automatically remaking any separated contacts which needed connecting for unbalanced operation. Of course, this would preclude the use of certain types of *electronic output* balancing arrangements, though, (the ones which will *not* accept the cold [−ve] being connected to ground), but as these can often create more problems than they solve, their demise may be no great loss. However, such is still not the case.

26.4 Avoiding Chaos

It is true that for many smaller, low budget studios, jackfields, plus 40 or 50 patch-cords, are not cheap to buy, but nor would it be cheap to buy all the necessary permutations of other connectors and leads which would be necessary to be able to make the optimum interconnections between *all* likely combinations of analogue equipment patching. If errors were to be avoided when concentration was being applied elsewhere (such as on the musical aspects of the recording), then each piece of equipment would need to be clearly labelled with which configuration of cable it should be connected to. All two- and three-pole inputs and outputs would need to be clearly labelled. Furthermore, each *lead* would need to be clearly labelled at *each end*, explaining the pin connections, and a large stock of leads would be needed to allow for all likely combinations of equipment. The biggest problem would come in remembering the optimal grounding system to be used with each set up, and consequently whether the leads should be ones with the shields/screens connected at both ends, or just the input or the output end. In all cases, only balanced-type cable, with a good screen, should be used, and *never* guitar-lead type cable, with just a single core and screen, as there would be no way in this case of keeping the signal negative separate from the screen. (Incidentally, for non-native English speakers, the words ‘shield’ and ‘screen’ are totally synonymous in this use of the words, screen being the more common word in the UK and shield in the USA.)

Clearly, to do things properly with a discrete lead system is impractical, though the practice is widespread in many smaller studios. Furthermore, it would be unreasonable to expect even an experienced technical person to remember all the differing permutations, especially if work was rapid and/or pressurised. Nevertheless, many mixing console manufacturers produce 30 or 40-channel consoles entirely terminated by individual connectors on the rear panels. This is an unashamed marketing exercise, as they are producing consoles which in almost any serious operation need to be connected to an appropriate jackfield. Their argument is often that if the greater part of the customers for such consoles are too ignorant to realise what is truly necessary, and if the other console manufacturers are selling ‘stripped down’ versions, then they would be unable to compete (and hence to survive) if they produced more expensive products. They do have a valid point; but the real nonsense is that it probably costs the customers twice as much (at least) to fit an external jackfield than it would cost to fit one at the factory during production.

Ben Duncan, in his book *High Performance Audio Power Amplifiers* (Newnes, 1996, p. 331), referred to the following quotation, which is so apt for this situation. It was from John Ruskin (1819–1900): 'There is hardly anything in the world that some man can't make just a little worse and sell just a little cheaper, and the people who buy on price alone are this man's lawful prey'.

26.5 Multiple Signal Path Considerations

What should also always be remembered is that a studio system is not just a big hi-fi system. A hi-fi system usually consists of a pair of signal channels, with the serial connection of no more than about three active devices. Typically, this would be a source (CD, tape, radio, etc.) a pre-amplifier and a power amplifier: there are usually no auxiliary sends and returns which are in simultaneous use with the main signal channels. There are therefore no parallel signal paths, except the left/right pair itself. Despite this apparent simplicity, many hi-fi enthusiasts go to great lengths to ensure the good quality of interconnect cables and plugs. It may seem strange to many people when they hear of hi-fi enthusiasts paying huge amounts of money for their inter-connect cables for their simple systems whilst many recording studios use only standard cable. Without doubt, there are some nonsenses involved in much of this cable mania, but there is at least *one* good engineering reason why hi-fi enthusiasts may be driven to pay vast amounts for interconnect cables is that much hi-fi *equipment* does not have output stages which can adequately drive the poorer quality cables. Top quality studio equipment tends to have output stages of much lower impedance and higher output capability, which are less prone to the limitations that can occur with much domestic hi-fi equipment. Good quality studio equipment is *designed* to drive long cables of standard quality, hence the 'benefit' from using esoteric hi-fi interconnects may not be relevant. Good quality standard cable will usually suffice. This is due to the fact that once we leave the realms of the seemingly simple serial signal paths, and enter the realms of multiple channels and parallel signal paths, a completely different approach needs to be taken to the system interfacing. A studio set-up is *much* more complicated than a domestic hi-fi system, and the opportunities for interference from mains-borne noise, radio frequency interference and interference induced from digital equipment are enormously greater. The wiring practices for hi-fi and studio systems are thus not necessarily directly applicable to each other. The circumstances of use and the characteristics of the equipment are often quite different.

26.6 Grounding of Signal Screens

Basically, for good interfacing, all equipment should use three-contact connectors, with the screen connection made nowhere else other than the chassis. Neve were doing this with their mixing consoles in the 1960s, and even then the practice was based on 20-year-old literature, so it seems that there are some slow learners around because so much equipment *still* does not respect this protocol. Neil Muncy outlined what he called 'The pin 1 problem' in a classic AES paper,[1] in which he clearly threw down the gauntlet to manufacturers. In another paper in the same journal, Stephen Macatee very concisely reinforced the same point, and on some aspects of bad interfacing went even further.[2] From that paper the following quotation has been taken:

The 'pin 1 problem'

> *Many audio manufacturers, consciously or unconsciously, connect balanced shields (screens) to audio signal ground – pin 1 for three-pin (XLR-type) connectors; the sleeve on ¼ inch (6.35 mm) jacks. Any currents induced into the shield modulate the signal reference to that ground. Normally, great pains are taken by circuit designers to ensure 'clean and quiet' audio signal grounds. It is surprising that the practice of draining noisy shield currents to audio signal ground is so widespread. Amazingly enough, acceptable performance in some systems is achievable, further providing confidence for the manufacturers to continue this improper practice – unfortunately for the unwitting user. The hum and buzz problems inherent in the balanced system with signal grounded shields have given balanced equipment a bad reputation. This has created great confusion and apprehension among users, system designers as well as equipment designers.*
>
> *Similar to the 'pin 2 is hot' issue, manufacturers have created the need for users to solve this design inconsistency. Until manufacturers provide a proper form of interconnect uniformity, users will have to continue their struggle for hum-free systems, incorporating previously unthinkable practices.*

The 'unthinkable' practices to which Stephen Macatee refers no doubt include the 'lifting' or removal of safety grounds on equipment for which they are an essential part of the design. Experience has shown that there have been a greater number of studios which violate this legal and sensible requirement than those who completely comply with it. The problem is that with many equipment combinations, even of very well known manufacturers, there can simply be no other way to achieve a hum-free system. Well, it *may* be possible by fitting correctly configured input and output stages, or by adding separate transformers or line termination amplifiers, but this could double the cost of the installation. The reality is that, human nature and economics being what they are, if the cost is going to significantly increase then the ground comes off; dangerous or illegal as it may be. There is absolutely no doubt where the blame lies for this situation; it is rooted in the ignorance and marketing of equipment manufacturers, together with the users who demand ever-cheaper equipment.

26.7 Balanced versus Unbalanced – No Obvious Choice

Balanced systems have two inherent advantages over unbalanced systems: the ability to reject more noise and interference, and the ability to completely free the signals from a noisy ground. Unfortunately, good balancing arrangements tend to be expensive, and this does not fit in well with the cost-conscious philosophies of most of the more competitively priced recording studios. There are situations where nominally balanced inputs are not very well balanced at all, and their performance can be problematical. There are other situations where the *un*balanced inputs are preferable on equipment which has dual inputs, simply because it is easier to make a good unbalanced input than a good balanced one. Frequently, it seems that balanced inputs of dubious quality are installed merely to make equipment *look* more professional, and hence they are probably no more than a marketing ploy. In many cases, unless there are great induced noise problems, the use of a good unbalanced input will be preferable to the use of a poor balanced one.

Good balancing requires the use of multiple chips, discrete components, or high quality transformers. If an inexpensive piece of equipment offers the option of a balanced input using a single chip, then its quality of performance is to be questioned, and it should not be used as an automatic preference. (Ben Duncan has written much about this subject, and it is worth consulting his work.[3]) Furthermore, balanced to unbalanced terminations, (in almost all cases except for high-quality, expensive, transformer balanced, floated outputs) do not make happy partnerships. Bearing this fact in mind, it obviously becomes difficult to generalise about which input to a piece of equipment with dual inputs is the 'best'. It may be that the balanced input is sonically the best one to use with balanced outputs, but the unbalanced input should be used with unbalanced outputs. It is difficult to give absolute advice here. The answer may also depend on cable length or numerous other factors.

The toughest problem to solve is usually the connection of certain types of electronically balanced outputs to unbalanced inputs. In fact, to avoid the possibility of mis-termination when some of the outputs may be connected from time to time to various different inputs, it may be wiser to find the optimum way to unbalance the outputs, then leave them that way. To do this, electronically non-floating outputs should leave the out-of-polarity (−ve, cold) connection disconnected. Electronically pseudo-floating outputs should have the out-of-polarity connection shorted to ground at the output terminals. It is important to ground them at the *output*, because if they are remotely grounded the low output impedance of the device can sometimes drive ground currents through any wiring loops, and can introduce distortion into the in-polarity side of the system. Equipment wrongly connected in this way can often operate inadequately, but with the problem going unnoticed because unless the clean sound has been heard, the degraded sound may well be taken as *the* sound, and the equipment may gain a poor reputation due solely to incorrect termination. And perhaps it may not even be reasonable to say 'gain an *unjustified* poor reputation', because the reputation may well *be* justified by the fact that the manufacturers have chosen an output topology which is prone to this sort of problem in its likely theatres of use. Once again, having balancing transformers available on the jackfield to interface incompatible inputs and outputs when patching-in effects is a very useful facility, though the use of good quality transformers is essential. In fact, this is sometimes the only answer to the problem.

26.8 Sixteen Options for One Cable

In Stephen Macatee's AES paper[2] he listed 16 possibilities for connecting different combinations of balanced and unbalanced inputs and outputs, the number being compounded by the possibilities of the screens (shields) being connected to *signal* ground, or *chassis* ground. Of the 16 possibilities only four could be optimally made with 'off-the-shelf' cables, and even one of those had exceptions. Of the other 12, one-to-one connection of the three contacts of the plugs on each end of the cable (or to the metal shells of the connectors) could not be made. None of this, incidentally, is referring to pin 2 or pin 3 hot anomalies. These sixteen possibilities all presumed a standard pin 2 hot. Figure 26.5 depicts the different possibilities for interconnections with signal or chassis grounds.

To quote from Macatee's paper 'The dashed lines in Figure [26.5] represent the units' chassis boundaries. Connections between the dashed lines are functions of the cable. Connections outside these lines are the manufacturers' choosing, whether conscious or unconscious … [They] include the two most common manufacturer shield grounding schemes – signal grounding the shield and chassis grounding the shield. Identifying these schemes for every unit in a system is essential to debug system hum and buzz. This is no simple task since chassis and signal grounds may be connected together.' This is all contrary to the expectation that one can buy equipment, take it out of its boxes, plug it together with standard cables and expect the system to work, and yet it would appear totally reasonable to *expect* to do so.

The other great problem with so many different possibilities for the connections is that even when a system is made to work noise-free, it only needs the substitution of one piece of equipment anywhere in the system, even remote from the direct signal chain, to upset the grounding. In such an instance it may be that it would be necessary to begin again, from square one, with the whole problem of noise solving. Even the concepts of 'source only' or 'destination only' screen grounding, each of which have their disciples, can be very system-specific in terms of which is best. The truly unfortunate thing is that all of these problems are loaded on to the owners and operators of the equipment due to the lack of understanding, care, or determination from the equipment manufacturers to conform to unified standards of wiring. It is actually a disgrace and shame upon the manufacturers who perpetrate much of this nonsense.

26.9 Some Comments

One aspect of a well-interconnected system is how relatively insensitive it is to the grounding (earthing) arrangements used for the power (mains) wiring. Good earthing is mandatory for the safety of the system, and 'clean' mains are always preferable to 'dirty' (noisy) mains. However, special grounding arrangements such as 'star' earthing are much less of a requirement for noise-free performance when the audio interconnections conform to the

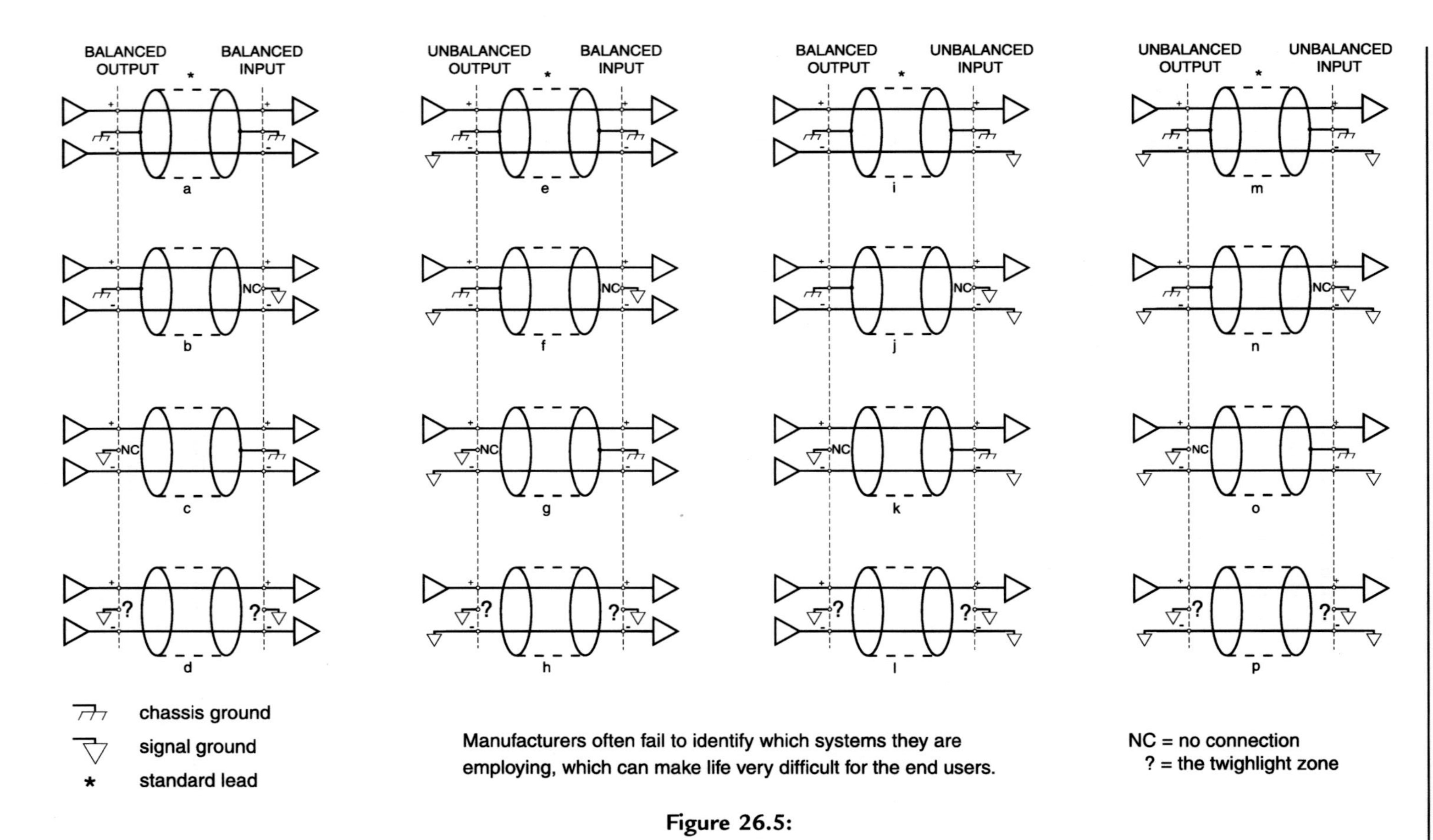

Figure 26.5:
Sixteen different balanced and unbalanced interconnections. Each are the optimised connections for the indicated situations. Note that only four, at the most, can be made with standard 1:1 leads (after Macatee).

preferred practices. The subject is huge, and is dealt with in great detail in the book by Philip Giddings, referred to in the Bibliography at the end of this chapter. Five hundred and fifty-one pages on audio systems wiring! The book is mandatory reading for anybody who is seriously studying the problems of audio wiring, but Giddings also calls for a more responsible attitude from the manufacturers in terms of making equipment to agreed standards, and lifting the burdens off the customers who provide their income. Manufacturers are receiving a lot of bad press on these points, and deservedly so.

The optimisation of sonic purity in many smaller studios is beyond the capacity of many owners, operators and installers of such facilities. This is a reality of the situation in which we find ourselves. Nevertheless, this chapter may not only give some guidance as to how to circumvent many of the common problems, but will also serve to relieve some of the frustration which people may feel when a system which is thought to be relatively simple obstinately refuses to behave correctly. One point to bear in mind at all times is that, in terms of their interconnections, many semi-professional/domestic studios present vastly more complex problems than ultra-expensive studios which use the finest equipment available. When a lesser system fails to operate to perfection, one should not put all of the blame on the technical personnel who installed it, because they may have inherited many intractable problems which have originated in the bad practices of many equipment manufacturers.

It is rather absurd that as a result of this, the most inexpensive equipment tends to require the most expensive installation, but that is what market forces have led to. So many corners have been cut during the manufacturing processes to keep the price of the equipment as low as possible at the point of sale that the problems and extra costs have been passed on to the purchasers in a rather clandestine way. Chaotic marketing free-for-alls are not conducive to good engineering solutions, but for the moment we simply have to live with this reality.

In fact, one of the great driving forces towards all-digital connections is that they leave so many of these problems behind. However, the analogue-to-digital and digital-to-analogue conversions are no easy subject to deal with either, at least where the highest standards of performance are required. The situation seems to be that for as long as microphones and loudspeakers use mechanical diaphragms, and for as long as electric guitars use strings, then these problems are still going to be with us.

26.10 Summary

The analogue audio interfacing of a large studio can be an expensive and complex process.

Much modern recording equipment is not optimally interconnectable.

In many cases, only the use of transformers in the signal path can solve the problems.

Standardisation of compatibility of all the inputs and outputs of a recording system via a 3-pole jackfield/patchbay is an essential feature for flexible and reliable operation.

The sleeve/ground of a 3-pole jack should *never* carry any signal current. The unbalancing of electronically balanced circuitry may require different treatments depending on circuit design.

The manufacturers should always be referred to before carrying out any unbalancing of electronically balanced circuitry.

Pin 1 of any XLR connector, or the ground/sleeve of a 3-pole jack, should ideally only be connected to the chassis of the equipment, and *not* to signal ground.

On equipment with dual, balanced and unbalanced inputs and outputs, the most sonically transparent option may not be obvious without experimentation.

Even a 'standard' XLR to XLR cable (3-conductors) has been shown to require various different wiring combinations for different grounding configurations. Optimally interfaced audio circuitry will probably be more tolerant of the earthing/grounding of the power supply system.

References

1 Muncy, Neil A., 'Noise Susceptibility in Analog and Digital Signal Processing Systems', *Journal of the Audio Engineering Society*, Vol. 43, No. 6, pp. 435–453, (June 1995)
2 Macatee, Stephen R., 'Considerations in Grounding and Shielding Audio Devices', *Journal of the Audio Engineering Society*, Vol. 43, No. 6, pp. 472–483, (June 1995)
3 Duncan, Ben, 'The New Age of Radio Defence', *Studio Sound*, Vol. 38, No. 10, pp. 85–88, (October 1996)

Bibliography

Giddings, P., *Audio System Design and Installation*, Focal Press, Boston, USA and Oxford, UK (1995)
Giddings, P., *Audio System Design and Installation*, Howard W. Sams, Indianapolis, IN, USA (1990)
Journal of the Audio Engineering Society, Vol. 43, No. 6, (June 1995). Issue dedicated to papers on audio system grounding and interconnecting
Davis, D. and Davis, C., *Sound System Engineering*, Second Edition, Focal Press, Boston, MA, USA (1997)
Davies, D. and Patronis, E., *Sound System Engineering*, Third Edition, Focal Press, Oxford, UK (2006)

APPENDIX 1

A More Detailed Analysis of the Non-Environment Control Room and its Absorption Systems

Introduction

As part of an investigation being undertaken at the Institute of Sound and Vibration Research (ISVR) at Southampton University, UK, into the effectiveness of various forms of low frequency absorption systems, advantage was taken of the opportunity to measure the progressive effects of the acoustic control treatments in a ground up studio during its construction.[1] (Tio Pete studios, Urduliz, Bilbao, in the Basque region of Spain.) The sound control room was the first part of the studio to be completed, in an *overall* shell (studio included) of 20 m × 12 m × 8 m high. During the construction of the control room, a series of 44 measurements were taken (11 at each of four different places) during the different phases of the work, beginning with a bare shell and ending with a finished control room. Measurement 1, the bare concrete shell, proved to be of little use, due to excessive reverberation, so only measurements 2 to 11 at the listening position are presented here. The general aim of the control room design was to create a monitoring environment with a decay time of around 0.2 s in a control room of about 200 m^3. If anything could be gleaned from the results that could reduce the proportion of overall volume consumed by the acoustic control system, then that would be useful. Indeed that was one of the prime objectives of the ISVR investigations – higher degrees of control in small rooms – because the great majority of recordings were, and still are, currently being made using inadequately controlled, small rooms.

A1.1 Room Philosophy

The room which was measured was of the Non-Environment, hard front-wall/highly absorbent rear-wall design concept, where the room decay time is source-position dependent, with the lowest decay time being for sources mounted in the hard front wall; the normal location for the loudspeakers. The rooms become progressively more live as the source moves towards the centre of the room, hence the rooms are not oppressive for the personnel working in them, and the small loudspeakers close to the mixing desk are in a rather more lively environment than the main monitors, which are located in positions which result in the greatest overall response flatness at the listening position.

The principal reflective surfaces are the front wall and the floor, and these are typically opposed by a highly absorbent rear wall and ceiling, respectively. Each side wall is reasonably absorbent, so they do not need to be as heavily treated as the rear wall and ceiling, which both face hard, reflective surfaces. Of obvious importance (if a maximally flat frequency response is to be achieved at the listening position and in a designated listening area around and behind it) is to have a relatively uniform and rather short decay time versus frequency. The means of achieving this in many such rooms has been somewhat empirically designed over many years. Nevertheless, the rooms have been developed to a state whereby they are widely considered to be excellent, and the wideband absorption is very effective. However, without a greater understanding of the complex interdependence of the absorption systems and their component materials, the prediction of their effectiveness in other circumstances, or the reduction of their size whilst maintaining the same absorption efficiency, has been something of a question of trial and error, and very time consuming. Hence the undertaking of the investigation programme at the ISVR.

A1.2 Systems of Absorption

In the room whose characteristic responses are presented here, the outer shell is of concrete, with an inner structure based on the old Camden partitioning principle. The 10 cm timber stud frame is covered on the outside with a double layer of 13 mm plasterboard, but the material sandwiched between the two layers is a 3.5 kg/m^2 plasticised deadsheet, instead of the insulation board used in the old designs. Between the studs, the cavities are filled with 10 cm of 40 kg/m^3 mineral wool (although frequently two 2 cm layers of a 60 to 80 kg/m^3 cotton-waste felt are used). The studs are then faced with deadsheet, followed by 2 cm of a 60 kg/m^3 cotton-waste felt. The ceiling structure is essentially similar to that of the walls, except for having a 22 cm depth due to the dimensions of the ceiling timbers.

The floor is a 10 cm, floated, reinforced concrete slab, and the front wall consists of a stud frame with multiple board layers with interspersed deadsheets, faced with 10 cm of irregular-faced granite and cement, to provide a massive baffle extension for the loudspeakers. The remaining treatment consists of an arrangement of plywood and chipboard panels, covered on *one* side with a damping layer of 3.5 kg/m^2 deadsheet, and on *both* sides with a 4 cm (2 × 2 cm) layer of cotton-waste felt. These are hung from chains, both parallel to the wall surfaces and also at various other angles, mounted in a manner which can be seen in Figure A.1.9(e).

Work by Alistair Walter, at the ISVR in 1998, showed the powerful waveguide effect of these panels, which had been thought for some time to be an important component part of the overall effectiveness of the absorption systems. (See also Figures 23.9 and 23.10.) In three investigations the solid panels themselves had been shown *not* to be responsible for any significant absorption, though the absorbent materials alone, suspended without the panels, had not been found to be even nearly as effective as with the panels inserted.[2,3,4]

The following is a list of the measurements taken at the listening position at each stage of the work. The photographs in Figure A.1.1 show the state of construction at the time that each measurement was taken, and the layout of that figure corresponds with the layout of Figures A.1.2 to A.1.6, for easy comparison. These Figures will be introduced section by section, as each measurement system is discussed in turn.

Description of Construction Work Prior to Each Measurement

Measurement No. 2

The empty shell, with only the framework for the vocal and machine rooms.

Measurement No. 3

Stud frames of the control room walls and ceiling, erected and covered on the outside with plasterboard/deadsheet/plasterboard sandwiches.

Measurement No. 4

Construction of concrete isolation wall around control room (see photograph of concrete block). See also Figure 4.28.

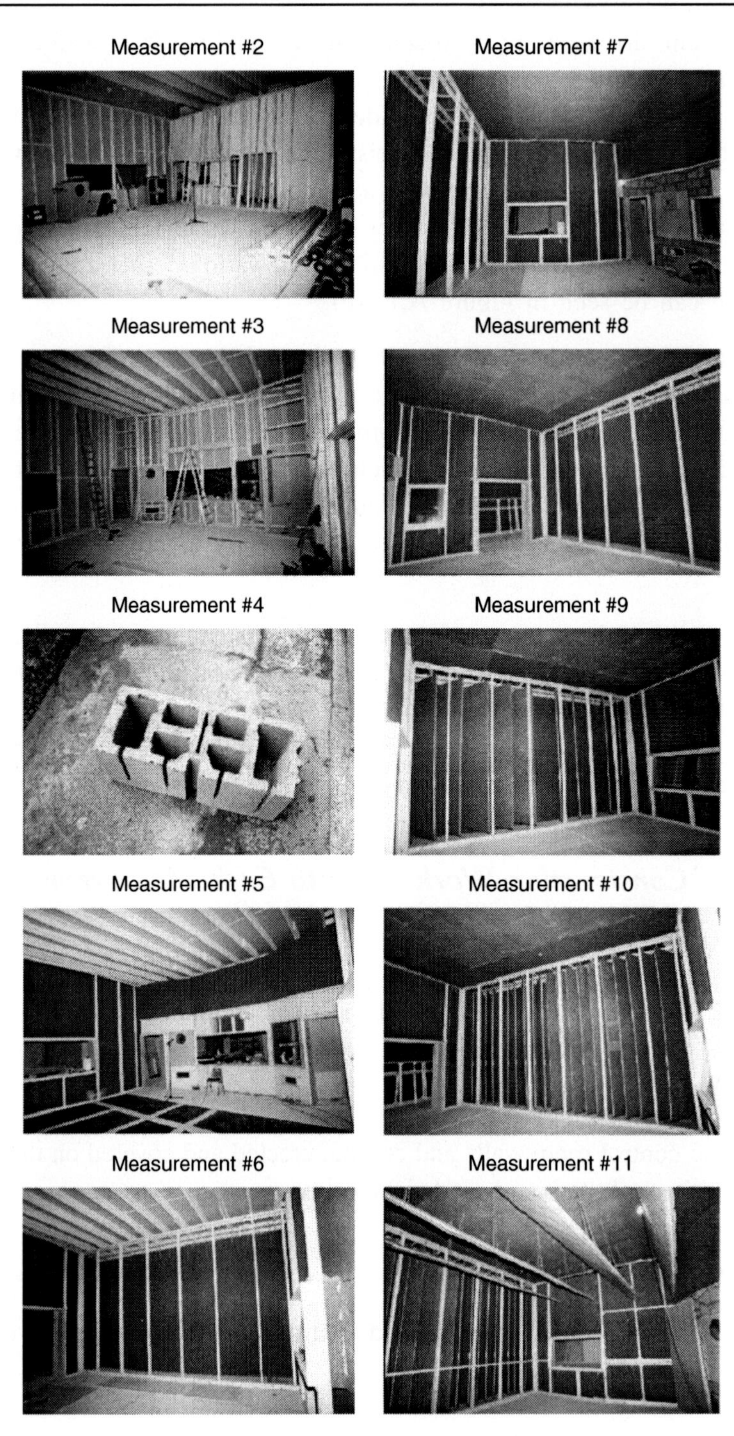

Figure A.1.1:
Photographs corresponding to measurement numbers.

Measurement No. 5

All wall frames filled with mineral wool and covered on inside with deadsheet and felt. Front wall boarded, and loudspeaker flush mounted. Vocal and machine rooms lined with deadsheet and mineral wool.

Measurement No. 6

Large (6.5 m × 4 m) absorbent-covered chipboard panel hung in front of back wall.

Measurement No. 7

Ceiling lined with deadsheet and felt.

Measurement No. 8

Absorbent-covered suspended panels installed next to both side walls.

Measurement No. 9

Ten waveguide panels mounted in front of back wall.

Measurement No. 10

Twenty waveguide panels mounted in front of back wall.

Measurement No. 11

Absorbent-covered suspended panels installed below the ceiling.

A1.3 Discussion of Results

A1.3.1 Waterfall Plots

Figure A.1.2 shows waterfall plots derived from the frequency response function measured at the intended listening position. Comparing measurements #2 and #3 of Figure A.1.2 shows that the introduction of the stud frames of the walls and ceiling, covered on the outside with plasterboard/deadsheet/plasterboard sandwiches, greatly reduces the decay time in the sub-100 Hz frequency range. The frequencies above 100 Hz remain broadly unaffected though. The introduction of the outer, concrete isolation shell between measurements #3 and #4 can be seen to increase the low frequency energy in the room, and not until measurement #11 does this fall back to that shown for measurement #3. These measurements are interesting in that they clearly show the effect of an outer concrete isolation shell on the internal acoustics. As was described at the beginning of Chapter 4, and in Sections 6.3 and 13.7.1, the heavier and more effective the isolation shell, the more difficult becomes the internal low frequency control.

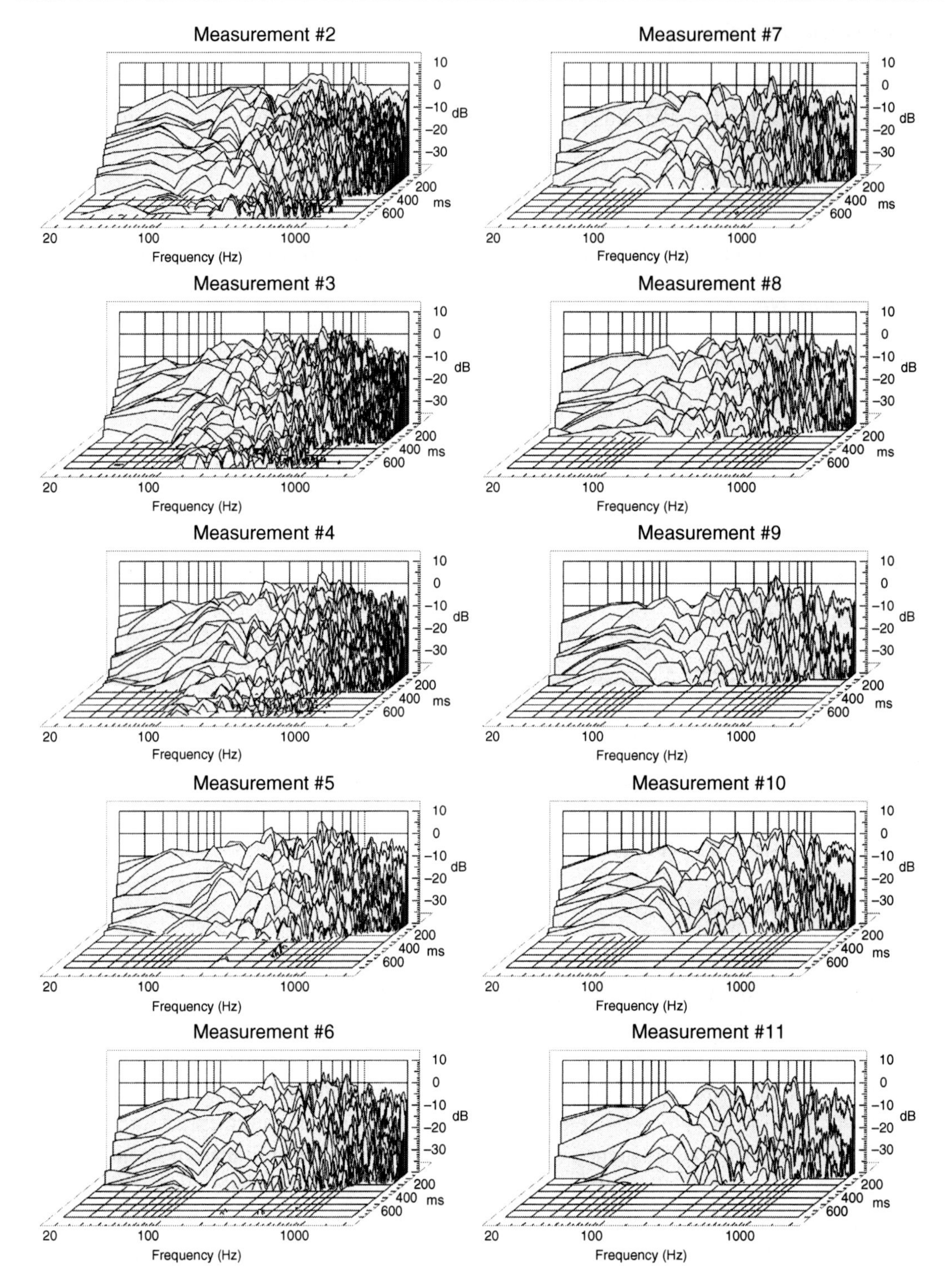

Figure A.1.2:
Waterfall plots.

Measurement #5 was taken after all the wall frames had been filled with mineral wool, and covered with deadsheet and felt. The front wall had also been boarded, resulting in an increase in low-frequency energy being radiated by the loudspeaker (compared to measurements #2 and #4) due to it being flush mounted at this stage. Despite the ceiling still being hard-surfaced, the effect of the treatment on the frequencies above 100 Hz is drastic, having the same order of effect that the plasterboard sandwiches had on the lower frequencies. At this point in the construction, a floor-to-ceiling 'chatter' was quite subjectively pronounced on speech within the room, as all the other reflexions had been very much reduced. This is a typical effect at this stage of a construction because of the ear's sensitivity to sounds occurring in isolated directions. However, it is interesting to note that very little of this significant subjective effect is evident on the plots, presumably because of the position of the source in the hard front wall.

Measurement #6 was taken after a large (6.5 m × 4 m) absorbent-covered chipboard panel was hung in front of the back wall. This produced an unexpected result – *very little* except a small *increase* in decay time in the 100 Hz region. Experience over many years had shown that rooms that have these large panels have exhibited much more isolation to the outside, without the internal acoustics suffering, as compared to rooms without the panels. It seems rather odd that such a large item could be introduced into a room with so little effect on the response (the slight rise in decay time at 100 Hz is presumably due to its masking effect on the deadsheet-covered back wall), but it seems to be the case. (Nevertheless, after the data presented in this set of measurements, the use of these panels was suspended in rooms where effective isolation shells already existed. Evidence which will be presented in following sections showed that the suspected manner in which they could work as extra panel absorbers in fact did not operate to any significant degree.)

Measurement #7 was taken after the lining of the ceiling with absorbent material. All trace of any low-level activity beyond 400 ms, evident in the earlier measurements, has been removed from the plot. The floor-to-ceiling 'chatter', previously heard on speech in the room, disappeared, but a low-level resonance was still audible. Prior to measurement #8, the suspended panels were installed in the side walls, and the waterfall plot shows little effect except for a decrease in decay time in the 100 – 500 Hz range.

Measurement #9 was taken after the mounting of 10 of the angled 'waveguide' panels in front of the large panel parallel to the rear wall. The waterfall plot shows that installing these panels (each 1.2 m × 4 m) reduced the decay time in the 200 – 1500 Hz range by a noticeable amount. Measurement #10 is as measurement #9, but with *20* panels occupying the same space as the previous 10. The increase in the number of panels appears from the waterfall plots to have had almost no effect at the listening position. The spacing of these panels, together with their optimum angle, was, at the time of analysing the measurements, the subject of research by Stuart Colam at the ISVR.

Finally, the ceiling 'trap' system, consisting of five panels, up to 1.2 m deep and 6 m long, was installed prior to measurement #11. The panels were set at angles such that they all pointed their lower edges towards the centres of the monitor loudspeakers. In the past, much importance has been attached to these panels, and it had been noticeable that rooms with insufficient ceiling space for them have been subjectively inferior to rooms with adequate ceiling traps. Measurement #11 clearly shows a very significant reduction in the decay time of the lowest octave, 20 – 40 Hz. This bears out the subjectively 'tighter' bass attributed to the rooms with adequate ceiling traps. The word 'trap' is perhaps a good choice here, because, being angled edge-on to the expanding wavefront, the panels would appear to offer little obstruction to the *entry* of the sound waves, but the exit route via the heavily-lined ducts which they form is extremely difficult to negotiate. The arrangement of these panels also serves to interrupt the parallel surfaces of the floor and ceiling, removing all traces of any remaining 'chatter'.

A1.3.2 Reverberation Time Plots

The reverberation time (RT) plots shown in Figure A.1.3 are actually point-to-point diagrams with each point determined by one-third octave Schroeder integration. As well as the usual −60 dB decay times, the figures also include the decay time to −30 dB as these can prove a useful guide to many subjective effects. One limitation of these plots is demonstrated in Figure A.1.7, which shows the equivalent RT for the set of filters combined with a theoretical model of the loudspeaker. This 'additional' reverberation, which is not a property of the room, is present in all of the plots and cannot, in general, be corrected for. In order to minimise the decay in the loudspeaker response, the sound source was a 15 in low resonance (20 Hz) drive unit in a 500 litre sealed cabinet, which was capable of giving a useful output down to 10 Hz. A comparison between the RTs for the later measurements in Figure A.1.3 and those in Figure A.1.7 reveals that much of the apparent reverberation at very low frequencies is due to the third-octave filters and the loudspeaker, as well as the room. The waterfall plots in Figure A.1.2 suggest that much low frequency absorption has taken place by the time of measurement #5; this is less apparent in the RT plots of Figure A.1.3. Nevertheless, the trend of the RT plots is generally downwards from measurement #3 onwards (measurement #3 was taken after the studwork shell introduced six relatively hard and plane surfaces around the microphone).

Two points worth observing from Figure A.1.3 are the reduction in RT due to the construction of the two remaining concrete outer isolation walls before measurement #4, and the very noticeable effect that the ceiling panels have at low frequencies (#11). The completion of the concrete isolation wall, tightly fitted round the rear and left wall of the control room, with fibrous material in the gap, actually appears to reduce the RT of the room. This may be due to a damping effect on the stud-framed walls. As previously mentioned, the addition of the large rear suspended panel also caused a small increase in RT to be apparent in measurement #6, although its effect is not obvious in the pressure amplitude response.

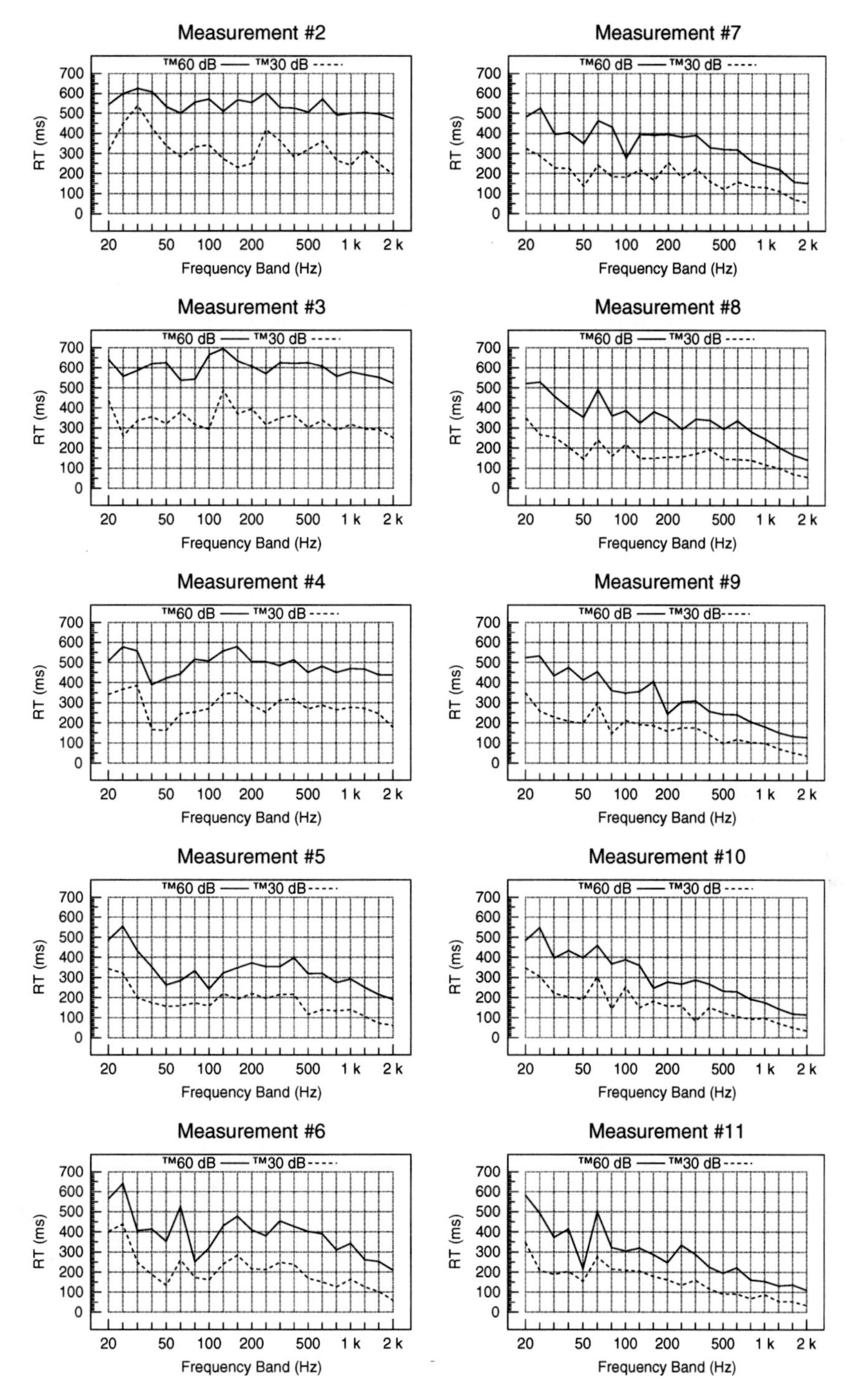

Figure A.1.3:
Reverberation time plots to −30 dB and −60 dB.

A1.3.3 Energy/Time Curves

Not surprisingly, the energy/time curves (ETC) in Figure A.1.4 show details which are not apparent in the waterfall or RT plots, such as discrete reflexions. One such reflexion is that due to the floor. This reflexion appears about 10 dB below the direct signal and about 5 milliseconds later, and is present in all the measurements (although it arrives more rapidly in measurement #2 because the loudspeaker cabinet was still on the floor). In the finished room, its independent nature would be broken up by the presence of the mixing console. Measurement #3 shows a rise in the level of reflected energy due to the completion of the smaller, inner case of the control room bringing the reflective surfaces nearer to the microphone. However, from then on, the trend of the plots is downwards (except for the previously mentioned effect on measurement #6, and some minor effects due to the concrete outer isolation shell before measurement #4).

Table A.1 shows the decay times to −20, −30, −40 and −50 dB for the last five plots. Ten milliseconds have been deducted from the time scales to allow for the arrival of the direct signal at 10 ms.

Only three reversals take place in the downward trend. Two are of only 1 ms, but the third is after the strangely low figure for measurement #9 crossing the −20 dB line. This is only due to the spike around the 28 ms mark on the plot being on the threshold of the line. If it were 1 dB higher, then the anomaly would not exist. (But this is well within the bounds of normal measurement error.)

If a diagonal line is drawn from the peak of the direct signal to the −60 dB/200 ms point, it becomes apparent that the reflected energy *has* been reduced in measurement #10 from that in #9. This suggests that the closer panel spacing is absorbing more energy. This fact is not apparent from the waterfall or RT plots. The most clear indication of the effectiveness of the ceiling panels is shown by the plot of measurement #11, with significantly reduced energy in the 100 − 200 ms region.

Table A.1: Decay times (milliseconds) for measurements #7 to #11

Measurement	−20 dB	−30 dB	−40 dB	−50 dB
#7	33	72	129	190
#8	18	73	127	185
#9	9	58	105	161
#10	18	55	106	149
#11	17	50	90	139

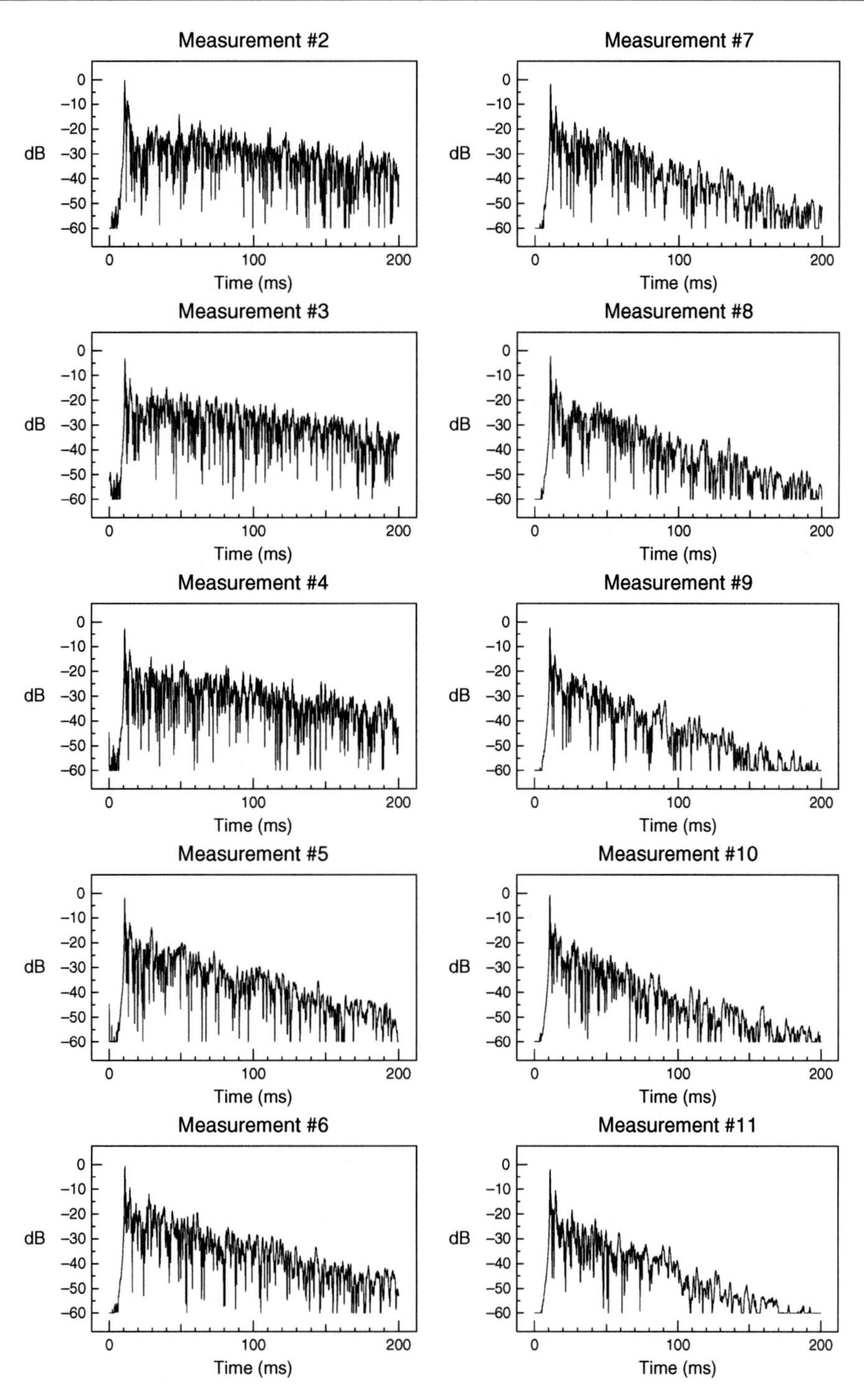

Figure A.1.4:
Energy/time plots (ETCs).

A1.3.4 Schroeder Integration Plots

Figure A.1.5 shows a sequence of Schroeder plots which again clearly show a significant downward overall trend in the decay times as the construction progressed, resulting in a final figure in measurement #11 of about 350 ms. An increase in early reverberant energy is apparent in measurements #3 and #4, after the progressive enclosure of the control room space, and an increase in decay time can be seen in measurement #6, after the installation of the large, rear suspended panel. (As mentioned earlier, the use of such panels was abandoned soon after these measurements were analysed, except in cases where the extra isolation which they afforded was required.)

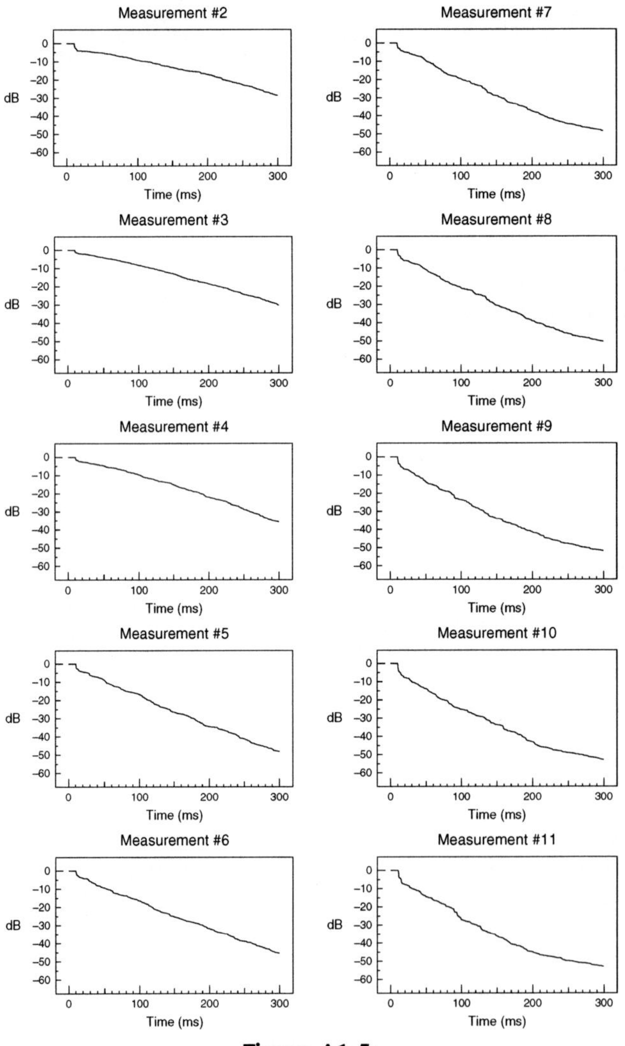

Figure A1.5:
Schroeder integration plots.

A1.3.5 Pressure Amplitude Response Plots

Once again, a clear trend is apparent in the pressure amplitude plots of Figure A.1.6; a progressive flattening of the response as the room treatment is added. The severe peak and dip in the sub-100 Hz region can be seen to change frequency during the early measurements, especially due to the additions to the structure before measurements #3, #4 and #6, but beyond measurement #7 there is a stabilisation of their frequencies, which is accompanied by a reduction in their amplitudes. The addition of the ceiling panels, as shown in measurement #11, deals most effectively with these irregularities, and significantly flattens the low-frequency response.

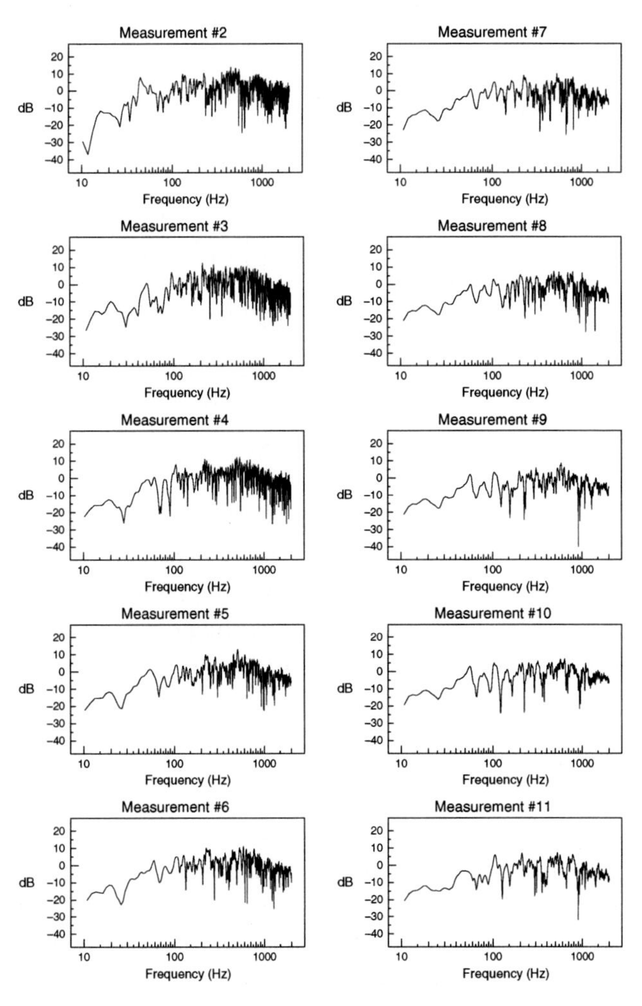

Figure A.1.6:
Pressure amplitude plots.

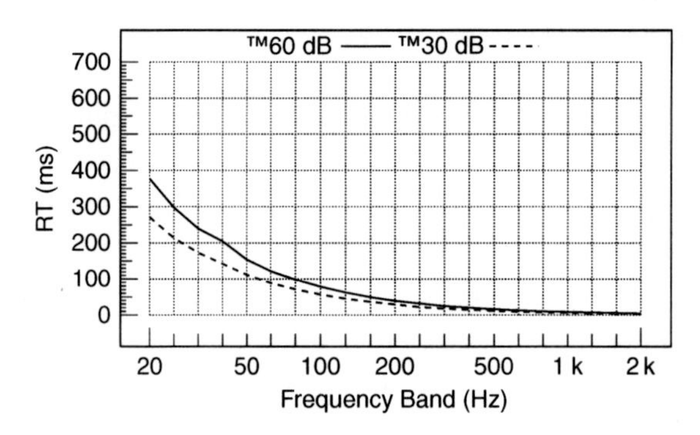

Figure A.1.7:
Equivalent reverberation time of one-third octave filters and loudspeaker combined. These will tend to cause the low frequency decay times to over-read on the plots in Figure A.1.3.

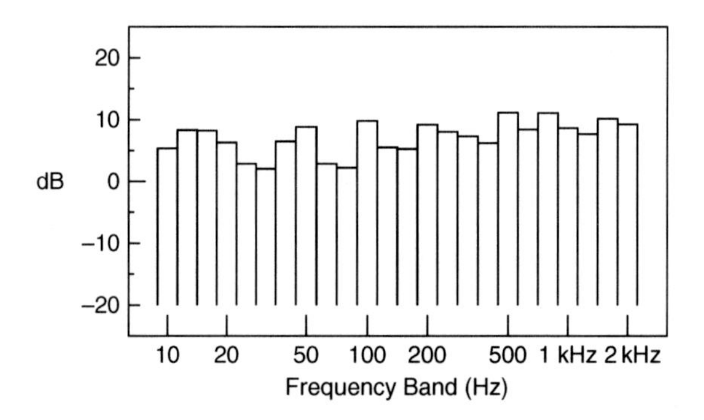

Figure A.1.8:
One-third octave response of completed room (Measurement No. 11) with correction for loudspeaker roll-off.

Figure A.1.8 shows the pressure amplitude plot for measurement #11, but this time in one-third octave bands, after having been corrected for the low frequency roll-off of the loudspeaker using a suitable theoretical model. The response is seen to lie within ±4 dB from 10 Hz to 2 kHz; a demonstration of the effectiveness of the acoustic treatment.

A1.4 Summary

The measurements presented here outline the acoustic control available from a system of membranes, deadsheets, waveguides and fibrous absorbers. It has been shown how a short and well-controlled decay time can be created in a concrete isolation 'bunker'. The effect of the introduction of each group of elements can be seen stage-by-stage. The techniques used are simple, require no special skills to install, and are flexible in their application. The major purpose of the exercise described was to provide a definitive proof of the concept.

It can be seen that in all of the series of measurements, each element of the acoustic control system, as it has been installed, has had a worthwhile effect on the overall room response (or rather the suppression of it). Despite the fact that no single display of the results has been capable of indicating all of the beneficial effects, the cumulative evidence from the whole presentation leaves little doubt that this method of room control is highly effective. The only exception that might apply is the effect of the large, rear, suspended panel. The origin of the concept of this panel was that it was discovered that it could be used to increase the isolation to and from the outside of the room without noticeably compromising the internal acoustics, but its beneficial effect in terms of low frequency absorption *within* the room has been shown to be insignificant.

Figures A.1.9 (a–e) to A.1.13 show various aspects of the construction of this type of room.

A1.5 The Scientific Findings

During the final stages of the production of the first edition of this book, Stuart Colam presented his doctoral thesis on the means by which the so-called 'trap' systems operate. Following on from work by Alistair Walter (see Figures 23.9 and 23.10) which clearly showed the waveguide effect, and after his own MSc work[2] which concluded that the suspended panels, in themselves (i.e. the wooden cores of the panels), did not contribute to any significant absorption, he made significant discoveries about how the panels functioned in *arrays.* It has now been shown that the *arrays* of panels have absorbent properties which are attributable only to the arrays as a whole, and not to their individual component parts.

(a)

Figure A1.9:
Various stages of the room construction during the measurements described in Section A1.2. (a) The Tio Pete control room at the stage of Measurement No 2.

(b)

Figure A1.9: (continued)
(b) The loudspeaker wall framing at the stage of Measurement No 3.

(c)

Figure A1.9: (continued)
(c) The front of the room during Measurement No 5.

(d)

Figure A1.9: (continued)
(d) The rear of the room at the stage of Measurement No 6.

(e)

Figure A1.9: (continued)
(e) The 20-panel array in place during Measurement No 10.

Figure A.1.10:
The completed acoustic treatment, viewed from the monitor loudspeaker wall.

Figure A.1.11:
The completed acoustic control system at Area Master Studios (now Planta Sónica 2), Vigo, Spain; viewed from the rear, and showing the closed nature of the ceiling panels from this viewpoint. Compare this with the view in Figure A.1.10.

Figure A.1.12:
Discosette 3, Lisbon, Portugal; at the stage of mounting the first fabric panels over the absorber systems. The floor is pink marble. The covering on the waveguide panels is Dacron.

Figure A.1.13:
The completed control room at Tio Pete Studios, Bilbao, Spain, the subject of the measurements presented in this appendix.

It has become clear from this work[3] that the effects of the arrays differ with the angle of incidence of the acoustic wave, and that the side-wall arrays, the rear-wall arrays and the ceiling arrays do not all operate in the same manner. Furthermore, the manners in which they do work are not all equal with frequency. Take the arrays on the side walls, for example. At low frequencies, it has been found that a surface wave exists as the wavefront propagates along the face of the array. It can be modelled by means of an acoustic impedance at the front edge of the array. This would suggest that the inclination of the panels on the side walls, with their leading edges facing the front of the room in Figure 13.7, should be of no consequence to low frequencies, because the surface acoustic impedance does not depend on the direction of the incident wave. However, at somewhat higher frequencies, where the lined-duct effect is important, things would change dramatically if the panels were mounted with their leading edges pointing to the rear of the room.

The rear-wall trap systems receive what is essentially a plane wave, which strikes them perpendicularly. It has been shown that the absorption at lower frequencies can be improved when the panels are set at an angle of about 20° to the rear wall, but placing them at such a relatively flat angle tends to make the panels more reflective to the upper-bass and lower-mid frequencies. Experience has shown 45° to be a better compromise in terms of producing a flatter, overall room response.

Another aspect of the absorption mechanisms in the trap that is still under investigation is the 'horn effect', suggested by Professor Frank Fahy. In effect, the inclined panels may act like reversed horns, with the high particle velocity at the mouths of the openings giving rise to high pressure at the throat, where the panels *almost* touch the wall. The phase differences of the wavefronts on each side of the panels, due to their waveguide effect, lead to high particle velocities through the porous material between the panel and the wall as the pressure differences try to equalise themselves. In turn, this appears to give rise to resistive losses due to the tortuosity and friction in the path through the fibrous materials, and hence, in turn, gives rise to considerable absorption. In a highly tuned system, this would be very selective in the frequencies of maximum absorption, but in practice, the effect seems to be quite low Q. Nevertheless, this may be just one of many mechanisms at work in these complex absorber systems.

The ceiling panels have usually been inclined such that their leading edges pointed towards the loudspeakers, in the belief that this provided the easiest access to the labyrinth for the expanding waves leaving the loudspeakers. However, the work by Colam suggested that at low frequencies, the lower edges of the panels may again provide a surface acoustic impedance, which should not be significantly affected by the precise angling. Despite this, further work by the same investigator seemed to show that the beneficial effect at very low frequencies could also be due to the broader spread of the effect created by the different panel depths and angles of incidence with respect to the ceiling. (See Figure 16.2.)

That these absorption systems should have been empirically developed will come as no surprise once the true complexity of their operation is better understood. They have been designed and developed by engineers, as opposed to scientists.[4] The scientific studies are now beginning to better understand what the engineers have achieved, and this should be a big help to further development. Indeed, it is the complex nature of the overall systems which has seriously inhibited computer modelling, because the input data simply has not been available. This has also been a block to more traditional design approaches to the Non-Environment rooms. One is simply not going to find how to make the practical absorption systems by reading the established text books. However, the fact that the Wright brothers did not have any text book to confirm the aerodynamic principles of their first aeroplane did not prevent it from flying. The engineering led to more science, which in turn led to better engineering. The parallel with these absorber systems is very real, and more science *and* engineering will yield even better results in the future.

The current 'Holy Grail' is to improve efficiency in order to reduce the size of the arrays, and to improve their effect in smaller rooms. Progress by engineering 'cut-and-try' approaches is slow, which is why more effort is being made on the scientific front, so that ways can be found to better predict the results of changes, and hopefully speed up the development.

Until about 2000, everybody involved had thought that the suspended panels, themselves, contributed to the absorption. When one-tenth scale modelling was carried out in 1990,[5] it failed to detect absorption, and the suspicion was that the modelling technique, itself, was flawed. Ten years later, Colam's work confirmed the modelling to have been correct, and that the suspended wooden panels do *not* contribute to the absorption. They tend to re-radiate the great majority of whatever vibrational energy they receive, in a reactive manner. However, the deadsheets covering one side of the panels provides some damping, and so *will* give rise to a certain amount of additional absorption.

A1.6 More Recent Research and Development

Figures A.1.14 to A.1.17 show various views of the 'critical listening room' at the University of Vigo, Spain, which has also been used as a test-bed for much physical measurement of the absorber systems. Some of the results of this work have already been published,[6,7] and Figure A.1.18 shows the averaged absorption across the face of the rear absorber. From Figure A.1.19 it can be seen very clearly that the waveguide effect of the panels is significant, and that it must lead to phase differences either side of each panel where their felt layers make contact with the rear lining of the 'trap'. This highlights the need for the felt on the panels to touch the felt on the rear surface, because when the out-of-step pressure differences on either side of the panels try to equalise themselves, there will be a flow through this resistive material and so the frictional losses will give rise to low frequency absorption.

Figure A1.14:
The Non-Environment, multi-format listening room in Vigo University, Spain, 2004.

Figure A1.15:
The Vigo University room under construction.

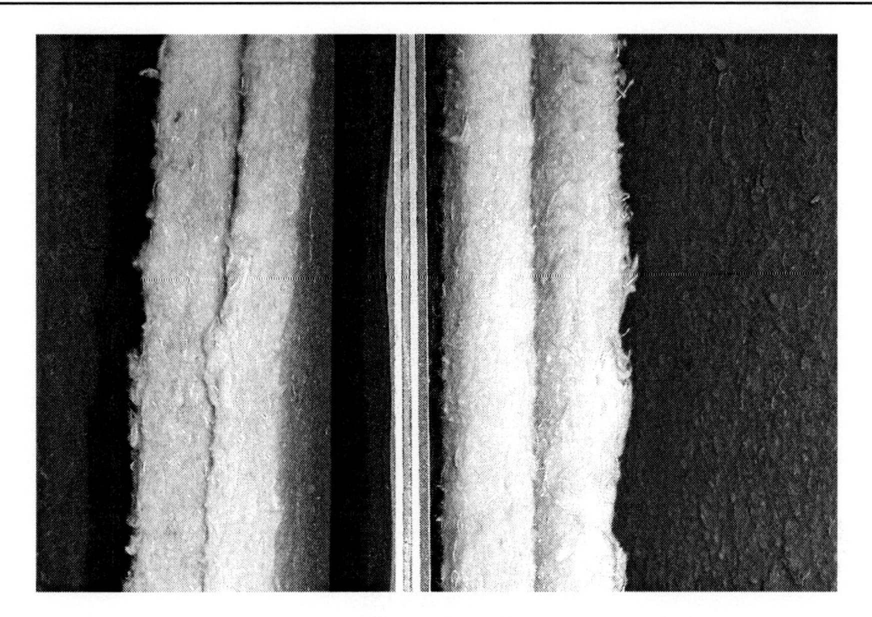

Figure A1.16:
Detail of the panel construction. The central core is of 12 mm plywood, with a layer of a 3.5 kg/m^2 deadsheet stapled to one side to damp any panel resonances. On either side, loosely fixed, are two 2 cm layers of a flame retardant cotton-waste felt of about 60 kg/m^3.

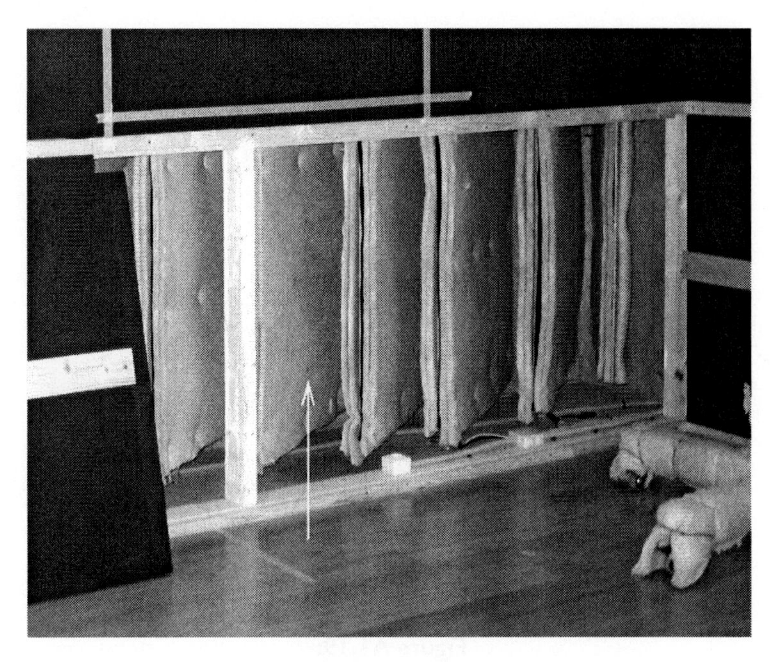

Figure A1.17:
The panel array behind the final, fabric covering.

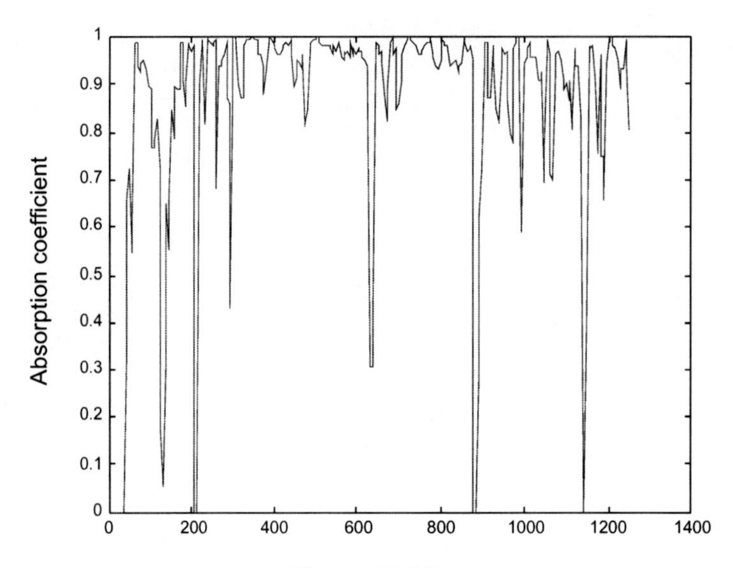

Figure A1.18:
Measured absorption coefficient of the rear wall absorption system. The deep, narrow dips are results of the measuring process.

(a)

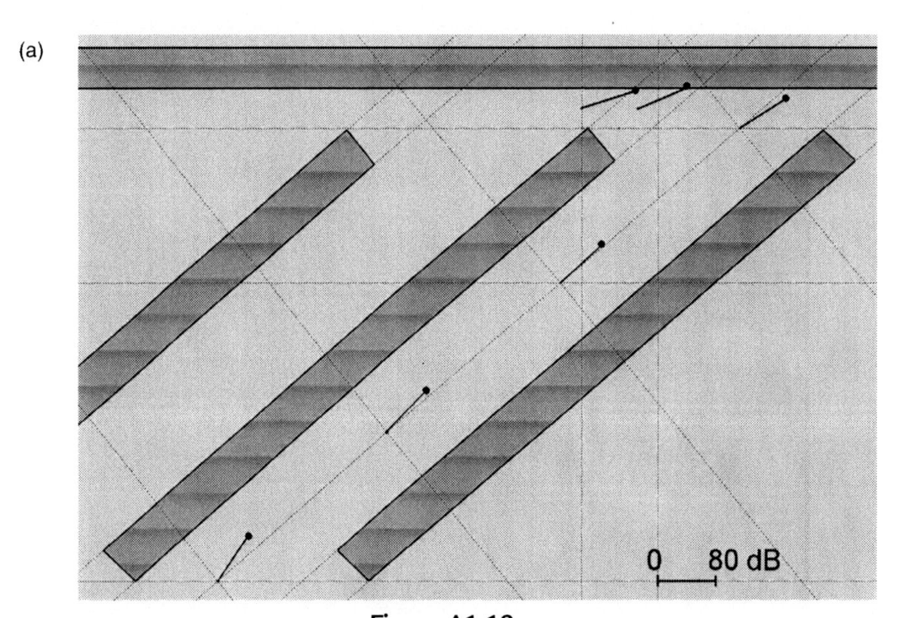

Figure A1.19:
Vector plots of the energy flow, clearly showing that the normally arriving (i.e. perpendicular to the wall) sound wave has been turned to flow parallel to the panels.(a) 63 Hz; (b) 500 Hz.

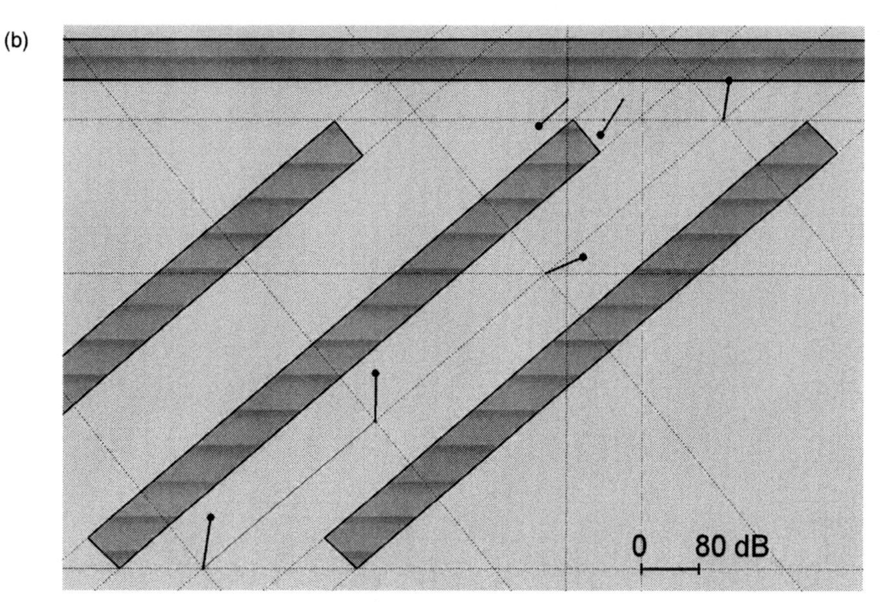

Figure A1.19:
(Continued) (b) 500 Hz.

Evidence has also been found of the cavities between the panels supporting quarter-wave resonances of a highly damped nature. Once again, it is the frictional losses due to the 4 cm fibrous lining that give rise to the absorption, due to resistive losses. Without doubt, there are multiple, overlapping mechanisms at work which all contribute to the general absorption of the 'traps'.

References

1 Newell, P. R., Holland, K. R., 'The measurement and analysis of a control room acoustic treatment during construction', *Proceedings of the Institute of Acoustics*, Vol. 22, Part 6, pp. 179–191, Reproduced Sound 16 conference, Stratford-upon-Avon, UK (2000)

2 Colam, S., Holland, K. R., 'An investigation into the performance of a chipboard panel as a low frequency sound absorber'. *Proceedings of the Institute of Acoustics*, Vol. 21, Part 8, pp 105–112, Reproduced Sound 15 conference, Stratford-upon-Avon, UK (1999)

3 Colam, Stuart, 'An investigation into an empirically designed passive sound absorber for use in recording studio control rooms'. Ph.D. thesis, ISVR, University of Southampton, UK (2002)

4 Newell, P. R., Holland, P. R., The acoustic 'trap' absorber systems: a review of recent research, *Proceedings of the Institute of Acoustics*, Vol. 25, Part 7, Autumn Conference, Oxford, UK (2003)

5 Soares, Luis, E. B. 'An experimental study of the responses of a recording studio control room at low frequencies using a 1:10 scale model', MSc thesis, ISVR, University of Southampton, UK (1991)

6 Torres-Guijarro, Soledad; Pena, Antonio; Rodrigez-Molares, Alfonso; and Degara-Quintela, Norberto, Sound field characterisation and absorption measurement of wideband absorbers, *Presented at the 126th convention of the AES*, Munich, Germany (May 2009)

7 Torres-Guijarro, Soledad; Pena, Antonio; Rodriguez-Molares, Alfonso; and Degara-Quintela, Norbero, A study of wideband absorbers in a Non-Environment control room: Characterisation of the sound field by means of p–p probe measurements, *Acta Acustica with Acustica*, Vol. 97, pp. 82–92, (2011)

APPENDIX 2

Measured Loudspeaker Responses

The purpose of this appendix is to emphasise the degree of difference in the responses of 38 loudspeakers, all of which were advertised by their manufacturers as small, professional, music-monitoring loudspeakers. All of the measurements were made in the anechoic chamber shown in Figure 4.2, and all measurements were made with the loudspeaker and microphone in the same position. They were all carried out by Dr Keith Holland, who is currently a senior lecturer at the Institute of Sound and Vibration Research. The measurements are incontrovertible.

None of the 38 plots of on-axis frequency response and harmonic distortion are similar to the eye, even when only viewed casually. Neither are the 38 plots of the step functions, nor the waterfalls. What is more, these are all *on-axis* responses taken in an anechoic chamber. When one considers that each loudspeaker will exhibit its own, unique, directivity characteristics, then when placed in a normal room it is totally inconceivable that any two of the loudspeakers would sound alike. What is shown in this appendix is a very powerful testimony to the fact that when we are monitoring via loudspeakers, the loudspeakers and rooms *will* significantly impose their characteristics on the overall sound unless careful measures are taken, such as those discussed in Chapters 15 to 20, and even then, perfection is unachievable.

Nevertheless, this appendix serves as a reminder of just how variable even *professional* monitoring loudspeakers are. When we delve into the world of domestic music system loudspeakers, the variability in responses becomes even more exaggerated.

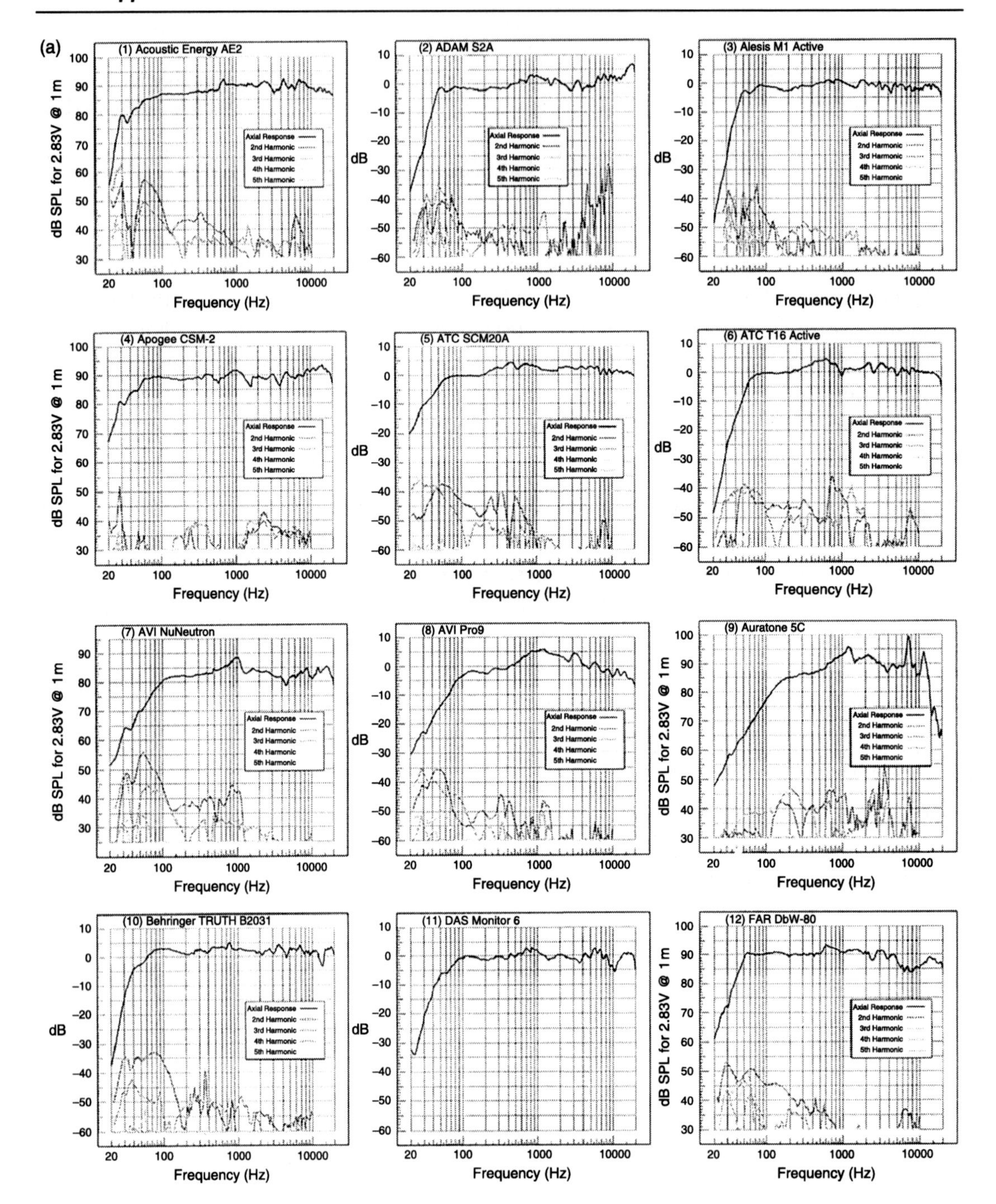

Figure A.2.1:
(a–c) On-axis frequency response and harmonic distortion (a) numbers 1–12

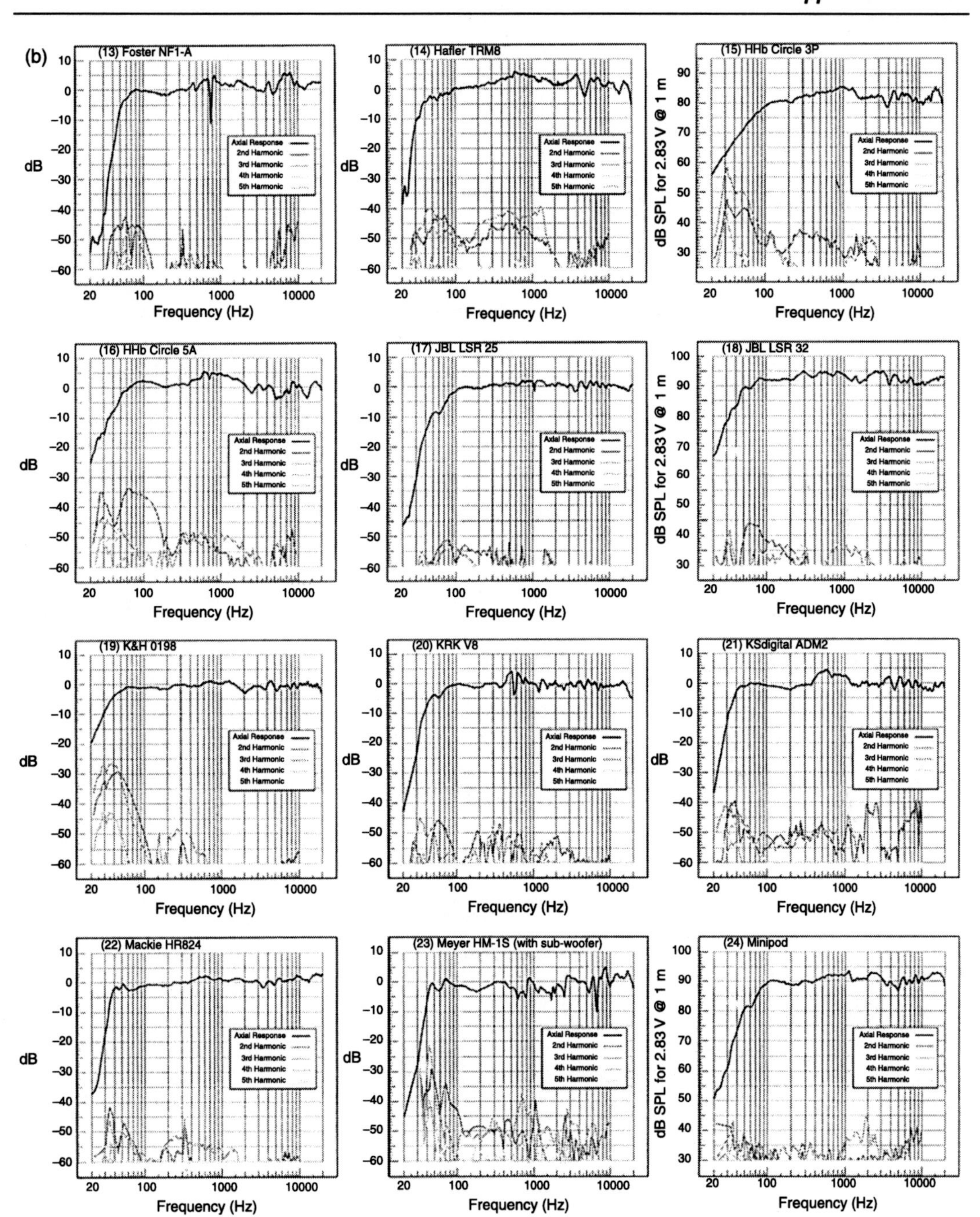

Figure A.2.1: (continued)
(b) numbers 13–24

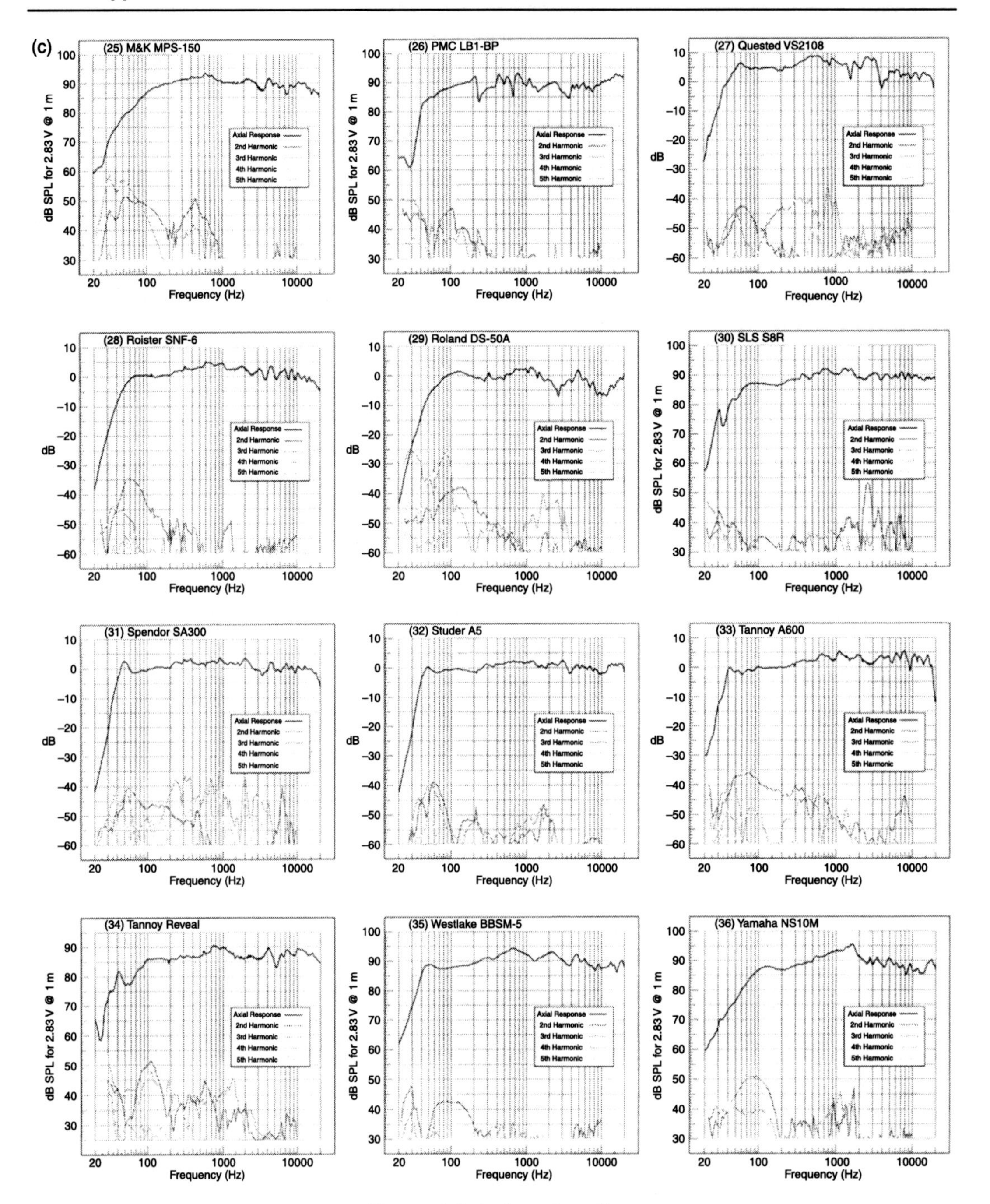

Figure A.2.1: (continued)
(c) numbers 25–36.

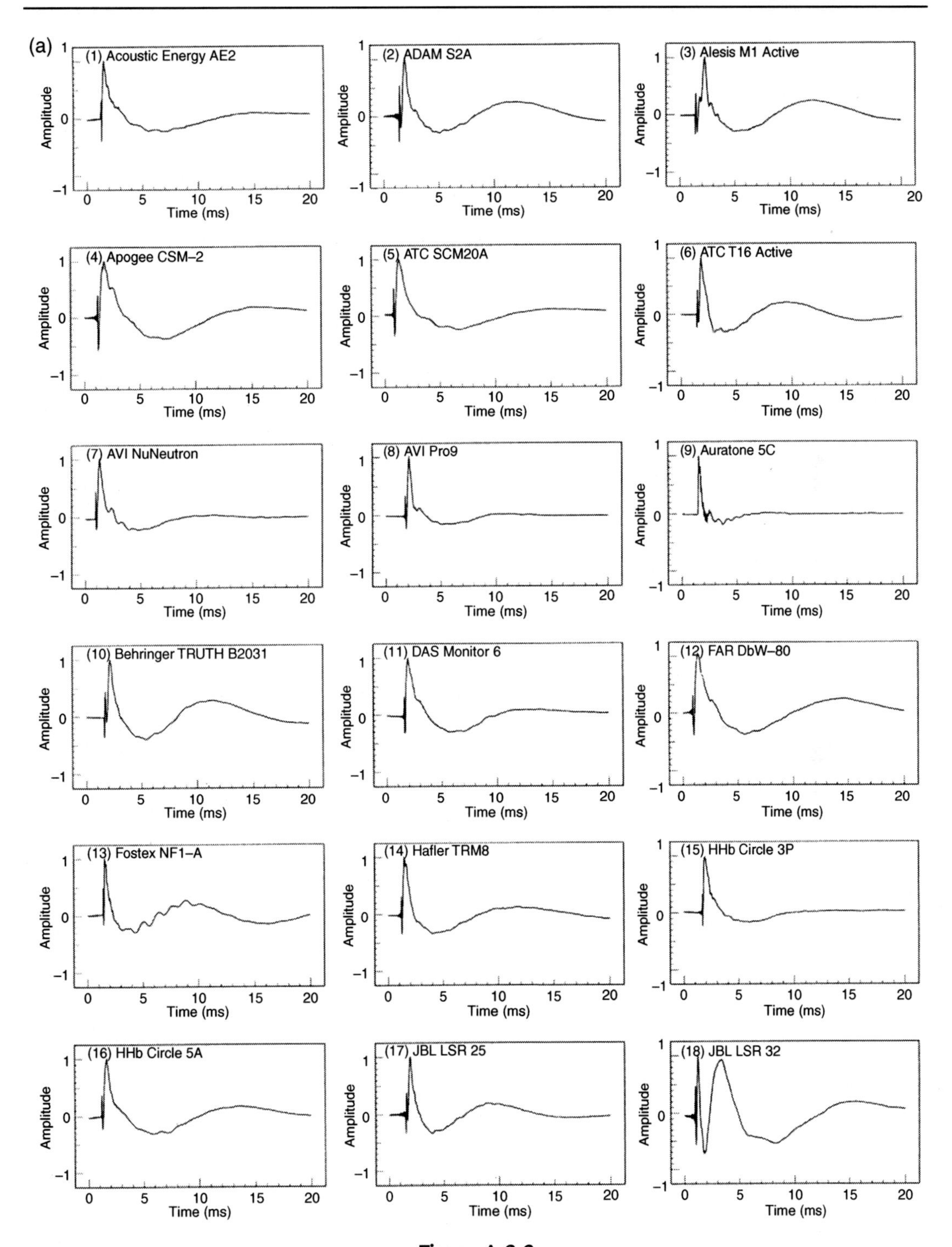

Figure A.2.2:
(a, b) Step responses (a) numbers 1–18

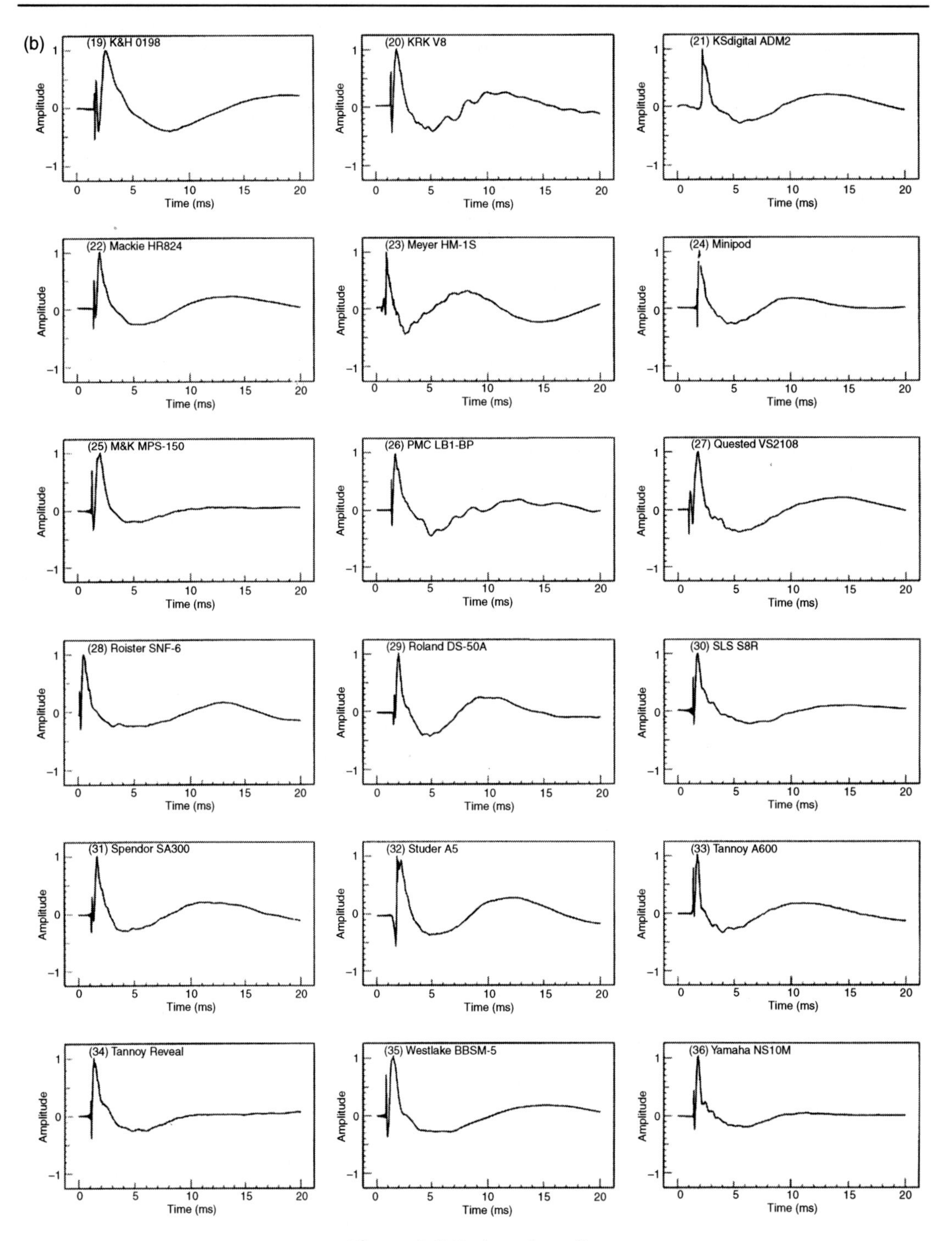

Figure A.2.2: (continued)
(b) 19–36.

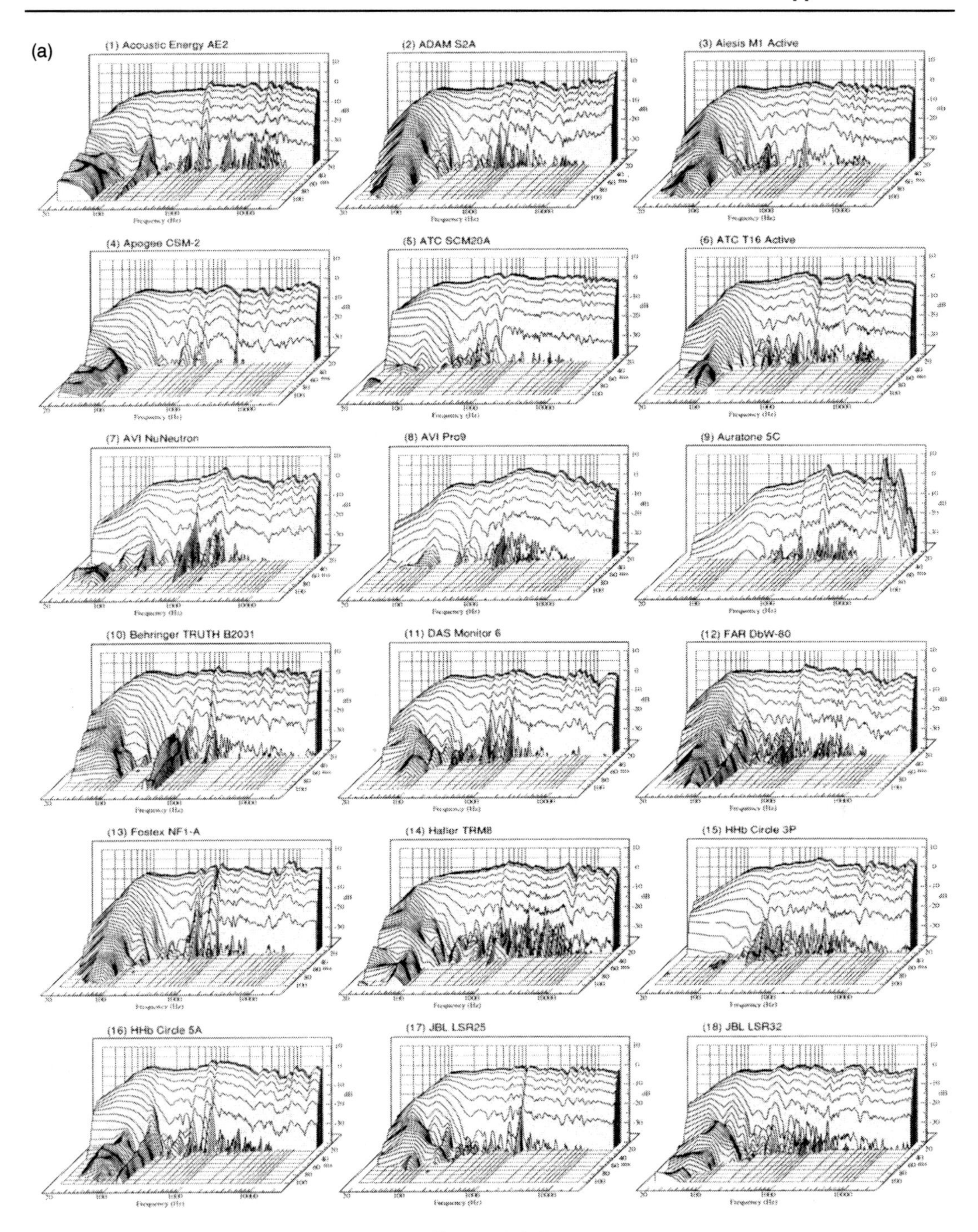

Figure A.2.3:
(a, b) Waterfall plots, (a) numbers 1–18

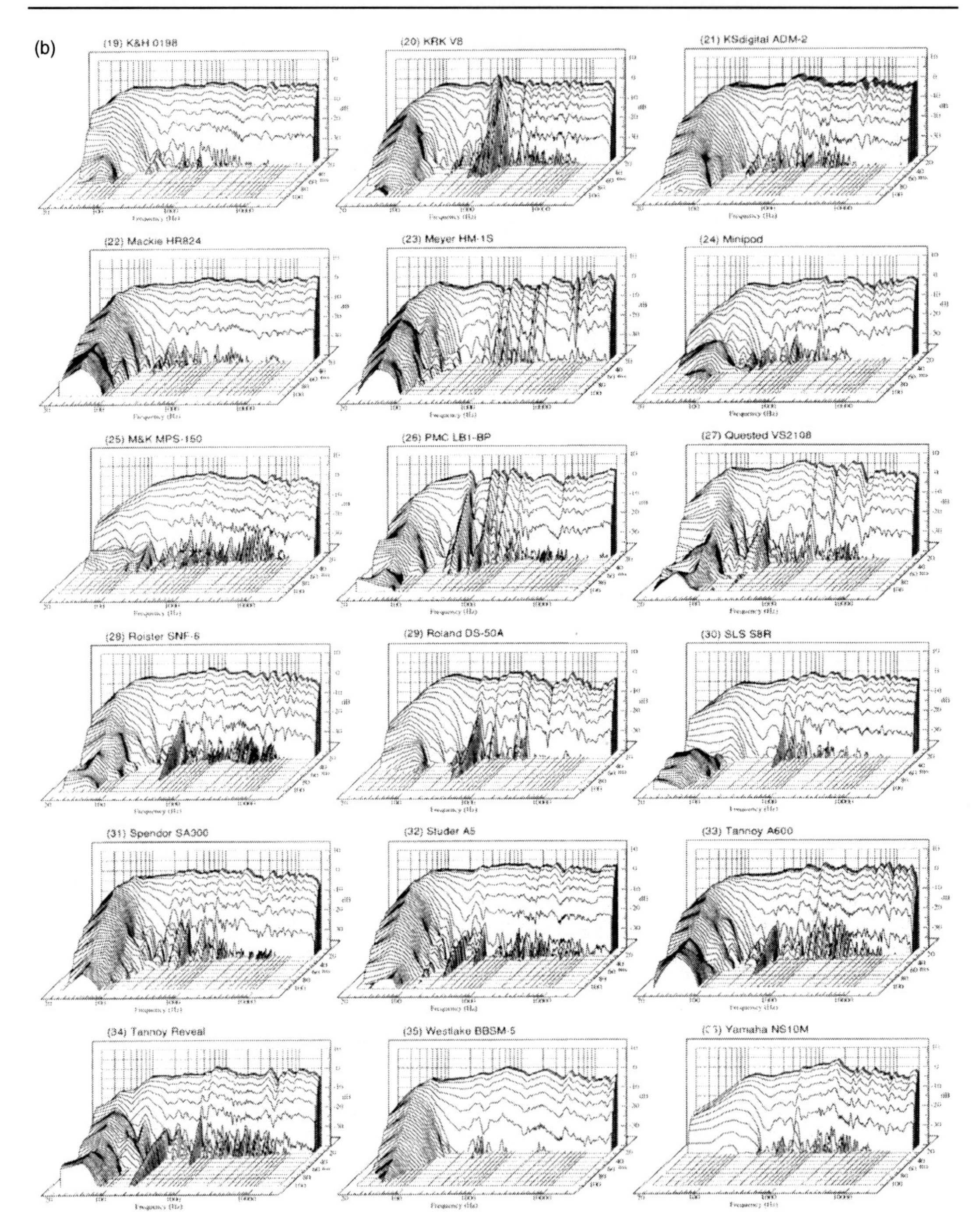

Figure A.2.3: (continued)
(b) 19–36.

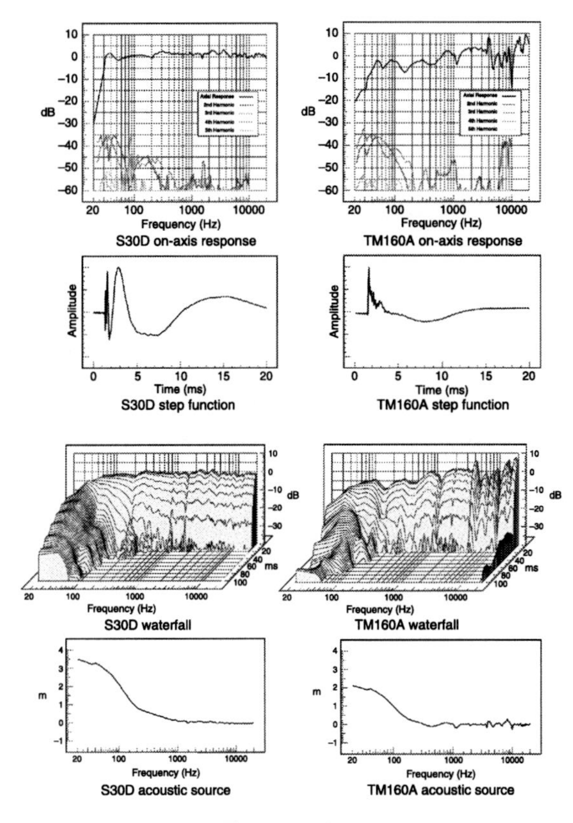

Figure A.2.4:
From the comparison of the various response characteristics of the Genelec S30D and the SAE TM160A loudspeakers, some interesting points can be highlighted. The S30D possesses a variety of switches for modifying the response according to mounting conditions or taste. When the switches are set to give a flat response in anechoic conditions, as shown above, the time response suffers badly, as can be seen from the waterfall and acoustic source plots. For the most natural sound, with faster transients, the most extended flat response is not always (and for small boxes it is rarely) the optimum response.[1] The side-by-side comparison of the responses of the two loudspeakers shown in this figure tends to raise the question as to just how it can be that two so dissimilar loudspeakers are ostensibly designed to perform similar tasks. Both loudspeakers were submitted as part of a loudspeaker test series for a recording industry magazine. They were tested under identical conditions to the 36 loudspeakers shown in the previous three figures. This once again highlights how the responses published by the manufacturers of almost all the loudspeakers are sadly lacking in details of their very important time responses.

Reference

1 Newell, P. R., Holland, K. R., and Mapp, P., 'The perception of the reception of a deception', *Proceedings of the Institute of Acoustics [Reproduced Sound 18 conference]*, Vol. 24, Part 8, UK (2002)

Glossary of Terms

Below is a brief description of some of the technical terms found throughout this text. Experience has shown that although some of the terms may be familiar to many readers, they have often been misunderstood or misused. This glossary attempts to clarify the definitions of the terms, at least as used in British English.

Acausal filter
A filter applied, usually digitally, which has advanced knowledge of the signal arriving via delaying the main signal on which it is intended to act – effect before cause through prior knowledge of the cause. See **Causal filter**.

Active systems
Filters or loudspeaker systems, for example, where an external power source needs to be applied as well as the drive signal. A filter based on semiconductors or valves, driven from an external battery or mains supply, and placed ahead of a power amplifier in a loudspeaker system is an example of an active filter. See **Passive systems**.

Anechoic chamber
A room that is designed to simulate free-field acoustic conditions by means of the placement of highly absorbent materials on all surfaces. The absorbent materials usually are in the form of wedges pointing into the room. This arrangement ensures that sound waves arriving at the room edges are maximally absorbed for all angles of incidence. The lower frequency limit of anechoic performance is set by the length of the wedges, which are effective down to a frequency where the wedge length is equal to one-quarter wavelength. However, what little reflexions still exist will return more weakly from the distant walls of the larger anechoic chambers than from the nearer walls of a smaller chamber using similar wedges.

Variously known as anechoic rooms or free-field rooms.

See also **Semi-anechoic chamber** and **Hemi-anechoic chamber**.

A to D
Analogue to **D**igital converter. A device using a highly stable internal clock which samples the audio voltage waveform at a rate higher than at least twice the highest frequency of interest,

and puts out a digital binary signal that represents the voltage level of each sample. For example, to sample a maximum frequency of 8 kHz, the sampling rate would need to be in excess of 16,000 samples per second. See also **D to A**.

Audio frequency range

The range of frequencies over which the human ear is sensitive is usually considered to be from 20 Hz to 20 kHz. A number of commonly used frequency ranges are listed below. The span of frequencies quoted for each range should not be treated as exact; they are included as an approximate guide only.

Name	Frequency range
Infrasonic	0–20 Hz
Very low	15–50 Hz
Low	20–250 Hz
Lower mid	200–1000 Hz
Mid	250 Hz–5 kHz
Upper mid	2–6 kHz
High	5–20 kHz
Very high	15–25 kHz
Ultrasonic	20 kHz–10^{13} Hz

Back e.m.f.'s

The electromotive forces (voltages) which are generated in a loudspeaker system by the mechano-magnetic interactions. They superimpose on the drive signal (forward e.m.f.) but are usually largely damped by the energy sinking (absorbing) action of the low output impedance of an amplifier, which thus provides a high damping factor. Excessive impedance in loudspeaker cables, for example, can reduce this damping effect, and hence in such systems the back e.m.f.'s would play a greater part in the overall response of the system.

Bookshelf loudspeaker

A genre of domestic orientated loudspeakers whose low frequency responses are aligned to try to achieve a flat far-field response when the loading provided by a wall is taken into account, as would typically be the case when a loudspeaker was mounted on a bookshelf.

Causal filter

A filter in which the effect takes place after the cause. See **Acausal filter**.

Cellular

Composed of cells, which can be either open or closed. In closed cell foams, for example, each cell acts like a small balloon. When compressed or distorted the air trapped inside cannot escape, and so the cell acts as a good spring but a poor sound absorber. Open cell foams are generally poorer springs but better acoustic absorbers. (For equal densities.)

Close field

The region close to a sound source, such as a loudspeaker, in which the sound-field is largely that due to the source, and is little affected by the room reflexions, resonances, or reverberation. See also **Near field**.

Codec (code-decode)

An algorithm for allowing data compression in digital systems, usually in accordance with psychoacoustic phenomena, which allows maximum data compression with minimum acceptable (perceptually/subjectively dependent) audible degradation of the reconstructed sound.

Damping

Damping refers to any mechanism that causes an oscillating system to lose energy. Damping of acoustic waves can result from the frictional losses associated with the propagation of sound through porous materials, the radiation of sound power, or causing a structure with internal losses to vibrate.

D to A

Digital to **A**nalogue converters receive digitally coded signals, representing voltage waveforms, and by means of clocking and filtering, produce an analogue output voltage that should be as close a representation as possible of the waveform represented by the digits. See also **A to D**.

dBA

A weighted (filtered) measurement scale where the filter curves are roughly the inverse of the 40 phon contour of equal loudness. The scale was principally developed to relate to the subjective annoyance of noise around the 30–50 dB SPL ear sensitivity at mid frequencies, by correcting the high and low frequency measured levels to the same subjective level as the mid frequencies. (See Figure 2.4.)

dBC

Similar to dBA but based on the inverse of the 80 phon contour of equal loudness, relating more to the subjective frequency balance at typical music listening levels. The low frequency response of the dBC curve is almost flat.

Deadsheets

Limp membranes having considerable inertia but little stiffness. They are widely used for the mechanical damping of acoustic waves. Those used in studio construction are normally between 3 and 15 kg/m^2.

Decibel

The standard unit of measure for level, or level difference. One tenth part of a bel. Multiplying a given quantity by 1.26 (to two significant figures) will give an increase in level of one decibel.

For example: 1 W×1.26 = 1.26 W (+1 dB relative to 1 W); 1.26 W×1.26 = 1.59 W (+2 dB relative to 1 W); 1.59 W×1.26 = 2.00 W (+3 dB relative to 1 W). It must be borne in mind that in *all* cases, a decibel represents a power ratio.

Decibels and sound pressure level (SPL)

Many observable physical phenomena cover a truly enormous dynamic range, and sound is no exception. The changes in pressure in the air due to the quietest of audible sounds are of the order of 20 μPa (20 micro-pascals), that is 0.00002 Pa, whereas those that are due to sounds on the threshold of ear-pain are of the order of 20 Pa, a ratio of one to one million. When the very loudest sounds, such as those generated by jet engines and rockets, are considered, this ratio becomes nearer to one to one thousand million! Clearly, the usual, linear number system is inefficient for an everyday description of such a wide dynamic range, so the concept of the bel was introduced to compress wide dynamic ranges into manageable numbers. The bel is simply *the logarithm of the ratio of two powers;* the decibel is one tenth of a bel.

Acoustic pressure is measured in pascals (newtons per square metre), which do not have the units of power. In order to express acoustic pressure in decibels it is therefore necessary to square the pressure and divide it by a squared reference pressure. For convenience, the squaring of the two pressures is usually taken outside the logarithm (a useful property of logarithms); the formula for converting from acoustic pressure to decibels can then be written:

$$\text{decibels} = 10 \times \log_{10} = \left\{\frac{p^2}{p_0^2}\right\} = 20 \times \log_{10}\left\{\frac{p}{p_0}\right\}$$

where p is the acoustic pressure of interest and p_0 is the reference pressure. When 20 μPa is used as the reference pressure, sound pressure expressed in decibels is referred to as *sound pressure level* (SPL). A sound pressure of 3 Pa is therefore equivalent to a sound pressure level of 103.5 dB, thus:

$$\text{SPL} = 20 \times \log_{10}\left\{\frac{3}{20 \times 10^{-6}}\right\} = 103.5\ \text{dB}$$

The acoustic dynamic range above can be expressed in decibels as sound pressure levels of 0 dB for the quietest sounds, through 120 dB for the threshold of pain, to 180 dB for the loudest (severely ear-damaging) sounds.

Decibels are also used to express electrical quantities, such as voltages and currents, in which case the reference quantity will depend upon the application (and should *always* be stated).

When dealing with quantities that already have the units of power, such as sound power or electrical power, the squaring inside the logarithm is unnecessary and the ratio of two powers, W_1 and W_2 expressed in decibels is then:

$$10 \times \log_{10}\left\{\frac{W_1}{W_2}\right\}$$

Distributed mode loudspeaker (DML)
A type of loudspeaker where the motor system excites a plate of high modal density. The source is diffuse and acts like neither a piston nor a point source. Radiation takes place from both sides of the panel, but the response is not typically that of a more conventional dipole, even when the panel is unbaffled. The stereo imaging is not as good as with conventional loudspeakers, but the response tends to be less disturbed by the room acoustics. Such loudspeakers can be advantageous as the ambient rear channels of a surround loudspeaker system for example.

Doppler distortion
Frequency modulation dependent on the speed with which a source of sound is either approaching or receding from a listening position. The most common example is perhaps that of a train whistle, or horn, which suddenly drops in pitch as the listening position is passed. The whistle exhibits a constant frequency of output, but when it is approaching the listening position, the period between the cycles is compressed as the arrival delay shortens with time. As the source approaches, effectively the wavelength is shortened. This gives rise to the impression that a higher frequency is being emitted. The time of arrival for each subsequent cycle is reduced as the train approaches because it is emitted from a point nearer to the listener than was the previous cycle. Once the listening point is passed, the opposite effect occurs, with each subsequent cycle emanating from a more distant position, lengthening the wavelength and hence suffering a greater arrival delay. The frequency is thus perceived to have lowered. The degree of pitch change is dependent upon the relative speed of the sound source and the listener. (C. J. Doppler, 1842.)

Eigenfrequency
Frequency of the **Eigentones (see below)**.

Eigentones
The natural resonant tones of a space (or any other resonant system, in fact), 'eigen' being German for 'own'. A room's own tones. If a room is driven (excited) by a noise signal containing all frequencies, then all the eigentones will be driven. When the drive signal is terminated, however, only the eigentones will continue to ring-on, the decay rate being a function of size and total absorption. If a room is driven by a *musical* signal, only the eigentones that correspond to frequencies in the musical signal will be driven, and then only those eigentones will ring on when the music stops.

The eigentones thus dictate which frequencies will ring-on when the drive signal is stopped, but the input signal determines which eigentones are driven. Eigentone is another term for resonant mode. (See **Mode**.) If a room is driven at a frequency that does *not* correspond with an eigentone (mode), then the room response will decay more rapidly, once the drive signal has been stopped, than the frequencies which correspond with the eigentones. This is why widely spaced room modes give rise to uneven overall responses in a room. Frequencies corresponding to eigentones (modes) will be reinforced or cancelled, depending on source and receiver position, but other frequencies will be unaffected. See **Standing waves and resonances**.

Euro (€)

The European currency unit, which is roughly equivalent to 1.35 US dollars in 2011. During the reading of the first draught of this book for assessment by the publishers, a referee requested that the currencies used in the text should be converted to one standard unit, such as the euro or the U.S. dollar. In reality, this proved difficult to achieve in any meaningful manner because the wild fluctuations of the currencies would lead to distorted senses of values outside of the local territory of each currency. In mid 2001, the euro would buy 87 cents of a U.S. dollar. By mid 2004 it would buy 1 dollar 34 cents, a rise in value of over 50%. Consequently, reference in this book to things manufactured in the USA is given in dollars, and to things made in Europe, in euros. Where any other currencies are stated, some means of assessing the relative value at the time referred to will be given in the text.

Far-field

The far-field is the region in which the radiation from a loudspeaker, or other acoustic source, can be considered equivalent to that of a point source, and in which the sound pressure *and* velocity reduce by 6 dB for each doubling of distance from the source.

Feedback

1. The instability that occurs when the output from a microphone is reproduced by a loudspeaker system, the output of which arrives back at the microphone. Regenerative feedback develops when the overall gain is greater than 1. Also known as 'howlround'.
2. The instability that occurs when an output of an electrical amplification system is reapplied to its input.
3. A term used by orchestral musicians to describe the way in which they hear the output from their own instruments reinforced via reflexions in their performance space. When fed to them via loudspeakers or headphones, the usual term is *foldback*.
4. *Negative* feedback (phase reversed) is used in electronic amplifier circuitry to reduce distortion.

Fourier Transform

The mathematical transform linking the time domain representation of a signal to its frequency domain representation. Application of the Fourier Transform to a signal (waveform) reveals the frequency components in terms of their magnitude and relative phase (the Spectrum). Application of the inverse Fourier Transform to the spectrum yields the original waveform.

Frequency
The rate of change of phase with time. The number of complete cycles per unit interval of time for a sinusoidally time-varying quantity. The repetition rate of any cyclic event. Unit: Hertz: (Hz); previously measured in units of cycles per second (c.p.s.), and even earlier in half-vibrations per second.

Frequency response
The response of a system in terms of its amplitude and phase response. See **(Pressure) amplitude response** and **Phase response**.

Group delay
The frequency dependent response delays through electrical or mechanical systems which are given rise to by phase distortions. The group delay is related to the degree of phase shift.

Haas effect
H. Haas, 1951. See **Precedence effect**.

Hemi-anechoic chamber
An anechoic chamber with one hard boundary (usually the floor) in which the sound sources can be embedded. Equates to 2π space (a hemisphere).

Hysteresis
The lagging of effect when an applied force varies in amount, and where the movement in one direction is not retraced when the force giving rise to that movement is reversed.

Impedance
A combination of resistance and reactance. Symbol Z. The ratio of pressure to velocity in acoustic systems, or voltage to current in electrical systems, expressed at a given frequency. See also **Resistance** and **Reactance**.

Infrasonic
Relating to frequencies below approximately 20 Hz. See **Subsonic**.

Intensity
'Sound intensity' is a very specific term, and represents the flow of acoustic energy. It is measured in units of watts per square metre, and should not be confused with either sound power level (SWL), sound pressure level (SPL) or loudness, but it *is* associated with the sound intensity level (SIL).

Interference field
See **Standing wave field**.

Kinetic energy
Energy of motion. Energy possessed by a body due to its motion, which can be converted to other forms of energy through the application of a braking force.

Linearity
A system can be said to be linear when the output contains only the frequencies applied to the input. A falling amplitude response is therefore a *linear* distortion. Such a response is *NOT* non-linear, as may be erroneously stated in many advertisements. See **Non-linearity**.

Loss factor
The reciprocal of the Q-factor. See **Q-factor**.

Magnetostriction
The property of some materials, for example iron, of expanding and contracting due to the influence of an applied magnetic field.

Microphone directivity patterns
Most microphones consist of a small diaphragm that moves in response to changes in the pressure exerted on it by a sound wave; the diaphragm motion is then detected and converted into an electrical signal.

The simplest form of microphone is one that has only one side of the diaphragm exposed to the sound-field. If the diaphragm is sufficiently small, such a microphone will respond equally to sounds from all directions, and is termed 'omni-directional'.

A microphone which has both sides of the diaphragm open to the sound-field will only detect the difference between the pressures on the two sides. When a sound wave is incident from a direction normal to the diaphragm, there will be a short delay between the pressure on the incident side and that on the far side, and the microphone responds to the resultant pressure difference. When a sound wave is incident from a direction in line with the diaphragm, the same pressure is exerted on both sides of the diaphragm and the sound is not detected. This arrangement results in a 'figure-of-eight' or dipole directivity pattern.

If an omni-directional microphone element and a figure-of-eight microphone element are mounted close together and their outputs summed, the resultant directivity pattern will lie between the two extremes of omni-directional and figure-of-eight patterns. If the sensitivities of the two elements are the same, the combined directivity pattern is known as cardioid, because of its heart shape. Various other patterns, such as hyper-cardioid and super-cardioid are achieved by varying the relative sensitivities of the omni-directional and figure-of-eight elements. Similar directivity patterns can be realised using only one microphone element. The microphone diaphragm is mounted at one end of a short tube, and the delay introduced to one side of the diaphragm by the tube gives rise to an approximation to a cardioid directivity pattern. More complex directivity patterns can be achieved by using more than two elements; the SoundField microphone, for example, has four elements that can be combined in a variety of ways.

Minimum phase and non-minimum phase

A minimum phase signal/system is one in which the phase shift associated with the amplitude response is the minimum that can be allowed whilst still exhibiting the properties of a *causal* system (one in which the output *never* arrives before the input). As there is a strict relationship between amplitude and phase in such systems, correcting either one will inevitably tend to correct the other. The low frequency response boost given rise to by the flush mounting of a loudspeaker in a wall is an example of a minimum phase response change, which therefore *can* be equalised to restore the free-field response in terms of both amplitude *and* phase (and hence it will also restore the time [transient] response). The essential factor is that no appreciable delay is involved between the generation of the signal and the effect of whatever is influencing it. If there is no appreciable delay, then there can be no appreciable phase-shifts, hence minimum-phase, or more fully, minimum phase-shift.

'Non-minimum phase' responses are those where amplitude correction, alone, cannot correct any phase disturbances. The far-field response of a loudspeaker in a reflective room is an example of a non-minimum phase effect. Here, there is a delay between the signal generation by the loudspeaker and the superimposition of the boundary reflexions on the overall response. Reflexion arrival times create phase irregularities, which are frequency and distance (time) dependent, so no simple manipulation of the amplitude response of the source can adequately compensate for the complex disturbances.

Another example of a non-minimum phase effect is in the combination of the various outputs of crossovers. In any filter circuits, either mechanical or electrical, there are inherent **Group delays** for any signal passing through them. The amount of group delay increases as the filter frequency lowers, and as its order (6, 12, 18, 24, etc. dB/octave) increases. A crossover will thus have different group delays associated with each section, and when the outputs are recombined, they will *not* produce an exact replica of the input signal. For this reason, conventional equalisation cannot be used to correct for response errors at crossover points. Amplitude correction will lead to further phase distortion, and hence time response errors. In practice, most crossovers above first-order are non-minimum phase devices.

Mode

A special pattern of vibration whose position remains invariant, as when a travelling wave and its reflexions superimpose themselves between two or more boundaries such that the peaks and troughs in the waveform coincide, and appear static. See **Standing wave field**.

Modes and resonances

Sound consists of tiny local changes in air density that propagate through the air as a wave motion at the speed of sound. The speed of sound is around 340 metres per second at normal room temperature and, although it is temperature dependent, it is independent of variations in the ambient pressure, and is the same at all frequencies. The frequency of a sound wave is measured in cycles per second (c/s or cps) known more usually these days as Hertz (Hz), and is usually represented by the symbol f. The distance that a sound wave travels in one cycle at any frequency is known as the wavelength, represented by the symbol λ, and has the units of metres. The speed of sound is represented by the symbol c. The relationship between wavelength, frequency and the speed of sound is simple; wavelength is equal to the speed of sound divided by the frequency, or $\lambda = c/f$. Therefore, for example, a sound wave at a frequency of 34 Hz has a wavelength of $340/34 = 10$ m.

As a sound wave propagates away from a source in a room, it will expand until it reaches a reflective room boundary, such as a wall, from which it will reflect back into the room. The reflected wave will continue to propagate until it reaches other boundaries from which it will also reflect. If there is nothing in either the room or the boundaries to absorb energy from the wave, the propagation and reflexion will continue indefinitely, but in practice some absorption is always present and the wave will decay with increasing time. The point on the cycle of a sound wave (the *phase* of the wave) when it reaches the boundary depends upon the distance to the boundary and the frequency of the wave. Figure 4.7 shows how waves at different frequencies, propagating from the same source, arrive at a boundary with different phases.

A rigid boundary will change the direction of propagation of an incident sound wave, but will maintain its phase, so the phase of a reflected wave can be calculated from the *total* distance propagated from the source. If this total distance is equal to a whole number of wavelengths, then the wave will have the same phase as it started with. When two boundaries are parallel to each other, a sound wave will reflect from one boundary towards the other, and then reflect back again to where it started, continuing back and forth until its energy is dissipated. If the distance between the boundaries is such that the 'round trip' from the source to the first boundary, on to the second boundary, and back to the source is a whole number of wavelengths, then the returning wave will have the same phase as the outgoing wave, and will serve to reinforce it. This situation is known as resonance. Resonances can also occur due to reflexions from multiple boundaries; the necessary requirement being that the sound wave eventually returns to a point with the same phase as when it left. One can imagine a whole set of possible combinations of reflexions in a typical room that allows the wave to return to its starting point, and therefore a whole set of frequencies for which resonance will occur. In fact, in theory, every room has an infinite number of possible resonances.

As stated above, if there is nothing in the room or the boundaries to absorb energy from a sound wave, a short duration sound pulse (or transient), emitted from a source, will propagate around the room indefinitely. Of the infinite number of possible paths that the wave can take, only those that correspond to resonances at frequencies contained in the pulse will be continually reinforced; all other paths will not be reinforced. After a short time, the resulting sound-field can be thought of as simply a sum of all of the resonances that have been excited. These resonant paths are known as the natural modes of the room, and the resonant frequencies are known as the natural frequencies, or '**eigentones**', of the room; both are determined uniquely by the room geometry. 'Eigen' is German for 'own', so the eigentones are a room's own particular, natural, resonance frequencies.

When sound absorption occurs within the room or boundaries, resonant modes still exist, but the wave will decay at a rate determined by the amount of absorption. To maintain a given sound level in a room in the presence of absorption, the source needs to be operated continuously, at a level dependent upon both whether or not resonant modes are being excited and the amount of absorption present. When the sound source emits a transient signal in the presence of absorption (for example, switching off a continuous signal), many different paths – not just resonant modes – will be excited, but after a short time, only the resonant modes will remain (because they tend to have less absorption); the room will 'ring' at the resonance frequencies until the modes decay. The reverberation time of a room is a measure of the average rate of decay of the sound in the room when an otherwise continuous sound source is switched off; it is the time taken for the sound level to fall to 60dB below its initial, continuously excited, level. As the amount of absorption is increased, the sound level at the resonant frequencies will reduce, but the bandwidth of each mode (the range of frequencies over which the mode can be excited to a significant degree) will increase. When the boundaries are fully absorbent, the room modes no longer exist (an anechoic chamber).

When sounds such as speech or music are heard in a room, the level of the continuous components of the sound will be determined by whether or not they coincide with any room resonances that are excited. The transient components will 'hang on' at the resonance frequencies after the transient has gone. It should also be understood that the perception of the modal activity is not uniform within the space, because of the spacial distribution of the nodes and antinodes.

Mutual coupling

Mutual coupling is the term used to describe the interaction between two or more sound sources radiating the same signal. If the diaphragms are receiving different, uncorrelated inputs, then the output *power* summation will be simply that of the output of the different diaphragms. However, if the diaphragms are receiving the *same* input, then for frequencies whose wavelengths are greater than eight times the distance between the diaphragms (i.e. the diaphragms are less than one-eighth of a wavelength apart) the outputs will be substantially in-phase at any point in the room. The radiated *pressures* (not powers) will then superimpose, giving rise to a 6 dB increase in SPL for each doubling of the radiating area if the diaphragms are radiating the same power. This implies *four times* the output power, yet when the radiated powers are equal, but the radiated signals are *uncorrelated* (i.e. totally different signals) only a 3 dB SPL increase (due to the simple power summation) would result. Where a close boundary reflects a wave back onto the radiating surface of a diaphragm, the effect is the same as if a second diaphragm were radiating the same signal – the mirrored room analogy – and so the diaphragm radiates twice the power that it would do if moved away from the boundary.

This seemingly $1 + 1 = 4$ situation is due to the fact that, for a given diaphragm velocity, the *power* output is proportional to the diaphragm radius raised to the fourth power. Doubling the diaphragm area therefore yields a four-fold increase in power. The increased low frequency radiating efficiency of large diaphragms can be thought of as being due to all the individual parts of the diaphragm mutually coupling. The pressure radiated by one part of the diaphragm resists the movement of the adjacent parts. The increase in radiation resistance on a mass-controlled diaphragm, typical of a heavy woofer (whose movement can be considered independent of the local air pressure), gives rise to increased work being done, by having a greater pressure of air to push against, so more power is radiated.

As the frequency of radiation rises, or the loudspeakers (or the loudspeaker and a reflective surface) are sited further apart (i.e. the diaphragms are separated by *more* than one-eighth of a wavelength), the coupling becomes less in-phase so the radiation boost reduces. As the frequency or separation distance continues to rise, regions of in and out-of-phase interference will result, giving rise to a combined output power response as shown in Figure 14.8. Only for a listener on the central plane between the loudspeakers will the 6dB pressure summation be maintained. (See also Figure 14.9.)

In a reverberant room, the output from a pair of stereo loudspeakers reproducing a central mono signal will sum by 6 dB on axis, at all frequencies, but the reverberant field will be driven by the combined power response, as shown in Figure 14.8(b). The *overall* sound, therefore, will be darker (with more low frequencies) than that from a central mono source radiating the same signal and receiving the same input power as the mutually coupling pair.

Much more on this subject can be found in Reference 1 at the end of the Glossary.

Near field
There are two quite distinct and separate definitions of the near field of a source of sound; one is related to the geometry of the source whilst the other has to do with the rate of expansion of radiating waves. The region beyond both near fields is known as the far field.

The **geometric near field** is defined as that region close to a source where the sound pressure does not vary as the inverse of the distance from a source. The extent of the geometric near field is dependent upon the detailed geometry of the sound source and is finite only at frequencies where the wavelength is shorter than a typical source dimension (for a circular piston this is when the wavelength is equal to the piston diameter); there is no geometric near field at lower frequencies. A point source does not have a geometric near field at any frequency.

The **acoustic near field** (or **hydrodynamic** near field) is defined as that region close to a source where the air motion (velocity field) does not vary as the inverse of the distance from a source, although the acoustic pressure may. The extent of the acoustic near field is inversely proportional to frequency: it is large at low frequencies and small at high frequencies. The sound-field radiated by a point source has a sound pressure that varies as the inverse of distance from the source at all frequencies and distances. The air motion only varies as the inverse of distance in the far field. For practical sources, the extent of the acoustic near field is affected also by source geometry.

Decisions relating to the proximity of a listener to a close-field monitor should be made by considering the geometric near field only; the ear, being essentially a pressure sensitive organ, is insensitive to the presence of the acoustic near field.

Newton
A standard unit of force; symbol N. Easy to remember examples are that one newton is roughly equal to a *weight* (on the earth's surface) of 100 g, and that an apple of medium size is attracted to the earth by a force of around one newton. A force of one newton bearing down on a spring would be applied by a 100 g weight on the surface of the earth.

Non-linearity
A system is said to be non-linear when the output contains frequencies which were *not* present in the input signal, and are not due to system noise. Harmonic distortion, intermodulation distortion and rattles are sources of non-linearities. A system exhibiting any of these is said to be non-linear. See **Linearity**.

Noise weighting curves (dBA, etc.)
The human ear does not have a flat frequency response; a low frequency noise will generally sound quieter than a higher frequency noise having the same sound pressure level. A measurement of sound pressure level therefore does not yield an accurate measurement of perceived loudness unless the frequency content of the noise is taken into account. Noise

weighting curves are used to convert sound pressure level measurements into approximations of perceived loudness, by discriminating against low and high frequency noises. The most commonly used noise-weighting curve is known as A-weighting. An A-weighting curve is simply a filter with a response that rises with increasing frequency up to 2 kHz, above which it falls off gently.

The frequency response of the human ear changes with changes in sound pressure level (see Figure 2.1), so different weighting curves are required for different levels. The dBA curve was developed for signals having loudness below 40 phon, the dBB curve was intended for use around 60 phons. At levels over about 80 phon, the dBC curve should be used. Other curves are also in use, such as dBD, which can be used for high level industrial noise, and dBG, which is used for infrasonic and very low frequency noise assessments. The responses of some of the weighting curves are shown in Figure 2.4.

The widespread use of the dBA curve for the assessment of noise can give rise to poor results in situations when another weighting curve is more appropriate. For example, Figure 2.4 shows the dBA curve superimposed on the inverse of the equal loudness contours – similar to those in Figure 2.1, but upside down. Only at about 1 kHz and 6 kHz do all the curves agree. Between 3 and 4 kHz, errors of up to 10 dB can be seen, and at low frequencies, the A-weighting curve can over-assess or under-assess noise nuisance levels by up to 20 dB, depending upon level. The dBA curve is often used at relatively high levels; a purpose that it was never intended for, and is not suited to, but sometimes this needs to be done for comparison purposes.

In any case, noise weighting should only be applied when one requires an approximation to the perceived loudness of a sound; it is therefore of most use in noise assessment. Noise weighting should never be applied when absolute values of sound pressure are required; in the measurement of loudspeaker frequency response, for example. Here, a flat measurement (unweighted) should be used unless clear specifications demand otherwise.

Objective and subjective assessment

In acoustics in general, and in audio in particular, there is often some disagreement between that which our measurements tell us and that which we hear. In audio, objective assessment involves measuring the performance of a piece of equipment using instruments, and comparing this performance with a desired specification. Subjective assessment, however, involves auditioning the equipment under carefully controlled conditions and assessing particular aspects of the sounds that are heard. The successful assessment of the quality or suitability of a piece of audio equipment therefore, ideally, needs both approaches. Objective assessment is more easily carried out in the laboratory, or during production runs, than subjective assessment. To make a reliable and repeatable subjective assessment usually requires the ears of a number of subjects, and hence, often, a large amount of time.

Particulate
Relating to particles. Particulate motion = motion of the particles.

Pascal
A pressure of one newton per square metre. See **Newton**.

Passive systems
Systems, such as filters, without any source of external power other than the signal energy itself. An inductor/capacitor filter immediately before a loudspeaker drive unit is an example of a passive filter. See **Active systems**.

Phase response
The relative phase of the input and output signals as a function of frequency.

Phon
A unit of perceived loudness, such that a given change in the phon level would always produce an equal, subjective loudness change, irrespective of the actual SPL change. The contours in Figure 2.1 represent phon levels, which it can be seen do not relate directly to the physical sound pressure levels.

Pink noise
Filtered white noise (reducing 3 dB/octave with frequency) which yields equal energy per octave.

Pinna
Plural: pinnae. The outer part of the ear that projects outside the head. The ear flap.

Plasticine
The trade name of synthetic clay used for modelling. It is highly plasticised and does not set hard.

Potential energy
The energy of position; such as imparted on a body by raising it in a gravity field. The energy concentrated in a spring is also potential energy.

Precedence effect
Also referred to as the **Haas effect**, and the **law of the first wavefront**. When two short-duration sounds are heard in rapid succession, the tendency is for the second sound to be psychoacoustically suppressed. The pair of sounds is perceived as one sound, coming from the direction of the first arriving source. The precedence effect operates when the second sound arrives within approximately 0.7–30 ms after the first sound. The affect can often be overridden if the second sound is 7–15 dB higher in level than the first sound.

(Pressure) amplitude response
The ratio of the output amplitude of a system divided by the amplitude of the input as a function of frequency. When sound pressure is involved, the term 'pressure' is prefixed: for electrical or mechanical systems, the term 'amplitude response' suffices.

Psychoacoustics
Unlike the related discipline of acoustics, which is concerned with the physics of sound, psychoacoustics is the science of the perception of sound, particularly by humans. The stereo illusion, the cocktail party effect and the perception of pitch are all examples of psychoacoustic phenomena.

PVA
Poly-vinyl acetate. A water-based adhesive that is water resistant once dry.

Q-factor
A measure of the sharpness of the peak in a resonant system. It is defined as:

$$Q = \frac{f\mathrm{res}}{\mathrm{Bh}}$$

where:
$f$res = resonant frequency
Bh = the half-power bandwidth of the resonance

Reactance
A reactive system is one which stores and releases energy without loss. Inductors and capacitors are reactive. Symbol X. The phase quadrature part of **Impedance**, q.v. Reactance can be thought of as the resistance to the flow of AC, as exhibited by inductors and capacitors, and is highly frequency dependent.

Resistance
Anything which impedes the flow of energy in such a way that the losses are dissipated (more usually into heat, but also into work). Symbol R. The (in-phase) part of an **impedance** q.v. Resistance acts equally on AC and DC currents, independent of frequency.

Semi-anechoic chamber
A room in which the absorption is incomplete, and contains a residual reflected component that can be corrected for during measurement analysis. See **Anechoic chamber**.

Shuffler
A type of circuit used in headphone reproduction to try to create inter-aural cross-correlation to simulate the effect of loudspeaker listening. This is done in an attempt to produce a frontal

sound stage, because the stereo sound stage is generally inside or above the head of a listener when using headphones. Pre-World War II, the word was also used for other purposes relating to middle/side (M & S) microphone matrixing.

Sine wave (and its frequency content)
A sine wave is a graph of the value of a single frequency signal against time. Strictly speaking, for a signal to consist of a single frequency, the sine wave must have existed for all time, as any change to the amplitude of the signal, such as during a switch-on or switch-off, gives rise to the generation of other frequencies: this has important implications for audio. Most audio signals contain pseudo-steady state sounds, such as notes played on an instrument. When these sounds are reproduced by an imperfect audio system, the excitation of any resonances in the reproduction chain will depend upon the frequency content of the signal. During a long note, the signal may be dominated by a few discrete frequencies, such as a set of harmonics, and the chances of resonances being excited are slim. However, during the start and stop of the note, a range of frequencies is produced, above and below those of the steady state signal, and the chances of resonances being excited are increased. This phenomenon leads to the apparent pitch of the note being 'pulled' towards the frequency of any nearby resonance during the start, and particularly the end, of the note.

Sound Power Level (SWL)
The level of sound power, expressed in decibels, relative to a stated reference value. The unit is the decibel referenced to 1 picowatt (1pW).

Sound Pressure Level (SPL)
The unit is the decibel, referenced to 0 dB SPL at 20 micro-pascals (20 μPa). It is defined by $20 \log_{10} (p_{rms}/p_{ref})$. Sound Pressure Level, or SPL, doubles or halves with every 6 dB change, unlike the sound power, which doubles and halves with 3 dB change, because the power relates to the square of the pressure. In the acoustic and electrical domains, sound power equates to electrical power and SPL to voltage. Subjective loudness tends to double or halve with 10 dB changes: 10 dB higher being twice as loud, and 10 dB lower being half as loud. See also **Intensity**. Ten decibels relates to a ten times *power* change.

Squawker
Term of American origin for a mid-range loudspeaker – onomatopoeically relating to the sound of a large bird such as a seagull or macaw.

Standing wave field
The pattern of wave superposition that occurs in a reflective environment, whereby the distribution of peaks and troughs in the response throughout the space appear to be stationary. See **Mode**.

Standing waves and resonances

Standing waves occur whenever two or more waves having the same frequency and type pass through the same point. The resultant spacial interference pattern, which consists of regions of high and low amplitude, is 'fixed' in space, even though the waves themselves are travelling.

Resonant standing waves only occur when a standing wave pattern is set up by interference between a wave and its reflexions from two or more surfaces. *And*, when the wave travels from a point, via the surfaces, back to that point, it is travelling in the original direction. *And*, when the distance travelled by this wave is equal to an exact number of wavelengths. The returning wave then reinforces itself, and if losses are low, the standing wave field becomes resonant.

The simplest resonant standing wave to visualise is that set up between two parallel walls spaced half a wavelength apart. A wave travelling from a point towards one of the walls is reflected back towards the other wall, from which it is reflected back again in the original direction. As the distance between the walls is one half of a wavelength, the total distance travelled by the wave on return to the point is one wavelength; the wave then travels away from the point with exactly the right phase to reinforce the next cycle of the wave. If the frequency of the wave or the distance between the walls is changed, a standing wave pattern will still exist between the walls, but resonance will not occur.

It should be stressed that standing waves *always* exist when like waves interfere, whether a resonance situation occurs or not, and that the common usage of the term 'standing wave' to describe only resonant conditions is both erroneous and misleading. See also, **Eigentones**.

Step function

Alternative names are 'unit step function' 'Heaviside step function' and 'Heaviside function'. (O. Heaviside, 1892.)

$$\begin{aligned} \mathrm{H}(x) &= 0 \text{ for } x < 0 \\ \mathrm{H}(x) &= 1 \text{ for } x > 0 \end{aligned}$$

Its value at $x = 0$ is not defined.

The alternative notation $\mathrm{u}(x)$ is more common in signal processing.

Subsonic

This, in British English at least, is an aerodynamic term meaning below the speed of sound (as opposed to **Supersonic**). Its use implying below 20 Hz is incorrect.

Compare:
Infrared with ultraviolet

Latin: sub – under, super – on top of, above
Infra – below, ultra – beyond

Conventionally, 'sub' usually pairs with 'super' and 'infra' with 'ultra'. See **Infrasonic**.

Supersonic
An aerodynamic term meaning *above* the *speed* of sound. Its use relating to beyond the frequency range of hearing is archaic. **Ultrasonic** is the term now used for frequencies above the range of human hearing.

Transfer function
Alternative term used (at least in electroacoustics, although not in all subjects) for the frequency response. What you get out relative to what you put in. A flat frequency response implies a flat transfer function.

Transient response
The response of a system to an impulsive input signal. An accurate time (transient) response requires an extended frequency response and a smooth phase response. A low frequency amplitude response roll-off, for example, will give rise to the lengthened time (transient) responses, as clearly shown in Figures 19.11 and 19.12. The more the frequency response is curtailed, either in terms of frequency of turnover or steepness of slope, the more the transient response will be smeared in time.

Tweeter
Term of American origin for a high-frequency loudspeaker, onomatopoeically imitating the high-pitched 'tweet-tweet' sound made by small birds.

Ultrasonic
Relating to frequencies above approximately 20 kHz. Some authorities limit the term to a maximum of 10^{13} Hz, beyond which the term 'hypersonic' is used. Hypersonic is also used in aerodynamics, relating to *speeds* beyond five times the speed of sound.

Weighting
The pre-multiplication of data by a set of weighting factors. A bias applied to improve measurement compatibility with subjective assessment.

White noise
A random noise signal containing all frequencies. Statistically the response has equal energy per bandwidth in hertz. For example, 20 Hz to 25 Hz (5 Hz bandwidth) would have equal energy to the band from 1000 Hz to 1005 Hz (also a 5 Hz bandwidth), and hence on a spectrum analyser shows a response rising 3 dB per octave as the frequency rises.

Woofer
Term of American origin for a low frequency loudspeaker – onomatopoeically relating to the deep bark, or 'woof' of a large dog.

Thanks to Dr Keith Holland for his contributions towards this Glossary, and to Professor Jamie Angus for verifying its accuracy.

Reference

1 Borwick, John, *Loudspeaker and Headphone Handbook*, 3rd ed, Chapters 1 and 9, Focal Press, Oxford, UK and Boston, USA (2001)

Index

Some of the items in this index refer to subject matter, and not to the precise wording, in order to help to guide readers to other relevant pages.

A

B

C

I

J

K

L

M

N

O

P

X

Y

Z